T0397036

Advanced Mathematical Analysis and its Applications

Advanced Mathematical Analysis and its Applications presents state-of-the-art developments in mathematical analysis through new and original contributions and surveys, with a particular emphasis on applications in engineering and mathematical sciences. New research directions are indicated in each of the chapters, and while this book is meant primarily for graduate students, there is content that will be equally useful and stimulating for faculty and researchers.

The readers of this book will require minimum knowledge of real, complex and functional analysis, and topology.

Features

- Suitable as a reference for graduate students, researchers, and faculty
- Contains the most up-to-date developments at the time of writing.

Advanced Mathematical Analysis and its Applications

Edited by

Pradip Debnath
Assam University, India

Delfim F. M. Torres
University of Aveiro, Portugal

Yeol Je Cho
Gyeongsang National University, Korea

CRC Press is an imprint of the
Taylor & Francis Group, an **informa** business
A CHAPMAN & HALL BOOK

Designed cover image: ©ShutterStock Images

First edition published 2024
by CRC Press
6000 Broken Sound Parkway NW, Suite 300, Boca Raton, FL 33487-2742

and by CRC Press
4 Park Square, Milton Park, Abingdon, Oxon, OX14 4RN

CRC Press is an imprint of Taylor & Francis Group, LLC

ISBN: 978-1-032-48151-7 (hbk)
ISBN: 978-1-032-48369-6 (pbk)
ISBN: 978-1-003-38867-8 (ebk)

DOI: 10.1201/9781003388678

Typeset in Nimbus
by codeMantra

Contents

Preface

This book is a collection of chapters from eminent contemporary mathematicians across the countries working on the advances of mathematical analysis. As suggested by the title, this book particularly focuses on the recent advances and applications of mathematical analysis in engineering and mathematical sciences. The first half of the book (Chapters 1–11) are dedicated to the study of fixed point theory and applications. Chapters 12–17 present recent advancements in fractional calculus. Periodic Dirichlet series and special functions are studied in Chapter 18, whereas Lotka–Volterra dynamical system and its discretization are presented in Chapter 19. Chapter 20 contains a study on pseudomonotone equilibrium problems. In Chapter 21, convergence analysis of an inertial alternating minimization algorithm is discussed. Ball convergence has been studied in Chapter 22. An interesting investigation on inner product trapezoid type inequalities in Hilbert spaces is available in Chapter 23. Finally, Chapters 24 and 25 present theory and applications of some generalized random variables.

This book is meant for graduate students, faculties and researchers working in mathematical analysis and its applications. New research directions are suggested within most of the chapters to enable the researchers to further advance their research. The readers of this book will require minimum prerequisites of real, complex and functional analysis and topology.

Editor Biographies

Pradip Debnath is an Assistant Professor (in Mathematics) at the Department of Applied Science and Humanities, Assam University, Silchar, India. He earned his Ph.D. in Mathematics from the National Institute of Technology Silchar, India. His research interests include fixed point theory, nonlinear functional analysis, soft computing and mathematical statistics.

He has published more than 60 papers in various journals of international repute and is an active reviewer for more than 50 international journals. He is also a reviewer for *Mathematical Reviews* published by the American Mathematical Society. He is the Lead Editor of the books *Metric Fixed Point Theory - Applications in Science, Engineering and Behavioural Sciences* (Springer Nature, 2021), *Soft Computing Techniques in Engineering, Health, Mathematical and Social Sciences* (CRC Press, 2021), *Fixed Point Theory and Fractional Calculus: Recent Advances and Applications* (Springer Nature, 2022), *Soft Computing: Recent Advances and Applications in Engineering and Mathematical Sciences* (CRC Press, 2023) and *Advances in Number Theory and Applied Analysis* (World Scientific, 2023). He is a topical advisory panel member of the journals "Axioms" and "Fractal and Fractional" and guest editor of several special issues for different journals.

He has successfully guided Ph.D. students in the areas of nonlinear analysis, soft computing and fixed point theory. He has recently completed a major Basic Science Research Project in fixed point theory funded by the UGC, the Government of India. Having been an academic gold medalist during his post-graduation studies from Assam University, Silchar, Dr. Debnath has qualified several national-level examinations in Mathematics in India.

Delfim F. M. Torres is a Portuguese Mathematician born on 16 August 1971 in Mozambique. He obtained a Ph.D. in Mathematics from the University of Aveiro (UA) in 2002, and habilitation in Mathematics, UA, in 2011. He is a full professor of Mathematics since 9 March 2015. He has been the Director of the R&D Unit CIDMA, the largest Portuguese research center for mathematics, and Coordinator of its Systems and Control Group. His main research areas are calculus of variations and optimal control; optimization; fractional derivatives and integrals; dynamic equations on time scales; and mathematical biology.

Torres has written outstanding scientific and pedagogical publications. In particular, he is the author of two books with Imperial College Press; three books with Springer; and editor of several other books. Professor Torres has been recognized four times as one of the top 1% of mathematicians on the prestigious global Clarivate Web of Science list and is the only Portuguese mathematician to be so honored. He has strong experience in graduate and post-graduate student supervision and teaching in Mathematics. Twenty-four Ph.D. theses in Mathematics have successfully been finished under his supervision. Moreover, he has been the leading member in several national and international R&D projects, including EU projects and networks. Professor Torres has been, since 2013, the Director of the Doctoral Programme Consortium in Mathematics and Applications (MAP-PDMA) of Universities of Minho, Aveiro, and Porto.

Delfim married in 2003 and has one daughter and two sons.

Yeol Je Cho is Emeritus Professor at the Department of Mathematics Education, Gyeongsang National University, Jinju, Korea, and Distinguished Professor at the School of Mathematical Sciences, the University of Electronic Science and Technology of China, Chengdu, Sichuan, China. In 1984, he received his Ph.D. in Mathematics from Pusan National University, Pusan, Korea. He is a fellow of the Korean Academy of Science and Technology, Seoul, Korea, since 2006, and a member of several mathematical societies. He has organized international conferences on nonlinear functional analysis and applications, fixed point theory and applications and workshops and symposiums on nonlinear analysis and applications.

He has published over 400 papers, 20 monographs and 12 books with renowned publishers from around the world. His research areas are nonlinear analysis and applications, especially fixed point theory and applications, some kinds of nonlinear problems, that is, equilibrium problems, variational inequality problems, saddle point problems, optimization problems, inequality theory and applications, and stability of functional equations and applications. He has delivered several invited talks at international conferences on nonlinear analysis and applications and is on the editorial boards of ten international journals of mathematics.

List of Contributors

Christopher Argyros
Department of Computing and Technology
Cameron University
Lawton, Oklahoma, USA

Ioannis K. Argyros
Department of Computing and Mathematical Sciences
Cameron University
Lawton, Oklahoma, USA

Zohreh Bagheri
Department of Mathematics, Ardabil Branch
Islamic Azad University
Ardabil, Iran

Benaoumeur Bayour
University of Mascara
Mascara, Algeria

Samira Hadi Bonab
Department of Mathematics, Ardabil Branch
Islamic Azad University
Ardabil, Iran

Shih-sen Chang
School of Mathematics
Sichuan University
Chengdu, China

Ilwoo Cho
Department of Mathematics and Statistics
Iowa, USA

Yeol Je Cho
Department of Mathematics Education
Gyeongsang National University
Jinju, Korea

Pradip Debnath
Department of Applied Science and Humanities
Assam University
Silchar, Cachar-788011, India

Hacen Dib
Department of Mathematics
Aboubekr Belkaid University
Tlemcen, Algeria

Silvestru Sever Dragomir
College of Engineering and Science
Victoria University
Melbourne, Australia

El Alaoui Fatima-Zahrae
TSI Team, Department of Mathematics
Moulay Ismail University
Meknes, Morocco

Santhosh George
Department of Mathematical and Computational Sciences
NIT Karnataka, India

Shigeru Kanemitsu
Kerala School of Mathematics
Kerala, India

Imre Kátai
Department of Computer Algebra
Eotvos Lorand University
Budapest, Hungary

S. K. Katiyar
Department of Mathematics
NIT Jalandhar, India

Seth Kermausuor
Department Mathematics and Computer Science
Alabama State University
Alabama, United States

Zguaid Khalid
TSI Team, Department of Mathematics
Moulay Ismail University
Meknes, Morocco

Leila Khitri-Kazi-Tani
Department of Mathematics
Aboubekr Belkaid University
Tlemcen, Algeria

Dae San Kim
Department of Mathematics
Sogang University
Seoul, Republic of Korea

Taekyun Kim
School of Science
Xian Technological University
Xi'an, China

Nabanita Konwar
Department of Mathematics
Birjhora Mahavidyalaya
Bongaigaon, Assam

Jongkyum Kwon
Department of Mathematics Education
Gyeongsang National University
Republic of Korea

Hyunseok Lee
Department of Mathematics
Kwangwoon University
Seoul, Republic of Korea

Márcia Lemos-Silva
Department of Mathematics
University of Aveiro
Aveiro, Portugal

Jay Mehta
Department of Mathematics
Sardar Patel University
Gujarat, India

Slobodanka Mitrović
Faculty of Forestry
University of Belgrade
Beograd, Serbia

Zoran D. Mitrović
Faculty of Electrical Engineering
University of Banja Luka Patre 5
Bosnia and Herzegovina

S. A. Mohiuddine
Department of General Required Courses, Mathematics,
Faculty of Applied Studies, King Abdulaziz University, Jeddah, Saudi Arabia
and
Operator Theory and Applications Research Group, Department of Mathematics, Faculty of Science, King Abdulaziz University, Jeddah, Saudi Arabia

Le Dung Muu
Thang Long University
Hanoi, Vietnam

Nimit Nimana
Department of Mathematics
Khon Kaen University
Khon Kaen, Thailand

Sehie Park
The National Academy of Sciences
Seoul, Republic of Korea

Vahid Parvaneh
Department of Mathematics
Gilan-E-Gharb Branch
Islamic Azad University
Gilan-E-Gharb, Iran

Jigen Peng
School of Mathematics and Information Science
Guangzhou University
GuangZhou, China

Mootta Prangprakhon
Department of Mathematics
Khon Kaen University
Khon Kaen, Thailand

B. V. Prithvi
Department of Mathematics
SRM Institute of Science and Technology
Chennai, India

Stojan Radenović
Faculty of Mechanical Engineering
University of Belgrade
Beograd, Serbia

N. Revathi
Department of Computer Science and Applications,
SRM Institute of Science and Technology, Ramapuram Campus
Chennai, Tamil Nadu, India

Priyadharsini Sivaraj
Department of Mathematics
Sri Krishna Arts and Science College
Coimbatore, India

K. Tamilvanan
Department of Mathematics, Faculty of Science & Humanities, R.M.K. Engineering College
Kavaraipettai, Tamil Nadu, India

Yuchao Tang
School of Mathematics and Information Science
Guangzhou University
Guangzhou, China

Nihal Taş
Department of Mathematics
Balikesir University
Balıkesir, Turkey

Tran Van Thang
Faculty of Natural Sciences
Electric Power University
Hanoi, Vietnam

Delfim F. M. Torres
Center for Research and Development in Mathematics and Applications (CIDMA)
Department of Mathematics
University of Aveiro
Aveiro, Portugal

Juan E. Nápoles Valdés
Notheast National University
FaCENA, Argentina

Sandra Vaz
Center of Mathematics and Applications, Department of Mathematics
University of Beira Interior
Covilhã, Portugal

Miguel Vivas-Cortez
Physical and Mathematics Science School
Pontifical Catholic University of Ecuador
Quito, Ecuador

Jelena Vujaković
Department of Mathematics
University of Pristina-Kosovska Mitrovica
Serbia

Yang Yang
School of Computer Science and Engineering
Sun Yat-sen University
Guangzhou, China

George Xianzhi Yuan
Business School
Sun Yat-Sen University
Guangzhou, China

Nedjoua Zine
University of Mascara
Mascara, Algeria

CHAPTER 1

Generalized Boyd-Wong-Type Contractions and Related Fixed-Figure Results

Pradip Debnath

Assam University Silchar

CONTENTS

1.1 INTRODUCTION AND PRELIMINARIES

Kannan [14, 15] initiated the study of fixed points for discontinuous mappings. Most of the mappings studied initially in this connection were continuous at their respective fixed points with points of discontinuity within their domains [2, 3, 4, 9]. A comparative study of contractive definitions – many of which did not guarantee the continuity of the mapping – can be found in the work of Rhoades [26, 27] and Pant [24].

Debnath and Srivastava [11] proved new extensions of Kannan's and Reich's theorems. Another Kannan-type contraction for multivalued asymptotic regular maps was presented by Debnath et al. [10].

Fixed point results for mappings with discontinuity found a wide variety of applications in different fields of science [5, 12, 16, 17, 18, 19, 30]. Recently, several authors have attempted to provide solutions to such problems from different points of view (see Bisht and Rakocević [1], Pant et al. [25], Tas and Ozgur [29], Ozgur and Tas [22], Debnath [8]).

The study of non-unique fixed points has also gained immense importance both in terms of theory and applications. The investigation of geometric properties of non-unique fixed points has given rise to a new branch of study called fixed-figure

DOI: 10.1201/9781003388678-1

problems. Such a problem primarily deals with establishing new contractive conditions which guarantee that a geometric figure is a subset of the set of fixed points of a given self-mapping.

Let $(\mathcal{W}, \eta)$ be a MS and $\Phi : \mathcal{W} \to \mathcal{W}$ a self-map. The set of fixed points of Φ is defined as $Fix(\Phi) = \{\theta \in \mathcal{W} : \Phi\theta = \theta\}$. The circle $C^r_{\theta_0} = \{\theta \in \mathcal{W} : \eta(\theta, \theta_0) = r\}$ (respectively, the disc $D^r_{\theta_0} = \{\theta \in \mathcal{W} : \eta(\theta, \theta_0) \leq r\}$) is called a fixed-circle (respectively, fixed disc) of Φ if $\Phi\theta = \theta$ for all $\theta \in C^r_{\theta_0}$ (respectively, for all $\theta \in D^r_{\theta_0}$). In general, a geometric figure $\mathcal{T}$ in a MS is said to be a fixed-figure of Φ if $\mathcal{T} \subseteq Fix(\Phi)$.

Solution to the fixed-figure problems was initiated by Ozgur and Tas [20, 21] and subsequently the major developments in this direction followed [28]. Geometric interpretations of theoretic fixed point results found applications in neural networks [13, 23].

In this chapter, we introduce some generalized Boyd-Wong-type contractive conditions and establish fixed-figure theorems such as fixed-point, fixed-circle and fixed-ellipse results in metric spaces. Some examples are provided to validate the results. The mappings for which these results are applicable are not necessarily continuous, and the completeness of the metric space under consideration is not necessary.

1.2 GENERALIZED BOYD-WONG-TYPE FIXED-FIGURE RESULTS

In this section, we present the main results. First, the concept of Boyd and Wong-type θ_0-contraction is defined.

Definition 1.1 *Let $(\mathcal{W}, \eta)$ be a MS and $\Phi : \mathcal{W} \to \mathcal{W}$ a self-map. Further, let the function $\zeta : [0, \infty) \to [0, \infty)$ be upper semi-continuous from right with $0 \leq \zeta(s) < s$ for $s > 0$ such that $\zeta(0) = 0$. If there exists $\theta_0 \in \mathcal{W}$ satisfying*

$$\eta(\Phi\theta, \theta) > 0 \implies \eta(\Phi\theta, \theta) \leq \zeta(\eta(\theta, \theta_0))$$

for all $\theta \in \mathcal{W}$, then Φ is called a Boyd and Wong type (BW-type) θ_0-contraction.

The following lemma will be used in the sequel.

Lemma 1.1 *Let $(\mathcal{W}, \eta)$ be a MS and $\Phi : \mathcal{W} \to \mathcal{W}$ a BW-type θ_0-contraction for $\theta_0 \in \mathcal{W}$. Then θ_0 is a fixed point of Φ.*

Proof. Let $\eta(\Phi\theta_0, \theta_0) > 0$, i.e., $\Phi\theta_0 \neq \theta_0$. Since Φ is a BW-type θ_0-contraction, we have

$$\eta(\Phi\theta_0, \theta_0) \leq \zeta(\eta(\theta_0, \theta_0)) = \zeta(0) = 0,$$

i.e., $\eta(\Phi\theta_0, \theta_0) = 0$, which is a contradiction.

Hence, we must have $\eta(\Phi\theta_0, \theta_0) = 0$, i.e. $\Phi\theta_0 = \theta_0$. □

Theorem 1.1 *Let $(\mathcal{W}, \eta)$ be a MS and $\Phi : \mathcal{W} \to \mathcal{W}$ a BW-type θ_0-contraction for $\theta_0 \in \mathcal{W}$. Define μ by*

$$\mu = \inf\{\eta(\Phi\theta, \theta) : \Phi\theta \neq \theta, \theta \in \mathcal{W}\}.$$

Then Φ fixes the circle $C^{\mu}_{\theta_0}$.

Proof. If $\mu = 0$, we have $C^{\mu}_{\theta_0} = \{\theta_0\}$. Using Lemma 17.1, clearly Φ fixes $C^{\mu}_{\theta_0} = \{\theta_0\}$.

Hence, suppose that $\mu > 0$ and let $\theta \in C^{\mu}_{\theta_0}$ such that $\eta(\Phi\theta, \theta) > 0$. Since Φ is a BW-type θ_0-contraction, we have that

$$\eta(\Phi\theta, \theta) \leq \zeta(\eta(\theta, \theta_0)) = \zeta(\mu) < \mu \leq \eta(\Phi\theta, \theta),$$

which is a contradiction.

Hence, we must have $\eta(\Phi\theta, \theta) = 0$, i.e., $\Phi\theta = \theta$, i.e., Φ fixes the circle $C^{\mu}_{\theta_0}$. □

Corollary 1.1 *Let $(\mathcal{W}, \eta)$ be a MS and $\Phi : \mathcal{W} \to \mathcal{W}$ a BW-type θ_0-contraction for $\theta_0 \in \mathcal{W}$ and μ as defined in Theorem 9.2. Then Φ fixes the disc $D^{\mu}_{\theta_0}$.*

Proof. Like in the previous case, if $\mu = 0$, we have $D^{\mu}_{\theta_0} = \{\theta_0\}$ and obviously Φ fixes $D^{\mu}_{\theta_0} = \{\theta_0\}$.

Hence, suppose that $\mu > 0$ and let $\theta \in D^{\mu}_{\theta_0}$ such that $\eta(\Phi\theta, \theta) > 0$. Since Φ is a BW-type θ_0-contraction, we have that

$$\eta(\Phi\theta, \theta) \leq \zeta(\eta(\theta, \theta_0)) < \eta(\theta, \theta_0) \leq \mu \leq \eta(\Phi\theta, \theta),$$

which leads to a contradiction.

Thus, we must have $\eta(\Phi\theta, \theta) = 0$, i.e., $\Phi\theta = \theta$, i.e., Φ fixes the disc $D_{\theta_0,\mu}$. □

Below we provide an example to validate Theorem 9.2.

Example 1.1 *Consider $\mathcal{W} = \mathbb{R}$ with the usual metric η. Define for all $\theta \in \mathcal{W}$, $\Phi : \mathcal{W} \to \mathcal{W}$ by*

$$\Phi\theta = \begin{cases} \theta, & \text{if } |\theta| \leq 7 \\ \theta + 2, & \text{if } |\theta| > 7. \end{cases}$$

Then Φ (see Figure 15.1) is a BW-type θ_0-contraction with $\theta_0 = 0$ and the function $\zeta : [0, \infty) \to [0, \infty)$ defined by $\zeta(t) = \frac{t}{3}$ for $t > 0$.

Then, since $\eta(\Phi\theta, \theta) = 2$ and $\zeta(\eta(\theta, \theta_0)) > 2$ for all θ such that $\eta(\Phi\theta, \theta) > 0$, we have $\mu = 2$ and Φ fixes the circle $C^2_0 = \{-2, 2\}$ and the disc $D^2_0 = [-2, 2]$.

We can observe that Φ is discontinuous at -2 and 2.

The next result can be thought of as a fixed-ellipse theorem. We define an ellipse $E_{\mu}(\theta_1, \theta_2)$ as

$$E_{\mu}(\theta_1, \theta_2) = \{\theta \in \mathcal{W} : \eta(\theta, \theta_1) + \eta(\theta, \theta_2) = \mu\}.$$

Theorem 1.2 *Let $(\mathcal{W}, \eta)$ be a MS and $\Phi : \mathcal{W} \to \mathcal{W}$ a self-mapping and μ defined like earlier by*

$$\mu = \inf\{\eta(\Phi\theta, \theta) : \Phi\theta \neq \theta, \theta \in \mathcal{W}\}.$$

If there exist $\theta_1, \theta_2 \in \mathcal{W}$ such that $\eta(\Phi\theta, \theta) > 0$ implies $\eta(\Phi\theta, \theta) \leq \zeta(\eta(\theta, \theta_1) + \eta(\theta, \theta_2))$, for all $\theta \in \mathcal{W}$, where $\zeta : [0, \infty) \to [0, \infty)$ is defined as earlier.

Then Φ fixes the ellipse $E_{\mu}(\theta_1, \theta_2)$.

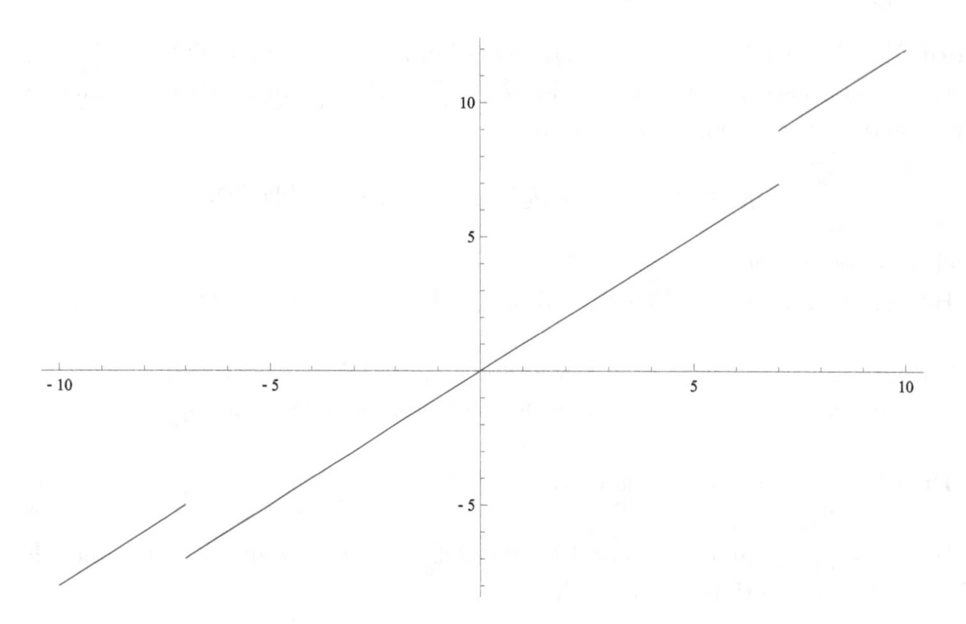

Figure 1.1 Plot of the function Φ.

Proof. We consider two cases. If $\mu = 0$, then $E_\mu(\theta_1, \theta_2) = C^\mu_{\theta_1} = C^\mu_{\theta_2} = \{\theta_1\} = \{\theta_2\}$, i.e., the two points θ_1 and θ_2 are coincident.

If $\eta(\theta_1, \Phi\theta_1) > 0$, then

$$\eta(\theta_1, \Phi\theta_1) \leq \zeta(\eta(\theta_1, \theta_1) + \eta(\theta_1, \theta_1)) = \zeta(0) = 0,$$

which is a contradiction. Hence, we must have $\eta(\theta_1, \Phi\theta_1) = 0$, i.e., $\Phi\theta_1 = \theta_1$.

Next, assume that $\mu > 0$ and let $\theta \in E_\mu(\theta_1, \theta_2)$ be such that $\theta \neq \Phi\theta$, i.e., $\eta(\theta, \Phi\theta) > 0$. Using the hypothesis, we have

$$\begin{aligned}\eta(\theta, \Phi\theta) &\leq \zeta(\eta(\theta, \theta_1) + \eta(\theta, \theta_2)) \\ &= \zeta(\mu) \\ &< \mu \\ &< \eta(\theta, \Phi\theta),\end{aligned}$$

which is again a contradiction. Hence, we must have $\eta(\theta, \Phi\theta) = 0$. □

The next example validates Theorem 9.3.

Example 1.2 *Consider $\mathcal{W} = \{-6, -5, -2, -1, 1, 2, 3, 4, 8\}$ endowed with the usual distance metric η on it. Define for all $\theta \in \mathcal{W}$, $\Phi : \mathcal{W} \to \mathcal{W}$ by*

$$\Phi\theta = \begin{cases} \theta + 6, & \text{if } \theta = 4 \\ \theta, & \text{otherwise.} \end{cases}$$

Further, define $\zeta : [0, \infty) \to [0, \infty)$ by $\zeta(t) = \frac{9}{10}t, t > 0$. Then Φ satisfies all conditions of Theorem 9.3 with $\theta_1 = -1, \theta_2 = 1$. We observe that $\mu = 6$ and $E_6(-1, 1) = \{-3, 3\}$.

Thus, it is easy to see that Φ fixes the ellipse $E_6(-1, 1)$.

Next, we define the concept of a generalized BW-type θ_0-contraction.

Definition 1.2 *Let $(\mathcal{W}, \eta)$ be a MS and $\Phi : \mathcal{W} \to \mathcal{W}$ a self-mapping. If there exists $\theta_0 \in \mathcal{W}$ such that $\eta(\theta, \Phi\theta) > 0$ implies $\eta(\theta, \Phi\theta) \leq \zeta(M(\theta, \theta_0))$, for all $\theta \in \mathcal{W}$, where $\zeta : [0, \infty) \to [0, \infty)$ has the property that $0 \leq \zeta(t) < t$ for $t > 0$ and $\zeta(0) = 0$, and*

$$M(\theta, \psi) = \max\left\{\eta(\theta, \psi), \eta(\theta, \Phi\theta), \eta(\psi, \Phi\psi), \frac{\eta(\theta, \Phi\psi) + \eta(\psi, \Phi\theta)}{2}\right\},$$

then Φ is called a generalized BW-type θ_0-contraction.

Lemma 1.2 *Let $(\mathcal{W}, \eta)$ be a MS and $\Phi : \mathcal{W} \to \mathcal{W}$ a generalized BW-type θ_0-contraction. Then $\theta_0 \in Fix(\Phi)$.*

Proof. Suppose that $\eta(\theta_0, \Phi\theta_0) > 0$. Using the hypothesis, we have that $M(\theta_0, \theta_0) = \eta(\theta_0, \Phi\theta_0)$.

Further,

$$\eta(\theta_0, \Phi\theta_0) \leq \zeta(M(\theta_0, \theta_0)) = \zeta(\eta(\theta_0, \Phi\theta_0)) < \eta(\theta_0, \Phi\theta_0),$$

which is a contradiction. Hence, $\eta(\theta_0, \Phi\theta_0) = 0$. □

Next, we establish a fixed-circle result using the notion of generalized BW-type θ_0-contraction.

Theorem 1.3 *Let $(\mathcal{W}, \eta)$ be a MS and $\Phi : \mathcal{W} \to \mathcal{W}$ a generalized BW-type θ_0-contraction for $\theta_0 \in \mathcal{W}$. Define μ by*

$$\mu = \inf\{\eta(\Phi\theta, \theta) : \Phi\theta \neq \theta, \theta \in \mathcal{W}\}.$$

If $\eta(\theta_0, \Phi\theta) \leq \mu$ for all $\theta \in \mathcal{W}$ with $\eta(\theta, \Phi\theta) > 0$, then Φ fixes the circle $C^{\mu}_{\theta_0}$.

Proof. If $\mu = 0$, then $C^{\mu}_{\theta_0} = \{\theta_0\}$. Thus, by Lemma 17.2, we have $\Phi\theta_0 = \theta_0$.

Suppose that $\mu > 0$ and let $\theta \in C^{\mu}_{\theta_0}$ with $\eta(\theta, \Phi\theta) > 0$. We have that

$$\begin{aligned} M(\theta, \theta_0) &= \max\left\{\eta(\theta, \theta_0), \eta(\theta, \Phi\theta), \eta(\theta_0, \Phi\theta_0), \frac{\eta(\theta, \Phi\theta_0) + \eta(\theta_0, \Phi\theta)}{2}\right\} \\ &= \max\left\{\eta(\theta, \theta_0), \eta(\theta, \Phi\theta), 0, \frac{\eta(\theta, \theta_0) + \eta(\theta_0, \Phi\theta)}{2}\right\} \\ &= \max\left\{\mu, \eta(\theta, \Phi\theta), \frac{\mu + \eta(\theta_0, \Phi\theta)}{2}\right\}. \end{aligned}$$

Hence, if $\eta(\theta_0, \Phi\theta) \leq \mu$, we have that $M(\theta, \theta_0) \leq \eta(\theta, \Phi\theta)$.

Further,

$$\begin{gathered} \eta(\theta, \Phi\theta) \leq \zeta(M(\theta, \theta_0)) \\ M(\theta, \theta_0) \end{gathered}$$

$$\leq \eta(\theta, \Phi\theta),$$

which is a contradiction. Hence, we must have $\eta(\theta, \Phi\theta) = 0$. Thus, Φ fixes the circle $C^{\mu}_{\theta_0}$. □

We have the following corollary which can be proved using similar arguments as in Theorem 9.4.

Corollary 1.2 *Let $(\mathcal{W}, \eta)$ be a MS and $\Phi : \mathcal{W} \to \mathcal{W}$ a generalized BW-type θ_0-contraction. Define μ by*

$$\mu = \inf\{\eta(\Phi\theta, \theta) : \Phi\theta \neq \theta, \theta \in \mathcal{W}\}.$$

If $\eta(\theta_0, \Phi\theta) \leq \mu$ for all $\theta \in \mathcal{W}$ with $\eta(\theta, \Phi\theta) > 0$, then Φ fixes the disc $D^{\mu}_{\theta_0}$.

Theorem 1.4 *Let $(\mathcal{W}, \eta)$ be a MS and $\Phi : \mathcal{W} \to \mathcal{W}$ a self-mapping and $M(\theta, \psi)$ be defined as in Definition and also μ defined like earlier by*

$$\mu = \inf\{\eta(\Phi\theta, \theta) : \Phi\theta \neq \theta, \theta \in \mathcal{W}\}.$$

If there exist $\theta_1, \theta_2 \in \mathcal{W}$ such that $\eta(\Phi\theta, \theta) > 0$ implies $\eta(\Phi\theta, \theta) \leq \zeta(M(\theta, \theta_1) + M(\theta, \theta_2))$, for all $\theta \in \mathcal{W}$, where $\zeta : [0, \infty) \to [0, \infty)$ has the property that $0 \leq \zeta(t) < t$ for $t > 0$. Further, if $\eta(\theta_1, \Phi\theta) \leq \mu$ and $\eta(\theta_2, \Phi\theta) \leq \mu$ for $\theta \in E_{\mu}(\theta_1, \theta_2)$, then Φ fixes the ellipse $E_{\mu}(\theta_1, \theta_2)$.

Proof. If $\mu = 0$, then $E_{\mu}(\theta_1, \theta_2) = C^{\mu}_{\theta_1} = C^{\mu}_{\theta_2} = \{\theta_1\} = \{\theta_2\}$.

Suppose that $\eta(\theta_1, \Phi\theta_1) > 0$. We have that

$$\begin{aligned}
\eta(\theta_1, \Phi\theta_1) &= \zeta\left(\eta\left(\frac{M(\theta_1, \theta_1) + M(\theta_1, \theta_1)}{2}\right)\right) \\
&= \zeta(M(\theta_1, \theta_1)) \\
&< M(\theta_1, \theta_1) \\
&= \eta(\theta_1, \Phi\theta_1),
\end{aligned}$$

which is a contradiction.

Hence, $\eta(\theta_1, \Phi\theta_1) = 0$.

Now suppose that $\mu > 0$ and $\theta \in E_{\mu}(\theta_1, \theta_2)$ such that $\eta(\theta, \Phi\theta) > 0$. We have that

$$\begin{aligned}
\eta(\theta, \Phi\theta) &\leq \zeta\left(\frac{M(\theta, \theta_1) + M(\theta, \theta_2)}{2}\right) \\
&< \frac{M(\theta, \theta_1) + M(\theta, \theta_2)}{2} \\
&= \eta(\theta, \Phi\theta),
\end{aligned}$$

a contradiction. Thus, we must have $\eta(\theta, \Phi\theta) = 0$.

Hence, in either case, Φ fixes the ellipse $E_{\mu}(\theta_1, \theta_2)$. □

Example 1.3 *Let $\mathcal{W} = \{-6, -3, -2, -1, 1, 2, 3, 4, 7\}$ be endowed with the usual distance metric η on it. Define for all $\theta \in \mathcal{W}$, $\Phi : \mathcal{W} \to \mathcal{W}$ by*

$$\Phi\theta = \begin{cases} \theta + 4, & \text{if } \theta = 3 \\ \theta, & \text{otherwise.} \end{cases}$$

Further, define $\zeta : [0, \infty) \to [0, \infty)$ by $\zeta(t) = \frac{19}{20}t, t > 0$. Then Φ satisfies all conditions of Theorem 1.4 with $\theta_1 = -1, \theta_2 = 1$. We observe that $\mu = 4$ and $E_4(-1, 1) = \{-2, 2\} \subseteq Fix(\Phi) = \mathcal{W} \setminus \{3\}$.

Thus, it is easy to see that Φ fixes the ellipse $E_4(-1, 1)$.

1.3 CONCLUSION AND FUTURE WORK

In this chapter, we have established a generalized Boyd-Wong-type contraction and established some fixed-figure results. In [28], Tas provided a geometric approach to Proinov type contractions and suggested several generalizations. Obtaining multi-valued analogues of the current results using the framework as in [6, 7] are some suggested future works.

Author contributions

The author contributed solely in writing this article.

Acknowledgments

The author is immensely thankful to the learned referees for their careful reading and constructive comments which resulted in the improvement of the manuscript. Wolfram Mathematica 7.0 has been used to generate the graph(s).

BIBLIOGRAPHY

[1] R. K. Bisht and V. Rakočević. Generalized Meir-Keeler type contractions and discontinuity at fixed point. *Fixed Point Theory*, 19(1):57–64, 2018.

[2] D. W. Boyd and J. S. Wong. On nonlinear contractions. *Proc. Amer. Math. Soc.*, 20:458–464, 1969.

[3] L. Ćirić. On contraction type mappings. *Math. Balkanica*, 1:52–57, 1971.

[4] L. Ćirić. A generalization of Banach's contraction principle. *Proc. Math. Amer. Math. Soc.*, 45(2):267–273, 1974.

[5] A. Das, S. A. Mohiuddine, A. Alotaibi, and B. C. Deuri. Generalization of Darbo-type theorem and application on existence of implicit fractional integral equations in tempered sequence spaces. *Alexandria Eng. J.*, 61:2010–2015, 2022.

[6] P. Debnath. Optimization through best proximity points for multivalued F-contractions. *Miskolc Math. Notes*, 22(1):143–151, 2021.

[7] P. Debnath. Banach, Kannan, Chatterjea, and Reich-type contractive inequalities for multivalued mappings and their common fixed points. *Math. Meth. Appl. Sci.*, 45(3):1587–1596, 2022.

[8] P. Debnath. Common fixed-point and fixed-circle results for a class of discontinuous F-contractive mappings. *Mathematics*, 10(9):Article ID: 1605, 2022.

[9] P. Debnath, N. Konwar, and S. Radenović. *Metric Fixed Point Theory: Applications in Science, Engineering and Behavioural Sciences.* Springer, Singapore, 2021.

[10] P. Debnath, Z. D. Mitrović, and H. M. Srivastava. Fixed points of some asymptotically regular multivalued mappings satisfying a Kannan-type condition. *Axioms*, 10(1): Article ID: 24, 2021.

[11] P. Debnath and H. M. Srivastava. New extensions of Kannan's and Reich's fixed point theorems for multivalued maps using Wardowski's technique with application to integral equations. *Symmetry*, 12(7):Article ID: 1090, 2020.

[12] M. Forti and P. Nistri. Global convergence of neural networks with discontinuous neuron activations. *IEEE Trans. Circuits Syst. I, Fundam. Theory Appl.*, 50(11):1421–1435, 2003.

[13] C. Guo, D. O'Regan, F. Deng, and R. P. Agarwal. Fixed points and exponential stability for a stochastic neural cellular neural network. *Appl. Math. Lett.*, 26(8):849–853, 2013.

[14] R. Kannan. Some results on fixed points. *Bull. Calc. Math. Soc.*, 60(1):71–77, 1968.

[15] R. Kannan. Some results on fixed points–II. *Amer. Math. Monthly*, 76(4):405–408, 1969.

[16] S. A. Mohiuddine, A. Das, and A. Alotaibi. Existence of solutions for nonlinear integral equations in tempered sequence spaces via generalized Darbo-type theorem. *J. Funct. Spaces*, 2022:Article ID: 4527439, 2022.

[17] X. Nie and W. X. Zheng. On multistability of competitive neural networks with discontinuous activation functions. In *4th Australian Control Conference (AUCC)*, Canberra, Australia, pages 245–250, 2014.

[18] X. Nie and W. X. Zheng. Multistability of neural networks with discontinuous non-monotonic piecewise linear activation functions and time-varying delays. *Neural Netw.*, 65:65–79, 2015.

[19] X. Nie and W. X. Zheng. Multistability of neural networks with discontinuous non-monotonic piecewise linear activation functions and time-varying delays. *IEEE Trans. Cybernat*, 46(3):679–693, 2015.

[20] N. Ozgur and N. Tas. Fixed-circle problem on s-metric spaces with a geometric viewpoint. *Facta Univ. Ser. Math. Inf.*, 34(3):459–472, 2019.

[21] N. Ozgur and N. Tas. Some fixed-circle theorems on metric spaces. *Bull. Malays. Math. Sci. Soc.*, 42(4):1433–1449, 2019.

[22] N. Ozgur and N. Tas. New discontinuity results at fixed point on metric spaces. *J. Fixed Point Theory Appl.*, 23(28):doi: 10.1007/s11784-021-00863-3, 2021.

[23] N. Ozgur, N. Tas, and J. F. Peters. New complex-valued activation functions. *Int. J. Optim. Control. Theor. Appl.*, 10(1):66–72, 2020.

[24] R. Pant. Discontinuity and fixed points. *J. Math. Anal. Appl.*, 240:284–289, 1999.

[25] R. P. Pant, N. Ozgur, and N. Tas. On discontinuity problem at fixed point. *Bull. Malays. Math. Sci. Soc.*, 43:499–517, 2020.

[26] B. E. Rhoades. A comparison of various definitions of contractive mappings. *Trans. Amer. Math. Soc.*, 226:257–290, 1977.

[27] B. E. Rhoades. Contractive definitions and continuity. *Contemp. Math.*, 42:233–245, 1988.

[28] N. Tas. A geometric approach to the Proinov type contractions. *Math. Moravica*, 26(1):123–132, 2022.

[29] N. Tas and N. Ozgur. A new contribution to discontinuity at fixed point. *Fixed Point Theory*, 20(2):715–728, 2019.

[30] H. Wu and C. Shan. Stability analysis for periodic solution of BAM neural networks with discontinuous neuron activations and impulses. *Appl. Math. Modell.*, 33(6):2564–2574, 2017.

CHAPTER 2

Remarks on the Metatheorem in Ordered Fixed Point Theory

Sehie Park

Seoul National University

CONTENTS

2.1 INTRODUCTION

Since the appearances of the Ekeland variational principle [6, 7] in 1972–1974 and the Caristi fixed point theorem [4] in 1976, nearly one thousand works followed on their equivalents, generalizations, imitations, applications, and related topics. Many of them are related to new spaces extending complete metric spaces, new metrics or topologies on them, and new order relations extending the so-called Caristi order.

DOI: 10.1201/9781003388678-2

While the author was working on the same subject from 1985 to 2000, in order to give some equivalents of the Ekeland principle, we obtained a Metatheorem in 1985–1987 [13, 14, 15, 16] on fixed point theorems related to the order theory. It claims that certain order theoretic maximal element statements can be equivalently reformulated to theorems on fixed points, stationary points, common fixed points, common stationary points of families of maps or multimaps. As usual in the mathematical community, our Metatheorem was not attracted for a long period.

Unknowing this, Fierro [9] in 2017 obtained an extended version of our particular form of Metatheorem in 2000 [17] for arbitrary preorderings, without metric considerations. He added two additional propositions to our particular form of equivalences. Against this, Boros, Iqbal, and Száz [1, 2] in 2022 claimed that an implication of Fierro's theorem is not adequate and obtained some related results in [1, 2, 11].

In such a situation, we returned to our Metatheorem in 2022 and obtained a large number of new applications [18, 19, 20, 21, 22, 23, 24, 25, 26, 27]. These are added to the traditional order theoretic results and, consequently, there appeared the so-called ordered fixed point theory [25]. This can be comparable to traditional several fields in the fixed point theory, that is, analytical fixed point theory is originated from Brouwer in 1912 and concerned mainly with topological vector spaces; metric fixed point theory is originated from Banach in 1922 and deals mainly with generalizations of contractions and nonexpansive maps; and topological fixed point theory relates mainly to pioneering works of Lefschetz, Nielsen, and Reidemeister.

In our previous work entitled "Foundations of Ordered Fixed Point Theory" [25], we established a large number of improved versions of historically well-known maximal element theorems and fixed point theorems related to order structure. It is based on our new 2023 Metatheorem and the Brøndsted-Jachymski Principle established by ourselves in 2022.

Recently, in 2022, we obtained extended versions of our Metatheorem in [18, 19, 20, 21, 22, 23, 24]. Motivated by this, our previous article [25] is based on the 2023 Metatheorem and the Brøndsted-Jachymski Principle. Actually [25] contains a large number of new improved results.

In this chapter, we introduce the 2023 Metatheorem, its short history, and the papers of Fierro [9], Boros-Iqbal-Száz [1, 2] and Iqbal-Száz [9] by comparing them with our works. This chapter is a historical supplement of [25] and organized as follows.

Section 2.2 is to introduce the 2023 Metatheorem and its practical particular forms in [25]. In Section 2.3, we introduce our traditional 2022 Metatheorem with some history. Section 2.4 deals with four examples of Metatheorem based on Brézis-Browder [3] and Fierro [8, 9, 10]. In Section 2.5, we give comments on the works of Fierro [9], Boros-Iqbal-Száz [1, 2], and Iqbal-Száz [11]. Section 2.6 is to resolve certain conflict related to our previous works and to give a proof of the extended Caristi fixed point theorem due to Chen-Cho-Yang [5] as an application of our Metatheorem.

Finally, Section 2.7 devotes to epilogue. For preliminaries, we can use [24, 25].

2.2 THE 2023 METATHEOREM

To give some equivalents of the well-known central result of Ekeland [6, 7] in 1972–1979 on the variational principle for approximate solutions of minimization problems, we obtained a metatheorem in 1985–1987 [13, 14, 15, 16] consisting of several equivalent statements. Later we found more additional statements and, consequently, we introduced several extended versions of the metatheorem in 2022 [18, 19, 20, 21]. Finally, we obtained the following form called the 2023 Metatheorem:

2023 Metatheorem. *Let X be a set, A its nonempty subset, $G(x, y)$ a sentence formula for $x, y \in X$, and $\neg$ denote the negation.*

Then the following statements are equivalent:

(α) *There exists an element $v \in A$ such that $G(v, w)$ for any $w \in X \backslash \{v\}$.*

($\beta 1$) *If $f : A \to X$ is a map such that, for any $x \in A$ with $x \neq f(x)$, there exists a $y \in X \backslash \{x\}$ satisfying $\neg G(x, y)$, then f has a fixed element $v \in A$, that is, $v = f(v)$.*

($\beta 2$) *If $\mathfrak{F}$ is a family of maps $f : A \to X$ such that, for any $x \in A$ with $x \neq f(x)$, there exists a $y \in X \backslash \{x\}$ satisfying $\neg G(x, y)$, then $\mathfrak{F}$ has a common fixed element $v \in A$, that is, $v = f(v)$ for all $f \in \mathfrak{F}$.*

($\gamma 1$) *If $f : A \to X$ is a map such that $\neg G(x, f(x))$ for any $x \in A$ with $x \neq f(x)$, then f has a fixed element $v \in A$, that is, $v = f(v)$.*

($\gamma 2$) *If $\mathfrak{F}$ is a family of maps $f : A \to X$ satisfying $\neg G(x, f(x))$ for all $x \in A$ with $x \neq f(x)$, then $\mathfrak{F}$ has a common fixed element $v \in A$, that is, $v = f(v)$ for all $f \in \mathfrak{F}$.*

($\delta 1$) *If $F : A \multimap X$ is a multimap such that, for any $x \in A \backslash F(x)$, there exists $y \in X \backslash \{x\}$ satisfying $\neg G(x, y)$, then F has a fixed element $v \in A$, that is, $v \in F(v)$.*

($\delta 2$) *Let $\mathfrak{F}$ be a family of multimaps $F : A \multimap X$ such that, for any $x \in A \backslash F(x)$ there exists $y \in X \backslash \{x\}$ satisfying $\neg G(x, y)$. Then $\mathfrak{F}$ has a common fixed element $v \in A$, that is, $v \in F(v)$ for all $F \in \mathfrak{F}$.*

($\epsilon 1$) *If $F : A \multimap X$ is a multimap satisfying $\neg G(x, y)$ for any $x \in A$ and any $y \in F(x) \backslash \{x\}$, then F has a stationary element $v \in A$, that is, $\{v\} = F(v)$.*

($\epsilon 2$) *If $\mathfrak{F}$ is a family of multimaps $F : A \multimap X$ such that $\neg G(x, y)$ holds for any $x \in A$ and any $y \in F(x) \backslash \{x\}$, then $\mathfrak{F}$ has a common stationary element $v \in A$, that is, $\{v\} = F(v)$ for all $F \in \mathfrak{F}$.*

($\zeta 1$) *If a multimap $F : A \multimap X$ satisfies that, for all $x \in A$ with $F(x) \neq \emptyset$, there exists $y \in X \backslash \{x\}$ satisfying $\neg G(x, y)$, then there exists $v \in A$ such that $F(v) = \emptyset$.*

(ζ2) *Let $\mathfrak{F}$ be a family of multimaps $F : A \multimap X$ such that, for all $x \in A$ with $F(x) \neq \emptyset$, there exists $y \in X \backslash \{x\}$ satisfying $\neg G(x, y)$. Then there exists $v \in A$ such that $F(v) = \emptyset$ for all $F \in \mathfrak{F}$.*

(η) *If Y is a subset of X such that, for each $x \in A \backslash Y$, there exists a $z \in X \backslash \{x\}$ satisfying $\neg G(x, z)$, then there exists a $v \in A \cap Y$.*

For the proof, see Park [25] and others. This metatheorem guarantees the truth of all items when one of them is true. Since 1985, we have shown nearly one hundred cases of such situation.

A preorder is the one satisfying reflexivity and transitivity.

As an application of metatheorem, we apply it to preordered sets when $G(x, y)$ means $x \not\preceq y$ (resp. $y \not\preceq x$). Since $(\beta 2) - (\zeta 2)$ implies $(\beta 1) - (\zeta 1)$, respectively, we take the following form as the prototype of maximal (resp. minimal) element principles:

Theorem A. *Let $(X, \preceq)$ be a preordered set and A be a nonempty subset of X. Then the following statements are equivalent:*

(α) *There exists a maximal (resp. minimal) element $v \in A$; that is, $v \not\preceq w$ (resp. $w \not\preceq v$) for any $w \in X \backslash \{v\}$.*

(β) *If $\mathfrak{F}$ is a family of maps $f : A \to X$ such that, for any $x \in A$ with $x \neq f(x)$, there exists a $y \in X \backslash \{x\}$ satisfying $x \preceq y$ (resp. $y \preceq x$), then $\mathfrak{F}$ has a common fixed element $v \in A$, that is, $v = f(v)$ for all $f \in \mathfrak{F}$.*

(γ) *If $\mathfrak{F}$ is a family of maps $f : A \to X$ satisfying $x \preceq f(x)$ (resp. $f(x) \preceq x$) for all $x \in A$ with $x \neq f(x)$, then $\mathfrak{F}$ has a common fixed element $v \in A$, that is, $v = f(v)$ for all $f \in \mathfrak{F}$.*

(δ) *Let $\mathfrak{F}$ be a family of multimaps $F : A \multimap X$ such that, for any $x \in A \backslash F(x)$, there exists $y \in X \backslash \{x\}$ satisfying $x \preceq y$ (resp. $y \preceq x$). Then $\mathfrak{F}$ has a common fixed element $v \in A$, that is, $v \in F(v)$ for all $F \in \mathfrak{F}$.*

(ϵ) *If $\mathfrak{F}$ is a family of multimaps $F : A \multimap X$ such that $x \preceq y$ (resp. $y \preceq x$) holds for any $x \in A$ and any $y \in F(x) \backslash \{x\}$, then $\mathfrak{F}$ has a common stationary element $v \in A$, that is, $\{v\} = F(v)$ for all $F \in \mathfrak{F}$.*

(ζ) *Let $\mathfrak{F}$ be a family of multimaps $F : A \multimap X$ such that, for all $x \in A$ with $F(x) \neq \emptyset$, there exists $y \in X \backslash \{x\}$ satisfying $x \preceq y$ (resp. $y \preceq x$). Then there exists $v \in A$ such that $F(v) = \emptyset$ for all $F \in \mathfrak{F}$.*

(η) *If Y is a subset of X such that, for each $x \in A \backslash Y$, there exists a $z \in X \backslash \{x\}$ satisfying $x \preceq z$ (resp. $z \preceq x$), then there exists a $v \in A \cap Y$.*

Remark 2.1 (1) Note that we claimed that $(\alpha) - (\eta)$ are equivalent in Theorem A and did not say that they are true. For a counterexample, consider the real line $\mathbb{R}$ with the usual order. However, we gave nearly one hundred examples such that they are true based on their original sources; see the articles mentioned in our [25].

(2) All of the elements v's in Theorem A are the same as we have seen in the proof of Metatheorem in [25].

Let $(X, \preceq)$ be a preordered set and $F : X \multimap X$ a multimap. For every $x \in X$, we denote

$$SF(x) := \{z \in X : u \preceq z \text{ for some } u \in F(x)\}.$$

From Theorem A, we have several consequences:

Theorem A1 *Let $(X, \preceq)$ be a preordered set, $F : X \multimap X$ be a map, and $x_0 \in X$ such that $A = (SF(x_0), \preceq)$ has an upper bound $v \in A$. Then the statements $(\alpha) - (\eta)$ of Theorem A for maximal case are equivalent.*

For the identity map $F = 1_X$, we have

$$S(x) := \{y \in X : x \preceq y\}.$$

Then Theorem A1 reduces to the following:

Theorem A2 *Let $(X, \preceq)$ be a preordered set, and $x_0 \in X$ such that $A = (S(x_0), \preceq)$ has an upper bound $v \in A$. Then the statements $(\alpha) - (\eta)$ of Theorem A1 are equivalent.*

Remark 2.2 When $(X, \preceq)$ is a partially ordered set, we obtained more concrete results as follows [25]:

(1) The statements $(\alpha) - (\eta)$ in Theorems A1 and A2 are true.

(2) Moreover, the same things hold for the minimal case of Theorems A1 and A2.

2.3 THE 2022 METATHEOREM

In this section, we introduce the early forms of the current 2023 Meta theorem which we used in 2022 [20–23, 25, 26]:

2.3.1 2022 Metatheorem

Let X be a set, A its nonempty subset, and $G(x, y)$ a sentence formula for $x, y \in X$. Then the following are equivalent:

(i) *[α] There exists an element $v \in A$ such that $G(v, w)$ for any $w \in X \backslash \{v\}$.*

(ii) *[δ1] If $T : A \multimap X$ is a multimap such that, for any $x \in A \backslash T(x)$, there exists a $y \in X \backslash \{x\}$ satisfying $\neg G(x, y)$, then T has a fixed element $v \in A$, that is, $v \in T(v)$.*

(iii) *[γ1] If $f : A \to X$ is a map such that. for any $x \in A$ with $x \neq f(x)$, there exists a $y \in X \backslash \{x\}$ satisfying $\neg G(x, y)$, then f has a fixed element $v \in A$, that is, $v = f(v)$.*

(iv) *[β1] If $f : A \to X$ is a map such that $\neg G(x, f(x))$ for any $x \in A$ with $x \neq f(x)$, then f has a fixed element $v \in A$, that is, $v = f(v)$.*

(v) *[ϵ1] If $T : A \multimap X$ is a multimap such that $\neg G(x, y)$ holds for any $x \in A$ and any $y \in T(x) \backslash \{x\}$, then T has a stationary element $v \in A$, that is, $\{v\} = T(v)$.*

(vi) [γ2] *If* $\mathfrak{F}$ *is a family of maps* $f : A \to X$ *satisfying* $\neg G(x, f(x))$ *for all* $x \in A$ *with* $x \neq f(x)$, *then* F *has a common fixed element* $v \in A$, *that is,* $v = f(v)$ *for all* $f \in \mathfrak{F}$.

(vii) [ϵ2] *If* $\mathfrak{F}$ *is a family of multimaps* $T : A \multimap X$ *such that* $\neg G(x, y)$ *holds for any* $x \in A$ *and any* $y \in T(x) \backslash \{x\}$, *then* $\mathfrak{F}$ *has a common stationary element* $v \in A$, *that is,* $\{v\} = T(v)$ *for all* $T \in \mathfrak{F}$.

(viii) [η] *If* Y *is a subset of* X *such that, for each* $x \in A \backslash Y$, *there exists a* $z \in X \backslash \{x\}$ *satisfying* $\neg G(x, z)$, *then there exists a* $v \in A \cap Y$.

(ix) [ς2] *Let* $\mathfrak{F}$ *be a family of multimaps* $T : A \multimap X$ *such that, for all* $x \in A$ *with* $T(x) \neq \emptyset$, *there exists* $y \in X \backslash \{x\}$ *satisfying* $d(x, y) \leq \varphi(x) - \varphi(y)$. *Then there exists* $v \in A$ *such that* $T(v) = \emptyset$ *for all* $T \in \mathfrak{F}$.

Here we recall some history of our Metatheorem. We show the items of the Metatheorem in each of our previous articles:

[13] in 1985—(i), (ii), (iii), (v) in the 2022 Metatheorem

[14, 15] in 1986, 1987—(i), (ii), (iii), (v), (vi) in the 2022 Metatheorem

[17] in 2000—(i), (ii), (iv), (v), (vi), (viii) in the 2022 Metatheorem

[18]—(i), (ii), (iii), (v), (vi), (vii), (viii) in the 2022 Metatheorem

[19, 20, 21]—(i)-(viii) in the 2022 Metatheorem

[22]—(i)-(ix) in the 2022 Metatheorem

[23, 24, 25, 26, 27]—(α) – (η) in the 2023 Metatheorem

2.4 EXAMPLES OF METATHEOREM

In our previous works in 2022–2023, we gave a large number of examples of our Metatheorems. In this chapter, we give only four examples as follows:

2.4.1 Brézis-Browder in 1976 [3]

Let X be an ordered set and $S(x) = \{y \in X : x \preccurlyeq y\}$ for each $x \in X$. The following is ([3], Corollary 1):

Example 4.1 *Let* $\phi : X \to \mathbb{R}$ *be a function, bounded above, and satisfying*

(1) $x \preccurlyeq y$ *and* $x \neq y$ *imply* $\phi(x) < \phi(y)$,

(4) *for any increasing sequence* $\{x_n\}$ *in* X, *there exists some* $y \in X$ *such that* $x_n \preccurlyeq y$ *for all* n.

Then, for each $a \in X$, *there exists* $\bar{a} \in X$ *such that* $a \preccurlyeq \bar{a}$ *and* $\bar{a}$ *is maximal (i.e.,* $S(\bar{a}) = \{\bar{a}\}$*).*

Applying Theorem A2 or the 2023 Metatheorem, Example 4.1 can be equivalently formulated to a large number of true propositions.

2.4.2 Fierro in 2015 [8]

Let E be a topological vector space with θ as zero. Given a cone P of E, a partial order is defined on E as $x \preccurlyeq y$ iff $y - x \in P$. Assume P is a cone of E such that E is a Riesz space. Additionally, E is assumed order complete (Dedekind), which means that every decreasing bounded from below net has a supremum.

A cone metric space is a pair (X, d), where X is a nonempty set and $d : X \times X \to E$ is a function satisfying the following two conditions: (i) for all $x,\ y \in X,\ d(x, y) = \theta$, iff $x = y$, and (ii) for all $x,\ y,\ z \in X,\ d(x, y) \preccurlyeq d(x, z) + d(y, z)$.

Let $\varphi : X \to E$ be a function. We say φ is lower semicontinuous, iff, for any $\alpha \in E$, the set $\{x \in X : \varphi(x) \preccurlyeq \alpha\}$ is closed. For this function, a Brøndsted type order $\preccurlyeq_\varphi$ is defined on X as follows: $x \preccurlyeq_\varphi y$ iff $d(x, y) \preccurlyeq \varphi(x) - \varphi(y)$.

In the sequel, $\mathcal{LS}(X)$ stands for the space of all lower semicontinuous and bounded below functions from X to E.

After such long preparation, Fierro ([8], Theorem 5) obtained the following extension of the well-known results by Bishop-Phelps lemma:

Example 4.2 *Suppose X is d-complete. Then, for each $\varphi \in \mathcal{LS}(X)$ and $x_0 \in X$, there exists a maximal element $x^* \in X$ such that $x_0 \preccurlyeq_\varphi x^*$.*

Applying Theorem A2 or the 2023 Metatheorem, Example 4.2 can be equivalently formulated to Theorems in [8] such as Theorem 6, Corollary 7, Theorem 8, Corollaries 15 and 16, and possibly more.

2.4.3 Fierro in 2017 [9]

Let $\preceq$ be a preordering on a nonempty set X. For each $x \in X$, we denote $S(x, \preceq) = \{y \in X : x \preceq y\}$.

The following is ([9], Theorem 2.1):

Example 4.3 *Let $x_0 \in X$. The following eight conditions are equivalent:*

(2.1.1) *there exists a maximal element $x^* \in X$ such that $x_0 \preceq x^*$.*

(2.1.2) *there exists $x_1 \in S(x_0, \preceq)$ such that, for each chain C in $S(x_1, \preceq)$, $\bigcap_{x \in C} S(x, \preceq) \neq \emptyset$;*

(2.1.3) *there exist $x_1 \in S(x_0, \preceq)$ and a maximal chain C^* in $S(x_1, \preceq)$ such that $\bigcap_{x \in C^*} S(x, \preceq) \neq \emptyset$;*

(2.1.4) *for each $T : S(x_0, \preceq) \to 2^X$ such that, for each $x \in S(x_0, \preceq) \backslash Tx$, there exists $y \in X \backslash \{x\}$ satisfying $x \preceq y$, there exists $z \in S(x_0, \preceq)$ such that $z \in Tz$;*

(2.1.5) *any function $f : S(x_0, \preceq) \to X$ such that $x \preceq f(x)$ for all $x \in S(x_0, \preceq)$, has a fixed point;*

(2.1.6) *for each $T : S(x_0, \preceq) \to 2^X \backslash \{\emptyset\}$ such that $x \preceq y$ for all $x \in S(x_0, \preceq)$ and $y \in Tx$, there exists $z \in S(x_0, \preceq)$ such that $Tz = \{z\}$;*

(2.1.7) *any family F of functions $f : S(x_0, \preceq) \to X$ such that $x \preceq f(x)$ for all $x \in S(x_0, \preceq)$, has a common fixed point;*

(2.1.8) *for any subset Y of X such that $S(x_0, \preceq) \cap Y = \emptyset$, there exists $x \in S(x_0, \preceq) \backslash Y$. satisfying $S(x, \preceq) = \{x\}$.*

This follows from our 2022 Metatheorem except (2.1.2) and (2.1.3). Note that (2.1.8) is incorrectly stated. Further comments on this example will be given in the next section.

2.4.4 Fierro in 2021 [10]

Let E be a Hausdorff t.v.s. and $\mathcal{U}$ denote the family of all of its balanced neighborhoods of 0. Let $\mathcal{F}$ be a family of subsets of E and $\langle E \rangle$ denote the family of finite subsets of E. We assume the following condition:

(C) For each $U \in \mathcal{U}$, there exist $B \in \mathcal{F}$ and $F \in \langle E \rangle$ such that $B \subset F + U$.

The following is ([10], Theorem 4.1):

Example 4.4 *Let X be a complete subset of E, $x_0 \in X$ and $\preceq$ be a preordering on X. Suppose the following conditions hold:*

(1) *$S(x, \preceq)$ is closed for each $x \in A = S(x_0, \preceq)$, and*

(2) *for each totally ordered subset C of A, $\mathcal{F}_C = \{\overline{S}(x, \preceq) \cap C\}_{x \in C}$ satisfies condition (C).*

Then $(\alpha) - (\eta)$ in Theorem A hold, for example,

(α) there exists a maximal element x^* of X such that $x^* \in A$.

Similar statements for ([10], Theorem 4.2) and Fierro's abstract Caristi theorem ([10], Example 4.1) can be established.

2.5 COMMENTS ON WORKS OF OTHER AUTHORS

In this section, we introduce some works related to Fierro [9].

2.5.1 Fierro [9]

In 2017, Fierro [9] noted that: "Park [17] (in 2000) states five equivalent conditions to maximality with respect to a specific preordering defined on a metric space. In this paper, we prove that these equivalences hold for arbitrary preorderings, without metric considerations, and two additional conditions are added to this set of equivalences."

From the beginning, Fierro made wrong statements since ours in [17] consists of six statements for quasi-metric spaces. Moreover, our earlier versions of Metatheorem in 1985–1987 are simply for sets as in the current ones without any order. Fierro failed to recognize our earlier Metatheorem in [13, 14, 15, 16] in 1985–1987. Note that the main part of his Theorem 2.1 (that is, (2.1.1) and (2.1.4)–(2.1.7)) follows from our previous versions of Theorem A2 or the 2022 Metatheorem (i), (ii), (iv). (v), (vi), respectively, in Section 2. He could not have a chance to see our original Metatheorem and to recognize the uniqueness of the point v in Metatheorem. Note that his condition (2.1.8) is an incorrect statement of our (vi) in [17] originated from Oettli-Théra [12].

Let X be a set and $G(x, y)$ a sentence formula for $x, y \in X$. A *chain* C in X is defined as follows:

(1) C is a nonempty subset of X;

(2) $G(x, x)$ holds for all $x \in C$;

(3) $G(x, y)$ and $G(y, x)$ imply $x = y$; and

(4) for any $x, y \in C$, $G(x, y)$ or $G(y, x)$ holds.

Now we prove the following:

Metatheorem* *Let X be a set, $G(x, y)$ a sentence formula for $x, y \in X$, and $S(x) = \{y \in X : G(x, y)\}$. Let $x_0 \in X$ and $A = S(x_0)$.*

Consider the following:

(α) *There exists an element $v \in A$ such that $G(v, w)$ for any $w \in X \backslash \{v\}$.*

($\theta 1$) *For $v \in A$ and for each chain C in $S(v)$, we have $\bigcap_{x \in C} S(x) \neq \emptyset$.*

($\theta 2$) *For $v \in A$ and a maximal chain C^* in $S(v)$, we have $\bigcap_{x \in C^*} S(x) \neq \emptyset$.*

Then $(\alpha) \Longrightarrow (\theta 1) \Longrightarrow (\theta 2)$.

Proof. $(\alpha) \Longrightarrow (\theta 1)$: By (α), $v \in S(x_0)$ implies $G(x_0, v)$ and $G(v, x_0)$. Hence $v = x_0$ by (3) in the definition of a chain. Now $C = \{v\}$ is the unique chain in $S(v)$, and $\bigcap_{x \in C} S(x) = S(v) \neq \emptyset$, which proves $(\theta 1)$.

$(\theta 1) \Longrightarrow (\theta 2)$: Let v be as in $(\theta 1)$. By Hausdorff maximal principle (which can be established), there exists a maximal chain C^* in $S(v)$ and from $(\theta 1)$, we have $\bigcup_{x \in C^*} S(x) \neq \emptyset$. Thus $(\theta 1)$ implies $(\theta 2)$. □

Let $\preceq$ be a *preorder* on a nonempty set X. For each $x \in X$, we denote

$$S_+(x) = \{y \in X : x \preceq y\}, \quad S_-(x) = \{y \in X : y \preceq x\},$$

and $G(x, y)$ means $y \preceq x$ (resp. $x \preceq y$).

Theorem A2* *Let $(X, \preceq)$ be a partially ordered set, $x_0 \in X$, and $A = S_+(x_0)$ have an upper bound (resp. $A = S_-(x_0)$ have a lower bound). Then the following equivalent statements hold:*

(α) *There exists a maximal (resp. minimal) element $v \in A$ such that $v \npreceq w$ (resp. $w \npreceq v$) for any $w \in X \backslash \{v\}$.*

(θ1) *There exists $v \in A$ such that, for each chain C in $S_+(v)$ (resp. $S_-(v)$) , we have $\bigcap_{x\in C} S_+(x) \neq \emptyset$ (resp. $\bigcap_{x\in C} S_-(x) \neq \emptyset$).*

(θ2) *There exist $v \in A$ and a maximal chain C^* in $S_+(v)$ (resp. $S_-(v)$), we have $\bigcap_{x\in C^*} S_+(x) \neq \emptyset$ (resp. $\bigcap_{x\in C^*} S_-(x) \neq \emptyset$).*

The maximal case was actually proved by Boros-Iqbal-Szás [2].

We introduced Metatheorem* and Theorem A.2* as correct restatements of Theorem 2.1 of Fierro [9] for (2.1.1)–(2.1.3) as follows:

(2.1.1) *there exists a maximal element $x^* \in X$ such that $x_0 \preceq x^*$:*

(2.1.2) *there exists $x_1 \in S(x_0, \preceq)$ such that for each chain C in $S(x_1, \preceq)$, we have $\bigcap_{x\in C} S(x, \preceq) \neq \emptyset$;*

(2.1.3) *there exist $x_1 \in S(x_0, \preceq)$ and a maximal chain C^* in $S(x_1, \preceq)$ such that $\bigcap_{x\in C^*} S(x, \preceq) \neq \emptyset$;*

(2.1.8) *for any subset Y of X such that $S(x_0, \preceq) \cap Y = \emptyset$, there exists $x \in S(x_0, \preceq)\backslash Y$ satisfying $S(x, \preceq) = \{x\}$.*

The statement (2.1.8) is incorrectly stated since $S(x_0, \preceq)$ has a maximal element by (2.1.1) or our original form of the 2022 Metatheorem (i); compare (2.1.8) with our (viii) there.

Fierro [9] also stated that "Due to Corollary 3.4, when (X, d) is a quasi-metric space, Theorem 1 in [17] (by Park in 2000) follows in the more general form. This result is stated as Corollary 3.8."

This statement is incorrect. Anyway, although [9] is informative, it needs serious corrections.

2.5.2 Boros, Iqbal, and Száz [1]

In 2022, Boros, Iqbal, and Szás [1, 2] and Iqbal and Szás [11] applied some of our results on Metatheorem. Boros, Iqbal, and Szás [1, 2] showed some incorrect statements of Fierro [9] and investigated details on preordered sets. This caused our return to Metatheorem after 22 years have passed. In 2022, we found some additional statements and, consequently, we obtained new extended versions of the Metatheorem, and their final version was called the 2023 Metatheorem in [25].

We begin with the following in [1]:

ABSTRACT. "We intensively investigate a very particular situation when

$$X = \{x \in \mathbb{R}^2 : x_1 + x_2 \leq 0\};$$
$$\varphi(x) = x_1 + x_2, \qquad f(x) = (x_1 + 1,\ x_2 - 1);$$
$$S(x) = \{y \in X : \varphi(x) \leq \varphi(y)$$

for all $x \in X$.

This example shows, in particular, that a maximality theorem published by Raúl Fierro in 2017 is not true without assuming the antisymmetry of the corresponding preorder.

A true particular case of this theorem improves and supplements a former similar theorem of Sehie Park, and has to be proved just after Zorn's lemma and a maximality principle of H. Brézis and F. Browder."

In [1], "Raúl Fierro tried to prove the following generalization of a theorem of Park [17] (in 2000)". The authors stated Theorem 2.1 of Fierro (Example 4.3 of our present chapter).

Then they gave the aforementioned example in Abstract, "this example will show, in particular, that implication (2.1.3) $\Longrightarrow$ (2.1.4) (in Theorem 2.1 of Fierro in 2017) is not true without assuming the antisymmetry of the relation $\preceq$.

A relational improvement of a true particular case of the theorem can be found in a subsequent paper [2], where the curious assertion (2.1.8) will also be reformulated.

This improvement generalizes and supplements a former similar theorem of Park [[17], Theorem 1]. (See also [14, 15] for some more general settings.)"

2.5.3 Boros, Iqbal, and Száz [2]

"In this paper, by using relational notations, we improve and supplement a true particular case of an inaccurate maximality theorem of Rául Fierro from 2017, which has to be proved in addition to Zorn's lemma and a famous maximality principle of H. Brézis and F. Browder."

"Fierro tried to prove the following closely related maximality theorem by generalizing and supplementing a similar theorem of Park ([18], Theorem 1). (See also [14, 15] for some more general settings.)"

Let $\preceq$ be a *partial order* on a nonempty set X, i.e. $\preceq$ is reflexive, antisymmetry and transitive.

To have a true, improved particular case of Theorem 1.2 of Fierro, suggested by ([17], Theorem 1) of Park, Boros-Iqbal- Száz [2] could prove ([2], Theorem 5.1) as follows:

Theorem 5.1 *If S is a partial order on a nonvoid set X, then for each $a \in X$ the following assertions are equivalent:*

(1) *there exists $b \in S(a)$ such that b is a strongly maximal element of $X(S)$;*

(2) *there exists $b \in S(a)$ such that for each chain C in $S(b)$ we have $\bigcap_{x \in C} S(x) \neq \emptyset$;*

(3) *there exist $b \in S(a)$ and a maximal chain C in $S(b)$ such that $\bigcap_{x \in C} S(x) \neq \emptyset$;*

(4) *for every relation T on X such that $S(x) \setminus \{x\} \neq \emptyset$ for all $x \in X$ with $x \in S(a) \setminus T(x)$, there exists $b \in S(a)$ such that b is a fixed point of T;*

(5) *if $S(x) \setminus \{x\} \neq \emptyset$ for all $x \in S(a)$, then for every relation T on X there exists $b \in S(a)$ such that b is a fixed point of T;*

(6) *for every extensive function f of X to itself there exists $b \in S(a)$ such that b is a fixed point of f;*

(7) *for every non-partial relation T on X such that, $y \in S(x)$ for all $x \in S(a)$ and $y \in T(x)\backslash\{x\}$, there exists $b \in S(a)$ such that b is a strong fixed point of T;*

(8) *for every nonvoid family $\mathcal{F}$ of extensive functions of X to itself there exists $b \in S(a)$ such that b is a fixed point of each element f of $\mathcal{F}$;*

(9) *for any $Y \subseteq X$, such that $S(x)\backslash\{x\} \neq \emptyset$ for all $x \in S(a)\backslash Y \neq \emptyset$, we have $S(a) \cap Y \neq \emptyset$.*

We show that this follows from our Metatheorems and the new Metatheorem*:

(1) — (i) = [α] in the 2022 Metatheorem

(2) — (2) in Metatheorem*

(3) — (3) in Metatheorem*

(4) — (ii) = [δ1] in the 2022 Metatheorem

(5) — A variant of (4)

(6) — (iv) = [β1] in the 2022 Metatheorem

(7) — (v) = [ϵ1] in the 2022 Metatheorem

(8) — (vi) = [γ1] in the 2022 Metatheorem

(9) — (viii) = [η] in the 2022 Metatheorem

Let $G(x, y)$ means $y \preceq x$. Then Theorem 5.1 can be improved as follows by our usage of terminology:

Theorem 5.1* *Let $(X, \preceq)$ be a partially ordered set, and A its nonempty subset. Then the following statements are equivalent:*

(1) *There exists an element $v \in A$ such that $v \npreceq w$ for any $w \in X\backslash\{v\}$.*

(2) *There exists $v \in A$ such that, for each chain C in $S(v)$, we have $\bigcap_{x \in C} S(x) \neq \emptyset$.*

(3) *There exist $v \in A$ and a maximal chain C^* in $S(v)$, we have $\bigcap_{x \in C^*} S(x) \neq \emptyset$;*

(4) *for every $T : X \multimap X$ such that $S(x)\backslash\{x\} \neq \emptyset$ for all $x \in X$ with $x \in A\backslash T(x)$, there exists $v \in A$ such that v is a fixed point of T;*

(5) *if $S(x)\backslash\{x\} \neq \emptyset$ for all $x \in A$, then for every multimap $T : X \multimap X$ there exists $v \in A$ such that v is a fixed point of T;*

(6) *for every progressive map $f : X \to X$ there exists $v \in A$ such that v is a fixed point of f.*

Proof. (1) $\Longrightarrow$ (2) $\Longrightarrow$ (3): Theorem A2* or Metatheorem*.

(3) $\Longrightarrow$ (4) $\Longrightarrow$ (5) $\Longrightarrow$ (6): ([2], p.8).

(6) $\Longrightarrow$ (1): the 2022 Metatheorem. □

From the 2023 Metatheorem, we can add up more equivalent statements to Theorem 4.1* and Theorem 5.1 of [2]. Note that only the implication (3) $\Longrightarrow$ (4) requires the antisymmetricity of partially ordered sets as Boros et al. [1, 2] used to claim.

In the previous two papers, the authors relied on unusually very detailed arguments in the order theory. This way is also followed by the following article.

2.5.4 Iqbal and Száz [11]

In this paper, the maximality principles of Brézis and Browder [3] in 1976 slightly generalized and improved. The authors made 15 critical remarks on [3], in several respects, not completely satisfactory for them. Moreover, they seem to establish a kind of relation theory. Among the contents of [3], consider the following:

Theorem 5.3 [11] *For an element x of $X(S)$, the following assertions are equivalent:*

(1) *x is a strong fixed point of S;*

(2) *x is a strongly maximal element of $X(S)$.*

Note that this corresponds to the equality of v in our Metatheorem as maximal elements, collectively fixed points, collectively stationary points, and other critical points.

Similarly, Corollary 8.9, Theorem 8.10, Corollary 8.11, and others in [11] can be regarded as variants of statements in our Metatheorem.

2.6 OUR RESOLUTION AND NEW CARISTI THEOREM

In a recent work, we obtained the following practical form of Theorems A2 and 5.1*:

Theorem 6.1 *Let $(X, \preccurlyeq)$ be a partially ordered set, $x_0 \in X$, and $A = S_+(x_0) = \{y \in X : x_0 \preccurlyeq y\}$ (resp. $A = S_-(x_0) = \{y \in X : y \preccurlyeq x_0\}$) have an upper bound (resp. a lower bound) $v \in A$.*

Then the equivalent statements $(\alpha) - (\eta)$ of Theorem A and $(\theta 1, 2)$ of Theorem A2 hold:*

(α) *$v \in A$ is a maximal (resp. minimal) element, that is, $v \not\preccurlyeq w$ (resp. $w \not\preccurlyeq v$) for any $w \in X \backslash \{v\}$.*

(θ1) *There exists $v \in A$ such that, for each chain C in $S_+(v)$ (resp. $S_-(v)$), we have $\bigcap_{x \in C} S_+(x) \neq \emptyset$ (resp. $\bigcap_{x \in C} S_-(x) \neq \emptyset$).*

(θ2) *There exist $v \in A$ and a maximal chain C^* in $S_+(v)$ (resp. $S_-(v)$), we have $\bigcap_{x \in C^*} S_+(x) \neq \emptyset$ (resp. $\bigcap_{x \in C^*} S_-(x) \neq \emptyset$).*

Proof. (α): Since A has an upper bound $v \in A$, for each $x \in A$, we have $x_0 \preccurlyeq x \preccurlyeq v$. If $v \preccurlyeq w$ for some $w \in X$, then $w \in S(x_0) = A$ and $w \preccurlyeq v$. Since $(X, \preccurlyeq)$ is partially ordered, we have $w = v$. Hence v is maximal.

The equivalency of (α) – ($\theta 2$) is routine from Theorem A2*. □

In view of this theorem, we have the following conclusion:

(1) Fierro's Theorem 2.1 (Example 4.3 in this chapter) holds for partially ordered sets except (2.1.8), whose correct form is originated from Oettli-Théra [12] and appeared in Park [17]. Fierro began with our [17], but could not recognize this fact. Moreover, he could not realize that the maximal point, collectively fixed point, collectively stationary point, and others in our Metatheorem are same.

(2) Theorem 5.1 of Boros, Iqbal, and Száz [2] states only equivalency of (1)–(9) for partially ordered sets. This can be improved as our Theorems 5.1* and 6.1. Moreover, such equivalency has far-reaching extensions as our 2023 Metatheorem.

This article originates from statements (2.1.2) and (2.1.3) of Fierro. But no other one practically applied them yet. Here we show that an elementary proof can be given to an extension of the Caristi fixed point theorem by (2.1.2).

In 2002, Chen-Cho-Yang [5] introduced the following concept of lower semicontinuity from above:

Definition 6.2 [5] Let X be a metric space. A function $f : X \to \mathbb{R} \cup \{+\infty\}$ is said to be *lower semicontinuous from above* at a point $x \in X$ if $x_n \to x$ as $n \to \infty$ and $f(x_1) \geq f(x_2) \geq \cdots \geq f(x_n) \geq \cdots$ imply that $f(x) \leq \lim_{n\to\infty} f(x_n)$.

Recall the following in [5]:

Theorem 6.3 (Caristi's Fixed Point Theorem) *Let (D, d) be a complete metric space and a function $\phi : D \to \mathbb{R}^+$ be lower semicontinuous from above. Suppose that a mapping $f : D \to D$ satisfies the following:*

$$d(x, f(x)) \leq \phi(x) - \phi(f(x)) \quad \textit{for all } x \in D.$$

Then there exists $x_0 \in D$ such that $f(x_0) = x_0$.

Note that (D, d) can be made into a partially ordered set by defining

$$x \preccurlyeq y \iff \phi(y) \leq \phi(x)$$

for $x,\ y \in D$.

Proof. Since $\phi : D \to \mathbb{R}^+$ is l.s.c. from below at any z, for any $\{x_n\}$ converging to z such that

$$\phi(x_1) \geq \phi(x_2) \geq \cdots \geq \phi(x_n) \geq \cdots \Longrightarrow \lim_{n\to\infty} \phi(x_n) \geq \phi(z)$$

and hence $x_1 \preccurlyeq x_2 \preccurlyeq \cdots \preccurlyeq x_n \preccurlyeq \cdots \preccurlyeq z$. Note that $C = \{z\} \subset S(x_1)$ is a chain in $S(z)$. Let $v = z \in C$. Then $C = \{v\} \subset \bigcup_{x \in C} S(x) \neq \emptyset$. Hence, in Theorem A2*, $(\theta 1)$ holds, v is maximal by (α), and our Caristi theorem holds by Theorem A2(γ). □

2.7 EPILOGUE

As we have seen in Metatheorem, the maximal element v plays various roles as fixed point, stationary point, common fixed point, common stationary point, etc. Almost all other authors seem to be not recognized this fact yet.

In this article, we extend our earlier Metatheorem and Theorem A by adding more equivalent statements. We showed that the maximal elements in certain preordered sets can be reformulated to fixed points or stationary points of maps or multimaps and to common fixed points or common stationary points of a family of maps or multimaps, and conversely. Actually such points are same as we have seen in the proof of Metatheorem. Therefore, if we have a theorem on any of such points, it can be converted automatically to almost twenty equivalent theorems on other types of points without any serious argument.

In many fields of mathematical sciences, there are a plentiful number of theorems concerning maximal points or various fixed points that can be applicable to our Metatheorem. Some of such theorems can be seen in our previous works and the present article. Therefore, a metatheorem like Theorem A is a machine to expand our knowledge easily. In this article, we presented relatively old and new such examples.

BIBLIOGRAPHY

[1] Z. Boros, M. Iqbal, A. Száz, *An instructive counterexample to a maximality theorem of Raúl Fierro* (2022), manuscript.

[2] Z. Boros, M. Iqbal, A. Száz, *A relational improvement of a true particular case of Fierro's maximality theorem* (2022), manuscript.

[3] H. Brézis, F.E. Browder, *A general principle on ordered sets in nonlinear functional analysis*, Adv. Math. 21 (1976), 355–364.

[4] J. Caristi, *Fixed point theorems for mappings satisfying inwardness conditions*, Trans. Amer. Math. Soc. 215 (1976), 241–251.

[5] Y. Chen, Y.J. Cho, L. Yang, *Note on the results with lower semicontinuity*, Bull. Kor. Math. Soc. 39 (2002), 535–541.

[6] I. Ekeland, *Sur les problèmes variationnels*, C.R. Acad. Sci. Paris 275 (1972), 1057–1059; 276 (1973), 1347–1348.

[7] I. Ekeland, *On the variational principle*, J. Math. Anal. Appl. 47 (1974), 324–353.

[8] R. Fierro, *Fixed point theorems for set-valued mappings on TVS-cone metric spaces*, Fixed Point Theory and Appl. 2015 (2015), 221. DOI 10.1186/s13663-015-0468-1

[9] R. Fierro, *Maximality, fixed points and variational principles for mappings on quasi-uniform spaces*, Filomat 31:16 (2017), 5345–5355. DOI: 10.2298/FIL1716345F

[10] R. Fierro, *An intersection theorem for topological vector spaces and applications*, J. Optim. Theory Appl. 191 (2021), 118–133. DOI: 10.1007/s10957-021-01927-7

[11] M. Iqbal, A. Száz, *An instructive treatment of the Brézis-Browder ordering and maximality principles* (2022), manuscript.

[12] W. Oettli, M. Thera, *Equivalents of Ekeland's principle*, Bull. Austral. Math. Soc. 48 (1993), 385–392.

[13] S. Park, *Some applications of Ekeland's variational principle to fixed point theory*, Approximation Theory and Applications (S.P. Singh, ed.), Pitman Res. Notes Math. 133 (1985), 59–172.

[14] S. Park, *Countable compactness, l.s.c. functions, and fixed points*, J. Korean Math. Soc. 23 (1986), 61–66.

[15] S. Park, *Equivalent formulations of Zorniś lemma and other maximmumm principles*, J. Korean Soc. Math. Edu. 25 (1986), 19–24.

[16] S. Park, *Partial orders and metric completeness*, Proc. Coll. Natur. Sci. SNU 12 (1987), 11–17.

[17] S. Park, *On generalizations of the Ekeland-type variational principles*, Nonlinear Anal. 39 (2000), 881–889.

[18] S. Park, *Equivalents of various maximum principles*, Results Nonlinear Anal. 5(2) (2022), 169–174.

[19] S. Park, *Applications of various maximum principles*, J. Fixed Point Theory 2022–2023, 1–23. ISSN:2052–5338.

[20] S. Park, *Equivalents of maximum principles for several spaces*, Top. Algebra Appl. 10 (2022), 68–76. 10.1515/taa-2022-0113

[21] S. Park, *Equivalents of ordered fixed point theorems of Kirk, Caristi, Nadler, Banach, and others*, Adv. Th. Nonlinear Anal. Appl. 6(4) (2022), 420–432.

[22] S. Park, *On the order-theoretic Cantor theorem of Granas and Horvath*, to appear. DOI: 10.13140/RG.2.2.28367.76962

[23] S. Park, *Applications of generalized Zorn's Lemma*, J. Nonlinear Anal. Optim. 13(2) (2022), 75–84. ISSN: 1906-9605. DOI: 10.13140/RG.2.2.26690.04801

[24] S. Park, *Generalizations of the Tarski type fixed point theorems*, Nonlinear Convex Anal. Optim. 1(2) (2022) DOI: 10.13140/RG.2.2.20462.48965

[25] S. Park, *Foundations of ordered fixed point theory*, J. Nat. Acad. Sci., ROK, Nat. Sci. Ser. 61(2) (2022), 41pp.

[26] S. Park, *Applications of several minimum principles*, Adv. Th. Nonlinear Anal. Appl. 7(1) (2023), 52–60. ISSN: 2567-2648.s

[27] S. Park, *Equivalents of certain minimum principles*, DOI: 10.13140/RG.2.2.19449.95843

CHAPTER 3

On Wardowski Type Results in the Framework of G-Metric Spaces

Jelena Vujaković

University of Pristina-Kosovska Mitrovica

Slobodanka Mitrović and Stojan Radenović

University of Belgrade

Zoran D. Mitrović

University of Banja Luka

CONTENTS

3.1 INTRODUCTION AND PRELIMINARIES

This year marks exactly 100 years since Stefan Banach proved his famous theorem on the fixed point of contractile mapping in a complete metric space. His result has become very popular among thousands of researchers in various fields of mathematics (see [22]–[34]). Since 1922, many mathematicians have tried to generalize Banach's famous result. These attempts took place in three main directions.

(1) The authors corrupted the axioms of the standard metric space (non-negativity of the metric $d : X \times X \to \mathbb{R}^{+}$; $d(x, y) = 0$ if and only if $x = y$; symmetry $d(x, y) = d(y, x)$ for all $x, y \in X$; and relation of the triangle $d(x, z) \leq d(x, y) + d(y, z)$ for all $x, y, z \in X$).

(2) Instead of the right side in the formulation of Banach contractive condition

$$d(Tx, Ty) \leq \lambda \cdot d(x, y) \tag{3.1}$$

of mapping $T : X \to X$, some authors take for example $\phi(d(x, y)) = (\phi \circ d)(x, y)$ where $\phi : \mathbb{R}^{+} \to \mathbb{R}^{+}$.

DOI: 10.1201/9781003388678-3

(3) In the formulation of Banach contraction principle instead for all $x, y \in X$ they write for all $(x, y) \in R$ where R is some subset of $X \times X$.

Using approach (1) various types of generalized standard metric spaces have emerged such as partial metric space, metric-like space, b-metric space, partial b-metric space, b-metric like spaces, S-metric space, S_b-metric space, G-metric space, G_b-metric space, and in the last few years, various researchers-mathematicians have paid great attention to the mentioned types of obtained spaces. In the previously mentioned types of generalizations of metric spaces, mappings from the product X^2 in $\mathbb{R}^+$ were used, i.e, from X^3 in $\mathbb{R}^+$. With the approach (2) we get well-known fixed point results in literature as for example Meir-Keeler, F. Browder, Boyd-Wong type results. The case (3) gives us some new generalizations of famous Banach contraction theorem as Ran-Reuring fixed point theorem in partially ordered sets ([40]).

The special generalization of Banach's result, which was introduced in 2012 by the Polish mathematician Darius Wardowski [33]. This generalization is of the second direction.

Namely, Wardowski considered mappings $F : (0, +\infty) \to \mathbb{R}$ that satisfy the following conditions:

(**F1**): F is a strictly increasing function;

(**F2**): A sequence $x_n \in (0, +\infty)$ converges to zero if and only if $F(x_n) \to -\infty$ as $n \to +\infty$;

(**F3**): $\lim_{x \to 0^+} x^k F(x) = 0$ for some $k \in (0, 1)$.

D.Wardowski denoted by $\mathcal{F}$ the collection of mappings $F : (0, +\infty) \to \mathbb{R}$ that satisfy the conditions (**F1**), (**F2**) and (**F3**). Using such functions, he introduced a new type of contraction in a given metric space in the following way:

Definition 3.1 *Let $F \in \mathcal{F}$ and let T be a mapping from a metric space (X, d) into itself. If there is a positive number τ such that for all $x, y \in X$ for which $d(Tx, Ty) > 0$,*

$$\tau + F(d(Tx, Ty)) \leq F(d(x, y)) \tag{3.2}$$

holds, then the mapping T is called an F-contraction.

The main result of D. Wardowski was the following.

Theorem 3.1 *Each F-contraction T on a complete metric space (X, d) has a unique fixed point. Moreover, for each $x \in X$, the corresponding Picard sequence $\{T^n x\}_{n \in \mathbb{N}}$ converges to that fixed point.*

Obviously, taking $F(x) = logx$ and $\tau = log\frac{1}{\lambda}, \lambda \in (0, 1)$ the condition (3.2) reduces to (3.1), i.e., Theorem 3.1. is a generalization of Banach's famous result from 1922.

This nice result inspired dozens of mathematicians to try to obtain new results by: applying similar ideas in various other spaces (for more details, see example [12]);

modifying the contractive condition (3.2) in various ways; or modifying conditions (**F1**)–(**F3**) for the function F.

Next we list some properties of a function F that follow just from property (**F1**):

1. F is continuous almost everywhere.
2. At each point $r \in (0, +\infty)$ there exist its left and right limits $\lim_{x \to r^-} F(r) = F(r^-)$ and $\lim_{x \to r^+} F(r) = F(r^+)$. Moreover, for the function F one of the following two properties holds: $F(0^+) = \mu \in \mathbb{R}$ or $F(0^+) = -\infty$.

Property (**F2**) is equivalent to
(**F2'**): $F(0^+) = -\infty$, as well as to
(**F2"**): $\inf_{x \in (0,+\infty)} F(x) = -\infty$.
For more details see [12].

Remark 3.1 *Note that the previous case (3.2) can only take place if $F(0^+) = -\infty$. Indeed, if $d \to 0^+$, then it follows from (3.2) that $\tau + F(0^+) \leq F(0^+)$, which is impossible if $F(0^+)$ is finite and F is a strictly increasing function. It means that condition (**F2**) for the function F is implicitly contained in the formulation of Theorem 3.1. In other words, there is no mapping T which satisfies Wardowski's condition (3.2) with function F satisfying (**F1**) and not satisfying (**F2**).*

In order to be able to consider F-contractions within G-metric spaces, let us first state some known things about them. Now, we present the necessary definitions and results in G-metric spaces, which will be useful for the rest of the paper. However, for more details, we refer to ([22]–[6]).

Definition 3.2 *Let X be a nonempty set. Suppose that $G : X^3 \to [0, +\infty)$ is a function satisfying the following conditions:*

(**G1**) $G(x, y, z) = 0$ if $x = y = z$;

(**G2**) $0 < G(x, x, y)$ for all $x, y \in X$ with $x \neq y$

(**G3**) $G(x, x, y) \leq G(x, y, z)$ for all $x, y, z \in X$ with $y \neq z$;

(**G4**) $G(x, y, z) = G(y, x, z) = G(x, z, y) = G(z, x, y) = G(z, y, x) = G(y, z, x)'$

(**G5**) $G(x, y, z) \leq G(x, a, a) + G(a, y, z)$ for all $x, y, z, a \in X$.

Then G is called a G-metric on X and (X, G) is called a G-metric space.

We note that from $G(x, y, z) = 0$ it follows that $x = y = z$. Actually, if for example x is different from y for two elements from X, then from (**G3**) follows $G(x, x, y) = 0$ what is according to the (**G2**) a contradiction.

Definition 3.3 *A G-metric space (X, G) is said to be symmetric if $G(x, y, y) = G(y, x, x)$ for all $x, y \in X$.*

Definition 3.4 *Let (X, G) be a G-metric space. We say that $\{x_n\}$is*

(1) a G-Cauchy sequence if, for any $\varepsilon > 0$, there is $n_0 \in \mathbb{N}$ such that for all $n, m, l \geq n_0, G(x_n, x_m, x_l) < \varepsilon$;

(2) a G-convergent sequence to $x \in X$ if, for any $\varepsilon > 0$, there is $n_0 \in \mathbb{N}$ such that for all $n, m \geq n_0, G(x, x_n, x_m) < \varepsilon$.

A G-metric space (X, G) is said to be $G-$complete if every G-Cauchy sequence in X is G-convergent in X.

Proposition 3.1 *Let (X, G) be a G-metric space. The following are equivalent*

(1) $\{x_n\}$ is G-convergent to x;
(2) $G(x_n, x_n, x) \to 0$ as $n \to +\infty$;
(3) $G(x_n, x, x) \to 0$ as $n \to +\infty$;
(4) $G(x_n, x_m, x) \to 0$ as $n, m \to +\infty$.

Proposition 3.2 *Let (X, G) be a G-metric space. The following are equivalent:*

(1) the sequence $\{x_n\}$ is G-Cauchy;
(2) $G(x_n, x_m, x_m) \to 0$ as $n, m \to +\infty$.

An interesting observation is that any G-metric of space (X, G) induces a metric d_G on X given by

$$d_G(x, y) = G(x, y, y) + G(y, x, x), \text{ for all } x, y \in X. \tag{3.3}$$

Moreover, (X, G) is G-complete if and only if (X, d_G) is complete.
It was observed that in the symmetric case ((X, G) is symmetric), many fixed point theorems on G-metric spaces are particular cases of existing fixed point theorems in metric spaces.

As in the context of metric spaces ([12], [29]) we recall the following two lemmas that we will use in the proofs of our main results. These both lemmas are important and are used to prove the Cauchyness of the sequence $x_n = fx_{n-1}, n \in \mathbb{N}$.

Lemma 3.1 *Let $\{x_n\}$ be a Picard sequence in G-metric space (X, G) such that*

$$G(x_{n+1}, x_{n+1}, x_n) < G(x_n, x_n, x_{n-1}), \tag{3.4}$$

for all $n \in \mathbb{N}$. Then $x_n \neq x_m$ whenever $n \neq m$.

Proof. Suppose $x_n = x_m$ for some two n, m from $\mathbb{N}$ with $n < m$. Then because $x_{n+1} = fx_n = fx_m = x_{m+1}$ we have

$$\begin{aligned} G(x_{n+1}, x_{n+1}, x_n) = G(x_{m+1}, x_{m+1}, x_m) < G(x_m, x_m, x_{m-1}) < \cdots \\ < G(x_{n+2}, x_{n+2}, x_{n+1}) < G(x_{n+1}, x_{n+1}, x_n), \end{aligned} \tag{3.5}$$

which is a contradiction. □

Lemma 3.2 *[22] Let (X,G) be a G-metric space and let $\{x_n\}$ be a sequence in X such that $\{G(x_n,x_n,x_{n+1})\}$ is non-increasing and $\lim_{n\to+\infty} G(x_n,x_n,x_{n+1})=0$. If $\{x_n\}$ is not a G-Cauchy sequence in (X,G), then there exist $\varepsilon>0$ and two sequences $\{m_k\}$ and $\{n_k\}$ of positive integers such that $n_k>m_k>k$ and the following sequences tend to ε^+ when $k\to+\infty$:*

$$G(x_{2m_k},x_{2n_k},x_{2n_k}),G(x_{2m_k},x_{2n_k-1},x_{2n_k-1}),G(x_{2m_k+1},x_{2n_k},x_{2n_k}),$$

$$G(x_{2m_k-1},x_{2n_k+1},x_{2n_k+1}),G(x_{2m_k+1},x_{2n_k+1},x_{2n_k+1}),\dots. \tag{3.6}$$

Remark 3.2 *This Lemma is true without the assumption that the sequence $\{G(x_n,x_n,x_{n+1})\}$ is decreasing. Then we have the following sequences*

$$G(x_{m_k},x_{n_k},x_{n_k}),G(x_{m_k},x_{n_k-1},x_{n_k-1}),G(x_{m_k+1},x_{n_k},x_{n_k}),$$

$$G(x_{m_k-1},x_{n_k+1},x_{n_k+1}),G(x_{m_k+1},x_{n_k+1},x_{n_k+1}),\dots \tag{3.7}$$

tend to ε^+ when $k\to+$.

We now list a few characteristic examples of G-metric spaces:

(1) [3] Let (X,d) be any metric space. Let us define the mapping $G:X^3\to[0,+\infty)$ as

$$G(x,y,z)=d(x,y)+d(y,z)+d(z,x). \tag{3.8}$$

Then it is easy to check that (X,G) is a symmetric G-metric space, where $G(x,x,y)=G(x,y,y)=2d(x,y)$.

(2) [3] If (X,G) is a given $G-$metric space, then with $G_1(x,y,z)=\frac{G(x,y,z)}{1+G(x,y,z)}$ new G-metric space is given. If G is symmetric, then so is G_1.

(3) [26] Let $X=\{a,b\}$ where $a\neq b$ and let $G:X^3\to[0,+\infty)$ be the function defined as:

$G(a,a,a)=G(b,b,b)=0, G(a,a,b)=G(a,b,a)=G(b,a,a)=1, G(a,b,b)=G(b,a,b)=G(b,b,a)=2$. Then G is G-metric on X, but it is not symmetric because $G(a,a,b)=1\neq 2=G(a,b,b)$.

3.2 MAIN RESULTS

Recently, Kumar and Aurora [14] introduced two different types of mappings from $(0,+\infty)$ to $\mathbb{R}$ called S_G and M_G mappings.

They denoted with S_G the class of all functions $F:(0,+\infty)\to\mathbb{R}$ such that

- **(S1)** F is strictly increasing, that means $x<y$ implies that $Fx<Fy$, where x,y are positive real.
- **(S2)** $\lim_{n\to+\infty} a_n=0$ if and only if $\lim_{n\to+\infty} F(a_n)=-\infty$, for every sequence $\{a_n\}$of positive numbers.
- **(S3)** F is continuous on $(0,+\infty)$.

Further with M_G is notation for class of all maps $F:(0,+\infty)\rightarrow\mathbb{R}$ such that

(M1) F is strictly increasing, that means $x<y$ implies that $Fx<Fy$, where x,y are positive real.

(M2) $\lim_{n\rightarrow+\infty}a_n=0$ if and only if $\lim_{n\rightarrow+\infty}F(a_n)=-\infty$, for every sequence $\{a_n\}$of positive numbers..

(M3) There exists $m\in(0,1)$ such that $\lim_{a\rightarrow 0+}a^m F(a)=0$.

In the same paper, they defined modified F-contractions of type S_G and M_G as follows:

Definition 3.5 *[14] Let (X,G) be a G-metric space and $f:X.\rightarrow X$ be a mapping. Then, f is known as modified generalized F-contraction of type S_G (resp.M_G), if there is $F\in S_G$ (resp.M_G) and $\lambda>0$ such that $G(fx,fy,fz)>0$, then*

$$\lambda+F(G(fx,fy,fz))\leq F(S_f(x,y,z)), \tag{3.9}$$

where

$$\begin{aligned} S_f(x,y,z) &= \max\{G(x,fy,fy),G(x,fz,fz),G(y,fx,fx),G(y,fz,fz),\\ &\quad G(z,fx,fx),G(z,fy,fy)\}. \end{aligned}$$

Then they formulated and proved their two theorems in the framework of G-metric spaces as follows:

Theorem 3.2 *([14], Theorem 2.4. resp. 2.5.) Let (X,G) be a $G-$complete G-metric space and $f:X\rightarrow X$ be a (resp. continuous) modified generalized F-contraction of type S_G (resp. M_G). Then f has a unique fixed point $u\in X$ and the sequence $\{f^n(x_0)\}$, where $n\in\mathbb{N}$, converges to u for each $u\in X$.*

To prove both theorems, the authors used **(F1)**, **(F2)** and the continuity of the mappings F. It is easy to see that the property **(F2)** is not necessary. And the existence of a unique fixed point is possible if (X,G) is a symmetric $G-$metric space (which is not given in the assumption of their theorems).

Note that the two formulated theorems on F-contractions within G-metric spaces within G-metric spaces differ only in the property of the function F. A careful consideration concludes that parts of the proof of the first and then the second theorem are doubtful. The problem is in defining the set S_f. In this work, we improved that and obtained natural results about F-contractions within G-metric spaces. In our work, we define S_f as:

$$S_f(x,y,z)=\max\{G(x,y,z),G(x,fx,fx),G(y,fy,fy),G(z,fz,fz)\}. \tag{3.10}$$

Therefore, in this part of this chapter, we will first reformulate the statement of Theorem 2.4 (resp. 2.5.) from [14] more precisely and give its proof. In our approach, we will not mention the types of S_G and M_G mappings $F:(0,+\infty)\rightarrow\mathbb{R}$, but only the properties that the function F should fulfill.

Theorem 3.3 *(Our version of the formulations): Let (X, G) be a $G-$complete G-metric space and f a mapping from X to itself where F satisfies only the condition (**F1**). If one of the mappings F, f is continuous then f has a unique fixed point and for every $x \in X$ the corresponding Picard sequence $\{f^n x\}_{n\in\mathbb{N}}$ converges to that unique fixed point.*

Proof. First, let x_0 be an arbitrary point in X. Let us further from the corresponding Picard sequence generated by the starting point x_0. Then $x_n = f(x_{n-1})$ for every $n \in \mathbb{N}$. If for some $p \in \mathbb{N}, x_p = x_{p-1}$ then x_{p-1} is a fixed point of f and the proof of the first part of the theorem is completed. Now suppose that x_n is a different from x_{n-1} for every $n \in \mathbb{N}$. In that case, we put in the given contractive condition (2.1) $x = x_{n-1}, y = z = x_n$ which is possible because then $G(fx, fy, fz) = G(x_n, x_{n+1}, x_{n+1}) > 0$ and then according to the (**G2**) we get

$$\tau + F(G(x_n, x_{n+1}, x_{n+1})) \leq F(S_f(x_{n-1}, x_n, x_n)). \tag{3.11}$$

By carefully calculating the right side of the inequality (2.3), we get

$$\begin{aligned} S_f(x_{n-1}, x_n, .x_n) &= \max\{G(x_{n-1}, x_n, x_n), G(x_{n-1}, x_n, x_n), \\ &\quad G(x_n, x_{n+1}, x_{n+1}), G(x_n, x_{n+1}, x_{n+1})\} \\ &= \max\{G(x_{n-1}, x_n, x_n), G(x_n, x_{n+1}, x_{n+1})\} \end{aligned}$$

that is., we get that

$$\tau + F(G(x_n, x_{n+1}, x_{n+1})) \leq F(G(x_{n-1}, x_n, x_n)), \tag{3.12}$$

since $\max\{G(x_{n-1}, x_n, x_n), G(x_n, x_{n+1}, x_{n+1})\} = G(x_{n-1}, x_n, x_n)$. Otherwise, if $\max\{G(x_{n-1}, x_n, x_n), G(x_n, x_{n+1}, x_{n+1})\} = G(x_n, x_{n+1}, x_{n+1})$ we get a contradiction with $\tau > 0$.

Whence from the previous relation (2.4) it follows $G(x_n, x_{n+1}, x_{n+1}) < G(x_{n-1}, x_n, x_n)$. This shows that there is a limit value of the sequence $G(x_n, x_{n+1}, x_{n+1})$ which is non-negative. By taking the limit in (2.4) when n tends $+\infty$ and using the property (**F1**) of the mapping F we obtain

$$\tau + F(\delta^+) \leq F(\delta^+) \tag{3.13}$$

where $\delta = \lim_{n\to+\infty} G(x_n, x_{n+1}, x_{n+1}) \geq 0$. Now, we see that (2.5) represents a contradiction with $\tau > 0$. This means that $\lim_{n\to+\infty} G(x_n, x_{n+1}, x_{n+1}) = \delta = 0$. From the strict decreasing of the sequence $G(x_n, x_{n+1}, x_{n+1})$ and the Lemma 3.4. we obtain that all members of the sequence $\{x_n\}$ are mutually different.

Now, to show that the sequence $\{x_n\}$ is G-Cauchy, we can apply Lemma 3.5. (similar to the works on metric spaces) putting in the contractive condition $x = x_{2m_k}, y = z = x_{2n_k}$ (because this is possible due to the difference of x with $y = z$). Thus we get

$$\tau + F(G(x_{2m_k+1}, x_{2n_k+1}, x_{2n_k+1})) \leq F(S_f(x_{2m_k}, x_{2n_k}, x_{2n_k})), \tag{3.14}$$

where

$$S_f(x_{2m_k}, x_{2n_k}, x_{2n_k}) = \max\{G(x_{2m_k}, x_{2n_k}, x_{2n_k}),\\ G(x_{2m_k}, x_{2m_k+1}, x_{2m_k+1}),\\ G(x_{2n_k}, x_{2n_k+1}, x_{2n_k+1})\}.$$

Now, switching to limes when $k \to +\infty$, we get

$$\tau + F(\varepsilon^+) \leq F(\max\{\varepsilon^+, 0, 0\}) = F(\varepsilon^+), \tag{3.15}$$

which is a contradiction with $\tau > 0$. Therefore, the sequence $\{x_n\}$ is G-Cauchy and then due to the G-completeness of the space (X, G) it converges to some point $u \in X$.

Let us first assume that the mapping f from X to itself is $G-$continuous. Then we get that $fu = u$. Since the sequence $\{x_n\}$ tends to u when $n \to +\infty$ it follows that $G(u, x_n, x_n) \to 0$ when $n \to +\infty$. Then the point of G-continuity of the mapping f (see the definition of G-continuity) we get that $G(fu, fx_n, fx_n) = G(fu, x_{n+1}, x_{n+1}) \to 0$ when $n \to +\infty$, i.e., $x_n \to fu$. Due to the uniqueness of limes in G-metric spaces, it follows that $fu = u$. The proof in the case that f is continuous mapping is complete.

Now suppose that F is a continuous mapping. Putting in the contractive condition $x = x_n, y = z = u$ we get

$$\tau + F(G(fx_n, fu, fu)) \leq F(S_f(x_n, u, u)), \tag{3.16}$$

where $S_f(x_n, u, u) = \max\{G(x_n, u, u), G(x_n, x_{n+1}, x_{n+1}), G(u, fu, fu)\}$. Moving to limes in the previous relation and using the continuity of the mapping F we get

$$\begin{aligned}
&\tau + F\left(\lim_{n\to+\infty} G(x_{n+1}, fu, fu)\right)\\
&\quad\leq F\left(\lim_{n\to+\infty} \max\{G(x_n, u, u), G(x_n, x_{n+1}, x_{n+1}), G(u, fu, fu)\}\right)\\
&\quad= F\left(\max\left\{\lim_{n\to+\infty} G(x_n, u, u), \lim_{n\to+\infty} G(x_n, x_{n+1}, x_{n+1}), \lim_{n\to+\infty} G(u, fu, fu)\right\}\right)\\
&\quad= F(\max\{0, 0, G(u, fu, fu)\}) = F(G(u, fu, fu)), \text{ i.e.,}
\end{aligned}$$

$$\tau + F(G(u, fu, fu)) \leq F(G(u, fu, fu)). \tag{3.17}$$

If $u \neq fu$, then due to $G(u, fu, fu) > 0$, a relation of contradiction with $\tau > 0$ is obtained. It remains to show the uniqueness of a fixed point. Let v be another fixed point of the mapping f different from the already obtained u. By putting $x = u, y = z = v$ in the contractive condition we get

$$\begin{aligned}
\tau + F(G(u, v, v)) &\leq F(\max\{G(u, v, v), G(u, u, u), G(v, v, v)\})\\
&= F(\max\{G(u, v, v), 0, 0\}) = F(G(u, v, v)).
\end{aligned}$$

From the obtained relation follows a contradiction with $\tau > 0$. Therefore, it must be $u = v$, i.e., f has a unique fixed point. The proof of the theorem is complete. □

We now state some consequences and comments on the last theorem.

Corollary 3.1 *Let (X, G) be a G-complete G-metric space, f a mapping from X to itself, and F a mapping from $(0, +\infty)$ to $\mathbb{R}$ that is strictly increasing. If there exists $\tau > 0$ such that for every $x, y, z \in X$ for which $G(fx, fy, fz) > 0$ the inequality*

$$\tau + F(G(fx, fy, fz)) \leq F(G(x, y, z)) \text{ fulfulled.} \tag{3.18}$$

Then f has a unique fixed point (say u) in X and for every $x \in X$ the corresponding Picard sequence $\{f^n x\}_{n \in \mathbb{N}}$ converges to that unique fixed point u.

Proof. According to the previous theorem, it is enough to show that from the given inequality (the contractive condition) the continuity of the mapping f follows. Let x_n be a sequence in X that converges to a point $x \in X$, i.e., $G(x_n, x, x) \to 0$ as $n \to +\infty$. It should be shown that $G(fx_n, fx, fx) \to 0$ as $n \to +\infty$. From the given inequality we have that $G(fx_n, fx, fx) \leq G(x_n, x, x) \to 0$ as $n \to +\infty$, that is., we obtain that $G(fx_n, fx, fx) \to 0$ as $n \to +\infty$. This proves the G-continuity of the mapping f. □

Corollary 3.2 *Let (X, G) be a G-complete G-metric space, f a mapping from X to itself, and F a mapping from $(0, +\infty)$ to $\mathbb{R}$ that is strictly increasing. If there exists $\tau > 0$ such that for every $x, y, z \in X$ for which $G(fx, fy, fz) > 0$ the inequality*

$$\begin{aligned} &\tau + F(G(fx, fy, fz)) \\ &\quad \leq F(\max\{G(x, fx, fx), G(y, fy, fy), G(z, fz, fz)\}) \text{ fulfilled.} \end{aligned} \tag{3.19}$$

If one of the mappings F, f is continuous, then f has a unique fixed point (say u) in X and for every $x \in X$ the corresponding Picard sequence $\{f^n x\}_{n \in \mathbb{N}}$ converges to that unique fixed point u.

Corollary 3.3 *Let (X, G) be a G-complete G-metric space, f a mapping from X to itself, and F a mapping from $(0, +\infty)$ to $\mathbb{R}$ that is strictly increasing. If there exists $\tau > 0$ such that for every $x, y, z \in X$ for which $G(fx, fy, fz) > 0$ the inequality*

$$\begin{aligned} &\tau + F(G(fx, fy, fz)) \\ &\quad \leq F(aG(x, y, z) + bG(x, fx, fx) + cG(y, fy, fy) + eG(z, fz, fz)) \end{aligned} \tag{3.20}$$

fulfilled, where $a, b, c, e \in [0, +\infty)$ and $a + b + c + e < 1$.

If one of the mappings F, f is continuous, then f has a unique fixed point (say u) in X and for every $x \in X$ the corresponding Picard sequence $\{f^n x\}_{n \in \mathbb{N}}$ converges to that unique fixed point u.

Remark 3.3 *All these consequences represent respectively Wardowski-Banach, Wardowski-Kannan and Wardowski-Reich types of results within G-metric spaces.*

At the end of our discussion, we will refer to two examples from the work of M. Kumar and S. Aurora. We will improve some parts of those two examples for the sake of young researchers who will eventually read the said paper.

Example 3.1 *([14], Example 2.3. Our repairs) Let* $X = [0, 2]$ *and let the function* $G : X^3 \to [0, +\infty)$ *defined by* $G(x, y, z) = |x - y| + |y - z| + |z - x|$ *and let* $f : X \to X$ *be given by* $fx = 1$ *if* $x \in [0, 2)$ *and* $f2 = \frac{1}{2}$*. We get that* (X, G) *is a symmetric G-complete G-metric space. We will show that the function* f *thus introduced on the set* $X = [0, 2]$ *is the modified generalized F-contraction of type* S_G *or type* M_G*. Due to the structure of the number set* $S_f(x, y, z)$ *and the area of definition of the function F, we have only the following two different cases:*

1. $x, y \in [0, 2), z = 2$ and **2.** $x \in [0, 2), y = z = 2$.
In the first case we get $G(fx, fy, fz) = G\left(1, 1, \frac{1}{2}\right) = 1$, while

$$\begin{aligned}
S_f(x, y, 2) &= \max\{G(x, fy, fy), G(x, fz, fz), G(y, fx, fx),\\
&\quad G(y, fz, fz), G(z, fx, fx), G(z, fy, fz)\}\\
&= \max\left\{G(x, 1, 1), G\left(x, \frac{1}{2}, \frac{1}{2}\right), G(y, 1, 1), G\left(y, \frac{1}{2}, \frac{1}{2}\right),\right.\\
&\quad G(2, 1, 1), G(2, 1, 1)\}\\
&= \max\{2|x - 1|, |2x - 1|, 2|y - 1|, |2y - 1|, 2, 2\}\\
&= \max\{2, 3, 2, 3\} = 3.
\end{aligned}$$

For the second case, we have:
$G(fx, fy, fz) = G\left(1, \frac{1}{2}, \frac{1}{2}\right) = 1$, while

$$\begin{aligned}
S_f(x, 2, 2) &= \max\{G(x, f2, f2), G(x, f2, f2), G(2, fx, fx),\\
&\quad G(2, f2, f2)\, G(2, fx, fx), G(2, f2, f2)\}\\
&= \max\left\{G\left(x, \frac{1}{2}, \frac{1}{2}\right), G(2, 1, 1), G\left(2, \frac{1}{2}, \frac{1}{2}\right)\right\}\\
&= \max\{|2x - 1|, 2, 3\} = 3.
\end{aligned} \tag{3.21}$$

Now, we see that in both cases, the condition

$$\tau + F(G(fx, fy, fz)) \leq F(S_f(x, y, z))$$

i.e.,

$$\tau + F(1) \leq F(3) \tag{3.22}$$

is possible for any strictly increasing function F defined on $(0, +\infty)$ and any positive number $\tau \in (0, F(3) - F(1))$, that is., f is a modified generalized function of type S_G and M_G.

Example 3.2 *([14], Example 2.6. Our repairs) Let* $X = [0,5]$ *and define the mapping* $G : X^3 \to [0,+\infty)$ *as in the previous example. Let* $f : X \to X$ *as* $fx = 3$ *if* $x \in [0,5)$ *and* $f5 = \frac{1}{3}$*. Similar to the previous example, we will check whether* f *is a modified generalized* F*-contraction of type* S_G *i.e., type* M_G*. For the same reason as in the previous example, we will distinguish two separate cases:*

1. $x, y \in [0,5), z = 5$. **2.** $x \in [0,5), y = z = 5$.

For the first case, we have that $G(fx, fy, fz) = G\left(3, 3, \frac{1}{3}\right) = \frac{16}{3}$, while in the second case also obtain the same value: $G(fx, fy, fz) = G\left(3, \frac{1}{3}, \frac{1}{3}\right) = \frac{16}{3}$. Let us now find $S_f(x, y, z)$ in both cases. In the first case we have:

$$\begin{aligned} S_f(x,y,z) &= S_f(x,y,5) \\ &= \max\{G(x,fy,fy), G(x,fz,fz), G(y,fx,fx), \\ &\quad G(y,fz,fz), G(z,fx,fx), G(z,fy,fy)\} \\ &= \max\left\{G(x,3,3), G\left(x,\frac{1}{3},\frac{1}{3}\right), G(y,3,3), G\left(y,\frac{1}{3},\frac{1}{3}\right),\right. \\ &\quad \left. G(5,3,3), G(5,3,3)\right\} \\ &= \max\left\{2|x-3|, 2|y-3|, \frac{2}{3}|3x-1|, \frac{2}{3}|3y-1|, 4, 4\right\} \\ &= \max\left\{2|x-3|, \frac{2}{3}|3x-1|, 4\right\} = \frac{28}{3}. \end{aligned} \tag{3.23}$$

In the second case, we find that

$$\begin{aligned} S_f(x,y,z) &= S_f(x,5,5) \\ &= \max\{G(x,fy,fy), G(x,fz,fz), G(y,fx,fx), \\ &\quad G(y,fz,fz), G(z,fx,fx), G(z,fy,fy)\} \\ &= \max\left\{G\left(x,\frac{1}{3},\frac{1}{3}\right), G\left(x,\frac{1}{3},\frac{1}{3}\right), G(5,3,3),\right. \\ &\quad \left. G\left(5,\frac{1}{3},\frac{1}{3}\right), G(5,3,3), G\left(5,\frac{1}{3},\frac{1}{3}\right)\right\} \\ &= \max\left\{2\left|x-\frac{1}{3}\right|, 4, \frac{28}{3}\right\} = \max\left\{4, \frac{28}{3}\right\} = \frac{28}{3}. \end{aligned} \tag{3.24}$$

So we got that in both cases $G(fx, fy, fz) = \frac{16}{3}$ and $S_f(x, y, z) = \frac{28}{3}$. Since F is a strictly increasing function, then the inequality

$$\begin{aligned} \tau + F(G(fx,fy,fz)) &\le F(S_f(x,y,z)), \text{ i.e.,} \\ \tau + F\left(\frac{16}{3}\right) &\le F\left(\frac{28}{3}\right) \end{aligned} \tag{3.25}$$

is possible for every such function F and $\tau \in \left(0, F\left(\frac{28}{3}\right) - F\left(\frac{16}{3}\right)\right]$. Hence, we have verified that the mapping f is a modified generalized F-contraction of type S_G and type M_G.

It is interesting to know if the function f from the previous two examples is a modified generalized F-contraction if $S_f(x,y,z)$ is as in our Theorem 7, i.e., if

$$S_f(x,y,z) = \max\{G(x,y,z), G(x,fx,fx), G(y,fy,fy), G(z,fz,fz)\}. \quad (3.26)$$

First, let f from the Example 3.1. For the case $x, y \in [0,2), z = 2$ we have:

$$\begin{aligned} S_f(x,y,z) &= S_f(x,y,2) \\ &= \max\left\{G(x,y,2), G(x,1,1), G(y,1,1), G\left(2,\frac{1}{2},\frac{1}{2}\right)\right\}, \end{aligned} \quad (3.27)$$

where $G(x,y,2) = |x-y| + 4 - (x+y), G(x,1,1) = 2|x-1|, G(y,1,1) = 2|y-1|, G\left(2,\frac{1}{2},\frac{1}{2}\right) = 3$. Further, we get

$$\begin{aligned} S_f(x,y,2) &= \max\{|x-y| + 4 - (x+y), 2|x-1|, 3\} \\ &= \max\{4,2,3\} = 4. \end{aligned} \quad (3.28)$$

Also, $G(fx,fy,fz) = G\left(1,1,\frac{1}{2}\right) = 1$.

If $x \in [0,2), y = z = 2$ we have:

$$\begin{aligned} S_f(x,y,z) &= S_f(x,2,2) \\ &= \max\left\{G(x,2,2), G(x,1,1), G\left(2,\frac{1}{2},\frac{1}{2}\right), G\left(2,\frac{1}{2},\frac{1}{2}\right)\right\} \\ &= \max\{2|x-2|, 2|x-1|, 3\} = \max\{4,2,3\} = 4, \end{aligned} \quad (3.29)$$

while $G(fx,fy,fz) = G(1,f2,f2) = G\left(1,\frac{1}{2},\frac{1}{2}\right) = 1$.

Hence, in both separated cases, we obtain $G(fx,fy,fz)=1$ and $S_f(x,y,z)=4$.

Now, it is obvious that the inequality

$$\tau + F(1) \leq F(4), \quad (3.30)$$

is possible whenever $\tau \in (0, F(4) - F(1)]$, that is., the function f is a modified generalized of type S_G and type M_G.

Now let f be the function from Example 3.2. and let $S_f(x,y,z)$ be as in our Theorem 7. By considering the two cases $x, y \in [0,5), z = 5$ and $x \in [0,5), y = z = 5$ it is not difficult to check (as in the previous one) that in both cases $G(fx,fy,fz) = \frac{16}{3}, S_f(x,y,z) = 10$. Since $G(fx,fy,fz) < S_f(x,y,z)$ we get obviously that the function f from Example 3.2. is a modified generalized F-contraction.

Remark 3.4 *Taking* $F(r) = \log r, \tau \in \left(0, \log \frac{10}{\frac{16}{3}}\right) = (0, \log \frac{15}{8}]$ *we get that Example 2. support our Theorem 7. Indeed, for* $G(fx,fy,fz) > 0$ *where* $x, y \in [0,5), z = 5$ *or* $x \in [0,5), y = z = 5$ *we have that there exists* $\tau \in (0, \log \frac{15}{8}], F(r) = \log r$ *so that*

$$\tau + F(G(fx,fy,fz)) \leq F(S_f(x,y,z)), \quad (3.31)$$

where
$S_f(x,y,z) = \max\{G(x,y,z), G(x,fx,fx), G(y,fy,fy), G(z,fz,fz)\}$.

3.3 CONCLUSION

With a note that the work of the mentioned authors contains a lot of errors and ambiguities, we have tried to correct them in our work and thus make the work available to many researchers who deal with F-contractions within generalized metric space. Some errors are typographical, but there are also essential ones. Such is, for example, Definition 1.2. for which the authors refer to the famous work of Banach from 1922. In examples 2.3 and 2.6, there are also typos and oversights such as $\lambda = \log \frac{1}{2}$ and $\lambda = \log \frac{1}{5}$. Since the proofs of Theorems 2.4 and 2.5 from [14] doubtful, it is the claims in them that are open and interesting for us. The structure of the number set $S_f(x, y, z)$ does not allow to prove that the sequence $G(x_n, x_{n+1}, x_{n+1})$ decreases in the usual way. In the mentioned paper, the proof of this is wrong (see page 4, line 8 from top). Correcting typos in Example 2.3. from [14], i.e., defining that $fx = 1$ for $x \in [0, 2)$ and $f2 = \frac{1}{2}$, we get that it supports our Theorem 7. As such it also supports Theorem 2.4. from [14], but her proof is doubtful.

BIBLIOGRAPHY

[1] M. Abbas, A. Hussain, B. Popović and S. Radenović, *Istratescu-Susuki-Ćirić-type fixed points results in the framework of G-metric spaces,* J. Nonlinear Sci. Appl., 9 (**2016**), 6077–6095

[2] R.P. Agarwal, Z. Kadelburg and S. Radenović, *On coupled fixed point results in asymmetric G-metric spaces,* J. Inequal. Appl., 2013 (**2013**), 528

[3] R.P. Agarwal, E. Karapinar, D. O'Regan, A.F.R.L.-de Hierro, *Fixed Point Theory in Metric Type Spaces,* Springer International Publishing, Switzerland, **2015**

[4] T. V. An, N. V. Dung, Z. Kadelburg, and S. Radenović, *Various generalizations of metric spaces and fixed point theorems,* Revista de la Real Academia de Ciencias Exactas, Fisicas y Naturales. Serie A. Matematicas, 109(1) (**2015**), 175–198

[5] H. Aydi, W. Shatanawi and C. Vetro, *On generalized weak G-contraction mapping in G-metric spaces,* Comp. Math. Appl., 62 (**2011**), 4223–4229.

[6] H. Aydi, D. Rakić, A. Aghajani, T. Došenović, M. S. Md. Noorani and H. Qawaqneh, *On fixed point results in G_b-metric spaces,* Mathematics, 7 (**2019**), 617. https://doi.org/10.3390/math7070617

[7] S. Banach, *Sur les operations dans les ensembles abstracts et leur application aux equations integrals,* Fundamenta Mathematicae, 3 (**1922**), 133–181

[8] Lj. Ćirić, *Some Recent Results in Metrical Fixed Point Theory,* University of Belgrade, Serbia, **2003**

[9] P. Debnath, N. Konwar, and S. Radenović, *Metric Fixed Point Theory, Applications in Science, Engineering and Behavioural Sciences,* Springer, Singapore, **2021**.

[10] N. V. Dung, and V. L. Hang, *A fixed point theorem for generalized F-contractions on complete metric spaces,* Vietnam J. Math., 43 (**2015**), 743–753.

[11] E. Karapinar, A. Fulga, and R. P. Agarwal, *A survey: F-contractions with related fixed point results,* J. Fixed Point Theory Appl., 22 (**2020**), 69.

[12] N. Fabiano, Z. Kadelburg, N. Mirkov, V. Š. Čavić, and S. Radenović, *On F-contractions: A Survey,* Contemporary Mathematics, 3(3) (**2022**), 327. http://ojs.wiserpub.com/index.php/CM/

[13] W. Kirk and N. Shahzad, *Fixed Point Theory in Distances Spaces,* Springer, Berlin, Germani, **2014**.

[14] M. Kumar and S. Arora, *Fixed point theorems for modified generalized F-contraction in G-metric spaces,* Bol. Soc. Paran. Mat. (3s.) 40 (**2022**), 1–8

[15] S. Mitrović, V. Parvaneh, M. De La Sen, J. Vujaković, S. Radenović, *Some new results for Jaggi-F-contraction type mappings on b-metric-like spaces,* Mathematics, 9 (**2021**), 2021. https://doi.org/10.3390/math9161921

[16] Z. Mustafa and B. Sims, *Some remarks concerning D-metric spaces,* Proc. Int. Conf. on Fixed Point Theory and Appl., Valencia (Spain), **2003**, 189–198.

[17] Z. Mustafa and B. Sims, *A new approach to generalized metric spaces,* J. Nonlinear and Convex Anal., 7(2) (**2006**), 289–297.

[18] Z. Mustafa, H. Obiedat and F. Awawdehand, *Some fixed point theorem for mapping on complete G-metric spaces,* Fixed Point Theory Appl., 2008, (**2008**), Article ID 189870, 12 pages.

[19] Z. Mustafa and B. Sims, *Fixed point theorems for contractive mappings in complete G-metric spaces,* Fixed Point Theory Appl., 2009 (**2009**), Article ID 917175.

[20] Z. Mustafa, W. Shatanawi and M. Bataineh, *Existence of fixed point results in G-metric spaces,* Int. J. Math. Math. Sci., 2009 (**2009**), Article ID 283028.

[21] Z. Mustafa and H. Obiedat, *A fixed point theorem of Reich in G-metric spaces,* CUBO-A Math. J., 12(1) (**2010**), 83–93.

[22] H. K. Nashine, Z. Kadelburg, and S.Radenović, *Coincidence and fixed point results under generalized weakly contractive condition in partially ordered G-metric spaces,* Filomat, 27(7) (**2013**), 1333–1343.

[23] H. Piri and P. Kumam, *Some fixed point theorems concerning F-contraction in complete metric spaces,* Fixed Point Theory Appl., 2014 (**2014**), Article ID 210, 1–13.

[24] H. Piri and P. Kumam, *Wardowski type fixed point theorems in complete metric spaces,* Fixed Point Theory Appl., 45 (**2016**), 1–12.

[25] A.C.M. Ran, M.C.B. Reuring, *A fixed point theorem in partially ordered sets and some applications to matrix equations,* Proc. Am. Math. Soc., 132 (**2004**), 1435–1443.

[26] S. Radenović, *Remarks on some recent coupled coincidence point results in symmetric G-metric spaces,* J. Oper. (**2013**), Article ID 290525, 8 pages

[27] S. Radenović, Z. Kadelburg, D. Jandrlić, and A. Jandrlić, *Some results on weakly contractive maps,* Bull. Iran. Math. Soc., 38(3) (**2012**), 625–645

[28] T. Rasham, M. Nazam, H. Aydi, and R.P. Agarwal, *Existence of common fixed points of generalized D-implicit locally contractive mappings on closed ball in multiplicative G-metric spaces with applications,* Mathematics, 10 (**2022**), 3369. https://doi.org/10.3390/math10183369

[29] W. Shatanawi, S. Chauhan, M. Postolache, M. Abbas and S. Radenović, *Common fixed points for contractive mappings of integral type in G-metric spaces,* J. Adv. Math. Stud., 6(1) (**2013**), 53–72

[30] R. Saadati, S. M. Vaezpour, P. Vetro and B. E. Rhoades, *Fixed point theorems in generalized partially ordered G-metric spaces,* Math. Comp. Modelling, 52(5–6) (**2010**), 797–801.

[31] J.Vujaković, N. Kontrec, M. Tošić, N. Fabiano, S. Radenović, *Some new results on F-contractions in complete metric spaces,* Mathematics, 10 (**2022**), 12. https://doi.org/10.3390/math10010012

[32] J. Vujaković, S. Radenović, *On some F -contraction of Piri-Kumam-Dung-type mappings in metric spaces,* Vojnotehnički glasnik/Military Technical Courier, 68(4) (**2020**), 697–714. https://doi.org/10.5937/vojtehg68-27385.

[33] D. Wardowski, *Fixed points of a new type of contractive mappings in complete metric spaces,* Fixed Point Theory Appl., 2012 (**2012**), Article ID 94, 1–6.

[34] D. Wardowski and N. V. Dung, *Fixed points of F-weak contractions on complete metric spaces,* Demonstr. Math., XLVII (**2014**), 146–155.

CHAPTER 4

Some New Fixed Point Results in Archimedean Type Intuitionistic Fuzzy b-Metric Space

Nabanita Konwar

Birjhora Mahavidyalaya

CONTENTS

4.1 INTRODUCTION

In 1965, Zadeh [32] put forward the concept of fuzzy set theory to represent and explain the mathematical configuration for the circumstances where the information is imprecise or vague. After this in 1986, Atanassaov [2] introduced a new generalization and called it as intuitionistic fuzzy set where the imprecise information is categorized on the basis of their belongingness property and non-belongingness property within a set. The belongingness property is named as the degree of membership and the non-belongingness property is named as the degree of non-membership function. Initially, Kaleva and Seikkala [25] developed the concept of fuzzy metric space. Later on Kramosil and Michalek [26], George and Veeramani [16] modified the definition of fuzzy metric space, respectively. Bakhtin [4] and Czerwik [7] put forward the concept of b-metric space where the condition of

DOI: 10.1201/9781003388678-4

triangular inequality was generalized. The study of fixed point theory in fuzzy metric spaces was developed by Heilpern [19]. Some extended generalizations appear in [1, 3, 5, 6, 7, 8, 9, 10, 11, 17, 23, 24, 21, 22, 20, 27, 28, 29, 30, 31].

In this chapter, we introduce the concept of Archimedean type intuitionistic fuzzy b metric space(in short, IFbMS). This work also contains the definition of the Caristi-Krirk balls in the setting of Archimedean type IFbMS. The main aim of the current work is to establish a few new fixed point theorem and to introduce the common fixed point for Archimedean type IFbMS. We also provide some generalizations of the resultant work.

4.2 SOME BASIC DEFINITIONS

Some preliminary definitions are listed below:

Definition 4.1 *[2] Let X be a non-empty set. Let us define a set F in X such that $F = \{(x, \hbar_F(x), \wp_F(x)) : x \in X, 0 \leq \hbar_I + \wp_I \leq 1\}$ where the mappings $\hbar_F : X \to [0,1] \subseteq \mathbb{R}$ implies the degree of membership and $\wp_F : X \to [0,1] \subseteq \mathbb{R}$ implies the degree of non-membership function of the element $x \in X$, then F is called intuitionistic fuzzy set(in short "IFS").*

Definition 4.2 *Let X be a non-empty set and $*$ is a continuous t-norm. Consider a fuzzy set $\hbar$ on $X^2 \times (0,\infty)$. Then the three-tuple $(X, \hbar, *)$ is known an fuzzy b-metric space if for a given number $b \geq 1$ and $\forall\ a_1, a_2, a_3 \in X$ and $s, t > 0$ following conditions are satisfied:*

1. $\hbar(a_1, a_2, t) > 0$,
2. $\hbar(a_1, a_2, t) = 1$ *iff* $a_1 = a_2$,
3. $\hbar(a_1, a_2, t) = \hbar(a_2, a_1, t)$,
4. $\hbar(a_1, a_3, t+s) \geq (\hbar(a_1, a_2, \frac{t}{b})) * (\hbar(a_2, a_3, \frac{s}{b}))$,
5. $\hbar(a_1, a_2, \cdot) : (0,\infty) \to [0,1]$ *is continuous.*

Definition 4.3 *Let X be a non-empty set, $*$ be a continuous t-norm, $\circ$ be a continuous t-conorm and $\hbar, \wp$ be the fuzzy sets on $X^2 \times (0,\infty)$. Then a five-tuple $(X, \hbar, \wp, *, \circ)$ is known as an intuitionistic fuzzy b metric space(in short, IFbMS) if for a given number $b \geq 1$ and $\forall\ a_1, a_2, a_3 \in X$ and $s, t > 0$ following conditions are satisfied:*

1. $\hbar(a_1, a_2, t) + \wp(a_1, a_2, t) \leq 1$,
2. $\hbar(a_1, a_2, t) > 0$,
3. $\hbar(a_1, a_2, t) = 1$ *iff* $a_1 = a_2$,
4. $\hbar(a_1, a_2, t) = \hbar(a_2, a_1, t)$,
5. $\hbar(a_1, a_3, t+s) \geq (\hbar(a_1, a_2, \frac{t}{b})) * (\hbar(a_2, a_3, \frac{s}{b}))$,
6. $\hbar(a_1, a_2, \cdot)$ *is nondecreasing function of* $\mathbb{R}^+$ *and* $\lim_{t\to\infty} \hbar(a_1, a_2, t) = 1$,

7. $\wp(a_1, a_2, t) < 1$,
8. $\wp(a_1, a_2, t) = 0$ *iff* $a_1 = a_2$,
9. $\wp(a_1, a_2, t) = \wp(a_2, a_1, t)$,
10. $\wp(a_1, a_3, t+s) \leq (\wp(a_1, a_2, \frac{t}{b})) \circ (\wp(a_2, a_3, \frac{s}{b}))$,
11. $\wp(a_1, a_2, \cdot)$ *is non-increasing function of* $\mathbb{R}^+$ *and* $\lim_{t\to\infty} \wp(a_1, a_2, t) = 0$.

Proposition 4.1 *[18] Consider a sequence of numbers* $\{x_n\} \in [0,1]$ *such that* $\lim_{n\to\infty} x_n = 1$ *and consider a t-norm* F *of H-type. Then* $\lim_{n\to\infty} T_{i=1}^{\infty} x_i = \lim_{n\to\infty} T_{i=1}^{\infty} x_{n+i} = 1$.

4.3 SOME FIXED POINT THEOREMS AND RELATED PROPOSITIONS

This section consists of some fixed point theorems and related properties in Archimedean type IFbMS.

4.3.1 Definitions

Definition 4.4 *Consider* $\zeta : [0,1] \to [0,1]$ *such that*

1. *If* $\zeta^{-1}(\{1\}) = \{1\}$ *then* ζ *is called amenable .*
2. *For all* $t, s \in [0,1]$, *if* $\zeta(t * s) \geq \zeta(t) * \zeta(s)$ *then* ζ *is called* $*$*-superadditive.*
3. *For all* $t, s \in [0,1]$, *if* $\zeta(t \circ s) \leq \zeta(t) \circ \zeta(s)$ *then* ζ *is called* $\circ$*-co-superadditive.*

Lemma 4.1 *Consider a nondecreasing continuous function* $\zeta : [0,1] \to [0,1]$. *If for some* $t \in (0,1)$, $\zeta(t) = 1$ *and* $(*, \circ)$ *is Archimedean, then* $\forall\ s \in [0,1]$, $\zeta(s) = 1$.

Definition 4.5 *Consider a IFbMS* $(X, \hbar, \wp, *, \circ)$ *and also consider* $\varphi : X \to [0,1]$ *and* $\zeta : [0,1] \to [0,1]$. *Then the Caristi-Krirk balls for any* $v \in X$ *such that* $\varphi(v) \neq 0$ *is defined as*

$$\begin{aligned} C_k(v) =\{u \in X : \zeta(\hbar(v, u, \frac{t}{b})) * \varphi(u) \geq \varphi(v) \text{ and} \\ \zeta(\wp(v, u, \frac{t}{b})) \circ \varphi(u) \leq 1 - \varphi(v), \forall t > 0\}. \end{aligned} \tag{4.1}$$

4.3.2 Theorems

Theorem 4.1 *Suppose* $(X, \hbar, \wp, *, \circ)$ *is a complete IFbMS and* $(*, \circ)$ *is Archimedean and continuous. Let* $T, S : X \longrightarrow X$ *be two self-maps. Consider a upper semi-continuous function* $\varphi : X \to [0,1]$ *such that there exist* $x \in X$, $\varphi(Sx) \neq 0$.

Assume that $\zeta : [0,1] \to [0,1]$ is a nondecreasing continuous function such that $\zeta(t * s) \geq \zeta(t) * \zeta(s)$ and $\zeta(t \circ s) \leq \zeta(t) \circ \zeta(s)$ along with $\zeta^{-1}(\{1\}) = \{1\}$ and satisfying the condition:

$$\zeta(\hbar(Sx, Tx, \frac{t}{b})) * \varphi(Tx) \geq \varphi(Sx) \text{ and}$$
$$\zeta(\wp(Sx, Tx, \frac{t}{b})) \circ \varphi(Tx) \leq 1 - \varphi(Sx), \forall x \in X, t > 0. \tag{4.2}$$

If $S(X)$ is complete, then T and S have a common fixed point in X.

Proof. For some $x \in X$ satisfying the condition $\varphi(Sx) \neq 0$, we have

$$C_k(x) = \{y \in X : \zeta(\hbar(x, y, \frac{t}{b})) * \varphi(y) \geq \varphi(x) \text{ and}$$
$$\zeta(\wp(x, y, \frac{t}{b})) \circ \varphi(y) \leq 1 - \varphi(x), \forall t > 0\}.$$

and

$$\underbrace{\sup}_{y \in C_k(x)} \varphi(y) = \alpha(x).$$

This implies that $1 \geq \alpha(x) \geq \varphi(y)$, for each $y \in C_k(x)$.

Since, for all x, $Tx \in C_k(Sx)$, therefore $C_k(Sx) \neq \phi$.

Let $x_1 = x$ and consider for all $t \geq 0$ x_{n+1} such that $Sx_{n+1} \in C_k(Sx_n)$ and $\varphi(Sx_{n+1}) \geq \alpha(Sx_n) - \frac{1}{n}$.

Since $Sx_{n+1} \in C_k(Sx_n)$,

$$\varphi(Sx_{n+1}) \geq \zeta(\hbar(Sx_n, Sx_{n+1}, \frac{t}{b})) * \varphi(Sx_{n+1}) \geq \varphi(Sx_n)$$

and

$$1 - \varphi(Sx_{n+1}) \leq \zeta(\wp(Sx_n, Sx_{n+1}, \frac{t}{b})) \circ \varphi(Sx_{n+1}) \leq 1 - \varphi(Sx_n),$$

for all $t > 0$.

Therefore $\{\varphi(Sx_n)\}$ is an increasing sequence and is convergent.

Since, $\alpha(Sx_n) \geq \varphi(Sx_{n+1}) \geq \alpha(Sx_n) - \frac{1}{n}$.

Therefore, $\lim_{n\to\infty} \alpha(Sx_n) = \lim_{n\to\infty} \varphi(Sx_n)$ exists.

Let

$$k = \lim_{n\to\infty} \alpha(Sx_n) = \lim_{n\to\infty} \varphi(Sx_n) \tag{4.3}$$

Next consider for any $n \in \mathbb{N}$,

$$\zeta(\hbar(Sx_n, Sx_m, \frac{t}{b})) * \varphi(Sx_m) \geq \varphi(Sx_n) \text{ and}$$
$$\zeta(\wp(Sx_n, Sx_m, \frac{t}{b})) \circ \varphi(Sx_m) \leq 1 - \varphi(Sx_n), \forall t > 0. \tag{4.4}$$

Now by using mathematical induction we have,

For $m = n+1$, since $Sx_{n+1} \in C_k(Sx_n)$ therefore the above inequality (4.4) is true.

Suppose that the inequality (4.4) is true for $m > n$.

Next verify it for $m+1$,

$$\begin{aligned}\zeta(\hbar(Sx_n, Sx_{m+1}, \frac{t}{b})) * \varphi(Sx_{m+1}) &\geq \zeta(\hbar(Sx_n, Sx_m, \frac{t}{2b}))* \\ &\quad \zeta(\hbar(Sx_m, Sx_{m+1}, \frac{t}{2b})) * \varphi(Sx_{m+1}) \\ &\geq \zeta(\hbar(Sx_n, Sx_m, \frac{t}{2b})) * \varphi(Sx_m) \\ &\geq \varphi(Sx_n).\end{aligned}$$

and

$$\begin{aligned}\zeta(\wp(Sx_n, Sx_{m+1}, \frac{t}{b})) \circ \varphi(Sx_{m+1}) &\leq \zeta(\wp(Sx_n, Sx_m, \frac{t}{2b}))\circ \\ &\quad \zeta(\wp(Sx_m, Sx_{m+1}, \frac{t}{2b})) \circ \varphi(Sx_{m+1}) \\ &\leq \zeta(\wp(Sx_n, Sx_m, \frac{t}{2b})) \circ \varphi(Sx_m) \\ &\leq 1 - \varphi(Sx_n).\end{aligned}$$

Therefore the inequality (4.4) is true for $m+1$. Hence the inequality (4.4) holds for all $m > n$.

Next we have to show that Sx_n is Cauchy sequence.

If possible suppose that Sx_n is not a Cauchy sequence. Therefore $\exists\ 0 < \delta < 1$ and $t > 0$ such that $\forall\ n \in \mathbb{N}\ \exists\ m \in \mathbb{N}$ such that

$$\hbar(Sx_n, Sx_m, \frac{t}{b}) \leq 1 - \epsilon \text{ and } \wp(Sx_n, Sx_m, \frac{t}{b}) \geq \epsilon.$$

Again for each $0 < \epsilon' < 1\ \exists\ N \in \mathbb{N}$ such that $k \geq \varphi(Sx_n) \geq k(1-\epsilon')$ and $k \leq \varphi(Sx_n) \leq k(\epsilon')$, for all $n > N$.

From (4.4) we can conclude that

$$\begin{aligned}k * \zeta((1-\epsilon)) &\geq \zeta(\hbar(Sx_n, Sx_m, \frac{t}{b})) * k \\ &\geq \zeta(\hbar(Sx_n, Sx_m, \frac{t}{b})) * \varphi(Sx_m) \\ &\geq \varphi(Sx_n) \\ &\geq k(1-\epsilon')\end{aligned}$$

and

$$\begin{aligned}k \circ \zeta((\epsilon)) &\leq \zeta(\wp(Sx_n, Sx_m, \frac{t}{b})) \circ k \\ &\leq \zeta(\wp(Sx_n, Sx_m, \frac{t}{b})) \circ \varphi(Sx_m)\end{aligned}$$

$$\leq \varphi(Sx_n)$$
$$\leq k(\epsilon'),$$

valid $\forall\ m > n > N$.

It implies, $k * \zeta((1-\epsilon)) \geq k(1-\epsilon')$ and $k \circ \zeta((\epsilon)) \leq k(\epsilon')$. But this statement is a contradiction with the Archimedean condition.

Hence Sx_n is a Cauchy sequence.

Since $S(X)$ is complete, the sequence Sx_n converges to $v = Su \in S(X)$ and as φ is upper semi-continuous $k = \lim_{n\to\infty} \sup \varphi(Sx_n) \leq \varphi(Su)$.

Now applying limit on (4.4), we have

$$\begin{aligned}\varphi(Sx_n) &\leq \lim_{m\to\infty} \sup(\zeta(\hbar(Sx_n, Sx_m, \frac{t}{b})) * \varphi(Sx_m)) \\ &\leq \zeta(\hbar(Sx_n, u, \frac{t}{b})) * \varphi(Su)\end{aligned}$$

and

$$\begin{aligned}1-\varphi(Sx_n) &\geq \lim_{m\to\infty} \inf(\zeta(\wp(Sx_n, Sx_m, \frac{t}{b})) \circ \varphi(Sx_m)) \\ &\geq \zeta(\wp(Sx_n, u, \frac{t}{b})) \circ \varphi(Su), \forall t > 0.\end{aligned}$$

Then $Su \in C_k(Sx_n)$. Hence $\alpha(Sx_n) \geq \varphi(Su)$.

Now from the inequality (4.4), $k \geq \varphi(Su)$ implies $k = \varphi(Su) = \varphi(v)$. Since $Su \in C_k(Sx_n)$ and $Tu \in C_k(Su)$, we have

$$\begin{aligned}\zeta(\hbar(Sx_n, Tu, \frac{t}{2b})) * \varphi(Tu) &\geq \zeta(\hbar(Sx_n, Su, \frac{t}{2b})) * \\ &\quad \zeta(\hbar(Su, Tu, \frac{t}{2b})) * \varphi(Tu) \\ &\geq \zeta(\hbar(Sx_n, Su, \frac{t}{2b})) * \varphi(Su) \\ &\geq \varphi(Sx_n)\end{aligned}$$

and

$$\begin{aligned}\zeta(\wp(Sx_n, Tu, \frac{t}{2b})) \circ \varphi(Tu) &\leq \zeta(\wp(Sx_n, Su, \frac{t}{2b})) \circ \\ &\quad \zeta(\wp(Su, Tu, \frac{t}{2b})) \circ \varphi(Tu) \\ &\leq \zeta(\wp(Sx_n, Su, \frac{t}{2b})) \circ \varphi(Su) \\ &\leq 1-\varphi(Sx_n), \forall t > 0.\end{aligned}$$

Hence $Tu \in C_k(Sx_n)\ \forall\ n \in \mathbb{N}$.

Then we have, $\varphi(Tu) \leq \alpha_n(x_n)\ \forall\ n \in \mathbb{N}$.

Thus we have $\varphi(Tu) \leq k$.

Therefore, $\varphi(Su) = k \geq \varphi(Tu) \geq \varphi(Su)$.
Thus $\varphi(Su) = \varphi(Tu) = k$ and for all $t > 0$,
$k * \zeta(\hbar(Su, Tu, \frac{t}{b})) \geq k$ and $k \circ \zeta(\wp(Su, Tu, \frac{t}{b})) \leq 1 - k$.
It implies that $\zeta(\hbar(Su, Tu, \frac{t}{b})) = 1$ and $\zeta(\wp(Su, Tu, \frac{t}{b})) = 1$.
Hence, for all $t > 0$, $\hbar(Su, Tu, \frac{t}{b}) = 1$ and $\wp(Su, Tu, \frac{t}{b}) = 1$.
And we have $Su = Tu$. □

Corollary 4.1 *Suppose $(X, \hbar, \wp, *, \circ)$ is a complete IFbMS and $(*, \circ)$ is Archimedean and continuous. Let $T, S : X \longrightarrow X$ be two self-maps where S is identity and $\varphi : X \to [0,1]$ be upper semi-continuous function such that there exist $x \in X$, $\varphi(Sx) \neq 0$. Assume that $\zeta : [0,1] \to [0,1]$ is a nondecreasing continuous function such that $\zeta(t * s) \geq \zeta(t) * \zeta(s)$ and $\zeta(t \circ s) \leq \zeta(t) \circ \zeta(s)$ and $\zeta^{-1}(\{1\}) = \{1\}$ and satisfying the condition:*

$$\zeta(\hbar(x, Tx, \frac{t}{b})) * \varphi(Tx) \geq \varphi(x) \text{ and}$$

$$\zeta(\wp(x, Tx, \frac{t}{b})) \circ \varphi(Tx) \leq 1 - \varphi(x), \forall x \in X, t > 0.$$

Then T has a fixed point in X.

Corollary 4.2 *Suppose $(X, \hbar, \wp, *, \circ)$ is a complete IFbMS and $(*, \circ)$ is Archimedean and continuous. Let $T, S : X \longrightarrow X$ be two self-maps where S is identity and $\varphi : X \to [0,1]$ be upper semi-continuous function such that there exist $x \in X$, $\varphi(Sx) \neq 0$. Assume that $\zeta : [0,1] \to [0,1]$ is also identity map satisfying the condition:*

$$\hbar(x, Tx, \frac{t}{b}) * \varphi(Tx) \geq \varphi(x) \text{ and}$$

$$\zeta(\wp(x, Tx, \frac{t}{b})) \circ \varphi(Tx) \leq 1 - \varphi(x), \forall x \in X, t > 0.$$

Then T has a fixed point in X.

Next we generalize the Theorem 9.2.

Theorem 4.2 *Suppose $(X, \hbar, \wp, *, \circ)$ is a complete IFbMS and $(*, \circ)$ is Archimedean and continuous. Let $T : X \longrightarrow X$ be a k-continuous self-map. Consider a upper semi-continuous function $\varphi : X \to [0,1]$ such that there exist $x \in X$, $\varphi(Sx) \neq 0$ satisfying the condition:*

$$\hbar(x, Tx, \frac{t}{b}) * \varphi(Tx) \geq \varphi(x) \text{ and } \wp(x, Tx, \frac{t}{b}) \circ \varphi(Tx) \leq 1 - \varphi(x), \tag{4.5}$$

for all $x \in X$ and $t > t_0$, for some $t_0 > 0$.
Then T has a fixed point in X.

Proof. Proceeding with the similar arguments as the proof of the Theorem 9.2 and considering that ζ is an identity map we obtain that $\{x_n\}$ is a Cauchy sequence.

Since X is complete, therefore for each $p \geq 1$, $\exists$ an element $y_0 \in X$, such that $\lim_{n\to\infty}(x_n) = y_0$ and $\lim_{n\to\infty}(T^p x_n) = y_0$. Then from the k continuity of T we have $\lim_{n\to\infty}(T^k x_n) \to Ty_0$. Thus as $\lim_{n\to\infty}(T^k x_n) \to y_0$, $Ty_0 = y_0$. This implies that y_0 is a fixed point of T. □

Next we provide one theorem which characterizes the completeness property of an Archimedean type IFbMS.

Theorem 4.3 *Suppose $(X, \hbar, \wp, *, \circ)$ is an IFbMS and $(*, \circ)$ is Archimedean and continuous. Consider that $\forall\ x \neq Tx$ and $t > 0$, each k-continuous self-map of X having a fixed point satisfying all conditions of Theorem 4.2 also satisfy the condition:*

$$\hbar(Tx, T^2x, \frac{t}{b}) > \hbar(x, Tx, \frac{t}{b}) \Rightarrow \hbar(Tx, T^2x, \frac{t}{b})^2 \geq \hbar(x, Tx, \frac{t}{b})$$

and

$$\wp(Tx, T^2x, \frac{t}{b}) < \wp(x, Tx, \frac{t}{b}) \Rightarrow \wp(Tx, T^2x, \frac{t}{b})^2 \leq \wp(x, Tx, \frac{t}{b})$$

In this case, X is complete.

Proof. Let us consider that each k-continuous self-map of X having a fixed point satisfying all conditions of Theorem 4.2.

If possible assume that X is not complete. Then $\exists$ a non-convergence Cauchy sequence in X consisting of distinct points, say $\{x_n\} = \{u_1, u_2, u_3, \ldots\}$.

Suppose $v \in X$ which is not a limit point of $\{x_n\}$. Then $\exists$ a least positive integer v_0 such that $v \neq u_{v_0}$.

Now for each $m \geq v_0$ and $t > 0$, we have

$$\hbar(v, u_{v_0}, \frac{t}{b}) < \hbar(u_{v_0}, v_m, \frac{t}{b})$$

and

$$\wp(v, u_{v_0}, \frac{t}{b}) > \wp(u_{v_0}, v_m, \frac{t}{b})$$

Construct a function $T : X \to X$ such that $T(v) = u_{v_0}$. Then for each v, $Tv \neq v$. From (4.5), for any $t > 0$ and $v \in X$, we have

$$\begin{aligned}\hbar(Tv, T^2v, \frac{t}{b}) &= \hbar(u_{v_0}, u_{T(v_0)}, \frac{t}{b})\\ &> \hbar(u_{v_0}, v_m, \frac{t}{b})\\ &= \hbar(v, Tv, \frac{t}{b})\end{aligned}$$

and

$$\begin{aligned}\wp(Tv, T^2v, \frac{t}{b}) &= \wp(u_{v_0}, u_{T(v_0)}, \frac{t}{b})\\ &< \wp(u_{v_0}, v_m, \frac{t}{b})\end{aligned}$$

$$= \wp(v, Tv, \frac{t}{b})$$

then $\hbar(Tv, T^2v, \frac{t}{b})^2 \geq \hbar(v, Tv, \frac{t}{b})$ and $\wp(Tv, T^2v, \frac{t}{b})^2 \leq \wp(v, Tv, \frac{t}{b})$

Consider $\varphi(v) = \hbar(v, Tv, \frac{t_0}{b})^2$ and $1 - \varphi(v) = \wp(v, Tv, \frac{t_0}{b})^2$, we have

$$\begin{aligned}\hbar(v, Tv, \frac{t_0}{b}) * \varphi(Tv) &= \hbar(v, Tv, \frac{t_0}{b}) * \hbar(Tv, T^2v, \frac{t_0}{b})^2 \\ &\geq \hbar(v, Tv, \frac{t_0}{b}) * \hbar(v, Tv, \frac{t_0}{b}) \\ &= \varphi(v)\end{aligned}$$

and

$$\begin{aligned}\wp(v, Tv, \frac{t_0}{b}) \circ \varphi(Tv) &= \wp(v, Tv, \frac{t_0}{b}) \circ \wp(Tv, T^2v, \frac{t_0}{b})^2 \\ &\leq \wp(v, Tv, \frac{t_0}{b}) \circ \wp(v, Tv, \frac{t_0}{b}) \\ &= 1 - \varphi(v)\end{aligned}$$

Moreover $\hbar(v, Tv, \frac{t}{b}) * \varphi(Tv) \geq \hbar(v, Tv, \frac{t_0}{b}) * \varphi(Tv) \geq \varphi(Tv)$ and $\wp(v, Tv, \frac{t}{b}) \circ \varphi(Tv) \leq \wp(v, Tv, \frac{t_0}{b}) \circ \varphi(Tv) \leq 1 - \varphi(Tv), \forall t > t_0$.

Hence, the function T fulfills all conditions of Theorem 4.2 along with the contractive one. Also T is a function having no fixed points, and its range is included in the Cauchy sequence $\{x_n\}$ which is not convergent. Therefore, $\exists$ no sequence $\{x_n\}$ in X for which $\{Tx_n\}$ converges i.e. $\exists$ no sequence $\{x_n\}$ in X for which the condition $(Tx_n) \to z \Longrightarrow (T^2x_n) \to Tz$ is violated. Hence, T is a 2-continuous function. Therefore, we obtain a self-mapping T of X which does not have a fixed point but satisfies all the conditions of Theorem 4.2, which is a contradiction.

Hence X is complete. □

Next we provide another generalization of the above theorem.

Corollary 4.3 *Suppose $(X, \hbar, \wp, *, \circ)$ is an IFbMS and $(*, \circ)$ is Archimedean and continuous. Let $T, S : X \longrightarrow X$. Consider that $\exists$ a function $\varphi : X \to [0, 1]$ such that*

1. $\hbar(Sx, Tx, \frac{t}{b}) * \varphi(Tx) \geq \varphi(Sx)$ *and* $\wp(Sx, Tx, \frac{t}{b}) \circ \varphi(Tx) \leq 1 - \varphi(Sx)$, *for all* $t \geq 0$ *and* $x \in X$.
2. $\hbar(Tx, Ty, \frac{t}{b})^2 > \min\{\hbar(Sx, Sy, \frac{t}{b})^2, \hbar(Sx, Ty, \frac{t}{b}) * \hbar(Tx, Sy, \frac{t}{b})\}$ *and* $\wp(Tx, Ty, \frac{t}{b})^2 < \max\{\wp(Sx, Sy, \frac{t}{b})^2, \wp(Sx, Ty, \frac{t}{b}) \circ \wp(Tx, Sy, \frac{t}{b})\}$, *for all* $x \neq y$ *and* $t \geq 0$.
3. $T(X) \subset S(X)$.
4. $T(X)$ *or* $S(X)$ *is complete.*

Then T and S have a common fixed point in X.

4.4 APPLICATION

One important use of Archimedean type IFbMS is Archimedean Compensatory Intuitionistic Fuzzy Logic(ACIFL) [13]. Initially ACFL [15] adds to the t-norm and t-conorm and it extends the Archimedean t-norm to Archimedean Logic. It also integrates its dual nature in the form of t-conorm and the negation operator. According to some mathematician [14], ACIFL has different properties and interpretations, which are involved in the truth values. Simultaneously, by using the ACFL generator function, many mathematicians generalized some fuzzy concepts. Some generalized s-shaped function models of this generator function are "Generalized Sigmoidal Function" and "Generalized Linguistic Modifier". Again a "Generalized Continuous Linguistic Variable" is defined as the parameterized family of different types of shape functions which are categorized as an increasing sigmoidal, decreasing sigmoidal, convex function, etc. Another application of Archimedean type IFbMS is Knowledge Discovery (KD) models. Such types of models are used to work out an optimization problem over the space of continuous parameters.

4.5 CONCLUSION

In this chapter, we have established the notion of Archimedean type IFbMS and introduced some common fixed point theorems in order to verify the nature of fixed points in this space. The work of this chapter is the generalized form of fuzzy metric space. Some more results are generalized with the help of the concept of k-continuous self-mapping. We have also provided some corollaries of the newly developed theorem. These new results impart a great deal of knowledge to the researchers to improve the field of fixed point theory in a new approach.

BIBLIOGRAPHY

[1] C. Alaca, D. Turkoglu, and C. Yildiz. Fixed points in intuitionistic fuzzy metric spaces. *Chaos, Solitons & Fractals*, 29:1073–1078, 2006.

[2] K. T. Atanassov. Intuitionistic fuzzy sets. *Fuzzy Sets Syst.*, 20:87–96, 1986.

[3] G. Babu, V. Ravindranadh, and D. T. Mosissa. Fixed points in b-metric spaces via simulation function. *Novi Sad J. Math*, 47(2):133–147, 2017.

[4] I.A. Bakhtin. The contraction principle in quasimetric spaces. *Funct. Anal.*, 30:26–37, 1989.

[5] D. Butnariu. Fixed points for fuzzy mappings. *Fuzzy Sets Syst.*, 7(2):191–207, 1982.

[6] I. Cristiana, R. Shahram, and S. E. Mohamad. Fixed points of some new contractions on intuitionistic fuzzy metric spaces. *Fixed Point Theory Appl.*, DOI: 10.1186/1687-1812-2013-168.

[7] S. Czerwik. Contraction mappings in b metric spaces. *Acta Math. Inform. Univ. Ostrav.*, 1(1):5–11, 1993.

[8] P. Debnath. Lacunary ideal convergence in intuitionistic fuzzy normed linear spaces. *Comput. Math. Appl.*, 63(3):708–715, 2012.

[9] P. Debnath. Results on lacunary difference ideal convergence in intuitionistic fuzzy normed linear spaces. *J. Intell. Fuzzy Syst.*, 28(3):1299–1306, 2015.

[10] P. Debnath. Results on lacunary difference ideal convergence in intuitionistic fuzzy normed linear spaces. *Ann. Fuzzy Math. Inform.*, 12(4):559–572, 2016.

[11] P. Debnath. Set-valued Meir-Keeler, Geraghty and Edelstein type fixed point results in b-metric spaces. *Rend. Circ. Mat. Palermo. Serie II*, 70(3):1389–1398, 2021.

[12] P. Debnath. Banach, Kannan, Chatterjea, and Reich-type contractive inequalities for multivalued mappings and their common fixed points. *Math Methods Appl. Sci.*, 45(3):1587–1596, 2022.

[13] R.A. Espin-Andrade, L. Cruz-Reyes, C. Llorente-Peralta, E. González-Caballero, W. Pedrycz, and S. Ruiz. Archimedean compensatory fuzzy logic as a pluralist contextual theory useful for knowledge discovery. *Int. J. Fuzzy Syst.*, 24(1):474–494, 2022.

[14] R.A. Espin-Andrade, E.R. Fernaińdez Gonzaiĺez, and E. Gonzalez. Compensatory fuzzy logic: a frame for reasoning and modeling preference knowledge in intelligent systems. *Soft Comput. Bus. Intell. Stud. Comput. Intell.*, 537:3–23, 2014.

[15] R.A. Espin-Andrade, E. Gonzalez, W. Pedrycz, and E.R. Fernaińdez Gonzaiĺez. Archimedean-compensatory fuzzy logic systems. *Int. J. Comput. Intell. Syst.*, 8(2):54–62, 2015.

[16] A. George and P. Veeramani. On some result in fuzzy metric spaces. *Fuzzy Sets Syst.*, 64:359–399, 1994.

[17] M. Grabiec. Fixed points in fuzzy metric spaces. *Fuzzy Sets Syst.*, 27:385–389, 1988.

[18] O. Hadźić. Fixed point theory in probalistic metric space. *Kluwer Academic, Dordrecht,* 2001.

[19] S. Heilpern. Fuzzy mappings and fixed point theorem. *J. Math. Anal. Appl.*, 83(2):566–569, 1981.

[20] N. Konwar and P. Debnath. Continuity and Banach contraction principle in intuitionistic fuzzy n-normed linear spaces. *J. Intell. Fuzzy Syst.*, 33(4):2363–2373, 2017.

[21] N. Konwar and P. Debnath. Intuitionistic fuzzy n-normed algebra and continuous product. *Proyecciones (Antofagasta)*, 37(1):68–83, 2018.

[22] N. Konwar and P. Debnath. Some new contractive conditions and related fixed point theorems in intuitionistic fuzzy n-Banach spaces. *J. Intell. Fuzzy Syst.*, 34(1):361–372, 2018.

[23] N. Konwar and P. Debnath. Some results on coincidence points for contractions in intuitionistic fuzzy-normed linear space. *Thai J. Math.*, 17(1):43–62, 2019.

[24] N. Konwar and P. Debnath. Fixed point results for a family of interpolative F-contractions in b-metric spaces. *Axioms*, 11(11):621, 2022.

[25] O. Kaleva and S. Seikkala. A Banach contraction theorem in fuzzy metric space. *Fuzzy Sets Syst.*, 12:215–229, 1984.

[26] I. Kramosil and J. Michalek. Fuzzy metric and statistical metric spaces. *Kybernetica*, 11:326–334, 1975.

[27] S. Manro and A. Tomar. Faintly compatible maps and existence of common fixed point in fuzzy metric space. *Ann. Fuzzy Math. Inform.*, 8(2):223–230, 2014.

[28] D. Rakić, A. Mukheimer, T. Došenović, Z. A. Mitrović, and S. Radenović. On some new fixed point results in fuzzy b metric spaces. *J. Inequal. Appl.*, 1:1–14, 2020.

[29] M. Saheli. A contractive mapping on fuzzy normed linear spaces. *Iran. J. Numer. Anal. Optim.*, 6(1):121–136, 2016.

[30] S. Shukla, D. Gopal, and A. F. R-L-de Hierro. Some fixed point theorems in 1-M-complete fuzzy metric-like spaces. *Int. J. Gen. Syst.*, DOI: 10.1080/03081079.2016.1153084, 2016.

[31] T. Suzuki. Basic inequality on a b-metric space and its applications. *J. Inequal. Appl.*, 1:256, 2017.

[32] L. A. Zadeh. Fuzzy sets. *Inform. Cont.*, 8:338–353, 1965.

CHAPTER 5

Fixed Point Theorems for Quasi Upper Semicontinuous Set-valued Mappings in p-Vector and Locally p-Convex Spaces

Shih-sen Chang

Sichuan University

Yeol Je Cho

Gyeongsang National University

Sehie Park

The National Academy of Sciences

George Xianzhi Yuan

Chengdu University
Sun Yat-Sen University
East China University of Science and Technology
College of Science, Chongqing University of Technology

CONTENTS

DOI: 10.1201/9781003388678-5

5.1 INTRODUCTION

It is known that the class of p-seminorm spaces $(0 < p \leq 1)$ is an important generalization of usual normed spaces with rich topological and geometrical structures, and related study has received a lot of attention (e.g., see Alghamdi et al. [3], Balachandran [4], Bayoumi [5], Ennassik and Taoudi [16], Ennassik et al. [17], Gholizadeh et al. [21], Granas and Dugundji [24], Jarchow [26], Kalton [27], [28], Kalton et al. [29], Park [37], Rolewicz [42], Xiao and Lu [48], Xiao and Zhu [49], Yuan [51]–[54], and many others). However, to the best of our knowledge, the corresponding basic tools and associated results in the category of nonlinear functional analysis have not been well developed, thus the goal of this chapter is to develop some fundamental fixed point theorems for quasi upper semicontinuous set-valued mappings including upper semicontinuous set-valued (USC) mappings as a special class under the framework of locally p-convex spaces which include locally convex spaces as a special case (for $p = 1$).

We all know that Schauder's fixed point theorem [43] (see also Leray and Schauder [32]) in normed spaces is one of the most powerful tools in dealing with nonlinear problems in analysis. Most notably, it has played a major role in the development of fixed point theory and related nonlinear analysis and mathematical theory of partial and differential equations and others. A generalization of Schauder's theorem from normed space to general topological vector spaces is an old conjecture in fixed point theory which is explained by Problem 54 of the book "The Scottish Book" by Mauldin [33] as stated as Schauder's conjecture: "**Every nonempty compact convex set in a topological vector space has the fixed point property**; or in its analytic way, **does a continuous function defined on a compact convex subset of a topological vector space to itself have a fixed point?**"

Based on the discussion given by Ennassik and Taoudi [16], Cauty [9], [10] tried to solve the Schauder conjecture (see also the comment given by Dobrowolski [14] on Professor Cauty's work), and they (Ennassik and Taoudi [16]) gave the solution to the Schauder conjecture by a different method for single-valued continuous mappings. From the respective development on the study of fixed point theory and related topics in nonlinear analysis, a number of works have been contributed, just mention a few of them, including Agarwal et al. [1], Ben-El-Mechaiekh [6], Ben-El-Mechaiekh and Saidi [7], Browder [8], Cellina [11], Chang [12], Chang et al. [13] Ennassik et al. [17], Fan [19], [20], Górniewicz [22], Granas and Dugundji [24], Guo et al. [23], Nhu [35], Park [37], Reich [39], Smart [44], Tychonoff [45], Weber [46], [47], Xiao and Lu [48], Xiao and Zhu [49], Xu [50], Yuan [51]–[54], Zeidler [55] and many other scholars, and also see the comprehensive references and related discussion under the general framework of topological vector space, or p-vector spaces for non-self set-valued mappings $(0 < p \leq 1)$.

The goal of this chapter is to establish a general fixed point theorem for compact single-valued continuous mapping in Hausdorff p-vector spaces, and the fixed point theorem for quasi upper semicontinuous set-valued mappings in Hausdorff locally p-convex for $p \in (0, 1]$. These new results provide an answer to Schauder conjecture in the affirmative under the setting of general p-vector spaces for compact single-valued continuous and also give the fixed point theorems for quasi upper semicontinuous set-valued mappings defined on s-convex subsets in Hausdorff locally p-convex spaces, which would be fundamental for nonlinear functional analysis in mathematics, where $s, p \in (0, 1]$.

This chapter has three sections as follows. Section 5.1 is the introduction. Section 5.2 describes general concepts for the p-convex subsets of topological vector spaces $(0 < p \leq 1)$. In Section 5.3, as the application of our approximation lemma in locally p-convex spaces, we establish general fixed point theorems for quasi upper semicontinuous self-mappings defined on nonempty s-convex subsets in locally p-convex spaces for $s, p \in (0, 1]$, which cover the upper semicontinuous set-valued (USC) mappings as a special class.

For the convenience of our discussion, throughout this chapter, all p-convex vector spaces are assumed to be Hausdorff, and p satisfying the condition for $0 < p \leq 1$ unless specified, and also we denote by $\mathbb{N}$ the set of all positive integers, i.e., $\mathbb{N} := \{1, 2, \cdots, \}$. For a set X, the 2^X denotes the family of all subsets of X.

5.2 SOME BASIC NOTIONS AND RESULTS OF P-VECTOR SPACES

Definition 5.1 *A set A in a vector space X is said to be p-convex for $0 < p \leq 1$ if, for any $x, y \in A$, $0 \leq s, t \leq 1$ with $s^p + t^p = 1$, we have $sx + ty \in A$; and if A is 1-convex, it is simply called convex (for $p = 1$) in general vector spaces; the set A is said to be absolutely p-convex if $sx + ty \in A$ for $0 \leq |s|, |t| \leq 1$ with $|s|^p + |t|^p \leq 1$.*

Definition 5.2 *If A is a subset of a topological vector space X, the closure of A is denoted by $\overline{A}$, then the p-convex hull of A and its closed p-convex hull are denoted by $C_p(A)$ and $\overline{C}_p(A)$, respectively, which are the smallest p-convex set containing A and the smallest closed p-convex set containing A, respectively.*

Definition 5.3 *Let A be p-convex and $x_1, \ldots, x_n \in A$, and $t_i \geq 0$, $\sum_1^n t_i^p = 1$. Then $\sum_1^n t_i x_i$ is called a p-convex combination of $\{x_i\}$ for $i = 1, 2, \ldots, n$. If $\sum_1^n |t_i|^p \leq 1$, then $\sum_1^n t_i x_i$ is called an absolutely p-convex combination. It is easy to see that $\sum_1^n t_i x_i \in A$ for a p-convex set A.*

Definition 5.4 *A subset A of a vector space X is called balanced (circled) if $\lambda A \subset A$ holds for all scalars λ satisfying $|\lambda| \leq 1$. By its definition, a balanced set A is symmetric, i.e., $A = -A$. We also say that the set A is absorbing if, for each $x \in X$, there is a real number $\rho_x > 0$ such that $\lambda x \in A$ for all $\lambda > 0$ with $|\lambda| \leq \rho_x$.*

By the Definition 5.4 above, it is easy to see that the system of all circled subsets of X is easily seen to be closed under the formation of linear combinations, arbitrary unions, and arbitrary intersections. In particular, every set $A \subset X$ determines the smallest circled subset $\hat{A}$ of X in which it is contained: $\hat{A}$ is called the circled hull of A. It is clear that $\hat{A} = \cup_{|\lambda| \leq 1} \lambda A$ holds, so that A is circled if and only if (in short, iff) $\hat{A} = A$. We use $\overline{\hat{A}}$ to denote the closed circled hull of $A \subset X$. In addition, if X is a topological vector space, we use $\text{int}(A)$ to denote the interior of set $A \subset X$, and if $0 \in \text{int}(A)$, then $\text{int}(A)$ is also circled, and we use ∂A to denote the boundary of A in X unless specified otherwise.

Definition 5.5 *A topological vector space is said to be locally p-convex if the origin has a fundamental set of absolutely p-convex 0-neighborhoods. This topology can be determined by p-seminorms, which are defined in the obvious way (see p. 52 of Bayoumi [5], Jarchow [26], or Rolewicz [42]). If $p = 1$, X is a usual locally convex space.*

Remark 5.1 *It is well known that a given p-seminorm P is said to be a p-norm if $x = 0$ whenever $P(x) = 0$. A vector space with a specific p-norm is called a p-normed space. Specifically, a Hausdorff topological vector space is locally bounded if and only if it is a p-normed space for some p-norm $\|\cdot\|_p$, where $0 < p \leq 1$ (see pp. 114 of Jarchow [26]).*

We also note that examples of p-normed spaces include $L^p(\mu)$-spaces and Hardy spaces H_p, $0 < p < 1$, endowed with their usual p-norms. Moreover, we would like to make the following important two points:

(1) First, by the fact that (e.g., see Kalton et al. [29] Roberts [41]) there is no open convex nonvoid subset in $L^p[0,1]$ (for $0 < p < 1$) except $L^p[0,1]$ itself. This means that p-normed paces with $0 < p < 1$ are not necessarily locally convex. Moreover, we know that every p-normed space is locally p-convex; and incorporating Lemma 5.2 below, it seems that p-vector spaces (for $0 < p \leq 1$) are nicer and bigger as we can use a p-convex subset in locally p-convex spaces to approximate convex subsets in topological vector spaces (TVS) by Lemma 5.1(ii), and also Lemma 5.3 in Section 5.3. In this way, it seems that p-vector spaces have better properties in terms of p-convexity than the usually (1−) convex subsets used in TVS with $p = 1$.

(2) Second, it is worthwhile noting that a 0-neighborhood in a topological vector space is always absorbing by Lemma 2.1.16 of Balachandran [4] or Proposition 2.2.3 of Jarchow [26].

The following result is very important and useful, which allows us to make the approximation for convex subsets in topological vector spaces by p-convex subsets in p-convex vector spaces. For the reader's convenience, we state the following result (see Lemma 2.1 of Ennassik and Taoudi [17], Remark 2.1 of Qiu and Rolewicz [38]).

Lemma 5.1 *Let A be a subset of a vector space X, then we have:*

(i) If A is p-convex with $0 < p < 1$, then $\alpha x \in A$ for any $x \in A$ and any $0 < \alpha \leq 1$.

(ii) If A is convex and $0 \in A$, then A is p-convex for any $p \in (0, 1]$.

(iii) If A is p-convex for some $p \in (0, 1)$, then A is s-convex for any $s \in (0, p]$.

Proof. See Lemma 2.1 of Ennassik and Taoudi [17], or, the Remark 2.1 of Qiu and Rolewicz [38]. □

Remark 5.2 *We would like to point out that results (i) and (iii) of Lemma 5.1 do not hold for $p = 1$. Indeed, any singleton $\{x\} \subset X$ is convex in topological vector spaces; but if $x \neq 0$, then it is not p-convex for any $p \in (0, 1)$ (see also Lemma 5.2 below).*

We also need the following proposition, which is Proposition 6.7.2 of Jarchow [26].

Proposition 5.1 *Let K be compact in a topological vector X and $(1 < p \leq 1)$. Then the closure $\overline{C}_p(K)$ of the p-convex hull and the closure $\overline{AC}_p(K)$ of absolutely p-convex hull of K are compact if and only if $\overline{C}_p(K)$ and $\overline{AC}_p(K)$ are complete, respectively.*

Before we close this section, we would like to point out that the structure of p-convexity when $p \in (0, 1)$ is really different from what we normally have for the concept of "convexity" used in topological vector spaces (TVS). In particular, maybe the following fact is one of the reasons for us to use better (p-convex) structures in p-vector spaces to approximate the corresponding structure of the convexity used in TVS (i.e., the p-vector space when $p = 1$).

Based on the discussion in p. 1740 by Xiao and Zhu [49], we have the following fact, which indicates that each p-convex subset is "bigger" than the convex subset in topological vector spaces for $0 < p < 1$.

Lemma 5.2 *Let x be a point of a p-vector space E, where assume $0 < p < 1$, then the p-convex hull and the closure of $\{x\}$ are given by*

$$C_p(\{x\}) = \begin{cases} \{tx : t \in (0, 1]\}, & \text{if } x \neq 0, \\ \{0\}, & \text{if } x = 0; \end{cases} \tag{5.1}$$

and

$$\overline{C_p(\{x\})} = \begin{cases} \{tx : t \in [0, 1]\}, & \text{if } x \neq 0, \\ \{0\}, & \text{if } x = 0. \end{cases} \tag{5.2}$$

Before the ending of this section, we note that if x is a given one point in p-vector space E, when $p = 1$, we have that $\overline{C_1(\{x\})} = C_1(\{x\}) = \{x\}$. This shows to be significantly different for the structure of p-convexity between $p = 1$ and $p \neq 1$! Thus it is necessary to establish fixed point theorems for single-valued mappings or the set-valued mappings with convex values, instead of p-convex values as zero is always contained by any p-closed subset when $p \in (0, 1)$ under general p-vector spaces or locally p-convex spaces.

Throughout this chapter, without loss of generality unless specified otherwise, for a given p-vector space E, where $p \in (0, 1]$, we always denote by $\mathfrak{U}$ the base of the p-vector space E's topology structure, which is the family of its 0-neighborhoods, and we assume that all p-vector spaces E are Hausdorff unless specified for $p \in (0, 1]$.

5.3 FIXED POINT THEOREMS IN p-VECTOR SPACES AND LOCALLY p-CONVEX SPACES

In this section, we establish a general fixed point theorem for compact single-valued continuous mapping in Hausdorff p-vector spaces, and fixed point theorem for quasi upper semicontinuous set-valued (USC) mappings in locally p-convex for $p \in (0, 1]$. These new results provide an answer to Schauder conjecture in the affirmative under the setting of general p-vector spaces for single-valued continuous, and upper semicontinuous set-valued mappings defined on s-convex subsets in locally p-convex spaces, which would be fundamental for nonlinear functional analysis in mathematics, where $s, p \in (0, 1]$.

Here, we first gather together necessary definitions, notations, and known facts needed in this section.

Definition 5.6 *Let X and Y be two topological spaces. A set-valued mapping (also called multifunction) $T : X \longrightarrow 2^Y$ is a point to set function such that for each $x \in X$, $T(x)$ is a subset of Y. The mapping T is said to be upper semicontinuous (USC) if the subset $T^{-1}(B) := \{x \in X : T(x) \cap B \neq \emptyset\}$ (equivalently, the set $\{x \in X : T(x) \subset B\}$) is closed (equivalently, open) for any closed (resp., open) subset B in Y. The function $T : X \to 2^Y$ is said to be lower semicontinuous (LSC) if the set $T^{-1}(A)$ is open for any open subset A in Y.*

Definition 5.7 *We recall that for two given topological spaces X and Y, and a set-valued mapping $T : X \to 2^Y$ is said to be compact if there is compact subset set C in Y such that $T(X)(:= \{y \in T(x), x \in X\})$ is contained in C, i.e., $F(X) \subset C$. Now we have the following non-compact versions of fixed point theorems for compact single-valued mappings defined in locally p-convex and topological vector spaces for $0 < p \leq 1$.*

We now state the following result which is a non-compact version of Theorem 3.1 and Theorem 3.3 by Ennassik and Taoudi [16].

Theorem 5.1 *If K is a nonempty closed p-convex subset of either a Hausdorff locally p-convex space (for $0 < p \leq 1$) or a Hausdorff topological vector space X, then the compact single-valued continuous mapping $T : K \to K$ has at least a fixed point.*

Proof. As T is compact, there exists a compact subset A in K such that $T(K) \subset A$. Let $K_0 := \overline{C}_p(A)$ be the closure of the p-convex hull of the subset A in K. Then K_0 is compact p-convex by Proposition 5.1, and the mapping $T : K_0 \to K_0$ is continuous. Then we can prove the conclusion by considering the self-mapping T on K_0 as applications of Theorem 3.1 and Theorem 3.3 given by Ennassik and Taoudi [16]. □

Now we are going to discuss how to establish the main results for the existence of fixed point theorem for quasi upper semicontinuous set-valued mappings defined on s-convex subsets under the framework of Hausdorff locally p-convex spaces, where $s, p \in (0, 1]$.

By following Repovš et al. [40] (see also Ewert and Neubrunn [18], Neubrunn [34], and Holá and Mirmostafaee [25]), we recall the following definition for quasi upper semicontinuous (QUSC) mappings which are a generalization of upper semicontinuous (USC) mappings.

Definition 5.8 *Let X and Y be two topological spaces and $T : X \longrightarrow 2^Y$ is a set-valued mapping. The mapping T is said to be quasi upper semicontinuous (QUSC) at $x \in X$ if, for each of its (x') neighborhood $W(x)$ and for each neighborhood V of the origin in Y, there exists a point $q(x) \in W(x)$ such that $x \in IntT_{-1}(T(q(x)) + V))$, where $T_{-1}(T(q(x)) + V)) = \{z \in X : T(z) \subset T(q(x)) + V\}$, and the notation $IntT_{-1}(T(q(x))+V))$ denotes the (topological) interior of the set $T_{-1}(T(q(x))+V))$ in X. The mapping T is said to be quasi supper semicontinuous if it is quasi upper semicontinuous at each point of its domain.*

Remark 5.3 *It is clear that in Definition 5.3 for QUSC mappings, for each $x \in X$, by taking $q(x)$ just being x itself, then it is just the definition for upper semicontinuous mappings given by Definition 5.1. Therefore, an USC mapping is QUSC, but a QUSC mapping may not be USC as shown by the example in pp. 1094 due to Repovš et al. [40]. But in this paper, we will focus on the study of fixed point theorem for upper semicontinuous set-valued mappings in locally p-convex spaces, where $p \in (0, 1]$. In addition, interesting readers can find more from Ewert and Neubrunn [18], Holá and Mirmostafaee [25], and Neubrunn [34] for the comprehensive study on the quasi continuity for both single and set-valued mappings and related application, and related reference wherein.*

By following the idea used by Repovš et al. [40] for the graph approximation of quasi upper semicontinuous set-valued mappings with the concept of the "p-convexity" used in locally p-convex spaces to replace the usual concept of "convexity" used in topological vector spaces (see also Ben-El-Mechaiekh [6], Ben-El-Mechaiekh and Saidi [7], Cellina [11], Kryszewski [31], Repovš et al. [40] and related references), we have the following Lemma 5.1 which is then used as a tool to establish a general fixed point theorem for upper semicontinous set-valued mappings in Hausdorff locally p-convex spaces for $p \in (0, 1]$.

Here we also recall that if X and Y are two topological spaces and $F : X \to 2^Y$ is a set-valued mapping, and we denote by either $Graph(F)$ or Γ_F for the graph of F in $X \times Y$, and α is a given open cover of Γ_F in $X \times Y$, then a (single- or set-valued) mapping $G : X \to Y$ is said to be an α-approximation (also called α-graph approximation) of F if for each point $p \in \Gamma_G$, there exists a point $q \in \Gamma_F$ such that p and q lie in some common elements of the over α. In the case Y is a topological vector space, if Ω is the open cover of X and V is an open neighborhood of their origin in Y, then $\Omega \times \{y + V\}_{y \in Y}$ is one open cover of $X \times Y$, which is denoted by $\Omega \times V$ in this section. We also refer the readers to the reference books by Dugungji [15] and Kelly [30] for the corresponding notations and concepts used in general topology below.

Lemma 5.3 *Let X be a paracompact space and Y be a topological vector space and $p \in (0, 1]$. If $F : X \to 2^Y$ is a quasi upper semicontinuous mapping with p-convex* *values, then for each open cover* Ω *of* X*, and each* p*-convex open neighborhood* V *of the origin in* Y*, there exists a continuous single-valued* $(\Omega \times V)$*-approximation for the set-valued mapping* F*. In particular, the conclusion holds if* V *is any convex open neighborhood of the origin in* Y*.*

Proof. Let Ω be an open covering of X, and let V be a p-convex open neighborhood of the origin in Y. For each $x \in X$, fix an arbitrary element $W(x) \in \Omega$ such that $x \in W(x)$, then we first claim the following statements:

(1) By the quasi upper semicontinuity (QUSC) of the mapping F, for each $x \in X$, there exists point $q(x) \in W(x)$ and an open neighborhood $U(x) \subset W(x)$ such that $F(z) \subset F(q(x)) + V$ for all $z \in U(x)$;

(2) As X is paracompact, by Theorem 3.5 of Dugundji [15] (see also Theorem 28 in Chapter 5 of Kelly [30]), without loss of the generality, let the family $\{G(x)\}_{x \in X}$ be a covering which is a star refinement of the covering $\{U(x)\}_{x \in X}$ of X (and see also the discussion on pp. 167–168 by Dugundji [15] for the concept of the star refinement for a given covering);

(3) Using the quasi upper semicontinuity property again for the mapping F, for each $x \in X$, there exists $q'(x) \in G(x)$ and a neighborhood $U'(x) \subset G(x)$ such that $F(z) \subset F(q'(x)) + V$ for all $z \in U'(x)$;

(4) Let $\{e_\alpha\}_{\alpha \in A}$ be a locally finite continuous partition of unity inscribed into the covering $\{U'(x)\}_{x \in X}$ of X, where A is the index set, with $\Sigma_{\alpha \in A}\, e_{\alpha(x)} = 1$ for each $x \in X$; and for each $\alpha \in A$, we can choose $x_\alpha \in X$ such that $supp\ e_\alpha \subset U'(x_\alpha)$, and choosing one point $y_\alpha \in F(q'(x_\alpha))$, where $supp\ e_\alpha$ is the support of e_α (defined by $supp\ e_\alpha := \overline{\{x \in X : e_\alpha(x) \neq 0\}}$); and

(5) Finally, define a mapping $f : X \to Y$ by $f(x) := \Sigma_{\alpha \in A} e_\alpha^{\frac{1}{p}}(x) y_\alpha$ for each $x \in X$, where $y_\alpha \in F(q'(x_\alpha))$ as given by (5.4) above, then f is well-defined, where the sum is taken over all $\alpha \in A$ with $e_\alpha(x) > 0$. By (5.3), it follows that $\Sigma_{\alpha \in A}(e_\alpha^{\frac{1}{p}}(x))^p = \Sigma_{\alpha \in A} e_\alpha(x) = 1$.

Now we show that f is indeed the desired single-valued continuous mapping, which is the $(\Omega \times V)$-approximation for the mapping F. Indeed for any given $x_0 \in X$, we have that

$$
\begin{aligned}
x_0 \in St\{x_0, \{supp\ e_\alpha\}_{\alpha \in A}\} \subset St\{x_0, \{U'(x)\}_{x \in X}\} \subset St\{x_0, \{G(x)\}_{x \in X}\} \\
\subset U(x') \subset W(x')
\end{aligned}
$$

for some $x' \in X$, where $St\{x_0, \{supp\ e_\alpha\}_{\alpha \in A}\}$ denotes the Star of the point $\{x_0\}$ with respect to the family $\{supp\ e_\alpha\}_{\alpha \in A}$ and defined by $St\{x_0, \{supp\ e_\alpha\}_{\alpha \in A}\} := \cup\{U : x_0 \in U, U \in \{supp\ e_\alpha\}_{\alpha \in A}\}$ (see also the corresponding discussion for the notation and concept on pp. 349 given by Ageev and Repovš [2]).

By the definition of quasi upper semi continuity, we have that $q(x') \in W(x')$. Hence the points x_0 and $q(x')$ are Ω-close.

Secondly, if $e_\alpha(x_0) > 0$ for $\alpha \in A$, then $x_0 \in G(x_\alpha)$ and $q'(x_\alpha) \in G(x_\alpha)$ by (5.3) above. Thus $q'(x_\alpha) \in St\{x_0, \{G(x)\}_{x\in X}\} \subset U(x')$. Therefore, $y_\alpha \in F(q'(x_\alpha)) \subset F(q(x')) + V$, i.e., $y_\alpha - v_\alpha \in V$ for some $v_\alpha \in F(q(x'))$ for $\alpha \in A$. But then, for $v := \Sigma_\alpha e_\alpha^{\frac{1}{p}}(x_0)v_\alpha \in F(q(x'))$ as F is p-convex valued and we know that $\Sigma_{\alpha\in A}(e_\alpha^{\frac{1}{p}}(x))^p = \Sigma_{\alpha\in A} e_\alpha(x) = 1$ as shown by (5.5) above, and $y_\alpha - v_\alpha \in V$, too for $\alpha \in A$, thus we have that $f(x_0) - v = \Sigma e_\alpha^{\frac{1}{p}}(x_0)(y_\alpha - v_\alpha) \in V$ as V is p-convex. Hence, the point $(x_0, f(x_0)) \in Graph(f)$ is $(\Omega \times V)$-close to the point $(q(x'), v) \in Graph(F)$.

In particular, as each convex neighborhood of the origin in Y is also p-convex for each $p \in (0, 1]$, thus the conclusion holds. The proof is complete. □

Now we have the following main result for quasi upper semicontinuous set-valued mappings in Hausdorff locally p-convex spaces.

Theorem 5.2 *Let K be a nonempty compact s-convex subset of a Hausdorff locally p-convex space X, where $p, s \in (0, 1]$. If $T : K \to 2^K$ is a quasi upper semicontinuous set-valued mapping with nonempty closed p-convex values, and the graph of T is closed, then T has a fixed point in K.*

Proof. We give the proof by using the graph approximation approach for quasi upper semicontinuous set-valued mappings established in this section above. Let $\mathfrak{U}$ be the family of absolutely p-convex open neighborhoods of the origin in X. By the fact the family $\{x + u\}_{x\in K}$ is an open covering of K, and we denote the family $\{x + u\}_{x\in K}$ by Ω. Now by Lemma 5.3, it follows that there exists one (single-valued) continuous mapping $f_u : K \to K$, which is $(\Omega \times u)$-approximation of the mapping T. By Theorem 5.1, f_u has a fixed point $x_u = f_u(x_u)$ in K for each $u \in \mathfrak{U}$. Note that $(x_u, f_u(x_u)) = (x_u, x_u) \in Graph(f_u)$, which is $(\Omega \times u)$-approximation of the Graph(T), and the graph of T is closed due to the assumption, we will go to prove T has a fixed point x^* which is indeed the limit of the family $\{x_u\}_{u\in\mathfrak{U}}$'s some sub-net in K, i.e., $x^* \in T(x^*)$, by using notations of language in general topology (for related references, see also Cellina [11], Ben-El-Mechaiekh [6] and Fan [19]).

Indeed, for any given open p-convex member u in $\mathfrak{U}$, as the set $\{x + u\}_{x\in K} \times \{y + u\}_{y\in K}$ is an open cover of $K \times K$, by Lemma 5.3, there exists a single-valued continuous mapping $f_u : K \to K$, which is $(\Omega \times u)$-approximation of the Graph(T), where $\Omega := \{x + u\}_{x\in K}$ as mentioned above. By Theorem 5.1, f_u has a fixed point $x_u = f_u(x_u)$ in K for each $u \in \mathfrak{U}$. Now for $x_u \in K$, by following the proof of Lemma 5.3, we observe that firstly, there exist $x'_u \in K$ and $q(x'_u) \in K$ such that $x_u \in x'_u + u$, and also $q(x'_u) \in x'_u + u$; and secondly, there also exists $v_u \in F(q(x'_u))$ such that $f_u(x_u) - v_u \in u$ which means that $f_u(x_u) \in v_u + u$.

In summary, for any given $u \in \mathfrak{U}$, there exist a continuous mapping $f_u : K \to K$, which has at least one fixed point $x_u \in K$ such that $x_u = f_u(x_u)$ with $(x_u, x_u) = (x_u, f_u(x_u)) \in Graph(f_u)$, and we also have the following statements:

(1) there exist $x'_u \in K$ and $q(x'_u) \in K$ such that $q(x'_u) \in x'_u + u$, and $x_u \in x'_u + u$; and

(2) there exist $v_u \in F(q(x'_u))$ such that $f_u(x_u) - v_u \in u$, which means $f_u(x_u) \in v_u + u$.

Since K is compact, without loss of the generality, we may assume that there exists a sub-net $(x_{u_i})_{u_i \in \mathfrak{U}}$ converges to x^* in K. Now we will show that x^* is the fixed point of T, i.e., $x^* \in T(x^*)$.

As K is compact, without loss of the generality, we may assume that three nets $\{x_u\}_{u\in\mathfrak{U}}$, $\{x'_u\}_{u\in\mathfrak{U}}$ and $\{q(x'_u)\}_{u\in\mathfrak{U}}$ in K have three sub-nets $\{x_{u_i}\}_{u_i\in\mathfrak{U}}$ converges to x^*, $\{x'_{u_i}\}_{u_i\in\mathfrak{U}}$ converges to x'^*, and $\{q(x'_{u_i})\}_{u_i\in\mathfrak{U}}$ converges to $q(x'^*)$ in K, respectively, in K. By the statement of (5.1) above, it is clear that we must have $x^* = x'^* = q(x'^*)$, as the family $\mathfrak{U}$ is the base of absolutely p-convex open neighborhoods for the origin in X; otherwise, by (1) we will have the contradiction, and thus our claim that $x^* = x'^* = q(x'^*)$ is true in locally p-convex space X.

Now we prove that x^* is a fixed point of T by using the statement of (5.2) for all $u \in \mathfrak{U}$. As the net $\{v_u\}_{u\in\mathfrak{U}} \subset K$, we may also assume its sub-net $\{v_{u_i}\}_{u_i\in\mathfrak{U}}$ converges to v^*. Then by the statement given by (2), it is clear that we have that $\lim_{u_i\in\mathfrak{U}} v_{u_i} = v^* = \lim_{u_i\in\mathfrak{U}} f_{u_i}(x_{u_i}) = \lim_{u_i\in\mathfrak{U}} x_{u_i} = x^*$. By the fact that $(v_{u_i}, q(x'_{u_i})) \in Graph(T)$, and the graph of T is closed by the assumption, it follows that $x^* = v^* \in T(x^*)$, which means x^* is a fixed point of T. The proof is complete. □

Remark 5.4 *Here we are not sure if the assumption "T(x) is with nonempty closed p-convex values" could be replaced by the condition "T(x) is with nonempty closed s-convex values" in Theorem 5.2. In fact, it seems that the proof of Theorem 4.3 given by Ennassik et al. [17] only goes through for the case $s \leq p$, not for the general case when both $s, p \in (0, 1]$ (please note that the letter p is used as the letter r by Ennassik et al. [17]). Thus, we are still looking for a proper way to prove if the conclusion of the Theorem 5.2 is true under Hausdorff topological vector spaces instead of locally p-convex spaces.*

By following the same idea used in the proof of Theorem 5.1, the conclusion of Theorem 5.2 still holds for compact quasi upper semicontinuous set-valued mappings as stated by Theorem 5.3 below and thus we omit its proof here.

Theorem 5.3 *If K is a nonempty closed s-convex subset of a Hausdorff locally p-convex space X, where $s, p \in (0, 1]$, then any compact quasi upper semicontinuous set-valued mapping $T : K \to 2^K$ with nonempty closed p-convex values, the graph of T is closed, has at least one fixed point.*

As an immediate consequence of Theorem 5.2, we have the following fixed point result for upper semicontinuous set-valued mappings in Hausdorff locally convex spaces for compact s-convex subsets, which include the common compact convex sets as a special class.

Corollary 5.1 *If K is a nonempty closed s-convex subset of a Hausdorff locally convex space X, where $s \in (0, 1]$, then any compact upper semicontinuous set-valued* *mapping $T : K \to 2^K$ with nonempty closed convex values, has at least one fixed point.*

Proof. Let $p = 1$ in Theorem 5.2, then the conclusion follows by Theorem 5.2. This completes the proof. □

As a special case of Theorem 5.3 or Corollary 5.1, we also have the following corollary.

Corollary 5.2 *If K is a nonempty compact s-convex subset of a Hausdorff locally convex space X, where $s \in (0, 1]$, then any upper semicontinuous set-valued mapping $T : K \to 2^K$ with nonempty closed convex values, has at least one fixed point.*

Remark 5.5 *Theorem 5.1 says that each compact single-valued mapping defined on a closed p-convex subsets $(0 < p \leq 1)$ in topological vector spaces has the fixed point property, which does not only include or improve most available results for fixed point theorems in the existing literature as special cases (just mention a few, Ben-El-Mechaiekh [6], Ben-El-Mechaiekh and Saidi [7], Ennassik and Taoudi [16], Mauldin [33], Granas and Dugundji [24], O'Regan and Precup [36], Reich [39], Park [37] and references wherein). In particular, we note that the answer to Schauder conjecture in the affirmative for a single-valued continuous mapping recently was obtained by Ennaassik and Taoudi [16] defined on nonempty compact p-convex subset in Hausdorff topological vector spaces, where $p \in (0, 1]$.*

We first note that Theorem 5.3 improve or unifies corresponding results given by Cauty [9], Cauty [10], Dobrowolski [14], Nhu [35], Park [37], Reich [39], Smart [44], Xiao and Lu [48], Xiao and Zhu [49], Yuan [52]–[54] under the framework of compact single-valued or upper semicontinuous set-valued mappings.

We also like to mention that by comparing with topological degree approach or related other methods used or developed by Cauty [9],[10], Nhu [35] and others, the arguments used in this section actually provide an accessible way for the study of nonlinear analysis for p-convex vector spaces $(0 < p \leq 1)$. The results given in this paper are new, and may easily understand by general readers in mathematical community, and see more by Yuan [53],[54] and related references on the study of nonlinear analysis and related applications in both p-vector and locally p-convex spaces for $0 < p \leq 1$.

ACKNOWLEDGMENT

This research is partially supported by the National Natural Science Foundation of China [grant numbers 71971031 and U1811462].

COMPLIANCE WITH ETHICAL STANDARDS

The author declares that there is no conflict of interest.

BIBLIOGRAPHY

[1] R.P. Agarwal, M. Meehan, and D. O'Regan, *Fixed Point Theory and Applications, Cambridge Tracts in Mathematics*, vol. 141, Cambridge University Press, Cambridge, 2001.

[2] S.M. Ageev, and D. Repovš, A selection theorem for strongly regular multivalued mappings, *Set-Valued Anal.*, 1998, 6, no. 4, 345–362.

[3] M.A. Alghamdi, D. O'Regan, and N. Shahzad, Krasnosel'skii type fixed point theorems for mappings on nonconvex sets, *Abstr. Appl. Anal. Article*, 2020, ID 267531, 1–23 (2012).

[4] V.K. Balachandran, *Topological Algebras*, vol. 185, Elsevier, Amsterdam, 2000.

[5] A. Bayoumi, *Foundations of Complex Analysis in Non Locally Convex Spaces. Function Theory without Convexity Condition, North-Holland Mathematics Studies*, Vol. 193. Elsevier Science B.V., Amsterdam, 2003.

[6] H. Ben-El-Mechaiekh. *Approximation and Selection Methods for Set-Valued Maps and Fixed Point Theory. Fixed Point Theory, Variational Analysis, and Optimization*, 77–136, CRC Press, Boca Raton, FL, 2014.

[7] H. Ben-El-Mechaiekh, and F.B. Saidi. On the continuous approximation of upper semi-continuous set-valued maps. *Questions Answers Gen. Topology*, 2013, 31, no. 2, 71–78.

[8] F.E. Browder, The fixed point theory of multi-valued mappings in topological vector spaces, *Math. Ann.*, 1968, 177, 283–301.

[9] R. Cauty, Rétractès absolus de voisinage algébriques.(French) [Algebraic absolute neighborhood retracts], *Serdica Math. J.*, 2005, 31, no. 4, 309–354.

[10] R. Cauty, Le théorème de Lefschetz - Hopf pour les applications compactes des espaces ULC. (French) [The Lefschetz - Hopf theorem for compact maps of uniformly locally contractible spaces], *J. Fixed Point Theory Appl.*, 2007, 1, no. 1, 123–134.

[11] A. Cellina. Approximation of set valued functions and fixed point theorems. *Ann. Mat. Pura Appl.*, 1969, 82, no. 4, 17–24.

[12] S.S. Chang, Some problems and results in the study of nonlinear analysis, Proceedings of the Second World Congress of Nonlinear Analysts, Part 7 (Athens, 1996), *Nonlinear Anal.*, 1997, 30, no. 7, 4197–4208.

[13] S.S. Chang, Y.J. Cho, and Y.Zhang, The topological versions of KKM theorem and Fan's matching theorem with applications, *Topol. Methods Nonlinear Anal.*, 1993, 1, no. 2, 231–245.

[14] T. Dobrowolski, Revisiting Cauty's proof of the Schauder conjecture, *Abstr. Appl. Anal.*, 2003, 7, 407–433.

[15] J. Dugundji, *Topology*, Allyn and Bacon, Inc.,Boston, 1978.

[16] M. Ennassik, and M.A. Taoudi, On the conjecture of Schauder, *J. Fixed Point Theory Appl.*, 2021, 23, no. 4, Paper No. 52, 15pp.

[17] M. Ennassik, L. Maniar and M.A. Taoudi, Fixed point theorems in r-normed and locally r-convex spaces and applications, *Fixed Point Theory*, 2021, 22, no. 2, 625–644.

[18] J. Ewert, and T. Neubrunn. On quasi-continuous multivalued maps. *Demonstration Math.*, 1988, 21, no. 3, 697–711.

[19] K. Fan, Fixed-point and minimax theorems in locally convex topological linear spaces, *Proc. Nat. Acad. Sci. U.S.A.*, 1952, 38, 121–126.

[20] K. Fan, A generalization of Tychonoff's fixed point theorem, *Math. Ann.*, 1960/1961, 142, 305–310.

[21] L. Gholizadeh, E. Karapinar, and M. Roohi, Some fixed point theorems in locally p-convex spaces, *Fixed Point Theory Appl.*, 2013, 2013, no. 312, 10 pp.

[22] L. Górniewicz, *Topological Fixed Point Theory of Multivalued Mappings. Mathematics and Its Applications*, vol. 495, Kluwer Academic Publishers, Dordrecht, 1999.

[23] T.X. Guo, R.X.Zhang, Y.C. Wang, and Z.C. Guo. Two fixed point theorems in complete random normed modules and their applications to backward stochastic equations, *J. Math. Anal. Appl.*, 2020, 483, no. 2, 123644, 30 pp.

[24] A. Granas and J. Dugundji, *Fixed Point Theory*, Springer-Verlag, New York, 2003.

[25] L. Holá, and A.K. Mirmostafaee, On continuity of set-valued mappings, *Topology Appl.*, 2022, 320, Paper No. 108200, 11 pp. (https://doi.org/10.1016/j.topol.2022.108200)

[26] H. Jarchow, *Locally Convex Spaces*, B.G. Teubner, Stuttgart, 1981.

[27] N.J. Kalton, Compact p-convex sets, *Q.J.Math.Oxf. Ser.*, 1977, 28, no.2, 301–308.

[28] N.J. Kalton, Universal spaces and universal bases in metric linear spaces, *Studia Math.*, 1977, 61, 161–191.

[29] N.J. Kalton, N.T. Peck, and J.W. Roberts, *An F-Space Sampler, London Mathematical Society Lecture Note Series*, vol. 89. Cambridge University Press, Cambridge, 1984.

[30] J.L. Kelly, *General Topology*, Van Nostrand, Princeton, NJ, 1957.

[31] W. Kryszewski. Graph-approximation of set-valued maps on noncompact domains. *Topology Appl.*, 1998, 83, no. 1, 1–21.

[32] J. Leray, and J. Schauder, Topologie et equations fonctionnelles, *Ann. Sci. Ecole Normale Sup.*, 1934, 51, 45–78.

[33] R.D. Mauldin, *The Scottish Book, Mathematics from the Scottish Café with Selected Problems from the New Scottish Book*, Second Edition, Birkhauser, Basel, 2015.

[34] T. Neubrunn. Quasi-continuity. *Real Anal. Exchange*, 1988/1989, 14, no. 2, 259–306. (https://doi.org/10.2307/44151947)

[35] N.T. Nhu, The fixed point property for weakly admissible compact convex sets: searching for a solution to Schauder's conjecture, *Topology Appl.*, 1996, 68, no. 1, 1–12.

[36] D. O'Regan, and R. Precup, *Theorems of Leray-Schauder Type and Applications*, Gordon and Breach Science Publishers, Philadelphia, 2001.

[37] S. Park, One hundred years of the Brouwer fixed point theorem, *J. Nat. Acad. Sci., ROK, Nat. Sci. Ser.*, 2021, 60, no. 1, 1–77.

[38] J. Qiu, and S. Rolewicz, Ekeland's variational principle in locally p-convex spaces and related results, *Stud. Math.*, 2008, 186, no. 3, 219–235.

[39] S. Reich, Fixed points in locally convex spaces. *Math. Z.*, 1972, 125, 17–31.

[40] D. Repovš, P.V. Semenov, and E.V. Ščepin. Approximation of upper semicontinuous maps on paracompact spaces. *Rocky Mountain J. Math.*, 1998, 28, no. 3, 1089–1101.

[41] J.W. Roberts, A compact convex set with no extreme points, *Studia Math.*, 1977, 60, no. 3, 255–266.

[42] S. Rolewicz, *Metric Linear Spaces*, PWN-Polish Scientific Publishers, Warszawa, 1985.

[43] J. Schauder, Der Fixpunktsatz in Funktionalraumen, *Stud. Math.*, 1930, 2, 171–180.

[44] D.R. Smart, *Fixed Point Theorems*, Cambridge University Press, Cambridge, 1980.

[45] A. Tychonoff, Ein Fixpunktsatz, *Math. Ann.*, 1935, 111, 767–776.

[46] H. Weber, Compact convex sets in non-locally convex linear spaces. Dedicated to the memory of Professor Gottfried Köthe, *Note Mat.*, 1992, 12, 271–289.

[47] H. Weber, Compact convex sets in non-locally convex linear spaces, Schauder-Tychonoff fixed point theorem, In, *Topology, Measures, and Fractals* (Warnemunde, 1991), 1992, 37–40, *Math. Res.*, 66, Akademie-Verlag, Berlin.

[48] J.Z. Xiao, and Y. Lu, Some fixed point theorems for s-convex subsets in p-normed spaces based on measures ofnoncompactness, *J. Fixed Point Theory Appl.*, 2018, 20, no.2, Paper No. 83, 22 pp.

[49] J.Z. Xiao, and X.H. Zhu, Some fixed point theorems for s-convex subsets in p-normed spaces, *Nonlinear Anal.*, 2011, 74, no. 5, 1738–1748.

[50] H.K. Xu, Metric fixed point theory for multivalued mappings, Dissertationes Math. *Rozprawy Mat.*, 2000, 389, 39 pp.

[51] G.X.Z. Yuan, The study of minimax inequalities and applications to economies and variational inequalities, *Mem. Amer. Math. Soc.*, 1998, 132, no. 625, 1–146.

[52] G.X.Z. Yuan, *KKM Theory and Applications in Nonlinear Analysis, Monographs and Textbooks in Pure and Applied Mathematics*, vol. 218. Marcel Dekker, Inc., New York, 1999.

[53] G.X.Z. Yuan, Nonlinear analysis by applying best approximation method in p-vector spaces. *Fixed Point Theory Algorithms Sci. Eng.*, 2022, 20 (2022) (https://doi.org/10.1186/s13663-022-00730-x).

[54] G.X.Z. Yuan, Nonlinear analysis in p-vector spaces for single-valued 1-set contractive mappings. *Fixed Point Theory Algorithms Sci. Eng.*, 2022, 26 (https://doi.org/10.1186/s13663-022-00735-6).

[55] E. Zeidler, *Nonlinear Functional Analysis and Its Applications, vol. I, Fixed-Point Theorems*, Springer Verlag, New York, 1986.

CHAPTER 6

Proinov E_S-Contraction Type Unique and Non-Unique Fixed-Point Results on S-Metric Spaces

Nihal Taş

Balıkesir University

CONTENTS

6.1 INTRODUCTION AND MOTIVATION

Fixed-point theory is a useful tool for solving some problems in mathematics and various research areas such as engineering, optimization, etc. This theory is studied with different approaches. One of them is to generalize the used contractive conditions such as Proinov type and Proinov type E-contractions (see [1, 19] and the references therein). Another approach is to generalize the used metric space. For example, the notion of S-metric space was introduced for this purpose (see [20] for more details). Recently, a new generalization is the fixed-figure problem given in [14] (for example,

DOI: 10.1201/9781003388678-6

see [2, 3, 5, 6, 10, 11, 12, 22, 23]). The last generalization is important to obtain some applications (for example, see [15, 16, 17] and the references therein).

Now we recall some basic notions as follows:

Definition 6.1 *[20] Let $\mathcal{Y} \neq \emptyset$ be any set and $\mathfrak{S} : \mathcal{Y} \times \mathcal{Y} \times \mathcal{Y} \to [0, \infty)$ be a function satisfying the following conditions for all $\mathfrak{q}, \mathfrak{w}, \mathfrak{t}, \mathfrak{a} \in \mathcal{Y}$.*

$(S1)$ $\mathfrak{S}(\mathfrak{q}, \mathfrak{w}, \mathfrak{t}) = 0$ if and only if $\mathfrak{q} = \mathfrak{w} = \mathfrak{t}$.

$(S2)$ $\mathfrak{S}(\mathfrak{q}, \mathfrak{w}, \mathfrak{t}) \leq \mathfrak{S}(\mathfrak{q}, \mathfrak{q}, \mathfrak{a}) + \mathfrak{S}(\mathfrak{w}, \mathfrak{w}, \mathfrak{a}) + \mathfrak{S}(\mathfrak{t}, \mathfrak{t}, \mathfrak{a})$.

Then $\mathfrak{S}$ is called an S-metric on $\mathcal{Y}$ and $(\mathcal{Y}, \mathfrak{S})$ is called an S-metric space.

Lemma 6.1 *[20] Let $(\mathcal{Y}, \mathfrak{S})$ be an S-metric space. Then we have*

$$\mathfrak{S}(\mathfrak{q}, \mathfrak{q}, \mathfrak{w}) = \mathfrak{S}(\mathfrak{w}, \mathfrak{w}, \mathfrak{q}). \tag{6.1}$$

The equality (6.1) can be considered as a symmetry property on an S-metric space.

Definition 6.2 *[20] Let $(\mathcal{Y}, \mathfrak{S})$ be an S-metric space.*

1. *A sequence $\{\mathfrak{q}_n\}$ in $\mathcal{Y}$ converges to $\mathfrak{q}$ if and only if $\mathfrak{S}(\mathfrak{q}_n, \mathfrak{q}_n, \mathfrak{q}) \to 0$ as $n \to \infty$. That is, there exists $n_0 \in \mathbb{N}$ such that for all $n \geq n_0$, $\mathfrak{S}(\mathfrak{q}_n, \mathfrak{q}_n, \mathfrak{q}) < \varepsilon$ for each $\varepsilon > 0$. We denote this by*

$$\lim_{n\to\infty} \mathfrak{q}_n = \mathfrak{q} \textit{ or } \lim_{n\to\infty} \mathfrak{S}(\mathfrak{q}_n, \mathfrak{q}_n, \mathfrak{q}) = 0.$$

2. *A sequence $\{\mathfrak{q}_n\}$ in $\mathcal{Y}$ is called a Cauchy sequence if $\mathfrak{S}(\mathfrak{q}_n, \mathfrak{q}_n, \mathfrak{q}_\mathfrak{d}) \to 0$ as $n, \mathfrak{d} \to \infty$. That is, there exists $n_0 \in \mathbb{N}$ such that for all $n, \mathfrak{d} \geq n_0$, $\mathfrak{S}(\mathfrak{q}_n, \mathfrak{q}_n, \mathfrak{q}_\mathfrak{d}) < \varepsilon$ for each $\varepsilon > 0$.*
3. *The S-metric space $(\mathcal{Y}, \mathfrak{S})$ is called complete if every Cauchy sequence is convergent.*

In the following, we see the relationship between a metric and an S-metric.

Lemma 6.2 *[4] Let $(\mathcal{Y}, \rho)$ be a metric space. Then the following properties are satisfied:*

1. *$\mathfrak{S}_\rho(\mathfrak{q}, \mathfrak{w}, \mathfrak{t}) = \rho(\mathfrak{q}, \mathfrak{t}) + \rho(\mathfrak{w}, \mathfrak{t})$ for all $\mathfrak{q}, \mathfrak{w}, \mathfrak{t} \in \mathcal{Y}$ is an S-metric on $\mathcal{Y}$.*
2. *$\mathfrak{q}_n \to \mathfrak{q}$ in $(\mathcal{Y}, \rho) \iff \mathfrak{q}_n \to \mathfrak{q}$ in $(\mathcal{Y}, \mathfrak{S}_\rho)$.*
3. *$\{\mathfrak{q}_n\}$ is Cauchy in $(\mathcal{Y}, \rho) \iff \{\mathfrak{q}_n\}$ is Cauchy in $(\mathcal{Y}, \mathfrak{S}_\rho)$.*
4. *$(\mathcal{Y}, \rho)$ is complete $\iff (\mathcal{Y}, \mathfrak{S}_\rho)$ is complete.*

$\mathfrak{S}_\rho$ is called the S-metric generated by the metric ρ. In the literature, there are some examples of an S-metric, which is not generated by any metric in [4, 9].

Recently, using the above definitions and Lemma 6.3, some fixed-point results have been given with different techniques on an S-metric space (see [4, 7, 8, 9, 20, 21]).

Lemma 6.3 *Let $(\mathcal{Y}, \mathfrak{S})$ be an S-metric spaceS-metric space and $\{\mathfrak{q}_n\}$ be a sequence in $\mathcal{Y}$ which is not Cauchy and*

$$\lim_{n\to\infty} \mathfrak{S}(\mathfrak{q}_n, \mathfrak{q}_n, \mathfrak{q}_{n+1}) = 0.$$

Then there are two subsequences $\{\mathfrak{q}_{n_k}\}$ and $\{\mathfrak{q}_{\mathfrak{d}_k}\}$ of $\{\mathfrak{q}_n\}$ and $e > 0$ such that

$$\lim_{k\to\infty} \mathfrak{S}\left(\mathfrak{q}_{n_{k+1}}, \mathfrak{q}_{n_{k+1}}, \mathfrak{q}_{\mathfrak{d}_{k+1}}\right) = \lim_{k\to\infty} \mathfrak{S}\left(\mathfrak{q}_{n_k}, \mathfrak{q}_{n_k}, \mathfrak{q}_{\mathfrak{d}_k}\right) \tag{6.2}$$

$$= \lim_{k\to\infty} \mathfrak{S}\left(\mathfrak{q}_{n_{k+1}}, \mathfrak{q}_{n_{k+1}}, \mathfrak{q}_{\mathfrak{d}_k}\right) \tag{6.3}$$

$$= \lim_{k\to\infty} \mathfrak{S}\left(\mathfrak{q}_{n_k}, \mathfrak{q}_{n_k}, \mathfrak{q}_{\mathfrak{d}_{k+1}}\right) = e.$$

Let $\mathfrak{N} : \mathcal{Y} \to \mathcal{Y}$ be a self-mapping and $\theta : \mathcal{Y} \times \mathcal{Y} \to [0, \infty)$ a mapping. $\mathfrak{N}$ is called triangular θ-orbital admissible (briefly, θ-t.o.a.) if the following statements are satisfied [18]:

(o) $\theta(\mathfrak{q}, \mathfrak{N}\mathfrak{q}) \geq 1 \Longrightarrow \theta(\mathfrak{N}\mathfrak{q}, \mathfrak{N}^2\mathfrak{q}) \geq 1$ for any $\mathfrak{q} \in \mathcal{Y}$,

(t_0) $\theta(\mathfrak{q}, \mathfrak{w}) \geq 1$ and $\theta(\mathfrak{w}, \mathfrak{N}\mathfrak{w}) \geq 1 \Longrightarrow \theta(\mathfrak{q}, \mathfrak{N}\mathfrak{w}) \geq 1$ for any $\mathfrak{q}, \mathfrak{w} \in \mathcal{Y}$.

Lemma 6.4 *[18] Let $\mathcal{Y} \neq \emptyset$ and $\{\mathfrak{q}_{\mathfrak{d}}\}$ be a sequence on $\mathcal{Y}$ by*

$$\mathfrak{q}_{\mathfrak{d}} = \mathfrak{N}\mathfrak{q}_{\mathfrak{d}-1},$$

for any $\mathfrak{d} \in \mathbb{N}$, where $\mathfrak{N}$ is a θ-t.o.a. If there is $\mathfrak{q}_0 \in \mathcal{Y}$ such that $\theta(\mathfrak{q}_0, \mathfrak{N}\mathfrak{q}_0) \geq 1$ then $\theta(\mathfrak{q}_n, \mathfrak{q}_{\mathfrak{d}}) \geq 1$ for all $n, \mathfrak{d} \in \mathbb{N}$.

6.2 MAIN RESULTS

In this section, we prove some fixed-point and fixed-figure results using the Proinov type E-contraction defined in [1] on S-metric spaces. Also, we give some illustrative examples.

6.2.1 Some Fixed-Point Results

In this subsection, suppose that $(\mathcal{Y}, \mathfrak{S})$ is a complete S-metric space, $T : \mathcal{Y} \to \mathcal{Y}$ is a self-mapping, $\alpha, \beta : (0, \infty) \to \mathbb{R}$ are two functions such that

$$\beta(\mathfrak{s}) < \alpha(\mathfrak{s}), \tag{6.4}$$

for all $\mathfrak{s} > 0$ and $\theta : \mathcal{Y} \times \mathcal{Y} \to [0, \infty)$ is a function.

Definition 6.3 *$\mathfrak{N}$ is called a (θ, α, β)-E_S-contraction if the inequality*

$$\theta(\mathfrak{q}, \mathfrak{w})\alpha\left[\mathfrak{S}(\mathfrak{N}\mathfrak{q}, \mathfrak{N}\mathfrak{q}, \mathfrak{N}\mathfrak{w})\right] \leq \beta\left[E_S(\mathfrak{q}, \mathfrak{w})\right], \tag{6.5}$$

holds for all $\mathfrak{q}, \mathfrak{w} \in \mathcal{Y}$ such that $\mathfrak{S}(\mathfrak{q}, \mathfrak{q}, \mathfrak{w}) > 0$ and $\mathfrak{S}(\mathfrak{N}\mathfrak{q}, \mathfrak{N}\mathfrak{q}, \mathfrak{N}\mathfrak{w}) > 0$, where

$$E_S(\mathfrak{q}, \mathfrak{w}) = \max\left\{ \begin{array}{c} \mathfrak{S}(\mathfrak{q}, \mathfrak{q}, \mathfrak{w}) + |\mathfrak{S}(\mathfrak{q}, \mathfrak{q}, \mathfrak{N}\mathfrak{q}) - \mathfrak{S}(\mathfrak{w}, \mathfrak{w}, \mathfrak{N}\mathfrak{w})|, \\ \frac{\mathfrak{S}(\mathfrak{q}, \mathfrak{q}, T\mathfrak{w}) + \mathfrak{S}(\mathfrak{w}, \mathfrak{w}, T\mathfrak{q})}{3} \end{array} \right\}.$$

Theorem 6.1 *Let $\mathfrak{N}$ be a (θ, α, β)-E_S-contraction such that*

(s_1) *α is non-decreasing and lower semi-continuous,*

(s_2) $\limsup\limits_{\mathfrak{s} \to \mathfrak{s}_0} \beta(\mathfrak{s}) < \alpha(\mathfrak{s}_0)$ *for any* $\mathfrak{s}_0 > 0$,

(s_3) *$\mathfrak{N}$ is θ-t.o.a. and there is $\mathfrak{q}_0 \in \mathcal{Y}$ such that $\theta(\mathfrak{q}_0, \mathfrak{N}\mathfrak{q}_0) \geq 1$,*

(s_4) *For any $\mathfrak{d} \in \mathbb{N}$, $\theta(\mathfrak{q}_\mathfrak{d}, \mathfrak{q}^*) \geq 1$ for any sequence $(\mathfrak{q}_\mathfrak{d})$ such that $\mathfrak{q}_\mathfrak{d} \to \mathfrak{q}^*$ and $\theta(\mathfrak{q}_\mathfrak{d}, \mathfrak{q}_{\mathfrak{d}+1}) \geq 1$.*

Then T has a fixed point.

Proof 6.1 *Let $\mathfrak{q}_0$ be any point in $\mathcal{Y}$ such that*

$$\theta(\mathfrak{q}_0, \mathfrak{N}\mathfrak{q}_0) \geq 1$$

and $\{\mathfrak{q}_\mathfrak{d}\}$ be the sequence on $\mathcal{Y}$ defined by

$$\mathfrak{q}_1 = \mathfrak{N}\mathfrak{q}_0 \text{ and } \mathfrak{q}_\mathfrak{d} = \mathfrak{N}\mathfrak{q}_{\mathfrak{d}-1}.$$

Suppose that

$$\mathfrak{S}(\mathfrak{q}_\mathfrak{d}, \mathfrak{q}_\mathfrak{d}, \mathfrak{q}_{\mathfrak{d}+1}) > 0,$$

for every natural number $\mathfrak{d}$. On the contrary, there is $\mathfrak{d}_i \in \mathbb{N}$ such that

$$\mathfrak{q}_{\mathfrak{d}_i} = \mathfrak{q}_{\mathfrak{d}_i+1}.$$

Using the definition of the sequence $\{\mathfrak{q}_{\mathfrak{d}_i}\}$, we have

$$\mathfrak{S}(\mathfrak{N}\mathfrak{q}_{\mathfrak{d}_i}, \mathfrak{N}\mathfrak{q}_{\mathfrak{d}_i}, \mathfrak{q}_{\mathfrak{d}_i}) = \mathfrak{S}(\mathfrak{q}_{\mathfrak{d}_i+1}, \mathfrak{q}_{\mathfrak{d}_i+1}, \mathfrak{q}_{\mathfrak{d}_i}) = 0$$

and

$$\mathfrak{N}\mathfrak{q}_{\mathfrak{d}_i} = \mathfrak{q}_{\mathfrak{d}_i},$$

that is, $\mathfrak{q}_{\mathfrak{d}_i} \in Fix(\mathfrak{N}) = \{\mathfrak{q} \in \mathcal{Y} : \mathfrak{q} = \mathfrak{N}\mathfrak{q}\}$. So, let us take $\mathfrak{q} = \mathfrak{q}_\mathfrak{d}$ and $\mathfrak{w} = \mathfrak{q}_{\mathfrak{d}+1}$ in (21.13). Using (21.12) and Lemma 6.4, we get

$$\begin{aligned}
\alpha\left[\mathfrak{S}(\mathfrak{q}_{\mathfrak{d}+1}, \mathfrak{q}_{\mathfrak{d}+1}, \mathfrak{q}_{\mathfrak{d}+2})\right] &= \alpha\left[\mathfrak{S}(\mathfrak{N}\mathfrak{q}_\mathfrak{d}, \mathfrak{N}\mathfrak{q}_\mathfrak{d}, \mathfrak{N}\mathfrak{q}_{\mathfrak{d}+1})\right] \\
&\leq \theta(\mathfrak{q}_\mathfrak{d}, \mathfrak{q}_{\mathfrak{d}+1})\alpha\left[\mathfrak{S}(\mathfrak{N}\mathfrak{q}_\mathfrak{d}, \mathfrak{N}\mathfrak{q}_\mathfrak{d}, \mathfrak{N}\mathfrak{q}_{\mathfrak{d}+1})\right] \\
&\leq \beta\left[E_S(\mathfrak{q}_\mathfrak{d}, \mathfrak{q}_{\mathfrak{d}+1})\right] < \alpha\left[E_S(\mathfrak{q}_\mathfrak{d}, \mathfrak{q}_{\mathfrak{d}+1})\right], \qquad (6.6)
\end{aligned}$$

where

$$\begin{aligned}
&E_S(\mathfrak{q}_\mathfrak{d}, \mathfrak{q}_{\mathfrak{d}+1}) \\
&= \max\left\{\begin{array}{c} \mathfrak{S}(\mathfrak{q}_\mathfrak{d}, \mathfrak{q}_\mathfrak{d}, \mathfrak{q}_{\mathfrak{d}+1}) + \left|\mathfrak{S}(\mathfrak{q}_\mathfrak{d}, \mathfrak{q}_\mathfrak{d}, \mathfrak{N}\mathfrak{q}_\mathfrak{d}) - \mathfrak{S}(\mathfrak{q}_{\mathfrak{d}+1}, \mathfrak{q}_{\mathfrak{d}+1}, \mathfrak{N}\mathfrak{q}_{\mathfrak{d}+1})\right|, \\ \frac{\mathfrak{S}(\mathfrak{q}_\mathfrak{d}, \mathfrak{q}_\mathfrak{d}, \mathfrak{N}\mathfrak{q}_{\mathfrak{d}+1}) + \mathfrak{S}(\mathfrak{q}_{\mathfrak{d}+1}, \mathfrak{q}_{\mathfrak{d}+1}, \mathfrak{N}\mathfrak{q}_\mathfrak{d})}{3} \end{array}\right\} \\
&= \max\left\{\begin{array}{c} \mathfrak{S}(\mathfrak{q}_\mathfrak{d}, \mathfrak{q}_\mathfrak{d}, \mathfrak{q}_{\mathfrak{d}+1}) + \left|\mathfrak{S}(\mathfrak{q}_\mathfrak{d}, \mathfrak{q}_\mathfrak{d}, \mathfrak{q}_{\mathfrak{d}+1}) - \mathfrak{S}(\mathfrak{q}_{\mathfrak{d}+1}, \mathfrak{q}_{\mathfrak{d}+1}, \mathfrak{q}_{\mathfrak{d}+2})\right|, \\ \frac{\mathfrak{S}(\mathfrak{q}_\mathfrak{d}, \mathfrak{q}_\mathfrak{d}, \mathfrak{q}_{\mathfrak{d}+2}) + \mathfrak{S}(\mathfrak{q}_{\mathfrak{d}+1}, \mathfrak{q}_{\mathfrak{d}+1}, \mathfrak{q}_{\mathfrak{d}+1})}{3} \end{array}\right\}
\end{aligned}$$

$$\leq \max\left\{ \begin{array}{c} \mathfrak{S}(\mathfrak{q}_{\mathfrak{d}},\mathfrak{q}_{\mathfrak{d}},\mathfrak{q}_{\mathfrak{d}+1}) + |\mathfrak{S}(\mathfrak{q}_{\mathfrak{d}},\mathfrak{q}_{\mathfrak{d}},\mathfrak{q}_{\mathfrak{d}+1}) - \mathfrak{S}(\mathfrak{q}_{\mathfrak{d}+1},\mathfrak{q}_{\mathfrak{d}+1},\mathfrak{q}_{\mathfrak{d}+2})|, \\ \frac{2\mathfrak{S}(\mathfrak{q}_{\mathfrak{d}},\mathfrak{q}_{\mathfrak{d}},\mathfrak{q}_{\mathfrak{d}+1})+\mathfrak{S}(\mathfrak{q}_{\mathfrak{d}+2},\mathfrak{q}_{\mathfrak{d}+2},\mathfrak{q}_{\mathfrak{d}+1})}{3} \end{array} \right\}.$$

If $\mathfrak{S}(\mathfrak{q}_{\mathfrak{d}+1},\mathfrak{q}_{\mathfrak{d}+1},\mathfrak{q}_{\mathfrak{d}+2}) > \mathfrak{S}(\mathfrak{q}_{\mathfrak{d}},\mathfrak{q}_{\mathfrak{d}},\mathfrak{q}_{\mathfrak{d}+1})$, *then we have*

$$E_S(\mathfrak{q}_{\mathfrak{d}},\mathfrak{q}_{\mathfrak{d}+1}) \leq \mathfrak{S}(\mathfrak{q}_{\mathfrak{d}+1},\mathfrak{q}_{\mathfrak{d}+1},\mathfrak{q}_{\mathfrak{d}+2}).$$

Using the inequality (21.14), we have

$$\alpha\left[\mathfrak{S}(\mathfrak{q}_{\mathfrak{d}+1},\mathfrak{q}_{\mathfrak{d}+1},\mathfrak{q}_{\mathfrak{d}+2})\right] < \alpha\left[\mathfrak{S}(\mathfrak{q}_{\mathfrak{d}+1},\mathfrak{q}_{\mathfrak{d}+1},\mathfrak{q}_{\mathfrak{d}+2})\right],$$

a contradiction. If $\mathfrak{S}(\mathfrak{q}_{\mathfrak{d}+1},\mathfrak{q}_{\mathfrak{d}+1},\mathfrak{q}_{\mathfrak{d}+2}) < \mathfrak{S}(\mathfrak{q}_{\mathfrak{d}},\mathfrak{q}_{\mathfrak{d}},\mathfrak{q}_{\mathfrak{d}+1})$, *then we have*

$$E_S(\mathfrak{q}_{\mathfrak{d}},\mathfrak{q}_{\mathfrak{d}+1}) \leq 2\mathfrak{S}(\mathfrak{q}_{\mathfrak{d}},\mathfrak{q}_{\mathfrak{d}},\mathfrak{q}_{\mathfrak{d}+1}) - \mathfrak{S}(\mathfrak{q}_{\mathfrak{d}+1},\mathfrak{q}_{\mathfrak{d}+1},\mathfrak{q}_{\mathfrak{d}+2}).$$

Using the inequality (21.14), we find

$$\alpha\left[\mathfrak{S}(\mathfrak{q}_{\mathfrak{d}+1},\mathfrak{q}_{\mathfrak{d}+1},\mathfrak{q}_{\mathfrak{d}+2})\right] < \alpha\left[2\mathfrak{S}(\mathfrak{q}_{\mathfrak{d}},\mathfrak{q}_{\mathfrak{d}},\mathfrak{q}_{\mathfrak{d}+1}) - \mathfrak{S}(\mathfrak{q}_{\mathfrak{d}+1},\mathfrak{q}_{\mathfrak{d}+1},\mathfrak{q}_{\mathfrak{d}+2})\right].$$

By (s_1), *we get*

$$\begin{aligned} \mathfrak{S}(\mathfrak{q}_{\mathfrak{d}+1},\mathfrak{q}_{\mathfrak{d}+1},\mathfrak{q}_{\mathfrak{d}+2}) &< 2\mathfrak{S}(\mathfrak{q}_{\mathfrak{d}},\mathfrak{q}_{\mathfrak{d}},\mathfrak{q}_{\mathfrak{d}+1}) - \mathfrak{S}(\mathfrak{q}_{\mathfrak{d}+1},\mathfrak{q}_{\mathfrak{d}+1},\mathfrak{q}_{\mathfrak{d}+2}) \\ &\Longrightarrow \mathfrak{S}(\mathfrak{q}_{\mathfrak{d}+1},\mathfrak{q}_{\mathfrak{d}+1},\mathfrak{q}_{\mathfrak{d}+2}) < \mathfrak{S}(\mathfrak{q}_{\mathfrak{d}},\mathfrak{q}_{\mathfrak{d}},\mathfrak{q}_{\mathfrak{d}+1}), \end{aligned}$$

so that, the sequence $\{\mathfrak{S}(\mathfrak{q}_{\mathfrak{d}},\mathfrak{q}_{\mathfrak{d}},\mathfrak{q}_{\mathfrak{d}+1})\}$ *is bounded below by* 0 *and a decreasing sequence. Therefore, there is* $\delta \geq 0$ *such that*

$$\lim_{\mathfrak{d}\to\infty} \mathfrak{S}(\mathfrak{q}_{\mathfrak{d}},\mathfrak{q}_{\mathfrak{d}},\mathfrak{q}_{\mathfrak{d}+1}) = \delta.$$

Then, we have

$$\lim_{\mathfrak{d}\to\infty} E_S(\mathfrak{q}_{\mathfrak{d}},\mathfrak{q}_{\mathfrak{d}+1}) = \delta.$$

We claim that $\delta = 0$. *On the contrary, we assume* $\delta > 0$. *Using the inequality (21.14), we have*

$$\alpha[\delta] = \lim_{\mathfrak{d}\to\infty} \alpha\left[\mathfrak{S}(\mathfrak{q}_{\mathfrak{d}+1},\mathfrak{q}_{\mathfrak{d}+1},\mathfrak{q}_{\mathfrak{d}+2})\right] \leq \limsup_{\mathfrak{d}\to\infty} \beta\left[E_S(\mathfrak{q}_{\mathfrak{d}},\mathfrak{q}_{\mathfrak{d}+1})\right] \leq \limsup_{\mathfrak{s}\to\delta} \alpha[\mathfrak{s}],$$

a contradiction with (s_2). *Consequently, it should be* $\delta = 0$, *that is,*

$$\lim_{\mathfrak{d}\to\infty} \mathfrak{S}(\mathfrak{q}_{\mathfrak{d}},\mathfrak{q}_{\mathfrak{d}},\mathfrak{q}_{\mathfrak{d}+1}) = \lim_{\mathfrak{d}\to\infty} \mathfrak{S}(\mathfrak{q}_{\mathfrak{d}},\mathfrak{q}_{\mathfrak{d}},\mathfrak{N}\mathfrak{q}_{\mathfrak{d}}) = 0. \tag{6.7}$$

Now we show that $\{\mathfrak{q}_{\mathfrak{d}}\}$ *is a Cauchy sequence. On the contrary, we assume that* $\{\mathfrak{q}_{\mathfrak{d}}\}$ *is not Cauchy. From Lemma 6.3, we can find two subsequences* $\{\mathfrak{q}_{\mathfrak{d}_j}\}$, $\{\mathfrak{q}_{k_j}\}$ *of the sequence* $\{\mathfrak{q}_{\mathfrak{d}}\}$ *such that (6.3) holds. Let us set*

$$a_j = \mathfrak{S}(\mathfrak{q}_{\mathfrak{d}_j},\mathfrak{q}_{\mathfrak{d}_j},\mathfrak{q}_{k_j})$$

and

$$b_j = E_S(\mathfrak{q}_{\mathfrak{d}_j}, \mathfrak{q}_{k_j}),$$

respectively. Using (6.3) and (21.15), we have

$$\lim_{j\to\infty} a_j = e,$$

$$\begin{aligned}
\lim_{j\to\infty} b_j &= \lim_{j\to\infty} E_S(\mathfrak{q}_{\mathfrak{d}_j}, \mathfrak{q}_{k_j}) \\
&= \lim_{j\to\infty} \max\left\{ \begin{array}{c} \mathfrak{S}(\mathfrak{q}_{\mathfrak{d}_j}, \mathfrak{q}_{\mathfrak{d}_j}, \mathfrak{q}_{k_j}) + \Big|\mathfrak{S}(\mathfrak{q}_{\mathfrak{d}_j}, \mathfrak{q}_{\mathfrak{d}_j}, \mathfrak{N}\mathfrak{q}_{\mathfrak{d}_j}) \\ -\mathfrak{S}(\mathfrak{q}_{k_j}, \mathfrak{q}_{k_j}, \mathfrak{N}\mathfrak{q}_{k_j})\Big|, \frac{\mathfrak{S}(\mathfrak{q}_{\mathfrak{d}_j}, \mathfrak{q}_{\mathfrak{d}_j}, \mathfrak{N}\mathfrak{q}_{k_j}) + \mathfrak{S}(\mathfrak{q}_{k_j}, \mathfrak{q}_{k_j}, \mathfrak{N}\mathfrak{q}_{\mathfrak{d}_j})}{3} \end{array} \right\} \\
&= \lim_{j\to\infty} \max\left\{ \begin{array}{c} \mathfrak{S}(\mathfrak{q}_{\mathfrak{d}_j}, \mathfrak{q}_{\mathfrak{d}_j}, \mathfrak{q}_{k_j}) + \Big|\mathfrak{S}(\mathfrak{q}_{\mathfrak{d}_j}, \mathfrak{q}_{\mathfrak{d}_j}, \mathfrak{q}_{\mathfrak{d}_j+1}) \\ -\mathfrak{S}(\mathfrak{q}_{k_j}, \mathfrak{q}_{k_j}, \mathfrak{q}_{k_j+1})\Big|, \frac{\mathfrak{S}(\mathfrak{q}_{\mathfrak{d}_j}, \mathfrak{q}_{\mathfrak{d}_j}, \mathfrak{q}_{k_j+1}) + \mathfrak{S}(\mathfrak{q}_{k_j}, \mathfrak{q}_{k_j}, \mathfrak{q}_{\mathfrak{d}_j+1})}{3} \end{array} \right\} \\
&= e.
\end{aligned}$$

For $\mathfrak{q} = \mathfrak{q}_{\mathfrak{d}_j}$ *and* $\mathfrak{w} = \mathfrak{q}_{k_j}$*, using (21.14) and Lemma 6.4, we find*

$$\begin{aligned}
\alpha\left[a_{j+1}\right] &= \alpha\left[\mathfrak{S}(\mathfrak{q}_{\mathfrak{d}_j+1}, \mathfrak{q}_{\mathfrak{d}_j+1}, \mathfrak{q}_{k_j+1})\right] \\
&\le \theta\left(\mathfrak{q}_{\mathfrak{d}_j}, \mathfrak{q}_{k_j}\right) \alpha\left[\mathfrak{S}(\mathfrak{N}\mathfrak{q}_{\mathfrak{d}_j}, \mathfrak{N}\mathfrak{q}_{\mathfrak{d}_j}, \mathfrak{N}\mathfrak{q}_{k_j})\right] \\
&\le \beta\left[E_S(\mathfrak{q}_{\mathfrak{d}_j}, \mathfrak{q}_{k_j})\right] = \beta\left[b_j\right].
\end{aligned} \tag{6.8}$$

Using (21.16), we find

$$\alpha\left[e\right] = \lim_{j\to\infty} \alpha\left[a_j\right] \le \limsup_{j\to\infty} \beta\left[b_j\right] \le \limsup_{t\to e} \beta\left[t\right],$$

a contradiction with (s_2)*. Consequently,* $\{\mathfrak{q}_{\mathfrak{d}}\}$ *is Cauchy on a complete metric space* $\mathcal{Y}$*. Let* $\mathfrak{t}$ *be the limit of the sequence* $\{\mathfrak{q}_{\mathfrak{d}}\}$*. Thereby,* $\mathfrak{t} \in Fix(\mathfrak{N})$ *under the condition* (s_4)*. If not, we have*

$$0 < \mathfrak{S}(\mathfrak{t}, \mathfrak{t}, \mathfrak{N}\mathfrak{t}) \le 2\mathfrak{S}(\mathfrak{t}, \mathfrak{t}, \mathfrak{N}\mathfrak{q}_{\mathfrak{d}}) + \mathfrak{S}(\mathfrak{N}\mathfrak{t}, \mathfrak{N}\mathfrak{t}, \mathfrak{N}\mathfrak{q}_{\mathfrak{d}}). \tag{6.9}$$

If $\mathfrak{S}(\mathfrak{N}\mathfrak{t}, \mathfrak{N}\mathfrak{t}, \mathfrak{N}\mathfrak{q}_{\mathfrak{d}}) = 0$ *for infinite natural numbers* $\mathfrak{d}$*, then using (21.17), we get*

$$0 < \mathfrak{S}(\mathfrak{t}, \mathfrak{t}, \mathfrak{N}\mathfrak{t}) \le 2\mathfrak{S}(\mathfrak{t}, \mathfrak{t}, \mathfrak{q}_{\mathfrak{d}+1}) \to 0$$

and so

$$\mathfrak{S}(\mathfrak{t}, \mathfrak{t}, \mathfrak{N}\mathfrak{t}) = 0.$$

If $\mathfrak{S}(\mathfrak{N}\mathfrak{t}, \mathfrak{N}\mathfrak{t}, \mathfrak{N}\mathfrak{q}_{\mathfrak{d}}) > 0$ *for any* $\mathfrak{d} \in \mathbb{N}$*, then using (21.14), we have*

$$\begin{aligned}
\alpha\left[\mathfrak{S}(\mathfrak{q}_{\mathfrak{d}+1}, \mathfrak{q}_{\mathfrak{d}+1}, \mathfrak{N}\mathfrak{t})\right] &\le \theta\left(\mathfrak{q}_{\mathfrak{d}}, \mathfrak{t}\right) \alpha\left[\mathfrak{S}(\mathfrak{N}\mathfrak{q}_{\mathfrak{d}}, \mathfrak{N}\mathfrak{q}_{\mathfrak{d}}, \mathfrak{N}\mathfrak{t})\right] \\
&\le \beta\left[E_S\left(\mathfrak{q}_{\mathfrak{d}}, \mathfrak{t}\right)\right] < \alpha\left[E_S\left(\mathfrak{q}_{\mathfrak{d}}, \mathfrak{t}\right)\right],
\end{aligned} \tag{6.10}$$

where

$$E_S(\mathfrak{q}_\mathfrak{d}, \mathfrak{t}) = \max\left\{ \begin{array}{c} \mathfrak{S}(\mathfrak{q}_\mathfrak{d}, \mathfrak{q}_\mathfrak{d}, \mathfrak{t}) + |\mathfrak{S}(\mathfrak{q}_\mathfrak{d}, \mathfrak{q}_\mathfrak{d}, \mathfrak{N}\mathfrak{q}_\mathfrak{d}) - \mathfrak{S}(\mathfrak{t}, \mathfrak{t}, \mathfrak{N}\mathfrak{t})|, \\ \frac{\mathfrak{S}(\mathfrak{q}_\mathfrak{d}, \mathfrak{q}_\mathfrak{d}, \mathfrak{N}\mathfrak{t}) + \mathfrak{S}(\mathfrak{t}, \mathfrak{t}, \mathfrak{N}\mathfrak{q}_\mathfrak{d})}{3} \end{array} \right\}$$
$$= \max\left\{ \begin{array}{c} \mathfrak{S}(\mathfrak{q}_\mathfrak{d}, \mathfrak{q}_\mathfrak{d}, \mathfrak{t}) + |\mathfrak{S}(\mathfrak{q}_\mathfrak{d}, \mathfrak{q}_\mathfrak{d}, \mathfrak{q}_{\mathfrak{d}+1}) - \mathfrak{S}(\mathfrak{t}, \mathfrak{t}, \mathfrak{N}\mathfrak{t})|, \\ \frac{\mathfrak{S}(\mathfrak{q}_\mathfrak{d}, \mathfrak{q}_\mathfrak{d}, \mathfrak{N}\mathfrak{t}) + \mathfrak{S}(\mathfrak{t}, \mathfrak{t}, \mathfrak{q}_{\mathfrak{d}+1})}{3} \end{array} \right\}$$
$$= \mathfrak{S}(\mathfrak{t}, \mathfrak{t}, \mathfrak{N}\mathfrak{t}),$$

for $\mathfrak{d}$ sufficiently large. Letting $\mathfrak{d} \to \infty$ in (21.18) and using the lower-semicontinuity of α, we find

$$\liminf_{\mathfrak{s} \to S(\mathfrak{t}, \mathfrak{t}, \mathfrak{N}\mathfrak{t})} \alpha[\mathfrak{s}] \leq \lim_{\mathfrak{d} \to \infty} \alpha[\mathfrak{S}(\mathfrak{N}\mathfrak{q}_\mathfrak{d}, \mathfrak{N}\mathfrak{q}_\mathfrak{d}, \mathfrak{N}\mathfrak{t})] \leq \beta[\mathfrak{S}(\mathfrak{t}, \mathfrak{t}, \mathfrak{N}\mathfrak{t})]$$
$$< \alpha[\mathfrak{S}(\mathfrak{t}, \mathfrak{t}, \mathfrak{N}\mathfrak{t})] < \liminf_{\mathfrak{s} \to \mathfrak{S}(\mathfrak{t}, \mathfrak{t}, \mathfrak{N}\mathfrak{t})} \alpha[\mathfrak{s}],$$

a contradiction. Hence we get

$$\mathfrak{N}\mathfrak{t} = \mathfrak{t},$$

that is,

$$\mathfrak{t} \in Fix(\mathfrak{N}).$$

Theorem 6.2 *If we add the condition*

$$(s_5)\ \theta(\mathfrak{q}, \mathfrak{w}) \geq 1 \ \textit{forany}\ \mathfrak{q}, \mathfrak{w} \in Fix(\mathfrak{N}),$$

to the hypothesis of Theorem 6.1, then $\mathfrak{N}$ has a unique fixed point.

Proof 6.2 *Assume $\mathfrak{w} \in Fix(\mathfrak{N})$ such that $\mathfrak{t} \neq \mathfrak{w}$. Using (21.14) and the symmetry property, we have*

$$\alpha[\mathfrak{S}(\mathfrak{t}, \mathfrak{t}, \mathfrak{w})] \leq \theta(\mathfrak{t}, \mathfrak{w})\alpha[\mathfrak{S}(\mathfrak{N}\mathfrak{t}, \mathfrak{N}\mathfrak{t}, \mathfrak{N}\mathfrak{w})] \leq \beta[E_S(\mathfrak{t}, \mathfrak{w})] < \alpha[E_S(\mathfrak{t}, \mathfrak{w})]$$
$$= \alpha\left[\max\left\{ \begin{array}{c} \mathfrak{S}(\mathfrak{t}, \mathfrak{t}, \mathfrak{w}) + |\mathfrak{S}(\mathfrak{t}, \mathfrak{t}, \mathfrak{N}\mathfrak{t}) - \mathfrak{S}(\mathfrak{w}, \mathfrak{w}, \mathfrak{N}\mathfrak{w})|, \\ \frac{\mathfrak{S}(\mathfrak{t}, \mathfrak{t}, \mathfrak{N}\mathfrak{w}) + \mathfrak{S}(\mathfrak{w}, \mathfrak{w}, \mathfrak{N}\mathfrak{t})}{3} \end{array} \right\}\right]$$
$$= \alpha\left[\max\left\{ \begin{array}{c} \mathfrak{S}(\mathfrak{t}, \mathfrak{t}, \mathfrak{w}) + |\mathfrak{S}(\mathfrak{t}, \mathfrak{t}, \mathfrak{t}) - \mathfrak{S}(\mathfrak{w}, \mathfrak{w}, \mathfrak{w})|, \\ \frac{\mathfrak{S}(\mathfrak{t}, \mathfrak{t}, \mathfrak{w}) + \mathfrak{S}(\mathfrak{w}, \mathfrak{w}, \mathfrak{t})}{3} \end{array} \right\}\right]$$
$$= \alpha[\mathfrak{S}(\mathfrak{t}, \mathfrak{t}, \mathfrak{w})],$$

a contradiction. Hence $\mathfrak{t} = \mathfrak{w}$. □

Corollary 6.1 *Let $\mathfrak{N}$ satisfies*

$$\alpha[\mathfrak{S}(\mathfrak{N}\mathfrak{q}, \mathfrak{N}\mathfrak{q}, \mathfrak{N}\mathfrak{w})] \leq \beta[E_S(\mathfrak{q}, \mathfrak{w})],$$

for every distinct $\mathfrak{q}, \mathfrak{w} \in \mathcal{Y}$ such that $\mathfrak{S}(\mathfrak{N}\mathfrak{q}, \mathfrak{N}\mathfrak{q}, \mathfrak{N}\mathfrak{w}) > 0$. Suppose that

(s_0) *$\beta(\mathfrak{s}) < \alpha(\mathfrak{s})$ for each $\mathfrak{s} > 0$,*

(s_1) *α is non-decreasing and lower semi-continuous,*

(s_2) *$\limsup_{\mathfrak{s} \to \mathfrak{s}_0} \beta(\mathfrak{s}) < \alpha(\mathfrak{s}_0)$ for any $\mathfrak{s}_0 > 0$.*

Then $\mathfrak{N}$ has a unique fixed point.

Proof 6.3 *If we take* $\theta(\mathfrak{q},\mathfrak{w})=1$ *in Theorem 6.1, it can be easily proved.*

Corollary 6.2 *Let* $\mathfrak{N}$ *satisfies*

$$\alpha\left[\mathfrak{S}(\mathfrak{N}\mathfrak{q},\mathfrak{N}\mathfrak{q},\mathfrak{N}\mathfrak{w})\right]\leq c\alpha\left[E_S(\mathfrak{q},\mathfrak{w})\right],$$

for every distinct $\mathfrak{q},\mathfrak{w}\in\mathcal{Y}$ *such that* $c\in[0,1)$, $\mathfrak{S}(\mathfrak{N}\mathfrak{q},\mathfrak{N}\mathfrak{q},\mathfrak{N}\mathfrak{w})>0$ *and* $\alpha:(0,\infty)\to(0,\infty)$ *is non-decreasing and left-continuous. Then* $\mathfrak{N}$ *has a unique fixed point.*

Proof 6.4 *If we take* $\beta(\mathfrak{s})=c\alpha(\mathfrak{s})$ *in Corollary 6.1, then it can be easily seen.*

Example 6.1 *Let* $\mathcal{Y}=[0,\infty)$ *be a complete* S*-metric space with the* S*-metric as*

$$\mathfrak{S}(\mathfrak{q},\mathfrak{w},z)=|\mathfrak{q}-\mathfrak{t}|+|\mathfrak{q}+\mathfrak{t}-2\mathfrak{w}|,$$

for all $\mathfrak{q},\mathfrak{w},\mathfrak{t}\in\mathcal{Y}$ *[9]. If we define a self-mapping* $\mathfrak{N}:\mathcal{Y}\to\mathcal{Y}$ *as*

$$\mathfrak{N}\mathfrak{q}=\frac{\mathfrak{q}}{2},$$

for all $\mathfrak{q}\in\mathcal{Y}$*, then* $\mathfrak{N}$ *satisfies the conditions of Theorems 6.1 and 6.2 with* $\alpha(\mathfrak{s})=\frac{\mathfrak{s}}{2}$, $\beta(\mathfrak{s})=\frac{\mathfrak{s}}{3}$ *and* $\theta(\mathfrak{q},\mathfrak{w})=1$*. Consequently,* $\mathfrak{N}$ *has a unique fixed point* $\mathfrak{q}=0$.

Definition 6.4 $\mathfrak{N}$ *is called a* (θ,α,β)-E_S^2*-contraction if the inequality*

$$\theta(\mathfrak{q},\mathfrak{w})\alpha\left[\mathfrak{S}(\mathfrak{N}^2\mathfrak{q},\mathfrak{N}^2\mathfrak{q},\mathfrak{N}^2\mathfrak{w})\right]\leq\beta\left[E_S^2(\mathfrak{q},\mathfrak{w})\right],$$

holds for all $\mathfrak{q},\mathfrak{w}\in\mathcal{Y}$ *such that* $\mathfrak{S}(\mathfrak{q},\mathfrak{q},\mathfrak{w})>0$ *and* $\mathfrak{S}(\mathfrak{N}\mathfrak{q},\mathfrak{N}\mathfrak{q},\mathfrak{N}\mathfrak{w})>0$*, where*

$$E_S^2(\mathfrak{q},\mathfrak{w})=\max\left\{\begin{array}{c}\mathfrak{S}(\mathfrak{w},\mathfrak{w},\mathfrak{N}\mathfrak{w})+|\mathfrak{S}(\mathfrak{q},\mathfrak{q},\mathfrak{w})-\mathfrak{S}(\mathfrak{q},\mathfrak{q},\mathfrak{N}\mathfrak{q})|,\\ \mathfrak{S}(\mathfrak{N}\mathfrak{q},\mathfrak{N}\mathfrak{q},\mathfrak{N}\mathfrak{w})+\left|\mathfrak{S}(\mathfrak{N}\mathfrak{q},\mathfrak{N}\mathfrak{q},\mathfrak{N}^2\mathfrak{q})\right.\\ \left.-\mathfrak{S}(\mathfrak{N}\mathfrak{w},\mathfrak{N}\mathfrak{w},\mathfrak{N}^2\mathfrak{w})\right|,\\ \mathfrak{S}(\mathfrak{N}\mathfrak{w},\mathfrak{N}\mathfrak{w},\mathfrak{N}^2\mathfrak{w})+\left|\mathfrak{S}(\mathfrak{N}\mathfrak{q},\mathfrak{N}\mathfrak{q},\mathfrak{N}^2\mathfrak{q})\right.\\ \left.-\mathfrak{S}(\mathfrak{w},\mathfrak{w},\mathfrak{N}\mathfrak{w})\right|\end{array}\right\}.$$

Theorem 6.3 *Let* $\mathfrak{N}$ *be a* (θ,α,β)-E_S^2*-contraction such that*

(s_1) α *is non-decreasing and lower semi-continuous,*

(s_2) $\limsup_{\mathfrak{s}\to\mathfrak{s}_0}\beta(\mathfrak{s})<\alpha(\mathfrak{s}_0)$ *for any* $\mathfrak{s}_0>0$,

(s_3) $\mathfrak{N}$ *is* θ*-t.o.a. and there is* $\mathfrak{q}_0\in\mathcal{Y}$ *such that* $\theta(\mathfrak{q}_0,\mathfrak{N}\mathfrak{q}_0)\geq 1$,

(s_6) $\mathfrak{N}^2$ *is continuous and* $\theta(\mathfrak{N}\mathfrak{w},\mathfrak{w})\geq 1$ *for any* $\mathfrak{w}\in Fix(\mathfrak{N}^2)$.

Then $\mathfrak{N}$ *has a fixed point.*

Proof 6.5 *Let* $\mathfrak{q}_0\in\mathcal{Y}$ *and the sequence* $\{\mathfrak{q}_\mathfrak{d}\}$ *be defined as in the proof of Theorem 6.1. By Lemma 6.4, we have*

$$\theta(\mathfrak{q}_\mathfrak{d},\mathfrak{q}_{\mathfrak{d}+1})\geq 1,$$

for any natural number $\mathfrak{d}$ *and so*

$$\alpha\left[\mathfrak{S}(\mathfrak{q}_{\mathfrak{d}+1},\mathfrak{q}_{\mathfrak{d}+1},\mathfrak{q}_{\mathfrak{d}+2})\right]$$

$$\begin{aligned} &\leq \theta(\mathfrak{q}_{\mathfrak{d}}, \mathfrak{q}_{\mathfrak{d}+1})\alpha\left[\mathfrak{S}(\mathfrak{N}^2\mathfrak{q}_{\mathfrak{d}}, \mathfrak{N}^2\mathfrak{q}_{\mathfrak{d}}, \mathfrak{N}^2\mathfrak{q}_{\mathfrak{d}+1})\right] \\ &\leq \beta\left[E_S^2(\mathfrak{q}_{\mathfrak{d}}, \mathfrak{q}_{\mathfrak{d}+1})\right] < \alpha\left[E_S^2(\mathfrak{q}_{\mathfrak{d}}, \mathfrak{q}_{\mathfrak{d}+1})\right]. \end{aligned} \tag{6.11}$$

By the monotonicity of α*, we have*

$$\mathfrak{S}(\mathfrak{q}_{\mathfrak{d}+1}, \mathfrak{q}_{\mathfrak{d}+1}, \mathfrak{q}_{\mathfrak{d}+2}) < E_S^2(\mathfrak{q}_{\mathfrak{d}}, \mathfrak{q}_{\mathfrak{d}+1}), \tag{6.12}$$

where

$$\begin{aligned} &E_S^2(\mathfrak{q}_{\mathfrak{d}},\mathfrak{q}_{\mathfrak{d}+1}) \\ &= \max\left\{\begin{array}{c} \mathfrak{S}(\mathfrak{q}_{\mathfrak{d}+1}, \mathfrak{q}_{\mathfrak{d}+1}, \mathfrak{N}\mathfrak{q}_{\mathfrak{d}+1}) + |\mathfrak{S}(\mathfrak{q}_{\mathfrak{d}}, \mathfrak{q}_{\mathfrak{d}}, \mathfrak{q}_{\mathfrak{d}+1}) - \mathfrak{S}(\mathfrak{q}_{\mathfrak{d}}, \mathfrak{q}_{\mathfrak{d}}, \mathfrak{N}\mathfrak{q}_{\mathfrak{d}})|, \\ \mathfrak{S}(\mathfrak{N}\mathfrak{q}_{\mathfrak{d}}, \mathfrak{N}\mathfrak{q}_{\mathfrak{d}}, \mathfrak{N}\mathfrak{q}_{\mathfrak{d}+1}) + \left|\mathfrak{S}(\mathfrak{N}\mathfrak{q}_{\mathfrak{d}}, \mathfrak{N}\mathfrak{q}_{\mathfrak{d}}, \mathfrak{N}^2\mathfrak{q}_{\mathfrak{d}})\right. \\ \left.-\mathfrak{S}(\mathfrak{N}\mathfrak{q}_{\mathfrak{d}+1}, \mathfrak{N}\mathfrak{q}_{\mathfrak{d}+1}, \mathfrak{N}^2\mathfrak{q}_{\mathfrak{d}+1})\right|, \\ \mathfrak{S}(\mathfrak{N}\mathfrak{q}_{\mathfrak{d}+1}, \mathfrak{N}\mathfrak{q}_{\mathfrak{d}+1}, \mathfrak{N}^2\mathfrak{q}_{\mathfrak{d}+1}) + \left|\mathfrak{S}(\mathfrak{N}\mathfrak{q}_{\mathfrak{d}}, \mathfrak{N}\mathfrak{q}_{\mathfrak{d}}, \mathfrak{N}^2\mathfrak{q}_{\mathfrak{d}})\right. \\ \left.-\mathfrak{S}(\mathfrak{q}_{\mathfrak{d}+1}, \mathfrak{q}_{\mathfrak{d}+1}, \mathfrak{N}\mathfrak{q}_{\mathfrak{d}+1})\right| \end{array}\right\} \\ &= \max\left\{\begin{array}{c} \mathfrak{S}(\mathfrak{q}_{\mathfrak{d}+1}, \mathfrak{q}_{\mathfrak{d}+1}, \mathfrak{q}_{\mathfrak{d}+2}) + |\mathfrak{S}(\mathfrak{q}_{\mathfrak{d}}, \mathfrak{q}_{\mathfrak{d}}, \mathfrak{q}_{\mathfrak{d}+1}) \\ -\mathfrak{S}(\mathfrak{q}_{\mathfrak{d}}, \mathfrak{q}_{\mathfrak{d}}, \mathfrak{q}_{\mathfrak{d}+1})|, \\ \mathfrak{S}(\mathfrak{q}_{\mathfrak{d}+1}, \mathfrak{q}_{\mathfrak{d}+1}, \mathfrak{q}_{\mathfrak{d}+2}) + |\mathfrak{S}(\mathfrak{q}_{\mathfrak{d}+1}, \mathfrak{q}_{\mathfrak{d}+1}, \mathfrak{q}_{\mathfrak{d}+2}) \\ -\mathfrak{S}(\mathfrak{q}_{\mathfrak{d}+2}, \mathfrak{q}_{\mathfrak{d}+2}, \mathfrak{q}_{\mathfrak{d}+3})|, \\ \mathfrak{S}(\mathfrak{q}_{\mathfrak{d}+2}, \mathfrak{q}_{\mathfrak{d}+2}, \mathfrak{q}_{\mathfrak{d}+3}) + |\mathfrak{S}(\mathfrak{q}_{\mathfrak{d}+1}, \mathfrak{q}_{\mathfrak{d}+1}, \mathfrak{q}_{\mathfrak{d}+2}) \\ -\mathfrak{S}(\mathfrak{q}_{\mathfrak{d}+1}, \mathfrak{q}_{\mathfrak{d}+1}, \mathfrak{q}_{\mathfrak{d}+2})| \end{array}\right\} \\ &= \max\left\{\begin{array}{c} \mathfrak{S}(\mathfrak{q}_{\mathfrak{d}+1}, \mathfrak{q}_{\mathfrak{d}+1}, \mathfrak{q}_{\mathfrak{d}+2}) + |\mathfrak{S}(\mathfrak{q}_{\mathfrak{d}+1}, \mathfrak{q}_{\mathfrak{d}+1}, \mathfrak{q}_{\mathfrak{d}+2}) \\ -\mathfrak{S}(\mathfrak{q}_{\mathfrak{d}+2}, \mathfrak{q}_{\mathfrak{d}+2}, \mathfrak{q}_{\mathfrak{d}+3})|, \\ \mathfrak{S}(\mathfrak{q}_{\mathfrak{d}+1}, \mathfrak{q}_{\mathfrak{d}+1}, \mathfrak{q}_{\mathfrak{d}+2}), \mathfrak{S}(\mathfrak{q}_{\mathfrak{d}+2}, \mathfrak{q}_{\mathfrak{d}+2}, \mathfrak{q}_{\mathfrak{d}+3}) \end{array}\right\}. \end{aligned}$$

If for some $\mathfrak{d} \in \mathbb{N}$, $\mathfrak{S}(\mathfrak{q}_{\mathfrak{d}+2}, \mathfrak{q}_{\mathfrak{d}+2}, \mathfrak{q}_{\mathfrak{d}+3}) \geq \mathfrak{S}(\mathfrak{q}_{\mathfrak{d}+1}, \mathfrak{q}_{\mathfrak{d}+1}, \mathfrak{q}_{\mathfrak{d}+2})$*, then we have*

$$E_S^2(\mathfrak{q}_{\mathfrak{d}}, \mathfrak{q}_{\mathfrak{d}+1}) = \mathfrak{S}(\mathfrak{q}_{\mathfrak{d}+2}, \mathfrak{q}_{\mathfrak{d}+2}, \mathfrak{q}_{\mathfrak{d}+3}).$$

Using (21.20), we get

$$\mathfrak{S}(\mathfrak{q}_{\mathfrak{d}+2}, \mathfrak{q}_{\mathfrak{d}+2}, \mathfrak{q}_{\mathfrak{d}+3}) < \mathfrak{S}(\mathfrak{q}_{\mathfrak{d}+2}, \mathfrak{q}_{\mathfrak{d}+2}, \mathfrak{q}_{\mathfrak{d}+3}),$$

a contradiction. If $\mathfrak{S}(\mathfrak{q}_{\mathfrak{d}+2}, \mathfrak{q}_{\mathfrak{d}+2}, \mathfrak{q}_{\mathfrak{d}+3}) \leq \mathfrak{S}(\mathfrak{q}_{\mathfrak{d}+1}, \mathfrak{q}_{\mathfrak{d}+1}, \mathfrak{q}_{\mathfrak{d}+2})$*, then we have*

$$E_S^2(\mathfrak{q}_{\mathfrak{d}}, \mathfrak{q}_{\mathfrak{d}+1}) = 2\mathfrak{S}(\mathfrak{q}_{\mathfrak{d}+1}, \mathfrak{q}_{\mathfrak{d}+1}, \mathfrak{q}_{\mathfrak{d}+2}) - \mathfrak{S}(\mathfrak{q}_{\mathfrak{d}+2}, \mathfrak{q}_{\mathfrak{d}+2}, \mathfrak{q}_{\mathfrak{d}+3}).$$

Using (21.20), we get

$$\mathfrak{S}(\mathfrak{q}_{\mathfrak{d}+2}, \mathfrak{q}_{\mathfrak{d}+2}, \mathfrak{q}_{\mathfrak{d}+3}) < \mathfrak{S}(\mathfrak{q}_{\mathfrak{d}+1}, \mathfrak{q}_{\mathfrak{d}+1}, \mathfrak{q}_{\mathfrak{d}+2}),$$

for all $\mathfrak{d} \in \mathbb{N}$*. Therefore, we have*

$$\begin{aligned} \mathfrak{S}(\mathfrak{q}_{\mathfrak{d}+2}, \mathfrak{q}_{\mathfrak{d}+2}, \mathfrak{q}_{\mathfrak{d}+3}) &< \mathfrak{S}(\mathfrak{q}_{\mathfrak{d}+1}, \mathfrak{q}_{\mathfrak{d}+1}, \mathfrak{q}_{\mathfrak{d}+2}) \\ &< \ldots < \mathfrak{S}(\mathfrak{q}_2, \mathfrak{q}_2, \mathfrak{q}_3) \\ &< \max\left\{\mathfrak{S}(\mathfrak{q}_1, \mathfrak{q}_1, \mathfrak{q}_2), \mathfrak{S}(\mathfrak{q}_0, \mathfrak{q}_0, \mathfrak{q}_1)\right\}. \end{aligned}$$

Hence, the sequence $\{\mathfrak{S}(\mathfrak{q}_{\mathfrak{d}}, \mathfrak{q}_{\mathfrak{d}}, \mathfrak{q}_{\mathfrak{d}+1})\}$ is bounded below by 0 and strictly decreasing. Therefore, there is $\delta \geq 0$ such that

$$\lim_{\mathfrak{d}\to\infty} \mathfrak{S}(\mathfrak{q}_{\mathfrak{d}+1}, \mathfrak{q}_{\mathfrak{d}+1}, \mathfrak{q}_{\mathfrak{d}+2}) = \delta.$$

Let $\delta > 0$. Since

$$\lim_{\mathfrak{d}\to\infty} E_S^2(\mathfrak{q}_{\mathfrak{d}}, \mathfrak{q}_{\mathfrak{d}+1}) = \lim_{\mathfrak{d}\to\infty} \mathfrak{S}(\mathfrak{q}_{\mathfrak{d}+1}, \mathfrak{q}_{\mathfrak{d}+1}, \mathfrak{q}_{\mathfrak{d}+2}) = \delta > 0$$

and using (21.19), (21.20), we get

$$\begin{aligned} \alpha\left[\delta\right] &= \lim_{\mathfrak{d}\to\infty} \alpha\left[\mathfrak{S}(\mathfrak{q}_{\mathfrak{d}+2}, \mathfrak{q}_{\mathfrak{d}+2}, \mathfrak{q}_{\mathfrak{d}+3})\right] \\ &\leq \limsup_{\mathfrak{d}\to\infty} \beta\left[E_S^2(\mathfrak{q}_{\mathfrak{d}}, \mathfrak{q}_{\mathfrak{d}+1})\right] \leq \limsup_{t\to\delta} \beta\left[t\right] < \alpha\left[\delta\right], \end{aligned}$$

a contradiction. So, we have

$$\lim_{\mathfrak{d}\to\infty} \mathfrak{S}(\mathfrak{q}_{\mathfrak{d}}, \mathfrak{q}_{\mathfrak{d}}, \mathfrak{q}_{\mathfrak{d}+1}) = 0. \tag{6.13}$$

We show that $\{\mathfrak{q}_{\mathfrak{d}}\}$ is a Cauchy sequence. On the contrary, we assume that $\{\mathfrak{q}_{\mathfrak{d}}\}$ is not Cauchy. From Lemma 6.3, we can find two subsequences $\{\mathfrak{q}_{\mathfrak{d}_j}\}$, $\{\mathfrak{q}_{k_j}\}$ of the sequence $\{\mathfrak{q}_{\mathfrak{d}}\}$ such that (6.3) holds. We have

$$\begin{aligned} &E_S^2(\mathfrak{q}_{\mathfrak{d}_j}, \mathfrak{q}_{k_j}) \\ &= \max\left\{\begin{array}{c} \mathfrak{S}(\mathfrak{q}_{k_j}, \mathfrak{q}_{k_j}, \mathfrak{N}\mathfrak{q}_{k_j}) + \left|\mathfrak{S}(\mathfrak{q}_{\mathfrak{d}_j}, \mathfrak{q}_{\mathfrak{d}_j}, \mathfrak{q}_{k_j}) - \mathfrak{S}(\mathfrak{q}_{\mathfrak{d}_j}, \mathfrak{q}_{\mathfrak{d}_j}, \mathfrak{N}\mathfrak{q}_{\mathfrak{d}_j})\right|, \\ \mathfrak{S}(\mathfrak{N}\mathfrak{q}_{\mathfrak{d}_j}, \mathfrak{N}\mathfrak{q}_{\mathfrak{d}_j}, \mathfrak{N}\mathfrak{q}_{k_j}) + \left|\mathfrak{S}(\mathfrak{N}\mathfrak{q}_{\mathfrak{d}_j}, \mathfrak{N}\mathfrak{q}_{\mathfrak{d}_j}, \mathfrak{N}^2\mathfrak{q}_{k_j}) \right. \\ \left. -\mathfrak{S}(\mathfrak{N}\mathfrak{q}_{k_j}, \mathfrak{N}\mathfrak{q}_{k_j}, \mathfrak{N}^2\mathfrak{q}_{k_j})\right|, \\ \mathfrak{S}(\mathfrak{N}\mathfrak{q}_{k_j}, \mathfrak{N}\mathfrak{q}_{k_j}, \mathfrak{N}^2\mathfrak{q}_{k_j}) + \left|\mathfrak{S}(\mathfrak{N}\mathfrak{q}_{\mathfrak{d}_j}, \mathfrak{N}\mathfrak{q}_{\mathfrak{d}_j}, \mathfrak{N}^2\mathfrak{q}_{k_j}) \right. \\ \left. -\mathfrak{S}(\mathfrak{q}_{k_j}, \mathfrak{q}_{k_j}, \mathfrak{N}\mathfrak{q}_{k_j})\right| \end{array}\right\} \\ &= \max\left\{\begin{array}{c} \mathfrak{S}(\mathfrak{q}_{k_j}, \mathfrak{q}_{k_j}, \mathfrak{q}_{k_j+1}) + \left|\mathfrak{S}(\mathfrak{q}_{\mathfrak{d}_j}, \mathfrak{q}_{\mathfrak{d}_j}, \mathfrak{q}_{k_j}) - \mathfrak{S}(\mathfrak{q}_{\mathfrak{d}_j}, \mathfrak{q}_{\mathfrak{d}_j}, \mathfrak{q}_{\mathfrak{d}_j+1})\right|, \\ \mathfrak{S}(\mathfrak{q}_{\mathfrak{d}_j+1}, \mathfrak{q}_{\mathfrak{d}_j+1}, \mathfrak{q}_{k_j+1}) + \left|\mathfrak{S}(\mathfrak{q}_{\mathfrak{d}_j+1}, \mathfrak{q}_{\mathfrak{d}_j+1}, \mathfrak{q}_{\mathfrak{d}_j+2}) \right. \\ \left. -\mathfrak{S}(\mathfrak{q}_{k_j+1}, \mathfrak{q}_{k_j+1}, \mathfrak{q}_{k_j+2})\right|, \\ \mathfrak{S}(\mathfrak{q}_{k_j+1}, \mathfrak{q}_{k_j+1}, \mathfrak{q}_{k_j+2}) + \left|\mathfrak{S}(\mathfrak{q}_{\mathfrak{d}_j+1}, \mathfrak{q}_{\mathfrak{d}_j+1}, \mathfrak{q}_{\mathfrak{d}_j+2}) \right. \\ \left. -\mathfrak{S}(\mathfrak{q}_{k_j}, \mathfrak{q}_{k_j}, \mathfrak{q}_{k_j+1})\right| \end{array}\right\} \end{aligned}$$

and so using (21.21), we obtain

$$\lim_{j\to\infty} E_S^2(\mathfrak{q}_{\mathfrak{d}_j}, \mathfrak{q}_{k_j}) = e.$$

Also, letting $j \to \infty$ in the following inequality, we get

$$\mathfrak{S}(\mathfrak{q}_{\mathfrak{d}_j+1}, \mathfrak{q}_{\mathfrak{d}_j+1}, \mathfrak{q}_{k_j+1}) - \mathfrak{S}(\mathfrak{q}_{\mathfrak{d}_j+1}, \mathfrak{q}_{\mathfrak{d}_j+1}, \mathfrak{q}_{\mathfrak{d}_j+2}) - \mathfrak{S}(\mathfrak{q}_{k_j+1}, \mathfrak{q}_{k_j+1}, \mathfrak{q}_{k_j+2})$$

$$\leq \mathfrak{S}(\mathfrak{q}_{\mathfrak{d}_j+2}, \mathfrak{q}_{\mathfrak{d}_j+2}, \mathfrak{q}_{k_j+2})$$
$$\leq \mathfrak{S}(\mathfrak{q}_{\mathfrak{d}_j+1}, \mathfrak{q}_{\mathfrak{d}_j+1}, \mathfrak{q}_{\mathfrak{d}_j+2}) + \mathfrak{S}(\mathfrak{q}_{\mathfrak{d}_j+1}, \mathfrak{q}_{\mathfrak{d}_j+1}, \mathfrak{q}_{k_j+1})$$
$$+ \mathfrak{S}(\mathfrak{q}_{k_j+1}, \mathfrak{q}_{k_j+1}, \mathfrak{q}_{k_j+2})$$

and

$$\lim_{j\to\infty} \mathfrak{S}(\mathfrak{q}_{\mathfrak{d}_j+2}, \mathfrak{q}_{\mathfrak{d}_j+2}, \mathfrak{q}_{k_j+2}) = e.$$

By Lemma 6.4 and $\theta\left(\mathfrak{q}_{\mathfrak{d}_j}, \mathfrak{q}_{k_j}\right) \geq 1$, *we find*

$$\alpha\left[\mathfrak{S}(\mathfrak{q}_{\mathfrak{d}_j+2}, \mathfrak{q}_{\mathfrak{d}_j+2}, \mathfrak{q}_{k_j+2})\right] \leq \beta\left[E_S^2(\mathfrak{q}_{\mathfrak{d}_j}, \mathfrak{q}_{k_j})\right].$$

So, we get

$$\alpha[e] = \lim_{j\to\infty} \alpha\left[\mathfrak{S}(\mathfrak{q}_{\mathfrak{d}_j+2}, \mathfrak{q}_{\mathfrak{d}_j+2}, \mathfrak{q}_{k_j+2})\right]$$
$$\leq \limsup_{j\to\infty} \beta\left[E_S^2(\mathfrak{q}_{\mathfrak{d}_j}, \mathfrak{q}_{k_j})\right] \leq \limsup_{\mathfrak{s}\to e} \beta[\mathfrak{s}],$$

a contradiction with (s_2). *Consequently,* $\{\mathfrak{q}_{\mathfrak{d}}\}$ *is Cauchy on a complete metric space* $\mathcal{Y}$. *Then there exists* $\mathfrak{t}$ *such that*

$$\lim_{\mathfrak{d}\to\infty} \mathfrak{S}(\mathfrak{q}_{\mathfrak{d}}, \mathfrak{q}_{\mathfrak{d}}, \mathfrak{t}) = 0.$$

Under the assumption (s_6), $\mathfrak{t}$ *is a fixed point of* $\mathfrak{N}$. *Since* $\mathfrak{N}^2$ *is continuous, we get*

$$\lim_{\mathfrak{d}\to\infty} \mathfrak{S}(\mathfrak{q}_{\mathfrak{d}}, \mathfrak{q}_{\mathfrak{d}}, \mathfrak{N}^2\mathfrak{t}) = \lim_{\mathfrak{d}\to\infty} \mathfrak{S}(\mathfrak{N}^2\mathfrak{q}_{\mathfrak{d}-2}, \mathfrak{N}^2\mathfrak{q}_{\mathfrak{d}-2}, \mathfrak{N}^2\mathfrak{t}) = 0$$

and so we have

$$\mathfrak{N}^2\mathfrak{t} = \mathfrak{t},$$

that is,

$$\mathfrak{t} \in Fix(\mathfrak{N}^2).$$

Assume $\mathfrak{t} \notin Fix(\mathfrak{N})$. *Using* (s_1) *and* (s_6), *we obtain*

$$\alpha[\mathfrak{S}(\mathfrak{N}\mathfrak{t}, T\mathfrak{t}, \mathfrak{t})] = \alpha\left[\mathfrak{S}(\mathfrak{N}^2\mathfrak{N}\mathfrak{t}, \mathfrak{N}^2\mathfrak{N}\mathfrak{t}, \mathfrak{N}^2\mathfrak{t})\right]$$
$$\leq \theta(\mathfrak{N}\mathfrak{t}, \mathfrak{t})\,\alpha\left[\mathfrak{S}(\mathfrak{N}^2\mathfrak{N}\mathfrak{t}, \mathfrak{N}^2\mathfrak{N}\mathfrak{t}, \mathfrak{N}^2\mathfrak{t})\right]$$
$$\leq \beta\left[E_S^2(\mathfrak{N}\mathfrak{t}, \mathfrak{t})\right] < \alpha\left[E_S^2(\mathfrak{N}\mathfrak{t}, \mathfrak{t})\right],$$

where

$$E_S^2(\mathfrak{N}\mathfrak{t}, \mathfrak{t}) = \mathfrak{S}(\mathfrak{t}, \mathfrak{t}, \mathfrak{N}\mathfrak{t}) = \mathfrak{S}(\mathfrak{N}\mathfrak{t}, \mathfrak{N}\mathfrak{t}, \mathfrak{t})$$

and so we get

$$\alpha[\mathfrak{S}(\mathfrak{N}\mathfrak{t}, \mathfrak{N}\mathfrak{t}, \mathfrak{t})] < \alpha[\mathfrak{S}(\mathfrak{N}\mathfrak{t}, \mathfrak{N}\mathfrak{t}, \mathfrak{t})],$$

a contradiction. Hence, $\mathfrak{S}(\mathfrak{N}\mathfrak{t}, \mathfrak{N}\mathfrak{t}, \mathfrak{t}) = 0$, *that is,*

$$\mathfrak{t} \in Fix(\mathfrak{N}).$$

Theorem 6.4 *If we add the condition*
 (s_5) $\theta(\mathfrak{q},\mathfrak{w}) \geq 1$ *for any* $\mathfrak{q},\mathfrak{w} \in Fix(\mathfrak{N})$,
to the hypothesis of Theorem 6.3, then $\mathfrak{N}$ *has a unique fixed point.*

Proof 6.6 *Let* $\mathfrak{t},\mathfrak{w} \in Fix(\mathfrak{N})$ *such that* $\mathfrak{t} \neq \mathfrak{w}$. *By* (s_5), *we get*

$$\begin{aligned} \alpha\left[\mathfrak{S}(\mathfrak{t},\mathfrak{t},\mathfrak{w})\right] &\leq \theta(\mathfrak{t},\mathfrak{w})\alpha\left[\mathfrak{S}(\mathfrak{N}^2\mathfrak{t},\mathfrak{N}^2\mathfrak{t},\mathfrak{N}^2\mathfrak{w})\right] \\ &\leq \beta\left[E_S^2(\mathfrak{t},\mathfrak{w})\right] < \alpha\left[E_S^2(\mathfrak{t},\mathfrak{w})\right] = \alpha\left[\mathfrak{S}(\mathfrak{t},\mathfrak{t},\mathfrak{w})\right], \end{aligned}$$

a contradiction. Consequently, T *has a unique fixed point.*

Corollary 6.3 *Let* $\mathfrak{N}$ *satisfies*

$$\alpha\left[\mathfrak{S}(\mathfrak{N}^2\mathfrak{q},\mathfrak{N}^2\mathfrak{q},\mathfrak{N}^2\mathfrak{w})\right] \leq \beta\left[E_S^2(\mathfrak{q},\mathfrak{w})\right],$$

for every distinct $\mathfrak{q},\mathfrak{w} \in \mathcal{Y}$ *such that* $\mathfrak{S}(\mathfrak{N}^2\mathfrak{q},\mathfrak{N}^2\mathfrak{q},\mathfrak{N}^2\mathfrak{w}) > 0$. *Suppose that*
 (s_0) $\beta(\mathfrak{s}) < \alpha(\mathfrak{s})$ *for each* $\mathfrak{s} > 0$,
 (s_1) α *is non-decreasing and lower semi-continuous,*
 (s_2) $\limsup\limits_{\mathfrak{s}\to\mathfrak{s}_0}\beta(\mathfrak{s}) < \alpha(\mathfrak{s}_0)$ *for any* $\mathfrak{s}_0 > 0$,
 (s_6) $\mathfrak{N}^2$ *is continuous.*
Then $\mathfrak{N}$ *has a unique fixed point.*

Proof 6.7 *If we take* $\theta(\mathfrak{q},\mathfrak{w}) = 1$ *in Theorem 6.3, then it can be easily proved.*

Corollary 6.4 *Let* $\mathfrak{N}$ *satisfies*

$$\alpha\left[\mathfrak{S}(\mathfrak{N}^2\mathfrak{q},\mathfrak{N}^2\mathfrak{q},\mathfrak{N}^2\mathfrak{w})\right] \leq c\alpha\left[E_S^2(\mathfrak{q},\mathfrak{w})\right],$$

for every distinct $\mathfrak{q},\mathfrak{w} \in \mathcal{Y}$ *such that* $c \in [0,1)$, $\mathfrak{S}(\mathfrak{N}\mathfrak{q},\mathfrak{N}\mathfrak{q},\mathfrak{N}\mathfrak{w}) > 0$ *and* $\alpha : (0,\infty) \to (0,\infty)$ *is non-decreasing and left-continuous. Then* $\mathfrak{N}$ *has a unique fixed point.*

Proof 6.8 *If we take* $\beta(\mathfrak{s}) = c\alpha(\mathfrak{s})$ *in Corollary 6.3, then it can be easily seen.*

Example 6.2 *Let* $\mathcal{Y} = [0,10]$ *be a complete* S*-metric space with the* S*-metric defined as in Example 6.1. Let us define a self-mapping* $\mathfrak{N} : \mathcal{Y} \to \mathcal{Y}$ *as*

$$\mathfrak{N}\mathfrak{q} = \begin{cases} 0, & \mathfrak{q} \in [0,2] \\ 2, & \mathfrak{q} \in (2,4] \\ \frac{\mathfrak{q}-4}{4}, & \mathfrak{q} \in (4,8] \\ \frac{\mathfrak{q}}{2}, & \mathfrak{q} \in (8,10] \end{cases},$$

for all $\mathfrak{q} \in \mathcal{Y}$. *Suppose that* $\alpha, \beta : (0,\infty) \to \mathbb{R}$ *are functions where* α *is non-decreasing and* $\beta(\mathfrak{s}) < \alpha(\mathfrak{s})$ *for any* $\mathfrak{s} > 0$. *Then* $\mathfrak{N}$ *satisfies the conditions of Theorem 6.1 for* $\mathfrak{q} = \frac{4}{3}$ *and* $\mathfrak{w} = \frac{7}{3}$. *On the other hand, we get*

$$\mathfrak{N}^2\mathfrak{q} = \begin{cases} 0, & \mathfrak{q} \in [0,8] \\ \frac{\mathfrak{q}-8}{8}, & \mathfrak{q} \in (8,10] \end{cases}$$

and $\mathfrak{N}^2$ is continuous. Therefore, $\mathfrak{N}$ satisfies the conditions Theorem 6.3 with

$$\theta(\mathfrak{q},\mathfrak{w}) = \begin{cases} \mathfrak{q}^3+\mathfrak{w}^3, & \mathfrak{q},\mathfrak{w}\in[0,8] \\ 1, & \mathfrak{q},\mathfrak{w}\in(8,10] \\ 2, & \mathfrak{q}\in(8,10],\mathfrak{w}\in[0,2] \\ 0, & otherwise \end{cases},$$

$$\alpha(\mathfrak{s}) = \frac{\mathfrak{s}}{4}$$

and

$$\beta(\mathfrak{s}) = \frac{\mathfrak{s}}{6}.$$

Consequently, $\mathfrak{N}$ has a unique fixed point $\mathfrak{q}=0$.

6.2.2 Some Fixed-Figure Results

In this subsection, suppose that $(\mathcal{Y}, S)$ is an S-metric space, $\mathfrak{N}:\mathcal{Y}\to\mathcal{Y}$ is a self-mapping, $\alpha,\beta:(0,\infty)\to\mathbb{R}$ are two functions such that

$$\beta(\mathfrak{s}) < \alpha(\mathfrak{s}),$$

for all $\mathfrak{s}>0$, α is non-decreasing and $\theta:\mathcal{Y}\times\mathcal{Y}\to[1,\infty)$ is a function.

Definition 6.5 *Let $\mathfrak{q}_0,\mathfrak{q}_1,\mathfrak{q}_2\in\mathcal{Y}$ and $r\in[0,\infty)$.*

(1) *[13] The circle centered at $\mathfrak{q}_0$ with radius r defined by*

$$C^{\mathfrak{S}}_{\mathfrak{q}_0,r} = \{\mathfrak{q}\in\mathcal{Y} : \mathfrak{S}(\mathfrak{q},\mathfrak{q},\mathfrak{q}_0) = r\}.$$

(2) *[20] The disc centered at $\mathfrak{q}_0$ with radius r defined by*

$$D^{\mathfrak{S}}_{\mathfrak{q}_0,r} = \{\mathfrak{q}\in\mathcal{Y} : \mathfrak{S}(\mathfrak{q},\mathfrak{q},\mathfrak{q}_0) \le r\}.$$

(3) *The ellipse $E^{\mathfrak{S}}_r(\mathfrak{q}_1,\mathfrak{q}_2)$ is defined by*

$$E^{\mathfrak{S}}_r(\mathfrak{q}_1,\mathfrak{q}_2) = \{\mathfrak{q}\in\mathcal{Y} : \mathfrak{S}(\mathfrak{q},\mathfrak{q},\mathfrak{q}_1)+\mathfrak{S}(\mathfrak{q},\mathfrak{q},\mathfrak{q}_2) = r\}.$$

(4) *The hyperbola $H^{\mathfrak{S}}_r(\mathfrak{q}_1,\mathfrak{q}_2)$ is defined by*

$$H^{\mathfrak{S}}_r(\mathfrak{q}_1,\mathfrak{q}_2) = \{\mathfrak{q}\in\mathcal{Y} : |\mathfrak{S}(\mathfrak{q},\mathfrak{q},\mathfrak{q}_1)-\mathfrak{S}(\mathfrak{q},\mathfrak{q},\mathfrak{q}_2)| = r\}.$$

(5) *The Cassini curve $C^S_r(\mathfrak{q}_1,\mathfrak{q}_2)$ is defined by*

$$C^S_r(\mathfrak{q}_1,\mathfrak{q}_2) = \{\mathfrak{q}\in\mathcal{Y} : \mathfrak{S}(\mathfrak{q},\mathfrak{q},\mathfrak{q}_1)\mathfrak{S}(\mathfrak{q},\mathfrak{q},\mathfrak{q}_2) = r\}.$$

(6) *The Apollonius circle $A^{\mathfrak{S}}_r(\mathfrak{q}_1,\mathfrak{q}_2)$ is defined by*

$$A^{\mathfrak{S}}_r(\mathfrak{q}_1,\mathfrak{q}_2) = \left\{\mathfrak{q}\in\mathcal{Y}-\{\mathfrak{q}_2\} : \frac{\mathfrak{S}(\mathfrak{q},\mathfrak{q},\mathfrak{q}_1)}{\mathfrak{S}(\mathfrak{q},\mathfrak{q},\mathfrak{q}_2)} = r\right\}.$$

Definition 6.6 *Let $\mathfrak{F}$ be a geometric figure. If $\mathfrak{F}\subset Fix(\mathfrak{N})$ then $\mathfrak{F}$ is called a fixed figure of $\mathfrak{N}$.*

Let the number r defined as

$$r = \inf\{\mathfrak{S}(\mathfrak{q},\mathfrak{q},\mathfrak{N}\mathfrak{q}) : \mathfrak{q}\notin Fix(\mathfrak{N})\}. \tag{6.14}$$

6.2.2.1 Some Fixed-Disc Results

In this section, we prove some fixed-disc (resp. fixed-circle) results on S-metric spaces. To do this, we give two new contractions.

Definition 6.7 *If there exists* $\mathfrak{q}_0 \in \mathcal{Y}$ *such that*

$$\theta(\mathfrak{q}, \mathfrak{q}_0)\alpha\left[\mathfrak{S}(\mathfrak{q}, \mathfrak{q}, \mathfrak{N}\mathfrak{q})\right] \leq \beta\left[\frac{E_S(\mathfrak{q}, \mathfrak{q}_0)}{2}\right],$$

for all $\mathfrak{q} \in \mathcal{Y} - Fix(\mathfrak{N})$, *then* $\mathfrak{N}$ *is called a* (θ, α, β)-E_{S_D}*-contraction.*

Theorem 6.5 *Let* $\mathfrak{N}$ *be a* (θ, α, β)-E_{S_D}*-contraction with* $\mathfrak{q}_0 \in \mathcal{Y}$. *If* $\mathfrak{S}(\mathfrak{q}_0, \mathfrak{q}_0, \mathfrak{N}\mathfrak{q}) \leq r$ *then* $\mathfrak{q}_0 \in Fix(\mathfrak{N})$ *and* $D^{\mathfrak{S}}_{\mathfrak{q}_0,r} \subset Fix(\mathfrak{N})$. *Especially,* $C^{\mathfrak{S}}_{\mathfrak{q}_0,r} \subset Fix(\mathfrak{N})$.

Proof 6.9 *At first, we prove* $\mathfrak{q}_0 \in Fix(\mathfrak{N})$. *On the contrary, we assume* $\mathfrak{q}_0 \notin Fix(\mathfrak{N})$. *Using the hypothesis, we get*

$$\alpha\left[\mathfrak{S}(\mathfrak{q}_0, \mathfrak{q}_0, \mathfrak{N}\mathfrak{q}_0)\right] \leq \theta(\mathfrak{q}_0, \mathfrak{q}_0)\alpha\left[\mathfrak{S}(\mathfrak{q}_0, \mathfrak{q}_0, \mathfrak{N}\mathfrak{q}_0)\right] \leq \beta\left[\frac{E_S(\mathfrak{q}_0, \mathfrak{q}_0)}{2}\right], \tag{6.15}$$

where

$$E_S(\mathfrak{q}_0, \mathfrak{q}_0) = \mathfrak{S}(\mathfrak{q}_0, \mathfrak{q}_0, \mathfrak{N}\mathfrak{q}_0).$$

From (21.22), we obtain

$$\alpha\left[\mathfrak{S}(\mathfrak{q}_0, \mathfrak{q}_0, \mathfrak{N}\mathfrak{q}_0)\right] \leq \beta\left[\frac{\mathfrak{S}(\mathfrak{q}_0, \mathfrak{q}_0, \mathfrak{N}\mathfrak{q}_0)}{2}\right] < \alpha\left[\frac{\mathfrak{S}(\mathfrak{q}_0, \mathfrak{q}_0, \mathfrak{N}\mathfrak{q}_0)}{2}\right],$$

a contradiction. Hence we get

$$\mathfrak{q}_0 \in Fix(\mathfrak{N}). \tag{6.16}$$

Now we show $D^{\mathfrak{S}}_{\mathfrak{q}_0,r} \subset Fix(\mathfrak{N})$ *under the following cases:*

Case 1: Let $r = 0$. *Then we have* $D^{\mathfrak{S}}_{\mathfrak{q}_0,r} = \{\mathfrak{q}_0\}$. *By (21.23), we get*

$$D^{\mathfrak{S}}_{\mathfrak{q}_0,r} \subset Fix(\mathfrak{N}).$$

Case 2: Let $r > 0$ *and* $\mathfrak{t} \in D^{\mathfrak{S}}_{\mathfrak{q}_0,r}$ *such that* $\mathfrak{q} \notin Fix(\mathfrak{N})$. *Using the hypothesis, we get*

$$\alpha\left[\mathfrak{S}(\mathfrak{q}, \mathfrak{q}, \mathfrak{N}\mathfrak{q})\right] \leq \theta(\mathfrak{q}, \mathfrak{q}_0)\alpha\left[\mathfrak{S}(\mathfrak{q}, \mathfrak{q}, \mathfrak{N}\mathfrak{q})\right] \leq \beta\left[\frac{E_S(\mathfrak{q}, \mathfrak{q}_0)}{2}\right], \tag{6.17}$$

where

$$E_S(\mathfrak{q}, \mathfrak{q}_0) \leq 2\mathfrak{S}(\mathfrak{q}, \mathfrak{q}, \mathfrak{N}\mathfrak{q}).$$

By (21.24), we find

$$\alpha\left[\mathfrak{S}(\mathfrak{q}, \mathfrak{q}, \mathfrak{N}\mathfrak{q})\right] \leq \beta\left[\frac{E_S(\mathfrak{q}, \mathfrak{q}_0)}{2}\right] < \alpha\left[\frac{E_S(\mathfrak{q}, \mathfrak{q}_0)}{2}\right] \leq \alpha\left[\mathfrak{S}(\mathfrak{q}, \mathfrak{q}, \mathfrak{N}\mathfrak{q})\right],$$

a contradiction. So $\mathfrak{q} \in Fix(\mathfrak{N})$. *Consequently, we have*

$$D^{\mathfrak{S}}_{\mathfrak{q}_0,r} \subset Fix(\mathfrak{N}).$$

Since the circle $C^{\mathfrak{S}}_{\mathfrak{q}_0,r}$ *is a boundary of the disc* $D^{\mathfrak{S}}_{\mathfrak{q}_0,r}$, *then we obtain*

$$C^{\mathfrak{S}}_{\mathfrak{q}_0,r} \subset Fix(\mathfrak{N}).$$

Definition 6.8 *If there exists* $\mathfrak{q}_0 \in \mathcal{Y}$ *such that*

$$\theta(\mathfrak{q},\mathfrak{q}_0)\alpha\left[\mathfrak{S}(\mathfrak{q},\mathfrak{q},\mathfrak{N}\mathfrak{q})\right] \le \beta\left[\frac{E_S^2(\mathfrak{q},\mathfrak{q}_0)}{2}\right],$$

for all $\mathfrak{q} \in \mathcal{Y} - Fix(\mathfrak{N})$, *then* $\mathfrak{N}$ *is called a* (θ,α,β)-$E^2_{S_D}$*-contraction.*

Theorem 6.6 *Let* $\mathfrak{N}$ *be a* (θ,α,β)-$E^2_{S_D}$*-contraction with* $\mathfrak{q}_0 \in \mathcal{Y}$. *If* $\mathfrak{S}(\mathfrak{q}_0,\mathfrak{q}_0,\mathfrak{N}\mathfrak{q}) \le r$, $\mathfrak{N}\mathfrak{q}_0 \in Fix(\mathfrak{N})$ *and* $\mathfrak{q} \in Fix(\mathfrak{N}^2)$, *then* $\mathfrak{q}_0 \in Fix(\mathfrak{N})$ *and* $D^{\mathfrak{S}}_{\mathfrak{q}_0,r} \subset Fix(\mathfrak{N})$. *Especially,* $C^{\mathfrak{S}}_{\mathfrak{q}_0,r} \subset Fix(\mathfrak{N})$.

Proof 6.10 *At first, we prove* $\mathfrak{q}_0 \in Fix(\mathfrak{N})$. *On the contrary, we assume* $\mathfrak{q}_0 \notin Fix(\mathfrak{N})$. *Using the hypothesis, we get*

$$\begin{aligned}\alpha\left[\mathfrak{S}(\mathfrak{q}_0,\mathfrak{q}_0,\mathfrak{N}\mathfrak{q}_0)\right] &\le \theta(\mathfrak{q}_0,\mathfrak{q}_0)\alpha\left[\mathfrak{S}(\mathfrak{q}_0,\mathfrak{q}_0,\mathfrak{N}\mathfrak{q}_0)\right]\\ &\le \beta\left[\frac{E_S^2(\mathfrak{q}_0,\mathfrak{q}_0)}{2}\right] = \beta\left[\mathfrak{S}(\mathfrak{q}_0,\mathfrak{q}_0,\mathfrak{N}\mathfrak{q}_0)\right]\\ &< \alpha\left[\mathfrak{S}(\mathfrak{q}_0,\mathfrak{q}_0,\mathfrak{N}\mathfrak{q}_0)\right],\end{aligned}$$

a contradiction. It should be

$$\mathfrak{q}_0 \in Fix(\mathfrak{N}). \tag{6.18}$$

Now we prove $D^{\mathfrak{S}}_{\mathfrak{q}_0,r} \subset Fix(\mathfrak{N})$ *under the following cases:*

Case 1: Let $r = 0$. *Then we have* $D^{\mathfrak{S}}_{\mathfrak{q}_0,r} = \{\mathfrak{q}_0\}$. *By (21.25), we get*

$$D^{\mathfrak{S}}_{\mathfrak{q}_0,r} \subset Fix(\mathfrak{N}).$$

Case 2: Let $r > 0$ *and* $\mathfrak{q} \in D^{\mathfrak{S}}_{\mathfrak{q}_0,r}$ *such that* $\mathfrak{q} \notin Fix(\mathfrak{N})$. *Using the hypothesis, we have*

$$\begin{aligned}\alpha\left[\mathfrak{S}(\mathfrak{q},\mathfrak{q},\mathfrak{N}\mathfrak{q})\right] \le \theta(\mathfrak{q},\mathfrak{q}_0)\alpha\left[\mathfrak{S}(\mathfrak{q},\mathfrak{q},\mathfrak{N}\mathfrak{q})\right] &\le \beta\left[\frac{E_S^2(\mathfrak{q},\mathfrak{q}_0)}{2}\right]\\ < \alpha\left[\frac{E_S^2(\mathfrak{q},\mathfrak{q}_0)}{2}\right] &\le \alpha\left[\mathfrak{S}(\mathfrak{q},\mathfrak{q},\mathfrak{N}\mathfrak{q})\right],\end{aligned}$$

a contradiction. Hence, it should be $\mathfrak{q} \in Fix(\mathfrak{N})$ *and we get*

$$D^{\mathfrak{S}}_{\mathfrak{q}_0,r} \subset Fix(\mathfrak{N}).$$

Since the circle $C^{\mathfrak{S}}_{\mathfrak{q}_0,r}$ *is a boundary of the disc* $D^{\mathfrak{S}}_{\mathfrak{q}_0,r}$, *then we obtain*

$$C^{\mathfrak{S}}_{\mathfrak{q}_0,r} \subset Fix(\mathfrak{N}).$$

Example 6.3 *Let $\mathcal{Y} = \mathbb{R}$ be an S-metric space with the S-metric defined as in Example 6.1. Let us define a self-mapping $\mathfrak{N} : \mathcal{Y} \to \mathcal{Y}$ as*

$$\mathfrak{N}\mathfrak{q} = \begin{cases} \mathfrak{q}, & \mathfrak{q} \in \mathcal{Y} - \{2\} \\ 3, & \mathfrak{q} = 2 \end{cases},$$

for all $\mathfrak{q} \in \mathcal{Y}$. Then $\mathfrak{N}$ is both (θ, α, β)-E_{S_D}-contraction and (θ, α, β)-$E^2_{S_D}$-contraction with $\mathfrak{q}_0 = 0$, $\theta(\mathfrak{q}, y) = 1$, $\alpha(\mathfrak{s}) = \frac{\mathfrak{s}}{2}$ and $\beta(\mathfrak{s}) = \frac{\mathfrak{s}}{3}$. Also, we have

$$r = 2.$$

Consequently, $\mathfrak{N}$ fixes the disc $D^{\mathfrak{S}}_{0,2} = [-1, 1]$ and the circle $C^{\mathfrak{S}}_{0,2} = \{-1, 1\}$.

6.2.2.2 Some Fixed-Ellipse Results

In this section, we prove two fixed-ellipse results on S-metric spaces.

Definition 6.9 *If there exists $\mathfrak{q}_1, \mathfrak{q}_2 \in \mathcal{Y}$ such that*

$$\theta(\mathfrak{q}_1, \mathfrak{q}_2)\alpha\left[\mathfrak{S}(\mathfrak{q}, \mathfrak{q}, \mathfrak{N}\mathfrak{q})\right] \leq \beta\left[\frac{E_S(\mathfrak{q}, \mathfrak{q}_1) + E_S(\mathfrak{q}, \mathfrak{q}_2)}{4}\right],$$

for all $\mathfrak{q} \in \mathcal{Y} - Fix(\mathfrak{N})$, then $\mathfrak{N}$ is called a (θ, α, β)-E_{S_E}-contraction.

Theorem 6.7 *Let $\mathfrak{N}$ be a (θ, α, β)-E_{S_E}-contraction with $\mathfrak{q}_1, \mathfrak{q}_2 \in \mathcal{Y}$. If $\mathfrak{q}_1, \mathfrak{q}_2 \in Fix(\mathfrak{N})$ and $\mathfrak{S}(\mathfrak{N}\mathfrak{q}, \mathfrak{N}\mathfrak{q}, \mathfrak{q}_1) + \mathfrak{S}(\mathfrak{N}\mathfrak{q}, \mathfrak{N}\mathfrak{q}, \mathfrak{q}_2) = r$ then $E^{\mathfrak{S}}_r(\mathfrak{q}_1, \mathfrak{q}_2) \subset Fix(\mathfrak{N})$.*

Proof 6.11 *Case 1: Let $r = 0$. Then we have $E^{\mathfrak{S}}_r(\mathfrak{q}_1, \mathfrak{q}_2) = \{\mathfrak{q}_1\} = \{\mathfrak{q}_2\}$. By the hypothesis, we get*

$$E^{\mathfrak{S}}_r(\mathfrak{q}_1, \mathfrak{q}_2) \subset Fix(\mathfrak{N}).$$

Case 2: Let $r > 0$ and $\mathfrak{q} \in E^{\mathfrak{S}}_r(\mathfrak{q}_1, \mathfrak{q}_2)$ such that $\mathfrak{q} \notin Fix(\mathfrak{N})$. Using the hypothesis, we have

$$\begin{aligned} \alpha\left[\mathfrak{S}(\mathfrak{q}, \mathfrak{q}, \mathfrak{N}\mathfrak{q})\right] &\leq \theta(\mathfrak{q}_1, \mathfrak{q}_2)\alpha\left[\mathfrak{S}(\mathfrak{q}, \mathfrak{q}, \mathfrak{N}\mathfrak{q})\right] \\ &\leq \beta\left[\frac{E_S(\mathfrak{q}, \mathfrak{q}_1) + E_S(\mathfrak{q}, \mathfrak{q}_2)}{4}\right] \\ &< \alpha\left[\frac{E_S(\mathfrak{q}, \mathfrak{q}_1) + E_S(\mathfrak{q}, \mathfrak{q}_2)}{4}\right] \\ &\leq \alpha\left[\mathfrak{S}(\mathfrak{q}, \mathfrak{q}, \mathfrak{N}\mathfrak{q})\right], \end{aligned}$$

a contradiction. So $\mathfrak{q} \in Fix(\mathfrak{N})$. Consequently, we have

$$E^{\mathfrak{S}}_r(\mathfrak{q}_1, \mathfrak{q}_2) \subset Fix(\mathfrak{N}).$$

Definition 6.10 *If there exists* $\mathfrak{q}_1, \mathfrak{q}_2 \in \mathcal{Y}$ *such that*

$$\theta(\mathfrak{q}_1, \mathfrak{q}_2)\alpha\left[\mathfrak{S}(\mathfrak{q}, \mathfrak{q}, \mathfrak{N}\mathfrak{q})\right] \leq \beta\left[\frac{E_S^2(\mathfrak{q}, \mathfrak{q}_1) + E_S^2(\mathfrak{q}, \mathfrak{q}_2)}{4}\right],$$

for all $\mathfrak{q} \in \mathcal{Y} - Fix(\mathfrak{N})$, *then* $\mathfrak{N}$ *is called a* (θ, α, β)-$E_{S_E}^2$*-contraction.*

Theorem 6.8 *Let* $\mathfrak{N}$ *be a* (θ, α, β)-$E_{S_E}^2$*-contraction with* $\mathfrak{q}_1, \mathfrak{q}_2 \in \mathcal{Y}$. *If* $\mathfrak{q}_1, \mathfrak{q}_2, \mathfrak{N}\mathfrak{q}_1, \mathfrak{N}\mathfrak{q}_2 \in Fix(\mathfrak{N})$, $\mathfrak{q} \in Fix(\mathfrak{N}^2)$ *and* $\mathfrak{S}(\mathfrak{N}\mathfrak{q}, \mathfrak{N}\mathfrak{q}, \mathfrak{q}_1) + \mathfrak{S}(\mathfrak{N}\mathfrak{q}, \mathfrak{N}\mathfrak{q}, \mathfrak{q}_2) = r$ *then* $E_r^{\mathfrak{S}}(\mathfrak{q}_1, \mathfrak{q}_2) \subset Fix(\mathfrak{N})$.

Proof 6.12 *Case 1: Let* $r = 0$. *Then we have* $E_r^{\mathfrak{S}}(\mathfrak{q}_1, \mathfrak{q}_2) = \{\mathfrak{q}_1\} = \{\mathfrak{q}_2\}$. *By the hypothesis, we get*

$$E_r^{\mathfrak{S}}(\mathfrak{q}_1, \mathfrak{q}_2) \subset Fix(\mathfrak{N}).$$

Case 2: Let $r > 0$ *and* $\mathfrak{q} \in E_r^{\mathfrak{S}}(\mathfrak{q}_1, \mathfrak{q}_2)$ *such that* $\mathfrak{q} \notin Fix(\mathfrak{N})$. *Using the hypothesis, we have*

$$\begin{aligned}
\alpha\left[\mathfrak{S}(\mathfrak{q}, \mathfrak{q}, \mathfrak{N}\mathfrak{q})\right] &\leq \theta(\mathfrak{q}_1, \mathfrak{q}_2)\alpha\left[\mathfrak{S}(\mathfrak{q}, \mathfrak{q}, \mathfrak{N}\mathfrak{q})\right] \\
&\leq \beta\left[\frac{E_S^2(\mathfrak{q}, \mathfrak{q}_1) + E_S^2(\mathfrak{q}, \mathfrak{q}_2)}{4}\right] \\
&< \alpha\left[\frac{E_S^2(\mathfrak{q}, \mathfrak{q}_1) + E_S^2(\mathfrak{q}, \mathfrak{q}_2)}{4}\right] \\
&\leq \alpha\left[\mathfrak{S}(\mathfrak{q}, \mathfrak{q}, \mathfrak{N}\mathfrak{q})\right],
\end{aligned}$$

a contradiction. So $\mathfrak{q} \in Fix(\mathfrak{N})$. *Consequently, we have*

$$E_r^{\mathfrak{S}}(\mathfrak{q}_1, \mathfrak{q}_2) \subset Fix(\mathfrak{N}).$$

Example 6.4 *Let* $\mathcal{Y} = [-1, 1] \cup \{2, 4\}$ *be an* S*-metric space with the* S*-metric defined as in Example 6.1. Let us define a self-mapping* $\mathfrak{N} : \mathcal{Y} \to \mathcal{Y}$ *as*

$$\mathfrak{N}\mathfrak{q} = \begin{cases} \mathfrak{q}, & \mathfrak{q} \in \mathcal{Y} - \{4\} \\ 2, & \mathfrak{q} = 4 \end{cases},$$

for all $\mathfrak{q} \in \mathcal{Y}$. *Then* $\mathfrak{N}$ *is both* (θ, α, β)-E_{S_E}*-contraction and* (θ, α, β)-$E_{S_E}^2$*-contraction with* $\mathfrak{q}_1 = -1$, $\mathfrak{q}_2 = 1$, $\theta(\mathfrak{q}, y) = 1$, $\alpha(\mathfrak{s}) = \frac{\mathfrak{s}}{2}$ *and* $\beta(\mathfrak{s}) = \frac{\mathfrak{s}}{3}$. *Also, we have*

$$r = 4.$$

Consequently, $\mathfrak{N}$ *fixes the ellipse* $E_4^{\mathfrak{S}}(-1, 1) = [-1, 1]$.

6.2.2.3 Some Fixed-Hyperbola Results

In this section, we prove two fixed-hyperbola results on S-metric spaces.

Definition 6.11 *If there exists* $\mathfrak{q}_1, \mathfrak{q}_2 \in \mathcal{Y}$ *such that*

$$\theta(\mathfrak{q}_1, \mathfrak{q}_2)\alpha\left[\mathfrak{S}(\mathfrak{q}, \mathfrak{q}, \mathfrak{N}\mathfrak{q})\right] \leq \beta\left[\left|E_S(\mathfrak{q}, \mathfrak{q}_1) - E_S(\mathfrak{q}, \mathfrak{q}_2)\right|\right]$$

and

$$\mathfrak{S}(\mathfrak{q}, \mathfrak{q}, \mathfrak{N}\mathfrak{q}) \geq \max\left\{\mathfrak{S}(\mathfrak{N}\mathfrak{q}, \mathfrak{N}\mathfrak{q}, \mathfrak{q}_1), \mathfrak{S}(\mathfrak{N}\mathfrak{q}, \mathfrak{N}\mathfrak{q}, \mathfrak{q}_2)\right\},$$

for all $\mathfrak{q} \in \mathcal{Y} - Fix(\mathfrak{N})$, *then* $\mathfrak{N}$ *is called a* (θ, α, β)-E_{S_H}*-contraction.*

Theorem 6.9 *Let* $\mathfrak{N}$ *be a* (θ, α, β)-E_{S_H}*-contraction with* $\mathfrak{q}_1, \mathfrak{q}_2 \in \mathcal{Y}$. *If* $\mathfrak{q}_1, \mathfrak{q}_2 \in Fix(\mathfrak{N})$ *and* $r > 0$ *then* $H_r^{\mathfrak{S}}(\mathfrak{q}_1, \mathfrak{q}_2) \subset Fix(\mathfrak{N})$.

Proof 6.13 *Let* $\mathfrak{q} \in H_r^{\mathfrak{S}}(\mathfrak{q}_1, \mathfrak{q}_2)$ *such that* $\mathfrak{q} \notin Fix(\mathfrak{N})$. *Using the hypothesis, we have*

$$\alpha\left[\mathfrak{S}(\mathfrak{q}, \mathfrak{q}, \mathfrak{N}\mathfrak{q})\right] \leq \theta(\mathfrak{q}_1, \mathfrak{q}_2)\alpha\left[\mathfrak{S}(\mathfrak{q}, \mathfrak{q}, \mathfrak{N}\mathfrak{q})\right] \leq \beta\left[\left|E_S(\mathfrak{q}, \mathfrak{q}_1) - E_S(\mathfrak{q}, \mathfrak{q}_2)\right|\right], \tag{6.19}$$

where

$$E_S(\mathfrak{q}, \mathfrak{q}_1) = \mathfrak{S}(\mathfrak{q}, \mathfrak{q}, \mathfrak{q}_1) + \mathfrak{S}(\mathfrak{q}, \mathfrak{q}, \mathfrak{N}\mathfrak{q})$$

and

$$E_S(\mathfrak{q}, \mathfrak{q}_2) = \mathfrak{S}(\mathfrak{q}, \mathfrak{q}, \mathfrak{q}_2) + \mathfrak{S}(\mathfrak{q}, \mathfrak{q}, \mathfrak{N}\mathfrak{q}).$$

By (21.26), we get

$$\begin{aligned}\alpha\left[\mathfrak{S}(\mathfrak{q}, \mathfrak{q}, \mathfrak{N}\mathfrak{q})\right] &\leq \beta\left[\left|\mathfrak{S}(\mathfrak{q}, \mathfrak{q}, \mathfrak{q}_1) - \mathfrak{S}(\mathfrak{q}, \mathfrak{q}, \mathfrak{q}_2)\right|\right] = \beta\left[r\right] \\ &< \alpha\left[r\right] \leq \alpha\left[\mathfrak{S}(\mathfrak{q}, \mathfrak{q}, \mathfrak{N}\mathfrak{q})\right],\end{aligned}$$

a contradiction. Hence we have

$$H_r^{\mathfrak{S}}(\mathfrak{q}_1, \mathfrak{q}_2) \subset Fix(\mathfrak{N}).$$

Definition 6.12 *If there exists* $\mathfrak{q}_1, \mathfrak{q}_2 \in \mathcal{Y}$ *such that*

$$\theta(\mathfrak{q}_1, \mathfrak{q}_2)\alpha\left[\mathfrak{S}(\mathfrak{q}, \mathfrak{q}, \mathfrak{N}\mathfrak{q})\right] \leq \beta\left[\left|E_S^2(\mathfrak{q}, \mathfrak{q}_1) - E_S^2(\mathfrak{q}, \mathfrak{q}_2)\right|\right],$$

for all $\mathfrak{q} \in \mathcal{Y} - Fix(\mathfrak{N})$, *then* $\mathfrak{N}$ *is called a* (θ, α, β)-$E_{S_H}^2$*-contraction.*

Theorem 6.10 *Let* $\mathfrak{N}$ *be a* (θ, α, β)-$E_{S_H}^2$*-contraction with* $\mathfrak{q}_1, \mathfrak{q}_2 \in \mathcal{Y}$. *If* $\mathfrak{q}_1, \mathfrak{q}_2, \mathfrak{N}\mathfrak{q}_1, \mathfrak{N}\mathfrak{q}_2 \in Fix(\mathfrak{N})$, $|\mathfrak{S}(\mathfrak{N}\mathfrak{q}, \mathfrak{N}\mathfrak{q}, \mathfrak{q}_1) - \mathfrak{S}(\mathfrak{N}\mathfrak{q}, \mathfrak{N}\mathfrak{q}, \mathfrak{q}_2)| = r$, $\mathfrak{q} \in Fix(\mathfrak{N}^2)$ *and* $r > 0$ *then* $H_r^{\mathfrak{S}}(\mathfrak{q}_1, \mathfrak{q}_2) \subset Fix(\mathfrak{N})$.

Proof 6.14 *Let* $\mathfrak{q} \in H_r^{\mathfrak{S}}(\mathfrak{q}_1, \mathfrak{q}_2)$ *such that* $\mathfrak{q} \notin Fix(\mathfrak{N})$. *Using the hypothesis, we have*

$$\alpha\left[\mathfrak{S}(\mathfrak{q}, \mathfrak{q}, \mathfrak{N}\mathfrak{q})\right] \leq \theta(\mathfrak{q}_1, \mathfrak{q}_2)\alpha\left[\mathfrak{S}(\mathfrak{q}, \mathfrak{q}, \mathfrak{N}\mathfrak{q})\right] \leq \beta\left[\left|E_S^2(\mathfrak{q}, \mathfrak{q}_1) - E_S^2(\mathfrak{q}, \mathfrak{q}_2)\right|\right], \tag{6.20}$$

where

$$E_S^2(\mathfrak{q}, \mathfrak{q}_1) = \mathfrak{S}(\mathfrak{N}\mathfrak{q}, \mathfrak{N}\mathfrak{q}, \mathfrak{q}_1) + \mathfrak{S}(\mathfrak{q}, \mathfrak{q}, \mathfrak{N}\mathfrak{q})$$

and

$$E_S^2(\mathfrak{q}, \mathfrak{q}_2) = \mathfrak{S}(\mathfrak{N}\mathfrak{q}, \mathfrak{N}\mathfrak{q}, \mathfrak{q}_2) + \mathfrak{S}(\mathfrak{q}, \mathfrak{q}, \mathfrak{N}\mathfrak{q}).$$

By (21.27), we get

$$\begin{aligned} \alpha\left[\mathfrak{S}(\mathfrak{q}, \mathfrak{q}, \mathfrak{N}\mathfrak{q})\right] &\leq \beta\left[\left|\mathfrak{S}(\mathfrak{N}\mathfrak{q}, \mathfrak{N}\mathfrak{q}, \mathfrak{q}_1) - \mathfrak{S}(\mathfrak{N}\mathfrak{q}, \mathfrak{N}\mathfrak{q}, \mathfrak{q}_2)\right|\right] \\ &= \beta\left[r\right] < \alpha\left[r\right] \leq \alpha\left[\mathfrak{S}(\mathfrak{q}, \mathfrak{q}, \mathfrak{N}\mathfrak{q})\right], \end{aligned}$$

a contradiction. Hence we get

$$H_r^{\mathfrak{S}}(\mathfrak{q}_1, \mathfrak{q}_2) \subset Fix(\mathfrak{N}).$$

Example 6.5 *Let $\mathcal{Y} = [-1, 1] \cup \{2, 3\}$ be an S-metric space with the S-metric defined as in Example 6.1. Let us define a self-mapping $\mathfrak{N} : \mathcal{Y} \to \mathcal{Y}$ as*

$$\mathfrak{N}\mathfrak{q} = \begin{cases} \mathfrak{q}, & \mathfrak{q} \in \mathcal{Y} - \{3\} \\ 2, & \mathfrak{q} = 3 \end{cases},$$

for all $\mathfrak{q} \in \mathcal{Y}$. Then $\mathfrak{N}$ is both (θ, α, β)-E_{S_H}-contraction and (θ, α, β)-$E_{S_H}^2$-contraction with $\mathfrak{q}_1 = -1$, $\mathfrak{q}_2 = 1$, $\theta(\mathfrak{q}, y) = 1$, $\alpha(\mathfrak{s}) = \frac{\mathfrak{s}}{2}$ and $\beta(\mathfrak{s}) = \frac{\mathfrak{s}}{3}$. Also, we have

$$r = 2.$$

Consequently, $\mathfrak{N}$ fixes the hyperbola $H_2^{\mathfrak{S}}(-1, 1) = \{-\frac{1}{2}, \frac{1}{2}\}$.

6.2.2.4 Some Fixed-Cassini Curve Results

In this section, we prove two fixed-Cassini curve theorems on S-metric spaces.

Definition 6.13 *If there exists $\mathfrak{q}_1, \mathfrak{q}_2 \in \mathcal{Y}$ such that*

$$\theta(\mathfrak{q}_1, \mathfrak{q}_2)\alpha\left[\mathfrak{S}(\mathfrak{q}, \mathfrak{q}, \mathfrak{N}\mathfrak{q})\right] \leq \beta\left[\sqrt{\frac{E_S(\mathfrak{q}, \mathfrak{q}_1)E_S(\mathfrak{q}, \mathfrak{q}_2)}{4}}\right]$$

and

$$\mathfrak{S}(\mathfrak{q}, \mathfrak{q}, \mathfrak{N}\mathfrak{q}) \geq \max\left\{\begin{array}{c} \mathfrak{S}(\mathfrak{q}, \mathfrak{q}, \mathfrak{q}_1), \mathfrak{S}(\mathfrak{q}, \mathfrak{q}, \mathfrak{q}_2), \\ \mathfrak{S}(\mathfrak{N}\mathfrak{q}, \mathfrak{N}\mathfrak{q}, \mathfrak{q}_1), \mathfrak{S}(\mathfrak{N}\mathfrak{q}, \mathfrak{N}\mathfrak{q}, \mathfrak{q}_2) \end{array}\right\},$$

for all $\mathfrak{q} \in \mathcal{Y} - Fix(\mathfrak{N})$, then $\mathfrak{N}$ is called a (θ, α, β)-E_{S_C}-contraction.

Theorem 6.11 *Let $\mathfrak{N}$ be a (θ, α, β)-E_{S_C}-contraction with $\mathfrak{q}_1, \mathfrak{q}_2 \in \mathcal{Y}$. If $\mathfrak{q}_1, \mathfrak{q}_2 \in Fix(\mathfrak{N})$ then $C_r^{\mathfrak{S}}(\mathfrak{q}_1, \mathfrak{q}_2) \subset Fix(\mathfrak{N})$.*

Proof 6.15 *Case 1: Let $r = 0$. Then we have $C_r^{\mathfrak{S}}(\mathfrak{q}_1, \mathfrak{q}_2) = \{\mathfrak{q}_1\} = \{\mathfrak{q}_2\}$. By the hypothesis, we get*

$$C_r^{\mathfrak{S}}(\mathfrak{q}_1, \mathfrak{q}_2) \subset Fix(\mathfrak{N}).$$

Case 2: Let $r > 0$ and $\mathfrak{q} \in C_r^{\mathfrak{S}}(\mathfrak{q}_1, \mathfrak{q}_2)$ such that $\mathfrak{q} \notin Fix(\mathfrak{N})$. Using the hypothesis, we find

$$\begin{aligned}
\alpha\left[\mathfrak{S}(\mathfrak{q}, \mathfrak{q}, \mathfrak{N}\mathfrak{q})\right] &\leq \theta(\mathfrak{q}_1, \mathfrak{q}_2)\alpha\left[\mathfrak{S}(\mathfrak{q}, \mathfrak{q}, \mathfrak{N}\mathfrak{q})\right] \\
&\leq \beta\left[\sqrt{\frac{E_S(\mathfrak{q}, \mathfrak{q}_1)E_S(\mathfrak{q}, \mathfrak{q}_2)}{4}}\right] \\
&< \alpha\left[\sqrt{\frac{E_S(\mathfrak{q}, \mathfrak{q}_1)E_S(\mathfrak{q}, \mathfrak{q}_2)}{4}}\right] \\
&\leq \alpha\left[\mathfrak{S}(\mathfrak{q}, \mathfrak{q}, \mathfrak{N}\mathfrak{q})\right],
\end{aligned}$$

a contradiction. Hence, we have

$$C_r^{\mathfrak{S}}(\mathfrak{q}_1, \mathfrak{q}_2) \subset Fix(\mathfrak{N}).$$

Definition 6.14 *If there exists $\mathfrak{q}_1, \mathfrak{q}_2 \in \mathcal{Y}$ such that*

$$\theta(\mathfrak{q}_1, \mathfrak{q}_2)\alpha\left[\mathfrak{S}(\mathfrak{q}, \mathfrak{q}, \mathfrak{N}\mathfrak{q})\right] \leq \beta\left[\sqrt{\frac{E_S^2(\mathfrak{q}, \mathfrak{q}_1)E_S^2(\mathfrak{q}, \mathfrak{q}_2)}{4}}\right]$$

and

$$\mathfrak{S}(\mathfrak{q}, \mathfrak{q}, \mathfrak{N}\mathfrak{q}) \geq \max\left\{\mathfrak{S}(\mathfrak{N}\mathfrak{q}, \mathfrak{N}\mathfrak{q}, \mathfrak{q}_1), \mathfrak{S}(\mathfrak{N}\mathfrak{q}, \mathfrak{N}\mathfrak{q}, \mathfrak{q}_2)\right\},$$

for all $\mathfrak{q} \in \mathcal{Y} - Fix(\mathfrak{N})$, then $\mathfrak{N}$ is called a (θ, α, β)-$E_{S_C}^2$-contraction.

Theorem 6.12 *Let $\mathfrak{N}$ be a (θ, α, β)-$E_{S_C}^2$-contraction with $\mathfrak{q}_1, \mathfrak{q}_2 \in \mathcal{Y}$. If $\mathfrak{q}_1, \mathfrak{q}_2, \mathfrak{N}\mathfrak{q}_1, \mathfrak{N}\mathfrak{q}_2 \in Fix(T)$ and $\mathfrak{q} \in Fix(\mathfrak{N}^2)$ then $C_r^{\mathfrak{S}}(\mathfrak{q}_1, \mathfrak{q}_2) \subset Fix(\mathfrak{N})$.*

Proof 6.16 *Case 1: Let $r = 0$. Then we have $C_r^{\mathfrak{S}}(\mathfrak{q}_1, \mathfrak{q}_2) = \{\mathfrak{q}_1\} = \{\mathfrak{q}_2\}$. By the hypothesis, we get*

$$C_r^{\mathfrak{S}}(\mathfrak{q}_1, \mathfrak{q}_2) \subset Fix(\mathfrak{N}).$$

Case 2: Let $r > 0$ and $\mathfrak{q} \in C_r^{\mathfrak{S}}(\mathfrak{q}_1, \mathfrak{q}_2)$ such that $\mathfrak{q} \notin Fix(\mathfrak{N})$. Using the hypothesis, we find

$$\begin{aligned}
\alpha\left[\mathfrak{S}(\mathfrak{q}, \mathfrak{q}, \mathfrak{N}\mathfrak{q})\right] &\leq \theta(\mathfrak{q}_1, \mathfrak{q}_2)\alpha\left[\mathfrak{S}(\mathfrak{q}, \mathfrak{q}, \mathfrak{N}\mathfrak{q})\right] \\
&\leq \beta\left[\sqrt{\frac{E_S^2(\mathfrak{q}, \mathfrak{q}_1)E_S^2(\mathfrak{q}, \mathfrak{q}_2)}{4}}\right] \\
&< \alpha\left[\sqrt{\frac{E_S^2(\mathfrak{q}, \mathfrak{q}_1)E_S^2(\mathfrak{q}, \mathfrak{q}_2)}{4}}\right] \\
&\leq \alpha\left[\mathfrak{S}(\mathfrak{q}, \mathfrak{q}, \mathfrak{N}\mathfrak{q})\right],
\end{aligned}$$

a contradiction. Hence, we have

$$C_r^{\mathfrak{S}}(\mathfrak{q}_1, \mathfrak{q}_2) \subset Fix(\mathfrak{N}).$$

Example 6.6 *Let $\mathcal{Y} = [-1, 1] \cup \{-\sqrt{2}, \sqrt{2}, 2, 3\}$ be an S-metric space with the S-metric defined as in Example 6.1. Let us define a self-mapping $\mathfrak{N} : \mathcal{Y} \to \mathcal{Y}$ as*

$$\mathfrak{N}\mathfrak{q} = \begin{cases} \mathfrak{q} & , \quad \mathfrak{q} \in \mathcal{Y} - \{3\} \\ 2 & , \quad \mathfrak{q} = 3 \end{cases},$$

for all $\mathfrak{q} \in \mathcal{Y}$. Then $\mathfrak{N}$ is both (θ, α, β)-E_{S_C}-contraction and (θ, α, β)-$E^2_{S_C}$-contraction with $\mathfrak{q}_1 = -1$, $\mathfrak{q}_2 = 1$, $\theta(\mathfrak{q}, y) = 1$, $\alpha(\mathfrak{s}) = \frac{\mathfrak{s}}{2}$ and $\beta(\mathfrak{s}) = \frac{\mathfrak{s}}{3}$. Also, we have

$$r = 2.$$

Consequently, $\mathfrak{N}$ fixes the Cassini curve $C_2^{\mathfrak{S}}(-1, 1) = \{-\sqrt{2}, 0, \sqrt{2}\}$.

6.2.2.5 Some Fixed-Apollonius Circle Results

In this section, we prove two fixed-Apollonius circle theorems on S-metric spaces.

Definition 6.15 *If there exists $\mathfrak{q}_1, \mathfrak{q}_2 \in \mathcal{Y}$ such that*

$$\theta(\mathfrak{q}_1, \mathfrak{q}_2)\alpha\left[\mathfrak{S}(\mathfrak{q}, \mathfrak{q}, \mathfrak{N}\mathfrak{q})\right] \leq \beta\left[\frac{E_S(\mathfrak{q}, \mathfrak{q}_1)}{E_S(\mathfrak{q}, \mathfrak{q}_2)}\right]$$

and

$$\mathfrak{S}(\mathfrak{q}, \mathfrak{q}, \mathfrak{N}\mathfrak{q}) \geq \max\left\{ \begin{array}{c} 1, \mathfrak{S}(\mathfrak{q}, \mathfrak{q}, \mathfrak{q}_1), \mathfrak{S}(\mathfrak{q}, \mathfrak{q}, \mathfrak{q}_2), \\ \mathfrak{S}(\mathfrak{N}\mathfrak{q}, \mathfrak{N}\mathfrak{q}, \mathfrak{q}_1), \mathfrak{S}(\mathfrak{N}\mathfrak{q}, \mathfrak{N}\mathfrak{q}, \mathfrak{q}_2) \end{array} \right\},$$

for all $\mathfrak{q} \in \mathcal{Y} - Fix(\mathfrak{N})$, then $\mathfrak{N}$ is called a (θ, α, β)-E_{S_A}-contraction.

Theorem 6.13 *Let $\mathfrak{N}$ be a (θ, α, β)-E_{S_A}-contraction with $\mathfrak{q}_1, \mathfrak{q}_2 \in \mathcal{Y}$. If $\mathfrak{q}_1, \mathfrak{q}_2 \in Fix(\mathfrak{N})$ then $A_r^{\mathfrak{S}}(\mathfrak{q}_1, \mathfrak{q}_2) \subset Fix(\mathfrak{N})$.*

Proof 6.17 *Case 1: Let $r = 0$. Then we have $A_r^{\mathfrak{S}}(\mathfrak{q}_1, \mathfrak{q}_2) = \{\mathfrak{q}_1\}$. By the hypothesis, we get*

$$A_r^{\mathfrak{S}}(\mathfrak{q}_1, \mathfrak{q}_2) \subset Fix(\mathfrak{N}).$$

Case 2: Let $r > 0$ and $\mathfrak{q} \in A_r^{\mathfrak{S}}(\mathfrak{q}_1, \mathfrak{q}_2)$ such that $\mathfrak{q} \notin Fix(\mathfrak{N})$. Using the hypothesis, we find

$$\begin{aligned} \alpha\left[\mathfrak{S}(\mathfrak{q}, \mathfrak{q}, \mathfrak{N}\mathfrak{q})\right] &\leq \theta(\mathfrak{q}_1, \mathfrak{q}_2)\alpha\left[\mathfrak{S}(\mathfrak{q}, \mathfrak{q}, \mathfrak{N}\mathfrak{q})\right] \\ &\leq \beta\left[\frac{E_S(\mathfrak{q}, \mathfrak{q}_1)}{E_S(\mathfrak{q}, \mathfrak{q}_2)}\right] \\ &< \alpha\left[\frac{E_S(\mathfrak{q}, \mathfrak{q}_1)}{E_S(\mathfrak{q}, \mathfrak{q}_2)}\right] \leq \alpha\left[1\right], \end{aligned}$$

a contradiction. Hence, we have

$$A_r^{\mathfrak{S}}(\mathfrak{q}_1, \mathfrak{q}_2) \subset Fix(\mathfrak{N}).$$

Definition 6.16 *If there exists* $\mathfrak{q}_1, \mathfrak{q}_2 \in \mathcal{Y}$ *such that*

$$\theta(\mathfrak{q}_1, \mathfrak{q}_2)\alpha\left[\mathfrak{S}(\mathfrak{q}, \mathfrak{q}, \mathfrak{N}\mathfrak{q})\right] \leq \beta\left[\frac{E_S^2(\mathfrak{q}, \mathfrak{q}_1)}{E_S^2(\mathfrak{q}, \mathfrak{q}_2)}\right]$$

and

$$\mathfrak{S}(\mathfrak{q}, \mathfrak{q}, \mathfrak{N}\mathfrak{q}) \geq \max\left\{1, \mathfrak{S}(\mathfrak{N}\mathfrak{q}, \mathfrak{N}\mathfrak{q}, \mathfrak{q}_1), \mathfrak{S}(\mathfrak{N}\mathfrak{q}, \mathfrak{N}\mathfrak{q}, \mathfrak{q}_2)\right\},$$

for all $\mathfrak{q} \in \mathcal{Y} - Fix(\mathfrak{N})$*, then* $\mathfrak{N}$ *is called a* (θ, α, β)*-*$E_{S_A}^2$*-contraction.*

Theorem 6.14 *Let* $\mathfrak{N}$ *be a* (θ, α, β)*-*$E_{S_A}^2$*-contraction with* $\mathfrak{q}_1, \mathfrak{q}_2 \in \mathcal{Y}$*. If* $\mathfrak{q}_1, \mathfrak{q}_2, \mathfrak{N}\mathfrak{q}_1, \mathfrak{N}\mathfrak{q}_2 \in Fix(\mathfrak{N})$ *and* $\mathfrak{q} \in Fix(\mathfrak{N}^2)$ *then* $A_r^{\mathfrak{S}}(\mathfrak{q}_1, \mathfrak{q}_2) \subset Fix(\mathfrak{N})$*.*

Proof 6.18 *Case 1: Let* $r = 0$*. Then we have* $A_r^{\mathfrak{S}}(\mathfrak{q}_1, \mathfrak{q}_2) = \{\mathfrak{q}_1\}$*. By the hypothesis, we get*

$$A_r^{\mathfrak{S}}(\mathfrak{q}_1, \mathfrak{q}_2) \subset Fix(\mathfrak{N}).$$

Case 2: Let $r > 0$ *and* $\mathfrak{q} \in A_r^{\mathfrak{S}}(\mathfrak{q}_1, \mathfrak{q}_2)$ *such that* $\mathfrak{q} \notin Fix(\mathfrak{N})$*. Using the hypothesis, we find*

$$\begin{aligned}\alpha\left[\mathfrak{S}(\mathfrak{q}, \mathfrak{q}, \mathfrak{N}\mathfrak{q})\right] &\leq \theta(\mathfrak{q}_1, \mathfrak{q}_2)\alpha\left[\mathfrak{S}(\mathfrak{q}, \mathfrak{q}, \mathfrak{N}\mathfrak{q})\right] \\ &\leq \beta\left[\frac{E_S^2(\mathfrak{q}, \mathfrak{q}_1)}{E_S^2(\mathfrak{q}, \mathfrak{q}_2)}\right] \\ &< \alpha\left[\frac{E_S^2(\mathfrak{q}, \mathfrak{q}_1)}{E_S^2(\mathfrak{q}, \mathfrak{q}_2)}\right] \leq \alpha\left[1\right],\end{aligned}$$

a contradiction. Hence, we have

$$A_r^{\mathfrak{S}}(\mathfrak{q}_1, \mathfrak{q}_2) \subset Fix(\mathfrak{N}).$$

BIBLIOGRAPHY

[1] M. A. Alghamdi, S. G. Ozyurt, A. Fulga, Fixed points of Proinov E-contractions, *Symmetry* 13 (2021), 962.

[2] H. Aytimur, N. Taş, A geometric interpretation to fixed-point theory on S_b-metric spaces, *Electron. J. Math. Anal. Appl.* 10 (2) (2022), 95–104.

[3] G. Z. Erçınar, *Some geometric properties of fixed points*, Ph.D. Thesis, Eskişehir Osmangazi University, 2020.

[4] N. T. Hieu, N. T. Ly, N. V. Dung, A generalization of Ciric quasi-contractions for maps on S-metric spaces, *Thai J. Math.* 13 (2) (2015) 369-380.

[5] M. Joshi, A. Tomar, S. K. Padaliya, Fixed point to fixed ellipse in metric spaces and discontinuous activation function, *Appl. Math. E-Notes* 21 (2021), 225-237.

[6] N. Mlaiki, N. Özgür, N. Taş, New fixed-circle results related to F_c-contractive and F_c-expanding mappings on metric spaces, arXiv:2101.10770.

[7] N. Y. Özgür, N. Taş, *Some generalizations of fixed point theorems on S-metric spaces*, Essays in Mathematics and Its Applications in Honor of Vladimir Arnold, New York, Springer, 2016.

[8] N. Y. Özgür and N. Taş, Some fixed point theorems on S-metric spaces, *Mat. Vesnik* 69 (1) (2017), 39–52.

[9] N. Y. Özgür and N. Taş, Some new contractive mappings on S-metric spaces and their relationships with the mapping (**S25**), Math. Sci. 11 (1) (2017), 7–16.

[10] N. Y. Özgür and N. Taş, *Some fixed-circle theorems and discontinuity at fixed circle*, AIP Conference Proceedings 1926, 020048 (2018).

[11] N. Y. Özgür and N. Taş, Some fixed-circle theorems on metric spaces, *Bull. Malays. Math. Sci. Soc.* 42 (4) (2019), 1433–1449.

[12] N. Özgür, Fixed-disc results via simulation functions, *Turkish J. Math.* 43 (6) (2019), 2794–2805.

[13] N. Y. Özgür and N. Taş, Some fixed-circle problem on S-metric spaces with a geometric viewpoint, *Facta Univ. Ser. Math. Inform.* 34 (2019), 459–472.

[14] N. Özgür and N. Taş, Geometric properties of fixed points and simulation functions, arXiv:2102.05417.

[15] R. P. Pant, N. Y. Özgür and N. Taş, Discontinuity at fixed points with applications, *Bull. Belg. Math. Soc. - Simon Stevin* 26 (2019), 571–589.

[16] R. P. Pant, N. Y. Özgür and N. Taş, On discontinuity problem at fixed point, *Bull. Malays. Math. Sci. Soc.* 43 (2020), 499–517.

[17] R. P. Pant, N. Özgür, N. Taş, A. Pant and M. C. Joshi, New results on discontinuity at fixed point, *J. Fixed Point Theory Appl.* 22 (2020), 39.

[18] O. Popescu, Some new fixed point theorems for α-Geraghty contraction type maps in metric spaces, *Fixed Point Theory Appl.* 2014 (2014), 190.

[19] P. D. Proinov, Fixed point theorems for generalized contractive mappings in metric spaces, *J. Fixed Point Theory Appl.* 22 (2020), 21.

[20] S. Sedghi, N. Shobe, A. Aliouche, A generalization of fixed point theorems in S-metric spaces, *Mat. Vesnik* 64 (3) (2012), 258–266.

[21] S. Sedghi, N. V. Dung, Fixed point theorems on S-metric spaces, *Mat. Vesnik* 66 (1) (2014), 113–124.

[22] N. Taş, Bilateral-type solutions to the fixed-circle problem with rectified linear units application, *Turkish J. Math.* 44 (4) (2020), 1330–1344.

[23] N. Taş, N. Özgür, *New fixed-figure results on metric spaces,* Fixed Point Theory and Fractional Calculus - Recent Advances and Applications, Springer, Singapore, 2022, 33–62.

$\eta_{\mathcal{A}}$-Admissible Mappings for Four Maps in C^*-Algebra-Valued $\mathcal{MP}$-Metric Spaces with an Application

Samira Hadi Bonab, Vahid Parvaneh, and Zohreh Bagheri
Islamic Azad University

CONTENTS

7.1 INTRODUCTION

In this chapter, we first introduce the concept of $\eta_{\mathcal{A}}$-admissible mapping in C^*-algebra-valued $\mathcal{MP}$-metric spaces, which is a generalization and combination of modular metric space, parametric metric space and C^*-algebra-valued metric space. Then we prove some fixed point theorems for four mappings in these spaces. To confirm the new results, we provide an example and an application about the solvability of operator equations and integral equations, respectively.

In recent years, many researchers have generalized the Banach fixed point [2] in nonlinear analysis. This theory has many applications in mathematics, for example:

Cone metric spaces [11], G-metric space [12], vector-valued metric space [7, 13], b-metric space [13], b-rectangular metric space [24], generalized parametric metric

DOI: 10.1201/9781003388678-7

[26], modular b-metric space [6, 22] etc. Also, many mathematicians have presented different definitions of contraction mappings on complete metric spaces and developed them in different ways and turned them into a general rule. For more information, see [4, 5, 23, 28].

Nakano [20] introduced modular spaces in connection with the theory of ordered spaces, which was later generalized in [19]. In [12], Hossein et al. introduced the concept of parametric metric spaces. Several authors investigated fixed point theorems for the contraction of multivalued maps. Kotabi and Latif [14] studied fixed points of multivalued maps in modular function spaces. Also see [3, 15] for more information.

In 2014, using the set of all positive elements of unit C^*-algebra instead of the set of real numbers, the concept of C^*-algebra valued metric spaces was introduced [16]. Later, many authors studied in this field and presented results. See [15, 18, 24, 27].

In 2012, the concept of α-admissible mapping was presented by Samet et al [25]. In this chapter, we introduce this concept for a C^*-algebra valued modular parametric metric space (C^*-av$\mathcal{MP}_{\mathcal{A}}$ms) via four mappings, which is a generalization and combination of modular metric space, parametric metric space and C^*-algebra-valued metric space. In the following, these concepts are used to prove some fixed point theorems through C^*-contraction and also Kannan-Ćirić C^*-contraction.

Throughout this chapter, $\mathcal{A}$ is denoted as the unital algebra with unit I. θ is the zero element. An involution on $\mathcal{A}$ corresponds to the conjugate linear map $\kappa \mapsto \kappa^*$ on $\mathcal{A}$ so that $\jmath^{**} = \jmath$ and $(\jmath\wp)^* = \wp^*\jmath^*$ for all $\jmath, \wp \in \mathcal{A}$. The pair $(\mathcal{A}, *)$ is called an $*$-algebra. A Banach $*$-algebra is an $*$-algebra $\mathcal{A}$ with the complete submultiplicative norm so that $\| \jmath^* \| = \| \jmath \|$ for all $\jmath \in \mathcal{A}$. A C^*-algebra is a Banach $*$-algebra so that $\| \jmath^*\jmath \| = \| \jmath \|^2$ for all $\jmath \in \mathcal{A}$. Let $\mathfrak{H}$ be a Hilbert space and $\mathfrak{B}(\mathfrak{H})$ be the set of bounded linear operators on $\mathfrak{H}$, then $\mathfrak{B}(\mathfrak{H})$ is a C^*-algebra with the operator norm. Let $\mathcal{A}_{sa}$ be the family of all self-adjoint elements in $\mathcal{A}$. An element $\jmath \in \mathcal{A}$ is positive ($\jmath \succeq \theta$) if $\jmath \in \mathcal{A}_{sa}$ and spectrum $\sigma(\jmath) = \{\lambda \in C \mid \lambda I - \jmath \textit{ is not invertible}\} \subseteq \mathbb{R}_+$, where $\jmath \in \mathcal{A}$. Set $\mathcal{A}_+ = \{\jmath \in \mathcal{A} : \jmath \succeq \theta\}$, then $\mathcal{A}_+ = \{\jmath^*\jmath : \jmath \in \mathcal{A}\}$ [18] and $(\jmath^*\jmath)^{\frac{1}{2}} = |\jmath|$. We write $x \prec y$ if $x \preceq y$ and $x \neq y$. Note that a partial ordering $\preceq$ on $\mathcal{A}_{sa}$ as follows: $\jmath \preceq \wp \Leftrightarrow \wp - \jmath \succeq \theta$. If $\jmath, \wp \in \mathcal{A}_{sa}$ and $q \in \mathcal{A}$, then $\jmath \preceq \wp \Rightarrow q^*\jmath q \preceq q^*\wp q$, and if $\jmath, \wp \in \mathcal{A}_+$ are invertible, then $\jmath \preceq \wp \Longrightarrow \theta \preceq \wp^{-1} \preceq \jmath^{-1}$.

Definition 7.1 *[2]. Consider a nonempty set* Π. *A mapping* $\mho : \Pi^2 \longrightarrow [0, \infty)$ *is called a metric on* Π, *if:*

1. $\mho(\zeta, \ell) = 0$ *iff* $\zeta = \ell$;
2. $\mho(\zeta, \ell) = \mho(\ell, \zeta)$ *for each* $\zeta, \ell \in \Pi$;
3. $\mho(\zeta, \ell) \leq \mho(\zeta, \eta) + \mho(\eta, \ell)$ *for each* $\zeta, \ell, \eta \in \Pi$.

Then $(\Pi, \mho)$ *is called a metric space.*

Definition 7.2 *[16] If the function* $\mho : \Pi^2 \to \mathcal{A}$ *(* Π *is a nonempty set) verifies for all* $\zeta, \ell, \eta \in \Pi$:

(i) $\theta \preceq \mho(\zeta, \ell)$ *and* $\mho(\zeta, \ell) = \theta$ *iff* $\zeta = \ell$;

(ii) $\mho(\zeta,\ell)=\mho(\ell,\zeta)$*;*

(iii) $\mho(\zeta,\ell)\preceq\mho(\zeta,\eta)+\mho(\eta,\ell)$.

Then $(\Pi,\mathcal{A},\mho)$ *is called a* C^**-algebra-valued metric space.*

Definition 7.3 *[15] Consider the self-mapping of* $\mathcal{L}:\Pi\to\Pi$ *and* $\mathcal{G}:\Pi\to\Pi$*. If* $\Im=\mathcal{L}\zeta=\mathcal{G}\zeta$ *for some* $\zeta\in\Pi$*, then* ζ *is called a coincidence point of* $\mathcal{L}$ *and* $\mathcal{G}$*.* $\Im$ *is said to be a coincidence point of* $\mathcal{L}$ *and* $\mathcal{G}$*.*

Definition 7.4 *[15] Consider the self-mapping of* $\mathcal{L}:\Pi\to\Pi$ *and* $\mathcal{G}:\Pi\to\Pi$*. If* $\mathcal{L}$ *and* $\mathcal{G}$ *commute at their coincidence points, then they are called w-compatible.*

Definition 7.5 *[17] The function* $\mathcal{W}:(0,+\infty)\times\Pi^2\to\mathcal{A}$ *is said to be a* C^**-algebra-valued modular metric on nonempty set* Π*, if*

(1) $\mathcal{W}_\lambda(\zeta,\ell)=\theta$ *iff* $\zeta=\ell$ *for all* $\lambda>0$*;*

(2) $\mathcal{W}_\lambda(\zeta,\ell)=\mathcal{W}_\lambda(\ell,\zeta)$ *for all* $\lambda>0,\ \zeta,\ell\in\Pi$*;*

(3) $\mathcal{W}_{\lambda+\mu}(\zeta,\ell)\preceq\mathcal{W}_\lambda(\zeta,\eta)+\mathcal{W}_\mu(\eta,\ell)$ *for all* $\zeta,\ell,\eta\in\Pi$ *and all* $\lambda,\mu>0$.

Then the pair $(\Pi,\mathcal{W})$ *is called* C^**-algebra-valued modular metric space .*

Definition 7.6 *[26] Let* Π *be a nonempty set. Function* $\mathcal{P}:\Pi^2\times(0,+\infty)\to[0,+\infty)$ *is said to be a parametric metric on* Π*, if*

(1) $\mathcal{P}(\zeta,\ell,\iota)=0$ *iff* $\zeta=\ell$*;*

(2) $\mathcal{P}(\zeta,\ell,\iota)=\mathcal{P}(\ell,\zeta,\iota)$ *for all* $\iota>0$*;*

(3) $\mathcal{P}(\zeta,\ell,\iota)\leq\mathcal{P}(\zeta,\eta,\iota)+\mathcal{P}(\eta,\ell,\iota)$ *for all* $\zeta,\ell,\eta\in\Pi$ *and all* $\iota>0$.

Then the pair $(\Pi,\mathcal{P})$ *is called parametric metric space.*

Definition 7.7 *[23] Let* $\alpha:\Pi^2\to[0,\infty)$ *(* Π *is a nonempty set). A mapping* $T:\Pi\to\Pi$ *is said to be an* η*-admissible mapping, if*

$$\eta(\zeta,\ell)\geq 1\Rightarrow\eta(T\zeta,T\ell)\geq 1 \text{ for all } \zeta,\ell\in\Pi.$$

Definition 7.8 *[1] The max function on* C^**-algebra* $\mathcal{A}$ *with the partial order relation* $\preceq$ *is defined by:*

$$\max\{\zeta,\ell\}=\ell\Leftrightarrow\zeta\preceq\ell \text{ and } \|\zeta\|\leq\|\ell\|,\quad \text{for all } \zeta,\ell\in\mathcal{A}^+.$$

7.2 $\eta_{\mathcal{A}}$-ADMISSIBLE MAPPING IN C^*-CONTRACTION

In this section, we present the concept of $\eta_{\mathcal{A}}$-admissible mapping in C^*-av$\mathcal{MP}_{\mathcal{A}}$ms to achieve a common fixed point for four maps. First, we introduce the following definitions:

Definition 7.9 *The function $\mathcal{MP} : (0,\infty)\times\Pi^2\times(0,\infty) \to [0,\infty]$ is called a modular parametric metric ($\mathcal{MP}m$) on nonempty set Π, if*

(1) $\mathcal{MP}_\lambda(\zeta,\ell,\iota) = 0$ *iff* $\zeta = \ell$;

(2) $\mathcal{MP}_\lambda(\zeta,\ell,\iota) = \mathcal{MP}_\lambda(\ell,\zeta,\iota)$ *for all* $\lambda,\iota > 0$;

(3) $\mathcal{MP}_{\lambda+\mu}(\zeta,\ell,\iota) \leq \mathcal{MP}_\lambda(\zeta,\eta,\iota) + \mathcal{MP}_\mu(\eta,\ell,\iota)$ *for all* $\zeta,\ell,\eta \in \Pi$.

Then the pair $(\Pi,\mathcal{MP})$ *is called a* $\mathcal{MP}m$ *space.*

Definition 7.10 *The function $\mathcal{P} : \Pi^2 \times (0,+\infty) \to \mathcal{A}^+$ is said to be a C^*-algebra-valued parametric metric on nonempty set Π if,*

(1) $\mathcal{P}(\zeta,\ell,\iota) = \theta$ *iff* $\zeta = \ell$;

(2) $\mathcal{P}(\zeta,\ell,\iota) = \mathcal{P}(\ell,\zeta,\iota)$ *for all* $\iota > 0$;

(3) $\mathcal{P}(\zeta,\ell,\iota) \preceq \mathcal{P}(\zeta,\eta,\iota) + \mathcal{P}(\eta,\ell,\iota)$ *for all* $\zeta,\ell,\eta \in \Pi$ *and all* $\iota > 0$.

Then the pair $(\Pi,\mathcal{P})$ *is called* C^**-algebra-valued parametric metric space.*

Definition 7.11 *Let $\eta : (0,\infty) \times \Pi^2 \times (0,\infty) \to \mathcal{A}^+$ (Π is a nonempty set). the mappings $\mathcal{L},\mathcal{G} : \Pi \to \Pi$ are called common $\eta_\mathcal{A}$-admissible mappings, if*

$$\eta_{\mathcal{A}_\lambda}(\zeta_0,\ell,\iota) \succeq I_\mathcal{A} \Rightarrow \eta_{\mathcal{A}_\lambda}(\mathcal{L}\zeta_0,\mathcal{G}\ell,\iota) \succeq I_\mathcal{A} \Rightarrow \eta_{\mathcal{A}_\lambda}(\mathcal{L}^2\zeta_0,\mathcal{G}^2\ell,\iota) \succeq I_\mathcal{A},$$

for all $\zeta_0,\ell \in \Pi$ *and* $\lambda,\iota > 0$.

Definition 7.12 *The function $\mathcal{MP}_\mathcal{A} : (0,\infty) \times \Pi^2 \times (0,\infty) \to \mathcal{A}^+$ is called a C^*-av$\mathcal{MP}_\mathcal{A}m$ on nonempty set Π if*

(1) $\mathcal{MP}_{\mathcal{A}_\lambda}(\zeta,\ell,\iota) = 0$ *iff* $\zeta = \ell$;

(2) $\mathcal{MP}_{\mathcal{A}_\lambda}(\zeta,\ell,\iota) = \mathcal{MP}_{\mathcal{A}_\lambda}(\ell,\zeta,\iota)$ *for all* $\lambda,\iota > 0$;

(3) $\mathcal{MP}_{\mathcal{A}_{\lambda+\mu}}(\zeta,\ell,\iota) \preceq \mathcal{MP}_{\mathcal{A}_\lambda}(\zeta,\eta,\iota) + \mathcal{MP}_{\mathcal{A}_\mu}(\eta,\ell,\iota)$ *for all* $\zeta,\ell,\eta \in \Pi$.

Then $(\Pi,\mathcal{A},\mathcal{MP}_\mathcal{A})$ *is called a* C^**-av*$\mathcal{MP}_\mathcal{A}ms$*.*

Theorem 7.1 *Let $(\Pi,\mathcal{A},\mathcal{MP}_\mathcal{A})$ be a complete C^*-av$\mathcal{MP}_\mathcal{A}ms$ and $\mathcal{L}$, $\mathcal{G}$, $\mathcal{Q}$, $\mathcal{I}$ be self-mappings on Π, so that the following conditions are satisfied:*

(i)

$$\eta_{\mathcal{A}_\lambda}(\zeta,\ell,\iota)\mathcal{MP}_{\mathcal{A}_\lambda}(\mathcal{L}\zeta,\mathcal{G}\ell,\iota) \preceq \partial^*[\mathcal{MP}_{\mathcal{A}_\lambda}(\mathcal{Q}\zeta,\mathcal{I}\ell,\iota)]\partial, \quad \text{for all } \zeta,\ell \in \Pi,\ \lambda,\iota > 0, \tag{7.1}$$

where $\eta_{\mathcal{A}_\lambda}(\zeta,\ell,\iota) \succeq I_\mathcal{A}$, $\| \partial \| < 1$;

(ii) $(\mathcal{L},\mathcal{G})$ *are common* $\eta_\mathcal{A}$*-admisible on* Π;

(iii) $\mathcal{L}(\Pi) \subset \mathcal{I}(\Pi)$ *and* $\mathcal{G}(\Pi) \subset \mathcal{Q}(\Pi)$.

Then

(A) If one of $\mathcal{L}(\Pi)\cup\mathcal{G}(\Pi)$ and $\mathcal{Q}(\Pi)\cup\mathcal{I}(\Pi)$ is complete, then $\{\mathcal{L},\mathcal{Q}\}$ and $\{\mathcal{G},\mathcal{I}\}$ have a unique coincidence point in Π.

(B) if $\{\mathcal{L},\mathcal{Q}\}$ and $\{\mathcal{G},\mathcal{I}\}$ are w-compatible, then $\mathcal{L}$, $\mathcal{G}$, $\mathcal{Q}$ and $\mathcal{I}$ have a unique common fixed point in Π.

Proof 7.1 *Let* $\zeta_0 \in \Pi$ *(*ζ_0 *is an arbitrary point), define the sequences* $\{\zeta_\hbar\}$ *and* $\{\ell_\hbar\}$ *in* Π *so that*

$$\begin{cases} \mathcal{L}\zeta_{2\hbar} = \mathcal{I}\zeta_{2\hbar+1} = \ell_{2\hbar+1} \\ \mathcal{G}\zeta_{2\hbar+1} = \mathcal{Q}\zeta_{2\hbar+2} = \ell_{2\hbar+2}, \qquad \forall \hbar \geq 0. \end{cases}$$

From (8.1), we get

$$\eta_{\mathcal{A}_\lambda}(\zeta_0, \zeta_1, \iota) = \eta_{\mathcal{A}_\lambda}(\mathcal{L}\zeta_0, \mathcal{G}\zeta_1, \iota) \succeq I_{\mathcal{A}} \Rightarrow \eta_{\mathcal{A}_\lambda}(\mathcal{L}^2\zeta_0, \mathcal{G}^2\zeta_1, \iota) \succeq I_{\mathcal{A}}$$
$$\Rightarrow \eta_{\mathcal{A}_\lambda}(\zeta_2, \zeta_3, \iota) \succeq I_{\mathcal{A}},$$

by induction, we obtain $\eta_{\mathcal{A}_\lambda}(\zeta_{2\hbar}, \zeta_{2\hbar+1}, \iota) \succeq I_{\mathcal{A}}$*, for all* $\hbar \geq 0$*.*

$$\begin{aligned} \mathcal{MP}_{\mathcal{A}_\lambda}(\ell_{2\hbar+1}, \ell_{2\hbar+2}, \iota) &= \mathcal{MP}_{\mathcal{A}_\lambda}(\mathcal{L}\zeta_{2\hbar}, \mathcal{G}\zeta_{2\hbar+1}, \iota) \\ &\preceq \eta_{\mathcal{A}_\lambda}(\zeta_{2\hbar}, \zeta_{2\hbar+1}, \iota)\mathcal{MP}_{\mathcal{A}_\lambda}(\mathcal{L}\zeta_{2\hbar}, \mathcal{G}\zeta_{2\hbar+1}, \iota) \\ &\preceq \partial^*[\mathcal{MP}_{\mathcal{A}_\lambda}(\mathcal{Q}\zeta_{2\hbar}, \mathcal{I}\zeta_{2\hbar+1}, \iota)]\partial \\ &= \partial^*[\mathcal{MP}_{\mathcal{A}_\lambda}(\ell_{2\hbar}, \ell_{2\hbar+1}, \iota)]\partial. \end{aligned}$$

So, by induction, we get

$$\mathcal{MP}_{\mathcal{A}_\lambda}(\ell_{2\hbar+1}, \ell_{2\hbar+2}, \iota) \preceq (\partial^*)^{2\hbar+1}[\mathcal{MP}_{\mathcal{A}_\lambda}(\ell_0, \ell_1, \iota)]\partial^{2\hbar+1}.$$

Similarly, it can be shown that

$$\mathcal{MP}_{\mathcal{A}_\lambda}(\ell_{2\hbar}, \ell_{2\hbar+1}, \iota) \preceq (\partial^*)^{2\hbar}[\mathcal{MP}_{\mathcal{A}_\lambda}(\ell_0, \ell_1, \iota)]\partial^{2\hbar}.$$

Now we can get for every $\hbar \in \mathbb{N}$

$$\begin{aligned} \mathcal{MP}_{\mathcal{A}_\lambda}(\ell_\hbar, \ell_{\hbar+1}, \iota) &\preceq (\partial^*)^\hbar[\mathcal{MP}_{\mathcal{A}_\lambda}(\ell_0, \ell_1, \iota)]\partial^\hbar \\ &\preceq (\partial^*)^\hbar\varpi_0\partial^\hbar, \end{aligned}$$

where $\varpi_0 := \mathcal{MP}_{\mathcal{A}_\lambda}(\ell_0, \ell_1, \iota)$*.*

Step II. We show that $\{\ell_{2\hbar}\}$ *is a Cauchy sequence in* Π*.*

Then for any $\Re, \hbar \geq 1$ *with* $\Re > \hbar$*, it follows that*

$$\begin{aligned} \mathcal{MP}_{\mathcal{A}_\lambda}(\ell_\hbar, \ell_\Re, \iota) &\preceq \sum_{i=\hbar}^{\Re-1} \mathcal{MP}_{\mathcal{A}_\lambda}(\ell_i, \ell_{i+1}, \iota) \\ &= \sum_{i=\hbar}^{\Re-1} (\partial^*)^i\varpi_0\partial^i \\ &\preceq \sum_{i=\hbar}^{\Re-1} (\partial^*)^i\varpi_0^{\frac{1}{2}}\varpi_0^{\frac{1}{2}}\partial^i \\ &= \sum_{i=\hbar}^{\Re-1} (\varpi_0^{\frac{1}{2}}\partial^i)^*(\varpi_0^{\frac{1}{2}}\partial^i) \end{aligned}$$

$$
\begin{aligned}
&= \sum_{i=\hbar}^{\Re-1} |\varpi_0^{\frac{1}{2}} \partial^i|^2 \\
&\preceq \| \sum_{i=\hbar}^{\Re-1} |\varpi_0^{\frac{1}{2}} \partial^i|^2 \| \,.\, I_{\mathcal{A}} \\
&\preceq \sum_{i=\hbar}^{\Re-1} \| \varpi_0^{\frac{1}{2}} \|^2 \| \partial^i \|^2 \,.\, I_{\mathcal{A}} \\
&\preceq \| \varpi_0^{\frac{1}{2}} \|^2 \sum_{i=\hbar}^{\Re-1} \| \partial \|^{2i} \,.\, I_{\mathcal{A}} \\
&\preceq \| \varpi_0^{\frac{1}{2}} \|^2 \, \frac{\| \partial \|^{2\hbar}}{1- \| \partial \|^2} . I_{\mathcal{A}} \to \theta_{\mathcal{A}} \ (as\ \hbar \to \infty).
\end{aligned}
$$

Hence, $\{\ell_{2\hbar}\}$ *is a Cauchy sequence in* Π.

Assuming that $\mathcal{Q}(\Pi)\cup\mathcal{I}(\Pi)$ *be complete. In this case, there is* $\mathcal{U} \in \mathcal{Q}(\Pi)\cup\mathcal{I}(\Pi)$ *so that* $\ell_\hbar \to \mathcal{U}$ *as* $\hbar \to \infty$. *Further, the subsequences* $\{\mathcal{Q}\zeta_{2\hbar+2}\} = \{\mathcal{G}\zeta_{2\hbar+1}\} = \{\ell_{2\hbar+2}\}$ *and* $\{\mathcal{I}\zeta_{2\hbar+1}\} = \{\mathcal{L}\zeta_{2\hbar}\} = \{\ell_{2\hbar+1}\}$ *of* $\{\ell_\hbar\}$ *also converge to the point* $\mathcal{U}$. *Since* $\mathcal{U} \in \mathcal{Q}(\Pi) \cup \mathcal{I}(\Pi)$, *we have* $\mathcal{U} \in \mathcal{Q}(\Pi)$ *or* $\mathcal{U} \in \mathcal{I}(\Pi)$. *If* $\mathcal{U} \in \mathcal{Q}(\Pi)$, *then we can find* $\mathcal{V} \in \Pi$ *so that* $\mathcal{Q}\mathcal{V} = \mathcal{U}$. *So, we claim that* $\mathcal{L}\mathcal{V} = \mathcal{U}$. *To do this, we see that*

$$
\begin{aligned}
\mathcal{MP}_{\mathcal{A}_{2\lambda}}(\mathcal{LV},\mathcal{U},\iota) &\preceq \mathcal{MP}_{\mathcal{A}_\lambda}(\mathcal{LV},\mathcal{G}\zeta_{2\hbar+1},\iota) + \mathcal{MP}_{\mathcal{A}_\lambda}(\mathcal{G}\zeta_{2\hbar+1},\mathcal{U},\iota) \\
&\preceq \eta_{\mathcal{A}_\lambda}(\mathcal{V},\zeta_{2\hbar+1},\iota)\mathcal{MP}_{\mathcal{A}_\lambda}(\mathcal{LV},\mathcal{G}\zeta_{2\hbar+1},\iota) \\
&\quad + \mathcal{MP}_{\mathcal{A}_\lambda}(\mathcal{G}\zeta_{2\hbar+1},\mathcal{U},\iota) \\
&\preceq \partial^*[\mathcal{MP}_{\mathcal{A}_\lambda}(\mathcal{QV},\mathcal{I}\zeta_{2\hbar+1},\iota)]\partial + \mathcal{MP}_{\mathcal{A}_\lambda}(\mathcal{G}\zeta_{2\hbar+1},\mathcal{U},\iota).
\end{aligned}
$$

We get

$$
\begin{aligned}
\| \mathcal{MP}_{\mathcal{A}_{2\lambda}}(\mathcal{LV},\mathcal{U},\iota) \| \leq \| \partial \|^2 \| \mathcal{MP}_{\mathcal{A}_\lambda}(\mathcal{U},\mathcal{I}\zeta_{2\hbar+1},\iota) \| \\
+ \| \mathcal{MP}_{\mathcal{A}_\lambda}(\mathcal{G}\zeta_{2\hbar+1},\mathcal{U},\iota) \| .
\end{aligned}
$$

Since $\| \partial \| < 1$, *making* $\hbar \to \infty$, *we have a contradiction, so* $\mathcal{MP}_{\mathcal{A}_{\in\lambda}}(\mathcal{LV},\mathcal{U},\iota) \preceq \theta_{\mathcal{A}}$. *Consequently, we have* $\mathcal{LV} = \mathcal{QV} = \mathcal{U}$ *and since* $\mathcal{U} \in \mathcal{L}(\Pi) \subset \mathcal{I}(\Pi)$, *there is* $\Im \in \Pi$ *so that* $\mathcal{I}\Im = \mathcal{U}$.

Now, we show that $\mathcal{G}\Im = \mathcal{U}$. *In fact, we have*

$$
\begin{aligned}
\mathcal{MP}_{\mathcal{A}_{2\lambda}}(\mathcal{G}\Im,\mathcal{U},\iota) &\preceq \mathcal{MP}_{\mathcal{A}_\lambda}(\mathcal{G}\Im,\mathcal{L}\zeta_{2\hbar},\iota) + \mathcal{MP}_{\mathcal{A}_\lambda}(\mathcal{L}\zeta_{2\hbar},\mathcal{U},\iota) \\
&\preceq \eta_{\mathcal{A}_\lambda}(\Im,\zeta_{2\hbar},\iota)\mathcal{MP}_{\mathcal{A}_\lambda}(\mathcal{G}\Im,\mathcal{L}\zeta_{2\hbar},\iota) + \mathcal{MP}_{\mathcal{A}_\lambda}(\mathcal{L}\zeta_{2\hbar},\mathcal{U},\iota)) \\
&\preceq \partial^*[\mathcal{MP}_{\mathcal{A}_\lambda}(\mathcal{I}\Im,\mathcal{Q}\zeta_{2\hbar},\iota)]\partial + \mathcal{MP}_{\mathcal{A}_\lambda}(\mathcal{L}\zeta_{2\hbar},\mathcal{U},\iota).
\end{aligned}
$$

We get

$$
\| \mathcal{MP}_{\mathcal{A}_{2\lambda}}(\mathcal{G}\Im,\mathcal{U},\iota) \| \leq \| \partial \|^2 \| \mathcal{MP}_{\mathcal{A}_\lambda}(\mathcal{U},\mathcal{Q}\zeta_{2\hbar},\iota) \| + \| \mathcal{MP}_{\mathcal{A}_\lambda}(\mathcal{L}\zeta_{2\hbar},\mathcal{U},\iota) \| .
$$

If $\hbar \to \infty$, *since* $\| \partial \| < 1$, $\mathcal{MP}_{\mathcal{A}_{2\lambda}}(\mathcal{G}\Im,\mathcal{U},\iota) \preceq \theta_{\mathcal{A}}$. *So,* $\mathcal{G}\Im = \mathcal{I}\Im = \mathcal{U}$. *Thus, the pairs* $\{\mathcal{L},\mathcal{Q}\}$ *and* $\{\mathcal{G},\mathcal{I}\}$ *have a common coincidence point in* Π.

Now, if $\{\mathcal{L},\mathcal{Q}\}$ and $\{\mathcal{G},\mathcal{I}\}$ be w-compatible, $\mathcal{LU}=\mathcal{LQV}=\mathcal{QLV}=\mathcal{QU}:=\Im_1$ and $\mathcal{GU}=\mathcal{GI}\Im=\mathcal{IG}\Im=\mathcal{IU}:=\Im_2$. Now

$$\begin{aligned}\mathcal{MP}_{\mathcal{A}_\lambda}(\Im_1,\Im_2,\iota)&=\mathcal{MP}_{\mathcal{A}_\lambda}(\mathcal{LU},\mathcal{GU},\iota)\\&\preceq \eta_{\mathcal{A}_\lambda}(\mathcal{U},\mathcal{U},\iota)\mathcal{MP}_{\mathcal{A}_\lambda}(\mathcal{LU},\mathcal{GU},\iota)\\&\preceq \partial^*[\mathcal{MP}_{\mathcal{A}_\lambda}(\mathcal{QU},\mathcal{IU},\iota)]\partial\\&=\partial^*[\mathcal{MP}_{\mathcal{A}_\lambda}(\Im_1,\Im_2,\iota)]\partial.\end{aligned}$$

This implies that

$$\|\mathcal{MP}_{\mathcal{A}_\lambda}(\Im_1,\Im_2,\iota)\|\le\|\partial\|^2\|\mathcal{MP}_{\mathcal{A}_\lambda}(\Im_1,\Im_2,\iota)\|.$$

Since $\|\partial\|<1$, implies that $\Im_1=\Im_2$ and hence $\mathcal{LU}=\mathcal{GU}=\mathcal{QU}=\mathcal{IU}$, that is, the point $\mathcal{U}$ is a coincidence point of $\{\mathcal{L},\mathcal{Q}\}$ and $\{\mathcal{G},\mathcal{I}\}$. Now, we show that $\mathcal{U}=\mathcal{GU}$. Hence, we have

$$\begin{aligned}\mathcal{MP}_{\mathcal{A}_\lambda}(\mathcal{U},\mathcal{GU},\iota)&=\mathcal{MP}_{\mathcal{A}_\lambda}(\mathcal{LV},\mathcal{GU},\iota)\\&\preceq \eta_{\mathcal{A}_\lambda}(\mathcal{V},\mathcal{U},\iota)\mathcal{MP}_{\mathcal{A}_\lambda}(\mathcal{LV},\mathcal{GU},\iota)\\&\preceq \partial^*[\mathcal{MP}_{\mathcal{A}_\lambda}(\mathcal{QV},\mathcal{IU},\iota)]\partial,\end{aligned}$$

so, we get

$$\|\mathcal{MP}_{\mathcal{A}_\lambda}(\mathcal{U},\mathcal{GU},\iota)\|\le\theta_{\mathcal{A}}.$$

Hence $\mathcal{GU}=\mathcal{U}$ and therefore $\mathcal{U}$ is a common fixed point of $\mathcal{L},\mathcal{G},\mathcal{Q}$ and $\mathcal{I}$.

Finally, to show the uniqueness of point $\mathcal{U}$, suppose that $\mathcal{U}^*$ be another common fixed point of $\mathcal{L},\mathcal{G},\mathcal{Q}$ and $\mathcal{I}$. From (8.1), it follows that

$$\begin{aligned}\mathcal{MP}_{\mathcal{A}_\lambda}(\mathcal{U},\mathcal{U}^*,\iota)&=\mathcal{MP}_{\mathcal{A}_\lambda}(\mathcal{LU},\mathcal{GU}^*,\iota)\\&\preceq \eta_{\mathcal{A}_\lambda}(\mathcal{U},\mathcal{U}^*,\iota)\mathcal{MP}_{\mathcal{A}_\lambda}(\mathcal{LU},\mathcal{GU}^*,\iota)\\&\preceq \partial^*[\mathcal{MP}_{\mathcal{A}_\lambda}(\mathcal{QU},\mathcal{IU}^*,\iota)]\partial.\end{aligned}$$

This implies that

$$\|\mathcal{MP}_{\mathcal{A}_\lambda}(\mathcal{U},\mathcal{U}^*,\iota)\|\preceq\|\partial\|^2\|\mathcal{MP}_{\mathcal{A}_\lambda}(\mathcal{U},\mathcal{U}^*,\iota)\|.$$

Since $\|\partial\|<1$, implies that $\mathcal{U}=\mathcal{U}^*$.

Suppose that $\mathcal{L}(\Pi)\cup\mathcal{G}(\Pi)$ be complete and $\mathcal{U}\in\mathcal{I}(\Pi)$. In this case, the proof is similar to the completeness of $\mathcal{Q}(\Pi)\cup\mathcal{I}(\Pi)$ and $\mathcal{U}\in\mathcal{I}(\Pi)$.

Corollary 7.1 *Let $(\Pi,\mathcal{A},\mathcal{MP}_{\mathcal{A}})$ be a complete C^*-av$\mathcal{MP}_{\mathcal{A}}$ms and $\mathcal{L}$, $\mathcal{G}$, $\mathcal{Q}$, $\mathcal{I}$ be self-mappings on Π satisfying $\mathcal{L}(\Pi)\subset\mathcal{I}(\Pi)$, $\mathcal{G}(\Pi)\subset\mathcal{Q}(\Pi)$, so that for some $\Re,\hbar\ge1$*

$$\eta_{\mathcal{A}_\lambda}(\zeta,\ell,\iota)\mathcal{MP}_{\mathcal{A}_\lambda}(\mathcal{L}^{\Re}\zeta,\mathcal{G}^{\hbar}\ell,\iota)\preceq\partial^*[\mathcal{MP}_{\mathcal{A}_\lambda}(\mathcal{Q}^{\Re}\zeta,\mathcal{I}^{\hbar}\ell,\iota)]\partial,\ \lambda,\iota>0,$$

for all $\zeta,\ell\in\Pi$, where $\eta_{\mathcal{A}_\lambda}(\zeta,\ell,\iota)\succeq I_{\mathcal{A}}$, $\|\partial\|<1$.

There is $\zeta_0 \in \Pi$ *so that* $\eta_{\mathcal{A}_\lambda}(\zeta_0, \ell, \iota) \succeq I_{\mathcal{A}} \Rightarrow \eta_{\mathcal{A}_\lambda}(\mathcal{L}\zeta_0, \mathcal{G}\ell, \iota) \succeq I_{\mathcal{A}}$.

If one of $\mathcal{L}(\Pi) \cup \mathcal{G}(\Pi)$ *and* $\mathcal{Q}(\Pi) \cup \mathcal{I}(\Pi)$ *be a complete subspace of* Π, *then* $\{\mathcal{L}, \mathcal{Q}\}$ *and* $\{\mathcal{G}, \mathcal{I}\}$ *have a unique coincidence point in* Π. *Further, if* $\{\mathcal{L}, \mathcal{Q}\}$ *and* $\{\mathcal{G}, \mathcal{I}\}$ *be w-compatible, then the mappings* $\mathcal{L}$, $\mathcal{G}$, $\mathcal{Q}$ *and* $\mathcal{I}$ *have a unique common fixed point in* Π.

Proof 7.2 *By using Theorem 7.1, it follows that* $\{\mathcal{L}^{\mathfrak{R}}, \mathcal{Q}^{\mathfrak{R}}\}$ *and* $\{\mathcal{G}^{\hbar}, \mathcal{I}^{\hbar}\}$ *have a unique common fixed point* $\Im \in \Pi$. *Now, we have*

$$\begin{aligned} \mathcal{L}(\Im) &= \mathcal{L}(\mathcal{L}^{\mathfrak{R}}(\Im)) = \mathcal{L}^{\mathfrak{R}+1}(\Im) = \mathcal{L}^{\mathfrak{R}}(\mathcal{L}(\Im)), \\ \mathcal{Q}(\Im) &= \mathcal{Q}(\mathcal{Q}^{\mathfrak{R}}(\Im)) = \mathcal{Q}^{\mathfrak{R}+1}(\Im) = \mathcal{Q}^{\mathfrak{R}}(\mathcal{Q}(\Im)) \end{aligned}$$

and therefore $\mathcal{L}(\Im)$ *and* $\mathcal{Q}(\Im)$ *are also fixed points for the mappings* $\mathcal{L}^{\mathfrak{R}}$ *and* $\mathcal{Q}^{\mathfrak{R}}$. *Hence,* $\mathcal{L}(\Im) = \mathcal{Q}(\Im) = \Im$. *By using the same reasoning in the proof of Theorem 7.1, we get* $\mathcal{G}(\Im) = \mathcal{I}(\Im) = \Im$. *Therefore, the proof is complete.*

Corollary 7.2 *Let* $\mathcal{L}$, $\mathcal{G}$ *and* $\mathcal{I}$ *be self-mappings on complete* C^**-av*$\mathcal{MP}_{\mathcal{A}}$*ms* $(\Pi, \mathcal{A}, \mathcal{MP}_{\mathcal{A}})$, *satisfying* $\mathcal{L}(\Pi) \cup \mathcal{G}(\Pi) \subset \mathcal{I}(\Pi)$, *so that for all* $\zeta, \ell \in \Pi$,

$$\eta_{\mathcal{A}_\lambda}(\zeta, \ell, \iota)\mathcal{MP}_{\mathcal{A}_\lambda}(\mathcal{L}\zeta, \mathcal{G}\ell, \iota) \tag{7.2}$$
$$\preceq \partial^*[\mathcal{MP}_{\mathcal{A}_\lambda}(\mathcal{I}\zeta, \mathcal{I}\ell, \iota)]\partial, \ \lambda, \iota > 0, \tag{7.3}$$

where $\eta_{\mathcal{A}_\lambda}(\zeta, \ell, \iota) \succeq I_{\mathcal{A}}$, $\| \partial \| < 1$.

There is $\zeta_0 \in \Pi$ *such that* $\eta_{\mathcal{A}_\lambda}(\zeta_0, \ell, \iota) \succeq I_{\mathcal{A}} \Rightarrow \eta_{\mathcal{A}_\lambda}(\mathcal{L}\zeta_0, \mathcal{G}\ell, \iota) \succeq I_{\mathcal{A}}$.

If one of $\mathcal{L}(\Pi) \cup \mathcal{G}(\Pi)$ *or* $\mathcal{I}(\Pi)$ *be a complete subspace of* Π, *then* $\{\mathcal{L}, \mathcal{I}\}$ *and* $\{\mathcal{G}, \mathcal{I}\}$ *have a unique coincidence point in* Π. *Furthermore, if* $\{\mathcal{L}, \mathcal{I}\}$ *and* $\{\mathcal{G}, \mathcal{I}\}$ *be w-compatible, then the mapping* $\mathcal{L}$, $\mathcal{G}$ *and* $\mathcal{I}$ *have a unique common fixed point in* Π.

Corollary 7.3 *Let* $\mathcal{L}$ *and* $\mathcal{I}$ *be self-mappings on complete* C^**-av*$\mathcal{MP}_{\mathcal{A}}$*ms* $(\Pi, \mathcal{A}, \mathcal{MP}_{\mathcal{A}})$, *satisfying* $\mathcal{L}(\Pi) \subset \mathcal{I}(\Pi)$, *so that for all* $\zeta, \ell \in \Pi$,

$$\eta_{\mathcal{A}_\lambda}(\zeta, \ell, \iota)\mathcal{MP}_{\mathcal{A}_\lambda}(\mathcal{L}\zeta, \mathcal{L}\ell, \iota) \preceq \partial^*[\mathcal{MP}_{\mathcal{A}_\lambda}(\mathcal{I}\zeta, \mathcal{I}\ell, \iota)]\partial, \ \lambda, \iota > 0, \tag{7.4}$$

where $\eta_{\mathcal{A}_\lambda}(\zeta, \ell, \iota) \succeq I_{\mathcal{A}}$, $\| \partial \| < 1$.

There is $\zeta_0 \in \Pi$ *such that* $\eta_{\mathcal{A}_\lambda}(\zeta_0, \ell, \iota) \succeq I_{\mathcal{A}} \Rightarrow \eta_{\mathcal{A}_\lambda}(\mathcal{L}\zeta_0, \mathcal{G}\ell, \iota) \succeq I_{\mathcal{A}}$.

If one of $\mathcal{L}(\Pi)$ *or* $\mathcal{I}(\Pi)$ *be a complete subspace of* Π, *then* $\{\mathcal{L}, \mathcal{I}\}$ *have a unique point of coincidence in* Π. *Furthermore, if* $\{\mathcal{L}, \mathcal{I}\}$ *be w-compatible, then the mapping* $\mathcal{L}$ *and* $\mathcal{I}$ *have a unique common fixed point in* Π.

7.3 $\eta_{\mathcal{A}}$-ADMISSIBLE MAPPING IN KANNAN-ĆIRIĆ C^*-CONTRACTION

Now, we generalize the Kannan-Ćirić contraction condition [28] as follows.

Theorem 7.2 *Let* $(\Pi, \mathcal{A}, \mathcal{MP}_{\mathcal{A}})$ *be a complete* C^**-av*$\mathcal{MP}_{\mathcal{A}}$*ms and* $\mathcal{L}$, $\mathcal{G}$, $\mathcal{Q}$, $\mathcal{I}$ *be self-mappings on* Π, *so that the following conditions are satisfied:*

(i)

$$\eta_{\mathcal{A}_\lambda}(\zeta,\ell,\iota)\mathcal{MP}_{\mathcal{A}_\lambda}(\mathcal{L}\zeta,\mathcal{G}\ell,\iota) \preceq \partial^*[\mathbf{P}(\zeta,\ell,\iota)]\partial, \text{ for all } \zeta,\ell\in\Pi,\ \lambda,\iota>0, \tag{7.5}$$

where $\eta_{\mathcal{A}_\lambda}(\zeta,\ell,\iota)\succeq I_{\mathcal{A}}$, $\|\partial\|<1$ *and*

$$\mathbf{P}(\zeta,\ell,\iota)=\max\{\mathcal{MP}_{\mathcal{A}_\lambda}(\mathcal{Q}\zeta,\mathcal{L}\zeta,\iota),\mathcal{MP}_{\mathcal{A}_\lambda}(\mathcal{I}\ell,\mathcal{G}\ell,\iota)\};$$

(ii) $(\mathcal{L},\mathcal{G})$ *are common* $\eta_{\mathcal{A}}$*-admisible on* Π*;*

(iii) $\mathcal{L}(\Pi)\subset\mathcal{I}(\Pi)$ *and* $\mathcal{G}(\Pi)\subset\mathcal{Q}(\Pi)$.

Then

(A) If one of $\mathcal{L}(\Pi)\cup\mathcal{G}(\Pi)$ *and* $\mathcal{Q}(\Pi)\cup\mathcal{I}(\Pi)$ *is complete, then* $\{\mathcal{L},\mathcal{Q}\}$ *and* $\{\mathcal{G},\mathcal{I}\}$ *have a unique coincidence point in* Π.

(B) if $\{\mathcal{L},\mathcal{Q}\}$ *and* $\{\mathcal{G},\mathcal{I}\}$ *are w-compatible, then the mappings* $\mathcal{L}$, $\mathcal{G}$, $\mathcal{Q}$ *and* $\mathcal{I}$ *have a unique common fixed point in* Π.

Proof. Let $\zeta_0\in\Pi$ (ζ_0 is an arbitrary point), define the sequences $\{\zeta_\hbar\}$ and $\{\ell_\hbar\}$ in Π so that

$$\begin{cases}\mathcal{L}\zeta_{2\hbar}=\mathcal{I}\zeta_{2\hbar+1}=\ell_{2\hbar+1}\\ \mathcal{G}\zeta_{2\hbar+1}=\mathcal{Q}\zeta_{2\hbar+2}=\ell_{2\hbar+2}, \qquad \forall\hbar\geq 0.\end{cases}$$

From (8.1), we get

$$\begin{aligned}\eta_{\mathcal{A}_\lambda}(\zeta_0,\zeta_1,\iota)=\eta_{\mathcal{A}_\lambda}(\mathcal{L}\zeta_0,\mathcal{G}\zeta_1,\iota)\succeq I_{\mathcal{A}} &\Rightarrow \eta_{\mathcal{A}_\lambda}(\mathcal{L}^2\zeta_0,\mathcal{G}^2\zeta_1,\iota)\succeq I_{\mathcal{A}}\\ &\Rightarrow \eta_{\mathcal{A}_\lambda}(\zeta_2,\zeta_3,\iota)\succeq I_{\mathcal{A}},\end{aligned}$$

by induction, we have $\eta_{\mathcal{A}_\lambda}(\zeta_{2\hbar},\zeta_{2\hbar+1},\iota)\succeq I_{\mathcal{A}}$, for all $\hbar\geq 0$.

$$\begin{aligned}\mathcal{MP}_{\mathcal{A}_\lambda}(\ell_{2\hbar+1},\ell_{2\hbar+2},\iota)&=\mathcal{MP}_{\mathcal{A}_\lambda}(\mathcal{L}\zeta_{2\hbar},\mathcal{G}\zeta_{2\hbar+1},\iota)\\ &\preceq \eta_{\mathcal{A}_\lambda}(\zeta_{2\hbar},\zeta_{2\hbar+1},\iota)\mathcal{MP}_{\mathcal{A}_\lambda}(\mathcal{L}\zeta_{2\hbar},\mathcal{G}\zeta_{2\hbar+1},\iota)\\ &\preceq \partial^*[\mathbf{P}(\zeta_{2\hbar},\zeta_{2\hbar+1},\iota)]\partial,\end{aligned}$$

where

$$\begin{aligned}\mathbf{P}(\zeta_{2\hbar},\zeta_{2\hbar+1},\iota)&=\max\{\mathcal{MP}_{\mathcal{A}_\lambda}(\mathcal{Q}\zeta_{2\hbar},\mathcal{L}\zeta_{2\hbar},\iota),\mathcal{MP}_{\mathcal{A}_\lambda}(\mathcal{I}\zeta_{2\hbar+1},\mathcal{G}\zeta_{2\hbar+1},\iota)\\ &=\max\{\mathcal{MP}_{\mathcal{A}_\lambda}(\ell_{2\hbar},\ell_{2\hbar+1},\iota),\mathcal{MP}_{\mathcal{A}_\lambda}(\ell_{2\hbar+1},\ell_{2\hbar+2},\iota)\}.\end{aligned}$$

If $\mathbf{P}(\zeta_{2\hbar},\zeta_{2\hbar+1},\iota)=\mathcal{MP}_{\mathcal{A}_\lambda}(\ell_{2\hbar+1},\ell_{2\hbar+2},\iota)$, then we simply see that it is impossible. So, $\mathbf{P}(\zeta_{2\hbar},\zeta_{2\hbar+1},\iota)=\mathcal{MP}_{\mathcal{A}_\lambda}(\ell_{2\hbar},\ell_{2\hbar+1},\iota)$ for all $\hbar\in\mathbb{N}$, and

$$\begin{aligned}\mathcal{MP}_{\mathcal{A}_\lambda}(\ell_{2\hbar+1},\ell_{2\hbar+2},\iota)&\preceq\partial^*[\mathbf{P}(\zeta_{2\hbar},\zeta_{2\hbar+1},\iota)]\partial\\ &=\partial^*[\mathcal{MP}_{\mathcal{A}_\lambda}(\ell_{2\hbar},\ell_{2\hbar+1},\iota)]\partial.\end{aligned}$$

By induction, we get

$$\mathcal{MP}_{\mathcal{A}_\lambda}(\ell_{2\hbar+1},\ell_{2\hbar+2},\iota)\preceq(\partial^*)^{2\hbar+1}[\mathcal{MP}_{\mathcal{A}_\lambda}(\ell_0,\ell_1,\iota)]\partial^{2\hbar+1}.$$

Similarly, it can be shown that

$$\mathcal{MP}_{\mathcal{A}_\lambda}(\ell_{2\hbar}, \ell_{2\hbar+1}, \iota) \preceq (\partial^*)^{2\hbar}[\mathcal{MP}_{\mathcal{A}_\lambda}(\ell_0, \ell_1, \iota)]\partial^{2\hbar}.$$

Now, we can get for every $\hbar \in \mathcal{N}$

$$\begin{aligned}\mathcal{MP}_{\mathcal{A}_\lambda}(\ell_\hbar, \ell_{\hbar+1}, \iota) &\preceq (\partial^*)^\hbar[\mathcal{MP}_{\mathcal{A}_\lambda}(\ell_0, \ell_1, \iota)]\partial^\hbar \\ &\preceq (\partial^*)^\hbar \varpi_0 \partial^\hbar,\end{aligned}$$

where $\varpi_0 := \mathcal{MP}_{\mathcal{A}_\lambda}(\ell_0, \ell_1, \iota)$.

Step II. We show that $\{\ell_{2\hbar}\}$ is a Cauchy sequence in Π. Then for any $\Re, \hbar \geq 1$ with $\Re > \hbar$, it follows that

$$\begin{aligned}\mathcal{MP}_{\mathcal{A}_\lambda}(\ell_\hbar, \ell_\Re, \iota) &\preceq \sum_{i=\hbar}^{\Re-1} \mathcal{MP}_{\mathcal{A}_\lambda}(\ell_i, \ell_{i+1}, \iota) \\ &= \sum_{i=\hbar}^{\Re-1} (\partial^*)^i \varpi_0 \partial^i \\ &\preceq \sum_{i=\hbar}^{\Re-1} (\partial^*)^i \varpi_0^{\frac{1}{2}} \varpi_0^{\frac{1}{2}} \partial^i \\ &= \sum_{i=\hbar}^{\Re-1} (\varpi_0^{\frac{1}{2}} \partial^i)^* (\varpi_0^{\frac{1}{2}} \partial^i) \\ &= \sum_{i=\hbar}^{\Re-1} |\varpi_0^{\frac{1}{2}} \partial^i|^2 \\ &\preceq \| \sum_{i=\hbar}^{\Re-1} |\varpi_0^{\frac{1}{2}} \partial^i|^2 \| \, . I_\mathcal{A} \\ &\preceq \sum_{i=\hbar}^{\Re-1} \| \varpi_0^{\frac{1}{2}} \|^2 \| \partial^i \|^2 \, . I_\mathcal{A} \\ &\preceq \| \varpi_0^{\frac{1}{2}} \|^2 \sum_{i=\hbar}^{\Re-1} \| \partial \|^{2i} \, . I_\mathcal{A} \\ &\preceq \| \varpi_0^{\frac{1}{2}} \|^2 \frac{\| \partial \|^{2\hbar}}{1 - \| \partial \|^2} . I_\mathcal{A} \to \theta_\mathcal{A} \ (as\ \hbar \to \infty).\end{aligned}$$

Therefore, $\{\ell_{2\hbar}\}$ is a Cauchy sequence in Π.

Suppose that $\mathcal{Q}(\Pi) \cup \mathcal{I}(\Pi)$ be complete. In this case there is $\mathcal{U} \in \mathcal{Q}(\Pi) \cup \mathcal{I}(\Pi)$ so that $\ell_\hbar \to \mathcal{U}$ as $\hbar \to \infty$. Furthermore, the subsequences $\{\mathcal{Q}\zeta_{2\hbar+2}\} = \{\mathcal{G}\zeta_{2\hbar+1}\} = \{\ell_{2\hbar+2}\}$ and $\{\mathcal{I}\zeta_{2\hbar+1}\} = \{\mathcal{L}\zeta_{2\hbar}\} = \{\ell_{2\hbar+1}\}$ of $\{\ell_\hbar\}$ also converge to the point $\mathcal{U}$. Since $\mathcal{U} \in \mathcal{Q}(\Pi) \cup \mathcal{I}(\Pi)$, we have $\mathcal{U} \in \mathcal{Q}(\Pi)$ or $\mathcal{U} \in \mathcal{I}(\Pi)$. If $\mathcal{U} \in \mathcal{Q}(\Pi)$, then we can find $\mathcal{V} \in \Pi$ so that $\mathcal{Q}\mathcal{V} = \mathcal{U}$. So, we claim that $\mathcal{L}\mathcal{V} = \mathcal{U}$. For this, we see that

$$\mathcal{MP}_{\mathcal{A}_{2\lambda}}(\mathcal{L}\mathcal{V}, \mathcal{U}, \iota) \preceq \mathcal{MP}_{\mathcal{A}_\lambda}(\mathcal{L}\mathcal{V}, \mathcal{G}\zeta_{2\hbar+1}, \iota) + \mathcal{MP}_{\mathcal{A}_\lambda}(\mathcal{G}\zeta_{2\hbar+1}, \mathcal{U}, \iota)$$

$$\begin{aligned}&\preceq \eta_{\mathcal{A}_\lambda}(\mathcal{V}, \zeta_{2\hbar+1}, \iota)\mathcal{MP}_{\mathcal{A}_\lambda}(\mathcal{LV}, \mathcal{G}\zeta_{2\hbar+1}, \iota)\\ &+ \mathcal{MP}_{\mathcal{A}_\lambda}(\mathcal{G}\zeta_{2\hbar+1}, \mathcal{U}, \iota)\\ &\preceq \partial^*[\mathbf{P}(\mathcal{V}, \zeta_{2\hbar+1}, \iota)]\partial + \mathcal{MP}_{\mathcal{A}_\lambda}(\mathcal{G}\zeta_{2\hbar+1}, \mathcal{U}, \iota),\end{aligned}$$

where

$$\mathbf{P}(\mathcal{V}, \zeta_{2\hbar+1}, \iota) = \max\{\mathcal{MP}_{\mathcal{A}_\lambda}(\mathcal{QV}, \mathcal{LV}, \iota), \mathcal{MP}_{\mathcal{A}_\lambda}(\mathcal{I}\zeta_{2\hbar+1}, \mathcal{G}\zeta_{2\hbar+1}, \iota)\}.$$

Therefore, we get

$$\| \mathcal{MP}_{\mathcal{A}_{2\lambda}}(\mathcal{LV}, \mathcal{U}, \iota) \| \leq \| \partial \|^2 \| \mathbf{P}(\mathcal{V}, \zeta_{2\hbar+1}, \iota) \| + \| \mathcal{MP}_{\mathcal{A}_\lambda}(\mathcal{G}\zeta_{2\hbar+1}, \mathcal{U}, \iota) \| .$$

Since $\| \partial \| < 1$, making $\hbar \to \infty$, we have a contradiction, so $\mathcal{MP}_{\mathcal{A}_{\in\lambda}}(\mathcal{LV}, \mathcal{U}, \iota) \preceq \theta_{\mathcal{A}}$. Consequently, we have $\mathcal{LV} = \mathcal{QV} = \mathcal{U}$ and since $\mathcal{U} \in \mathcal{L}(\Pi) \subset \mathcal{I}(\Pi)$, there is $\Im \in \Pi$ so that $\mathcal{I}\Im = \mathcal{U}$.

Now, we show that $\mathcal{G}\Im = \mathcal{U}$. So, we have

$$\begin{aligned}\mathcal{MP}_{\mathcal{A}_{2\lambda}}(\mathcal{G}\Im, \mathcal{U}, \iota) &\preceq \mathcal{MP}_{\mathcal{A}_\lambda}(\mathcal{G}\Im, \mathcal{L}\zeta_{2\hbar}, \iota) + \mathcal{MP}_{\mathcal{A}_\lambda}(\mathcal{L}\zeta_{2\hbar}, \mathcal{U}, \iota)\\ &\preceq \eta_{\mathcal{A}_\lambda}(\Im, \zeta_{2\hbar}, \iota)\mathcal{MP}_{\mathcal{A}_\lambda}(\mathcal{G}\Im, \mathcal{L}\zeta_{2\hbar}, \iota) + \mathcal{MP}_{\mathcal{A}_\lambda}(\mathcal{L}\zeta_{2\hbar}, \mathcal{U}, \iota))\\ &\preceq \partial^*[\mathbf{P}(\Im, \zeta_{2\hbar}, \iota)]\partial + \mathcal{MP}_{\mathcal{A}_\lambda}(\mathcal{L}\zeta_{2\hbar}, \mathcal{U}, \iota),\end{aligned}$$

where

$$\mathbf{P}(\Im, \zeta_{2\hbar}, \iota) = \max\{\mathcal{MP}_{\mathcal{A}_\lambda}(\mathcal{Q}\Im, \mathcal{L}\Im, \iota), \mathcal{MP}_{\mathcal{A}_\lambda}(\mathcal{I}\zeta_{2\hbar}, \mathcal{G}\zeta_{2\hbar}, \iota)\}.$$

We get

$$\| \mathcal{MP}_{\mathcal{A}_{2\lambda}}(\mathcal{G}\Im, \mathcal{U}, \iota) \| \leq \| \partial \|^2 \| \mathcal{U}(\Im, \zeta_{2\hbar}, \iota) \| + \| \mathcal{MP}_{\mathcal{A}_\lambda}(\mathcal{L}\zeta_{2\hbar}, \mathcal{U}, \iota) \| .$$

If $\hbar \to \infty$, since $\| \partial \| < 1$, $\mathcal{MP}_{\mathcal{A}_{2\lambda}}(\mathcal{G}\Im, \mathcal{U}, \iota) \preceq \theta_{\mathcal{A}}$. So, $\mathcal{G}\Im = \mathcal{I}\Im = \mathcal{U}$. Hence, $\{\mathcal{L}, \mathcal{Q}\}$ and $\{\mathcal{G}, \mathcal{I}\}$ have a common coincidence point in Π.

Now, if $\{\mathcal{L}, \mathcal{Q}\}$ and $\{\mathcal{G}, \mathcal{I}\}$ be w-compatible, $\mathcal{LU} = \mathcal{LQV} = \mathcal{QLV} = \mathcal{QU} := \Im_1$ and $\mathcal{GU} = \mathcal{GI}\Im = \mathcal{IG}\Im = \mathcal{IU} := \Im_2$. Now

$$\begin{aligned}\mathcal{MP}_{\mathcal{A}_\lambda}(\Im_1, \Im_2, \iota) &= \mathcal{MP}_{\mathcal{A}_\lambda}(\mathcal{LU}, \mathcal{GU}, \iota)\\ &\preceq \eta_{\mathcal{A}_\lambda}(\mathcal{U}, \mathcal{U}, \iota)\mathcal{MP}_{\mathcal{A}_\lambda}(\mathcal{LU}, \mathcal{GU}, \iota)\\ &\preceq \partial^*[\mathbf{P}(\mathcal{U}, \mathcal{U}, \iota)]\partial,\end{aligned}$$

where

$$\mathbf{P}(\mathcal{U}, \mathcal{U}, \iota) = \max\{\mathcal{MP}_{\mathcal{A}_\lambda}(\mathcal{QU}, \mathcal{LU}, \iota), \mathcal{MP}_{\mathcal{A}_\lambda}(\mathcal{IU}, \mathcal{GU}, \iota)\}.$$

This implies that

$$\mathcal{MP}_{\mathcal{A}_\lambda}(\Im_1, \Im_2, \iota) \preceq \theta_{\mathcal{A}}.$$

So $\Im_1 = \Im_2$ and hence $\mathcal{LU} = \mathcal{GU} = \mathcal{QU} = \mathcal{IU}$, i.e., the point $\mathcal{U}$ is a coincidence point of $\{\mathcal{L}, \mathcal{Q}\}$ and $\{\mathcal{G}, \mathcal{I}\}$. Now, we show that $\mathcal{U} = \mathcal{GU}$. So, we have

$$\begin{aligned}\mathcal{MP}_{\mathcal{A}_\lambda}(\mathcal{U}, \mathcal{GU}, \iota) &= \mathcal{MP}_{\mathcal{A}_\lambda}(\mathcal{LV}, \mathcal{GU}, \iota)\\ &\preceq \eta_{\mathcal{A}_\lambda}(\mathcal{V}, \mathcal{U}, \iota)\mathcal{MP}_{\mathcal{A}_\lambda}(\mathcal{LV}, \mathcal{GU}, \iota)\\ &\preceq \partial^*[\mathbf{P}(\mathcal{V}, \mathcal{U}, \iota)]\partial,\end{aligned}$$

where

$$\begin{aligned}\mathbf{P}(\mathcal{V}, \mathcal{U}, \iota) &= \max\{\mathcal{MP}_{\mathcal{A}_\lambda}(\mathcal{QV}, \mathcal{LV}, \iota), \mathcal{MP}_{\mathcal{A}_\lambda}(\mathcal{IU}, \mathcal{GU}, \iota)\}\\ &= \mathcal{MP}_{\mathcal{A}_\lambda}(\mathcal{U}, \mathcal{GU}, \iota).\end{aligned}$$

So, we get

$$\| \mathcal{MP}_{\mathcal{A}_\lambda}(\mathcal{U}, \mathcal{GU}, \iota) \| \leq \| \partial \|^2 \| \mathcal{MP}_{\mathcal{A}_\lambda}(\mathcal{U}, \mathcal{GU}, \iota) \| .$$

Since $\| \partial \| < 1$, then $\mathcal{GU} = \mathcal{U}$ and hence $\mathcal{U}$ is a common fixed point of $\mathcal{L}, \mathcal{G}, \mathcal{Q}$ and $\mathcal{I}$.

Finally, to show the uniqueness of point $\mathcal{U}$, suppose that $\mathcal{U}^*$ be another common fixed point of $\mathcal{L}, \mathcal{G}, \mathcal{Q}$ and $\mathcal{I}$. From (8.1), it follows that

$$\begin{aligned}\mathcal{MP}_{\mathcal{A}_\lambda}(\mathcal{U}, \mathcal{U}^*, \iota) &= \mathcal{MP}_{\mathcal{A}_\lambda}(\mathcal{LU}, \mathcal{GU}^*, \iota)\\ &\preceq \eta_{\mathcal{A}_\lambda}(\mathcal{U}, \mathcal{U}^*, \iota)\mathcal{MP}_{\mathcal{A}_\lambda}(\mathcal{LU}, \mathcal{GU}^*, \iota)\\ &\preceq \partial^*[\mathbf{P}(\mathcal{U}, \mathcal{U}^*, \iota)]\partial,\end{aligned}$$

where

$$\mathbf{P}(\mathcal{U}, \mathcal{U}^*, \iota) = \max\{\mathcal{MP}_{\mathcal{A}_\lambda}(\mathcal{QU}, \mathcal{LU}, \iota), \mathcal{MP}_{\mathcal{A}_\lambda}(\mathcal{IU}^*, \mathcal{GU}^*, \iota)\}.$$

This implies that

$$\mathcal{MP}_{\mathcal{A}_\lambda}(\mathcal{U}, \mathcal{U}^*, \iota) \preceq \theta_{\mathcal{A}}.$$

Then $\mathcal{U} = \mathcal{U}^*$.

Suppose that $\mathcal{L}(\Pi) \cup \mathcal{G}(\Pi)$ be complete and $\mathcal{U} \in \mathcal{I}(\Pi)$. In this case, the proof is similar to the completeness of $\mathcal{Q}(\Pi) \cup \mathcal{I}(\Pi)$ and $\mathcal{U} \in \mathcal{I}(\Pi)$. □

Corollary 7.4 *Let $(\Pi, \mathcal{A}, \mathcal{MP}_{\mathcal{A}})$ be a complete C^*-av$\mathcal{MP}_{\mathcal{A}}$ms and $\mathcal{L}$, $\mathcal{G}$, $\mathcal{Q}$, $\mathcal{I}$ be self-mappings on Π satisfying $\mathcal{L}(\Pi) \subset \mathcal{I}(\Pi)$, $\mathcal{G}(\Pi) \subset \mathcal{Q}(\Pi)$, so that for some $\Re, \hbar \geq 1$*

$$\eta_{\mathcal{A}_\lambda}(\varsigma, \ell, \iota)\mathcal{MP}_{\mathcal{A}_\lambda}(\mathcal{L}^{\Re}\varsigma, \mathcal{G}^{\hbar}\ell, \iota) \preceq \partial^*[\mathbf{P}(\varsigma, \ell, \iota)]\partial, \ \lambda, \iota > 0, \tag{7.6}$$

for all $\varsigma, \ell \in \Pi$, where $\eta_{\mathcal{A}_\lambda}(\varsigma, \ell, \iota) \succeq I_{\mathcal{A}}$, $\| \partial \| < 1$ and

$$\mathbf{P}(\varsigma, \ell, \iota) = \max\{\mathcal{MP}_{\mathcal{A}_\lambda}(\mathcal{Q}^{\Re}\varsigma, \mathcal{L}^{\Re}\varsigma, \iota), \mathcal{MP}_{\mathcal{A}_\lambda}(\mathcal{I}^{\hbar}\ell, \mathcal{G}^{\hbar}\ell, \iota)\}.$$

There is $\varsigma_0 \in \Pi$ so that $\eta_{\mathcal{A}_\lambda}(\varsigma_0, \ell, \iota) \succeq I_{\mathcal{A}} \Rightarrow \eta_{\mathcal{A}_\lambda}(\mathcal{L}\varsigma_0, \mathcal{G}\ell, \iota) \succeq I_{\mathcal{A}}$.

If one of $\mathcal{L}(\Pi) \cup \mathcal{G}(\Pi)$ and $\mathcal{Q}(\Pi) \cup \mathcal{I}(\Pi)$ be a complete subspace of Π, then $\{\mathcal{L}, \mathcal{Q}\}$ and $\{\mathcal{G}, \mathcal{I}\}$ have a unique coincidence point in Π. Furthermore, if $\{\mathcal{L}, \mathcal{Q}\}$ and $\{\mathcal{G}, \mathcal{I}\}$ be w-compatible, then the mappings $\mathcal{L}$, $\mathcal{G}$, $\mathcal{Q}$ and $\mathcal{I}$ have a unique common fixed point in Π.

Corollary 7.5 *Let $\mathcal{L}$, $\mathcal{G}$ and $\mathcal{I}$ be self-mappings on complete C^*-av$\mathcal{MP}_{\mathcal{A}}$ms $(\Pi, \mathcal{A}, \mathcal{MP}_{\mathcal{A}})$, satisfying $\mathcal{L}(\Pi) \cup \mathcal{G}(\Pi) \subset \mathcal{I}(\Pi)$, so that for all $\varsigma, \ell \in \Pi$,*

$$\eta_{\mathcal{A}_\lambda}(\varsigma, \ell, \iota)\mathcal{MP}_{\mathcal{A}_\lambda}(\mathcal{L}\varsigma, \mathcal{G}\ell, \iota) \preceq \partial^*[\mathbf{P}(\varsigma, \ell, \iota)]\partial, \ \lambda, \iota > 0, \tag{7.7}$$

where $\eta_{\mathcal{A}_\lambda}(\varsigma, \ell, \iota) \succeq I_{\mathcal{A}}$, $\| \partial \| < 1$ and

$$\mathbf{P}(\varsigma, \ell, \iota) = \max\{\mathcal{MP}_{\mathcal{A}_\lambda}(\mathcal{I}\varsigma, \mathcal{L}\varsigma, \iota), \mathcal{MP}_{\mathcal{A}_\lambda}(\mathcal{I}\ell, \mathcal{G}\ell, \iota)\}.$$

There is $\varsigma_0 \in \Pi$ such that $\eta_{\mathcal{A}_\lambda}(\varsigma_0, \ell, \iota) \succeq I_{\mathcal{A}} \Rightarrow \eta_{\mathcal{A}_\lambda}(\mathcal{L}\varsigma_0, \mathcal{G}\ell, \iota) \succeq I_{\mathcal{A}}$.

If one of $\mathcal{L}(\Pi) \cup \mathcal{G}(\Pi)$ or $\mathcal{I}(\Pi)$ be a complete subspace of Π, then $\{\mathcal{L}, \mathcal{I}\}$ and $\{\mathcal{G}, \mathcal{I}\}$ have a unique coincidence point in Π. Furthermore, if $\{\mathcal{L}, \mathcal{I}\}$ and $\{\mathcal{G}, \mathcal{I}\}$ be w-compatible, then the mapping $\mathcal{L}$, $\mathcal{G}$ and $\mathcal{I}$ have a unique common fixed point in Π.

Corollary 7.6 *Let $\mathcal{L}$ and $\mathcal{I}$ be self-mappings on complete C^*-av$\mathcal{MP}_{\mathcal{A}}$ms $(\Pi, \mathcal{A}, \mathcal{MP}_{\mathcal{A}})$, satisfying $\mathcal{L}(\Pi) \subset \mathcal{I}(\Pi)$, so that for all $\varsigma, \ell \in \Pi$,*

$$\eta_{\mathcal{A}_\lambda}(\varsigma, \ell, \iota)\mathcal{MP}_{\mathcal{A}_\lambda}(\mathcal{L}\varsigma, \mathcal{L}\ell, \iota) \preceq \partial^*[\mathbf{P}(\varsigma, \ell, \iota)]\partial, \ \lambda, \iota > 0, \tag{7.8}$$

where $\eta_{\mathcal{A}_\lambda}(\varsigma, \ell, \iota) \succeq I_{\mathcal{A}}$, $\| \partial \| < 1$ and

$$\mathbf{P}(\varsigma, \ell, \iota) = \max\{\mathcal{MP}_{\mathcal{A}_\lambda}(\mathcal{I}\varsigma, \mathcal{L}\varsigma, \iota), \mathcal{MP}_{\mathcal{A}_\lambda}(\mathcal{I}\ell, \mathcal{L}\ell, \iota)\}.$$

There is $\varsigma_0 \in \Pi$ so that $\eta_{\mathcal{A}_\lambda}(\varsigma_0, \ell, \iota) \succeq I_{\mathcal{A}} \Rightarrow \eta_{\mathcal{A}_\lambda}(\mathcal{L}\varsigma_0, \mathcal{G}\ell, \iota) \succeq I_{\mathcal{A}}$.

If one of $\mathcal{L}(\Pi)$ or $\mathcal{I}(\Pi)$ be a complete subspace of Π, then $\{\mathcal{L}, \mathcal{I}\}$ have a unique coincidence point in Π. Furthermore, if $\{\mathcal{L}, \mathcal{I}\}$ be w-compatible, then the mapping $\mathcal{L}$ and $\mathcal{I}$ have a unique common fixed point in Π.

7.4 EXAMPLE

Example 7.1 Let $\mathfrak{H}$ be a Hilbert space and $\mathfrak{L}(\mathfrak{H})$ be the set of linear bounded operators on $\mathfrak{H}$. Let $\mathfrak{D}_1, \mathfrak{D}_2, ..., \mathfrak{D}_\hbar \in \mathfrak{L}(\mathfrak{H})$, with $\sum_{\hbar=1}^{\infty} \parallel \mathfrak{D}_\hbar \parallel^2 < 1$ and $\Pi \in \mathfrak{L}(\mathfrak{H})$, $\mathfrak{P} \in \mathfrak{L}(\mathfrak{H})_+$. Then the operator equation

$$\mathfrak{X} - \sum_{\hbar=1}^{\infty} \mathfrak{D}_\hbar^* \mathfrak{X} \mathfrak{D}_\hbar = \mathfrak{P},$$

has a unique solution in $\mathfrak{L}(\mathfrak{H})$.

Proof. Set $\varrho = \sum_{\hbar=1}^{\infty} \parallel \mathfrak{D}_\hbar \parallel^2$. Clear that if $\varrho = 0$, then the $\mathfrak{D}_\hbar = \theta$ $(\hbar \in \mathbb{N})$, and the equation has a unique solution in $\mathfrak{L}(\mathfrak{H})$. Without loss of generality, one can suppose that $\varrho > 0$.

Choose a positive operator $\mathcal{T} \in \mathfrak{L}(\mathfrak{H})$. For $\mathfrak{X}, \mathfrak{Y} \in \mathfrak{L}(\mathfrak{H})$, set

$$\mathcal{MP}_{\mathcal{A}_\lambda}(\mathfrak{X}, \mathfrak{Y}, \iota) = \frac{\iota \mho(\mathfrak{X}, \mathfrak{Y})}{\lambda} \mathcal{T}.$$

It is clear that $(\mathfrak{L}(\mathfrak{H}), \mathcal{A}, \mathcal{MP}_\mathcal{A})$ is a C^*-av$\mathcal{MP}_\mathcal{A}$m and is complete since $\mathfrak{L}(\mathfrak{H})$ is a Banach space. We defined the mapping $\mathcal{F} : \mathfrak{L}(\mathfrak{H}) \to \mathfrak{L}(\mathfrak{H})$ by

$$\mathcal{F}(\mathfrak{X}) = \sum_{\hbar=1}^{\infty} \mathfrak{D}_\hbar^* \mathfrak{X} \mathfrak{D}_\hbar + \mathfrak{P}$$

and $\eta_{\mathcal{A}_\lambda}(\zeta, \ell, \iota) = I_\mathcal{A}$. Then

$$\begin{aligned}
I_\mathcal{A}\mathcal{MP}_{\mathcal{A}_\lambda}(\mathcal{F}\mathfrak{X}, \mathcal{F}\mathfrak{Y}, \iota) &= \frac{\iota \parallel \mathcal{F}\mathfrak{X} - \mathcal{F}\mathfrak{Y} \parallel}{\lambda} \mathcal{T} \\
&= \frac{\iota \parallel \sum_{\hbar=1}^{\infty} \mathfrak{D}_\hbar^* (\mathfrak{X} - \mathfrak{Y}) \mathfrak{D}_\hbar \parallel}{\lambda} \mathcal{T} \\
&\preceq \frac{\iota \sum_{\hbar=1}^{\infty} \parallel \mathfrak{D}_\hbar \parallel^2 \parallel (\mathfrak{X} - \mathfrak{Y}) \parallel}{\lambda} \mathcal{T} \\
&= \frac{\varrho \iota \mho(\mathfrak{X} - \mathfrak{Y})}{\lambda} \mathcal{T} \\
&= (\varrho^{\frac{1}{2}} I_\mathcal{A})^* [\frac{\iota \mho(\mathfrak{X}, \mathfrak{Y})}{\lambda} \mathcal{T}] (\varrho^{\frac{1}{2}} I_\mathcal{A}) \\
&= (\varrho^{\frac{1}{2}} I_\mathcal{A})^* [\mathcal{MP}_{\mathcal{A}_\lambda}(\mathfrak{X}, \mathfrak{Y}, \iota)] (\varrho^{\frac{1}{2}} I_\mathcal{A}).
\end{aligned}$$

By using Theorem 7.1 for mapping $\mathcal{F}$, there is a unique fixed point $\mathfrak{X} \in \mathfrak{L}(\mathfrak{H})$. Moreover, since $\sum_{\hbar=1}^{\infty} \mathfrak{D}_\hbar^* \mathfrak{X} \mathfrak{D}_\hbar + \mathfrak{P}$ is a positive operator, the solution is a Hermitian operator. □

7.5 APPLICATION

Let $\mathfrak{X} = \mathfrak{L}^\infty(\mathfrak{S})$ the set of essentially bounded measurable functions on $\mathfrak{S}$. Consider the Hilbert space $\mathfrak{H} = \mathfrak{L}^2(\mathfrak{S})$ and $\mathfrak{L}(\mathfrak{H}) = \mathcal{A}$, where $\mathfrak{S}$ is a Lebesgue measurable set and $m(\mathfrak{S}) < \infty$. We consider Fredholm's integral equation as follows:

$$\zeta(\iota) = \int_{\mathfrak{S}} \Upsilon(\iota, \varsigma, \zeta(\varsigma))d\varsigma + \mathfrak{h}(\iota), \text{ for all } \varsigma, \iota \in \mathfrak{S}, \tag{7.9}$$

where $\Upsilon : \mathfrak{S}^2 \times \mathbb{R} \longrightarrow \mathbb{R}$ and $\mathfrak{h} \in \mathfrak{L}^\infty(\mathfrak{S})$. Define $\mathcal{MP}_\mathcal{A} : (0,\infty) \times \mathfrak{X}^2 \times (0,\infty) \to \mathcal{A}^+$ by:

$$\mathcal{MP}_{\mathcal{A}_\lambda}(\zeta, \ell, \iota) = M_{\frac{\iota|\zeta-\ell|}{\lambda}},$$

for all $\zeta, \ell \in \mathfrak{X}$ and $\lambda, \iota > 0$, where $M_{\frac{\iota|\zeta-\ell|}{\lambda}}$ is the multiplication operator on $\mathfrak{L}^2(\mathfrak{S})$.

Then $(\mathfrak{X}, \mathcal{A}, \mathcal{MP}_\mathcal{A})$ is a complete C^*-av$\mathcal{MP}_\mathcal{A}$ms. Now we consider the following assumption:

there is $\kappa \in (0, \frac{1}{2})$ so that for all $\zeta, \ell \in \mathfrak{X}$, suppose the following condition holds:

$$|\Upsilon(\varsigma, \iota, \zeta(\varsigma)) - \Upsilon(\varsigma, \iota, \ell(\varsigma))| \leq \kappa(|\zeta - \ell|).$$

Theorem 7.3 *Suppose that the aforementioned assumptions hold. Then the integral equation ((8.23)) has a unique solution in $\mathfrak{X}$.*

Proof. We define $\mathcal{L} : \mathfrak{X} \to \mathfrak{X}$ by

$$\mathcal{L}(\zeta)(\iota) = \int_{\mathfrak{S}} \Upsilon(\iota, \varsigma, \zeta(\varsigma))d\varsigma + \mathfrak{h}(\iota), \quad \forall \varsigma, \iota \in \mathfrak{S},$$

and $\eta : (0,\infty) \times \mathfrak{X}^2 \times (0,\infty) \to \mathcal{A}^+$ by

$$\eta_{\mathcal{A}_\lambda}(\zeta, \ell, \iota) = I_\mathcal{A}.$$

Set $\varrho = \kappa I_\mathcal{A}$, then $\varrho \in \mathfrak{L}(\mathfrak{H})_+$ and $\| \varrho \| = \kappa < 1$. For every $\varphi \in \mathfrak{H}$, we have

$$\begin{aligned}
\| \mathcal{MP}_{\mathcal{A}_\lambda}(\mathcal{L}\zeta, \mathcal{L}\ell, \iota) \| &= \| M_{\frac{\iota|\mathcal{L}\zeta-\mathcal{L}\ell|}{\lambda}} \| \\
&= \sup_{\|\varphi\|=1} (M_{\frac{\iota|\mathcal{L}\zeta-\mathcal{L}\ell|}{\lambda}}\varphi, \varphi) \\
&\leq \sup_{\|\varphi\|=1} \int_{\mathfrak{S}} (\frac{\iota}{\lambda}| \int_{\mathfrak{S}} \Upsilon(\iota, \varsigma, \zeta(\varsigma)) \\
&\quad - \Upsilon(\iota, \varsigma, \ell(\varsigma))d\varsigma|)\varphi(r)\overline{\varphi(r)}dr \\
&\leq \sup_{\|\varphi\|=1} \int_{\mathfrak{S}} (\frac{\iota}{\lambda} \int_{\mathfrak{S}} |\Upsilon(\iota, \varsigma, \zeta(\varsigma)) \\
&\quad - \Upsilon(\iota, \varsigma, \ell(\varsigma))|d\varsigma)|\varphi(r)|^2 dr \\
&\leq \sup_{\|\varphi\|=1} \int_{\mathfrak{S}} |\varphi(r)|^2 dr \,.\, \| \kappa(\frac{\iota|\zeta-\ell|}{\lambda}) \|_\infty \\
&\leq \| \kappa(\frac{\iota|\zeta-\ell|}{\lambda}) \|_\infty \\
&= \| \varrho \| \| \mathcal{MP}_{\mathcal{A}_\lambda}(\zeta, \ell, \iota) \| .
\end{aligned}$$

This implies that

$$\| \eta_\lambda(\zeta,\ell,\iota)\mathcal{MP}_{\mathcal{A}_\lambda}(\mathcal{L}\zeta,\mathcal{L}\ell,\iota) \| \leq \| \varrho \| \| \| \mathcal{MP}_{\mathcal{A}_\lambda}(\zeta,\ell,\iota) \|,$$

for $\lambda, \iota > 0$. Since $\| \varrho \| < 1$, so $\mathcal{L}$ is an $(\eta_{\mathcal{A}}, \varphi_{\mathcal{A}})$-contractive mapping and Theorem 7.1 hold for a mapping $\mathcal{L}$. Therefore, the equation ((8.23)) has a unique solution, that is, $\mathcal{L}$ has a unique fixed point.

□

7.6 CONCLUSION

In this chapter, we stated the concept of $\eta_{\mathcal{A}}$-admissible mapping in C^*-algebra-valued $\mathcal{MP}$-metric spaces for C^*-contraction and also Kannan-Ćirić C^*-contraction. We have combined the three spaces of modular metric, parametric metric and C^*-algebra-valued metric space. Using this new space, we presented a new development of the Banach contraction principle. To confirm the new results, we provide an example and an application about the solvability of operator equations and integral equations, respectively. Our results extend and generalize the relevant results in [12, 14, 16, 20, 25, 28].

BIBLIOGRAPHY

[1] M. Asim, M. Imdad, C^*-algebra valued symmetric spaces and fixed point results with an application, *Korean J. Math.* 28 (1) (2020), 17–30

[2] S. Banach, Sur les opérations dans les ensembles abstraits et leur application aux équations intégrales, *Fund. Math.* 3 (1922), 133–181.

[3] P. Debnath, Banach, Kannan, Chatterjea, and Reich-type contractive inequalities for multivalued mappings and their common fixed points, *Math. Methods Appl. Sci.*, 45 (3) (2022), 1587–1596.

[4] P. Debnath, Optimization through best proximity points for multivalued F-contractions, *Miskolc Math Notes*, 22 (1) (2021), 143–151.

[5] P. Debnath, A new extension of Kannan's fixed point theorem via F-contraction with application to integral equations, *Asian-Eur J Math*, 15 (07) (2022), 2250123.

[6] M. E. Ege, C. Alaca, Some results for modular b-metric spaces and an application to system of linear equations, *Azerb. J. Math.* 8 (2018), 3–14.

[7] A.D. Filip, Petruşel, Fixed point theorems on spaces endowed with vector-valued metrics, *J. Fixed Point Theory Appl.* 2010 (2010), 1–15.

[8] S. Hadi Bonab, R. Abazari, A. Bagheri Vakilabad, Partially ordered cone metric spaces and coupled fixed point theorems via α-series, *Math. Anal. Contemporary Appl.* 1 (1) (2019), 50–61.

[9] S. Hadi Bonab, R. Abazari1, A. Bagheri Vakilabad, H. Hosseinzadeh, Coupled fixed point theorems on G-metric spaces via α-series, *Global Anal. Discrete Math.* 6 (1) (2021) 1–12.

[10] S. Hadi Bonab, R. Abazaril, A. Bagheri Vakilabad, H. Hosseinzadeh, Generalized metric spaces endowed with vector-valued metrics and matrix equations by tripled fixed point theorems, *J. Inequal. Appl.* 2014 (2020) 1–16.

[11] H. Hosseinzadeh, S. Hadi Bonab. Kh. Amini Sefidab, Some common fixed point theorems for four mapping in generalized metric spaces, *Thai J. Math.* 20 (1) (2022), 425–437.

[12] N. Hussain, S. Khaleghizadeh, P. Salimi, A. A. N. Abdou, A new approach to fixed point results in triangular intuitionistic fuzzy metric spaces, *Abstr. Appl. Anal.* 2014 (2014), 16 pages.

[13] T. Kamran, M. Samreen, Q.U. Ain, A generalization of b-metric space and some fixed point theorems, *Mathematics*, 5 (19) 2017).

[14] M.A. Kutbi, A. Latif, Fixed points of multivalued mappings in modular function spaces, *Fixed Point Theory Appl.* 2009, Article ID 786357 (2009).

[15] Z.H. Ma, L.N. Jiang, C^*-Algebra valued b-metric spaces and related fixed point theorems, *Fixed Point Theory Appl.* 2015 (222) (2015), 1–12.

[16] Z. Ma, L. Jiang, H. Sun, C^*-Algebra-valued metric spaces and related fixed point theorems, *Fixed Point Theory Appl.* 2014 (1) (2014).

[17] B. Moeini, A. H. Ansari, C. Park, JHR-Operator pairs in C^*-algebra-valued modular metric spaces and related fixed point results with application, *Numer. Funct. Anal. Optim.* 2014 (2018), 1785–1805.

[18] G. J. Murphy, *C^*-Algebras and Operator Theory*, Academic Press, Inc., Boston, MA, 1990.

[19] J. Musielak, W. Orlicz, On modular spaces, *Stud. Math.* 18 (1959), 591–597.

[20] H. Nakano, *Modulared Semi-Ordered Spaces*, Maruzen, Tokyo (1950).

[21] V. Parvaneh , S. Hadi Bonab, H. Hosseinzadeh, H. Aydi, A tripled fixed point theorem in C^*-algebra-valued metric spaces and application in integral equations, *Adv. Math. Phys.* 2021 (2021) 1–6.

[22] V. Parvaneh, N. Hussain, M. A. Kutbi, M. Khorshdi, Some fixed point results in extended modular b-metric spaces with application to integral equations, *J. Math. Anal.* 10 (5) (2019), 14–33.

[23] B. Samet, C. Vetro, P. Vetro, Fixed point theorems for $\alpha-\psi$-contractive type mapping, *Nonlinear Anal.* 75 (2012), 2154–2165.

[24] J. R. Roshan, V. Parvaneh, Z. Kadelburg, N. Hussain, New fixed point results in b-rectangular metric spaces, *Nonlinear Anal. Model. Control.* 21 (5) (2016), 614–634.

[25] B. Samet, C. Vetro, P. Vetro, Fixed point theorems for $\alpha-\psi$-contractive type mappings, *Nonlinear Anal.* 75 (2012), 2154–2165.

[26] T. Stephen, Y. Rohen, N. Mlaiki, M. Bina, N. Hussain, D. Rizk, On fixed points of rational contractions in generalized parametric metric and fuzzy metric spaces, *J. Inequal. Appl.* 2021 (125) (2021), 1–15.

[27] J. Uma Maheswari, A. Anbarasan, M. Gunaseelan, V. Parvaneh, S. Hadi Bonab. Solving an integral equation via C^*-algebra-valued partial b-metrics, *Fixed Point Theory Algorithms Sci. Eng.* 18 (2022), 1–14.

[28] J. V. Morales, Subordinate semimetric spaces and fixed point theorems, *J. Math.* 12 (2018), 1–5.

CHAPTER 8

Summarized Proofs to Find Common Fixed Points of Prešić Contractions for Four Maps

Samira Hadi Bonab

*Department of Mathematics, Ardabil Branch,
Islamic Azad University, Ardabil, Iran*

Vahid Parvaneh

*Department of Mathematics, Gilan-E-Gharb Branch,
Islamic Azad University, Gilan-E-Gharb, Iran*

Zohreh Bagheri

*Department of Mathematics, Azadshahr Branch,
Islamic Azad University, Azadshahr, Iran*

CONTENTS

8.1 INTRODUCTION

One of the more intriguing generalizations of Banach's fixed point theorem (BFPT) [2] was made by Hardy and Rogers [14] in 1973 when they generalized the BFPT to the Hardy-Rogers contraction (HRC).

DOI: 10.1201/9781003388678-8

As is well known, there are numerous fields and branches of mathematics where the BFPT can be used. This traditional result has been generalized by other authors (for further information, see [1, 4, 7, 8, 9, 11, 15] and [20–24]).

We generalize the PHRC discussed in [28] to the Prešić-Hardy-Rogers contraction (PHRC) for four maps in this chapter. We'll employ a technique to condense some fixed point theorems' proofs.

Definition 8.1 *[2] Let $\Re \neq \emptyset$. A mapping $\chi : \Re^2 \longrightarrow [0, \infty)$ is called a metric on $\Re$, if:*

1. $\chi(\iota, \varsigma) = 0$ *if and only if* $\iota = \varsigma$;
2. $\chi(\iota, \varsigma) = \chi(\varsigma, \iota)$ *for each* $\iota, \varsigma \in \Re$;
3. $\chi(\iota, \varsigma) \leq \chi(\iota, \kappa) + \chi(\kappa, \varsigma)$ *for each* $\iota, \varsigma, \kappa \in \Re$.

A set $\Re$ will be a metric space whenever it is equipped with a metric χ and is denoted by $(\Re, \chi)$.

Definition 8.2 *Let $\mathcal{Q}, \mathcal{Q}' : \Re \to \Re$ be self-mappings. If $\mathbf{q} = \mathcal{Q}\iota = \mathcal{Q}'\iota$ for some $\iota \in \Re$, then ι is called a coincidence point of $\mathcal{Q}$ and $\mathcal{Q}'$. $\mathbf{q}$ is called a point of coincidence of $\mathcal{Q}$ and $\mathcal{Q}'$.*

Definition 8.3 *Let $\mathcal{Q}, \mathcal{Q}' : \Re \to \Re$ be self-mappings. If $\mathcal{Q}$ and $\mathcal{Q}'$ commute at their coincidence points, then they are called w-compatible.*

Definition 8.4 *[14] Let $(\Re, \chi)$ be a metric space and $\mathcal{G} : \Re \to \Re$. $\mathcal{G}$ will be a HRC if we have some nonnegative coefficients γ_i, such that $\mathcal{G}_{i=1}^{5}\gamma_i < 1$ and for each $\iota, \varsigma \in \Re$,*

$$\chi(\mathcal{G}\iota, \mathcal{G}\varsigma) \leq \gamma_1\chi(\iota, \varsigma) + \gamma_2\chi(\iota, \mathcal{G}\iota) + \gamma_3\chi(\varsigma, \mathcal{G}\varsigma) + \gamma_4\chi(\iota, \mathcal{G}\varsigma) + \gamma_5\chi(\varsigma, \mathcal{G}\iota).$$

For further details, the readers are referred to [18, 22, 27, 29] and the references therein.

This chapter will go over the HRCs and PHRCs in order to arrive at a common fixed point for four maps.

8.2 MAIN RESULTS

Our main theorem serves as the basis for the debate. This section states and demonstrates our main finding after first generalizing the HRC for four self-mappings $\mathcal{Q}$, $\mathcal{Q}'$, $\mathcal{G}$ and $\mathcal{G}'$ on $\Re$.

Theorem 8.1 *Let $(\Re, \chi)$ be a complete metric space (c.m.s.) and $\mathcal{Q}$, $\mathcal{Q}'$, $\mathcal{G}$ and $\mathcal{G}'$ be self-mappings on $\Re$, such that*

(i) for all $\iota, \varsigma \in \Re$,

$$\chi(\mathcal{Q}\iota, \mathcal{Q}'\varsigma) \leq \gamma_1\chi(\mathcal{G}\iota, \mathcal{G}'\varsigma) + \gamma_2\chi(\mathcal{G}\iota, \mathcal{Q}\iota) + \gamma_3\chi(\mathcal{G}'\varsigma, \mathcal{Q}'\varsigma)$$

$$+ \gamma_4\chi(\mathcal{G}\iota, \mathcal{Q}'\varsigma) + \gamma_5\chi(\mathcal{G}'\varsigma, \mathcal{Q}\iota) \tag{8.1}$$

where $\gamma_1 + \gamma_2 + \gamma_3 + \gamma_4 + \gamma_5 < 1$, $\gamma_i \geq 0$ *and* $\gamma_4 \leq \gamma_5$,

(ii) $\mathcal{Q}(\Re) \subset \mathcal{G}'(\Re)$ *and* $\mathcal{Q}'(\Re) \subset \mathcal{G}(\Re)$.

Then

(A) If one of $\mathcal{Q}(\Re) \cup \mathcal{Q}'(\Re)$ *and* $\mathcal{G}(\Re) \cup \mathcal{G}'(\Re)$ *is complete, then there is a unique point of coincidence for the pairs* $\{\mathcal{Q}, \mathcal{G}\}$ *and* $\{\mathcal{Q}', \mathcal{G}'\}$ *in* $\Re$.

(B) If the pairs $\{\mathcal{Q}, \mathcal{G}\}$ *and* $\{\mathcal{Q}', \mathcal{G}'\}$ *are w-compatible, then there is a unique common fixed point for the mappings* $\mathcal{Q}$, $\mathcal{Q}'$, $\mathcal{G}$ *and* $\mathcal{G}'$ *in* $\Re$.

Proof 8.1 *For constant point* $\iota_0 \in \Re$, *build the sequences* $\{\iota_\mathbf{n}\}$ *and* $\{\varsigma_\mathbf{n}\}$ *in* $\Re$ *such that*

$$\begin{cases} \mathcal{Q}\iota_{2\mathbf{n}} = \mathcal{G}'\iota_{2\mathbf{n}+1} = \varsigma_{2\mathbf{n}+1} \\ \mathcal{Q}'\iota_{2\mathbf{n}+1} = \mathcal{G}\iota_{2\mathbf{n}+2} = \varsigma_{2\mathbf{n}+2}, \qquad \forall \mathbf{n} \geq 0. \end{cases}$$

From 8.1, we have

$$\begin{aligned} \chi(\varsigma_{2\mathbf{n}+1}, \varsigma_{2\mathbf{n}+2}) = \chi(\mathcal{Q}\iota_{2\mathbf{n}}, \mathcal{Q}'\iota_{2\mathbf{n}+1}) &\leq \gamma_1\chi(\mathcal{G}\iota_{2\mathbf{n}}, \mathcal{G}'\iota_{2\mathbf{n}+1}) + \gamma_2\chi(\mathcal{G}\iota_{2\mathbf{n}}, \mathcal{Q}\iota_{2\mathbf{n}}) \\ &\quad + \gamma_3\chi(\mathcal{G}'\iota_{2\mathbf{n}+1}, \mathcal{Q}'\iota_{2\mathbf{n}+1}) + \gamma_4\chi(\mathcal{G}\iota_{2\mathbf{n}}, \mathcal{Q}'\iota_{2\mathbf{n}+1}) \\ &\quad + \gamma_5\chi(\mathcal{G}'\iota_{2\mathbf{n}+1}, \mathcal{Q}\iota_{2\mathbf{n}}) \\ &= \gamma_1\chi(\varsigma_{2\mathbf{n}}, \varsigma_{2\mathbf{n}+1}) + \gamma_2\chi(\varsigma_{2\mathbf{n}}, \varsigma_{2\mathbf{n}+1}) \\ &\quad + \gamma_3\chi(\varsigma_{2\mathbf{n}+1}, \varsigma_{2\mathbf{n}+2}) + \gamma_4\chi(\varsigma_{2\mathbf{n}}, \varsigma_{2\mathbf{n}+2}) + \gamma_5\chi(\varsigma_{2\mathbf{n}+1}, \varsigma_{2\mathbf{n}+1}) \\ &\leq \gamma_1\chi(\varsigma_{2\mathbf{n}}, \varsigma_{2\mathbf{n}+1}) + \gamma_2\chi(\varsigma_{2\mathbf{n}}, \varsigma_{2\mathbf{n}+1}) + \gamma_3\chi(\varsigma_{2\mathbf{n}+1}, \varsigma_{2\mathbf{n}+2}) \\ &\quad + \gamma_4\chi(\varsigma_{2\mathbf{n}}, \varsigma_{2\mathbf{n}+1}) + \gamma_4\chi(\varsigma_{2\mathbf{n}+1}, \varsigma_{2\mathbf{n}+2}) \\ &= (\gamma_1 + \gamma_2 + \gamma_4)\chi(\varsigma_{2\mathbf{n}}, \varsigma_{2\mathbf{n}+1}) + (\gamma_3 + \gamma_4)\chi(\varsigma_{2\mathbf{n}+1}, \varsigma_{2\mathbf{n}+2}). \end{aligned}$$

Then, $\gamma_1 + \gamma_2 + \gamma_3 + 2\gamma_4 < 1$ *implies that* $1 - \gamma_3 - \gamma_4 > 0$. *Hence,*

$$\chi(\varsigma_{2\mathbf{n}+1}, \varsigma_{2\mathbf{n}+2}) \leq \Big(\frac{\gamma_1 + \gamma_2 + \gamma_4}{1 - \gamma_3 - \gamma_4}\Big)\chi(\varsigma_{2\mathbf{n}}, \varsigma_{2\mathbf{n}+1}).$$

As a result

$$\begin{aligned} \chi(\varsigma_{2\mathbf{n}+1}, \varsigma_{2\mathbf{n}+2}) &\leq \Big(\frac{\gamma_1 + \gamma_2 + \gamma_4}{1 - \gamma_3 - \gamma_4}\Big)\chi(\varsigma_{2\mathbf{n}}, \varsigma_{2\mathbf{n}+1}) \\ &\leq \Big(\frac{\gamma_1 + \gamma_2 + \gamma_4}{1 - \gamma_3 - \gamma_4}\Big)^2 \chi(\varsigma_{2\mathbf{n}-1}, \varsigma_{2\mathbf{n}}) \\ &\;\;\vdots \\ &\leq \Big(\frac{\gamma_1 + \gamma_2 + \gamma_4}{1 - \gamma_3 - \gamma_4}\Big)^{\mathbf{2n+1}} \chi(\varsigma_0, \varsigma_1) \\ &= \mathbf{C}^{\mathbf{2n+1}}\chi(\varsigma_0, \varsigma_1), \end{aligned}$$

where, $\mathbf{C} = \Big(\frac{\gamma_1 + \gamma_2 + \gamma_4}{1 - \gamma_3 - \gamma_4}\Big)$.

Similarly, it can be shown that

$$\chi(\varsigma_{2n+2}, \varsigma_{2n+3}) \leq \mathbf{C}^{\mathbf{2n+2}}\chi(\varsigma_0, \varsigma_1). \tag{8.2}$$

Hence,

$$\chi(\varsigma_k, \varsigma_{k+1}) \leq \mathbf{C}^{\mathbf{k}}\chi(\varsigma_0, \varsigma_1), \tag{8.3}$$

for all k.

We will show now that $\{\varsigma_{2n}\}$ is a Cauchy sequence in X.

We begin by demonstrating that $\{\varsigma_{\mathbf{n}}\}$ is a Cauchy sequence. If $\mathbf{m}, \mathbf{n} \geq 1$ and $\mathbf{m} > \mathbf{n}$, then

$$\begin{aligned}
\chi(\varsigma_{\mathbf{n}}, \varsigma_{\mathbf{m}}) &\leq \mathcal{G}_{i=\mathbf{n}}^{\mathbf{m}-1}\chi(\varsigma_i, \varsigma_{i+1}) \leq \mathcal{G}_{i=\mathbf{n}}^{\mathbf{m}-1}\mathbf{C}^i\chi(\varsigma_0, \varsigma_1) \\
&= (\mathbf{C}^{\mathbf{n}} + \cdots + \mathbf{C}^{\mathbf{m}-1})\chi(\varsigma_0, \varsigma_1) \\
&\leq \mathbf{C}^{\mathbf{n}}(1 + \mathbf{C}^{\mathbf{n}} + \mathbf{C}^{\mathbf{n}+1} + \cdots)\chi(\varsigma_0, \varsigma_1) \\
&\leq \frac{\mathbf{C}^{\mathbf{n}}}{1-\mathbf{C}}\chi(\varsigma_0, \varsigma_1) \longrightarrow 0 \quad \text{as } \mathbf{n} \longrightarrow \infty.
\end{aligned}$$

So, $\{\varsigma_{\mathbf{n}}\}$ is a Cauchy sequence in $\Re$.

Let $\mathcal{G}(\Re) \cup \mathcal{G}'(\Re)$ be complete. Then $\varsigma_{\mathbf{n}} \to \ell$ as $\mathbf{n} \to \infty$ for some $\ell \in \mathcal{G}(\Re) \cup \mathcal{G}'(\Re)$. Moreover, the subsequences $\{\mathcal{G}\iota_{2\mathbf{n}+2}\} = \{\mathcal{Q}'\iota_{2\mathbf{n}+1}\} = \{\varsigma_{2\mathbf{n}+2}\}$ and $\{\mathcal{G}'\iota_{2\mathbf{n}+1}\} = \{\mathcal{Q}\iota_{2\mathbf{n}}\} = \{\varsigma_{2\mathbf{n}+1}\}$ of $\{\varsigma_{\mathbf{n}}\}$ also converge to the point ℓ. Now, since $\ell \in \mathcal{G}(\Re) \cup \mathcal{G}'(\Re)$, we have $\ell \in \mathcal{G}(\Re)$ or $\ell \in \mathcal{G}'(\Re)$. If $\ell \in \mathcal{G}(\Re)$, then we can find $\mathbf{v} \in \Re$ such that $\mathcal{G}\mathbf{v} = \ell$ and we claim that $\mathcal{Q}\mathbf{v} = \ell$. For this purpose, it can be seen that for all $k \geq 1$,

$$\begin{aligned}
\chi(\mathcal{Q}\mathbf{v}, \ell) &\leq \chi(\mathcal{Q}\mathbf{v}, \mathcal{Q}'\iota_{2\mathbf{n}+1}) + \chi(\mathcal{Q}'\iota_{2\mathbf{n}+1}, \ell) \\
&\leq \gamma_1\chi(\mathcal{G}\mathbf{v}, \mathcal{G}'\iota_{2\mathbf{n}+1}) + \gamma_2\chi(\mathcal{G}\mathbf{v}, \mathcal{Q}\mathbf{v}) + \gamma_3\chi(\mathcal{G}'\iota_{2\mathbf{n}+1}, \mathcal{Q}'\iota_{2\mathbf{n}+1}) \\
&\quad + \gamma_4\chi(\mathcal{G}\mathbf{v}, \mathcal{Q}'\iota_{2\mathbf{n}+1}) + \gamma_5\chi(\mathcal{G}'\iota_{2\mathbf{n}+1}, \mathcal{Q}\mathbf{v}) + \chi(\mathcal{Q}'\iota_{2\mathbf{n}+1}, \ell).
\end{aligned}$$

Making $n \to \infty$, we have $(1 - \gamma_2 - \gamma_5)\chi(\mathcal{Q}\mathbf{v}, \ell) \leq 0$. Consequently, we have $\mathcal{Q}\mathbf{v} = \mathcal{G}\mathbf{v} = \ell$ and so, since $\ell \in \mathcal{Q}(\Re) \subset \mathcal{G}'(\Re)$, there exists $\mathbf{q} \in \Re$ such that $\mathcal{G}'\mathbf{q} = \ell$.

Now, we can show that $\mathcal{Q}'\mathbf{q} = \ell$. In fact, for all $k \geq 1$, we have

$$\begin{aligned}
\chi(\mathcal{Q}'\mathbf{q}, \ell) &\leq \chi(\mathcal{Q}'\mathbf{q}, \mathcal{Q}\iota_{2\mathbf{n}}) + \chi(\mathcal{Q}\iota_{2\mathbf{n}}, \ell) \\
&= \chi(\mathcal{Q}\iota_{2\mathbf{n}}, \mathcal{Q}'\mathbf{q}) + \chi(\mathcal{Q}\iota_{2\mathbf{n}}, \ell) \\
&\leq \gamma_1\chi(\mathcal{G}\iota_{2\mathbf{n}}, \mathcal{G}'\mathbf{q}) + \gamma_2\chi(\mathcal{G}\iota_{2\mathbf{n}}, \mathcal{Q}\iota_{2\mathbf{n}}) + \gamma_3\chi(\mathcal{G}'\mathbf{q}, \mathcal{Q}'\mathbf{q}) \\
&\quad + \gamma_4\chi(\mathcal{G}\iota_{2\mathbf{n}}, \mathcal{Q}'\mathbf{q}) + \gamma_5\chi(\mathcal{G}'\mathbf{q}, \mathcal{Q}\iota_{2\mathbf{n}}) + \chi(\mathcal{Q}\iota_{2\mathbf{n}}, \ell).
\end{aligned}$$

If $n \to \infty$, $(1 - \gamma_3 - \gamma_4)\chi(\mathcal{Q}'\mathbf{q}, \ell) \leq 0$. So, $\mathcal{Q}'\mathbf{q} = \mathcal{G}'\mathbf{q} = \ell$. Thus, the pairs $\{\mathcal{Q}, \mathcal{G}\}$ and $\{\mathcal{Q}', \mathcal{G}'\}$ possess a common point of coincidence in $\Re$.

Now, if $\{\mathcal{Q}, \mathcal{G}\}$ and $\{\mathcal{Q}', \mathcal{G}'\}$ be weakly compatible, $\mathcal{Q}\ell = \mathcal{Q}\mathcal{G}\mathbf{v} = \mathcal{G}\mathcal{Q}\mathbf{v} = \mathcal{G}\ell := \mathbf{q}_1$ and $\mathcal{Q}'\ell = \mathcal{Q}'\mathcal{G}'\mathbf{q} = \mathcal{G}'\mathcal{Q}'\mathbf{q} = \mathcal{G}'\ell := \mathbf{q}_2$. Now

$$\begin{aligned}
\chi(\mathbf{q}_1, \mathbf{q}_2) = \chi(\mathcal{Q}\ell, \mathcal{Q}'\ell) &\leq \gamma_1\chi(\mathcal{G}\ell, \mathcal{G}'\ell) + \gamma_2\chi(\mathcal{G}\ell, \mathcal{Q}\ell) + \gamma_3\chi(\mathcal{G}'\ell, \mathcal{Q}'\ell) \\
&\quad + \gamma_4\chi(\mathcal{G}\ell, \mathcal{Q}'\ell) + \gamma_5\chi(\mathcal{G}'\ell, \mathcal{Q}\ell)
\end{aligned}$$

$$= (\gamma_1 + \gamma_4 + \gamma_5)\chi(\mathbf{q}_1, \mathbf{q}_2)$$

where by $\gamma_1 + \gamma_4 + \gamma_5 \leq 1$ *implies that* $\mathbf{q}_1 = \mathbf{q}_2$ *and hence* $\mathcal{Q}\ell = \mathcal{Q}'\ell = \mathcal{G}\ell = \mathcal{G}'\ell$, *i.e., the point* ℓ *is a coincidence point of the pairs* $\{\mathcal{Q}, \mathcal{G}\}$ *and* $\{\mathcal{Q}', \mathcal{G}'\}$. *Further, we show that* $\ell = \mathcal{Q}'\ell$. *In fact, we have*

$$\begin{aligned}\chi(\ell, \mathcal{Q}'\ell) = \chi(\mathcal{Q}\mathbf{v}, \mathcal{Q}'\ell) &\leq \gamma_1\chi(\mathcal{G}\mathbf{v}, \mathcal{G}'\ell) + \gamma_2\chi(\mathcal{G}\mathbf{v}, \mathcal{Q}\mathbf{v}) + \gamma_3\chi(\mathcal{G}'\ell, \mathcal{Q}'\ell)\\ &+ \gamma_4\chi(\mathcal{G}\mathbf{v}, \mathcal{Q}'\ell) + \gamma_5\chi(\mathcal{G}'\ell, \mathcal{Q}\mathbf{v})\\ &= (\gamma_3 + \gamma_4)\chi(\ell, \mathcal{Q}'\ell)\end{aligned}$$

which implies that $\mathcal{Q}'\ell = \ell$ *and hence* ℓ *is a common fixed point of* $\mathcal{Q}, \mathcal{Q}', \mathcal{G}$ *and* $\mathcal{G}'$.

Consider that ℓ^* *is yet another common fixed point of* $\mathcal{Q}, \mathcal{Q}', \mathcal{G}$ *and* $\mathcal{G}'$ *to further support the uniqueness.*

From 8.1, it follows that

$$\begin{aligned}\chi(\ell, \ell^*) = \chi(\mathcal{Q}\ell, \mathcal{Q}'\ell^*) &\leq \gamma_1\chi(\mathcal{G}\ell, \mathcal{G}'\ell^*) + \gamma_2\chi(\mathcal{G}\ell, \mathcal{Q}\ell) + \gamma_3\chi(\mathcal{G}'\ell^*, \mathcal{Q}'\ell^*)\\ &+ \gamma_4\chi(\mathcal{G}\ell, \mathcal{Q}'\ell^*) + \gamma_5\chi(\mathcal{G}'\ell^*, \mathcal{Q}\ell)\\ &= (\gamma_3 + \gamma_4)\chi(\ell, \mathcal{Q}'\ell^*)\end{aligned}$$

which implies that $\ell = \ell^*$.

Suppose that $\mathcal{Q}(\mathfrak{R}) \cup \mathcal{Q}'(\mathfrak{R})$ *is complete and* $\ell \in \mathcal{G}'(\mathfrak{R})$. *Then the proof lines are similar to those of the completeness of* $\mathcal{G}(\mathfrak{R}) \cup \mathcal{G}'(\mathfrak{R})$ *and so, we omit it here. This completes the proof.*

Corollary 8.1 *Let* $(\mathfrak{R}, \chi)$ *be a c.m.s. and* $\mathcal{Q}$, $\mathcal{Q}'$, $\mathcal{G}$ *and* $\mathcal{G}'$ *be self-mappings on* $\mathfrak{R}$ *satisfying* $\mathcal{Q}(\mathfrak{R}) \subset \mathcal{G}'(\mathfrak{R})$, $\mathcal{Q}'(\mathfrak{R}) \subset \mathcal{G}(\mathfrak{R})$ *and, for some* $\mathbf{m}, \mathbf{n} \geq 1$

$$\begin{aligned}\chi(\mathcal{Q}^{\mathbf{m}}\iota, \mathcal{Q}'^{\mathbf{n}}\varsigma) &\leq \gamma_1\chi(\mathcal{G}^{\mathbf{m}}\iota, \mathcal{G}'^{\mathbf{n}}\varsigma) + \gamma_2\chi(\mathcal{G}^{\mathbf{m}}\iota, \mathcal{Q}^{\mathbf{m}}\iota) + \gamma_3\chi(\mathcal{G}'^{\mathbf{n}}\varsigma, \mathcal{Q}'^{\mathbf{n}}\varsigma)\\ &+ \gamma_4\chi(\mathcal{G}^{\mathbf{m}}\iota, \mathcal{Q}'^{\mathbf{n}}\varsigma) + \gamma_5\chi(\mathcal{G}'^{\mathbf{n}}\varsigma, \mathcal{Q}^{\mathbf{m}}\iota)\end{aligned} \tag{8.4}$$

for all $\iota, \varsigma \in \mathfrak{R}$, *where* $\gamma_1 + \gamma_2 + \gamma_3 + \gamma_4 + \gamma_5 < 1$, $\gamma_i \geq 0$ *and* $\gamma_4 \leq \gamma_5$.

If one of $\mathcal{Q}(\mathfrak{R}) \cup \mathcal{Q}'(\mathfrak{R})$ *and* $\mathcal{G}(\mathfrak{R}) \cup \mathcal{G}'(\mathfrak{R})$ *be a complete in* $\mathfrak{R}$, *then there is a unique point of coincidence for the pairs* $\{\mathcal{Q}, \mathcal{G}\}$ *and* $\{\mathcal{Q}', \mathcal{G}'\}$ *in* $\mathfrak{R}$. *Moreover, if the pairs* $\{\mathcal{Q}, \mathcal{G}\}$ *and* $\{\mathcal{Q}', \mathcal{G}'\}$ *be weakly compatible, then there exists a unique common fixed point for the mappings* $\mathcal{Q}$, $\mathcal{Q}'$, $\mathcal{G}$ *and* $\mathcal{G}'$ *in* $\mathfrak{R}$.

Proof 8.2 *From Theorem 8.1* $\{\mathcal{Q}^{\mathbf{m}}, \mathcal{G}^{\mathbf{m}}\}$ *and* $\{\mathcal{Q}'^{\mathbf{m}}, \mathcal{G}'^{\mathbf{n}}\}$ *possess a unique common fixed point* $\mathbf{q} \in \mathfrak{R}$. *Now, we have*

$$\begin{aligned}\mathcal{Q}(\mathbf{q}) = \mathcal{Q}(\mathcal{Q}^{\mathbf{m}}(\mathbf{q})) = \mathcal{Q}^{\mathbf{m}+1}(\mathbf{q}) = \mathcal{Q}^{\mathbf{m}}(\mathcal{Q}(\mathbf{q})),\\ \mathcal{G}(\mathbf{q}) = \mathcal{G}(\mathcal{G}^{\mathbf{m}}(\mathbf{q})) = \mathcal{G}^{\mathbf{m}+1}(\mathbf{q}) = \mathcal{G}^{\mathbf{m}}(\mathcal{G}(\mathbf{q}))\end{aligned}$$

and so $\mathcal{Q}(\mathbf{q})$ *and* $\mathcal{G}(\mathbf{q})$ *are also fixed points for the mappings* $\mathcal{Q}^{\mathbf{m}}$ *and* $\mathcal{G}^{\mathbf{m}}$. *Thus,* $\mathcal{Q}(\mathbf{q}) = \mathcal{G}(\mathbf{q}) = \mathbf{q}$.

We obtain $\mathcal{Q}'(\mathbf{q}) = \mathcal{G}'(\mathbf{q}) = \mathbf{q}$ *by using the same argument as in the proof of the theorem 8.1. This completes the proof.*

Corollary 8.2 *Let $\mathcal{Q}$, $\mathcal{Q}'$ and $\mathcal{G}'$ be self-mappings on c.m.s. $(\Re, \chi)$, satisfying $\mathcal{Q}(\Re) \cup \mathcal{Q}'(\Re) \subset \mathcal{G}'(\Re)$ and for all $\iota, \varsigma \in \Re$,*

$$\chi(\mathcal{Q}\iota, \mathcal{Q}'\varsigma) \leq \gamma_1\chi(\mathcal{G}'\iota, \mathcal{G}'\varsigma) + \gamma_2\chi(\mathcal{G}'\iota, \mathcal{Q}\iota) + \gamma_3\chi(\mathcal{G}'\varsigma, \mathcal{Q}'\varsigma) + \gamma_4\chi(\mathcal{G}'\iota, \mathcal{Q}'\varsigma) + \gamma_5\chi(\mathcal{G}'\varsigma, \mathcal{Q}\iota) \tag{8.5}$$

where $\gamma_1 + \gamma_2 + \gamma_3 + \gamma_4 + \gamma_5 < 1$, $\gamma_i \geq 0$ and $\gamma_4 \leq \gamma_5$.

If one of $\mathcal{Q}(\Re) \cup \mathcal{Q}'(\Re)$ or $\mathcal{G}'(\Re)$ be complete in $\Re$, then there exists a unique point of coincidence for the pairs $\{\mathcal{Q}, \mathcal{G}'\}$ and $\{\mathcal{Q}', \mathcal{G}'\}$ in $\Re$. Moreover, if the pairs $\{\mathcal{Q}, \mathcal{G}'\}$ and $\{\mathcal{Q}', \mathcal{G}'\}$ be weakly compatible, then there is a common fixed point for the mappings $\mathcal{Q}$, $\mathcal{Q}'$ and $\mathcal{G}'$ in $\Re$.

Corollary 8.3 *Let $\mathcal{Q}$ and $\mathcal{G}'$ be self-mappings on c.m.s. $(\Re, \chi)$ such that $\mathcal{Q}(\Re) \subset \mathcal{G}'(\Re)$ and for all $\iota, \varsigma \in \Re$,*

$$\chi(\mathcal{Q}\iota, \mathcal{Q}\varsigma) \leq \gamma_1\chi(\mathcal{G}'\iota, \mathcal{G}'\varsigma) + \gamma_2\chi(\mathcal{G}'\iota, \mathcal{Q}\iota) + \gamma_3\chi(\mathcal{G}'\varsigma, \mathcal{Q}\varsigma) + \gamma_4\chi(\mathcal{G}'\iota, \mathcal{Q}\varsigma) + \gamma_5\chi(\mathcal{G}'\varsigma, \mathcal{Q}\iota) \tag{8.6}$$

where $\gamma_1 + \gamma_2 + \gamma_3 + \gamma_4 + \gamma_5 < 1$, $\gamma_i \geq 0$ and $\gamma_4 \leq \gamma_5$.

If one of $\mathcal{Q}(\Re)$ or $\mathcal{G}'(\Re)$ be complete in $\Re$, then there is a unique point of coincidence for the pair $\{\mathcal{Q}, \mathcal{G}'\}$ in $\Re$. Moreover, if the pair $\{\mathcal{Q}, \mathcal{G}'\}$ be weakly compatible, then there is a unique common fixed point for the mappings $\mathcal{Q}$ and $\mathcal{G}'$ in $\Re$.

8.3 EXAMPLE

Example 8.1 Let $\Re = [0, \infty)$ and $\chi : \Re \times \Re \to \mathbb{R}$ be defined by $\chi(\iota, \varsigma) = |\iota - \varsigma|$. Then $(\Re, \chi)$ is a c.m.s.. Consider four mappings $\mathcal{Q}, \mathcal{Q}', \mathcal{G}, \mathcal{G}' : \Re \to \Re$ defined by

$$\mathcal{Q}\iota = \frac{3\iota}{7}, \qquad \mathcal{Q}'\iota = \frac{2\iota}{7}, \qquad \mathcal{G}'\iota = \frac{7\iota}{3}, \qquad \mathcal{G}\iota = \frac{7\iota}{2}, \text{ for all } \iota \in \Re.$$

Clearly, $\mathcal{Q}(\Re) \subset \mathcal{G}'(\Re)$ and $\mathcal{Q}'(\Re) \subset S(\Re)$. Also, the pairs $\{\mathcal{Q}, \mathcal{G}\}$ and $\{\mathcal{Q}', \mathcal{G}'\}$ have a unique point of coincidence in $\Re$. Moreover, the pairs $\{\mathcal{Q}, \mathcal{G}\}$ and $\{\mathcal{Q}', \mathcal{G}'\}$ are w-compatible, that is, $\mathcal{Q}\mathcal{G}\iota = \mathcal{G}\mathcal{Q}\iota = \iota$ and $\mathcal{Q}'\mathcal{G}'\iota = \mathcal{G}'\mathcal{Q}'\iota = \iota$ in $\iota = 0$.

Now, for all $\iota, \varsigma \in \Re$ and $\gamma_i = \frac{1}{7}$, $i = \{1, \cdots, 5\}$,

$$\begin{aligned}
&\chi(\mathcal{Q}\iota, \mathcal{Q}'\varsigma) = \frac{1}{7}(|3\iota - 2\varsigma|) \\
&\chi(\mathcal{G}\iota, \mathcal{G}'\varsigma) = (|\frac{7\iota}{2} - \frac{7\varsigma}{3}|), \quad \chi(\mathcal{Q}\iota, \mathcal{G}\iota) = (\frac{43\iota}{14}), \quad \chi(\mathcal{Q}'\varsigma, \mathcal{G}'\varsigma) = (\frac{43y}{21}) \\
&\chi(\mathcal{Q}'\varsigma, \mathcal{G}\iota) = (|\frac{2\varsigma}{7} - \frac{7\iota}{2}|), \quad \chi(\mathcal{Q}\iota, \mathcal{G}'\varsigma) = (|\frac{3\iota}{7} - \frac{7\varsigma}{3}|) \\
&M = \chi(\mathcal{G}\iota, \mathcal{G}'\varsigma) + \chi(\mathcal{Q}\iota, \mathcal{G}\iota) + \chi(\mathcal{Q}'\varsigma, \mathcal{G}'\varsigma) + \chi(\mathcal{Q}'\varsigma, \mathcal{G}\iota) + \chi(\mathcal{Q}\iota, \mathcal{G}'\varsigma) \\
&\quad = \frac{147\iota}{14}
\end{aligned}$$

$$N = \chi(\mathcal{G}\iota, \mathcal{G}'\varsigma) + \chi(\mathcal{Q}\iota, \mathcal{G}\iota) + \chi(\mathcal{Q}'\varsigma, \mathcal{G}'\varsigma) + \chi(\mathcal{Q}'\varsigma, \mathcal{G}\iota) + \chi(\mathcal{Q}\iota, \mathcal{G}'\varsigma)$$
$$= \frac{147\varsigma}{21}.$$

If $\iota \geq \varsigma$; then

$$\begin{aligned}\chi(\mathcal{Q}\iota, \mathcal{Q}'\varsigma) &= \frac{1}{7}(|3\iota - 2\varsigma|) \leq (\frac{3\iota}{7}) \leq \frac{1}{7}M \\ &= \gamma_1\chi(\mathcal{G}\iota, \mathcal{G}'\varsigma) + \gamma_2\chi(\mathcal{Q}\iota, \mathcal{G}\iota) + \gamma_3\chi(\mathcal{Q}'\varsigma, \mathcal{G}'\varsigma) + \gamma_4\chi(\mathcal{Q}'\varsigma, \mathcal{G}\iota) \\ &\quad + \gamma_5\chi(\mathcal{Q}\iota, \mathcal{G}'\varsigma)\end{aligned}$$

and if $\iota \leq \varsigma$ then

$$\begin{aligned}\chi(\mathcal{Q}\iota, \mathcal{Q}'\varsigma) &= \frac{1}{7}(|3\iota - 2\varsigma|) \leq (\frac{2\varsigma}{7}) \leq \frac{1}{7}N \\ &= \gamma_1\chi(\mathcal{G}\iota, \mathcal{G}'\varsigma) + \gamma_2\chi(\mathcal{Q}\iota, \mathcal{G}\iota) + \gamma_3\chi(\mathcal{Q}'\varsigma, \mathcal{G}'\varsigma) + \gamma_4\chi(\mathcal{Q}'\varsigma, \mathcal{G}\iota) \\ &\quad + \gamma_5\chi(\mathcal{Q}\iota, \mathcal{G}'\varsigma)\end{aligned}$$

Thus, all the conditions of Theorem 8.1 are satisfied with $\gamma_1 + \gamma_2 + \gamma_3 + \gamma_4 + \gamma_5 = \frac{5}{7} < 1$. Then the mappings $\mathcal{Q}$, $\mathcal{Q}'$, $\mathcal{G}$ and $\mathcal{G}'$ have a unique common fixed point.

8.4 PREŠIĆ-HARDY-ROGERS TYPE FIXED POINT RESULTS

In this section, we integrate Prešić and Hardy-Rogers' results to create a novel extension of the BFPT. First we recall some key results.

Theorem 8.2 *[2] Let $(\Re, \chi)$ be a c.m.s. and let $\mathcal{G} : \Re \to \Re$ so that*

$$\chi(\mathcal{G}\iota, \mathcal{G}\varsigma) \leq \gamma\chi(\iota, \varsigma) \text{ for all } \iota, \varsigma \in \Re,$$

where $\gamma \in [0, 1)$. Then $\sigma = \mathcal{G}\sigma$ for a unique $\sigma \in \Re$. Also, $\iota_{\mathbf{n}+1} = \mathcal{G}\iota_{\mathbf{n}} \to \sigma$, for each $\iota_0 \in \Re$.

Many different techniques have been used to broaden and generalize the BFPT (see, for instance, [23]).

The following is the outcome of Prešić [25]:

Theorem 8.3 *[25] Let $(\Re, \chi)$ be a c.m.s., let $\mathcal{G} : \Re^k \to \Re$ (k is a positive integer), and let*

$$\chi(\mathcal{G}(\iota_1, ..., \iota_k), \mathcal{G}(\iota_2, ..., \iota_{k+1})) \leq \sum_{i=1}^{k} \alpha_i\chi(\iota_i, \iota_{i+1}) \tag{8.7}$$

for all $\iota_1, ..., \iota_{k+1}$ in $\Re$, where $\alpha_i \geq 0$ and $\sum_{i=1}^{k} \alpha_i \in [0, 1)$. Then there is a unique fixed point ι^* for $\mathcal{G}$ (that is, $\mathcal{G}(\iota^*, ..., \iota^*) = \iota^*$). Moreover, for all arbitrary points $\iota_1, ..., \iota_{k+1}$ in $\Re$, the sequence $\{\iota_{\mathbf{n}}\} \to \iota^*$ where $\iota_{\mathbf{n}+k} = \mathcal{G}(\iota_{\mathbf{n}}, \iota_{\mathbf{n}+1}, ..., \iota_{\mathbf{n}+k-1})$.

It is clear that Theorem 8.3 coincides with the BFPT for $k = 1$. The following is how Ćirić and Prešić generalized the aforementioned theorem [6].

Theorem 8.4 *[6] Let $(\Re, \chi)$ be a c.m.s., $\mathcal{G} : \Re^k \to \Re$ (k is a positive integer) and let*

$$\chi(\mathcal{G}(\iota_1, ..., \iota_k), \mathcal{G}(\iota_2, ..., \iota_{k+1})) \leq \alpha \max\{\chi(\iota_i, \iota_{i+1}) : 1 \leq i \leq k\}, \tag{8.8}$$

for all $\iota_1, ..., \iota_{k+1}$ in $\Re$, where $\alpha \in [0, 1)$. Then there is a fixed point $\iota^* \in \Re$ for $\mathcal{G}$. Also, for all points $\iota_1, ..., \iota_{k+1} \in \Re$, the sequence $\{\iota_{\mathbf{n}}\} \to \iota^*$ where $\iota_{\mathbf{n}+k} = \mathcal{G}(\iota_{\mathbf{n}}, \iota_{\mathbf{n}+1}, ..., \iota_{\mathbf{n}+k-1})$. The fixed point of $\mathcal{G}$ is unique if

$$\chi(\mathcal{G}(\rho, ..., \rho), \mathcal{G}(\varrho, ..., \varrho)) < \chi(\rho, \varrho),$$

for all $\rho, \varrho \in \Re$ with $\rho \neq \varrho$.

We recommend [4, 25, 23] for further information on Prešić type contractions.

Definition 8.5 *Let $(\Re, \chi)$ be a c.m.s., k be a positive integer, $\mathcal{Q}, \mathcal{Q}' : \Re^k \to \Re$ and $\mathcal{G}, \mathcal{G}' : \Re \to \Re$ be self-mappings. Then $\mathcal{Q}, \mathcal{Q}', \mathcal{G}, \mathcal{G}'$ are said to satisfy the Prešić-Kannan type contractive condition if:*

$$\begin{aligned}\chi(\mathcal{Q}(\iota_1, \cdots, \iota_k), \mathcal{Q}'(\iota_2, \cdots, \iota_{k+1})) \leq \delta\Big[&\sum_{i=1}^{k} \chi(\mathcal{G}\iota_i, \mathcal{Q}(\iota_i, \cdots, \iota_i)) \\ &+ \sum_{j=2}^{k+1} \chi(\mathcal{G}'\iota_j, \mathcal{Q}'(\iota_j, \cdots, \iota_j))\Big],\end{aligned} \tag{8.9}$$

for all $\iota_1, \cdots, \iota_k, \iota_{k+1}$, where $0 \leq 2k\delta < 1$.

Definition 8.6 *Let $(\Re, \chi)$ be a c.m.s., k be a positive integer, $\mathcal{Q}, \mathcal{Q}' : \Re^k \to \Re$ and $\mathcal{G}, \mathcal{G}' : \Re \to \Re$ be self-mappings. Then $\mathcal{Q}, \mathcal{Q}', \mathcal{G}, \mathcal{G}'$ are said to satisfy the Prešić-Reich type contractive condition if:*

$$\begin{aligned}\chi(\mathcal{Q}(\iota_1, \cdots, \iota_k), \mathcal{Q}'(\iota_2, \cdots, \iota_{k+1})) \leq &\sum_{i=1}^{k} \gamma_i \chi(\mathcal{G}\iota_i, \mathcal{G}'\iota_{i+1}) \\ &+ \sum_{i=1}^{k} \delta_i \chi(\mathcal{G}\iota_i, \mathcal{Q}(\iota_i, \cdots, \iota_i)) \\ &+ \sum_{j=2}^{k+1} \delta_j \chi(\mathcal{G}'\iota_j, \mathcal{Q}'(\iota_j, \cdots, \iota_j)),\end{aligned} \tag{8.10}$$

for all $\iota_1, \cdots, \iota_k, \iota_{k+1}$, where

$$\sum_{i=1}^{k} \gamma_i + \sum_{i=1}^{k} \delta_i + \sum_{j=2}^{k+1} \delta_j < 1.$$

Remark 8.1 *If in the aforementioned definition,* $\gamma_i = 0$*, for all* $i \in \{1, 2, ..., k\}$*, then we reach the Prešić-Kannan type contraction. Also, if* $\delta_i = 0$*, then we reach the Prešić type contraction.*

Definition 8.7 *Let* $(\Re, \chi)$ *be a c.m.s.,* k *be a positive integer,* $\mathcal{Q}, \mathcal{Q}' : \Re^k \to \Re$ *and* $\mathcal{G}, \mathcal{G}' : \Re \to \Re$ *be self-mappings. Then* $\mathcal{Q}, \mathcal{Q}', \mathcal{G}, \mathcal{G}'$ *are said to satisfy the Prešić-Chatterjea type contractive condition provided that:*

$$\chi(\mathcal{Q}(\iota_1, \cdots, \iota_k), \mathcal{Q}'(\iota_2, \cdots, \iota_{k+1})) \leq \alpha\Big[\sum_{i=1}^{k}\sum_{j=2}^{k+1}\chi(\mathcal{G}\iota_i, \mathcal{Q}'(\iota_j, \cdots, \iota_j)) + \sum_{j=2}^{k+1}\sum_{i=1}^{k}\chi(\mathcal{G}'\iota_j, \mathcal{Q}(\iota_i, \cdots, \iota_i))\Big] \tag{8.11}$$

for all $\iota_1, \cdots, \iota_k, \iota_{k+1}$*, where* $0 \leq 2\alpha k^2 < 1$.

Definition 8.8 *Let* $(\Re, \chi)$ *be a c.m.s.,* $\mathcal{Q}, \mathcal{Q}' : \Re^k \to \Re$ *and* $\mathcal{G}, \mathcal{G}' : \Re \to \Re$ *be self-mappings. Then* $\mathcal{Q}, \mathcal{Q}', \mathcal{G}, \mathcal{G}'$ *are said to satisfy the generalized-Prešić type contractive condition provided that:*

$$\begin{aligned}\chi(\mathcal{Q}(\iota_1, \cdots, \iota_k), \mathcal{Q}'(\iota_2, \cdots, \iota_{k+1})) \leq{}& \sum_{i=1}^{k}\gamma_i\chi(\mathcal{G}\iota_i, \mathcal{G}'\iota_{i+1}) \\ &+ \sum_{i=1}^{k}\delta_i\chi(\mathcal{G}\iota_i, \mathcal{Q}(\iota_i, \cdots, \iota_i)) \\ &+ \sum_{j=2}^{k+1}\delta_j\chi(\mathcal{G}'\iota_j, \mathcal{Q}'(\iota_j, \cdots, \iota_j)) \\ &+ \delta\Big[\sum_{i=1}^{k}\sum_{j=2}^{k+1}\chi(\mathcal{G}\iota_i, \mathcal{Q}'(\iota_j, \cdots, \iota_j)) \\ &+ \sum_{j=2}^{k+1}\sum_{i=1}^{k}\chi(\mathcal{G}'\iota_j, \mathcal{Q}(\iota_i, \cdots, \iota_i))\Big]\end{aligned} \tag{8.12}$$

for all $\iota_1, \cdots, \iota_k, \iota_{k+1}$*, where*

$$\sum_{i=1}^{k}\gamma_i + k\sum_{i=1}^{k}\delta_i + k\sum_{j=2}^{k+1}\delta_j + \delta[2k^2] < 1.$$

Remark 8.2 *If in the aforementioned definition,* $\gamma_i = 0$*, for all* $i \in \{1, 2, ..., k\}$*,* $\delta_i = 0$*, for all* $i \in \{1, 2, ..., k+1\}$ *and* $\delta = \alpha$*, then we reach the Prešić-Chatterjea type contraction. Also, if* $\delta = 0$*, then we reach the Prešić-Reich type contraction.*

Definition 8.9 *Let $(\Re, \chi)$ be a c.m.s., k be a positive integer, $\mathcal{Q}, \mathcal{Q}' : \Re^k \to \Re$ and $\mathcal{G}, \mathcal{G}' : \Re \to \Re$ be self-mappings. Then $\mathcal{Q}, \mathcal{Q}', \mathcal{G}, \mathcal{G}'$ are said to satisfy the Prešić-Hardy-Rogers type contractive condition if:*

$$\begin{aligned}\chi(\mathcal{Q}(\iota_1, \cdots, \iota_k), \mathcal{Q}'(\iota_2, \cdots, \iota_{k+1})) \leq & \sum_{i=1}^{k} \gamma_i \chi(\mathcal{G}\iota_i, \mathcal{G}'\iota_{i+1}) \\ & + \sum_{i=1}^{k} \delta_i \chi(\mathcal{G}\iota_i, \mathcal{Q}(\iota_i, \cdots, \iota_i)) \\ & + \sum_{i=1}^{k} \sum_{j=2}^{k+1} \delta_{i,j} \chi(\mathcal{G}\iota_i, \mathcal{Q}'(\iota_j, \cdots, \iota_j)) \\ & + \sum_{j=2}^{k+1} \sum_{i=1}^{k} \delta_{i,j} \chi(\mathcal{G}'\iota_j, \mathcal{Q}(\iota_i, \cdots, \iota_i)) \\ & + \sum_{j=2}^{k+1} \delta_j \chi(\mathcal{G}'\iota_j, \mathcal{Q}'(\iota_j, \cdots, \iota_j)) \qquad (8.13)\end{aligned}$$

for all $\iota_1, \cdots, \iota_k, \iota_{k+1}$, where

$$\sum_{i=1}^{k} \gamma_i + \sum_{i=1}^{k} \delta_i + \sum_{i=1}^{k} \sum_{j=2}^{k+1} \delta_{i,j} + \sum_{j=2}^{k+1} \sum_{i=1}^{k} \delta_{i,j} + \sum_{j=2}^{k+1} \delta_j < 1.$$

Remark 8.3 *If in the aforementioned definition, $\delta_{i,j} = \delta$, for all $i \in \{1, 2, ..., k\}$ and $j \in \{2, ..., k+1\}$, then we reach the generalized-Prešić type contraction.*

Lemma 8.1 *Suppose that $\chi_1, \ldots, \chi_{\mathbf{n}}$ be some complete metrics on nonempty sets $\mathbf{Q}_1, \ldots, \mathbf{Q}_{\mathbf{n}}$, respectively, and let $\upsilon : [0, \infty)^{\mathbf{n}} \longrightarrow [0, \infty)$ so that $\upsilon(\sigma_1, \ldots, \sigma_{\mathbf{n}}) = 0$ if and only if $\sigma_i = 0$ for all $i = 1, 2, 3, \ldots, \mathbf{n}$ and*

$$\begin{aligned}\upsilon((a_{11} + a_{12}), (a_{21} + a_{22}), ..., (a_{n1} + a_{n2})) \leq & \ \upsilon(a_{11}, a_{21}, \ldots, a_{n1}) \\ & + \upsilon(a_{12}, a_{22}, \ldots, a_{n2}), \qquad (8.14)\end{aligned}$$

for all $a_{ij} \in [0, \infty)$. Then

$$\widetilde{\chi}((\iota_{11}, \iota_{12}, ..., \iota_{1\mathbf{n}}), (\iota_{21}, \iota_{22}, ..., \iota_{2\mathbf{n}})) = \upsilon(\chi_1(\iota_{11}, \iota_{21}), \chi_2(\iota_{12}, \iota_{22}), \ldots, \chi_{\mathbf{n}}(\iota_{1\mathbf{n}}, \iota_{2\mathbf{n}})),$$

is a metric in $[\mathbf{Q}_1 \times \mathbf{Q}_2 \times \ldots \times \mathbf{Q}_{\mathbf{n}}]^2$.

Proof 8.3 *The triangle inequality is only demonstrated in metric spaces. Let $\varsigma_j \in \mathbf{Q}_j$ for all $1 \leq j \leq \mathbf{n}$. So,*

$$\begin{aligned}\widetilde{\chi}((\iota_{11}, \iota_{12}, ..., \iota_{1\mathbf{n}}), (\iota_{21}, \iota_{22}, ..., \iota_{2\mathbf{n}})) &= \upsilon(\chi_1(\iota_{11}, \iota_{21}), \chi_2(\iota_{12}, \iota_{22}), \ldots, \chi_{\mathbf{n}}(\iota_{1\mathbf{n}}, \iota_{2\mathbf{n}})) \\ &\leq \upsilon\Big[\big((\chi_1(\iota_{11}, \varsigma_1) + \chi_1(\varsigma_1, \iota_{21})\big), \big(\chi_2(\iota_{12}, \varsigma_2) + \chi_2(\varsigma_2, \iota_{22})\big), , ..., \big(\chi_{\mathbf{n}}(\iota_{1\mathbf{n}}, \varsigma_{\mathbf{n}})\end{aligned}$$

$$+ \chi_{\mathbf{n}}(\varsigma_{\mathbf{n}}, \iota_{2\mathbf{n}}))\Big]$$
$$\leq \upsilon\Big[(\chi_1(\iota_{11}, \varsigma_1), \chi_2(\iota_{12}, \varsigma_2), \ldots, \chi_{\mathbf{n}}(\iota_{1\mathbf{n}}, \varsigma_{\mathbf{n}})\Big] + \upsilon\Big[(\chi_1(\varsigma_1, \iota_{21}), \chi_2(\varsigma_2, \iota_{22}), \chi_{\mathbf{n}}(\varsigma_{\mathbf{n}}, \iota_{2\mathbf{n}})\Big]$$
$$\leq \widetilde{\chi}((\iota_{11}, \iota_{12}, ..., \iota_{1\mathbf{n}}), (\varsigma_1, \varsigma_2, \ldots, \varsigma_{\mathbf{n}})) + \widetilde{\chi}((\varsigma_1, \varsigma_2, \ldots, \varsigma_{\mathbf{n}}), (\iota_{21}, \iota_{22}, \ldots, \iota_{2\mathbf{n}})).$$

Theorem 8.5 *Let $(\Re, \chi)$ be a c.m.s., k be a positive integer, $\mathcal{Q}, \mathcal{Q}' : \Re^k \to \Re$ and $\mathcal{G}, \mathcal{G}' : \Re \to \Re$ as continuous functions such that:*

(i) $\mathcal{Q}(\Re^k) \subseteq \mathcal{G}'(\Re)$ and $\mathcal{Q}'(\Re^k) \subseteq \mathcal{G}(\Re)$,

(ii)

$$\chi\Big(\mathcal{Q}(\iota_1, \iota_2, ..., \iota_k), \mathcal{Q}'(\iota_2, \iota_3, ..., \iota_{k+1})\Big) \leq \sum_{i=1}^{k} \gamma_i \chi(\mathcal{G}\iota_i, \mathcal{G}'\iota_{i+1})$$
$$+ \sum_{i=1}^{k} \delta_i \chi(\mathcal{G}\iota_i, \mathcal{Q}(\iota_i, \cdots, \iota_i))$$
$$+ \sum_{i=1}^{k} \sum_{j=2}^{k+1} \delta_{i,j} \chi(\mathcal{G}\iota_i, \mathcal{Q}'(\iota_j, \cdots, \iota_j))$$
$$+ \sum_{j=2}^{k+1} \sum_{i=1}^{k} \delta_{i,j} \chi(\mathcal{G}'\iota_j, \mathcal{Q}(\iota_i, \cdots, \iota_i))$$
$$+ \sum_{j=2}^{k+1} \delta_j \chi(\mathcal{G}'\iota_j, \mathcal{Q}'(\iota_j, \cdots, \iota_j)), \quad (8.15)$$

for all $\iota_i \subseteq \Re$ where

$$\sum_{i=1}^{k} \gamma_i + \sum_{i=1}^{k} \delta_i + \sum_{i=1}^{k} \sum_{j=2}^{k+1} \delta_{i,j} + \sum_{j=2}^{k+1} \sum_{i=1}^{k} \delta_{i,j} + \sum_{j=2}^{k+1} \delta_j < 1.$$

(iii) If one of $\mathcal{Q}(\Re^k) \cup \mathcal{Q}'(\Re^k)$ and $\mathcal{G}(\Re) \cup \mathcal{G}'(\Re)$ is complete, then there is a unique point of coincidence for the pairs $\{\mathcal{Q}, \mathcal{G}\}$ and $\{\mathcal{Q}', \mathcal{G}'\}$ in $\Re$.

(iv) If the pairs $\{\mathcal{Q}, \mathcal{G}\}$ and $\{\mathcal{Q}', \mathcal{G}'\}$ are k-w-compatible, then $\mathcal{Q}, \mathcal{Q}', \mathcal{G}, \mathcal{G}'$ have at least a Prešić-type common fixed point.

Proof 8.4 *We define the mappings $\widetilde{\mathcal{Q}}, \widetilde{\mathcal{Q}'}, \widetilde{\mathcal{G}}, \widetilde{\mathcal{G}'} : \Re^k \to \Re^k$ by*

$$\widetilde{\mathcal{Q}}(\sigma_1, ..., \sigma_k) = (\mathcal{Q}(\sigma_1, ..., \sigma_k), ..., \mathcal{Q}(\sigma_1, ..., \sigma_k))$$

and

$$\widetilde{\mathcal{Q}'}(\sigma_1, ..., \sigma_k) = (\mathcal{Q}'(\sigma_1, ..., \sigma_k), ..., \mathcal{Q}'(\sigma_1, ..., \sigma_k)),$$

and

$$\widetilde{\mathcal{G}}(\sigma_1, ..., \sigma_k) = (\mathcal{G}(\sigma_1), ..., \mathcal{G}(\sigma_k)), \text{ and } \widetilde{\mathcal{G}}'(\sigma_1, ..., \sigma_k) = (\mathcal{G}'(\sigma_1), ..., \mathcal{G}'(\sigma_k)).$$

Clearly, $\widetilde{\mathcal{Q}}$ *and* $\widetilde{\mathcal{Q}'}$ *are continuous. We show that* $\widetilde{\mathcal{Q}}$ *and* $\widetilde{\mathcal{Q}'}$ *meets all requirement of the theorem 8.1, where*

$$\widetilde{\chi}((\iota_1, \iota_2, ..., \iota_k), (\sigma_1, \sigma_2, ..., \sigma_k)) = \frac{\chi(\iota_1, \sigma_1) + \chi(\iota_2, \sigma_2) + \cdots + \chi(\iota_k, \sigma_k)}{k}.$$

From (8.15) we have

$$\begin{aligned}
&\widetilde{\chi}(\widetilde{\mathcal{Q}}(\iota_1,\iota_2, ..., \iota_k), \widetilde{\mathcal{Q}'}(\iota_2, \iota_3, ..., \iota_{k+1}))\\
&\quad = \widetilde{\chi}\Big((\mathcal{Q}(\iota_1, \iota_2, ..., \iota_k), \mathcal{Q}(\iota_1, \iota_2, ..., \iota_k), \ldots, \mathcal{Q}(\iota_1, \iota_2, ..., \iota_k)),\\
&\qquad (\mathcal{Q}'(\iota_2, \iota_3, ..., \iota_{k+1}), \mathcal{Q}'(\iota_2, \iota_3, ..., \iota_{k+1}), \ldots, \mathcal{Q}'(\iota_2, \iota_3, ..., \iota_{k+1}))\Big)\\
&\quad = \chi(\mathcal{Q}(\iota_1, \iota_2, ..., \iota_k), \mathcal{Q}'(\iota_2, \iota_3, ..., \iota_{k+1}))\\
&\quad \le \sum_{i=1}^{k} \gamma_i \chi(\mathcal{G}\iota_i, \mathcal{G}'\iota_{i+1}) + \sum_{i=1}^{k} \delta_i \chi(\mathcal{G}\iota_i, \mathcal{Q}(\iota_i, \cdots, \iota_i))\\
&\qquad + \sum_{i=1}^{k}\sum_{j=2}^{k+1} \delta_{i,j} \chi(\mathcal{G}\iota_i, \mathcal{Q}'(\iota_j, \cdots, \iota_j))\\
&\qquad + \sum_{j=2}^{k+1}\sum_{i=1}^{k} \delta_{i,j} \chi(\mathcal{G}'\iota_j, \mathcal{Q}(\iota_i, \cdots, \iota_i))\\
&\qquad + \sum_{j=2}^{k+1} \delta_j \chi(\mathcal{G}'\iota_j, \mathcal{Q}'(\iota_j, \cdots, \iota_j))
\end{aligned}$$

$$\begin{aligned}
&\le k[\Gamma_1 \widetilde{\chi}(\widetilde{\mathcal{G}}(\iota_1, \iota_2, ..., \iota_k), \widetilde{\mathcal{G}}'(\iota_2, \iota_3, ..., \iota_{k+1})) + \Gamma_2 \widetilde{\chi}(\widetilde{\mathcal{G}}(\iota_1, \iota_2, ..., \iota_k),\\
&\quad \widetilde{\mathcal{Q}}(\iota_1, \iota_2, ..., \iota_k)) + \Gamma_3 \widetilde{\chi}(\widetilde{\mathcal{G}}'(\iota_2, \iota_3, ..., \iota_{k+1}), \widetilde{\mathcal{Q}'}(\iota_2, \iota_3, ..., \iota_{k+1}))\\
&\quad + \Gamma_4 \widetilde{\chi}(\widetilde{\mathcal{G}}(\iota_1, \iota_2, ..., \iota_k), \widetilde{\mathcal{Q}'}(\iota_2, \iota_3, ..., \iota_{k+1}))\\
&\quad + \Gamma_5 \widetilde{\chi}(\widetilde{\mathcal{G}}'(\iota_2, \iota_3, ..., \iota_{k+1}), \widetilde{\mathcal{Q}}(\iota_1, \iota_2, ..., \iota_k))],
\end{aligned}$$

where $\Gamma_1 = \sum_{i=1}^{k} \gamma_i$, $\Gamma_2 = \sum_{i=1}^{k} \delta_i$, $\Gamma_3 = \sum_{i=1}^{k}\sum_{j=2}^{k+1} \delta_{i,j}$, $\Gamma_4 = \sum_{i=1}^{k}\sum_{j=2}^{k+1} \delta_{i,j}$, $\Gamma_5 = \sum_{j=2}^{k+1} \delta_j$.

Now, with the same process of Theorem 8.1, we deduce that there is at least a common fixed point for $\widetilde{\mathcal{Q}}$ *and* $\widetilde{\mathcal{Q}'}$ *which implies that there exist* $\sigma_1, ..., \sigma_{\mathbf{n}}$ *such that* $\mathcal{Q}(\sigma_1, ..., \sigma_{\mathbf{n}}) = \mathcal{Q}'(\sigma_1, ..., \sigma_{\mathbf{n}}) = \sigma_1 = ... = \sigma_{\mathbf{n}}$*, that is, there exists at least a Prešić-Hardy-Rogers type fixed point of* $\mathcal{Q}$ *and* $\mathcal{Q}'$*.*

Since the pairs $\{\mathcal{Q}, \mathcal{G}\}$ *and* $\{\mathcal{Q}', \mathcal{G}'\}$ *are k-w-compatible, then again by Theorem 8.1, the mappings there is at least a Prešić-Hardy-Rogers type common fixed point for* $\widetilde{\mathcal{Q}}, \widetilde{\mathcal{Q}'}, \widetilde{\mathcal{G}}$ *and* $\widetilde{\mathcal{G}}'$*.*

Remark 8.4 *If in the aforementioned theorem, $k = 1$, then we obtain the result of Hardy-Rogers [28]. If $\delta_i = \delta_{i,j} = 0$, for all $i \in \{1, 2, ..., k\}, j \in \{2, ..., k+1\}$ and the mapping $\mathcal{Q} : \Re^k \to \Re$, then we obtain the fixed point result of Prešić [25].*

Corollary 8.4 *Let $(\Re, \chi)$ be a c.m.s., k be a positive integer, $\mathcal{Q}, \mathcal{Q}' : \Re^k \to \Re$ and $\mathcal{G}, \mathcal{G}' : \Re \to \Re$ be continuous functions such that $\mathcal{Q}(\Re^k) \subseteq \mathcal{G}'(\Re)$ and $\mathcal{Q}'(\Re^k) \subseteq \mathcal{G}(\Re)$ and, for some $\mathbf{m}, \mathbf{n} \geq 1$*

$$\begin{aligned}\chi\Big(\mathcal{Q}^{\mathbf{m}}(\iota_1, \iota_2, ..., \iota_k), \mathcal{Q}'^{\mathbf{n}}(\iota_2, \iota_3, ..., \iota_{k+1})\Big) \leq & \sum_{i=1}^{k} \gamma_i \chi(\mathcal{G}^{\mathbf{m}}\iota_i, \mathcal{G}'^{\mathbf{n}}\iota_{i+1}) \\ & + \sum_{i=1}^{k} \delta_i \chi(\mathcal{G}^{\mathbf{m}}\iota_i, \mathcal{Q}^{\mathbf{m}}(\iota_i, \cdots, \iota_i)) \\ & + \sum_{i=1}^{k}\sum_{j=2}^{k+1} \delta_{i,j} \chi(\mathcal{G}^{\mathbf{m}}\iota_i, \mathcal{Q}'^{\mathbf{n}}(\iota_j, \cdots, \iota_j)) \\ & + \sum_{j=2}^{k+1}\sum_{i=1}^{k} \delta_{i,j} \chi(\mathcal{G}'^{\mathbf{n}}\iota_j, \mathcal{Q}^{\mathbf{m}}(\iota_i, \cdots, \iota_i)) \\ & + \sum_{j=2}^{k+1} \delta_j \chi(\mathcal{G}'^{\mathbf{n}}\iota_j, \mathcal{Q}'^{\mathbf{n}}(\iota_j, \cdots, \iota_j)), \quad (8.16)\end{aligned}$$

for all $\iota_i, \iota_j \subseteq \Re$ where

$$\sum_{i=1}^{k} \gamma_i + \sum_{i=1}^{k} \delta_i + \sum_{i=1}^{k}\sum_{j=2}^{k+1} \delta_{i,j} + \sum_{j=2}^{k+1}\sum_{i=1}^{k} \delta_{i,j} + \sum_{j=2}^{k+1} \delta_i < 1.$$

(iii) If one of $\mathcal{Q}(\Re^k) \cup \mathcal{Q}'(\Re^k)$ and $\mathcal{G}(\Re) \cup \mathcal{G}'(\Re)$ is complete, then there is a unique point of coincidence for the pairs $\{\mathcal{Q}, \mathcal{G}\}$ and $\{\mathcal{Q}', \mathcal{G}'\}$ in $\Re$.

(iv) If the pairs $\{\mathcal{Q}, \mathcal{G}\}$ and $\{\mathcal{Q}', \mathcal{G}'\}$ are k-w-compatible, then there is at least a Prešić-type common fixed point for $\mathcal{Q}, \mathcal{Q}', \mathcal{G}$ and $\mathcal{G}'$.

Corollary 8.5 *Let $(\Re, \chi)$ be a c.m.s., k be a positive integer, $\mathcal{Q}, \mathcal{Q}' : \Re^k \to \Re$ and $\mathcal{G} : \Re \to \Re$ are continuous functions such that $\mathcal{Q}(\Re^k) \cup \mathcal{Q}'(\Re^k) \subseteq \mathcal{G}(\Re)$, and*

$$\begin{aligned}\chi\Big(\mathcal{Q}(\iota_1, \iota_2, ..., \iota_k), \mathcal{Q}'(\iota_2, \iota_3, ..., \iota_{k+1})\Big) \leq & \sum_{i=1}^{k} \gamma_i \chi(\mathcal{G}\iota_i, \mathcal{G}\iota_{i+1}) \\ & + \sum_{i=1}^{k} \delta_i \chi(\mathcal{G}\iota_i, \mathcal{Q}(\iota_i, \cdots, \iota_i)) \\ & + \sum_{i=1}^{k}\sum_{j=2}^{k+1} \delta_{i,j} \chi(\mathcal{G}\iota_i, \mathcal{Q}'(\iota_j, \cdots, \iota_j))\end{aligned}$$

$$+\sum_{j=2}^{k+1}\sum_{i=1}^{k}\delta_{i,j}\chi(\mathcal{G}\iota_j, \mathcal{Q}(\iota_i,\cdots,\iota_i))$$
$$+\sum_{j=2}^{k+1}\delta_j\chi(\mathcal{G}\iota_j, \mathcal{Q}'(\iota_j,\cdots,\iota_j)), \quad (8.17)$$

for all $\iota_i, \iota_j \subseteq \Re$ *where*

$$\sum_{i=1}^{k}\gamma_i+\sum_{i=1}^{k}\delta_i+\sum_{i=1}^{k}\sum_{j=2}^{k+1}\delta_{i,j}+\sum_{j=2}^{k+1}\sum_{i=1}^{k}\delta_{i,j}+\sum_{j=2}^{k+1}\delta_j<1.$$

(iii) If one of $\mathcal{Q}(\Re^k)\cup\mathcal{Q}'(\Re^k)$ *and* $\mathcal{G}(\Re)$ *is complete, then there exists a unique point of coincidence for the pairs* $\{\mathcal{Q},\mathcal{G}\}$ *and* $\{\mathcal{Q}',\mathcal{G}\}$ *in* $\Re$.

(iv) If the pairs $\{\mathcal{Q},\mathcal{G}\}$ *and* $\{\mathcal{Q}',\mathcal{G}\}$ *are k-w-compatible, then there is at least a Prešić-type common fixed point for* $\mathcal{Q}$, $\mathcal{Q}'$ *and* $\mathcal{G}$.

Corollary 8.6 *Let* $(\Re,\chi)$ *be a c.m.s., k be a positive integer,* $\mathcal{Q}:\Re^k\to\Re$ *and* $\mathcal{G}:\Re\to\Re$ *be continuous functions such that* $\mathcal{Q}(\Re^k)\subseteq\mathcal{G}(\Re)$, *and*

$$\chi\Big(\mathcal{Q}(\iota_1,\iota_2,...,\iota_k),\mathcal{Q}(\iota_2,\iota_3,...,\iota_{k+1})\Big)\leq\sum_{i=1}^{k}\gamma_i\chi(\mathcal{G}\iota_i,\mathcal{G}\iota_{i+1})$$
$$+\sum_{i=1}^{k}\delta_i\chi(\mathcal{G}\iota_i, \mathcal{Q}(\iota_i,\cdots,\iota_i))$$
$$+\sum_{i=1}^{k}\sum_{j=2}^{k+1}\delta_{i,j}\chi(\mathcal{G}\iota_i, \mathcal{Q}(\iota_j,\cdots,\iota_j))$$
$$+\sum_{j=2}^{k+1}\sum_{i=1}^{k}\delta_{i,j}\chi(\mathcal{G}\iota_j, \mathcal{Q}(\iota_i,\cdots,\iota_i))$$
$$+\sum_{j=2}^{k+1}\delta_j\chi(\mathcal{G}\iota_j, \mathcal{Q}(\iota_j,\cdots,\iota_j)), \quad (8.18)$$

for all $\iota_i, \iota_j \subseteq \Re$ *where*

$$\sum_{i=1}^{k}\gamma_i+\sum_{i=1}^{k}\delta_i+\sum_{i=1}^{k}\sum_{j=2}^{k+1}\delta_{i,j}+\sum_{j=2}^{k+1}\sum_{i=1}^{k}\delta_{i,j}+\sum_{j=2}^{k+1}\delta_j<1.$$

(iii) If one of $\mathcal{Q}(\Re^k)$ *and* $\mathcal{G}(\Re)$ *is complete, then there is a unique point of coincidence for the pair* $(\mathcal{Q},\mathcal{G})$ *in* $\Re$.

(iv) If the pair $\{\mathcal{Q},\mathcal{G}\}$ *is k-w-compatible, then there is at least a Prešić-type common fixed point for* $\mathcal{Q}$ *and* $\mathcal{G}$.

According to Remarks 8.1, 8.2 and 8.3, we have the following results:

Corollary 8.7 *Let $(\Re, \chi)$ be a c.m.s., k be a positive integer, $\mathcal{Q}, \mathcal{Q}' : \Re^k \to \Re$ and $\mathcal{G}, \mathcal{G}' : \Re \to \Re$ be continuous functions such that*

(i) $\mathcal{Q}(\Re^k) \subseteq \mathcal{G}'(\Re)$ and $\mathcal{Q}'(\Re^k) \subseteq \mathcal{G}(\Re)$,

(ii)

$$\begin{aligned}\chi\Big(\mathcal{Q}(\iota_1, \iota_2, ..., \iota_k), \mathcal{Q}'(\iota_2, \iota_3, ..., \iota_{k+1})\Big) \leq & \sum_{i=1}^{k} \gamma_i \chi(\mathcal{G}\iota_i, \mathcal{G}'\iota_{i+1}) \\ & + \sum_{i=1}^{k} \delta_i \chi(\mathcal{G}\iota_i, \mathcal{Q}(\iota_i, \cdots, \iota_i)) \\ & + \sum_{j=2}^{k+1} \delta_j \chi(\mathcal{G}'\iota_j, \mathcal{Q}'(\iota_j, \cdots, \iota_j)) \\ & + \delta\Big[\sum_{i=1}^{k}\sum_{j=2}^{k+1} \chi(\mathcal{G}\iota_i, \mathcal{Q}'(\iota_j, \cdots, \iota_j)) \\ & + \sum_{j=2}^{k+1}\sum_{i=1}^{k} \chi(\mathcal{G}'\iota_j, \mathcal{Q}(\iota_i, \cdots, \iota_i))\Big], \quad (8.19)\end{aligned}$$

for all $\iota_i, \iota_j \subseteq \Re$ where

$$\sum_{i=1}^{k} \gamma_i + \sum_{i=1}^{k} \delta_i + 2\delta k^2 + \sum_{j=2}^{k+1} \delta_j < 1.$$

(iii) If one of $\mathcal{Q}(\Re^k) \cup \mathcal{Q}'(\Re^k)$ and $\mathcal{G}(\Re) \cup \mathcal{G}'(\Re)$ is complete, then there is a unique point of coincidence for the pairs $(\mathcal{Q}, \mathcal{G})$ and $(\mathcal{Q}', \mathcal{G}')$ in $\Re$.

(iv) If the pairs $\{\mathcal{Q}, \mathcal{G}\}$ and $\{\mathcal{Q}', \mathcal{G}'\}$ are k-w-compatible, then there is at least a Prešić-type common fixed point for $\mathcal{Q}, \mathcal{Q}', \mathcal{G}$ and $\mathcal{G}'$.

In the aforementioned corollary, for $k = 1$ and $\mathcal{Q} : \Re^k \to \Re$, we obtain the Ćirić fixed point result [5].

Corollary 8.8 *Let $(\Re, \chi)$ be a c.m.s., k be a positive integer, $\mathcal{Q}, \mathcal{Q}' : \Re^k \to \Re$ and $\mathcal{G}, \mathcal{G}' : \Re \to \Re$ be continuous functions such that*

(i) $\mathcal{Q}(\Re^k) \subseteq \mathcal{G}'(\Re)$ and $\mathcal{Q}'(\Re^k) \subseteq \mathcal{G}(\Re)$,

(ii)

$$\begin{aligned}\chi\Big(\mathcal{Q}(\iota_1, \iota_2, ..., \iota_k), \mathcal{Q}'(\iota_2, \iota_3, ..., \iota_{k+1})\Big) \leq & \sum_{i=1}^{k} \gamma_i \chi(\mathcal{G}\iota_i, \mathcal{G}'\iota_{i+1}) \\ & + \sum_{i=1}^{k} \delta_i \chi(\mathcal{G}\iota_i, \mathcal{Q}(\iota_i, \cdots, \iota_i)) \\ & + \sum_{j=2}^{k+1} \delta_j \chi(\mathcal{G}'\iota_j, \mathcal{Q}'(\iota_j, \cdots, \iota_j)), \quad (8.20)\end{aligned}$$

for all $\iota_i, \iota_j \subseteq \Re$ *where*

$$\sum_{i=1}^{k}\gamma_i + \sum_{i=1}^{k}\delta_i + \sum_{j=2}^{k+1}\delta_j < 1.$$

(iii) If one of $\mathcal{Q}(\Re^k) \cup \mathcal{Q}'(\Re^k)$ *and* $\mathcal{G}(\Re) \cup \mathcal{G}'(\Re)$ *is complete, then there is a unique point of coincidence for the pairs* $\{\mathcal{Q}, \mathcal{G}\}$ *and* $\{\mathcal{Q}', \mathcal{G}'\}$ *in* $\Re$.

(iv) If the pairs $\{\mathcal{Q}, \mathcal{G}\}$ *and* $\{\mathcal{Q}', \mathcal{G}'\}$ *are k-w-compatible, then there exists at least a Prešić-type common fixed point for* $\mathcal{Q}, \mathcal{Q}', \mathcal{G}, \mathcal{G}'$.

In the aforementioned corollary, for $k = 1$ and $\mathcal{Q} : \Re^k \to \Re$, we obtain the Reich fixed point result [26].

Corollary 8.9 *Let* $(\Re, \chi)$ *be a c.m.s., k be a positive integer,* $\mathcal{Q}, \mathcal{Q}' : \Re^k \to \Re$ *and* $\mathcal{G}, \mathcal{G}' : \Re \to \Re$ *as continuous functions such that*

(i) $\mathcal{Q}(\Re^k) \subseteq \mathcal{G}'(\Re)$ *and* $\mathcal{Q}'(\Re^k) \subseteq \mathcal{G}(\Re)$,

(ii)

$$\chi\Big(\mathcal{Q}(\iota_1, \iota_2, ..., \iota_k), \mathcal{Q}'(\iota_2, \iota_3, ..., \iota_{k+1})\Big) \leq \delta\Big(\sum_{i=1}^{k}\chi(\mathcal{G}\iota_i, \mathcal{Q}(\iota_i, \cdots, \iota_i)) + \sum_{j=2}^{k+1}\chi(\mathcal{G}'\iota_j, \mathcal{Q}'(\iota_j, \cdots, \iota_j))\Big), \quad (8.21)$$

for all $\iota_i, \iota_j \subseteq \Re$ *where* $2k\delta < 1$.

(iii) If one of $\mathcal{Q}(\Re^k) \cup \mathcal{Q}'(\Re^k)$ *and* $\mathcal{G}(\Re) \cup \mathcal{G}'(\Re)$ *is complete, then there is a unique point of coincidence for the pairs* $\{\mathcal{Q}, \mathcal{G}\}$ *and* $\{\mathcal{Q}', \mathcal{G}'\}$ *in* $\Re$.

(iv) If the pairs $\{\mathcal{Q}, \mathcal{G}\}$ *and* $\{\mathcal{Q}', \mathcal{G}'\}$ *are k-w-compatible, then there is at least a Prešić-type common fixed point for* $\mathcal{Q}, \mathcal{Q}', \mathcal{G}, \mathcal{G}'$.

In the aforementioned corollary, for $k = 1$ and $\mathcal{Q} : \Re^k \to \Re$, we obtain the Kannan fixed point result [16].

Corollary 8.10 *Let* $(\Re, \chi)$ *be a c.m.s., k be a positive integer,* $\mathcal{Q}, \mathcal{Q}' : \Re^k \to \Re$ *and* $\mathcal{G}, \mathcal{G}' : \Re \to \Re$ *be continuous functions such that*

(i) $\mathcal{Q}(\Re^k) \subseteq \mathcal{G}'(\Re)$ *and* $\mathcal{Q}'(\Re^k) \subseteq \mathcal{G}(\Re)$,

(ii)

$$\chi\Big(\mathcal{Q}(\iota_1, \iota_2, ..., \iota_k), \mathcal{Q}'(\iota_2, \iota_3, ..., \iota_{k+1})\Big) \leq \alpha\Big(\sum_{i=1}^{k}\sum_{j=2}^{k+1}\chi(\mathcal{G}\iota_i, \mathcal{Q}'(\iota_j, \cdots, \iota_j)) + \sum_{j=2}^{k+1}\sum_{i=1}^{k}\chi(\mathcal{G}'\iota_j, \mathcal{Q}(\iota_i, \cdots, \iota_i))\Big), \quad (8.22)$$

for all $\iota_i, \iota_j \subseteq \Re$ *where* $2k^2\alpha < 1$.

(iii) If one of $\mathcal{Q}(\Re^k) \cup \mathcal{Q}'(\Re^k)$ and $\mathcal{G}(\Re) \cup \mathcal{G}'(\Re)$ is complete, then there is a unique point of coincidence for the pairs $\{\mathcal{Q}, \mathcal{G}\}$ and $\{\mathcal{Q}', \mathcal{G}'\}$ in $\Re$.

(iv) If the pairs $\{\mathcal{Q}, \mathcal{G}\}$ and $\{\mathcal{Q}', \mathcal{G}'\}$ are k-w-compatible, then there is at least a Prešić-type common fixed point for $\mathcal{Q}, \mathcal{Q}', \mathcal{G}, \mathcal{G}'$.

In the aforementioned corollary, for $k = 1$ and $\mathcal{Q} : \Re^k \to \Re$, we obtain the of Chatterjea fixed point result [3].

8.5 EXAMPLE

Example 8.2

Let $X = [0, 1]$. Consider the metric space and γ_i presented in Example 8.2, we define mappings $\mathcal{Q}, \mathcal{Q}' : \Re^2 \to \Re$ and $\mathcal{G}, \mathcal{G}' : \Re \to \Re$ by

$$\mathcal{Q}(\iota, \varsigma) = \frac{3\iota + \varsigma}{15}, \quad \mathcal{Q}'(\iota, \varsigma) = \frac{\iota + 2\varsigma}{15}, \quad \mathcal{G}'\iota = \frac{10\iota}{3}, \quad \mathcal{G}\iota = \frac{15\iota}{2}, \text{ for all } \iota \in \Re.$$

Clearly, $\mathcal{Q}(\Re^2) \subset \mathcal{G}'(\Re)$ and $\mathcal{Q}'(\Re^2) \subset S(\Re)$. Also, the pairs $\{\mathcal{Q}, \mathcal{G}\}$ and $\{\mathcal{Q}', \mathcal{G}'\}$ have a unique point of coincidence in $\Re$. Moreover, the pairs $\{\mathcal{Q}, \mathcal{G}\}$ and $\{\mathcal{Q}', \mathcal{G}'\}$ are 2-w-compatible, that is, $\mathcal{G}\mathcal{Q}(\iota, \iota) = \mathcal{Q}(\mathcal{G}\iota, \mathcal{G}\iota)$ and $\mathcal{G}'\mathcal{Q}'(\iota, \iota) = \mathcal{Q}'(\mathcal{G}'\iota, \mathcal{G}'\iota)$.

Now, for all $\iota, \varsigma, \kappa \in \Re$

$$\chi(\mathcal{Q}(\iota, \varsigma), \mathcal{Q}'(\varsigma, \kappa)) = \frac{1}{15}(|3\iota - 2\kappa|),$$

$$\sum_{i=1}^{2} \chi(\mathcal{G}\iota_i, \mathcal{G}'\iota_{i+1}) = (|\frac{15\iota}{2} - \frac{10\varsigma}{3}| + |\frac{15\varsigma}{2} - \frac{10\kappa}{3}|),$$

$$\sum_{i=1}^{2} \chi(\mathcal{G}\iota_i, \mathcal{Q}(\iota_i, \iota_i) = (\frac{217\iota}{30} + \frac{217\varsigma}{30}),$$

$$\sum_{i=1}^{2}\sum_{j=2}^{3} \chi(\mathcal{G}\iota_i, \mathcal{Q}'(\iota_j, \iota_j)) = (|\frac{15\iota}{2} - \frac{3\varsigma}{15}| + |\frac{15\iota}{2} - \frac{3\kappa}{15}| + |\frac{219\varsigma}{30}| + |\frac{15\varsigma}{2} - \frac{3\kappa}{15}|),$$

$$\sum_{j=2}^{3}\sum_{i=1}^{2} \chi(\mathcal{G}'\iota_j, \mathcal{Q}(\iota_i, \iota_i)) = (|\frac{10\varsigma}{3} - \frac{4\iota}{15}| + |\frac{10\varsigma}{3} - \frac{4\varsigma}{15}| + |\frac{10\kappa}{3} - \frac{4\iota}{15}| + |\frac{10\kappa}{3} - \frac{4\varsigma}{15}|),$$

$$\sum_{j=2}^{3} \chi(\mathcal{G}'\iota_j, \mathcal{Q}'(\iota_j, \iota_j)) = (|\frac{47\varsigma}{15}| + \frac{47\kappa}{15}),$$

$$\mathbf{A} = \sum_{i=1}^{2} \gamma_i \chi(\mathcal{G}\iota_i, \mathcal{G}'\iota_{i+1}) + \sum_{i=1}^{2} \chi(\mathcal{G}\iota_i, \mathcal{Q}(\iota_i, \iota_i)) + \sum_{j=2}^{3}\sum_{i=1}^{2} \chi(\mathcal{G}'\iota_j, \mathcal{Q}(\iota_i, \iota_i))$$
$$+ \sum_{i=1}^{2}\sum_{j=2}^{3} \chi(\mathcal{G}\iota_i, \mathcal{Q}'(\iota_j, \iota_j)) + \sum_{j=2}^{3} \chi(\mathcal{G}'\iota_j, \mathcal{Q}'(\iota_j, \iota_j)) = \frac{908\iota}{30},$$

$$\mathbf{B} = \sum_{i=1}^{2} \gamma_i \chi(\mathcal{G}\iota_i, \mathcal{G}'\iota_{i+1}) + \sum_{i=1}^{2} \chi(\mathcal{G}\iota_i, \mathcal{Q}(\iota_i, \iota_i)) + \sum_{j=2}^{3}\sum_{i=1}^{2} \chi(\mathcal{G}'\iota_j, \mathcal{Q}(\iota_i, \iota_i))$$

$$+\sum_{i=1}^{2}\sum_{j=2}^{3}\chi(\mathcal{G}\iota_i, \mathcal{Q}'(\iota_j, \iota_j)) + \sum_{j=2}^{3}\chi(\mathcal{G}'\iota_j, \mathcal{Q}'(\iota_j, \iota_j)) = \frac{203\kappa}{15}.$$

If $\iota \geq \kappa$, then

$$\chi(\mathcal{Q}(\iota,\varsigma), \mathcal{Q}'(\varsigma,\kappa)) = \frac{1}{15}(|3\iota - 2\kappa|) \leq (\frac{3\iota}{15}) \leq \frac{1}{15}(\frac{908\iota}{30}) = \frac{1}{15}\mathbf{A},$$

and if $\iota \leq \kappa$, then

$$\chi(\mathcal{Q}(\iota,\varsigma), \mathcal{Q}'(\varsigma,\kappa)) = \frac{1}{15}(|3\iota - 2\kappa|) \leq (\frac{2\kappa}{15}) \leq \frac{1}{15}(\frac{203\kappa}{15}) = \frac{1}{15}\mathbf{B}.$$

Thus, all the conditions of Theorem 8.5 for $k = 2$ are satisfied with

$$\sum_{i=1}^{2}\gamma_i + \sum_{i=1}^{2}\delta_i + \sum_{i=1}^{2}\sum_{j=2}^{3}\delta_{i,j} + \sum_{j=2}^{3}\sum_{i=1}^{2}\delta_{i,j} + \sum_{j=2}^{3}\delta_j = \frac{14}{15} < 1.$$

Then the mapping $\mathcal{Q}$, $\mathcal{Q}'$, $\mathcal{G}$ and $\mathcal{G}'$ have a unique Prešić-Hardy-Rogers type common fixed point.

8.6 APPLICATION

Let $\Re = C([a,b],\mathbb{R})$ be the set of real continuous functions defined on $[a,b]$. Consider the Fredholm integral equations system below:

$$\begin{cases} \iota_1(t) = \int_a^b \mathbf{A}(t,s,\iota_1(s),...,\iota_k(s))ds + g(t) \\ \iota_2(t) = \int_a^b \mathbf{A}(t,s,\iota_1(s),...,\iota_k(s))ds + g(t) \\ \vdots \\ \iota_k(t) = \int_a^b \mathbf{A}(t,s,\iota_1(s),...,\iota_k(s))ds + g(t) \end{cases} \tag{8.23}$$

for all $s,t \in [a,b]$, where $\mathbf{A} : [a,b]^2 \times \mathbb{R}^k \to \mathbb{R}$ and $g : [a,b] \to \mathbb{R}$. Let

$$\chi(\iota,\varsigma) = \max_{t\in[a,b]} |\iota(t) - \varsigma(t)|.$$

Then $(\Re, \chi)$ is a metric space. We now take this presumption into consideration:

Let for all $\iota, \varsigma \in \Re$ there exist the coefficients γ_i, δ_i and $\delta_{i,j}$ so that

$$\sum_{i=1}^{k}\gamma_i + \sum_{i=1}^{k}\delta_i + \sum_{i=1}^{k}\sum_{j=2}^{k+1}\delta_{i,j} + \sum_{j=2}^{k+1}\sum_{i=1}^{k}\delta_{i,j} + \sum_{j=2}^{k+1}\delta_j < 1$$

and

$$|\mathbf{A}(s,t,\iota_1(s),...,\iota_k(s)) - \mathbf{A}(s,t,\iota_2(s),...,\iota_{k+1}(s))|$$

$$\leq \frac{1}{b-a}\Big[\sum_{i=1}^{k}\gamma_i|\iota_i(s)-\iota_{i+1}(s)|$$

$$+\sum_{i=1}^{k}\delta_i|\iota_i(s)-\int_a^b \mathbf{A}(t,s,\iota_i(s),...,\iota_i(s))ds-g(t)|$$

$$+\sum_{i=1}^{k}\sum_{j=2}^{k+1}\delta_{i,j}|\iota_i(s)-\int_a^b \mathbf{A}(t,s,\iota_j(s),...,\iota_j(s))ds-g(t)|$$

$$+\sum_{j=2}^{k+1}\sum_{i=1}^{k}\delta_{i,j}|\iota_j(s)-\int_a^b \mathbf{A}(t,s,\iota_i(s),...,\iota_i(s))ds-g(t)|$$

$$+\sum_{j=2}^{k+1}\delta_j|\iota_j(s)-\int_a^b \mathbf{A}(t,s,\iota_j(s),...,\iota_j(s))ds-g(t)|\Big].$$

Theorem 8.6 *Assume that the aforementioned presumptions are true. The integral equation ((8.23)) then has a unique solution in* $\Re$.

Proof. We define $\mathcal{Q}:\Re^k\to\Re$ by

$$\mathcal{Q}(\iota_1,...,\iota_k)(t)=\int_a^b \mathbf{A}(t,s,\iota_1(s),...,\iota_k(s))ds+g(t),\quad \forall s,t\in[a,b].$$

So, we have

$$|\mathcal{Q}(\iota_1,...,\iota_k)-\mathcal{Q}(\iota_2,...,\iota_{k+1})|$$

$$=\Big(|\int_a^b \mathbf{A}(t,s,\iota_1(s),...,\iota_k(s))-\mathbf{A}(t,s,\iota_2(s),...,\iota_{k+1}(s))ds|\Big)$$

$$\leq\Big(\int_a^b |\mathbf{A}(t,s,\iota_1(s),...,\iota_k(s))-\mathbf{A}(t,s,\iota_2(s),...,\iota_{k+1}(s))|ds\Big)$$

$$\leq\Big(\frac{1}{b-a}\int_a^b ds.\Big[\sum_{i=1}^{k}\gamma_i|\iota_i(s)-\iota_{i+1}(s)|$$

$$+\sum_{i=1}^{k}\delta_i|\iota_i(s)-\mathcal{Q}(\iota_i,\cdots,\iota_i)(s)|$$

$$+\sum_{i=1}^{k}\sum_{j=2}^{k+1}\delta_{i,j}|\iota_i(s)-\mathcal{Q}(\iota_j,\cdots,\iota_j)(s)|$$

$$+\sum_{j=2}^{k+1}\sum_{i=1}^{k}\delta_{i,j}|\iota_j(s)-\mathcal{Q}(\iota_i,\cdots,\iota_i)(s)|$$

$$+\sum_{j=2}^{k+1}\delta_j|\iota_j(s)-\mathcal{Q}(\iota_j,\cdots,\iota_j)(s)|\Big]$$

$$
\begin{aligned}
&\leq \sum_{i=1}^{k} \gamma_i \chi(\iota_i, \iota_{i+1}) + \sum_{i=1}^{k} \delta_i \chi(\iota_i, \mathcal{Q}(\iota_i, \cdots, \iota_i)) \\
&+ \sum_{i=1}^{k} \sum_{j=2}^{k+1} \delta_{i,j} \chi(\iota_i, \mathcal{Q}(\iota_j, \cdots, \iota_j)) \\
&+ \sum_{j=2}^{k+1} \sum_{i=1}^{k} \delta_{i,j} \chi(\iota_j, \mathcal{Q}(\iota_i, \cdots, \iota_i)) + \sum_{j=2}^{k+1} \delta_j \chi(\iota_j, \mathcal{Q}(\iota_j, \cdots, \iota_j)).
\end{aligned}
$$

It follows that the mapping $\mathcal{Q}$ estimates each and every requirement of Corollary 8.6. As a result, the Fredholm integral equation ((8.23)) has a solution, since there is a unique common fixed point for the mapping $\mathcal{Q}$. □

8.7 CONCLUSION

In this chapter, the idea of Hardy-Rogers type contractions for four maps is first introduced. Additionally, we expand the idea of Hardy-Rogers type contractions for four mappings to Prešić-Hardy-Rogers type contractions for four mappings. Since the proof of the fixed point results for such mappings is very long, we summarize the proofs using a previously unexplored technique. An example and an application regarding the solvability of a class of integral equations are given to demonstrate the new methodology.

BIBLIOGRAPHY

[1] M. Abbas, V. Parvaneh, A. Razani, Periodic points of T-Ćirić generalized contraction mappings in ordered metric spaces, *Georgian Math. J.*, 19 (4) (2012), 597–610.

[2] S. Banach, Sur les opérations dans les ensembles abstraits et leur application auxéquations intégrales, *Fund. Math.*, 3 (1922), 133–181.

[3] S.K. Chatterjea, Fixed-point theorems, *Comptes Rendus de l'Académie Bulgare des Sciences*, 25 (1972), 727–730.

[4] Y.Z. Chen, A Prešić type contractive condition and its applications, *Nonlinear Anal.*, 71 (2009), 2012–2017.

[5] L.B. Ćirić, Generalized contractions and fixed-point theorems, *Publ. Inst. Math.*, 12 (26) (1971), 19–26.

[6] L.B. Ćirić, S.B. Prešić, On Prešić type generalisation of Banach contraction mapping principle, *Acta. Math. Univ. Com.*, LXXVI (2) (2007), 143–147.

[7] P. Debnath, Optimization through best proximity points for multivalued F-contractions. *Miskolc Mathematical Notes*, 22 (1) (2021), 143–151.

[8] P. Debnath, A new extension of Kannan's fixed point theorem via F-contraction with application to integral equations. *Asian-Eur. J. Math.*, 15 (07) (2022), 2250123.

[9] P. Debnath, Banach, Kannan, Chatterjea, and Reich-type contractive inequalities for multivalued mappings and their common fixed points. *Math. Meth. Appl. Sci.*, 45 (3) (2022), 1587–1596.

[10] P. Debnath, Common fixed-point and fixed-circle results for a class of discontinuous F-contractive mappings, *Mathematics*, 10 (9) (2022), 1605.

[11] S. Hadi Bonab, R. Abazari, A. Bagheri Vakilabad, Partially ordered cone metric spaces and coupled fixed point theorems via α-series, *Mathematical Anal. Contemporary Appl.*, 1 (1) (2019), 50–61.

[12] S. Hadi Bonab, R. Abazari, A. Bagheri Vakilabad, H. Hosseinzadeh, Coupled fixed point theorems on G-metric spaces via α-series, *Global Anal. Discrete Math.*, 6 (1) (2021), 1–12.

[13] S. Hadi Bonab, R. Abazari, A. Bagheri Vakilabad, H. Hosseinzadeh, Generalized metric spaces endowed with vector-valued metrics and matrix equations by tripled fixed point theorems, *J. Inequal. Appl.*, 2014 (2020), 1–16.

[14] G. E. Hardy, T.D. Rogers, A generalization of a fixed point theorem of Reich, *Can. Math. Bull.*, 16 (1973), 201–206.

[15] H. Hosseinzadeh, S. Hadi Bonab, Kh. Amini Sefidab, Some common fixed point theorems for four mapping in generalized metric spaces, *Thai J. Math.*, 20 (1) (2022), 425–437.

[16] R. Kannan, Some results on fixed points. II, *Am. Math. Month.*, 76 (4) (1969), 405–408.

[17] N. Konwar and P. Debnath, Fixed point results for a family of interpolative F-contractions in b-metric spaces. *Axioms*, 11 (11) (2022), 621.

[18] A. Latif, T. Nazir, M. Abbas, Fixed point results for multivalued Prešic type weakly contractive mappings, *Mathematics*, 7 (7) (2019), 601.

[19] B. Mohammadi, V. Parvaneh, H. Aydi, On extended interpolative Ciric-Reich-Rus type F-contractions and an application, *J. Inequal. Appl.*, 290 (2019).

[20] Z. Mustafa, V. Parvaneh, M. Abbas, J.R. Roshan, Some coincidence point results for generalized (ψ, φ)-weakly contractive mappings in ordered G-metric spaces. *Fixed Point Theory Appl.*, 2013 (326) (2013). https://doi.org/10.1186/1687-1812-2013-326.

[21] Z. Mustafa, V. Parvaneh, J.R. Roshan, Z. Kadelburg, b_2-Metric spaces and some fixed point theorems, *Fixed Point Theory Appl.*, 2014 (144) (2014). https://doi.org/10.1186/1687-1812-2014-144.

[22] J. Olaleru, V. Olisama, M. Abbas, Coupled best proximity points of generalised Hardy-Rogers type cyclic (w)-contraction mappings, *Int. J. Math. Sci. Optim.: Theor. Appl.*, 2015 (2015), 33–54.

[23] V. Parvaneh, F. Golkarmanesh, R. George, Fixed points of Wardowski-ciric-Presic type contractive mappings in a partial rectangular b-metric space, *J. Math. Anal.*, 8 (1) (2017), 183–201.

[24] V. Parvaneh, S. Hadi Bonab, H. Hosseinzadeh, H. Aydi, A tripled fixed point theorem in C^*-algebra-valued metric spaces and application in integral equations, *Adv. Math. Phys.*, 2021 (2021), 1–6.

[25] S.B. Prešić, Sur une classe d'inèquations aux differences finies et sur la convergence de certaines suites, *Pub. de. l'Institut Math. Belgrade.*, 5 (19) (1965), 75–78.

[26] S. Reich, Some remarks concerning contraction mappings, *Canad. Math. Bull.*, 14 (1971), 121–124.

[27] S. Shukla, S. Radenović, Some generalizations of Prešić type mappings and applications, *An. Stiint. Univ.'Al. I. Cuza'Iasi, Mat.*, (2014), https://doi.org/10.1515/aicu-2015-0026.

[28] S. Shukla, S. Radenović, S. Pantelić, Some fixed point theorems for Prešić-Hardy-Rogers type contractions in metric spaces, *J. Math.*, 2013 (2013), 1–8.

[29] S. Shukla, S. Radojević, Z. A. Veljković, S. Radenović, Some coincidence and common fixed point theorems for ordered Prešić-Reich type contractions, *J. Inequal. Appl.*, 2013 (520) (2013).

CHAPTER 9

Fixed Point Method: Ulam Stability of Mixed Type Functional Equation in β-Banach Modules

K. Tamilvanan
R.M.K. Engineering College

N. Revathi
Periyar University Centre for PG and Research Studies

S. A. Mohiuddine
King Abdulaziz University

CONTENTS

9.1 INTRODUCTION

The study of stability problems for functional equations is one of the most significant research areas in mathematics, which originated in issues related to applied mathematics. The first question concerning the stability of homomorphisms was given by Ulam [24] as follows.

Given a group $(G, *)$, a metric group $(G', \cdot)$ with the metric d, and a mapping f from G and G', does $\delta > 0$ exist such that

$$d(f(x * y), f(x) \cdot f(y)) \leq \delta$$

DOI: 10.1201/9781003388678-9

for all $x, y \in G$? If such a mapping exists, then does a homomorphism $g : G \to G'$ exist such that

$$d(f(x), g(x)) \leq \epsilon$$

for all $x \in G$? Ulam defined such a problem in 1940 and solved it the following year for the Cauchy functional equation

$$f(x+y) = f(x) + f(y)$$

by the method of Hyers [16]. By permitting the Cauchy difference operator to be controlled by $\epsilon(\|u\|^p + \|v\|^p)$, Aoki [4] for approximate additive mappings and Th. M. Rassias [26] for approximate linear mappings generalized Hyers' results. A generalization of Rassias' theorem was achieved in 1994 by Găvruţa [14], who took the place of $\epsilon(\|u\|^p + \|v\|^p)$ is a general control function $\varphi(u, v)$.

The functional equations

$$f(v+w) = f(v) + f(w)$$

and

$$f(v+w) + f(v-w) = 2f(v) + 2f(w)$$

is called the additive functional equation and quadratic functional equation, respectively. In particular, each additive and quadratic solution of functional equations shall be an additive function and a quadratic function.

The stability characteristics of various functional equations can be used for domains that are unrelated, it is important to remember. For instance, Zhou [27] used the stability of the functional equation $f(u-v)+f(u+v) = 2f(u)+f(v)$ to demonstrate a Ditzian conjecture regarding the correlation between a mappings smoothness and the degree of its approximation by the related Bernstein polynomials.

These stability findings can be used in stochastic analysis, mathematical finance and statistics, as well as in sociology and psychology. There have recently been a number of additional, fascinating analyses, modifications, extensions, and generalizations of the original Ulam problem presented (see, for instance, [13, 17, 1, 24, 20, 21, 23, 25, 26]). For more recent results in this context, we refer to [7, 8, 9, 10, 11, 12].

In this work, we examine the Ulam stability results of the mixed type additive-quadratic functional equation

$$\begin{aligned} f\left(\sum_{1\leq a\leq m} as_a\right) + \sum_{1\leq a\leq m} f\left(-as_a + \sum_{b=1;a\neq b}^{m} bs_b\right) &= (m-3)\sum_{1\leq a<b\leq m} f\left(as_a + bs_b\right) \\ &\quad - \left(m^2 - 5m + 2\right)\sum_{1\leq a\leq m} a^2\left[\frac{f(s_a)+f(-s_a)}{2}\right] \end{aligned}$$

$$- (m^2 - 5m + 4) \sum_{1 \le a \le m} a \left[\frac{f(s_a) - f(-s_a)}{2} \right] \tag{9.1}$$

where $f(0) = 0$, and m is a non-negative integer with $m > 4$ in β-Banach modules with the help of fixed point approach. We here discuss our results for odd mapping (additive functional equation) and even mapping (quadratic functional equation) as well as discuss the stability of mixed case (quadratic-additive functional equation). We provide some applications in which the stability of mixed type quadratic-additive functional equation can be controlled by sums and products of powers of norms.

Here, using a fixed point approach for three cases, we examine the stability (in the sense of Ulam stability) of (9.1) in β-Banach modules. We can further divide this section into three subsections. We obtain the stability results for the odd case in Section 9.2.1, the even case in Section 9.2.2, and the major results of the function equation (9.1) for the mixed case in Section 9.2.3.

Let us assume Ψ^* is a unital Banach algebra with $\|\cdot\|_{\Psi^*}$, $\Psi_1^* = \{v \in \Psi^* | \ \|v\|_{\Psi^*} = 1\}$, M is a β-normed left Banach Ψ^*-module and B is a β-normed left Ψ^*-module. And also consider $\mathbb{K}$ refers either $\mathbb{R}$ or $\mathbb{C}$ and a real number β with $0 < \beta \le 1$. To proceed with our primary findings, we can instantly use the definition of β-normed space in [5].

Theorem 9.1 *[9, 2, 3, 22] Let (Λ, d) be a complete generalized metric space and a strictly contractive mapping $\Phi : \Lambda \to \Lambda$ with $0 < L < 1$,*

$$i.e., \ d(\Phi s_1, \Phi s_2) \le L d(s_1, s_2), \quad \forall \ s_1, s_2 \in \Lambda.$$

Then for each given $s \in \Lambda$, either

$$d(\Phi^m s, \Phi^{m+1} s) = \infty, \quad \forall \ m \ge 0,$$

or there exist an integer $m_0 > 0$ such that

(1) $d(\Phi^m s, \Phi^{m+1} s) < \infty, \quad \forall \ m \ge m_0$;

(2) the sequence $\{\Phi^m s\}$ converges to a fixed point t^* of Φ;

(3) t^* is a unique fixed point of Φ in $\Lambda^* = \{t \in \Lambda \ | d(\Phi^{m_0} s, t) < \infty\}$;

(4) $d(t, t^*) \le \frac{1}{1-L} d(t, \Phi t), \quad \forall \ t \in \Lambda^*$.

Theorem 9.2 *[20] If an odd mapping $f : B \to M$ satisfies the functional equation (9.1), then the mapping f is additive.*

Theorem 9.3 *[20] If an even mapping $f : B \to M$ satisfies the functional equation (9.1), then the mapping f is quadratic.*

Theorem 9.4 *[20] If a mapping $f : B \to M$ satisfies $f(0) = 0$ and the functional equation (9.1) for all $s_1, s_2, \cdots, s_m \in B$ if and only if there exists a mapping $Q : B \times B \to M$ which is symmetric bi-additive and a mapping $A : B \to M$ is additive such that $f(s) = Q(s, s) + A(s)$ for all s in B.*

9.2 MAIN RESULTS

We use the abbreviations for the mapping $f : B \to M$:

$$\begin{aligned}
&D_v f(s_1, s_2, \cdots, s_m) \\
&:= f\left(\sum_{1\leq a\leq m} as_a\right) + \sum_{1\leq a\leq m} f\left(-as_a + \sum_{b=1;a\neq b}^{m} bs_b\right) \\
&- (m-3) \sum_{1\leq a<b\leq m} f\left(as_a + bs_b\right) \\
&\qquad + \left(m^2 - 5m + 2\right) \sum_{1\leq a\leq m} a^2 \left[\frac{f(s_a) + f(-s_a)}{2}\right] \\
&+ \left(m^2 - 5m + 4\right) \sum_{1\leq a\leq m} a \left[\frac{f(s_a) - f(-s_a)}{2}\right]
\end{aligned}$$

for all $s_1, s_2, \cdots, s_m \in B$ and $v \in \Psi_1^*$.

9.2.1 Stability Results: When f Is Odd

Theorem 9.5 *Let a mapping $\chi : B^m \to [0, \infty)$ such that*

$$\lim_{l\to\infty} \frac{1}{|2|^{l\beta}} \psi\left(2^l s_1, 2^l s_2, \cdots, 2^l s_m\right) = 0, \tag{9.1}$$

for all $s_1, s_2, \cdots, s_m \in B$. Let an odd mapping $f : B \to M$ such that

$$\| D_v f\left(s_1, s_2, \cdots, s_m\right) \|_\beta \leq \chi\left(s_1, s_2, \cdots, s_m\right), \tag{9.2}$$

for all $s_1, s_2, \cdots, s_m \in B$, and $v \in \Psi_1^*$. If there exists $0 < L < 1$ such that

$$s \to \varrho(s) = \frac{\chi\left(0, s, 0, \cdots, 0\right)}{\left(m^2 - 5m + 4\right)}$$

and

$$\varrho(2s) \leq |2|^\beta L \varrho(s), \tag{9.3}$$

for all $s \in B$, then there exists a unique additive mapping $A : B \to M$ satisfying

$$\|A(s) - f(s)\|_\beta \leq \frac{\varrho(s)}{|2|^\beta - |2|^\beta L}, \tag{9.4}$$

for all $s \in B$. Moreover, if $f(ns)$ is continuous in $n \in \mathbb{R}$ for all $s \in B$, then A is Ψ^*-linear.

Proof. Take $v = 1$. Replacing $(s_1, s_2, \cdots, s_m)$ by $(0, s, 0, \cdots, 0)$ in (9.2), we have

$$\left\|2(m^2 - 5m + 4) f(s) - (m^2 - 5m + 4) f(2s)\right\|_\beta \ \leq \ \chi\left(0, s, 0, \cdots, 0\right)$$

$$\Rightarrow \left\| f(s) - \frac{f(2s)}{2} \right\|_{\beta} \leq L\varrho(s), \tag{9.5}$$

for all $s \in B$. Consider

$$\Lambda := \{u | u : B \to M, u(0) = 0\}$$

and define the generalized metric on Λ as below:

$$d(u, r) = \inf\{c \in [0, \infty) | \|u(s) - r(s)\|_{\beta} \leq c\varrho(s), \quad s \in B\}. \tag{9.6}$$

It is simple to demonstrate that (Λ, d) is a complete generalized metric space (see [5]).

Next, we define a mapping $\Phi : \Lambda \to \Lambda$ by

$$(\Phi u)(s) = \frac{1}{2} u(2s), \tag{9.7}$$

for all $u \in \Lambda$ and $s \in B$. Let $u, r \in \Lambda$ and c be an arbitrary constant with $c \in [0, \infty]$ with $d(u, r) < c$. By definition of d, we have

$$\|u(s) - r(s)\|_{\beta} \leq c\varrho(s), \tag{9.8}$$

for all $s \in B$. By the given hypothesis and the last inequality (9.8), one has

$$\left\| \frac{1}{2} u(2s) - \frac{1}{2} r(2s) \right\|_{\beta} \leq cL\varrho(s), \tag{9.9}$$

for all $s \in B$. Hence,

$$d(\Phi u, \Phi r) \leq L d(u, r).$$

From inequality (9.5), we get

$$d(\Phi f, f) \leq \frac{1}{|2|^{\beta}}.$$

By Theorem 9.1, Φ has a unique fixed point $A : B \to M$ in $\lambda^* = \{u \in \Lambda |\ d(u, r) < \infty\}$ satisfies

$$A(s) := \lim_{l \to \infty} (\Phi^l f)(s) = \lim_{l \to \infty} \frac{1}{2^l} f\left(2^l s\right) \tag{9.10}$$

and $A(2s) = 2A(s)$ for all $s \in B$. Also,

$$\begin{aligned} d(A, f) &\leq \frac{1}{1 - L} d(\Phi f, f) \\ &\leq \frac{1}{|2|^{\beta} - |2|^{\beta} L}. \end{aligned} \tag{9.11}$$

Thus, inequality (9.4) valid for all $s \in B$.

Now, we want to prove that A is an additive function. Using inequalities (9.1), (9.2) and (9.10), we have

$$\begin{aligned}\|D_1 A(s_1, s_2, \cdots, s_m)\|_\beta &= \lim_{l\to\infty} \frac{1}{|2|^{l\beta}} \|D_1 f(2^l s_1, 2^l s_2, \cdots, 2^l s_m)\|_\beta \\ &\leq \lim_{l\to\infty} \frac{1}{|2|^{l\beta}} \chi(2^l s_1, 2^l s_2, \cdots, 2^l s_m) = 0,\end{aligned}$$

that is,

$$\begin{aligned} f\left(\sum_{1\leq a\leq m} a s_a\right) + \sum_{1\leq a\leq m} f\left(-a s_a + \sum_{b=1; a\neq b}^{m} b s_b\right) &= (m-3) \sum_{1\leq a<b\leq m} f(a s_a + b s_b) \\ &\quad - (m^2 - 5m + 4) \sum_{1\leq a\leq m} a f(s_a)\end{aligned}$$

for all $s_1, s_2, \cdots, s_m \in B$. By Theorem 9.2, the mapping A is odd.

Next, we want to prove that the function A is unique. Let us consider an another odd mapping $A^{'} : B \to M$ such that (9.4). Since

$$d(f, A^{'}) \leq \frac{1}{|2|^\beta (1 - L)}$$

and the function $A^{'}$ is additive, we obtain $A^{'} \in \Lambda^*$ and $(\Phi A^{'})(s) = \frac{1}{2} A^{'}(2s) = A(s)$ for all $s \in B$. That is, $A^{'}$ is a fixed point of Φ in Λ^*. Hence, $A^{'} = A$.

Moreover, if $f(ns)$ is continuous in $n \in \mathbb{R}$ for all $s \in B$, then by the proof of [26], A is $\mathbb{R}$-linear. Replacing $(s_1, s_2, \cdots, s_m)$ by $(0, s, 0, \cdots, 0)$ in (9.2), we have

$$\|(m^2 - 5m + 4)) f(2vs) - 2(m^2 - 5m + 4) v f(s)\|_\beta \leq \psi(0, s, 0, \cdots, 0) \qquad (9.12)$$

for all $s \in B$ and all $v \in \Psi_1^*$. Thus, using definition of A, inequalities (9.1) and (9.12), we get

$$\begin{aligned}&\|(m^2 - 5m + 4) A(2vs) - 2(m^2 - 5m + 4) v A(s)\|_\beta \\ &= \lim_{l\to\infty} \frac{1}{|2|^{l\beta}} \|(m^2 - 5m + 4) f(2^{l+1} vs) - 2(m^2 - 5m + 4) v f(s)\|_\beta \\ &\leq \lim_{l\to\infty} \frac{1}{|2|^{l\beta}} \chi(0, 2^l s, 0, \cdots, 0) = 0\end{aligned}$$

for all $s \in B$ and all $v \in \Psi_1^*$. So,

$$(m^2 - 5m + 4) A(2vs) - 2(m^2 - 5m + 4) v A(s) = 0$$

for all $s \in B$ and all $v \in \Psi_1^*$. Since A is additive, we get $A(vs) = vA(s)$ for all $s \in B$ and all $v \in \Psi_1^* \cup \{0\}$. Since A is $\mathbb{R}$-linear, let $v \in \Psi^* \backslash \{0\}$.

$$A(vs) = A\left(\|v\|_{\Psi^*} \cdot \frac{v}{\|v\|_{\Psi^*}} s\right)$$

$$\begin{aligned} &= \|v\|_{\Psi^*} \cdot A\left(\frac{v}{\|v\|_{\Psi^*}} s\right) \\ &= \|v\|_{\Psi^*} \cdot \frac{v}{\|v\|_{\Psi^*}} A(s) \\ &= vA(s), \qquad s \in B, \quad v \in \Psi^*. \end{aligned}$$

Hence, A is Ψ^*-linear. □

Corollary 9.1 *If an odd mapping $f : B \to M$ such that*

$$\|D_v f(s_1, s_2, \cdots, s_m)\|_\beta \leq \delta + \epsilon\left(\sum_{i=1}^{m} \|s_i\|_\beta^t\right), \tag{9.13}$$

for all $s_1, s_2, \cdots, s_m \in B$, and $v \in \Psi_1^*$, then there exists a unique additive mapping $A : B \to M$ satisfies

$$\|f(s) - A(s)\|_\beta \leq \frac{\left(\delta + \epsilon\|s\|_\beta^t\right)}{(m^2 - 5m + 4)\left(|2|^\beta - |2|^{\beta t}\right)},$$

where $0 < t < 1$, $\delta, \epsilon \in [0, \infty)$ and for all $s \in \Psi$. Moreover, if $f(ns)$ is continuous in $n \in \mathbb{R}$ for all $s \in B$, then A is Ψ^*-linear.

Proof. By letting

$$\chi(s_1, s_2, \cdots, s_m) = \delta + \epsilon\left(\sum_{i=1}^{m} \|s\|_\beta^t\right)$$

and $L = |2|^{\beta(t-1)}$ in Theorem 9.5, we have needed result. □

Corollary 9.2 *Let $t > 0$ such that $mt < 1$ and $\delta, \epsilon \in \mathbb{R}^+$, and an odd mapping $f : B \to M$ such that*

$$\|D_v f(s_1, s_2, \cdots, s_m)\|_\beta \leq \delta + \epsilon\left[\prod_{i=1}^{m} \|s_i\|_\beta^t + \sum_{i=1}^{m} \|s_i\|_\beta^{mt}\right], \quad s_1, s_2, \cdots, s_m \in B,$$

and $v \in \Psi_1^*$, then there exists a unique additive mapping $A : B \to M$ satisfying

$$\|f(s) - A(s)\|_\beta \leq \frac{\left(\delta + \epsilon\|s\|_\beta^{mt}\right)}{(m^2 - 5m + 4)\left(|2|^\beta - |2|^{\beta mt}\right)}, \tag{9.14}$$

for all $s \in B$. Moreover, if $f(ns)$ is continuous in $n \in \mathbb{R}$ for all $s \in B$, then A is Ψ^*-linear.

Proof. By letting

$$\chi(s_1, s_2, \cdots, s_m) = \delta + \epsilon \left[\prod_{i=1}^{m} \|s_i\|_\beta^t + \sum_{i=1}^{m} \|s_i\|_\beta^{mt} \right]$$

and $L = |2|^{\beta(mt-1)}$ in Theorem 9.5, we obtain our desired result. □

Theorem 9.6 *Let an odd mapping* $\chi : B^m \to [0, \infty)$ *such that*

$$\lim_{l \to \infty} |2|^{l\beta} \chi\left(2^{-l}s_1, 2^{-l}s_2, \cdots, 2^{-l}s_m\right) = 0 \tag{9.15}$$

for all $s_1, s_2, \cdots, s_m \in B$. *If an odd mapping* $f : B \to M$ *satisfies* (9.2). *If there exists* $0 < L < 1$ *such that*

$$s \to \varrho(s) = \frac{\chi(0, s, 0, \cdots, 0)}{(m^2 - 5m + 4)}$$

and

$$\varrho(s) \leq |2|^{-\beta} L \varrho(2s) \tag{9.16}$$

for all $s \in B$, *then there exists a unique additive mapping* $A : B \to M$ *satisfying*

$$\|f(s) - A(s)\|_\beta \leq \frac{L}{|2|^\beta - |2|^\beta L} \varrho(s), \tag{9.17}$$

for all $s \in B$. *Moreover, if* $f(ns)$ *is continuous in* $n \in \mathbb{R}$ *for all* $s \in B$, *then* A *is* Ψ^**-linear.*

Proof. Take $v = 1$ and replacing $(s_1, s_2, \cdots, s_m)$ by $(0, s, 0, \cdots, 0)$ in (9.2), we have

$$\left\|2(m^2 - 5m + 4)f(s) - (m^2 - 5m + 4)f(2s)\right\|_\beta \leq \chi(0, s, 0, \cdots, 0), \tag{9.18}$$

for all $s \in B$. Replacing s by $\frac{s}{2}$ in (9.18), we obtain

$$\left\|2f\left(\frac{s}{2}\right) - f(s)\right\|_\beta \leq L\varrho(s) \tag{9.19}$$

for all $s \in B$. Let us define the set

$$\Lambda := \{u | u : B \to M, u(0) = 0\}$$

and the generalized metric on Λ as

$$d(u, r) = \inf\{c \in [0, \infty) | \|u(s) - r(s)\|_\beta \leq c\varrho(s), \quad \forall\, s \in B\}. \tag{9.20}$$

Clearly, (Λ, d) is a complete generalized metric space (see [5]).

Next, we can define a mapping $\Phi : \Lambda \to \Lambda$ by

$$(\Phi u)(s) = 2u\left(\frac{s}{2}\right), \tag{9.21}$$

for all $u \in \Lambda$ and all $s \in B$. Let $u, r \in \Lambda$ and an arbitrary constant $c \in [0, \infty]$ with $d(u, r) < c$. Using the definition of d, we obtain

$$\|u(s) - r(s)\|_\beta \le c\varrho(s), \tag{9.22}$$

for all $s \in B$. By the given hypothesis and the above inequality, we have

$$\left\|2u\left(\frac{s}{2}\right) - 2r\left(\frac{s}{2}\right)\right\|_\beta \le cL\varrho(s), \tag{9.23}$$

for all $s \in B$. Hence,

$$d(\Phi u, \Phi r) \le Ld(u, r).$$

From inequality (9.19), we have

$$d(\Phi f, f) \le \frac{L}{|2|^\beta}.$$

From Theorem 9.1, F has a unique fixed point $A : B \to M$ in $\Lambda^* = \{u \in \Lambda | \; d(u, r) < \infty\}$ satisfies

$$A(s) := \lim_{l\to\infty} (\Phi^l f)(s) = \lim_{l\to\infty} 2^l f\left(\frac{s}{2^l}\right) \tag{9.24}$$

and $A\left(\frac{s}{2}\right) = \frac{1}{2}A(s)$ for all $s \in B$. Also,

$$\begin{aligned} d(A, f) &\le \frac{1}{1-L} d(\Phi f, f) \\ &\le \frac{L}{|2|^\beta - |2|^\beta L}. \end{aligned} \tag{9.25}$$

Thus, condition (9.17) hold for all $s \in B$. Next, we want to prove that the function A is additive. Using inequalities (9.15), (9.2) and (9.24), we obtain

$$\begin{aligned} \|D_1A(s_1, s_2, \cdots, s_m)\|_\beta &= \lim_{l\to\infty} |2|^{l\beta} \|D_1 f\left(\frac{s_1}{2^l}, \frac{s_2}{2^l}, \cdots, \frac{s_m}{2^l}\right)\|_\beta \\ &\le \lim_{l\to\infty} |2|^{l\beta} \chi\left(\frac{s_1}{2^l}, \frac{s_2}{2^l}, \cdots, \frac{s_m}{2^l}\right) = 0, \end{aligned}$$

for all $s_1, s_2, \cdots, s_m \in B$. Thus, by Theorem 9.2, the function A is odd. Now, we need to prove that the function A is unique. Consider an another additive mapping $A' : B \to M$ satisfies (9.17). As a result,

$$d(f, A') \le \frac{L}{(1-L)|2|^\beta}$$

and the function A' is additive, we obtain $A' \in \Lambda^*$ and $(\Phi A')(s) = 2A'\left(\frac{s}{2}\right) = A(s)$ for all $s \in B$. That is, A' is a fixed point of Φ in Λ^*. Clearly, $A' = A$.

Moreover, if $f(ns)$ is continuous in $n \in \mathbb{R}$ for all $s \in B$, then by using proof of [26], A is $\mathbb{R}$-linear. Replacing $(s_1, s_2, \cdots, s_m)$ by $\left(0, \frac{s}{2}, 0, \cdots, 0\right)$ in (9.2), we have

$$\|(m^2 - 5m + 4)f(vs) - 2(m^2 - 5m + 4)vf\left(\frac{s}{2}\right)\|_\beta$$

$$\leq \chi\left(0, \frac{s}{2}, 0, \cdots, 0\right) \tag{9.26}$$

for all $s \in B$ and all $v \in \Psi_1^*$. By using definition of A, inequalities (9.15) and (9.26), we have

$$\begin{aligned}
&\|(m^2 - 5m + 4)A(vs) - 2(m^2 - 5m + 4)vA\left(\frac{s}{2}\right)\|_\beta \\
&= \lim_{l\to\infty} |2|^{l\beta}\|((m^2 - 5m + 4)f\left(\frac{vs}{2^l}\right) - 2(m^2 - 5m + 4)vf\left(\frac{s}{2^{l+1}}\right)\|_\beta \\
&\leq \lim_{l\to\infty} |2|^{l\beta}\chi\left(0, \frac{s}{2^{l+1}}, 0, \cdots, 0\right) = 0
\end{aligned}$$

for all $s \in B$ and all $v \in \Psi_1^*$. So,

$$(m^2 - 5m + 4)A(vs) - 2(m^2 - 5m + 4)vA\left(\frac{s}{2}\right) = 0$$

for all $s \in B$ and all $v \in \Psi_1^*$. Since, the function A is additive, we obtain $A(vs) = vA(s)$, for all $s \in B$ and all $v \in \Psi_1^* \cup \{0\}$. Since A is $\mathbb{R}$-linear, let $v \in \Psi^*\backslash\{0\}$, then

$$A(vs) = vA_1(s),$$

for all $s \in B$ and $v \in \Psi^*$. Hence, the additive function A is Ψ^*-linear. □

As the applications of Theorem 9.6, one can get the following Corollaries 9.3 and 9.4

Corollary 9.3 *If an odd mapping $f : B \to M$ such that*

$$\|D_v f(s_1, s_2, \cdots, s_m)\|_\beta \leq \gamma\left(\sum_{i=1}^m \|s_i\|_\beta^t\right), \tag{9.27}$$

for all $s_1, s_2, \cdots, s_m \in B$ and $v \in \Psi_1^*$, then there exists a unique additive mapping $A : B \to M$ satisfies

$$\|f(s) - A(s)\|_\beta \leq \frac{\epsilon\|s\|_\beta^t}{(m^2 - 5m + 4)\left(|2|^{\beta t} - |2|^\beta\right)},$$

where $t > 1$ and $\epsilon \in \mathbb{R}^+$, for all $s \in B$. Moreover, if $f(ns)$ is continuous in $n \in \mathbb{R}$ for all $s \in B$, then the function A is Ψ^*-linear.

Proof. By letting

$$\chi(s_1, s_2, \cdots, s_m) = \epsilon\left(\sum_{i=1}^m \|s\|_\beta^t\right)$$

and $L = |2|^{\beta(1-t)}$ in Theorem 9.6, we obtain our needed result. □

Corollary 9.4 *If an odd mapping $f : B \to M$ such that*

$$\|D_v f(s_1, s_2, \cdots, s_m)\|_\beta \leq \epsilon \left[\prod_{i=1}^{m} \|s_i\|_\beta^t + \sum_{i=1}^{m} \|s_i\|_\beta^{mt}\right]$$

for all $s_1, s_2, \cdots, s_m \in B$ and $r \in \Psi_1^*$, then there exists a unique additive mapping $A : B \to M$ satisfying

$$\|f(s) - A(s)\|_\beta \leq \frac{\epsilon \|s\|_\beta^{mt}}{(m^2 - 5m + 4)\left(|2|^{\beta mt} - |2|^\beta\right)}, \tag{9.28}$$

where $t > 0$ and $\epsilon \in \mathbb{R}^+$ with $mt > 1$, for all $s \in B$. Moreover, if $f(ns)$ is continuous in $n \in \mathbb{R}$ for all $s \in B$, then the function A is Ψ^*-linear.

Proof. By letting

$$\chi(s_1, s_2, \cdots, s_m) = \epsilon \left[\prod_{i=1}^{m} \|s_i\|_\beta^t + \sum_{i=1}^{m} \|s_i\|_\beta^{mt}\right]$$

and $L = |2|^{\beta(1-mt)}$ in Theorem 9.6, we obtain our needed result. □

9.2.2 Stability Results: When f Is Even

Theorem 9.7 *Let a mapping $\chi : B^m \to [0, \infty)$ such that*

$$\lim_{l \to \infty} \frac{1}{|2|^{2l\beta}} \chi\left(2^l s_1, 2^l s_2, \cdots, 2^l s_m\right) = 0 \tag{9.29}$$

for all $s_1, s_2, \cdots, s_m \in B$. Let an even mapping $f : B \to M$ with $f(0) = 0$ such that (9.2). If there exists $0 < L < 1$ such that

$$s \to \varrho(s) = \frac{\chi(0, s, 0, \cdots, 0)}{(m^2 - 5m + 2)},$$

and

$$\varrho(2s) \leq |2|^{2\beta} L \varrho(s), \tag{9.30}$$

for all $s \in B$, then there exists a unique quadratic mapping $Q : B \to M$ satisfying

$$\|f(s) - Q(s)\|_\beta \leq \frac{\varrho(s)}{|2|^{2\beta} - |2|^{2\beta} L}, \tag{9.31}$$

for all $s \in B$. Moreover, if $f(ns)$ is continuous in $n \in \mathbb{R}$ for all $s \in B$, then the mapping Q is Ψ^*-quadratic.

i.e., $Q_2(sv) = s^2 Q_2(v)$ for all $v \in V$ and all $s \in B^*$.

Proof. Setting $v = 1$ and replacing $(s_1, s_2, \cdots, s_m)$ by $(0, s, 0, \cdots, 0)$ in (9.2), we have

$$\begin{aligned} \left\|(m^2-5m+2)f(2s) - 2^2(m^2-5m+2)f(s)\right\|_\beta &\leq \chi(0,s,0,\cdots,0) \\ \left\|\frac{f(2s)}{2^2} - f(s)\right\|_\beta &\leq L\varrho(s), \end{aligned} \tag{9.32}$$

for all $s \in B$. Consider the set $\Lambda := \{u | u : B \to M, u(0) = 0\}$ and define the generalized metric on Λ as below:

$$d(u,r) = \inf\{c \in [0,\infty) \mid \|u(s) - r(s)\|_\beta \leq c\varrho(s), \ \ \forall\, s \in B\}. \tag{9.33}$$

Clearly, (Λ, d) is a complete generalized metric space (see [5]). We can define the mapping $\Phi : \Lambda \to \Lambda$ by

$$(\Phi u)(s) = \frac{1}{2^2}u(2s), \tag{9.34}$$

for all $u \in \Lambda$ and $s \in B$. Let $u, r \in \Lambda$ and $c \in [0,\infty]$ be an arbitrary constant with $d(u,r) < c$. By the definition of d, we obtain

$$\|u(s) - r(s)\|_\beta \leq c\varrho(s), \tag{9.35}$$

for all $s \in B$. By the given hypothesis and the last inequality, one has

$$\left\|\frac{1}{2^2}u(2s) - \frac{1}{2^2}r(2s)\right\|_\beta \leq cL\varrho(s), \tag{9.36}$$

for all $s \in B$. Hence,

$$d(\Phi u, \Phi r) \leq Ld(u,r).$$

By using inequality (9.32) that

$$d(\Phi f, f) \leq \frac{1}{|2|^{2\beta}}.$$

By Theorem 9.1, Φ has a unique fixed point $Q : B \to M$ in $\Lambda^* = \{u \in \Lambda |\ d(u,r) < \infty\}$ such that

$$Q(s) := \lim_{l\to\infty} (\Phi^l f)(s) = \lim_{l\to\infty} \frac{1}{2^{2l}} f\left(2^l s\right) \tag{9.37}$$

and $Q(2s) = 2^2 Q(s)$ for all $s \in B$. Also,

$$\begin{aligned} d(Q,f) &\leq \frac{d(\Phi f, f)}{1-L} \\ &\leq \frac{1}{|2|^{2\beta} - |2|^{2\beta}L}. \end{aligned} \tag{9.38}$$

Thus, inequality (9.31) holds for all $s \in B$. Next, we want to prove that the function Q is quadratic. By inequalities (9.29), (9.2) and (9.37), we obtain

$$\|D_1 Q(s_1, s_2, \cdots, s_m)\|_\beta = \lim_{l\to\infty} \frac{1}{|2|^{2l\beta}} \|D_1 f\left(2^l s_1, 2^l s_2, \cdots, 2^l s_m\right)\|_\beta$$

$$\leq \lim_{l\to\infty} \frac{1}{|2|^{2l\beta}} \chi\left(2^l s_1, 2^l s_2, \cdots, 2^l s_m\right) = 0,$$

that is,

$$\begin{aligned} f\left(\sum_{1\leq a\leq m} as_a\right) + \sum_{1\leq a\leq m} f\left(-as_a + \sum_{b=1;a\neq b}^{m} bs_b\right) \\ = (m-3) \sum_{1\leq a<b\leq m} f\left(as_a + bs_b\right) \\ - \left(m^2 - 5m + 2\right) \sum_{1\leq a\leq m} a^2 f(s_a) \end{aligned}$$

for all $s_1, s_2, \cdots, s_m \in B$. By Theorem 9.3, the function Q is even. Next, we want to prove that the function Q is unique. Suppose an another quadratic mapping $Q^{'}: B \to M$ satisfies the inequality (9.31). Then,

$$d(f, Q^{'}) \leq \frac{1}{|2|^{2\beta} - |2|^{2\beta} L}$$

and the function $Q^{'}$ is quadratic, which implies $Q^{'} \in \Lambda^*$ and $(\Phi Q^{'})(s) = \frac{1}{2^2} Q^{'}(2s) = Q(s)$ for all $s \in B$, i.e., $Q^{'}$ is a fixed point of Φ in Λ^*. Hence, $Q^{'} = Q$.

Moreover, if $f(ns)$ is continuous in $n \in \mathbb{R}$ for all $s \in B$, then using the proof of [26], the function Q is $\mathbb{R}$-quadratic.

Replacing $(s_1, s_2, \cdots, s_m)$ by $(0, s, 0, \cdots, 0)$ in (9.2), we have

$$\begin{aligned} \|(m^2 - 5m + 2) f(2vs) - 2^2 (m^2 - 5m + 2) v^2 f(s)\|_\beta \\ \leq \chi(0, s, 0, \cdots, 0), \end{aligned} \tag{9.39}$$

for all $s \in B$ and all $v \in \Psi_1^*$. Using definition of Q, (9.29) and (9.39), we obtain

$$\begin{aligned} &\|(m^2 - 5m + 2) Q(2vs) - (m^2 - 5m + 2) v^2 Q(s)\|_\beta \\ &= \lim_{l\to\infty} \frac{1}{|2|^{l\beta}} \|(m^2 - 5m + 2) f(2^{m+1} vs) - (m^2 - 5m + 2) v^2 f(2^m s)\|_\beta \\ &\leq \lim_{l\to\infty} \frac{1}{|2|^{l\beta}} \chi\left(0, 2^l s, 0, \cdots, 0\right) = 0, \end{aligned}$$

for all $s \in B$ and all $v \in \Phi_1^*$. So,

$$(m^2 - 5m + 2) Q(2vs) - (m^2 - 5m + 2) v^2 Q(s) = 0,$$

for all $s \in B$ and all $v \in \Psi_1^*$. Since, the function Q is quadratic, we have $Q(vs) = v^2 Q(s)$, for all $s \in B$ and all $v \in \Psi_1^* \cup \{0\}$. Since the function Q is $\mathbb{R}$-quadratic, let $v \in \Psi^* \backslash \{0\}$,

$$Q(vs) = Q\left(\|v\|_{\Psi^*} \cdot \frac{v}{\|v\|_{\Psi^*}} s\right)$$

$$
\begin{aligned}
&= \|v\|^2_{\Psi^*} \cdot Q\left(\frac{v}{\|v\|_{\Psi^*}} s\right) \\
&= \|v\|^2_{\Psi^*} \cdot \left(\frac{v}{\|v\|_{\Psi^*}}\right)^2 Q(s) \\
&= v^2 Q(s),
\end{aligned}
$$

for all $v \in B$, and all $v \in \Psi^*$. Hence, the function Q is Ψ^*-quadratic. □

Corollary 9.5 *Let an even function $f : B \to M$ satisfies $f(0) = 0$ with*

$$\|D_s f(s_1, s_2, \cdots, s_m)\|_\beta \leq \delta + \epsilon\left(\sum_{i=1}^{m} \|s_i\|^t_\beta\right), \tag{9.40}$$

for all $s_1, s_2, \cdots, s_m \in B$ and $v \in \Psi^*_1$, then there exists a unique quadratic mapping $Q : B \to M$ satisfying

$$\|f(s) - Q(s)\|_\beta \leq \frac{\left(\delta + \epsilon\|s\|^t_\beta\right)}{(m^2 - 5m + 2)\left(|2|^{2\beta} - |2|^{\beta t}\right)},$$

where $0 < t < 2$, $\epsilon, \delta \in [0, \infty)$ and for all $s \in B$. Moreover, if $f(ns)$ is continuous in $n \in \mathbb{R}$ for all $s \in B$, then the function Q is Ψ^*-quadratic.

Proof. By setting

$$\chi(s_1, s_2, \cdots, s_m) = \delta + \epsilon\left(\sum_{i=1}^{m} \|s\|^t_\beta\right)$$

and $L = |2|^{\beta(t-2)}$ in Theorem 9.7, we obtain our needed results. □

Corollary 9.6 *Let $t > 0$ such that $mt < 2$ and $\delta, \epsilon \in \mathbb{R}^+$, and $f : B \to M$ be an even mapping and $f(0) = 0$ such that*

$$\|D_v f(s_1, s_2, \cdots, s_m)\|_\beta \leq \delta + \epsilon\left[\prod_{i=1}^{m} \|s_i\|^t_\beta + \sum_{i=1}^{m} \|s_i\|^{mt}_\beta\right]$$

for all $s_1, s_2, \cdots, s_m \in B$ and $v \in \Psi^*_1$, then there exists a unique quadratic mapping $Q : B \to M$ satisfies

$$\|f(s) - Q(s)\|_\beta \leq \frac{\left(\delta + \epsilon\|s\|^{mt}_\beta\right)}{(m^2 - 5m + 2)\left(|2|^{2\beta} - |2|^{\beta mt}\right)}, \tag{9.41}$$

for all $s \in B$. Moreover, if $f(ns)$ is continuous in $n \in \mathbb{R}$ for all fixed $s \in B$, then the function Q is Ψ^*-quadratic.

Proof. By setting

$$\chi(s_1, s_2, \cdots, s_m) = \delta + \epsilon\left(\sum_{i=1}^{m} \|s\|_\beta^t\right)$$

and $L = |2|^{\beta(mt-2)}$ in Theorem 9.7, we obtain our needed result.

□

Theorem 9.8 *Let $\chi : B^m \to [0, \infty)$ be an even mapping such that*

$$\lim_{l\to\infty} |2|^{2l\beta}\chi\left(2^{-l}s_1, 2^{-l}s_2, \cdots, 2^{-l}s_m\right) = 0 \tag{9.42}$$

for all $s_1, s_2, \cdots, s_m \in B$. Let $f : B \to M$ be an even mapping with $f(0) = 0$ such that (9.2). If there exists $0 < L < 1$ satisfies

$$s \to \varrho(s) = \frac{\chi\left(0, s, 0, \cdots, 0\right)}{(m^2 - 5m + 2)},$$

and

$$\varrho(s) \le |2|^{-2\beta}L\varrho(2s), \tag{9.43}$$

for all $s \in B$, then there exists a unique quadratic mapping $Q : B \to M$ satisfying

$$\|f(s) - Q(s)\|_\beta \le \frac{L}{|2|^{2\beta} - |2|^{2\beta}L}\varrho(s), \tag{9.44}$$

for all $s \in B$. Moreover, if $s(ns)$ is continuous in $n \in \mathbb{R}$ for all $s \in B$, then the function Q is Ψ^*-quadratic.

Proof. Take $v = 1$ and replacing $(s_1, s_2, s_3, \cdots, s_m)$ by $(0, s, 0, \cdots, 0)$ in (9.2), we obtain

$$\left\|(m^2 - 5m + 2)f(2s) - 2^2(m^2 - 5m + 2)f(s)\right\|_\beta \le \chi\left(0, s, 0, \cdots, 0\right), \tag{9.45}$$

for all $s \in B$. Replacing s by $\frac{s}{2}$ in (9.45), we obtain

$$\left\|2^2 f\left(\frac{s}{2}\right) - f(s)\right\|_\beta \le L\varrho(s), \tag{9.46}$$

for all $s \in B$. Consider the set $\Lambda := \{u | u : B \to M, u(0) = 0\}$ and define the generalized metric on Λ as

$$d(u, r) = \inf\{c \in [0, \infty) \mid \|u(s) - r(s)\|_\beta \le c\varrho(s), \ \ \forall\ s \in B\}. \tag{9.47}$$

Clearly, (Λ, d) is a complete generalized metric space (see [5]). Now, we define a mapping $\Phi : \Lambda \to \Lambda$ by

$$(\Phi u)(s) = 2^2 u\left(\frac{s}{2}\right), \quad \forall\ u \in \Lambda, \tag{9.48}$$

for all $s \in B$. Let $u, r \in \Lambda$ and an arbitrary constant $c \in [0, \infty]$ with $d(u, r) < c$. By the definition of d, we arrive

$$\|u(s) - r(s)\|_\beta \leq c\varrho(s), \tag{9.49}$$

for all $s \in B$. By the given hypothesis and the above inequality, we have

$$\left\|2^2 u\left(\frac{s}{2}\right) - 2^2 r\left(\frac{s}{2}\right)\right\|_\beta \leq cL\varrho(s), \tag{9.50}$$

for all $s \in B$. Hence,

$$d(\Phi u, \Phi r) \leq Ld(u, r).$$

By using inequality (9.46) that

$$d(\Phi f, f) \leq \frac{L}{|2|^{2\beta}}.$$

Thus, by Theorem 9.1, Φ has a unique fixed point $Q : B \to M$ in $\Lambda^* = \{u \in \Lambda |\ d(u, r) < \infty\}$ such that

$$Q(s) := \lim_{l \to \infty} (\Phi^l f)(s) = \lim_{l \to \infty} 2^{2l} f\left(\frac{s}{2^l}\right), \tag{9.51}$$

and $Q\left(\frac{s}{2}\right) = \frac{1}{2^2} Q(s)$, for all $s \in B$. Also,

$$\begin{aligned} d(Q, f) &\leq \frac{1}{1 - L} d(\Phi f, f) \\ &\leq \frac{L}{|2|^{2\beta} - |2|^{2\beta} L}. \end{aligned} \tag{9.52}$$

Thus, inequality (9.44) holds for all $s \in B$. Next, we prove that Q is quadratic. By inequalities (9.2), (9.42) and (9.51), we obtain

$$\begin{aligned} \|D_1 Q(s_1, s_2, \cdots, s_m)\|_\beta &= \lim_{l \to \infty} |2|^{2l\beta} \left\|D_1 f\left(\frac{s_1}{2^l}, \frac{s_2}{2^l}, \cdots, \frac{s_m}{2^l}\right)\right\|_\beta \\ &\leq \lim_{l \to \infty} |2|^{2l\beta} \chi\left(\frac{s_1}{2^l}, \frac{s_2}{2^l}, \cdots, \frac{s_m}{2^l}\right) = 0, \end{aligned}$$

Thus, by Theorem 9.3, the mapping Q is quadratic. Next, we want to prove that the function Q is unique. Suppose an another quadratic mapping $Q' : B \to M$ satisfies (9.44). Then,

$$d(f, Q') \leq \frac{L}{|2|^{2\beta} - |2|^{2\beta} L}$$

and Q' is quadratic, which implies $Q' \in \Lambda^*$ and $(\Phi Q')(s) = 2^2 Q'\left(\frac{s}{2}\right) = Q(s)$ for all $s \in B$, i.e., Q' is a fixed point of Φ in Λ^*. Hence, we obtain $Q' = Q$.

Moreover, if $f(ns)$ is continuous in $n \in \mathbb{R}$ for all $s \in B$, then using the proof of [26], Q is $\mathbb{R}$-quadratic. Replacing $(s_1, s_2, \cdots, s_m)$ by $\left(0, \frac{s}{2}, 0, \cdots, 0\right)$ in (9.2), we have

$$\left\|(m^2 - 5m + 2)f(vs) - 2^2(m^2 - 5m + 2)v^2 f\left(\frac{s}{2}\right)\right\|_\beta$$

$$\leq \chi\left(0, \frac{s}{2}, 0, \cdots, 0\right), \tag{9.53}$$

for all $s \in B$ and all $v \in \Phi_1^*$. Using definition of Q, (9.42) and (9.53), we obtain

$$\begin{aligned}
&\left\|(m^2 - 5m + 2)Q(vs) - 2^2(m^2 - 5m + 2)v^2 Q\left(\frac{s}{2}\right)\right\|_\beta \\
&= \lim_{l\to\infty} |2|^{2l\beta} \|(m^2 - 5m + 2)f\left(\frac{vs}{2^l}\right) - 2^2(m^2 - 5m + 2)v^2 f\left(\frac{s}{2^{l+1}}\right)\|_\beta \\
&\leq \lim_{l\to\infty} |2|^{2l\beta}\chi\left(0, \frac{s}{2^{l+1}}, 0, \cdots, 0\right) = 0,
\end{aligned}$$

for all $s \in B$ and all $v \in \Psi_1^*$. Thus,

$$(m^2 - 5m + 2)Q(vs) - 2^2(m^2 - 5m + 2)s^2 Q\left(\frac{s}{2}\right) = 0,$$

for all $s \in B$ and all $v \in \Psi_1^*$. Since, the function Q is quadratic, we obtain, $Q(vs) = v^2 Q(s)$, for all $s \in B$ and all $v \in \Psi_1^* \cup \{0\}$. Since, the function Q is $\mathbb{R}$-quadratic, let $v \in \Psi^* \backslash \{0\}$,

$$\begin{aligned}
Q(vs) &= Q\left(\|v\|_{\Psi^*} \cdot \frac{v}{\|v\|_{\Psi^*}} s\right) \\
&= \|v\|_{\Psi^*}^2 \cdot Q\left(\frac{v}{\|v\|_{\Psi^*}} s\right) \\
&= \|v\|_{\Psi^*}^2 \cdot \left(\frac{v}{\|v\|_{\Psi^*}}\right)^2 Q(s) \\
&= v^2 Q(s),
\end{aligned}$$

for all $s \in B$ and all $v \in \Psi^*$. Hence, then function Q is Ψ^*-quadratic. □

Corollary 9.7 *Let an even mapping $f : B \to M$ such that $f(0) = 0$ with*

$$\|D_v f(s_1, s_2, \cdots, s_m)\|_\beta \leq \epsilon\left(\sum_{i=1}^m \|s_i\|_\beta^t\right), \tag{9.54}$$

for all $s_1, s_2, \cdots, s_m \in B$, and all $v \in \Psi_1^*$, then there exists a unique quadratic mapping $Q : B \to M$ satisfying

$$\|f(s) - Q(s)\|_\beta \leq \frac{\epsilon\|s\|_\beta^t}{(m^2 - 5m + 2)\left(|2|^{\beta t} - |2|^{2\beta}\right)},$$

where $t > 0$ and $\epsilon \in \mathbb{R}^+$, for all $s \in B$. Moreover, if $f(ns)$ is continuous in $n \in \mathbb{R}$ for all $s \in B$, then the function Q is Ψ^*-quadratic.

Proof. Setting

$$\chi(s_1, s_2, \cdots, s_m) = \delta + \epsilon\left(\sum_{i=1}^m \|s\|_\beta^t\right)$$

and $L = |2|^{\beta(2-t)}$ in Theorem 9.8, we obtain our needed result. □

Corollary 9.8 *Let an even mapping $f : B \to M$ be such that $f(0) = 0$ with*

$$\|D_v f(s_1, s_2, \cdots, s_m)\|_\beta \leq \epsilon \left[\prod_{i=1}^{m} \|s_i\|_\beta^t + \sum_{i=1}^{m} \|s_i\|_\beta^{mt} \right],$$

for all $s_1, s_2, \cdots, s_m \in B$ and $r \in \Psi_1^*$, then there exists a unique quadratic mapping $Q : B \to M$ satisfying

$$\|f(s) - Q(s)\|_\beta \leq \frac{\epsilon \|s\|_\beta^{mt}}{(m^2 - 5m + 2)\left(|2|^{\beta mt} - |2|^{2\beta}\right)}, \tag{9.55}$$

where $t > 0$ such that $mt > 2$ and $\epsilon \in \mathbb{R}^+$, for all $s \in B$. Moreover, if $f(ns)$ is continuous in $n \in \mathbb{R}$ for all $s \in B$, then the function Q is Ψ^*-quadratic.

Proof. By letting

$$\chi(s_1, s_2, \cdots, s_m) = \delta + \epsilon \left(\sum_{i=1}^{m} \|s\|_\beta^t \right)$$

and $L = |2|^{\beta(2-mt)}$ in Theorem 9.8, we obtain our needed result. □

9.2.3 Stability Results for the Mixed Case

Theorem 9.9 *Let a mapping $\chi : B^m \to [0, \infty)$ such that*

$$\lim_{l \to \infty} \frac{1}{|2|^{l\beta}} \chi\left(2^l s_1, 2^l s_2, \cdots, 2^l s_m\right) = 0, \quad \lim_{l \to \infty} \frac{1}{|2|^{2l\beta}} \chi\left(2^l s_1, 2^l s_2, \cdots, 2^l s_m\right) = 0 \tag{9.56}$$

for all $s_1, s_2, \cdots, s_m \in B$. If a mapping $f : B \to M$ and $f(0) = 0$ such that (9.2). If there exists a constant $0 < L < 1$ such that

$$\begin{aligned} \chi(0, 2s, 0, \cdots, 0) &\leq |2|^\beta L \chi(0, s, 0, \cdots, 0) \quad \textit{and} \\ \chi(0, 2s, 0, \cdots, 0) &\leq |2|^{2\beta} L \chi(0, s, 0, \cdots, 0), \end{aligned} \tag{9.57}$$

for all $s \in B$, then there exists a unique additive mapping $A : B \to M$ and a unique quadratic mapping $Q : B \to M$ such that

$$\begin{aligned} \|f(s) - A(s) - Q(s)\|_\beta \leq & \frac{(\chi(0, s, 0, \cdots, 0) + \chi(0, -s, 0, \cdots, 0))}{|2|^{2\beta} - |2|^{2\beta} L} \\ & \cdot \left[\frac{|2|^\beta}{(m^2 - 5m + 4)} + \frac{1}{(m^2 - 5m + 2)} \right], \end{aligned}$$

for all $s \in B$. Moreover, if $f(ns)$ is continuous in $n \in \mathbb{R}$ for all $s \in B$, then the function A is Ψ^*-linear and the function Q is Ψ^*-quadratic.

Proof. If the function f is split into even and odd parts by letting

$$f_e(s) = \frac{f(s) + f(-s)}{2} \quad \text{and} \quad f_o(s) = \frac{f(s) - f(-s)}{2} \tag{9.58}$$

for all $s \in B$, then $f(s) = f_e(s) + f_o(s)$. Let

$$\phi(s_1, s_2, \cdots, s_m) = \frac{[\chi(s_1, s_2, \cdots, s_m) + \chi(-s_1, -s_2, \cdots, -s_m))}{2^\beta},$$

then by (9.56), (9.57) and (9.58), we have

$$\lim_{l\to\infty} \frac{1}{|2|^{l\beta}}\phi\left(2^l s_1, 2^l s_2, \cdots, 2^l s_m\right) = 0; \qquad \lim_{l\to\infty} \frac{1}{|2|^{2l\beta}}\phi\left(2^l s_1, 2^l s_2, \cdots, 2^l s_m\right) = 0,$$

$$\phi(0, 2s, 0, \cdots, 0) \leq |2|^\beta L\phi(0, s, 0, \cdots, 0), \text{ and } \quad \phi(0, 2s, 0, \cdots, 0) \leq |2|^{2\beta} L\phi(0, s, 0, \cdots, 0),$$

$$\|D_v f_o(s_1, s_2, \cdots, s_m)\|_\beta \leq \phi(s_1, s_2, \cdots, s_m), \qquad \|D_v f_e(s_1, s_2, \cdots, s_m)\|_\beta \leq \phi(s_1, s_2, \cdots, s_m).$$

Hence, by Theorem 9.5 and 9.7, there exists a unique additive mapping $A : B \to M$ and a unique quadratic mapping $Q : B \to M$ such that

$$\|f_o(s) - A(s)\|_\beta \leq \frac{1}{(m^2 - 5m + 4)|2|^\beta(1 - L)}\phi(0, s, 0, \cdots, 0),$$

and

$$\|f_e(s) - Q(s)\|_\beta \leq \frac{1}{(m^2 - 5m + 2)|2|^{2\beta}(1 - L)}\phi(0, s, 0, \cdots, 0)$$

for all $s \in B$. Thus,

$$\begin{aligned}
&\|f(s) - A(s) - Q(s)\|_\beta \leq \|f_o(s) - A(s)\|_\beta + \|f_e(s) - Q(s)\|_\beta \\
\leq\ & \left[\frac{1}{(m^2 - 5m + 4)|2|^\beta(1 - L)} + \frac{1}{(m^2 - 5m + 2)|2|^{2\beta}(1 - L)}\right] \\
&\cdot\phi(0, s, 0, \cdots, 0) \\
\leq\ & \frac{1}{|2|^{2\beta} - |2|^{2\beta}L}\left[\frac{|2|^\beta}{(m^2 - 5m + 4)} + \frac{1}{(m^2 - 5m + 2)}\right](\chi(0, s, 0, \cdots, 0) \\
&\cdot + \chi(0, -s, 0, \cdots, 0))
\end{aligned}$$

for all $s \in B$. □

Corollary 9.9 *Let a mapping $f : B \to M$ with $f(0) = 0$ such that*

$$\|D_v f\left(s_1, s_2, \cdots, s_m\right)\|_\beta \leq \delta + \epsilon \sum_{i=1}^{m} \|s_i\|_\beta^t, \tag{9.59}$$

for all $s_1, s_2, \cdots, s_m \in B$, and all $v \in \Psi_1^*$, then there exists a unique additive mapping $A : B \to M$ and a unique quadratic mapping $Q : B \to M$ such that

$$\begin{aligned}
\|f(s) - A(s) - Q(s)\|_\beta \leq\ & \frac{2\left(\delta + \epsilon\|s\|_\beta^t\right)}{\left(|2|^{2\beta} - |2|^{\beta(t+1)}\right)}\left[\frac{|2|^\beta}{(m^2 - 5m + 4)}\right. \\
&\left. + \frac{1}{(m^2 - 5m + 2)}\right],
\end{aligned}$$

where $0 < t < 1$ and $\delta, \epsilon \in \mathbb{R}^+$, for all $s \in B$. Moreover, if $f(ns)$ is continuous in $n \in \mathbb{R}$ for all $s \in B$, then the function A is Ψ^*-linear and the function Q is Ψ^*-quadratic.

Corollary 9.10 *Let a mapping* $f : B \to M$ *with* $f(0) = 0$ *such that*

$$\|D_v f(s_1, s_2, \cdots, s_m)\|_\beta \leq \epsilon \sum_{i=1}^{m} \|s_i\|_\beta^t, \tag{9.60}$$

for all $s_1, s_2, \cdots, s_m \in B$ *and* $v \in \Psi_1^*$, *then there exists a unique additive mapping* $A : B \to M$ *and a unique quadratic mapping* $Q : B \to M$ *such that*

$$\|f(s) - A(s) - Q(s)\|_\beta \leq \frac{2\epsilon\|s\|_\beta^t}{(|2|^{2\beta} - |2|^{\beta t})}\left[\frac{|2|^\beta}{(m^2 - 5m + 4)} + \frac{1}{(m^2 - 5m + 2)}\right],$$

for all $s \in B$, *where* $t > 2$ *and* $\epsilon \in \mathbb{R}^+$. *Moreover, if* $f(ns)$ *is continuous in* $n \in \mathbb{R}$ *for all* $s \in B$, *then the function* Q *is* Ψ^**-quadratic and the function* A *is* Ψ^**-linear.*

Theorem 9.10 *Let a mapping* $\chi : B^m \to [0, \infty)$ *such that*

$$\lim_{l\to\infty} |2|^{l\beta} \chi\left(\frac{s_1}{2^l}, \frac{s_2}{2^l}, \cdots, \frac{s_m}{2^l}\right) = 0, \lim_{l\to\infty} |2|^{2l\beta} \chi\left(\frac{s_1}{2^l}, \frac{s_2}{2^l}, \cdots, \frac{s_m}{2^l}\right) = 0 \tag{9.61}$$

for all $s_1, s_2, \cdots, s_m \in B$. *If a mapping* $f : B \to M$ *with* $f(0) = 0$ *such that* (9.2). *If there exists a constant* $0 < L < 1$ *such that*

$$\begin{aligned} \chi(0, s, 0, \cdots, 0) &\leq |2|^{-\beta} L\chi(0, 2s, 0, \cdots, 0) \quad \textit{and} \\ \chi(0, s, 0, \cdots, 0) &\leq |2|^{-2\beta} L\chi(0, 2s, 0, \cdots, 0), \end{aligned} \tag{9.62}$$

for all $s \in B$, *then there exists a unique additive mapping* $A : B \to M$ *and a unique quadratic mapping* $Q : B \to M$ *such that*

$$\begin{aligned} \|f(s) - A(s) - Q(s)\|_\beta &\leq \frac{(\chi(0, s, 0, \cdots, 0) + \chi(0, -s, 0, \cdots, 0))\, L}{|2|^{2\beta}(1 - L)} \\ &\quad \cdot \left[\frac{|2|^\beta}{(m^2 - 5m + 4)} + \frac{1}{(m^2 - 5m + 2)}\right], \end{aligned}$$

for all $s \in B$. *Moreover, if* $f(ns)$ *is continuous in* $n \in \mathbb{R}$ *for all* $s \in B$, *then the function* Q *is* Ψ^**-quadratic and the function* A *is* Ψ^**-linear.*

Corollary 9.11 *If a mapping* $f : B \to M$ *with* $f(0) = 0$ *such that*

$$\|D_v f(s_1, s_2, \cdots, s_m)\|_\beta \leq \epsilon \sum_{i=1}^{m} \|s_i\|_\beta^t, \tag{9.63}$$

for all $s_1, s_2, \cdots, s_m \in B$ *and* $v \in \Psi_1^*$, *then there exists a unique additive mapping* $A : B \to M$ *and a unique quadratic mapping* $Q : B \to M$ *such that*

$$\|f(s) - A(s) - Q(s)\|_\beta \leq \frac{2\epsilon\|s\|_\beta^t}{(|2|^{\beta t} - |2|^{2\beta})}\left[\frac{|2|^\beta}{(m^2 - 5m + 4)} + \frac{1}{(m^2 - 5m + 2)}\right],$$

for all $s \in B$, *where* $t > 2$ *and* $\epsilon \in \mathbb{R}^+$. *Moreover, if* $f(ns)$ *is continuous in* $n \in \mathbb{R}$ *for all* $s \in B$, *then the function* Q *is* Ψ^**-quadratic and the function* A *is* Ψ^**-linear.*

Corollary 9.12 *If a mapping $f : B \to M$ with $f(0) = 0$ such that*

$$\|D_v f(s_1, s_2, \cdots, s_m)\|_\beta \leq \delta + \epsilon \sum_{i=1}^{m} \|s_i\|_\beta^t, \tag{9.64}$$

for all $s_1, s_2, \cdots, s_m \in B$, and $v \in \Psi_1^*$, then there exists a unique additive mapping $A : B \to M$ and a mapping quadratic mapping $Q : B \to M$ such that

$$\|f(s) - A(s) - Q(s)\|_\beta \leq \frac{2\left(\delta + \epsilon\|s\|_\beta^t\right)}{\left(|2|^{\beta(t+1)} - |2|^{2\beta}\right)} \left[\frac{|2|^\beta}{(m^2 - 5m + 4)} + \frac{1}{(m^2 - 5m + 2)}\right],$$

for all $s \in B$, where $0 < t < 1$ and $\delta, \epsilon \in \mathbb{R}^+$. Moreover, if $f(ns)$ is continuous in $n \in \mathbb{R}$ for all $s \in B$, then the function Q is Ψ^*-quadratic and the function A is Ψ^*-linear.

9.3 CONCLUSION

The Ulam stability of (9.1) has been investigated in this work using the fixed point approach. It was obtained in Section 9.2.3 with the aid of Section 9.2.1, if the function f is odd, and Section 9.2.2, where the function f is even.

BIBLIOGRAPHY

[1] N. Alessa, K. Tamilvanan, G. Balasubramanian, K. Loganathan, Stability results of the functional equation deriving from quadratic function in random normed spaces, *AIMS Math.* 6 (2021), no. 3, 2385–2397.

[2] N. Alessa, K. Tamilvanan, K. Loganathan, T. S. Karthik, John Michael Rassias, Orthogonal stability and nonstability of a generalized quartic functional equation in quasi-β-normed spaces, *J. Funct. Spaces* 2021, Art. ID 5577833, 7 pp.

[3] N. Alessa, K. Tamilvanan, K. Loganathan, K. Kalai Selvi, Hyers-Ulam stability of functional equation deriving from quadratic mapping in non-Archimedean (n, β)-normed spaces, *J. Funct. Spaces* 2021, Art. ID 9953214, 10 pp.

[4] T. Aoki, On the stability of the linear transformation in Banach spaces, *J. Math. Soc. Japan* 2 (1950), 64–66.

[5] V. K. Balachandran, *Topological Algebras*, reprint of the 1999 original, North-Holland Mathematics Studies, 185, North-Holland Publishing Co., Amsterdam, 2000.

[6] L. Cădariu and V. Radu, Fixed point methods for the generalized stability of functional equations in a single variable, *Fixed Point Theory Appl.* 2008, Art. ID 749392, 15 pp.

[7] P. Debnath, Banach, Kannan, Chatterjea, and Reich-type contractive inequalities for multivalued mappings and their common fixed points, *Math. Methods Appl. Sci.* 45(3) (2022) 1587–1596.

[8] P. Debnath, A new extension of Kannan's fixed point theorem via F-contraction with application to integral equations, *Asian-Eur. J. Math.* 15(7) (2022) 2250123.

[9] P. Debnath, New common fixed point theorems for Gornicki-type mappings and enriched contractions, *Sao Paulo J. Math. Sci.* 16 (2022) 1401–1408.

[10] P. Debnath, Common fixed-point and fixed-circle results for a class of discontinuous F-contractive mappings, *Mathematics* 10(9) (2022) 1605.

[11] P. Debnath, N. Konwar and S. Radenović, *Metric Fixed Point Theory: Applications in Science, Engineering and Behavioural Sciences*, Springer, Singapore, 2021.

[12] P. Debnath, H. M. Srivastava, P. Kumam and B. Hazarika, *Metric Fixed Point Theory: Applications in Science, Engineering and Behavioural Sciences*, Springer, Singapore, 2022.

[13] J. B. Diaz and B. Margolis, A fixed point theorem of the alternative, for contractions on a generalized complete metric space, *Bull. Amer. Math. Soc.* 74 (1968), 305–309.

[14] P. Găvruţa, A generalization of the Hyers-Ulam-Rassias stability of approximately additive mappings, *J. Math. Anal. Appl.* 184 (1994), no. 3, 431–436.

[15] M. E. Gordji, H. Khodaei and Th. M. Rassias, Fixed points and stability for quadratic mappings in β-normed left Banach modules on Banach algebras, *Results Math.* 61 (2012), no. 3-4, 393–400.

[16] D. H. Hyers, On the stability of the linear functional equation, *Proc. Nat. Acad. Sci. U.S.A.* 27 (1941), 222–224.

[17] S. O. Kim, K. Tamilvanan, Fuzzy stability results of generalized quartic functional equations, *Mathematics* 2021; 9(2), Art. ID 120.

[18] C.-G. Park, On the stability of the linear mapping in Banach modules, *J. Math. Anal. Appl.* 275 (2002), no. 2, 711–720.

[19] T. M. Rassias, On the stability of the linear mapping in Banach spaces, *Proc. Amer. Math. Soc.* 72 (1978), no. 2, 297–300.

[20] K. Tamilvanan, R. T. Alqahtani, S. A. Mohiuddine, Stability results of mixed type quadratic-additive functional equation in β-Banach modules by using fixed-point technique, *Mathematics* 2022, 10, Art. ID. 493.

[21] K. Tamilvanan, Y. Almalki, S. A. Mohiuddine, R. P. Agarwal, Stability results of quadratic-additive functional equation based on Hyers technique in matrix paranormed spaces, *Mathematics* 2022, 10(11), Art. ID 1940.

[22] K. Tamilvanan, G. Balasubramanian, Nazek Alessa, K. Loganathan, Hyers-Ulam stability of additive functional equation using direct and fixed-point methods, *J. Math.* 2020, Art. ID 6678772, 9 pp.

[23] K. Tamilvanan, J. R. Lee and C. Park, Ulam stability of a functional equation deriving from quadratic and additive mappings in random normed spaces, *AIMS Math.* 6 (2021), no. 1, 908–924.

[24] S. M. Ulam, *A Collection of Mathematical Problems*, Interscience Tracts in Pure and Applied Mathematics, no. 8, Interscience Publishers, New York, 1960.

[25] N. Uthirasamy, K. Tamilvanan and M. J. Kabeto, Ulam stability and non-stability of additive functional equation in IFN-spaces and 2-Banach spaces by different methods, *J. Funct. Spaces* 2022, Art. ID 8028634, 14 pp.

[26] T. Xu, J. M. Rassias and W. Xu, A fixed point approach to the stability of a general mixed additive-cubic equation on Banach modules, *Acta Math. Sci. Ser. B (Engl. Ed.)* 32 (2012), no. 3, 866–892.

[27] D. X. Zhou, On a conjecture of Z. Ditzian, *J. Approx. Theory* 69 (1992), no. 2, 167–172.

CHAPTER 10

Hybrid Steepest Descent Methods for Solving Variational Inequalities with Fixed Point Constraints in a Hilbert Space: An Annotated Bibliography

Mootta Prangprakhon and Nimit Nimana

Khon Kaen University

CONTENTS

10.1 INTRODUCTION

Constrained convex optimization problems provide a unified and general framework in which a diverse range of limitations and specifications of objective functions are imposed on the required solution. These limitations and specifications are illustrated

DOI: 10.1201/9781003388678-10

as convex constraint sets. It is of interest to observe that many real-world problems arising in engineering, economics, architecture, management science, medicine, and so on are mostly constrained rather than unconstrained convex optimization problems.

To solve constrained convex optimization problems, various interesting approaches have been proposed. One of the classical and significant approaches to deal with these problems is to construct iterative methods by making use of some algorithmic operators, for example, metric projections, and subgradient projections. However, it is worth noting that the metric projection can only be computed when the constraint sets are simple enough in the sense that closed-form expressions of such sets do exist.

10.1.1 Hybrid Steepest Descent Method (HSDM)

As the constraint sets reflect limitations and specifications of the considered objective functions, the structure of these sets can be highly complicated in some situations in which the metric projection onto such sets might be difficult to compute. Therefore, some existing iterative methods that employ metric projections might not be suitable or applicable in this case. This obstacle leads to finding new approaches to cope with these problems. Fortunately, by exploring the benefit of fixed point theory, instead of finding closed-form expressions of metric projections, we can find those of nonexpansive operators whose fixed point sets coincide with the given complicated constraint sets. This fascinating idea leads to establishing the celebrating *hybrid steepest descent method* (in short, HSDM) which was proposed by Yamada [1] in the paper:

The method was proposed to solve the so-called variational inequality problem over the fixed point set of a nonexpansive operator. Namely, let H be a real Hilbert space and $F : H \to H$ be a κ-Lipschitz continuous and η-strongly monotone operator. That is, there exists a constant $\kappa > 0$ such that

$$\|Fx - Fy\| \leq \kappa\|x - y\|,$$

and there exists a constant $\eta > 0$ such that

$$\langle Fx - Fy, x - y\rangle \geq \eta\|x - y\|^2,$$

for all $x, y \in H$, respectively. Moreover, let $T : H \to H$ be a nonexpansive operator. That is

$$\|Tx - Ty\| \leq \|x - y\|,$$

for all $x, y \in H$, along with its fixed point set $FixT := \{x \in H : Tx = x\}$ which is nonempty. The variational inequality problem is to find a point $x^* \in FixT$ such that

$$\langle x - x^*, Fx^*\rangle \geq 0, \qquad \text{(VIP}(F, FixT))$$

for all $x \in FixT$. This problem is considered to be one of the most popular problems in constrained convex optimization problems according to its wide range of applications. Moreover, since the operator F is strongly monotone and Lipschitz continuous and the fixed point set $FixT$ of a nonexpansive operator is a closed and convex set, by the Banach contraction principle, the uniqueness and the existence of the solution of VIP$(F, FixT)$ are guaranteed.

To solve the VIP$(F, FixT)$, Yamada proposed the HSDM of the form:

$$\begin{cases} x^1 \in H \text{ is chosen}, \\ x^{n+1} = Tx^n - \mu\alpha_{n+1}F(Tx^n)), \end{cases} \tag{HSDM}$$

for all $n \in \mathbb{N}$. By letting $\mu \in (0, 2\eta/\kappa^2)$ and assuming that the step size $\{\alpha_n\}_{n=1}^{\infty} \subset (0, 1]$ is

1. a slowly diminishing sequence (in short, SDM), i.e., $\lim_{n\to\infty} \alpha_n = 0$ and $\sum_{n=1}^{\infty} \alpha_n = \infty$,
2. $\sum_{n=1}^{\infty} |\alpha_{n+1} - \alpha_n| < \infty$ (or equivalently, $\lim_{n\to\infty} \frac{\alpha_n - \alpha_{n+1}}{\alpha_{n+1}^2} = 0$),

Yamada proved that any sequence $\{x^n\}_{n=1}^{\infty}$ generated by HSDM converges strongly to the unique solution of VIP$(F, FixT)$. Yamada also considered the problem in the case when the constrained set is the nonempty common fixed point sets of a finite family of nonexpansive operators $\bigcap_{i=1}^{m} FixT_i$; however, we omit the details here.

An important particular situation of VIP$(F, FixT)$ is the smooth minimization problem over the fixed point constraint. To be precise, let $f : H \to \mathbb{R}$ be a continuously differentiable convex function, the smooth minimization problem is to solve

$$\begin{array}{ll} \text{minimize} & f(x) \\ \text{subject to} & x \in FixT. \end{array} \tag{S-MIN}$$

That is, the smooth minimization problem is to find a point $x^* \in FixT$ such that $f(x^*) \leq f(x)$ for all $x \in FixT$. Since the objective function f is smooth convex and it is well known that the fixed point set a nonexpansive operator T is closed and convex, S-MIN is equivalent to the following specific VIP$(F, FixT)$: find $x^* \in FixT$ such that

$$\langle x - x^*, \nabla f(x^*)\rangle \geq 0,$$

for all $x \in FixT$, where ∇f is the gradient of f. Therefore, S-MIN is nothing else but a special case of the VIP$(F, FixT)$ in the case when $F := \nabla f$. Moreover, the strong convergence of the sequence $\{x^n\}_{n=1}^{\infty}$ generated by HSDM to the unique solution of S-MIN is also guaranteed.

These ideas were the new beginning that paved the way for new research works in fixed point theory, variational inequalities, and other related fields. Nowadays, many of the advances in the development and applications of HSDM have penetrated virtually various types of constrained convex optimization problems aside from the VIP$(F, FixT)$. These gave rise to new iterative methods to solve problems such as minimization problems, hierarchical fixed point problems, and equilibrium problems. The goal of this chapter is to create a brief survey of some research works related to HSDM and its development in the context of an annotated bibliography of some research works, which were considered within fixed point constraints and to present a perspective on new research directions which could possibly be accomplished in the future.

10.1.2 Scope of the Paper and an Apology

Since there is a great amount of literature which constitutes HSDM and is inspired by such a method, to include every research work relating to that would be a gargantuan task beyond our original goal. Hence, to create this annotated bibliography, we collect only some of the research works associated with HSDM in the case when the considered constraint sets are the fixed point sets of some nonlinear operators in a Hilbert space. Moreover, we emphasize here that, in the development of HSDM, our attention will be mainly focused on some of the research works which can be used to solve VIP in the case when the operator F is a monotone operator only.

Even though we have scrupulously checked and taken every precaution throughout the process of writing this manuscript, some errors may inevitably occur due to lack of knowledge and general oversight which are unintentional. For this reason, we apologize for the negligence and the omission of information in this chapter. If the reader has any suggestions or any additional information which pertain to the scope of this chapter, we will be very grateful.

10.1.3 Organization of the Paper

This annotated bibliography is organized as follows. In Section 10.2, we present an annotated bibliography of some research works related to fixed point constraints which were proposed before HSDM. In Section 10.3, we demonstrate the development of HSDM through an annotated bibliography of some research works, which were proposed to solve VIP in the cases when the constraint set can either be illustrated as a fixed point set or the intersection of fixed point sets.

10.2 SOME WORKS OF FIXED POINT CONSTRAINTS BEFORE HSDM

HSDM has a long line of history. Below, we present an annotated bibliography of some of the interesting works considered within fixed point constraints that were investigated before the constitution of HSDM.

Wittmann [2] investigated the solving of the S-MIN in the case when the objective function f is defined by $f := \frac{1}{2}\| \cdot - x^1\|^2$, where the point x^1 is arbitrarily given. To solve this, Wittmann proposed the following method:

$$\begin{cases} x^1 \in H \text{ is chosen}, \\ x^{n+1} = Tx^n - \alpha_{n+1}(Tx^n - x^1), \end{cases}$$

for all $n \in \mathbb{N}$. By assuming that the step size $\{\alpha_n\}_{n=1}^{\infty} \subset [0, 1]$ is SDM and satisfies $\sum_{n=1}^{\infty} |\alpha_{n+1} - \alpha_n| < \infty$, Wittmann proved that any sequence $\{x^n\}_{n=1}^{\infty}$ generated by the proposed method converges strongly to the unique solution of S-MIN. It was noted that the problem considered by Wittmann can be seen as the best approximation problem of finding a point in $FixT$ which is close to the point x^1.

Bauschke [3] extended Wittmann's best approximation problem to the S-MIN in the case when the objective function is defined by $f := \frac{1}{2}\|\cdot - a\|^2$, where the point $a \in H$ is given and such a point is called the anchor point. Moreover, the constraint set of this work is assumed to be the nonempty intersection of a finite family of nonexpansive operators $T_i : H \to H$, for all $i = 1, 2, \ldots, m$. To solve this, Bauschke proposed the following method:

$$\begin{cases} x^1 \in H \text{ is chosen,} \\ x^{n+1} = T_{[n+1]}x^n - \alpha_{n+1}(T_{[n+1]}x^n - a), \end{cases}$$

for all $n \in \mathbb{N}$, where the function $[\cdot]$ is the modulo m function taking values in $\{1, \ldots, m\}$. By assuming that the constrained set satisfying

$$\begin{aligned} \bigcap_{i=1}^{m} FixT_i &= Fix(T_m \cdots T_1) \\ &= Fix(T_1 T_m \cdots T_3 T_2) \\ &= \cdots = Fix(T_{m-1}T_{m-2} \cdots T_1 T_m) \end{aligned} \tag{10.1}$$

and by supposing that the step size sequence $\{\alpha_n\}_{n=1}^{\infty} \subset [0, 1)$ is SDM and satisfies $\sum_{n=1}^{\infty} |\alpha_{n+m} - \alpha_n| < \infty$, Bauschke proved that the sequence $\{x^n\}_{n=1}^{\infty}$ generated by the proposed method converges strongly to the unique solution of S-MIN. Note that the starting point $x^1 \in H$ is possibly different from the anchor point a.

Yamada et al. [4] investigated the solving of the S-MIN in the case when the objective function f is a quadratic function defined by $f(x) := \frac{1}{2}\langle Bx, x\rangle - \langle b, x\rangle$, for all $x \in H$, in which the operator $B : H \to H$ is a strongly positive bounded self-adjoint linear operator with $\|Id - B\| < 1$ and b is a point on H. To solve this, they proposed the following method:

$$\begin{cases} x^1 \in H \text{ is chosen,} \\ x^{n+1} = Tx^n - \alpha_{n+1}(B(Tx^n) - b), \end{cases}$$

for all $n \in \mathbb{N}$. By assuming that the step size $\{\alpha_n\}_{n=1}^{\infty} \subset (0, 1]$ is SDM and satisfies $\sum_{n=1}^{\infty} |\alpha_{n+1} - \alpha_n| < \infty$, they proved that the sequence $\{x^n\}_{n=1}^{\infty}$ generated by the proposed method converges strongly to the unique solution of S-MIN. Moreover, it is worth mentioning that the case of common fixed point constraints of a finite family of nonexpansive operators with the condition (10.1) was also investigated in the paper.

Deutsch and Yamada [5] considered S-MIN in the case when the constrained set is the nonempty intersection of the fixed point sets of a finite family of nonexpansive operators $T_i : H \to H$, $i = 1, 2, \ldots, m$, with the condition (10.1). Denote the set $\Delta := \bigcup_{i=1}^{m} \text{conv}(T_i(H))$, where $\text{conv}(T_i(H))$ represents the convex hull of the image $T_i(H)$. They assumed that the objective function f is twice differentiable on some open set $U \supset \Delta$, and its Hessian $\nabla^2 f : U \to \mathcal{B}(H)$ satisfies the uniformly strongly positive and uniformly bounded over Δ, that is $\nabla^2 f(x)$ is self-adjoint for all $x \in \Delta$, and there exist real numbers A and a such that $A \geq a \geq 0$ and

$$a\|u\|^2 \leq \langle \nabla^2 f(x)u, u\rangle \leq A\|u\|^2,$$

for all $x \in \Delta$ and $u \in H$. To solve this, they proposed the following iterative method:

$$\begin{cases} x^1 \in H \text{ is chosen,} \\ x^{n+1} = T_{[n+1]}x^n - \mu\alpha_{n+1}\nabla f(T_{[n+1]}x^n), \end{cases}$$

for all $n \in \mathbb{N}$, where the function $[\cdot]$ is the modulo m function taking values in $\{1, \dots, m\}$. By letting $\mu \in (0, 2/A)$ and by assuming that the step size sequence $\{\alpha_n\}_{n=1}^{\infty} \subset [0,1]$ is SDM and satisfies $\sum_{n=1}^{\infty} |\alpha_{n+m} - \alpha_n| < \infty$, they proved that the sequence $\{x^n\}_{n=1}^{\infty}$ generated by the proposed method converges strongly to the unique solution of S-MIN.

10.3 THE DEVELOPMENT OF HSDM

The interest in the development of HSDM has been continuously increasing for over two decades. Numerous iterative methods and acceleration schemes inspired by HSDM have been proposed in order to deal with a variety of problems. In this section, we present an annotated bibliography of some related works motivated by HSDM in which the constraint set of these works was expressed as a fixed point set or the intersection of fixed point sets. These works are categorized by the type of considered problems and they utilized various algorithmic operators such as nonexpansive operators, cutter operators, quasi-nonexpansive operators, etc. To begin with, we start with an annotated bibliography of the works that presented some variant algorithms of HSDM to solve VIP($F, FixT$) in the following subsections.

10.3.1 VIP($F, FixT$)

Some of the interesting paper works on solving VIP($F, FixT$) where the considered operator T is a nonlinear operator are presented as follows:

Ogura and Yamada [6] considered VIP($F, FixT$) in the case when the whole space H is a finite dimensional real Hilbert space and the operator T is an attractive nonexpansive operator. That is, T is nonexpansive with $FixT \neq \emptyset$ and $\|Tx - y\| < \|x - y\|$ for all $x \notin FixT$ and $y \in FixT$. Furthermore, the boundedness of the fixed point set of T was assumed. For the operator F, the Lipschitz continuity over $T(H)$ and the paramonotonicity over $FixT$ were assumed. In view of HSDM with $\mu = 1$, the convergence of the sequence $\{x^n\}_{n=1}^{\infty}$ generated by HSDM was proved under the SDM of the step size sequence $\{\alpha_n\}_{n=1}^{\infty} \subset [0, \infty)$ and the boundedness of the sequence $\{x^n\}_{n=1}^{\infty}$. It should be noted that the bounded properties of $FixT$ and the sequence $\{x^n\}_{n=1}^{\infty}$ are satisfied if the operators T satisfy the asymptotically shrinking, i.e., there is some constant $R > 0$ such that $\sup_{\|u\| \geq R} \frac{\|Tu\|}{\|u\|} < 1$.

Ogura and Yamada [7] considered VIP$(F, FixT)$ with the same setting as in [6]. In this work, they obtained the strong convergence which stated that the sequence $\{x^n\}_{n=1}^{\infty}$ generated by HSDM converges strongly to a solution point without assuming that the sequence $\{x^n\}_{n=1}^{\infty}$ is bounded.

Xu and Kim [8] also investigated the solving of VIP$(F, FixT)$ with the same setting as in Yamada [1] and analyzed the convergence result of HSDM with some modifications. To be precise, in place of using the conditions of the step size $\{\alpha_n\}_{n=1}^{\infty} \subset (0,1]$ given by Yamada, they proposed the variant conditions of the step size $\{\alpha_n\}_{n=1}^{\infty} \subset (0,1]$ stating that $\{\alpha_n\}_{n=1}^{\infty} \subset (0,1]$ is SDM and satisfies $\lim_{n\to\infty} \frac{\alpha_n}{\alpha_{n+1}} = 1$. Strong convergence for the considered method was proved in the paper. Note that the second condition given by Xu and Kim is strictly weaker than the second condition of the original one. An important and simple example of the step size satisfying these conditions is $\alpha_n = \frac{1}{n}$ for all $n \geq 1$. This satisfies the second condition of Xu and Kim; however, it does not satisfy that of Yamada. Moreover, they also considered the case when the constraint set was assumed to be the nonempty intersection of a finite family of nonexpansive operators $T_i : H \to H$, for all $i = 1, 2, \ldots, m$. They also considered the following method:

$$\begin{cases} x^1 \in H \text{ is chosen}, \\ x^{n+1} = T_{[n+1]}x^n - \mu\alpha_{n+1}F(T_{[n+1]}x^n), \end{cases} \tag{10.2}$$

for all $n \in \mathbb{N}$, where the function $[\cdot]$ is the modulo m function taking values in $\{1, \ldots, m\}$. This method was proposed in [1]. By assuming that the constraint set satisfying the condition (10.1), by letting $\mu \in (0, 2\eta/\kappa^2)$ and by supposing that the step size sequence $\{\alpha_n\}_{n=1}^{\infty} \subset [0,1)$ is SDM and satisfies $\lim_{n\to\infty} \frac{\alpha_n}{\alpha_{n+m}} = 1$, they proved that the sequence $\{x^n\}_{n=1}^{\infty}$ generated by the proposed method converges strongly to the unique solution of VIP$(F, FixT)$ with a constraint set expressed as the common fixed point set.

Yamada and Ogura [9] extended VIP$(F, FixT)$ to the case when the operator T is quasi-nonexpansive with $FixT \neq \emptyset$, that is, $\|Tx - y\| \leq \|x - y\|$ for all $x \in H$ and $y \in FixT$, and the operator F is κ-Lipschitz continuous and η-strongly monotone over $T(H)$. They introduced the so-called quasi-shrinking operator and proved its useful properties. In view of HSDM, by assuming that the step size sequence $\{\alpha_n\}_{n=1}^{\infty} \subset [0, \infty)$ is SDM and assuming that there are some $y \in FixT$ and the parameter $\mu \in (0, 2\eta/\kappa^2)$ for which T is quasi-shrinking on a closed ball centered at y and some specific radius, they proved that the sequence $\{x^n\}_{n=1}^{\infty}$ generated by HSDM converges strongly to the considered problem.

Zeng, Wong and Yao [10] considered the solving of VIP$(F, FixT)$ in the same manner as [8]. In this work, they relaxed the constant parameter μ to the sequence $\{\mu_n\}_{n=1}^{\infty} \in (0, \frac{2\eta}{\kappa^2})$. By assuming that the condition (10.1) satisfies $\sum_{n=1}^{\infty} \alpha_n = \infty$ and $|\mu_n - \eta/\kappa^2| \leq \sqrt{\eta^2 - c\kappa^2}/\kappa^2$ for some $c \in (0, \eta^2/\kappa^2)$ and $\lim_{n\to\infty}(\mu_{n+1} - (\alpha_n/\alpha_{n+1})\mu_n) = 0$, they proved that the generated sequence converges strongly to the unique solution of the considered problem provided that $\limsup_{n\to\infty}\langle Tx_n - x_{n+1}, Tx_n - x_n\rangle \leq 0$.

Iiduka [11] also investigated the solving of VIP$(F, FixT)$ in the case when the operator T is firmly nonexpansive with $FixT \neq \emptyset$, and the operator F is extended to the class of monotone and hemicontinuous operators. In this situation, the nonemptiness of the solution of VIP$(F, FixT)$ needs to be assumed without the continuity of F. For simplicity, we may denote the solution set by VIP$(F, FixT)$. To solve this, Iiduka proposed the following method:

$$\begin{cases} x^1 \in H \text{ is chosen}, \\ y^n = T(x^n - \alpha_n F x^n), \\ x^{n+1} = \lambda_n x^n + (1 - \lambda_n) y^n, \end{cases} \tag{10.3}$$

for all $n \in \mathbb{N}$. By assumming the sequences $\{\alpha_n\}_{n=1}^{\infty} \subset (0,1)$ and $\{\lambda_n\}_{n=1}^{\infty} \subset [0,1)$ satisfy $\sum_{n=1}^{\infty} {\alpha_n}^2 < \infty$, $\limsup_{n\to\infty} \lambda_n < 1$, the generated sequence $\{x^n\}_{n=1}^{\infty}$ is bounded, and the following conditions hold true:

(I-1) there is $n_0 \in \mathbb{N}$ such that

$$\text{VIP}(F, FixT) \subset \bigcap_{n \geq n_0} \{x \in FixT : \langle x^n - x, Fx^n \rangle \geq 0\},$$

(I-2) $\|x^n - y^n\| = o(\alpha_n)$,

Iiduka proved a weak convergence of the generated sequence $\{x^n\}_{n=1}^{\infty}$ to a solution point in VIP$(F, FixT)$. Even if the conditions (I-1) and (I-2) seem to be restricted, a series of remarks and practical situations in which these two assumptions hold are discussed in the paper. Iiduka also presented some numerical examples of minimizing a quadratic function over the intersection of simple sets, and the solution of another minimization problem, respectively.

Ceng, Ansari and Yao [12] also considered VIP$(F, FixT)$ in a more general setting of the operators. Namely, let C be a nonempty closed and convex subset of H, the operator $T : C \to C$ is nonexpansive with $FixT \neq \emptyset$, and the operator F in the classical sense is extended to $\mu F - \gamma V$, where this operator $F : C \to H$ is given in the same manner as the one used in [1], whereas the operator $V : C \to H$ is L-Lipschitz continuous. They proposed both implicit and explicit schemes for solving VIP$(\mu F - \gamma V, FixT)$. Focusing on the explicit iteration, the sequence $\{x^n\}_{n=1}^{\infty}$ is generated as follow:

$$\begin{cases} x^1 \in C \text{ is chosen}, \\ x^{n+1} = P_C[\alpha_n \gamma V x^n + Tx^n - \mu \alpha_n F(Tx^n)], \end{cases} \tag{10.4}$$

for all $n \in \mathbb{N}$. By assuming the parameters $\mu \in (0, 2\eta/\kappa^2)$, $\gamma \in [0, (1 - \sqrt{1 - \mu(2\eta - \mu\kappa^2)})/L)$, the step size $\{\alpha_n\}_{n=1}^{\infty} \subset (0,1)$ and $\{\lambda_n\}_{n=1}^{\infty} \subset [0,1)$ satisfy the conditions proposed in [8], they proved a strong convergence result for the proposed method.

Iiduka [13] considered VIP$(F, FixT)$ in the case when the whole space H is a Euclidean space $\mathbb{R}^N$, the operator T is firmly nonexpansive with $FixT \neq \emptyset$, the operator F is continuous and the subset $C \subset \mathbb{R}^N$ is nonempty closed and convex with $FixT \subset C$. He proposed an iterative method in a similar fashion to (10.3) where x^1 is chosen in C and update $x^{n+1} = P_C[\lambda_n x^n + (1-\lambda_n)y^n]$ for all $n \in \mathbb{N}$. By assuming that the assumptions in [11] hold, the convergence of the proposed method is guaranteed. An interesting contribution of this paper is the application of the problem setting and its solving method to the power control problem for code-division multiple access (CDMA) data networks.

Cegielski and Grossmann [14] extended VIP$(F, FixT)$ to the case when the operator T is quasi-nonexpansive with $FixT \neq \emptyset$, and the operator F is κ-Lipschitz continuous and η-strongly monotone over $T(H)$. The main contribution of the paper is to approximate the infinite dimensional problem VIP$(F, FixT)$ by using some family of finite dimensional problems. Actually, for every $n \in \mathbb{N}$, let $H_n \subset H$ be a finite dimensional subspace which is nested, that is $H_n \subset H_{n+1}$. Let $T_n : H \to H_n$ be a quasi-nonexpansive operator with $FixT_n \subset FixT$ and $\bigcap_{n=1}^{\infty} FixT_n \neq \emptyset$. Furthermore, let $F_n : H \to H_n$ be a uniformly κ-Lipschitz continuous and η-strongly monotone over H_n. They considered the sequence of problems: Find $\bar{x}^n \in FixT_n$ such that

$$\langle x - \bar{x}^n, F_n\bar{x}^n \rangle \geq 0, \qquad (\text{VIP}(F_n, FixT_n))$$

for all $x \in FixT_n$. They constructed the following iteration-discretization scheme:

$$\begin{cases} x^1 \in H \text{ is chosen,} \\ x^{n+1} = T_n x^n - \mu\alpha_n F_n(T_n x^n)), \end{cases}$$

for all $n \in \mathbb{N}$. Denote by $\bar{x}^n \in FixT_n$ the unique solution of VIP$(F_n, FixT_n)$. By assuming that the sequence $\{F_n\bar{x}^n\}_{n=1}^{\infty}$ is bounded and it holds that $\sum_{n=1}^{\infty} \|\bar{x}^n - x^*\| < \infty$, the strong convergence of the generated sequence $\{x^n\}_{n=1}^{\infty}$ to the solution of VIP$(F, FixT)$ is guaranteed provided that the parameter $\mu \in (0, 2\eta/\kappa^2)$ and the step size $\{\alpha_n\}_{n=1}^{\infty} \subset (0,1]$ is SDM. Moreover, they also applied the proposed method to solve an optimal control problem.

Cegielski and Zalas [15] considered the approximation scheme for solving VIP$(F, FixT)$ in the case when the operator T is quasi-nonexpansive with $FixT \neq \emptyset$, and the operator F is κ-Lipschitz continuous and η-strongly monotone over $T(H)$. The key contribution of this paper is to utilize a family of certain quasi-nonexpansive operators with its fixed point containing $FixT$ for approximating the solution of VIP$(F, FixT)$. For every $n \in \mathbb{N}$, let $T_n : H \to H_n$ be a ρ_n-quasi-nonexpansive operator with $FixT_n \neq \emptyset$, that is, $\|T_n x - y\|^2 \leq \|x - y\| - \rho_n\|x - T_n x\|^2$ for all $x \in H$ and $y \in FixT_n$. Moreover, it was assumed that $FixT \subset \bigcap_{n=1}^{\infty} FixT_n$. They constructed the so-called generalized hybrid steepest descent method:

$$\begin{cases} x^1 \in H \text{ is chosen,} \\ x^{n+1} = T_n x^n - \mu\alpha_n F(T_n x^n)), \end{cases}$$

for all $n \in \mathbb{N}$. They proved the strong convergence of the generated sequence $\{x^n\}_{n=1}^{\infty}$ to the solution of VIP$(F, FixT)$ provided that the parameter $\mu \in (0, 2\eta/\kappa^2)$, the step size $\{\alpha_n\}_{n=1}^{\infty} \subset (0,1]$ is SDM and the s-approximately shrinking property with respect to $FixT$ (see, [15, Definition 11] for more details).

Cegielski et al. [16] considered VIP$(F, FixT)$ in the setting of a Euclidean space $\mathbb{R}^N$ in which the operator T is a cutter with $FixT \neq \emptyset$, that is, $x - Tx, Tx - y \geq 0$ for all $x \in \mathbb{R}^N$ and $y \in FixT$, and $Id - T$ is demi-closed at 0. Furthermore, the operator F is strongly monotone and continuous. Moreover, they assumed the following assumption: For some $y \in FixT$, there exist $\beta > 0$ and a bounded set $E \subset \mathbb{R}^N$ such that

$$\langle Fx, x - y\rangle \geq \beta\|F(x)\| \qquad \text{for all } x \notin E. \tag{10.5}$$

They proposed an iterative method as follows: Let $x^1 \in \mathbb{R}^N$ be a given point and $\{\alpha_n\}_{n=1}^{\infty}$ be a sequence of positive numbers. For a current iterate $x^n \in \mathbb{R}^N$, calculate the shifted point

$$z^n = \begin{cases} x^n - \alpha_n \dfrac{F(x^n)}{\|F(x^n)\|} & \text{if } F(x^n) \neq \emptyset, \\ x^n & \text{if } F(x^n) = \emptyset. \end{cases}$$

Furthermore, choose $\beta_n \in [\mu, 2-\mu]$ for some $\mu \in (0,1)$ and calculate the next iterate as follows:

$$x^{n+1} = P_{H_n}(z^n) = \begin{cases} z^n - \beta_n \dfrac{\langle z^n - Tx^n, x^n - Tx^n\rangle}{\|x^n - Tx^n\|^2}(x^n - Tx^n) & \text{if } z^n \notin H_n, \\ z^n & \text{if } z^n \in H_n, \end{cases} \tag{10.6}$$

where $H_n = H(x^n, Tx^n) = \{u \in \mathbb{R}^N : \langle u - Tx^n, x^n - Tx^n\rangle \leq 0\}$. By assuming the step size $\{\alpha_n\}_{n=1}^{\infty} \subset (0,1]$ is SDM, the convergence of the sequence $\{x^n\}_{n=1}^{\infty}$ to the solution of VIP$(F, FixT)$ was obtained. It should be noted that the assumption (10.5) seems to be a keystone of the convergence results instead of the Lipschitz continuity of F which is typically assumed when dealing with VIP$(F, FixT)$.

Iemoto, Hishinuma and Iiduka [17] extended the consideration in [11] to the case when T is strongly nonexpansive with $FixT \neq \emptyset$, that is, T is nonexpansive and if for any bounded sequences $\{u^n\}_{n=1}^{\infty}$ and $\{v^n\}_{n=1}^{\infty}$ in H such that $\|u^n - v^n\| - \|Tu^n - Tv^n\| \to 0$, it holds that $\|(u^n - v^n) - (Tu^n - Tv^n)\| \to 0$. The weak convergence results and some numerical examples were obtained in a similar fashion to [11].

Al-Musallam, Cegielski, and Grossmann [18] also extended VIP$(F, FixT)$ in a similar way to [14]. The main contribution is to consider the discretizations of the considered problem and to investigate the convergence of the solutions of discretized problems to the unique solution of the original problem. Moreover, they also relaxed the Lipschitz continuous properties of the operator T and the discretized operators F_n to the locally Lipschitz continuous property. In this situation, the convergence was also discussed in the paper.

Cegielski [19] considered VIP$(F, FixT)$ in the same setting as [15]. By using the idea of the approximating sequence in [15], the generalized version of the hybrid steepest descent method was proposed. To be specific, for every $n \in \mathbb{N}$, let $T_n : H \to$

H be ρ_n-strongly quasi-nonexpansive and $V_n : H \to H$ be quasi-nonexpansive with $FixT \subset \bigcap_{n=1}^{\infty} FixT_n \subset \bigcap_{n=1}^{\infty} FixV_n$. Furthermore, let $F_n : H \to H$ be κ_n-Lipschitz continuous and η_n-strongly monotone with $0 < \underline{\eta} \leq \eta_n \leq \overline{\kappa} < \infty$ for some $\underline{\eta}$ and $\overline{\kappa}$. The proposed method is in the form:

$$\begin{cases} x^1 \in H \text{ is chosen}, \\ x^{n+1} = (1-\beta_n)T_n x^n + \beta_n V_n x^n - (1-\beta_n)\mu\alpha_n F_n(T_n x^n)), \end{cases}$$

for all $n \in \mathbb{N}$. By assuming that $\liminf_{n\to\infty} \rho_n > 0$, $\beta_n \in [0, \beta]$ for some $\beta < 1$, $\lambda_n \in [0, \mu_n]$ where $\mu_n \in \left[\varepsilon, (2\eta_n - \varepsilon)/\kappa_n^2\right]$ for some small $\varepsilon > 0$, and the step size $\{\alpha_n\}_{n=1}^{\infty} \subset (0, 1]$ is SDM, the strong convergence of the generated sequence $\{x^n\}_{n=1}^{\infty}$ to the solution of VIP$(F, FixT)$ was obtained under further assumptions that $\lim_{n\to\infty} \|F_n x^* - Fx^*\| = 0$ and the sequence $\{T_n\}_{n=1}^{\infty}$ satisfies the DC principle. The latter means that, for every bounded sequence $\{x^n\}_{n=1}^{\infty}$ with $\|T_n x^n - x^n\| \to 0$ and for any its weak cluster point y^*, it holds $y^* \in \bigcap_{n=1}^{\infty} FixT_n$. Moreover, the convergence result in the case when the proposed method is approximately shrinking was also investigated to fulfill the case when the family $\{T_n\}_{n=1}^{\infty}$ does not satisfy the DC principle.

Gibali, Reich and Zalas [20] also considered VIP$(F, FixT)$ in the same setting as [16]. By making use of the sequence of cutter operators $\{T_n\}_{n=1}^{\infty}$ in which $FixT \subset \bigcap_{n=1}^{\infty} FixT_n$, and by replacing the operator T in (10.6) by T_n and by updating the next iterate x^{n+1} as the relaxation of $P_{H_n}(z^n)$, they proved the convergence of the proposed method in the same manner as [16].

10.3.2 VIP$\left(F, \bigcap_{i=1}^{m} FixT_i\right)$

Let $T_i : H \to H, i = 1, \ldots, m$, be operators with its nonempty common fixed point set $\bigcap_{i=1}^{m} FixT_i$, and let F be a monotone type operator. In this subsection, we are interested in the variational inequality problem of finding a point $x^* \in \bigcap_{i=1}^{m} FixT_i$ such that

$$\langle x - x^*, Fx^* \rangle \geq 0, \qquad \left(\text{VIP}\left(F, \bigcap_{i=1}^{m} FixT_i\right)\right)$$

for all $x \in \bigcap_{i=1}^{m} FixT_i$. Apart from the works of Yamada [1] and Xu and Kim [8], some interesting research works on solving the variational inequality problem when the constraint set is the intersection of the fixed point sets of nonlinear operators are presented as follows [21].

Zhang and Yang [22] extended VIP$\left(F, \bigcap_{i=1}^{m} FixT_i\right)$ to the case when the operator F is extended to $\mu F - \gamma V$ as in [12]. By replacing the operator T in (10.4) with the composition $T_{n,m}T_{n,m-1}\cdots T_{n,2}T_{n,1}$ where $T_{n,i} := (1-\beta_{n,i})Id + \beta_{n,i}T_i$ for all $i = 1,\ldots,m$. By assuming that $\{\alpha_n\}_{n=1}^{\infty}$ is SDM, $0 < \gamma < \tau/a$ with $\tau := \mu(\eta - 0.5\mu\kappa^2)$ and $\{\beta_{n,i}\}_{n=1}^{\infty} \subset (a,b)$ for some $a, b \in (0,1)$ with $\lim_{n\to\infty} |\beta_{n+1,i} - \beta_{n,i}| = 0$ for all $i = 1,\ldots,m$, the strong convergence of the proposed method to the unique solution of the considered problem was guaranteed.

Gibali, Reich and Zalas [23] extended the investigation of [20] by replacing the finite dimensional Euclidean space with the general real Hilbert space and by replacing the fixed point constraint set with the common fixed point constraint. Instead of assuming the assumption (10.5), they assumed that the operator F satisfied the κ-Lipschitz continuity and η-strongly monotone as a traditional direction. This yields that the boundedness of the generated sequence $\{x^n\}_{n=1}^{\infty}$ was obtained without the normalization of the estimate $F(x^n)$ as in [20]. The strong convergence result was obtained provided that the sequence $\{\alpha_n\}_{n=1}^{\infty}$ was SDM. Some numerical examples were also presented.

Hieu [24] considered VIP$\left(F, \bigcap_{i=1}^{m} FixT_i\right)$ in the case when the operators $T_i, i = 1\ldots,m$, are demi-contractive with the uniform constant $\beta \in [0,1)$ and the operator F is η-strongly monotone and κ-Lipschitz continuous. Hieu proposed the parallel iterative scheme in the following form:

$$\begin{cases} x^1 \in H \text{ is chosen,} \\ x^{n+1} = \sum_{i=1}^{m} \gamma_{n,i} T_{n,i}(Id - \mu\alpha_n F)x^n, \end{cases}$$

where $T_{n,i} := (1-\beta_{n,i})Id + \beta_{n,i}T_i$ for all $n \in \mathbb{N}, i = 1,\ldots,m$. By assuming that $\{\alpha_n\}_{n=1}^{\infty} \subset (0,1]$ is SDM, $\{\beta_{n,i}\}_{n=1}^{\infty} \subset [a,(1-b)/2)$ for some $a, b \in (0,1)$ and $\{\gamma_{n,i}\}_{n=1}^{\infty} \subset (0,1)$ such that $\sum_{i=1}^{m} \gamma_{n,i} = 1$ and $\liminf_{n\to\infty} \gamma_{n,i} > 0$ for all $i = 1,\ldots,m$, the strong convergence of the generated sequence $\{x^n\}_{n=1}^{\infty}$ to the solution of VIP$\left(F, \bigcap_{i=1}^{m} FixT_i\right)$ was obtained. Moreover, some numerical examples were also presented.

Prangprakhon, Nimana and Petrot [25] considered VIP$\left(F, \bigcap_{i=1}^{m} FixT_i\right)$ in the case when the operator F is η-strongly monotone and κ-Lipschitz continuous and the operators $T_i, i = 1,2,\ldots,m$, are firmly nonexpansive operators with $\bigcap_{i=1}^{m} FixT_i \neq \emptyset$.

They proposed an iterative method for solving the problem as follows:

$$\begin{cases} x^1 \in H \text{ is chosen}, \\ y_{n,0} = x^n - \mu\alpha_n F(x^n), \\ y_{n,i} = T_i y_{n,i-1} + e_{n,i}, \ i = 1, \dots, m, \\ x^{n+1} = (1 - \lambda_n) y_{n,0} + \lambda_n y_{n,m}, \end{cases}$$

for all $n \in \mathbb{N}$. An interesting thing about this method is the terms $e_i^n, i = 1 \dots, m$, which can be seen as the additional information when computing the operators T_i' values. The authors stated that, by adding the terms e_i^n, the estimates $y_{n,i}$ might get closer to $FixT_i$ which possibly yields the faster convergence. Moreover, the terms e_i^n can also be viewed as the numerical errors in computing the operators' values which may occur when the explicit form of each T_i is not known, or even when each operator's value is found approximately by solving a subproblem, for instance a metric projection onto a nonempty closed convex set. By assuming that the parameter $\mu \in (0, 2\eta/\kappa^2)$, $\{\alpha_n\}_{n=1}^\infty \subset (0, 1]$ is SDM, and $\{\lambda_n\}_{n=1}^\infty \subset [\varepsilon, 1-\varepsilon]$ for some constant $\varepsilon \in (0, 1/2]$, they proved that the sequence $\{x^n\}_{n=1}^\infty$ generated by the proposed method converges strongly to the unique solution of the considered problem provided that $\sum_{n=1}^\infty \|e_{n,i}\| < \infty$ for all $i = 1, 2, ..., m$.

Prangprakhon and Nimana [26] considered VIP$\left(F, \bigcap_{i=1}^m FixT_i\right)$ in the case when the operator F is η-strongly monotone and κ-Lipschitz continuous, and the operators T_i, $i = 1, 2, \dots, m$, are cutters. To solve this, they presented an iterative algorithm called the extrapolated sequential constraint method with conjugate gradient direction (ESCoM-CGD). This method is motivated by the idea of the accelerated version of HSDM known as the hybrid conjugate gradient method [28] and the idea of the extrapolated cyclic cutter method [30] for solving common fixed-point problems. Denote $T := T_m T_{m-1} \cdots T_1$, $S_0 := Id$, and $S_i := T_i T_{i-1} \cdots T_1$ for all $i = 1, 2, \dots, m$. The proposed method has the form:

$$\begin{cases} x^1 \in H \text{ is chosen}, \\ d^1 = -F(x^1), \\ y^n = x^n + \mu\beta_n d^n, \\ \sigma(y^n) = \begin{cases} \dfrac{\sum_{i=1}^n \langle Ty^n - S_{i-1}y^n, S_i y^n - S_{i-1}y^n\rangle}{\|Ty^n - y^n\|^2}, & \text{for } y^n \neq \bigcap_{i=1}^m FixT_i, \\ 1, \text{ otherwise}, \end{cases} \\ x^{n+1} = T_m(y^n + \lambda_n \sigma(y^n)(Ty^n - y^n)), \\ d^{n+1} = -F(x^{n+1}) + \varphi_{n+1} d^n, \end{cases}$$

for all $n \in \mathbb{N}$, where $\mu \in (0, 2\eta/\kappa^2)$, $\{\beta_n\}_{n=1}^\infty \subset (0, 1]$, $\{\varphi_n\}_{n=1}^\infty \subset [0, \infty)$ and $\{\lambda_n\}_{n=1}^\infty$ is a positive sequence. It can be seen that this method consists of two interesting acceleration schemes: one is the extrapolation function $\sigma(y^n)$ and the other one is the search direction d^n. To prove the convergence of ESCoM-CGD, they assumed that $\lim_{n\to\infty} \beta_n = 0$, $\sum_{n=1}^{\infty} \beta_k = \infty$, $\lim_{n\to\infty} \varphi_n = 0$, and $\{\lambda_n\}_{n=1}^{\infty} \subset [\varepsilon, 2-\varepsilon]$ for a constant $\varepsilon \in (0,1)$. Moreover, they assumed further that the sequence $\{x^n\}_{n=1}^{\infty}$ is bounded and that $\{T_i\}_{i=1}^{m}$ satisfies the DC principle. Under these conditions, the strong convergence was guaranteed. Apart from theoretical perspective, they also presented numerical results and numerical comparisons of the proposed method with some existing algorithms such as the hybrid conjugate gradient method and the hybrid three-term conjugate gradient method [29]. The results demonstrated that in some situation the proposed algorithm had better convergence than the other two methods.

Petrot et al. [27] investigated the solving of VIP$\left(F, \bigcap_{i=1}^{m} FixT_i\right)$ in the case when the operator F is η-strongly monotone and κ-Lipschitz continuous, and the operators T_i, $i = 1, 2, \ldots, m$, are firmly nonexpansive operators. To solve this, they presented an iterative algorithm called the dynamic distributed conjugate gradient method (DDCGM). This method is based on the idea of the hybrid three-term conjugate gradient method (HTCGM) [29] and the extrapolated simultaneous subgradient projection method (ESSPM) [32]. An interesting thing about this method is that, at each iteration, the method allows the computation of each firmly nonexpansive operator along with its dynamic weight to take place independently. This strategy was included in order to improve the convergence behavior of DDCGM by updating control factors. The proposed method is in the form:

$$\begin{cases} x^1 \in H \text{ is chosen,} \\ y^n = x^n + \mu\beta_n \dfrac{d^n}{\max\{1, \|d^n\|\}}, \\ x^{n+1} = y^n + \lambda_n \left(\displaystyle\sum_{i=1}^{m} \omega_i(y^n) T_i y^n - y^n \right), \\ d^{n+1} = -(1+\gamma_{n+1}) F(x^{n+1}) + \varphi_{n+1} \dfrac{d^n}{\max\{1, \|d^n\|\}}, \end{cases}$$

for all $n \in \mathbb{N}$, where $\mu \in (0, 2\eta/\kappa^2)$, $\{\varphi_k\}_{k=1}^{\infty} \subset [0,1]$, $\{\gamma_k\}_{k=1}^{\infty} \subset [0,1]$, $\{\beta_k\}_{k=1}^{\infty} \subset (0,1]$, and $\{\lambda_n\}_{n=1}^{\infty}$ is a nonnegative real sequence. In order to obtain the strong convergence of DDCGM, they supposed that $\lim_{k\to\infty} \varphi_k = 0$, $\lim_{k\to\infty} \gamma_k = 0$, $\lim_{k\to\infty} \beta_k = 0$, $\sum_{k=1}^{\infty} \beta_k = \infty$, and $\{\lambda_n\}_{n=1}^{\infty} \subset [\varepsilon, 2-\varepsilon]$ for a constant $\varepsilon \in (0,1)$. Moreover, they supposed further that the dynamic weight function $w : H \to \Delta_m$ is ρ-regular with respect to $\{T_i\}_{i=1}^{m}$. Unlike [26], the strong convergence of the method was guaranteed without assuming that the sequence $\{x^n\}_{n=1}^{\infty}$ is bounded. Apart from this, they also presented numerical results of DDCGM by applying the method to solve the image classification problem via support vector machines. The experiment demonstrated that the proposed method had better convergence than HTCGM and ESSPM in term of archiving less misclassification rate and using less computational runtime.

10.4 CONCLUSIONS

In this annotated bibliography we provided some research works related to the solving of the well known variational inequality over the fixed point constraints. We reviewed the so-called hybrid steepest descent method and its variances. In our opinions, there are many open research problems relating to the aforementioned topic. The simplest direction is to obtain the boundedness of the generated sequence $\{x^n\}_{n=1}^{\infty}$ as the direction investigated by Petrot et al. [27].

BIBLIOGRAPHY

[1] Yamada, I. (2001). The hybrid steepest descent method for the variational inequality problemover the intersection of fixed point sets of nonexpansive mappings. In D. Butnariu, Y. Censor, & S. Reich (Eds.), *Inherently Parallel Algorithms in Feasibility and Optimization and Their Applications* (pp. 473–504). Elsevier.

[2] Wittmann, R. (1992). Approximation of fixed points of nonexpansive mappings. *Archiv der Mathematik*, 58(5), 486–491.

[3] Bauschke, H. H. (1996). The approximation of fixed points of compositions of nonexpansive mappings in Hilbert space. *Journal of Mathematical Analysis and Applications*, 202(1), 150–159.

[4] Yamada, I., Ogura, N., Yamashita, Y., & Sakaniwa, K. (1998). Quadratic optimization of fixed points of nonexpansive mappings in Hilbert space. *Numerical Functional Analysis and Optimization*, 19(1–2), 165–190.

[5] Deutsch, F., & Yamada, I. (1998). Minimizing certain convex functions over the intersection of the fixed point sets of nonexpansive mappings. *Numerical Functional Analysis and Optimization*, 19(1–2), 33–56.

[6] Ogura, N., & Yamada, I. (2002). Non-strictly convex minimization over the fixed point set of the asymptotically shrinking nonexpansive mapping. *Numerical Functional Analysis and Optimization*, 23(1–2), 113–137.

[7] Ogura, N., & Yamada, I. (2003). Non-strictly convex minimization over the bounded fixed point set of nonexpansive mapping. *Numerical Functional Analysis and Optimization*, 24(1–2), 129–135.

[8] Xu, H. K., & Kim, T. H. (2003). Convergence of hybrid steepest-descent methods for variational inequalities. *Journal of Optimization Theory and Applications*, 119(1), 185–201.

[9] Yamada, I., & Ogura, N. (2005). Hybrid steepest descent method for variational inequality problem over the fixed point set of certain quasi-nonexpansive mappings. *Numerical Functional Analysis and Optimization*, 25(7–8), 619–655.

[10] Zeng, L. C. , Wong, N. C. & Yao, J. C. (2007). Convergence analysis of modified hybrid steepest-descent methods with variable parameters for variational inequalities. *Journal of Optimization Theory and Applications*, 132, 51–69.

[11] Iiduka, H. (2010). A new iterative algorithm for the variational inequality problem over the fixed point set of a firmly nonexpansive mapping. *Optimization*, 59(6), 873–885.

[12] Ceng, L. C., Ansari, Q. H., & Yao, J. C. (2011). Some iterative methods for finding fixed points and for solving constrained convex minimization problems. *Nonlinear Analysis: Theory, Methods & Applications*, 74(16), 5286–5302.

[13] Iiduka, H. (2012). Fixed point optimization algorithm and its application to power control in CDMA data networks. *Mathematical Programming*, 133(1), 227–242.

[14] Cegielski, A., & Grossmann, C. (2013). Iteration–discretization methods for variational inequalities over fixed point sets. *Nonlinear Analysis: Theory, Methods & Applications*, 85, 31–42.

[15] Cegielski, A., & Zalas, R. (2013). Methods for variational inequality problem over the intersection of fixed point sets of quasi-nonexpansive operators. *Numerical Functional Analysis and Optimization*, 34(3), 255–283.

[16] Cegielski, A., Gibali, A., Reich, S., & Zalas, R. (2013). An algorithm for solving the variational inequality problem over the fixed point set of a quasi-nonexpansive operator in Euclidean space. *Numerical Functional Analysis and Optimization*, 34(10), 1067–1096.

[17] Iemoto, S., Hishinuma, K., & Iiduka, H. (2014). Approximate solutions to variational inequality over the fixed point set of a strongly nonexpansive mapping. *Fixed Point Theory and Applications*, 2014(1), 1–14.

[18] Al-Musallam, F., Cegielski, A., & Grossmann, C. (2015). Contraction behaviour of iteration–discretization based on gradient type projections. *Optimization*, 64(1), 25–39.

[19] Cegielski, A. (2015). Application of quasi-nonexpansive operators to an iterative method for variational inequality. *SIAM Journal on Optimization*, 25(4), 2165–2181.

[20] Gibali, A., Reich, S., & Zalas, R. (2015). Iterative methods for solving variational inequalities in Euclidean space. *Journal of Fixed Point Theory and Applications*, 17(4), 775–811.

[21] Zhou, H., & Wang, P. (2014). A new iteration method for variational inequalities on the set of common fixed points for a finite family of quasi-pseudocontractions in Hilbert spaces. *Journal of Inequalities and Applications*, 2014(1), 1–12.

[22] Zhang, C., & Yang, C. (2014). A new explicit iterative algorithm for solving a class of variational inequalities over the common fixed points set of a finite family of nonexpansive mappings. *Fixed Point Theory and Applications*, 2014(1), 1–11.

[23] Gibali, A., Reich, S., & Zalas, R. (2017). Outer approximation methods for solving variational inequalities in Hilbert space. *Optimization*, 66(3), 417–437.

[24] Hieu, D. V. (2019). An explicit parallel algorithm for variational inequalities. *Bulletin of the Malaysian Mathematical Sciences Society*, 42(1), 201–221.

[25] Prangprakhon, M., Nimana, N., & Petrot, N. (2020). A seqential constraint method for solving variational inequality over the intersection of fixed point sets. *Thai Journal of Mathematics*, 18(3), 1105–1123.

[26] Prangprakhon, M., & Nimana, N. (2021). Extrapolated sequential constraint method for variational inequality over the intersection of fixed-point sets. *Numerical Algorithms*, 88(3), 1051–1075.

[27] Petrot, N., Prangprakhon, M., Promsinchai, P., & Nimana, N. (2022). A dynamic distributed conjugate gradient method for variational inequality problem over the common fixed-point constraints. *Numerical Algorithms*, 1–30. (Article in Press)

[28] Iiduka, H., & Yamada, I. (2009). A use of conjugate gradient direction for the convex optimization problem over the fixed point set of a nonexpansive mapping. *SIAM Journal on Optimization*, 19(4), 1881–1893.

[29] Iiduka, H. (2011). Three-term conjugate gradient method for the convex optimization problem over the fixed point set of a nonexpansive mapping. *Applied Mathematics and Computation*, 217(13), 6315–6327. Chicago

[30] Cegielski, A., & Censor, Y. (2012). Extrapolation and local acceleration of an iterative process for common fixed point problems. *Journal of Mathematical Analysis and Applications*, 394(2), 809–818.

[31] Cegielski, A., & Nimana, N. (2019). Extrapolated cyclic subgradient projection methods for the convex feasibility problems and their numerical behaviour. *Optimization*, 68(1), 145–161.

[32] Cegielski, A. (2014). Extrapolated simultaneous subgradient projection method for variational inequality over the intersection of convex subsets. *Journal of Nonlinear and Convex Analysis*, 15(2), 211–218.

CHAPTER 11

Generalized Kannan Maps with Application to Iterated Function System

B. V. Prithvi

SRM Institute of Science and Technology

S. K. Katiyar

Dr B R Ambedkar National Institute of Technology

CONTENTS

ABBREVIATIONS AND NOTATIONS

Abbreviations

- Pompeiu-Hausdorff (PH), Iterated function system (IFS).
- Fixed Point Theory (FPT), Banach contraction principle (BCP).

Notations

- $\emptyset$:= nonempty set; θ_f := fixed point; $\vee, \bigvee$:= max.
- $\mathbb{N}$:= The set of positive integers.
- $\mathbb{N}_N$:= The set of first N positive integers.
- $\mathbb{R}_+$:= The set of all positive real numbers.

DOI: 10.1201/9781003388678-11

- $\mathbb{R}^0_+$:= The set of all non-negative real numbers.
- $(\mathcal{M},\mu)$:= Metric space, $(\mathcal{H}(\mathcal{M}),\sigma(\mu))$:= PH metric space.
- $G\theta := G(\theta)$; $G^{\circ k}(\theta)$:= kth composition of map G at θ.
- $2^{\mathcal{M}} - \{\emptyset\}$:= The set of all nonempty subsets of $\mathcal{M}$.
- $\mathcal{H}(\mathcal{M})$:= The set of all nonempty compact subsets of $\mathcal{M}$.
- $\mu, \tau, \tau^*, \sigma, \tau_m, \tau_M, \mu^*$:= Distance functionals.

11.1 INTRODUCTION

Post the contraction map brought in by Banach [1], the one which received equal appraisal and attention among the fixed point theorists is a map described by Kannan [2], now called the Kannan map. The objective behind both these maps is to claim the existence and uniqueness of a fixed point for any map G that satisfies their respective conditions over a complete metric space $(\mathcal{M},\mu)$. However, while the contraction condition necessitates G to be continuous, the Kannan condition does not. This way, Kannan allowed continuous and discontinuous maps to show up for the uniqueness of fixed points. In addition to this, the Kannan map holds the property of metric completeness, a property which states – "a metric space $(\mathcal{M},\mu)$ is complete if and only if every Kannan map G on $\mathcal{M}$ has a fixed point" (cf. [9], [33]). Whereas, in the case of the contraction map, there exists a metric space $(\mathcal{M},\mu)$ that is not complete yet every contraction map G on $\mathcal{M}$ has a fixed point (see [10]). Also, a unique description of Kannan's condition during the 1960s allowed the Kannan map to stand out among various other variants ([4]–[8]) of the contraction map that appeared during the same decade. For details of maps that Kannan map inspired further, one may see [3], [34], [35], and [36].

It has been 100 years since Banach's breakthrough, popularly known as BCP, happened. Yet, its applications are still on the cards. One such application was brought out by Hutchinson [11] when he proved the existence of a unique fixed point for any finite union of contraction maps. It acts as a pathway to the generation of fractals such as the Cantor set, Sierpiński triangle, Weierstrass curve, etc. (see [12]). As a result, the fixed point associated with a "finite union of maps of the same class (called Hutchinson map)" is called a *fractal* in general. For example, consider contraction maps $P_1, P_2 : \mathbb{R} \to \mathbb{R}$ defined by

$$G_1\theta = \frac{\theta}{3}, G_2\theta = \frac{\theta+2}{3}, \ \forall\, \theta \in \mathbb{R}.$$

Then there exists an element $\mathcal{C} \in \mathcal{H}(\mathcal{M})$ such that

$$(G_1 \cup G_2)(\mathcal{C}) = \mathcal{C} \text{ and } \lim_{\mathbb{N}\ni k\to\infty} (G_1 \cup G_2)^{\circ k}([0,1]) = \mathcal{C}.$$

Here, $\mathcal{C}$ is the Cantor set (a fractal). Further, Barnsley [12] popularized this theory of fractals by introducing the notion of IFS. Ever since then, IFS got included in various applications of BCP. Classically, it is a finite collection of contraction maps defined on complete metric space $(\mathcal{M}, \mu)$, and denoted by $\{G_i; (\mathcal{M}, \mu) \mid i \in \mathbb{N}_N\}$. The terminology of IFS is a narrative of the action of the Hutchinson map that approaches (converges) fractal after several iterates (compositions) of it, taken at any arbitrary member of $\mathcal{M}$.

Since fractal is a fixed point of the Hutchinson map, it attracted fixed point researchers in the 21st century to replace contractions (in the classical IFS) with its variants and solve the fixed point problem. For example, see, countable contractions in [13], F-contractions in [14], weak θ-contractions in [15], cyclic ϕ-contractions in [16], cyclic Meir-Keeler contractions in [17], Kannan maps in [18], weakly contractive maps in [27], generalized enriched contractions in [28], interpolative δ-operators in [29] and cyclic weak ϕ-contractions in [30]. But, with regard to Kannan maps from Sahu et al. [18], the problem remained unsolved, as reported by Dung and Petruşel [20] via a counter-example. In addition, Dung and Petruşel provided a partial solution by solving for only a pair of Kannan maps with added commutativity assumptions. On the other hand, Miculescu and Mihail [26] presented a solution in the context of Reich-type IFS. However, they involved severe impositions that increase complexity of the system, as depicted by Georgescu et al. (see section 4, [19]) in their similar kind of system of Hardy-Rogers type. It marks the interest of this chapter wherein generalizations of the Kannan map, namely δ-Kannan and generalized δ-Kannan, are introduced via incorporation of constant $\delta \in [1, \infty)$ in Kannan's condition. Unlike any other generalization of the Kannan map in the literature, for example, see [21]–[25], the ones nurtured here hold the ability to replace the original Kannan map for it improves the range of the Kannan parameter from $(0, \frac{1}{2})$ to $[0, 1)$ (see Remark 11.1). Further, using a finite collection of generalized δ-Kannan maps, the unsolved fixed point problem associated with Kannan IFS is solved as an application to generalized Kannan IFS.

Section 11.2 introduces the improved version of the Kannan map with an example to illustrate the same. It also includes the fixed point theorem associated with it that shows it is a Picard operator. Section 11.3 takes the improvement of the Kannan map in Section 11.2 to a higher level with an aim to resolve the fixed point problem of Kannan IFS. Lemma 11.2 plays the head role in determining its solution along with Example 11.2 that resolves the issue raised in Example 2.4 of [20]. Finally, Section 11.3 concludes with a brief description of the work undertaken in this chapter.

11.2 GENERALIZED KANNAN MAP

Definition 11.1 *Let $(\mathcal{M}, \mu)$ be a metric space. Suppose there exists a $\delta \in [1, \infty)$ such that $G : (\mathcal{M}, \mu) \to (\mathcal{M}, \mu)$ satisfies the following*

$$\mu(G\theta, G\vartheta) \leq \beta. \left[\frac{\mu(\theta, G\theta)}{\delta} + \mu(\vartheta, G\vartheta)\right] \ \forall \ \theta, \vartheta \in \mathcal{M} : \theta \neq \vartheta, \tag{11.1}$$

where

$$\beta \in [0, \infty) : \beta < \frac{\delta}{1+\delta}.$$

*Then G is said to be a **δ-Kannan** map .*

Remark 11.1 $0 \leq \beta < \frac{\delta}{1+\delta} \Rightarrow \beta \in [0,1) \; \forall \; \delta \in [1, \infty)$.

Remark 11.2 *When $\delta = 1$, the Kannan [2] map is retrieved.*

Remark 11.3 *In the case of the Kannan map, the range varies from 0 to $\frac{1}{2}$. Whereas, in the case of the δ-Kannan map, the range of β varies from 0 to 1. It is due to the incorporation of the parameter δ that acts as a regulator for β in the interval $[0, 1)$.*

Let us revisit the example proposed by Reich [3].

Example 11.1 *Define $G : ([0,1].|.|) \to ([0,1].|.|)$ by*

$$G(\theta) = \begin{cases} \frac{\theta}{3}, & \theta \in [0,1), \\ \frac{1}{6}, & \theta = 1. \end{cases}$$

*Then G being discontinuous is not a contraction. But, G **is not a Kannan map** either. This is because for $\theta = 0, \vartheta = \frac{1}{3}$*

$$|G0 - G\frac{1}{3}| = |0 - \frac{1}{9}| \leq \beta(|0 - G0| + |\frac{1}{3} - G\frac{1}{3}|) = \beta(|0-0| + |\frac{1}{3} - \frac{1}{9}|).$$

It gives $\beta \geq \frac{1}{2}$. In general, we obtain $\beta \geq \frac{1}{2} . \frac{|\theta - \vartheta|}{|\theta| + |\vartheta|} \; \forall \; \theta \in [0,1)$. Now, in the light of the δ-Kannan condition we have

$$\frac{1}{3}|\theta - \vartheta| \leq \frac{2\beta}{3}(\frac{|\theta|}{\delta} + |\vartheta|) \Rightarrow \beta \geq \frac{1}{2}.\frac{|\theta - \vartheta|}{\frac{|\theta|}{\delta} + |\vartheta|} \; \forall \; \theta \in [0,1).$$

*In particular, if we consider the previous case $\theta = 0, \vartheta = \frac{1}{3}$, we obtain $\beta \geq \frac{1}{2}$ which is true for any $\delta \in (1, \infty)$. Thus, G **is a δ-Kannan map**, with $\delta \in (1, \infty)$.*

Definition 11.2 *[31] Let $(\mathcal{M}, \mu)$ be a metric space and $G : \mathcal{M} \to \mathcal{M}$ be its self-map. For each $\theta \in \mathcal{M}$, if the sequence of iterates $< G^{\circ n}(\theta) >$ converge to θ_f, i.e., if*

$$\exists \text{ unique } \theta_f \in \mathcal{M} : G(\theta_f) = \theta_f = \lim_{\mathbb{N} \ni n \to \infty} G^{\circ n}(\theta), \forall \; \theta \in \mathcal{M},$$

then G is called a Picard operator.

Theorem 11.1 *Let $(\mathcal{M}, \mu)$ be a complete metric space and let $G : \mathcal{M} \to \mathcal{M}$ be a δ-Kannan map. Then G is a Picard operator.*

Proof 11.1 *Let $\theta, \vartheta \in \mathcal{M} : \theta \neq \vartheta$. Using triangle inequality, we deduce*

$$\mu(\theta,\vartheta) \leq \mu(\theta, G\theta) + \beta\left[\frac{\mu(\theta, G\theta)}{\delta} + \mu(\vartheta, G\vartheta)\right] + \mu(P\vartheta, \vartheta).$$

Inequating, we get

$$\mu(\theta,\vartheta) \leq \left(1+\frac{\beta}{\delta}\right).\mu(\theta, P\theta) + (1+\beta).\mu(\vartheta, P\vartheta). \tag{11.2}$$

Now, consider a sequence of iterates $< G^{\circ n}(\theta) >_{\theta\in\mathcal{M}}$. From Definition 11.1, we see that, after iteration of any j steps

$$\mu\big(G^{\circ j}(\theta), G^{\circ(j+1)}(\theta)\big) \leq s^j.\mu\big(\theta, P\theta\big), j \in \mathbb{N}, \tag{11.3}$$

where $s := \frac{\frac{\beta}{\delta}}{1-\beta}$, and

$$\beta < \frac{\delta}{1+\delta} \Rightarrow \left(\frac{1+\delta}{\delta}\right)\beta < 1 \Rightarrow \frac{\beta}{\delta} + \beta < 1 \Rightarrow s < 1.$$

Replace θ by $G^{\circ r}(\theta)$ and v by $G^{\circ t}(\theta)$ respectively in inequation (11.2), then in view of inequation (11.3) we obtain,

$$\mu\big(G^{\circ r}(\theta), G^{\circ t}(\theta)\big) \leq \left[\left(1+\frac{\beta}{\delta}\right).s^r + (1-\beta).s^t\right]\mu(\theta, G\theta).$$

It implies that for any $n \in \mathbb{N}$, the sequence $< G^{\circ n}(\theta) >$ of G-iterates generated at any $\theta \in \mathcal{M}$, is a Cauchy sequence. Since $\mathcal{M}$ is complete, the Cauchy sequence possesses a limit say $\theta^* \in \mathcal{M}$, i.e., we have $\theta^* = \lim_{n\to\infty} G^{\circ n}(\theta)$. Claim that θ^* is a fixed point of G. To check, assume $\mu(\theta^*, G\theta^*) > 0$. From Definition 11.1 of the δ-Kannan map, it follows that

$$\mu\big(G^{\circ(n+1)}\theta, G\theta^*\big) \leq \beta.\left[\frac{\mu\big(G^{\circ(n+1)}\theta, G^{\circ(n)}\theta\big)}{\delta} + \mu\big(G\theta^*, \theta^*\big)\right].$$

As $n \to \infty$, we get

$$(1-\beta).\mu(\theta^*, G\theta^*) = 0 \Rightarrow \mu(\theta^*, G\theta^*) = 0,$$

which is a contradiction. Hence, $\mu(\theta^*, G\theta^*) = 0$. This infers that θ^* is a fixed point. For the uniqueness of fixed point, one may check it through the traditional approach of assuming existence of another fixed point say $\theta^{**}(\neq \theta^*)$. It shall follow by contradiction that $\theta_f := \theta^* = \theta^{**}$. This concludes the proof.

Remark 11.4 *In Theorem 11.1, if G is continuous, we have*

$$G(\theta^*) = G(\lim_{n\to\infty} G^{\circ n}(\theta)) = \lim_{n\to\infty} G^{\circ(n+1)}(\theta) = \theta^*,$$

to affirm that θ^* is a fixed point of G.

Note 11.1 *In Definition 11.1, if the condition (11.1) is replaced by the following*

$$\mu(G\theta, G\vartheta) \leq \beta.\left[\mu(\theta, G\theta) + \frac{\mu(\vartheta, G\vartheta)}{\delta}\right],$$

the previous results are still true.

11.3 APPLICATION TO ITERATED FUNCTION SYSTEM

Definition 11.3 *Let $(\mathcal{M}, \mu)$ be a metric space. Suppose there exists a $\delta \in [1, \infty)$ such that $G : (\mathcal{M}, \mu) \to (\mathcal{M}, \mu)$ satisfies the following*

$$\mu(G\theta, G\vartheta) \leq \beta.\max\left\{\left[\frac{\mu(\theta, G\theta)}{\delta} + \mu(\vartheta, G\vartheta)\right], \left[\mu(\theta, G\theta) + \frac{\mu(\vartheta, G\vartheta)}{\delta}\right]\right\},$$

for each $\theta, \vartheta \in \mathcal{M} : \theta \neq \vartheta$, where

$$\beta \in [0, \infty) : 0 \leq \beta < \frac{\delta}{1+\delta}.$$

*Then G is said to be a **generalized δ-Kannan** map.*

Lemma 11.1 *Let $(\mathcal{M}, \mu)$ be a complete metric space and let $G : \mathcal{M} \to \mathcal{M}$ be a generalized δ-Kannan map. Then G is a Picard operator.*

Proof 11.2 *Proof shall follow on similar lines as in that of Theorem 11.1, so omitted.*

Definition 11.4 *[32] If $\Theta, \Upsilon \in 2^{\mathcal{M}} - \{\emptyset\}$, the functionals $\tau, \tau^*, \sigma, \tau_m, \tau_M, \mu^*$: $(2^{\mathcal{M}} - \{\emptyset\}) \times (2^{\mathcal{M}} - \{\emptyset\}) \to \mathbb{R}^0_+$ are defined as follows –*

- $\tau(\Theta, \Upsilon) = \inf\{\mu(\theta, \upsilon) \mid \theta \in \Theta, \upsilon \in \Upsilon\}$. *In particular, if $\theta_0 \in \mathcal{M}$ then* $\tau(\theta_0, \Upsilon) := \tau(\{\theta_0\}, \Upsilon)$.

- $\tau^*(\Theta, \Upsilon) = \sup\{\tau(\theta, \Upsilon) \mid \theta \in \Theta\}$, $\tau^*(\Upsilon, \Theta) = \sup\{\tau(\upsilon, \Theta) \mid \upsilon \in \Upsilon\}$.

- $\sigma(\Theta, \Upsilon) = \tau^*(\Theta, \Upsilon) \vee \tau^*(\Upsilon, \Theta)$.

- $\tau_m(\Theta, G(\Theta)) = \min\limits_{\theta \in \Theta} \mu(\theta, G\theta), \tau_M(\Theta, G(\Theta)) = \max\limits_{\theta \in \Theta} \mu(\theta, G\theta)$.

- $\mu^*(\Theta, \Upsilon) = \sup\{\mu(\theta, \upsilon) \mid \theta \in \Theta, \upsilon \in \Upsilon\}$.

Here, τ is called the gap functional between U and V, τ^* is the excess functional (excess of U over V and V over U respectively), σ is the PH functional, and μ^* is called the diameter between U and V.

Remark 11.5 1. *For each $\theta_0 \in \Theta, \upsilon_0 \in \Upsilon$,*

$$\mu^*(\Theta, \Upsilon) \geq \sigma(\Theta, \Upsilon) \geq \begin{cases} \tau^*(\Theta, \Upsilon) \geq \tau(\theta_0, \Upsilon) \geq \tau(\Theta, \Upsilon), \\ \tau^*(\Upsilon, \Theta) \geq \tau(\upsilon_0, \Theta) \geq \tau(\Upsilon, \Theta). \end{cases}$$

2. $\tau_m(\Theta, G(\Theta)),\ \sigma(\Theta, G(\Theta)) \leq \tau_M(\Theta, G(\Theta)), \forall\ \Theta \in \mathcal{H}(\mathcal{M})$.

Remark 11.6 *[12] $(\mathcal{H}(\mathcal{M}), \sigma(\mu))$ forms a metric space, called Hausdorff space (in particular, Hausdorff metric space).*

Proposition 11.1 *[12] (Property of compactness) Consider two sets $\Theta, \Upsilon \in \mathcal{H}(\mathcal{M})$. Then*

$$\exists\ \hat{\upsilon} \in \Upsilon, \hat{\theta} \in \Theta : \tau(\theta, \Upsilon) = \mu(\theta, \hat{\upsilon}), \tau^*(\Theta, \Upsilon) = \tau(\hat{\theta}, \Upsilon).$$

Proposition 11.2 *Let $(\mathcal{M}, \mu)$ be a metric space and suppose $G : \mathcal{M} \to \mathcal{M}$ be continuous. Consider the induced map $G_H := G : \mathcal{H}(\mathcal{M}) \to \mathcal{H}(\mathcal{M})$, given by*

$$G_H(\Theta) = \bigcup_{\theta \in \Theta} \{G\theta\}, \forall\ \Theta \in \mathcal{H}(\mathcal{M}).$$

Then G_H is well-defined.

Proof 11.3 *Since G is a continuous self-map of $\mathcal{M}$, the G-image of any compact subset of $\mathcal{M}$ is compact. So, we may infer that*

$$\Theta \in \mathcal{H}(\mathcal{M}) \Rightarrow G_H(\Theta) \in \mathcal{H}(\mathcal{M}),$$

for each Θ, i.e., G_H is well-defined.

Proposition 11.3 *If $a, b \in \mathbb{R}^+ : a \leq b$, then $\exists\ c \in [1, \infty) : a \leq b \leq c.a$*

Note 11.2 *The constant c in the aforementioned proposition is the least upper bound of all constants in $[1, \infty)$ for which the result is true.*

Lemma 11.2 *If G is a* ***generalized δ-Kannan*** *map, i.e., if there exists $\beta \in [0, 1)$ for some $\delta \in [1, \infty)$ such that, for each $\theta, \vartheta \in \mathcal{M} : \theta \neq \vartheta$,*

$$\mu(G\theta, Gv) \leq \beta. \max\left\{ \left[\frac{\mu(\theta, G\theta)}{\delta} + \mu(v, Gv)\right], \left[\mu(\theta, G\theta) + \frac{\mu(v, Gv)}{\delta}\right] \right\},$$

then G_H (if well-defined) is also so, i.e., there exists $\beta^* \in [0,1)$ (for the same δ) so that, for each $\Theta, \Upsilon \in \mathcal{H}(\mathcal{M}) : \Theta \neq \Upsilon$,

$$\sigma(G_H(\Theta), G_H(\Upsilon)) \\ \leq \beta^*. \max\left\{ \left[\frac{\sigma(\Theta, G_H(\Theta))}{\delta} + \sigma(\Upsilon, G_H(\Upsilon))\right], \\ \left[\sigma(\Theta, G_H(\Theta)) + \frac{\sigma(\Upsilon, G_H(\Upsilon))}{\delta}\right] \right\}.$$

Proof 11.4 *Suppose G_H be well-defined. Let $\Theta, \Upsilon \in \mathcal{H}(\mathcal{M}) : \Theta \neq \Upsilon$ be arbitrary. Then, by the hypothesis,*

$$\mu(G\theta, G\vartheta) \leq \beta. \max\left\{ \left[\frac{\mu(\theta, G\theta)}{\delta} + \mu(\vartheta, G\vartheta)\right], \left[\mu(\theta, G\theta) + \frac{\mu(\vartheta, G\vartheta)}{\delta}\right] \right\}$$

is true, for each $\theta \in \Theta, \vartheta \in \Upsilon$. Consider (fixed) $\theta^* \in \Theta$ be arbitrary. It follows that

$$\tau(G\theta^*, G_H(\Upsilon)) \leq \mu(G\theta^*, G\vartheta), \\ \leq \beta. \max\left\{ \left[\frac{\mu(\theta^*, G\theta^*)}{\delta} + \mu(\vartheta, Gv)\right], \left[\mu(\theta^*, G\theta^*) + \frac{\mu(\vartheta, G\vartheta)}{\delta}\right] \right\}, \forall\, \vartheta \in \Upsilon.$$

Thereby,

$$\tau(G\theta^*, G_H(\Upsilon)) \\ \leq \beta. \max\left\{ \left[\frac{\mu(\theta^*, G\theta^*)}{\delta} + \tau_m(\Upsilon, G_H(\Upsilon))\right], \\ \left[\mu(\theta^*, G\theta^*) + \frac{\tau_m(\Upsilon, G_H(\Upsilon))}{\delta}\right] \right\}$$

is true. But, since θ^* is arbitrary, in general we have

$$\tau(G\theta, G_H(\Upsilon)) \\ \leq \beta. \max\left\{ \left[\frac{\mu(\theta, G\theta)}{\delta} + \tau_m(\Upsilon, G_H(\Upsilon))\right], \\ \left[\mu(\theta, G\theta) + \frac{\tau_m(\Upsilon, G_H(\Upsilon))}{\delta}\right] \right\}, \\ \leq \beta. \max\left\{ \left[\frac{\tau_M(\Theta, G_H(\Theta))}{\delta} + \tau_M(\Upsilon, G_H(\Upsilon))\right], \\ \left[\tau_M(\Theta, G_H(\Theta)) + \frac{\tau_M(\Upsilon, G_H(\Upsilon))}{\delta}\right] \right\},$$

true for each $\theta \in \Theta$. From Remark 11.5(2.), we have

$$\tau^*(A, G_H(A)) \leq \tau_M(A, G_H(A)), \forall\, A \in \mathcal{H}(\mathcal{M}).$$

Consequently, by Proposition 11.3, there exists a constant $e_A \in [1, \infty)$ *such that*

$$\tau^*(A, G_H(A)) \leq \tau_M(A, G_H(A)) \leq e_A.\tau^*(A, G_H(A)), \forall\ A \in \mathcal{H}(\mathcal{M}).$$

So, it follows that

$$\tau(G\theta, G_H(\Upsilon)) \leq \beta.\max\left\{ \left[\frac{e_\Theta.\tau^*(\Theta, G_H(\Theta))}{\delta} + e_\Upsilon.\tau^*(\Upsilon, G_H(\Upsilon))\right], \left[e_\Theta.\tau^*(\Theta, G_H(\Theta)) + \frac{e_\Upsilon.\tau^*(\Upsilon, G_H(\Upsilon))}{\delta}\right] \right\}, \forall\ \theta \in \Theta.$$

Then, by the compactness of Θ *(Proposition 11.1), we have*

$$\begin{aligned} &\tau^*(G_H(\Theta), G_H(\Upsilon)) \\ &\leq \beta(e_\Theta \vee e_\Upsilon).\max\left\{ \left[\frac{\tau^*(\Theta, G_H(\Theta))}{\delta} + \tau^*(\Upsilon, G_H(\Upsilon))\right], \right. \\ &\left. \left[\tau^*(\Theta, G_H(\Theta)) + \frac{\tau^*(\Upsilon, G_H(\Upsilon))}{\delta}\right] \right\}, \\ &\leq \beta(e_\Theta \vee e_\Upsilon).\max\left\{ \left[\frac{\sigma(\Theta, G_H(\Theta))}{\delta} + \sigma(\Upsilon, G_H(\Upsilon))\right], \right. \\ &\left. \left[\sigma(\Theta, G_H(\Theta)) + \frac{\sigma(\Upsilon, G_H(\Upsilon))}{\delta}\right] \right\}, \end{aligned}$$

in view of Remark 11.5(1.).
Similarly, one may witness the same for $\tau^*(G_H(\Upsilon), G_H(\Theta))$*. But, since* $\Theta, \Upsilon \in \mathcal{H}(\mathcal{M})$ *are arbitrary, let* $e := \sup_{A \in \mathcal{H}(\mathcal{M})} e_A$*. We obtain,*

$$\sigma(G(\Theta), G(\Upsilon)) \leq \beta^*.\max\left\{ \left[\frac{\sigma(\Theta, G_H(\Theta))}{\delta} + \sigma(\Upsilon, G_H(\Upsilon))\right], \left[\sigma(\Theta, G_H(\Theta)) + \frac{\sigma(\Upsilon, G_H(\Upsilon))}{\delta}\right] \right\},$$

true for each $\Theta, \Upsilon \in \mathcal{H}(\mathcal{M}) : \Theta \neq \Upsilon$*, with* $\beta^* := e\beta$*. Thus, whenever there exists* $\beta \in [0, 1)$ *corresponding to some* $\delta \in [1, \infty)$ *such that* G *is a generalized* δ*-Kannan map, there exists a* $\beta^* \in [0, 1)$ *so that* G_H *is also a generalized* δ*-Kannan map, for the same* δ*.*

Remark 11.7 *Indeed, such a* β^* *exists, if* δ *is properly chosen according to the value of* e *in the following way –*

$$\begin{cases} \text{Choose } \delta \in (1, \infty)\ \forall\ e \in (1, 2), & \text{so that } \exists\ \beta < \frac{1}{e} < 1. \\ \text{Choose } \delta = 1\ \forall\ e \in [2, \infty), & \text{so that } \exists\ \beta < \frac{1}{2e} < \frac{1}{2}. \end{cases}$$

Proof 11.5 *From Definition 11.3 we have,* $\beta < \frac{\delta}{1+\delta} < 1 \; \forall \; \delta \geq 1$*. So, if*

$$\beta < \frac{1}{e} \leq \frac{\delta}{1+\delta} < 1, \; for \; some \; \delta > 1, \; e > 1 \Rightarrow \; \exists \; \beta^* < 1,$$

for some $\delta > 1$*. But, we want to have a common value of* δ *for which the aforesaid is true. So, we choose* β *according to the equation* $\beta < \frac{\delta}{e(1+\delta)}$*. This suggests one may choose*

$$\delta \geq \frac{1}{e-1} \; \forall \; e \in (1,2) : \exists \; \beta < \frac{\delta}{e(1+\delta)} < \frac{\delta}{1+\delta} \Rightarrow \exists \; \beta^* < \frac{\delta}{1+\delta}.$$

Else if, for some $e \geq 1$*,*

$$\beta < \frac{\delta}{1+\delta} < \frac{1}{e} \leq 1, \; for \; some \; \delta > 1 \Rightarrow \; \exists \; \beta^* < \frac{e\delta}{1+\delta} < 1,$$

for some $\delta > 1$*. In fact,*

$$\begin{aligned} \frac{e\delta}{1+\delta} < 1 \; &\Rightarrow \delta < \frac{1}{e-1}\left(\Leftrightarrow 1 < e < 2 \right), \\ &\Rightarrow 1+\delta < \frac{e}{e-1}, \\ &\Rightarrow \frac{1}{1+\delta} > \frac{e-1}{e}, \\ &\Rightarrow \frac{\delta}{1+\delta} > \frac{\delta(e-1)}{e} = \beta, \\ &\Rightarrow \frac{e\delta}{1+\delta} > \delta(e-1) = \beta^*. \end{aligned}$$

But, again, we want to have a common value of δ *for which the aforesaid is true. So, we choose* β *according to the equation* $\beta < \frac{\delta}{e(1+\delta)}$*. In particular,* $\beta = \frac{\delta(e-1)}{e^2}$ *and* $\beta^* = \frac{\delta(e-1)}{e}$ *would suffice for a common value of* δ*. Thus, it suggests one may choose*

$$1 < \delta < \frac{1}{e-1} \; \forall \; e \in (1,2) : \exists \; \beta < \frac{\delta}{e(1+\delta)} < \frac{\delta}{1+\delta} \Rightarrow \exists \; \beta^* < \frac{\delta}{1+\delta}.$$

On the other hand, for each $e \geq 2$*,*

$$\beta^* < \frac{1}{2} \Leftrightarrow \beta < \frac{1}{2e} < \frac{1}{2} \Leftrightarrow \delta = 1.$$

To understand the previous result, let us revisit the example (Example 2.4., [20]) given by Dung and Petruşel to contradict results concerned with the Kannan map in [18]. The generalized δ-Kannan map becomes Kannan when $\delta = 1$, so in what follows, we shall prove the example in assertive for $\delta \in (1, \infty)$.

Example 11.2 *Define a metric space $(\mathcal{M}, \mu)$, where $\mathcal{M} = \{0, 1, 2\}$, under the distance function $\mu : \mathcal{M} \times \mathcal{M} \to \mathbb{R}^+$ defined by*

$$\begin{aligned} \mu(0,0) &= \mu(1,1) = \mu(2,2) = 0, \\ \mu(0,1) &= \mu(1,0) = 5, \\ \mu(1,2) &= \mu(2,1) = 2, \\ \mu(0,2) &= \mu(2,0) = 3. \end{aligned}$$

Let $G : \mathcal{M} \to \mathcal{M} : G0 = 1, G1 = G2 = 2$. Then there exists $\beta \in [0, 1)$ such that G is a generalized δ-Kannan map for some $\delta \in (1, \infty)$, and there exists $\beta^* \in [0, 1)$ so that G_H is also a generalized δ-Kannan map, for the same $\delta \in (1, \infty)$.

Proof 11.6 *Let us first see through the following cases to show that G is generalized δ-Kannan.*

Case 11.1 $(\theta, \vartheta) \in \{(0, 0), (1, 1), (2, 2), (1, 2)\}$

Here,

$$\mu(G\theta, G\vartheta) \leq \beta. \max\left\{ \left[\frac{\mu(\theta, G\theta)}{\delta} + \mu(\vartheta, G\vartheta)\right], \left[\mu(\theta, G\theta) + \frac{\mu(\vartheta, G\vartheta)}{\delta}\right] \right\}$$

is true, for each $\beta \in [0, 1)$.

Case 11.2 $(\theta, \vartheta) \in \{(0, 1), (0, 2)\}$

For $\theta = 0, \vartheta = 1$, we have

$$2 \leq \beta. \max\{\frac{5}{\delta} + 2, 5 + \frac{2}{\delta}\} \Rightarrow \beta \geq \frac{2}{\frac{2}{\delta} + 5} > \frac{2}{7} \ or \ \beta \geq \frac{2}{2 + \frac{5}{\delta}} > \frac{2}{7},$$

whereas for $\theta = 0, \vartheta = 2$, we have $2 \leq 5\beta$, i.e., $\beta \geq \frac{2}{5}$.

Thus, there exists $\beta \in [\frac{2}{5}, 1)$ such that G is a generalized δ-Kannan map for any $\delta \in (1, \infty)$. Here, $\mathcal{H}(\mathcal{M})$ is the power set of $\mathcal{M}$ without the member $\emptyset$, i.e., $\{\{0\}, \{1\}, \{2\}, \{0, 1\}, \{0, 2\}, \{1, 2\}, \mathcal{M}\}$. Since G is continuous, $G_H : \mathcal{H}(\mathcal{M}) \to \mathcal{H}(\mathcal{M})$ is well-defined (by Proposition 11.2). Next, to proceed further, to show that G_H is also generalized δ-Kannan through various cases like in the previous cases is a hideous task. So, instead, we try to find

$$e_A \geq \frac{\tau_M(A, G_H(A))}{\tau^*(A, G_H(A))}, \forall \ A \in \mathcal{H}(\mathcal{M}).$$

- *For singleton sets, $e_A \geq 0$.*

- *For $A = \{0, 1\}$, $\tau_M(A, G_H(A)) = 5, \tau^*(A, G_H(A)) = 3$. So, $e_A \geq \frac{5}{3}$.*

- *For* $A = \{0, 2\}$, $\tau_M(A, G_H(A)) = 5, \tau^*(A, G_H(A)) = 3$. *So,* $e_A \geq \frac{5}{3}$.
- *For* $A = \{1, 2\}$, $\tau_M(A, G_H(A)) = 2, \tau^*(A, G_H(A)) = 2$. *So,* $e_A \geq 1$.
- *For* $A = \mathcal{M}$, $\tau_M(A, G_H(A)) = 5, \tau^*(A, G_H(A)) = 3$. *So,* $e_A \geq \frac{5}{3}$.

In view of these derivations, we have $e \in [\frac{5}{3}, \infty)$. *Consider* $e = \frac{5}{3}$. *Then, in view of the proof of Lemma 11.2, for each* $\Theta, \Upsilon \in \mathcal{H}(\mathcal{M}) : \Theta \neq \Upsilon$, *we have*

$$\sigma(G_H(\Theta), G_H(\Upsilon)) \leq \beta^* . \max \left\{ \left[\frac{\sigma(\Theta, G_H(\Theta))}{\delta} + \sigma(\Upsilon, G_H(\Upsilon)) \right], \left[\sigma(\Theta, G_H(\Theta)) + \frac{\sigma(\Upsilon, G_H(\Upsilon))}{\delta} \right] \right\},$$

where $\beta^* = \frac{5\beta}{3}$. *Now, choose* $\beta \in [\frac{2}{5}, \frac{3}{5})$. *Then* $\beta^* \in [\frac{2}{3}, 1)$, *where*

$$\beta < \frac{3}{5} \leq \frac{\delta}{1+\delta} < 1 \Rightarrow \delta \geq \frac{3}{2}.$$

But, we want to have a common value (range) of δ. *So, choose*

$$\beta < \frac{3\delta}{5(1+\delta)} < \frac{3}{5} < \frac{\delta}{1+\delta} < 1 \Rightarrow \beta^* < \frac{\delta}{1+\delta}.$$

Check for $\delta = 3$. *We see that there exists* $\beta \in [\frac{2}{5}, \frac{9}{20})$ *and* $\beta^* \in [\frac{2}{3}, \frac{3}{4})$. *Therefore, we may infer that there exists* $\beta \in [0, 1)$ *such that if* G *is a generalized* δ*-Kannan map for some* $\delta \in (1, \infty)$, *there do also exists* $\beta^* \in [0, 1)$ *so that* G_H *is a generalized* δ*-Kannan map for the same* $\delta \in (1, \infty)$.

The following definition is an improvement to the one proposed by Barnsley [12].

Definition 11.5 *Let* $(\mathcal{M}, \mu)$ *be a complete metric space. A finite collection of generalized* δ*-Kannan maps* $\{G_k : \mathcal{M} \to \mathcal{M} \mid k \in \mathbb{N}_N\}$ *that corresponds to the finite collection of generalized* δ*-Kannan maps* $\{G_{H_k} : \mathcal{H}(\mathcal{M}) \to \mathcal{H}(\mathcal{M}) \mid k \in \mathbb{N}_N\}$, *is said to be a generalized Kannan IFS .*

Lemma 11.3 *[29] If* $\{M_i\}_{i \in \wedge}, \{Q_i\}_{i \in \wedge}$ *are two finite collections of sets in* $(\mathcal{H}(\mathcal{M}), \sigma(\mu))$, *then*

1. $\sigma(\bigcup_{i \in \wedge} M_i, \bigcup_{i \in \wedge} Q_i) \leq \max_{i \in \wedge} \sigma(M_i, Q_i)$.
2. $\sigma(M, \bigcup_{i \in \wedge} Q_i) \leq \bigvee_{i \in \wedge} \sigma(M, Q_i), \forall\, M \in \mathcal{H}(\mathcal{M})$.

Proof 11.7 *Let us understand the proof through a case* $i = 2$, *i.e., let* $M_1, M_2, Q_1, Q_2 \in \mathcal{H}(\mathcal{M})$. *Also, suppose* $M, N, Q \in \mathcal{H}(\mathcal{M}) : M \subset N$. *Then*

$$\begin{aligned}\tau^*(Q_1 \cup Q_2, M) = \sup_{q \in Q_1 \cup Q_2} \tau(q, M) &= \sup_{q \in Q_1} \tau(q, M) \vee \sup_{q \in Q_2} \tau(q, M) \\ &= \tau^*(Q_1, M) \vee \tau^*(Q_2, M).\end{aligned} \tag{11.4}$$

We note that, for each $q \in Q$, $\tau(q, N) \leq \mu(q, n)\ \forall\ n \in N$. *This implies,* $\tau(q, N) \leq \tau(q, M)\ \forall\ q \in Q$. *Hence, by Proposition 11.1,* $\tau^*(Q, N) \leq \tau^*(Q, M)$. *Keeping this fact in view, since* $Q_1, Q_2 \subset Q_1 \cup Q_2$, *we have*

$$\tau^*(M, Q_1 \cup Q_2) \leq \tau^*(M, Q_1) \vee \tau^*(M, Q_2). \tag{11.5}$$

Replace M *by* $M_1 \cup M_2$ *in the equations (11.4) and (11.5). Then the desired result follows for the case* $i = 2$. *For any finite case, the proof follows from induction.*

Lemma 11.4 *[12]* $(\mathcal{H}(\mathcal{M}), \sigma(\mu))$ *forms a complete (compact) metric space whenever* $(\mathcal{M}, \mu)$ *is complete (compact).*

Theorem 11.2 *Consider a complete metric space* $(\mathcal{M}, \mu)$. *Let*

$$\{G_k; \beta_k; \mathcal{M} \mid k \in \mathbb{N}_N\}$$

denotes a generalized δ*-Kannan IFS. Consider the Hutchinson map* $C_H : (\mathcal{H}(\mathcal{M}), \sigma) \to (\mathcal{H}(\mathcal{M}), \sigma)$ *given by*

$$C_H(\Theta) = \bigcup_{k=1}^{N} G_{Hk}(\Theta) = \bigcup_{k=1}^{N} \bigcup_{\theta \in \Theta} \{G_k \theta\}, \forall\ \Theta \in \mathcal{H}(\mathcal{M}).$$

There exists constant $c_1 \in [1, \infty)$ *such that, if* $\beta_k < \frac{1}{ec_1}$ *for each* $k \in \mathbb{N}_N$, *then* C_H *is a generalized* δ*-Kannan map, i.e.,* $\forall\ \Theta, \Upsilon \in \mathcal{H}(\mathcal{M}) : \Theta \neq \Upsilon$,

$$\begin{aligned}\sigma(C_H(\Theta), C_H(\Upsilon)) \leq \Lambda_2^*. \max\Bigg\{ &\left[\frac{\sigma(\Theta, C_H(\Theta))}{\delta} + \sigma(\Upsilon, C_H(\Upsilon))\right], \\ &\left[\sigma(\Theta, C_H(\Theta)) + \frac{\sigma(\Upsilon, C_H(\Upsilon))}{\delta}\right]\Bigg\}.\end{aligned}$$

where $\Lambda_2^* := \Lambda_2 c_1, \Lambda_2 := \max_{k \in \mathbb{N}_N} \beta_k^*$, *and* $\Delta := \min_{k \in \mathbb{N}_N} \delta_k$. *Moreover,* C_H *is a Picard operator as it obeys*

$$\boldsymbol{F} = C_H(\boldsymbol{F}) = \bigcup_{k=1}^{N} C_{Hk}(F) : \boldsymbol{F} = \lim_{k \to \infty} C_H^{\circ k}(S), \forall\ S \in \mathcal{H}(\mathcal{M}),$$

uniquely for the resulting fractal $\boldsymbol{F}$.

Proof 11.8 *Since finite union of compact sets is compact, the map C_H turns out to be well-defined. Let $\Theta, \Upsilon \in \mathcal{H}(\mathcal{M}) : \Theta \neq [\Upsilon$. Using Lemmas 11.3 and 11.2, we deduce*

$$\begin{aligned}
\sigma(C_H(\Theta), C_H(\Upsilon)) &= \sigma\Big(\bigcup_{k=1}^{N} G_{Hk}(\Theta), \bigcup_{k=1}^{N} G_{Hk}(\Upsilon)\Big), \\
&\leq \max\{\sigma(G_{Hk}(\Theta), G_{Hk}(\Upsilon)) \mid k \in \mathbb{N}_N\}, \\
&= \sigma(G_{Hk^*}(\Theta), G_{Hk^*}(\Upsilon)) \text{ for some } k^* \in \mathbb{N}_N, \\
&\leq \beta^*_{k^*}. \max\left\{ \left[\frac{\sigma(\Theta, G_{Hk^*}(\Theta))}{\delta} + \sigma(\Upsilon, G_{Hk^*}(\Upsilon))\right], \right. \\
&\left. \left[\sigma(\Theta, G_{Hk^*}(\Theta)) + \frac{\sigma(\Upsilon, G_{Hk^*}(\Upsilon))}{\delta}\right] \right\}, \\
&\leq \Lambda_2. \max\left\{ \left[\frac{\bigvee_{k=1}^{N} \sigma(\Theta, G_{Hk}(\Theta))}{\Delta} + \bigvee_{k=1}^{N} \sigma(\Upsilon, G_{Hk}(\Upsilon))\right], \right. \\
&\left. \left[\bigvee_{k=1}^{N} \sigma(\Theta, G_{Hk}(\Theta)) + \frac{\bigvee_{k=1}^{N} \sigma(\Upsilon, G_{Hk}(\Upsilon))}{\Delta}\right] \right\}, \\
&\leq \Lambda^*_2. \max\left\{ \left[\frac{\sigma(\Theta, C_H(\Theta))}{\delta} + \sigma(\Upsilon, C_H(\Upsilon))\right], [\sigma(\Theta, C_H(\Theta)) \right. \\
&\left. + \frac{\sigma(\Upsilon, C_H(\Upsilon))}{\delta}\right] \Big\},
\end{aligned}$$

in view of the Proposition 11.3. Hence, C is a generalized δ-Kannan map. Rest of the proof follows from Lemmas 11.4 and 11.1.

11.4 CONCLUSION

In this chapter, we introduce a novel map that generalizes the Kannan map. The generalization is due to an improvement in the range of the Kannan parameter. As an application to this improvement, we present sufficient conditions under which the unsolved problem about Kannan IFS, raised in [20], can be solved.

BIBLIOGRAPHY

[1] S. Banach, Sur les opérations dans les ensembles abstraits et leur application aux equations integrales, *Fund. Math.* 3 (1922) 133–181.

[2] R. Kannan, Some results on fixed points-II, *Am. Math. Month.* 76(4) (1969) 405–408.

[3] S. Reich, Some remarks concerning contraction mappings, *Canad. Math. Bull.* 14 (1971) 121–124.

[4] E. Rakotch, A note on contractive mappings, *Proc. Am. Math. Soc.* 13 (1962) 459–465.

[5] M. Edelstein, An extension of Banach's contraction principle, *Proc. Amer. Math. Soc.* 12 (1961) 7–10.

[6] M. Edelstein, On fixed and periodic points under contractive mappings, *J. Lond. Math. Soc.* 37 (1962) 74–79.

[7] D. W. Boyd, J. S. W. Wong, On nonlinear contractions, *Proc. Am. Math. Soc.* 20(2) (1969) 458–458.

[8] F. E. Browder, On the convergence of successive approximations for nonlinear functional equations, *Indagationes Mathematicae Proceedings*, 71 (1967) 27–35.

[9] P. V. Subrahmanyam, Completeness and fixed-points, *Monatsh. Math.* 80 (1975) 325–330.

[10] E. H. Connell, Properties of fixed point spaces, *Proc. Amer. Math. Soc.* 10 (1959) 974–979.

[11] J. Hutchinson, Fractals and self-similarity, *Indiana Univ. Math. J.* 30 (1981) 713–747.

[12] M. F. Barnsley, *Fractals everywhere*, Academic Press, Dublin, 1988. (2nd Edition, Morgan Kaufmann 1993; 3rd Edition, Dover Publications, 2012).

[13] N. A. Secelean, The existence of the attractor of countable iterated function systems, *Mediterr. J. Math.* 9 (2012) 61–79.

[14] N. A. Secelean, Iterated function systems consisting of F-contractions, *Fixed Point Theory Appl.* 277(1) (2013) 1–13.

[15] M. Imdad, W. M. Alfapih, I. A. Khan, Weak θ-contractions and some fixed point results with applications to fractal theory, *Adv. Differ. Equ.* 439 (2018) 1–18.

[16] R. Pasupathi, A. K. B. Chand, M. A. Navascueś, Cyclic iterated function systems, *J. Fixed Point Theory Appl.* 22(58) (2020) 1–17.

[17] R. Pasupathi, A. K. B. Chand, M. A. Navascueś, Cyclic Meir-Keeler contraction and its fractals, *Num. Func. Ana. Opt.* 42(9) (2021) 1053–1072.

[18] D. R. Sahu, A. Chakraborty, R. P. Dubey, K-iterated function system, *Fractals* 18(1) (2010) 139–144.

[19] F. Georgescu, R. Miculescu, A. Mihail, Hardy-Rogers type iterated function systems. *Qual. Theory of Dyn. Sys.* 19(37) (2020) 1–13.

[20] N. V. Dung, A. Petruşel, On iterated function systems consisting of Kannan maps, Reich maps, Chatterjea type maps, and related results, *J. Fixed Point Theory Appl.* 19 (2017) 2271–2285.

[21] P. Debnath, H. M. Srivastava, New extensions of Kannan's and Reich's fixed point theorems for multivalued maps using Wardowski's technique with application to integral equations, *Symmetry* 12(7), 1090 (2020) 1–9.

[22] P. Debnath, A new extension of Kannan's fixed point theorem via F -contraction with application to integral equations, *Asian-Eur. J. Math.* 15(1) (2021) 383–391.

[23] P. Debnath, Z. D. Mitrović, S. Y. Cho, Common fixed points of Kannan, Chatterjea and Reich type pairs of self-maps in a complete metric space, *São Paulo J. Math. Sci.* 15(7), 2250123 (2022) 1–9.

[24] P. Debnath, Banach, Kannan, Chatterjea, and Reich-type contractive inequalities for multivalued mappings and their common fixed points, *Math. Meth. Appl. Sci.* 45(3) (2022) 1587–1596.

[25] P. Debnath, Z. D. Mitrović, H. M. Srivastava, Fixed points of some asymptotically regular multivalued mappings satisfying a Kannan-type condition, *Axioms* 10(1), 24 (2021) 1–7.

[26] R. Miculescu, A. Mihail, Riech-type iterated function systems, *J. Fixed Point Theory Appl.* 18(2) (2016) 285–296.

[27] K. Leśniak, N. Snigireva, F. Strobin, Weakly contractive iterated function systems and beyond: a manual. *J. Differ. Eqn. Appl.* 26(8) (2020) 1114–1173.

[28] M. Abbas, R. Anjum, H. Iqbal, Generalized enriched cyclic contractions with application to generalized iterated function system, *Chaos, Solitons & Fractals* 154 (2022) 1–8.

[29] B. V. Prithvi, S. K. Katiyar, Interpolative operators: Fractal to multivalued fractal, *Chaos Solitons & Fractals* 164, 112449 (2022) 1–12.

[30] U. Kifayat, S. K. Katiyar, Cylic weak ϕ iterated function system, *Top. Algebra Appl.* 10 (2022) 161–166.

[31] I. A. Rus, Picard operators and applications, *Sci. Math. Jpn.* 58 (2003) 191–219.

[32] N. Hu Shouchuan, S. Papageorgiou, *Handbook of Multivalued Analysis*, vol. 1, Kluwer Academic Publishers, New York, 1997.

[33] P. V. Subrahmanyam, *Elementary Fixed Point Theorems* (© Springer Nature Singapore Pte Ltd. 2018), Forum for Interdisciplinary Mathematics, Springer.

[34] B. E. Rhoades, A comparison of various definitions of contractive mappings, *Trans. Am. Math. Soc.* 226 (1977) 257–290.

[35] P. Debnath, H. M. Srivastava, P. Kumam and B. Hazarika, *Fixed Point Theory and Fractional Calculus: Recent Advances and Applications*, Springer, Singapore, 2022.

[36] P. Debnath, H. M. Srivastava, K. Chakraborty and P. Kumam, *Advances in Number Theory and Applied Analysis*, World Scientific, Singapore, 2023.

CHAPTER 12

Stability Analysis of Lotka-Volterra Models: Continuous, Discrete and Fractional

Sandra Vaz
University of Beira Interior

Delfim F. M. Torres
University of Aveiro

CONTENTS

DOI: 10.1201/9781003388678-12

12.1 INTRODUCTION

The Lotka–Volterra model is a very well-known model discovered in the middle of the 20s of the 20th century by the mathematician Vito Volterra (1860–1940), who intended to explain the oscillatory levels of certain fish catches in the Adriatic sea [18], and, at the same time, by the biophysicist Alfred Lotka (1880–1949), who studied the same predator–prey interaction and published the book "Elements of Physical Biology" [26]. The Lotka model was similar to the Volterra one, and the predator–prey model is nowadays called the Lotka–Volterra model [8, 11].

The Lotka–Volterra model is a nonlinear population model that has several applications, e.g., in epidemiology [3, 23], biology [18], and economics [10]. For this reason, it has captivated researchers throughout the years, and numerous papers about Lotka–Volterra models and their dynamics are available: see [1, 27] and references therein.

Sumarti, Nurfitriyana and Nurwenda, in 2014, applied the continuous Lotka–Volterra model with several interaction laws to the banking system and studied the local stability of the equilibrium points [24]. Here we consider the case of the Lotka–Volterra model where the growth behavior of one of the variables is not exponential but logistic, and the functional response function is given by the Michaelis–Menten functional

$$\frac{p}{1+\gamma D}. \tag{12.1}$$

Precisely, the model that we study here is the following one:

$$\begin{cases} {}^{c}D^{\sigma}D(t) &= \alpha D(t)\left(1-D(t)\right) - \frac{pD(t)L(t)}{1+\gamma D(t)}, \\ {}^{c}D^{\sigma}L(t) &= \frac{pD(t)L(t)}{1+\gamma D(t)} - \beta L(t), \quad 0<\sigma\leq 1, \end{cases} \tag{12.2}$$

where $D(t)$ represents the deposits in the instant t and $L(t)$ represents the loans in time t. Both are fractions of the total. All the parameters of the system are positive, and α represents the interest rate of the deposits: if the value increases, then the deposits volume will also increase. The interest rate of the loans is represented by β: if the value increases, then the loans volume will decrease. There are several functional responses that can be applied to such models. In this work, we use the Michaelis–Menten equation adopted from biochemistry to describe enzyme kinetics. Following Sumarti's work [24], in our model, the loan functional response to the deposit approaches a constant as the volume of the deposit increases. The parameter p represents the maximum rate of the mixture between deposit and loan volumes; $\frac{p}{\gamma}$ is the maximum proportion of the deposit able to be used as a loan; and $\frac{1}{\gamma}$ is the volume of deposit necessary to achieve one half of the rate p.

In fractional calculus, several notions of differentiation are available, such as the Grünwald–Letnikov derivative, the Riemann–Liouville derivative, or the Caputo derivative, among others [5, 6]. Here we consider fractional differentiation in the Caputo sense: the model is expressed in terms of the Caputo fractional derivative ${}^{c}D^{\sigma}$. If $\sigma = 1$, then we recover the standard first-order derivative: ${}^{c}D^{1}D(t) = \dot{D}(t)$ [4].

For most nonlinear models, it is not possible to obtain an exact solution, and the use of numerical schemes is vital to obtain numerical approximations and represent them graphically. There are several numerical methods; some of them are standard, whereas others are nonstandard. For instance, Euler and Heun methods are standard discretization methods, whereas Mickens' method is an example of a nonstandard scheme [9, 13, 14, 15, 16].

In Section 12.2, we analyze the equilibrium points of (12.2) and their local stability when $\sigma = 1$. Afterward, in Section 12.3, we discretize the model using the Euler numerical scheme [9], and in Section 12.4 the Mickens numerical scheme [13, 14, 15, 16, 17], and analyze the dynamical consistency with the continuous model (12.2). Section 12.5 is dedicated to fractional calculus to study (12.2) for $0 < \sigma < 1$. We prove the existence and uniqueness of a positive solution (Proposition 12.2) and the well-posededness of the model (Theorem 12.4); we determine the local stability of the equilibrium points of the system (Section 12.5.5), and we represent the solutions graphically (Section 12.5.6). For the fractional derivative, it is necessary to use a suitable numerical scheme. Here we use the modified trapezoidal method that involves the modified Euler scheme [2, 19]. Throughout our work, we represent graphically some of the solutions for different methods by comparing them. We present our conclusions in Section 12.6. All our numerical simulations were done using Mathematica, version 13.1.

12.2 THE MODIFIED LOTKA–VOLTERRA MODEL

In this section, we start by describing the model. Then we determine the feasible region for model (12.2) with $\sigma = 1$, the equilibrium points, and their local stability.

12.2.1 Model Description

In the model, we have two populations. The variable $D(t)$ represents the deposit volume that, in the absence of the other variable, grows in a logistic way. The loan volume $L(t)$, in case of absence of deposits, tends to extinction with exponential decay. The functional response is the Michaelis–Menten equation. Using this equation, the loan functional response to the deposit approaches a constant as the volume of the deposit increases. The parameters α and β represent interest rates and p the maximum rate of the mixture between deposit and loan volumes. The model is reasonable because in the real world the deposits and the loans are not infinite:

$$\begin{cases} \dot{D} &= \alpha D\left(1 - D\right) - \dfrac{pDL}{1 + \gamma D}, \\ \dot{L} &= \dfrac{pDL}{1 + \gamma D} - \beta L. \end{cases} \tag{12.3}$$

Throughout the text, we consider the following two hypotheses:

(P_1) $0 < \alpha < \beta < \dfrac{p}{1+\gamma} < 1$;

(P_2) all the initial conditions are positive.

As we shall prove, the first condition is necessary for the existence of non-negative equilibrium points.

12.2.2 Non-Negativity and Boundedness of the Solutions

Because of the meaning of model (12.3), it has to satisfy some properties, namely the non-negativity of the solutions and their boundedness.

In general, Proposition 12.1 asserts that the deposit and loan volumes are always non-negative.

Proposition 12.1 *Under hypotheses (P_1) and (P_2), the solutions of system (12.3) are non-negative for all $t > 0$, that is, the solutions are in*

$$\Omega_+ = \left\{(D, L) \in (\mathbb{R}_0^+)^2 : D \geqslant 0, L \geqslant 0\right\}.$$

Proof. Considering the first equation of (12.3),

$$\frac{dD}{dt} = \alpha D\,(1-D) - \frac{pDL}{1+\gamma D} \Leftrightarrow \frac{dD}{D} = \left[\alpha\,(1-D) - \frac{pL}{1+\gamma D}\right] dt,$$

and by integrating both sides, we get

$$D(T) = D_1 \exp^{\displaystyle\int_0^T \alpha\,(1-D) - \frac{pL}{1+\gamma D}\,dt} \geqslant 0, \quad D_1 = e^{D_0}.$$

Analogously, considering the second equation of (12.3),

$$\frac{dL}{dt} = L\left(\frac{pD}{1+\gamma D} - \beta\right),$$

and we have

$$L(T) = L_1 \exp^{\displaystyle\int_0^T \frac{pD}{1+\gamma D} - \beta\,dt} \geqslant 0, \quad L_1 = e^{L_0}.$$

Therefore, $(D, T) \in \Omega_+$. □

Theorem 12.1 *Let $M = \max\{D(0), 1\}$. If hypotheses (P_1) and (P_2) hold, then the feasible region is given by*

$$\Omega = \left\{(D, L) \in (\mathbb{R}_0^+)^2 : 0 \leq D \leq M \text{ and } 0 \leq D + L \leq \left(\frac{\alpha + 4\beta}{4\beta}\right) M\right\},$$

that is, any solution that starts in Ω remains in Ω.

Proof. Considering the first equation of (12.3),

$$\frac{dD}{dt} = \alpha D(1-D) - \frac{pDL}{1+\gamma D} \Rightarrow \frac{dD}{dt} \leq \alpha D(1-D) \Leftrightarrow \frac{dD}{D(1-D)} \leq \alpha dt.$$

Integrating both sides,

$$\int \frac{1}{D}dD - \int \frac{-1}{1-D}dD \leq \alpha t + k \Leftrightarrow D(t) \leq \frac{C_1 \exp^{\alpha t}}{1+C_1 \exp^{\alpha t}}.$$

If $D(0) = D_0$, then $C_1 = \dfrac{D_0}{1-D_0}$. Thus,

$$D(t) \leqslant \frac{D_0 \exp^{\alpha t}}{1-D_0(1-\exp^{\alpha t})},$$

that is, $\lim_{t\to\infty} D(t) = 1$ and $\lim_{t\to 0} D(t) = D_0$. Let $W(t) = D(t) + L(t)$. Then,

$$\begin{aligned}\frac{dW}{dt} &= \frac{dD}{dt} + \frac{dL}{dt}\\ &= \alpha D(1-D) + \beta D - \beta D - \beta L\\ &= \alpha D(1-D) + \beta D - \beta W.\end{aligned}$$

If $f(D) = \alpha D(1-D)$, then $\overline{D} = \frac{1}{2}$, which is a critical point that is the maximum. Therefore, $f(\overline{D}) = \frac{\alpha 1}{4}$ and

$$\begin{aligned}\frac{dW}{dt} &\leq \frac{\alpha}{4} + \beta D - \beta W,\\ \frac{dW}{dt} &\leq \left(\frac{\alpha+4\beta}{4}\right) M - \beta W,\\ \frac{dW}{dt} + \beta W &\leq \left(\frac{\alpha+4\beta}{4}\right) M.\end{aligned}$$

Solving the differential equation,

$$W(t) \leq W_0 \exp^{-\beta t} + \left(\frac{\alpha+4\beta}{4\beta}\right) M(1-\exp^{-\beta t})$$

and we have $\lim_{t\to+\infty} W(t) = \left(\dfrac{\alpha+4\beta}{4\beta}\right) M.$ □

12.2.3 Stability Analysis

The equilibrium points of (12.3) are

$$\begin{cases}\dot{D} &= 0,\\ \dot{L} &= 0,\end{cases} \Leftrightarrow \begin{cases}\overline{D} &= 0,\\ \overline{L} &= 0,\end{cases} \vee \begin{cases}\overline{D} &= 1,\\ \overline{L} &= 0,\end{cases} \vee \begin{cases}\overline{D} &= \dfrac{\beta}{p-\gamma\beta},\\ \overline{L} &= \dfrac{\alpha}{p-\gamma\beta}\left(1-\dfrac{\beta}{p-\gamma\beta}\right).\end{cases}$$

We obtain three equilibrium points:

$$e_1 = (0,0), \quad e_2 = (1,0), \quad e_3 = \left(\frac{\beta}{p-\gamma\beta}, \frac{\alpha}{p-\gamma\beta}\left(1-\frac{\beta}{p-\gamma\beta}\right)\right).$$

The equilibrium e_1 means that both deposit and loan volumes tend to extinction; and e_2 means the deposit volume tends to 1 and the loan volume tends to extinction. It is important to note that the equilibrium point e_3 only exists if $p-\gamma\beta > 0$ and $1-\frac{\beta}{p-\gamma\beta} > 0$. It can be seen that if the previous inequality is true, then $p-\gamma\beta > 0$ is also true. The necessary condition for e_3 to exist is in (P_1) and the meaning of e_3 is: the deposit volume tends to $\frac{\beta}{p-\gamma\beta}$ and the loan volume tends to $\frac{\alpha}{p-\gamma\beta}\left(1-\frac{\beta}{p-\gamma\beta}\right)$.

To study the local stability, we linearize the system. The Jacobian matrix is

$$J(D,L) = \begin{bmatrix} \alpha(1-2D) - \dfrac{pL}{(1+\gamma D)^2} & -\frac{pD}{1+\gamma D} \\ \frac{pL}{(1+\gamma D)^2} & \frac{pD}{1+\gamma D} - \beta \end{bmatrix}.$$

In $e_1 = (0,0)$,

$$J(e_1) = \begin{bmatrix} \alpha & 0 \\ 0 & -\beta \end{bmatrix}$$

and so

$$P(\lambda) = \det(J(e_1) - \lambda Id) = (\alpha-\lambda)(-\beta-\lambda).$$

The eigenvalues are $\lambda_1 = \alpha$ and $\lambda_2 = -\beta$. By hypothesis (P_1), $\lambda_1 > 0$ and $\lambda_2 < 0$. This means e_1 is a *saddle point.*

In $e_2 = (1,0)$,

$$J(e_2) = \begin{bmatrix} -\alpha & -\dfrac{p}{1+\gamma} \\ 0 & \dfrac{p}{1+\gamma} - \beta \end{bmatrix}$$

and

$$P(\lambda) = \det(J(e_2) - \lambda Id) = (-\alpha-\lambda)\left(\frac{p}{1+\gamma} - \beta - \lambda\right).$$

The eigenvalues are $\lambda_1 = -\alpha$ and $\lambda_2 = \dfrac{p-\beta(1+\gamma)}{1+\gamma}$. From (P_1) it can be seen that $\lambda_1 < 0$ and $\lambda_2 > 0$. Once again, e_2 is a *saddle point.*

For $e_3 = \left(\dfrac{\beta}{p-\gamma\beta}, \dfrac{\alpha}{p-\gamma\beta}\left(1-\dfrac{\beta}{p-\gamma\beta}\right)\right)$,

$$J(e_3) = \begin{bmatrix} -\dfrac{\alpha\beta}{p(p-\gamma\beta)}(p(1-\gamma)+\gamma\beta(1+\gamma)) & -\beta \\ \dfrac{\alpha}{p}((p-\gamma\beta)-\beta) & 0 \end{bmatrix}$$

and

$$P(\lambda) = \det(J(e_3) - \lambda Id) = \lambda^2 + \frac{\alpha\beta}{p(p-\gamma\beta)}\left(p(1-\gamma) + \gamma\beta(1+\gamma)\right)\lambda + \frac{\alpha\beta}{p}\left((p-\gamma\beta) - \beta\right).$$

By the Routh–Hurwitz criterion, if all coefficients of $P(\lambda)$ are greater than zero, then we can conclude that all its roots are negative. The independent term is positive by hypothesis (P_1). For the same to happen to

$$\frac{\alpha\beta}{p(p-\gamma\beta)}\left(p(1-\gamma) + \gamma\beta(1+\gamma)\right), \tag{12.4}$$

the following condition must be satisfied:

$$\beta > \frac{p(\gamma-1)}{\gamma(\gamma+1)}.$$

Since all parameters are positive, if $\gamma < 1$, then e_3 is a *sink* or *asymptotically stable.*

12.2.4 Graphical Analysis

Throughout the work, we represent graphically the solutions, under different initial conditions, and compare different methods, observing similarities and differences. In our simulations, for comparison reasons, the parameters will always be $a = 0.1$, $b = 0.3$, $\gamma = 0.1$ and $p = 0.4$.

In Figure 12.1, the continuous model (12.2) is represented using different initial conditions. We see that if $(D_0, L_0) = (0.2, 0.3)$, then the loan volume tends to zero allowing the deposit volume to increase. When the deposit volume is almost 1, then the loan volume increases, converging to e_3. On the other hand, if

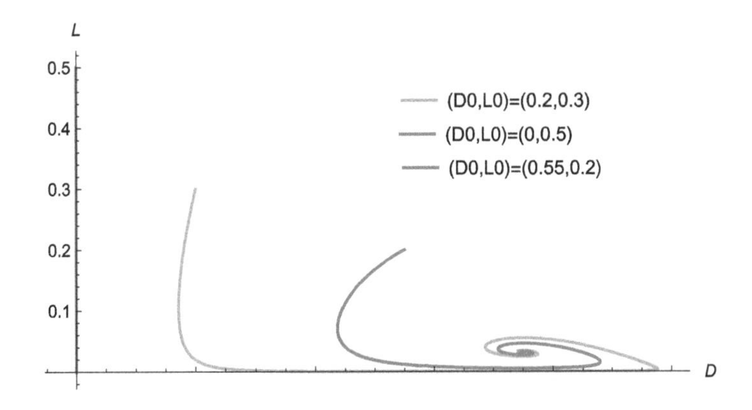

Figure 12.1 Continuous model (12.2) with different initial conditions.

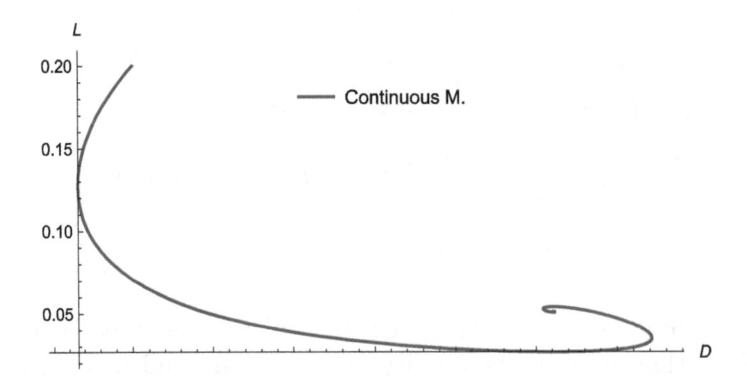

Figure 12.2 Continuous model (12.2) with $(D_0, L_0) = (0.55, 0.2)$.

$(D_0, L_0) = (0, 0.5)$, since the bank has no deposits, the loan volume tends to extinction. Finally, if $(D_0, L_0) = (0.55, 0.2)$, then the deposit and loan volumes tend to the equilibrium point e_3.

Figure 12.2 corresponds to initial conditions $(D_0, L_0) = (0.55, 0.2)$ and will be used for comparison with the other models in the sequel.

12.3 EULER'S NUMERICAL SCHEME

Leonhard Euler (1707–1783) presented his method, now known as the Euler method, in *Institutionum calculi integralis* (1768–1770). Euler's method is an iterative process for approximating the solution of a problem. The scheme (12.5) is precisely the Euler method. The step-size $h = t_{i+1} - t_i$, $i = 0, \ldots, n$, is uniform and $t_i = t_0 + ih$, $i = 0, \ldots, n$, where $h = \frac{b-a}{n}$:

$$\begin{aligned} y_0 &= y(t_0), \\ y(t_{i+1}) \approx y_{i+1} &= y_i + hf(t_i, y_i), \quad i = 0, \ldots, n. \end{aligned} \tag{12.5}$$

For physical models, this method is not frequently used because it may be difficult to prove the non-negativity of the solutions or it may lead to numerical instabilities, that is, to present dynamical inconsistency with the continuous model [12, 14, 22]. However, as we shall see, in the present context, we are able to prove, under some conditions, the non-negativity of the solutions as well as dynamical consistency.

12.3.1 Model Discretization

Using the concepts just introduced,

$$\begin{cases} \dot{D}(t) \approx \dfrac{D(t+h) - D(t)}{h}, \\ \dot{L}(t) \approx \dfrac{L(t+h) - L(t)}{h}, \end{cases}$$

and we get

$$\begin{cases} \dfrac{D_{i+1}-D_i}{h} = \alpha D_i\left(1-D_i\right) - \dfrac{pD_iL_i}{1+\gamma D_i}, \\ \dfrac{L_{i+1}-L_i}{h} = \dfrac{pD_iL_i}{1+\gamma D_i} - \beta L_i, \quad i \geq 0, \end{cases}$$

$$\Leftrightarrow \begin{cases} D_{i+1} = D_i\left(\alpha h\left(1-D_i\right) - \dfrac{phL_i}{1+\gamma D_i} + 1\right), \\ L_{i+1} = L_i\left(\dfrac{phD_i}{1+\gamma D_i} - \beta h + 1\right), \quad i \geq 0. \end{cases} \tag{12.6}$$

System (12.6) is the explicit discrete model obtained by the standard Euler's scheme.

12.3.2 Non-Negativity and Boundedness of the Solutions

Now we prove that the discrete system (12.6) satisfies the non-negativity condition and the boundedness of the solutions. First, we prove an auxiliary lemma under the following hypothesis:

(H_1) $1-\beta h > 0$.

Lemma 12.1 *Let h be the step-size and suppose that (P_1) and (H_1) hold. Then,*

$$\frac{1+\alpha h}{ph} \geq 1.$$

Proof. From hypotheses (P_1) and (H_1), $0 > 1+\alpha h - \beta h > 1 + \alpha h - ph$ because $\beta h < ph$. □

Theorem 12.2 *If hypotheses (P_1), (P_2) and (H_1) hold, then all solutions of system* (12.6) *are non-negative for all $n \geq 0$ with the feasible region being given by*

$$\Omega = \left\{(D,L) \in (\mathbb{R}_0^+)^2 : W_n = D_n + L_n \leq 1\right\}.$$

Proof. By hypotheses (P_1) and (H_1), we easily see that the second equation of (12.6),

$$L_{n+1} = L_n\left(1-\beta h + phD_n\right),$$

satisfy $L_{n+1} \geqslant 0$ for all n. Regarding the first equation, by (P_1) it follows that

$$\begin{aligned} D_{n+1} &= D_n\left(1+\alpha h - h\left(\alpha D_n + \frac{pL_n}{1+\gamma D_n}\right)\right) \\ &> D_n(1+\alpha h - ph(D_n+L_n)) \\ &> D_n(1+\alpha h - phW_n). \end{aligned}$$

Therefore, $D_{n+1} \geq 0$ if, and only if, $1+\alpha h - phW_n \geq 0$, that is,

$$W_n \leq \frac{1+\alpha h}{ph}.$$

Let $W_n = D_n + L_n$. By (12.6), we have

$$\begin{aligned}
&\frac{W_{n+1}-W_n}{h} = \alpha D_n(1-D_n) - \beta L_n \\
\Leftrightarrow &\frac{W_{n+1}-W_n}{h} = \alpha D_n(1-D_n) - \beta L_n - \beta D_n + \beta D_n \\
\Leftrightarrow &\frac{W_{n+1}-W_n}{h} = \alpha D_n(1-D_n) - \beta W_n + \beta D_n
\end{aligned}$$

and from (P_1) it follows that

$$\begin{aligned}
W_{n+1} &\leq (1-\beta h)W_n + \beta h D_n(1-D_n) + \beta h D_n \\
&= (1-\beta h)W_n + \beta h D_n(2-D_n).
\end{aligned}$$

Let $f(D) = \beta h D(2-D)$. Then, the critical point is $\overline{D} = 1$, which is a maximum, and $f(\overline{D}) = \beta h$. Moreover, $W_{n+1} \leq (1-\beta h)W_n + \beta h$, that is,

$$W_{n+1} \leq (1-\beta h)^{n+1} W_0 + (1-(1-\beta h)^n).$$

Thus, $\lim_{n\to+\infty} W_{n+1} = 1$, which means that $W_n \leq 1 \leq \dfrac{1+\alpha h}{ph}$ for all $n \geqslant 0$. We conclude that the feasible region is

$$\Omega = \left\{(D,L) \in \mathbb{R}_0^{+2} : 0 \leq W_n = D_n + L_n \leq 1\right\}$$

and that (12.6) satisfies the non-negativity condition and the boundedness condition if (H_1) is verified. □

12.3.3 Stability Analysis

The stationary points of the Euler discrete system (12.6) are obtained from

$$\begin{aligned}
&(D^*, L^*) = F(D^*, L^*) \\
&\Leftrightarrow \begin{cases} D^* = 0, \\ L^* = 0, \end{cases} \vee \begin{cases} D^* = 1, \\ L^* = 0, \end{cases} \vee \begin{cases} D^* = \frac{\beta}{p-\gamma\beta}, \\ L^* = \frac{\alpha}{p-\gamma\beta}\left(1-\frac{\beta}{p-\gamma\beta}\right), \end{cases}
\end{aligned}$$

that is,

$$e_1 = (0,0), \quad e_2 = (1,0), \quad e_3 = \left(\frac{\beta}{p-\gamma\beta}, \frac{\alpha}{p-\gamma\beta}\left(1-\frac{\beta}{p-\gamma\beta}\right)\right),$$

which are equal to the ones of the continuous model (12.3).

To study their local stability, we linearize the system. The Jacobian matrix is given by

$$J = \begin{bmatrix} \alpha h(1-2D) - \dfrac{phL}{(1+\gamma D)^2} + 1 & -\dfrac{phD}{1+\gamma D} \\ \dfrac{phL}{(1+\gamma D)^2} & 1 - \beta h + \dfrac{phD}{1+\gamma D} \end{bmatrix}.$$

For $e_1 = (0, 0)$, the Jacobian matrix is

$$J(e_1) = \begin{bmatrix} 1 + \alpha h & 0 \\ 0 & 1 - \beta h \end{bmatrix}$$

and

$$P(\lambda) = \det(J(e_1) - \lambda Id) = (1 + \alpha h - \lambda)(1 - \beta h - \lambda).$$

The eigenvalues are $\lambda_1 = 1 + \alpha h$ and $\lambda_2 = 1 - \beta h$. By (P_1), $\lambda_1 > 1$, and by (H_1), $\lambda_2 < 1$. We conclude that e_1 is a *saddle point.*

For $e_2 = (1, 0)$, the Jacobian matrix is

$$J(e_2) = \begin{bmatrix} 1 - \alpha h & -\dfrac{ph}{1+\gamma} \\ 0 & 1 - \beta h + \dfrac{ph}{1+\gamma} \end{bmatrix}$$

and

$$P(\lambda) = \det(J(e_2) - \lambda Id) = (1 - \alpha h - \lambda)\left(\frac{ph}{1+\gamma} - \beta h + 1 - \lambda\right).$$

The eigenvalues are $\lambda_1 = 1 - \alpha h$ and $\lambda_2 = 1 - h\left(\beta - \frac{p}{1+\gamma}\right)$. By (P_1), we can see that $\lambda_2 > 1$ and $\lambda_1 < 1$, and thus e_2 is a *saddle point.*

For $e_3 = \left(\frac{\beta}{p-\gamma\beta}, \frac{\alpha}{p-\gamma\beta}\left(1 - \frac{\beta}{p-\gamma\beta}\right)\right)$, the Jacobian matrix is

$$J(e_3) = \begin{bmatrix} 1 + \alpha\beta h\left(\dfrac{1+\gamma}{p} - \dfrac{2}{p-\gamma\beta}\right) & -\beta h \\ \dfrac{\alpha h(p-\gamma\beta)}{p}\left(1 - \dfrac{\beta}{p-\gamma\beta}\right) & 1 \end{bmatrix}.$$

Letting $a_{11} = 1 - \dfrac{\alpha\beta h}{p(p-\gamma\beta)}(p(1-\gamma) + \gamma\beta(1+\gamma))$, then

$$\begin{aligned} P(\lambda) &= \det(J(e_3) - \lambda Id) \\ &= (a_{11} - \lambda)(1 - \lambda) + \frac{\alpha\beta h^2(p-\gamma\beta)}{p}\left(1 - \frac{\beta}{p-\gamma\beta}\right). \end{aligned}$$

From the Schur–Cohn criterion for quadratic polynomials $P(x)$, if $P(1) > 0$, $P(-1) > 0$, and $\mid P(0) \mid < 1$, then all the roots are inside the unit circle (see [7, 12]). Here we have

$$P(1) = \frac{\alpha\beta h^2(p-\gamma\beta)}{p}\left(1 - \frac{\beta}{p-\gamma\beta}\right) = E^* h^2,$$

so by (P_1) we can conclude that $P(1) > 0$; while

$$P(-1) = 2(1 + a_{11}) + \frac{\alpha\beta h^2(p-\gamma\beta)}{p}\left(1 - \frac{\beta}{p-\gamma\beta}\right).$$

Let us recall that for e_3 to be a sink in the continuous model, condition (12.4) has to be satisfied, that is,

$$\frac{\alpha\beta}{p(p-\gamma\beta)}\left(p(1-\gamma)+\gamma\beta(1+\gamma)\right)>0,$$

and so we can conclude that $a_{11}<1$. Now we show that $a_{11}>0$. From the expression of a_{11} by (P_1) it can be seen that

$$D^*=\frac{\beta\left(p(1-\gamma)+\gamma\beta(1+\gamma)\right)}{p(p-\gamma\beta)}<\frac{p(1-\gamma)+\gamma\beta(1+\gamma)}{(1+\gamma)(p-\gamma\beta)}<1,$$

because

$$\frac{p(1-\gamma)+\gamma\beta(1+\gamma)}{(1+\gamma)(p-\gamma\beta)}<1 \Leftrightarrow 2\gamma(p-\beta(1+\gamma))>0.$$

Since

$$D^*<1 \Leftrightarrow 2-\alpha hD^*>2-\alpha h>1+(1-\beta h)>0, \tag{12.7}$$

one has $P(-1)>0$ by (P_1) and (H_1). Moreover,

$$P(0)=1-\alpha D^*h+E^*h^2.$$

First we need to show that $P(0)>-1$. From (12.7),

$$P(0)>-1 \Leftrightarrow 2-\alpha D^*h+E^*h^2>0.$$

For us to have $P(0)<1$ we need to show that

$$-\alpha D^*h+E^*h^2<0. \tag{12.8}$$

Because (12.8) is a polynomial $g(h)$ of degree 2 with roots

$$h_1=0 \quad \vee \quad h_2=\frac{\alpha D^*}{E^*}=\frac{p(1-\gamma)+\gamma\beta(1+\gamma)}{(p-\gamma\beta)^2-\beta(p-\gamma\beta)}, \tag{12.9}$$

we have $P(0)<1$ as long as $0<h<h_2$. Therefore, we have shown that $\mid P(0)\mid<1$ as long as $0<h<h_2$ and, in this case, e_3 is a *sink* or *asymptotically stable.*

Under Euler's numerical scheme, it is not straightforward to show the boundedness of the solutions, and we need to impose some conditions to have dynamical consistency with the continuous model. In order for e_3 to be asymptotically stable, condition (12.4) must be also satisfied, and the step size must be smaller than h_2, as given in (12.9). By (P_1), all parameters are less than 1, so it is not a difficult condition to be attained.

12.3.4 Graphical Analysis

For Euler's numerical scheme, we take the time interval to be $[0, 500]$ and the step-size as $h = 0.25$. In Figure 12.3, some solutions of (12.6), for different initial conditions, are plotted.

Figure 12.3 is similar to Figure 12.1.

In Figure 12.4, we compare the solution of the continuous model (12.3) with the one obtained by Euler's method with initial conditions $(D_0, L_0) = (0.55, 0.2)$. There are mild differences between the plots, which are, however, qualitatively the same.

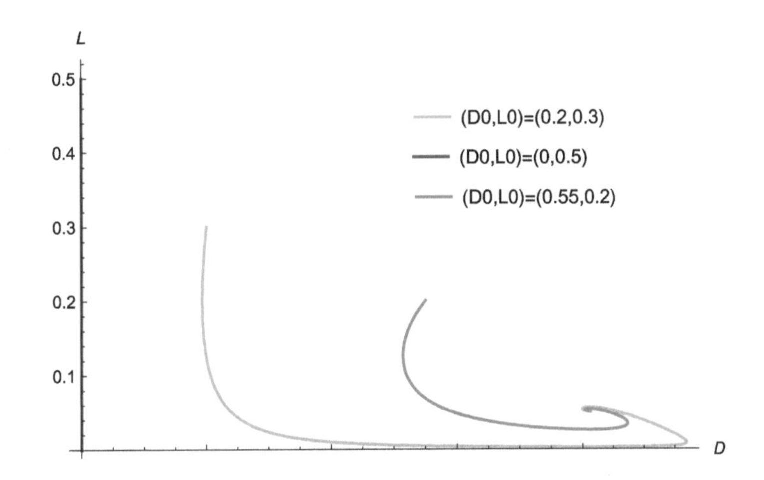

Figure 12.3 Solutions to Euler's discrete model (12.6) with different initial conditions.

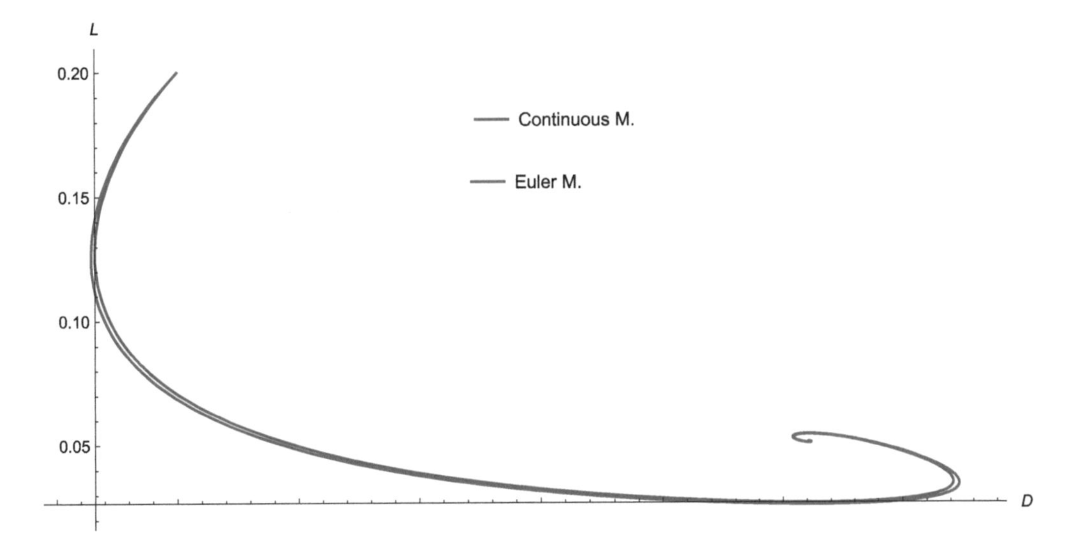

Figure 12.4 Euler's method compared with the continuous model.

12.4 MICKENS' NUMERICAL SCHEME

The nonstandard finite difference numerical scheme (NSFD), started by [13], is based on Mickens' work [14, 15, 16, 17]. Is was created with the goal of solving some problems produced by Euler's method, namely numerical instabilities and the difficulty of showing the non-negativity of the solutions for physical models, that is, the difficulty on showing dynamical consistency with the continuous model [14].

Mickens' numerical scheme has two main rules [14]. The first is: the derivative is approximated by

$$\frac{dx}{dt} \approx \frac{x_{k+1} - \varphi(h)x_k}{\psi(h)}, \qquad h = \Delta t,$$

where h is the step size, and $\varphi(h)$ and $\psi(h)$ satisfy

$$\varphi(h) = 1 + O(h^2) \quad \text{and} \quad \psi(h) = h + O(h^2).$$

The numerator function $\varphi(h)$ and the denominator function $\psi(h)$ may take different forms. Generally, $\varphi(h) = 1$, but it can be different, for instance, $\varphi(h) = \cos(\lambda h)$; while $\psi(h)$ can be, for example,

$$\psi(h) = \frac{1 - e^{-\lambda h}}{\lambda},$$

where λ is a parameter that appears in the model.

The second main rule of Mickens' numerical scheme is: the linear and nonlinear terms may require a non-local representation. For examples of such non-local representations see, e.g., [14, 15, 16].

12.4.1 Model Discretization

Applying Mickens' rules,

$$\begin{cases} \dot{D}(t) & \approx \dfrac{D((n+1)h) - D(nh)}{\phi(h)} = \dfrac{D_{n+1} - D_n}{\phi(h)}, \\ \dot{L}(t) & \approx \dfrac{L((n+1)h) - L(nh)}{\phi(h)} = \dfrac{L_{n+1} - L_n}{\phi(h)}. \end{cases}$$

By [17], if the populations have to satisfy a boundedness condition, then the denominator function is obtained by the condition of the continuous model (12.2). We have seen that

$$\dot{W} = \left(\frac{\alpha + 4\beta}{4\beta}\right) M - \beta W,$$

so the denominator function is chosen to be $\phi(h) = \dfrac{1 - e^{-\beta h}}{\beta}$. For simplification in writing, in this work we denote $\phi := \phi(h)$.

The discrete model obtained from Mickens method is then given by

$$\begin{cases} \dfrac{D_{n+1}-D_n}{\phi} &= \alpha D_n\left(1-D_{n+1}\right) - p\dfrac{D_{n+1}L_n}{1+\gamma D_n}, \\ \\ \dfrac{L_{n+1}-L_n}{\phi} &= p\dfrac{D_{n+1}L_n}{1+\gamma D_n} - \beta L_{n+1}, \end{cases}$$

$$\Leftrightarrow \begin{cases} D_{n+1} &= \dfrac{(1+\alpha\phi)D_n}{1+\alpha\phi D_n + \dfrac{p\phi L_n}{1+\gamma D_n}}, \\ \\ L_{n+1} &= \dfrac{(1+\gamma D_n + p\phi D_{n+1})L_n}{(1+\beta\phi)(1+\gamma D_n)}, \quad n \geq 0. \end{cases} \tag{12.10}$$

The explicit discrete model is

$$\begin{cases} D_{n+1} &= \dfrac{(1+\alpha\phi)D_n}{1+\alpha\phi D_n + \dfrac{p\phi L_n}{1+\gamma D_n}}, \\ \\ L_{n+1} &= \left(1+\dfrac{p\phi(1+\alpha\phi)D_n}{p\phi L_n + (1+\gamma D_n)\left(1+\alpha\phi D_n\right)}\right)\dfrac{L_n}{1+\beta\phi}, \end{cases} \tag{12.11}$$

for $n \geq 0$.

12.4.2 Non-Negativity and Boundedness of Solutions

If the initial conditions are non-negative, by (P_1), and the direct observation of (12.11), we have the non-negativity of the solutions.

Theorem 12.3 *Under hypotheses (P_1) and (P_2), if $\xi = 1+\alpha\phi$, then*

$$\Omega = \left\{(D,L) \in (\mathbb{R}_0^+)_+^2 : 0 \leq D_n \leq 1 \quad and \quad W_n \leq 1 \leq \frac{4\alpha^2+\xi\beta^2}{4\alpha\beta}\right\},$$

where $W_n = D_n + L_n$.

Proof. By the first equation of (12.11), we have

$$D_{n+1} = \frac{(1+\alpha\phi)D_n}{1+\alpha\phi D_n + \dfrac{p\phi L_n}{1+\gamma D_n}} < \frac{(1+\alpha\phi)D_n}{1+\alpha\phi D_n}$$

for some $n > 0$ and

$$\lim_{n\to+\infty} \frac{(1+\alpha\phi)D_n}{1+\alpha\phi D_n} = 1.$$

Thus, $0 \leq D_n \leq 1$. Adding the equations of (12.10) and considering the first equation of (12.10), we have

$$\begin{aligned}\frac{W_{n+1}-W_n}{\phi} &= \alpha D_n - \alpha D_n D_{n+1} - \beta L_{n+1} - \beta D_{n+1} + \beta D_{n+1}\\ &= \alpha D_n + D_{n+1}(\beta - \alpha D_n) - \beta W_{n+1}\\ &\leq \alpha D_n + \beta D_{n+1}\left(1 - \frac{\alpha}{\xi\beta} D_{n+1}\right) - \beta W_{n+1}.\end{aligned}$$

Let $f(D) = \beta D\left(1 - \frac{D}{K}\right)$, where $K = \frac{\xi\beta}{\alpha}$. Then, $\overline{D} = \frac{K}{2}$ is the critical point and maximizer, so that $f(D)$ attains its maximum at $f(\overline{D}) = \frac{K\beta}{4}$. Therefore,

$$\begin{aligned}\frac{W_{n+1}-W_n}{\phi} &\leq \alpha + \frac{K\beta^2}{4} - \beta W_{n+1},\\ W_{n+1} &\leq \frac{A}{1+\beta\phi} + \frac{W_n}{1+\beta\phi},\end{aligned}$$

where $A = \alpha + \frac{\beta K}{4} = \frac{4\alpha^2+\beta^2\xi}{4\alpha}$. We conclude that

$$\begin{aligned}W_{n+1} &\leq \left(\frac{1}{1+\beta\phi}\right)^{n+1} W_0 + \frac{A\phi}{1+\beta\phi}\left(\frac{1-\left(\frac{1}{1+\beta\phi}\right)^n}{1-\frac{1}{1+\beta\phi}}\right)\\ &\leq \left(\frac{1}{1+\beta\phi}\right)^{n+1} W_0 + \frac{A}{\beta}\left(1 - \left[\frac{1}{1+\beta\phi}\right]^n\right)\end{aligned}$$

and $\lim_{n\to+\infty} W_{n+1} = \dfrac{4\alpha^2+\xi\beta^2}{4\alpha\beta}$. □

Considering hypothesis (P_1), $\phi(h) = h + O(h^2)$ and $\xi = 1 + \alpha\phi$, and it is reasonable to consider $1 < \xi < 2$.

12.4.3 Stability Analysis

The stationary points of the discrete system (12.11) are:

$$(D^*, L^*) = F(D^*, L^*)$$

$$\Leftrightarrow \begin{cases} D^* = 0, \\ L^* = 0, \end{cases} \vee \begin{cases} D^* = 1, \\ L^* = 0, \end{cases} \vee \begin{cases} D^* = \dfrac{\beta}{p-\gamma\beta}, \\ L^* = \dfrac{\alpha}{p-\gamma\beta}\left(1 - \dfrac{\beta}{p-\gamma\beta}\right). \end{cases}$$

The stationary points are equal to the ones of the continuous model (12.2):

$$e_1 = (0,0), \quad e_2 = (1,0), \quad e_3 = \left(\frac{\beta}{p-\gamma\beta}, \frac{\alpha}{p-\gamma\beta}\left(1 - \frac{\beta}{p-\gamma\beta}\right)\right).$$

Once again, to study the local stability, we linearize the system. The Jacobian matrix $J(D,F)$ of (12.11) is given by

$$\begin{bmatrix} j_{1,1} & j_{1,2} \\ j_{2,1} & j_{2,2} \end{bmatrix},$$

where

$$j_{1,1} = \frac{(1+\alpha\phi)(1+\frac{p\phi L}{1+\gamma D})\left(1+\frac{\gamma D}{1+\gamma D}\right)}{\left(1+\frac{p\phi L}{1+\gamma D}+\alpha\phi D\right)^2},$$

$$j_{1,2} = -\frac{(1+\alpha\phi)p\phi D}{(1+\gamma D)\left(1+\frac{p\phi L}{1+\gamma D}+\alpha\phi D\right)^2},$$

$$j_{2,1} = \frac{p\phi(1+\alpha\phi)L(1+\phi(pL-\alpha\gamma D^2))}{(1+\beta\phi)\left(p\phi L+(1+\gamma D)(1+\alpha\phi D)\right)^2},$$

and

$$j_{2,2} = \frac{1}{1+\beta\phi}\left(1+\frac{p\phi(1+\alpha\phi)D(1+\gamma D)\left(1+\alpha\phi D\right)}{\left(p\phi L+(1+\gamma D)(1+\alpha\phi D)\right)^2}\right).$$

For $e_1=(0,0)$, the Jacobian matrix is

$$J(e_1) = \begin{bmatrix} 1+\alpha\phi & 0 \\ 0 & \dfrac{1}{1+\beta\phi} \end{bmatrix}$$

and

$$P(\lambda) = \det(J(e_1)-\lambda Id) = (1+\alpha\phi-\lambda)\left(\frac{1}{1+\beta\phi}-\lambda\right).$$

In this way, the eigenvalues are $\lambda_1 = 1+\alpha\phi$ and $\lambda_2 = \dfrac{1}{1+\beta\phi}$. By (P_1), we have $\lambda_1 > 1$ and $\lambda_2 < 1$, that is, e_1 is a *saddle point.*

For $e_2=(1,0)$, the Jacobian matrix is

$$J(e_2) = \begin{bmatrix} \dfrac{1}{1+\alpha\phi} & -\dfrac{p\phi}{(1+\gamma)(1+\alpha\phi)} \\ 0 & \dfrac{1+\gamma+p\phi}{(1+\gamma)(1+\beta\phi)} \end{bmatrix}$$

and the characteristic polynomial is

$$P(\lambda) = \det(J(e_2)-\lambda Id) = \left(\frac{1}{1+\alpha\phi}-\lambda\right)\left(\frac{1+\gamma+p\phi}{(1+\gamma)(1+\beta\phi)}-\lambda\right).$$

The eigenvalues are $\lambda_1 = \dfrac{1}{1+\alpha\phi}$ and $\lambda_2 = \dfrac{1+\gamma+p\phi}{(1+\gamma)(1+\beta\phi)}$. Once more, it can be seen that $\lambda_1 < 1$ and

$$\lambda_2 > 1 \Leftrightarrow p > \beta(1+\gamma),$$

which is true by (P_1). We conclude that e_2 is a *saddle point.*

For $e_3 = \left(\frac{\beta}{p-\gamma\beta}, \frac{\alpha}{p-\gamma\beta}\left(1-\frac{\beta}{p-\gamma\beta}\right)\right)$, let $k^* = \left(1-\frac{\beta}{p-\gamma\beta}\right)$. Then we have

$$J(e_3) = \begin{bmatrix} \frac{1+\alpha\phi k^*\left(1+\frac{\gamma\beta}{p}\right)}{1+\alpha\phi} & -\frac{\beta\phi}{1+\alpha\phi} \\ \frac{\alpha\phi k^*\left(1+\alpha\phi k^* - \frac{\gamma\beta}{p}\left(1+\frac{\alpha\beta\phi}{p-\gamma\beta}\right)\right)}{(1+\alpha\phi)(1+\beta\phi)} & \frac{1}{1+\beta\phi}\left(1+\frac{\beta\phi\left(1+\frac{\alpha\phi\beta}{p-\gamma\beta}\right)}{1+\alpha\phi}\right) \end{bmatrix}$$

and after some computations, it can be seen that

$$\begin{aligned} P(\lambda) &= \det(J(e_3) - \lambda Id) \\ &= \lambda^2 - \left(1 + \frac{1}{1+\alpha\phi} + \frac{\alpha\phi k^*\left(1+\frac{\gamma\beta}{p}(1+\beta\phi)\right)}{(1+\beta\phi)(1+\alpha\phi)}\right)\lambda \\ &\quad + \frac{1+\alpha\phi k^*}{1+\alpha\phi} + \frac{\alpha\beta\gamma\phi k^*}{p(1+\alpha\phi)(1+\beta\phi)}. \end{aligned}$$

By the Schur–Cohn criterion for quadratic polynomials, if $P(1) > 0$, $P(-1) > 0$, and $\mid P(0) \mid < 1$, then all of its roots are in the unit circle [7, 12]. After some computations, and by (P_1),

- $P(1) = \dfrac{\alpha\beta\phi^2 k^*\left(1-\frac{\gamma\beta}{p}\right)}{(1+\alpha\phi)(1+\beta\phi)} > 0;$

- $P(-1) = 2 + \frac{1}{1+\alpha\phi} + \frac{\alpha\phi k^*\left(1+\frac{\gamma\beta}{p}(1+\beta\phi)\right)}{(1+\beta\phi)(1+\alpha\phi)} + \frac{1+\alpha\phi k^*}{1+\alpha\phi} + \frac{\alpha\beta\gamma\phi k^*}{p(1+\alpha\phi)(1+\beta\phi)} > 0.$

Regarding

$$P(0) = \frac{1+\alpha\phi k^*}{1+\alpha\phi} + \frac{\alpha\beta\gamma\phi k^*}{p(1+\alpha\phi)(1+\beta\phi)} > 0,$$

we need to show that $P(0) < 1$. It can be seen that

$$\begin{aligned} P(0) < 1 &\Leftrightarrow \frac{1+\alpha\phi k^*}{1+\alpha\phi} + \frac{\beta\gamma\alpha\phi k^*}{(1+\alpha\phi)(1+\beta\phi)p} < 1 \\ &\Leftrightarrow \frac{(1+\alpha\phi k^*)p(1+\beta\phi) + \beta\gamma\alpha\phi k^*}{p(1+\alpha\phi)(1+\beta\phi)} < 1 \\ &\Leftrightarrow \alpha\beta\phi\left(-\frac{p(1+\beta\phi)}{p-\gamma\beta} + \frac{\phi(p-\gamma\beta)-\beta\phi}{p-\gamma\beta}\right) < 0 \\ &\Leftrightarrow \frac{\alpha\beta\phi}{p-\gamma\beta}(-p(1+\beta\phi) + \phi(p-\gamma\beta) - \beta\phi) < 0 \end{aligned}$$

$$\Leftrightarrow \beta > \frac{-p(1-\phi)}{1+p+\gamma}$$

since all parameters are positive. So, if $\phi < 1$, then $P(0) < 1$ and e_3 is a *sink* or an *asymptotically stable point.*

We proved the boundedness of the solutions to Mickens' numerical scheme and the dynamical consistency with the continuous model, without the need of considering the condition obtained for the continuous model. For that to happen we need to have $\phi < 1$, which means that the discretization step size has to be less than one.

12.4.4 Graphical Analysis

Figure 12.5 presents solutions to the Mickens discrete model (12.11) with different initial conditions.

Figure 12.5 is similar to Figures 12.1 and 12.3.

Figure 12.6 compares the three previous models (12.2), (12.6) and (12.10). As we can see, the differences between them are only mild.

12.5 FRACTIONAL CALCULUS (FC)

The calculus of non-integer order, known as the Fractional Calculus (FC), is the branch of mathematics that studies the extension of the derivative and integral concepts to an arbitrary order, not necessarily of a fraction order [20].

Considering $f(x) = \frac{1}{2}x^2$, the first and second derivatives are $f'(x) = x$ and $f''(x) = 1$, respectively. But what about the derivative of order $n = \frac{1}{2}$? This was the question that the father of fractional calculus, L'Hôpital, considered, asking it, by

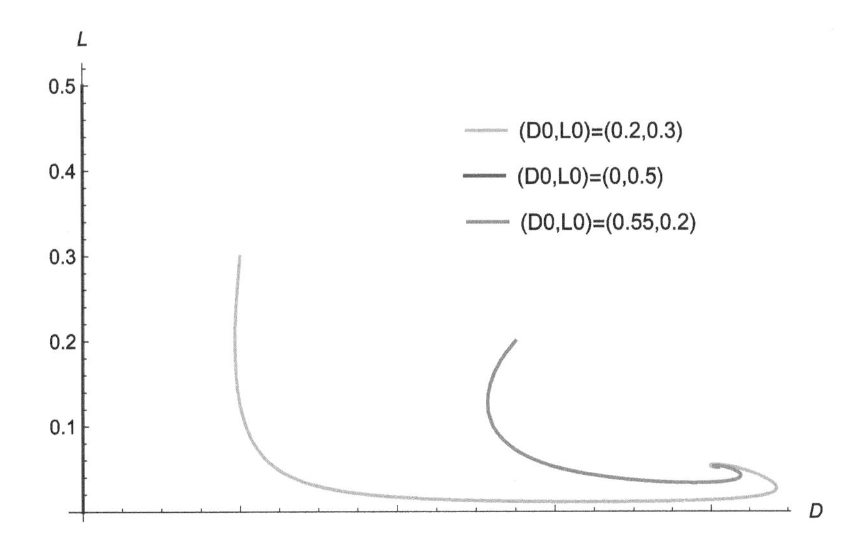

Figure 12.5 Solutions to Mickens discrete model (12.11) with different initial conditions.

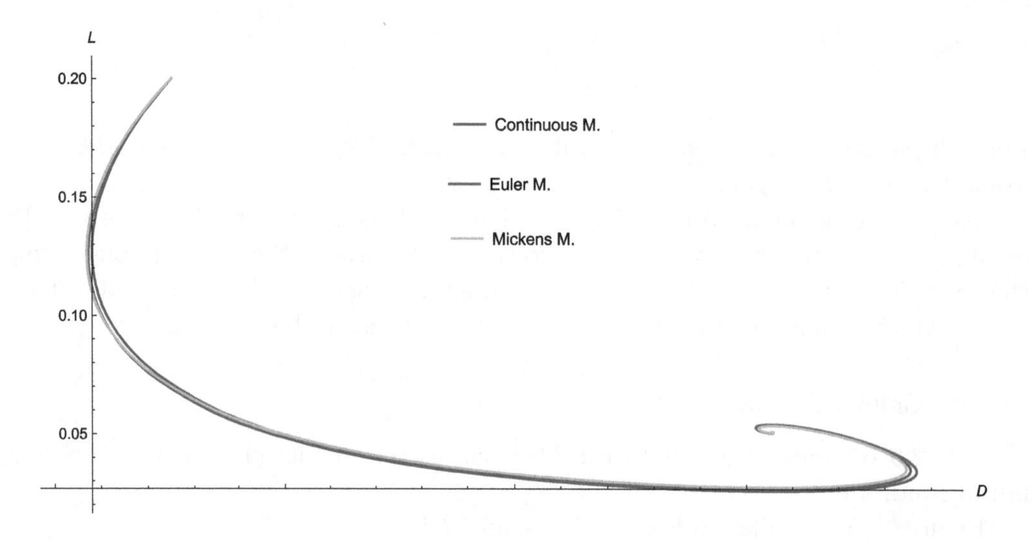

Figure 12.6 Solutions of Mickens' model (12.11) versus solutions of previous models (12.2) and (12.6).

letter, to Leibniz, in 1695. Since then, several mathematicians worked in such kind of calculus, namely Grünwald, Letnikov, Riemann, Liouville, Caputo, among many others [21, 25].

12.5.1 Preliminaries on FC

In fractional calculus, there are some functions, namely the Gamma, Beta and Mittag–Leffler functions, that play a crucial role.

Definition 12.1 (Gamma function) *Let $z \in \mathbb{C}$. The gamma function is defined by*

$$\Gamma(z) = \int_0^\infty e^{-t} t^{z-1} \, dt. \tag{12.12}$$

The integral (12.12) converges if $Re(z) > 0$. The Gamma function has the following important property:

$$\Gamma(z+1) = z\Gamma(z).$$

Definition 12.2 (Beta function) *Let $z, w \in \mathbb{C}$. The beta function is defined by*

$$B(z,w) = \int_0^1 \tau^{z-1}(1-\tau)^{w-1} \, d\tau \tag{12.13}$$

for $Re(w) > 0$.

Using the Laplace transform, we can rewrite (12.13) as

$$B(z,w) = \frac{\Gamma(z)\Gamma(w)}{\Gamma(z+w)}. \tag{12.14}$$

By (12.14), we conclude that $B(z,w) = B(w,z)$.

It is known that e^z,

$$e^z = \sum_{k=0}^{\infty} \frac{z^k}{\Gamma(k+1)}, \tag{12.15}$$

has an important role in the integration of ordinary differential equations. Similar role has the Mittag–Leffler function for fractional differential equations. The Mittag–Leffler function is a generalization of the exponential function (12.15).

Definition 12.3 (Mittag–Leffler function) *The Mittag–Leffler function of two parameters is defined by*

$$E_{\alpha,\beta} = \sum_{k=0}^{\infty} \frac{z^k}{\Gamma(\alpha k+\beta)},$$

where $\alpha, \beta \in \mathbb{C}$ and $Re(\alpha) > 0$.

The Mittag–Leffler function is uniformly convergent in every compact subset of $\mathbb{C}$. If $\alpha = \beta = 1$, then $E_{1,1}(z) = e^z$.

The fractional calculus has several formulations. The two most well-known are the Riemann–Liouville and Caputo approaches. Here we use fractional derivatives in Caputo's sense.

The fractional derivative of Riemann–Liouville is obtained using the arbitrary order integral of Riemann–Liouville and the integer order derivative.

Definition 12.4 (Riemann–Liouville fractional order derivative) *Let $\alpha > 0$, $m = \lceil \alpha \rceil$, and $v = m - \alpha$. The Riemann–Liouville fractional order derivative of order α of function f is defined by*

$$\begin{aligned}
D^\alpha f(x) &= D^m[J^v f(x)] \\
&= D^m \left(\frac{1}{\Gamma(\alpha)} \int_0^x (x-t)^{\alpha-1} f(t)\, dt \right) \\
&= \frac{d^m}{dt^m} \left(\frac{1}{\Gamma(\alpha)} \int_0^x (x-t)^{\alpha-1} f(t)\, dt \right) \\
&= \frac{1}{\Gamma(\alpha)} \frac{d^m}{dt^m} \int_0^x (x-t)^{\alpha-1} f(t)\, dt.
\end{aligned}$$

When one uses the Laplace transform of the Riemann–Liouville fractional order derivative, initial values with integer-order derivatives are not obtained. This fact has no physical meaning [6]. For this reason, in many applications, the Caputo fractional derivative is preferred.

The definition of Caputo's fractional derivative is similar to the Riemann–Liouville definition, the difference being the order one takes the operations of Riemann–Liouville integration and integer-order differentiation. Indeed, in Caputo's definition, first we compute the derivative of integer order and then the fractional order integral.

Definition 12.5 (Caputo fractional order derivative) *Let $\alpha > 0$, $m = \lceil \alpha \rceil$, and $v = m - \alpha$. The Caputo fractional order derivative of order α of function f is defined by*

$$\begin{aligned}{}^{c}D^{\alpha} f(x) &= J^{v}[D^{m} f(x)] = J^{v}\left[\frac{d^{m}}{dx^{m}} f(x)\right] = J^{v} f^{(m)}(x) \\ &= \frac{1}{\Gamma(v)} \int_{0}^{x} (x-t)^{v-1} f^{(m)}(t)\, dt.\end{aligned}$$

Caputo's definition has two advantages with respect to the Riemann–Liouville definition: (i) in applications that involve fractional differential equations, the presence of initial values is physically meaningful; (ii) the derivative of a constant is zero, in contrast with the Riemann–Liouville.

12.5.2 Model Description

Let $0 < \sigma < 1$, $(D(0), L(0)) \in (\mathbb{R}_0^+)^2$, and

$$\begin{cases} {}^{c}D^{\sigma} D(t) = \alpha D\,(1-D) - \frac{pDL}{1+\gamma D}, \\ {}^{c}D^{\sigma} L(t) = \frac{pDL}{1+\gamma D} - \beta L. \end{cases} \tag{12.16}$$

To write (12.16) in a compact way, let

$$(\mathbb{R}_0^+)^2 = \{X \in \mathbb{R}^2 : X \geqslant 0\},$$

$X(t) = (D(t) \quad L(t))^T$, and

$${}^{c}D^{\sigma} X(t) = F(X(t)), \quad X(0) = (D(0), L(0)) \in (\mathbb{R}_0^+)^2, \tag{12.17}$$

with

$$F(X) = \begin{pmatrix} \alpha D\,(1-D) - \frac{pDL}{1+\gamma D} \\ \frac{pDL}{1+\gamma D} - \beta L \end{pmatrix}. \tag{12.18}$$

12.5.3 Existence and Uniqueness

Before the stability analysis, it is necessary to show the existence and uniqueness of non-negative solutions. The following lemma is important.

Lemma 12.2 (See [23]) *Let us assume that F in (12.18) satisfies the following conditions:*

- *$F(X)$ and $\frac{\partial F(X)}{\partial X}$ are continuous in $X \in \mathbb{R}^n$;*
- *$\|F(X)\| \leqslant w + \lambda \|X\|$, $\forall X \in \mathbb{R}^n$, where w and λ are positive constants.*

Then a solution to (12.17)–(12.18) exists and is unique.

Proposition 12.2 (Existence and uniqueness) *There exists only one solution to the IVP* (12.17) *in*

$$(\mathbb{R}_0^+)^2 = \left\{(D,L) \in \mathbb{R}^2 : (D(t),L(t)) \geqslant 0, \forall t > 0\right\}.$$

Proof. The existence and uniqueness of solution follows from Lemma 12.2. The vector function (12.18) is a polynomial, so it is continuously differentiable. Let

$$Z_1 = \begin{pmatrix} -\alpha & -p \\ 0 & p \end{pmatrix}, \quad Z_2 = \begin{pmatrix} 1 & 0 \\ 0 & \frac{1}{1+\gamma D} \end{pmatrix}, \quad B = \begin{pmatrix} \alpha & 0 \\ 0 & -\beta \end{pmatrix}.$$

Then,

$$F(X) = DZ_1Z_2X + BX. \tag{12.19}$$

Note that

$$Z_2 = \begin{pmatrix} 1 & 0 \\ 0 & \frac{1}{1+\gamma D} \end{pmatrix} < \begin{pmatrix} 1 & 0 \\ 0 & 1 \end{pmatrix}.$$

Using the sup norm, $\|Z_2\| \leq 1$, so

$$\begin{aligned} \|F(X)\| &\leq \|Z_1X\| + \|BX\| \\ &= (\|Z_1\| + \|B\|)\,\|X\| \\ &< \epsilon + (\|Z_1\| + \|B\|)\,\|X\| \end{aligned}$$

for some $\epsilon > 0$. So, by Lemma 12.2, system (12.17) has a unique solution. The proof of the non-negative of the solutions follows the same idea of [3]. To prove that $(D_k^*(t), L_k^*(t)) \in (\mathbb{R}_0^+)^2$ for all $t \geqslant 0$, let us consider the following auxiliary fractional differential system:

$$\begin{cases} {}^cD^\sigma D(t) = \alpha D\,(1-D) - \frac{pDL}{1+\gamma D} + \dfrac{1}{k}, \\ {}^cD^\sigma L(t) = \frac{pDL}{1+\gamma D} - \beta L + \dfrac{1}{k}, \end{cases}$$

with $k \in \mathbb{N}$. By contradiction, let us assume that exists a time instant where the condition fails. Let

$$t_0 = \inf\left\{t > 0 : (D_k^*(t), L_k^*(t)) \notin (\mathbb{R}_0^+)^2\right\}.$$

Then $(D_k^*(t_0), L_k^*(t_0)) \in (\mathbb{R}_0^+)^2$ and one of the quantities $D_k^*(t_0)$ or $L_k^*(t_0)$ is zero. Let us suppose that $D_k^*(t_0) = 0$. Then,

$${}^cD^\sigma D_k^*(t_0) = 0 + \frac{1}{k} > 0.$$

By continuity of ${}^cD^\sigma D_k^*$, we conclude that ${}^cD^\sigma D_k^*([t_0, t_0 + \zeta]) \subseteq \mathbb{R}^+$ so D_k^* is non-negative. Analogously, we can do the same for ${}^cD^\sigma L_k^*$, obtaining the intended contradiction. It follows by Lemma 1 of [3], when $k \to \infty$, that $(D^*(t), L^*(t)) \in (\mathbb{R}_0^+)^2$ for all $t \geqslant 0$. □

12.5.4 The Boundedness Condition

From Proposition 12.2, there exists only one solution to the IVP (12.17). The proof of our next result uses some auxiliary results found in [3, 23].

Theorem 12.4 *The solution to the IVP* (12.17) *is in*

$$\Omega = \left\{(D, L) \in (\mathbb{R}_0^+)^2 : W(t) = D(t) + L(t) \leqslant W(0) + \frac{A}{\beta}\right\},$$

where $A = \dfrac{\alpha + 4\beta}{4} M$.

Proof. Let

$$\begin{aligned} {}^cD^\sigma W(t) &\leqslant \alpha D(t)(1 - D(t)) - \beta L(t) - \beta D(t) + \beta D(t) \\ &\leqslant \frac{\alpha}{4} M + \beta M - \beta W(t) \leqslant \left(\frac{\alpha + 4\beta}{4}\right) M - \beta W(t). \end{aligned}$$

It is known that

$${}^cD^\sigma W(t) = J^{1-\alpha}\dot{W}(t) \quad \text{and} \quad \phi_\alpha(t) = \frac{t^{\alpha-1}}{\Gamma(\alpha)}, \qquad \text{for } t > 0.$$

Then

$$\begin{aligned} \mathcal{L}\{{}^cD^\sigma W(t)\} &= \mathcal{L}\{\phi_{1-\alpha}(t) * \dot{W}(t)\} = \mathcal{L}\{\phi_{1-\alpha}(t)\} \cdot \mathcal{L}\{\dot{W}(t)\} \\ &= s^{\alpha-1}(sW(s) - W(0)) = s^\alpha W(s) - s^{\alpha-1}W(0), \end{aligned}$$

so that

$$\begin{aligned} {}^cD^\sigma W(t) + \beta W(t) \leq A &\Leftrightarrow \mathcal{L}\{{}^cD^\sigma W(t) + \beta W(t)\} \leq \mathcal{L}\{A\} \\ &\Leftrightarrow s^\alpha W(s) - s^{\alpha-1}W(0) + \beta W(s) \leq \frac{A}{s} \\ &\Leftrightarrow W(s) \leq \frac{A}{s(s^\alpha + \beta)} + \frac{s^{\alpha-1}}{s^\alpha + \beta} W(0). \end{aligned}$$

Therefore,

$$\begin{aligned} W(t) &= \mathcal{L}^{-1}\{W(s)\} \\ &\leq \mathcal{L}^{-1}\left\{\frac{A}{s(s^\alpha + \beta)}\right\} + \mathcal{L}^{-1}\left\{\frac{s^{\alpha-1}}{s^\alpha + \beta} W(0)\right\} \\ &= A\mathcal{L}^{-1}\left\{\frac{1}{s(s^\alpha + \beta)}\right\} + W(0)\mathcal{L}^{-1}\left\{\frac{s^{\alpha-1}}{s^\alpha + \beta}\right\} \\ &= \frac{A}{\beta}(1 - E_\alpha(-\beta t^\alpha)) + W(0)E_\alpha(-\beta t^\alpha). \end{aligned}$$

Since $0 \leqslant E_\alpha(-\beta t^\alpha) \leqslant 1$, we have $W(t) \leqslant W(0) + \frac{A}{\beta}$. □

12.5.5 Stability Analysis

The equilibrium points of the fractional system (12.16) are

$$\begin{cases} {}^cD^\alpha D(t) = 0 \\ {}^cD^\alpha L(t) = 0 \end{cases} \Leftrightarrow \begin{cases} D = 0, \\ L = 0, \end{cases} \vee \begin{cases} D = 1, \\ L = 0, \end{cases} \vee \begin{cases} D = \frac{\beta}{p-\gamma\beta}, \\ L = \frac{\alpha}{p-\gamma\beta}\left(1 - \frac{\beta}{p-\gamma\beta}\right). \end{cases}$$

They are similar to the ones of the continuous model (12.3):

$$e_1 = (0,0), \quad e_2 = (1,0), \quad e_3 = \left(\frac{\beta}{p-\gamma\beta}, \frac{\alpha}{p-\gamma\beta}\left(1 - \frac{\beta}{p-\gamma\beta}\right)\right),$$

and the study of their local stability is equal to the continuous model. Precisely,

- e_1 and e_2 are *saddle points*;
- e_3 is a *sink* or an *asymptotically stable point* as long as condition (12.4) holds.

12.5.6 Graphical Analysis

We now present some plots obtained by the modified trapezoidal method, for $\sigma = 0.95$, step-size $h = 0.25$, and the time interval $[0, 300]$.

In Figure 12.7, we observe that the orbit with initial conditions $(D_0, L_0) = (0.2, 0.3)$ tends to the equilibrium point faster than the other methods. The same happens with $(D_0, L_0) = (0.55, 0.2)$.

Figure 12.8 shows that the equilibrium point is attained faster and Figure 12.9 illustrates, graphically, that when σ tends to 1 the solution of (12.16) tends to the solution of the continuous model (12.3), showing that the modified trapezoidal method is a good discretization method for our model.

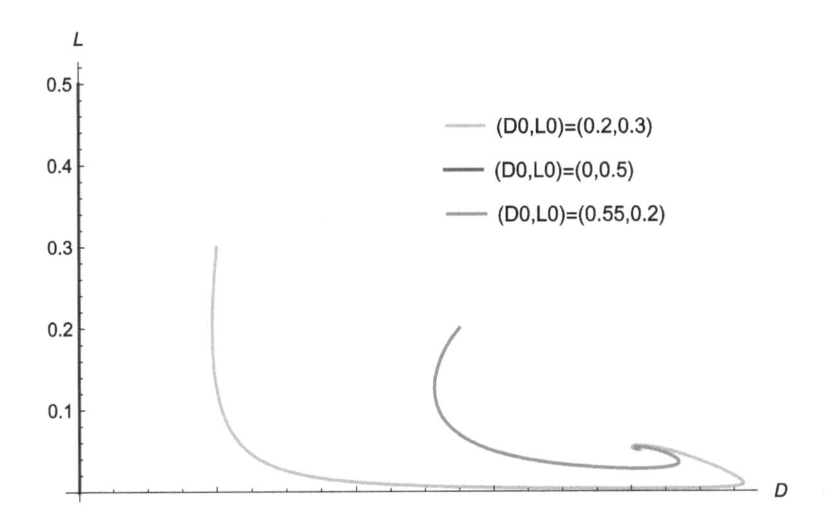

Figure 12.7 Solutions to the fractional model (12.16) with different initial conditions.

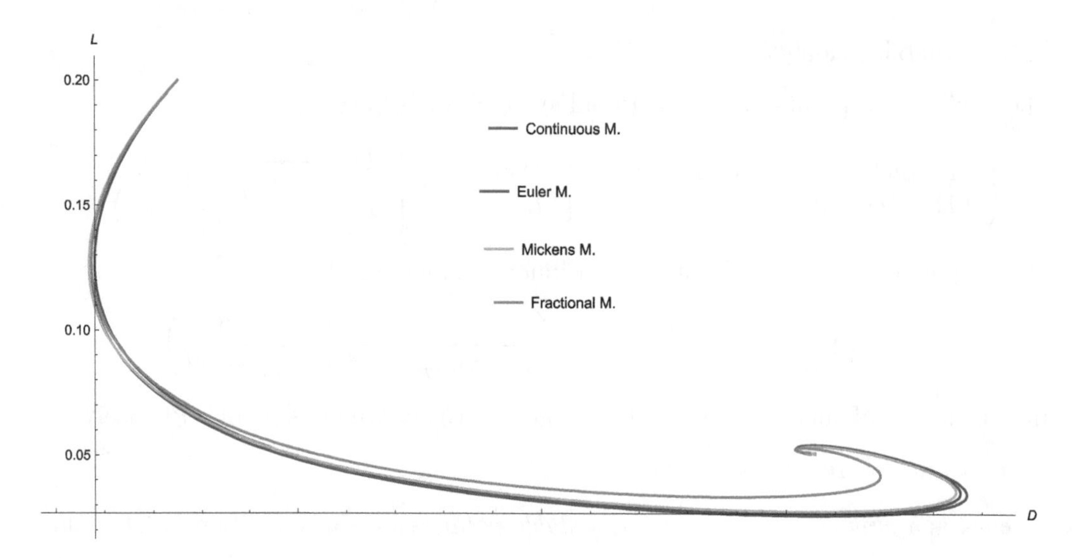

Figure 12.8 Solution to fractional model compared with solutions to continuous and discrete Euler's and Micken's models.

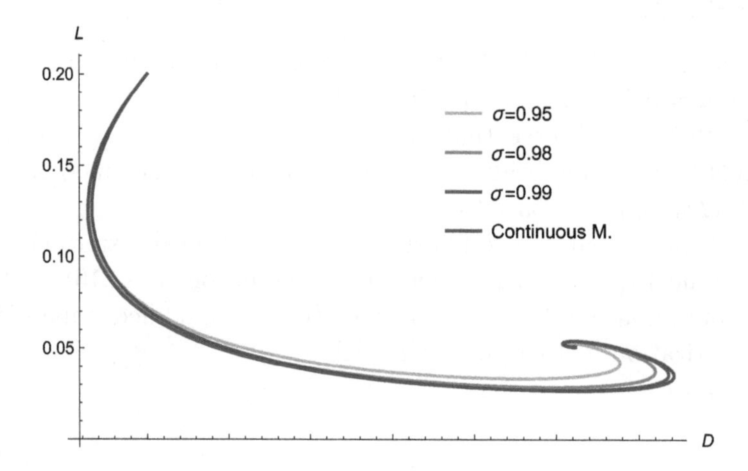

Figure 12.9 Solution to fractional system (12.16), for different values of σ, compared with the solution to the continuous model (12.3).

12.6 CONCLUSIONS

In this work, we considered a predator–prey Lotka–Volterra type model using the Michaelis–Menten equation as functional response. Financially, such functional response seems suitable. The use of the logistic function, to explain the growth of the deposit volume in the absence of a loan, is also realistic because the banks have a finite amount of deposit volume. We analyzed the proposed model through different methods and perspectives. Our study allows us to know if there are necessary conditions in order to achieve an asymptotic stable equilibrium point and what extra conditions are needed to be imposed by the numerical methods to be dynamically consistent. It is important to note that for the positive equilibrium point of the continuous model to be asymptotically stable, the condition (12.4) over the parameters has to be satisfied. For the Euler numerical scheme, some extra conditions need to be imposed, namely $1 - \beta h > 0$ and $h \in]0, h_2[$, where the expression of h_2 is in (12.9). Regarding Mickens' NSFD scheme, to prove that e_3 is asymptotically stable, it is necessary that $\phi(h) < 1$. Writing the model using fractional calculus, and using the same initial conditions, our graphical analysis allowed us to conclude that with an order of differentiation equal to $\sigma = 0.95$, the equilibrium point is attained faster and, as σ tends to one, the trajectories of the fractional system tend to be one of the continuous models.

ACKNOWLEDGMENTS

The authors were partially supported by the Portuguese Foundation for Science and Technology (FCT): Vaz through the Center of Mathematics and Applications of *Universidade da Beira Interior* (CMA-UBI), project UIDB/00212/2020; Torres through the Center for Research and Development in Mathematics and Applications (CIDMA), project UIDB/04106/2020.

BIBLIOGRAPHY

[1] A. Acharya, S. Bandyopadhyay, J. T. Cronin, J. Goddard, A. Muthunayake, and R. Shivaji. The diffusive Lotka-Volterra competition model in fragmented patches I: Coexistence. *Nonlinear Anal. Real World Appl.*, 70:Paper No. 103775, 2023.

[2] H. F. Ahmed. Fractional Euler method; an effective tool for solving fractional differential equations. *J. Egyptian Math. Soc.*, 26(1):38–43, 2018.

[3] R. Almeida, A. M. C. Brito da Cruz, N. Martins, and M. T. T. Monteiro. An epidemiological MSEIR model described by the Caputo fractional derivative. *Int. J. Dyn. Control*, 7(2):776–784, 2019.

[4] R. Almeida, S. Pooseh, and D. F. M. Torres. *Computational methods in the fractional calculus of variations.* Imperial College Press, London, 2015.

[5] M. Caputo. Linear models of dissipation whose q is almost frequency independent—ii. *Geophys J. Int.*, 13(5):529–539, 1967.

[6] M. Dias de Carvalho and J. E. Ottoni. Introdução ao cálculo fracionário com aplicações. *Rev. Mat. Ouro Preto*, 1:50–77, 2018.

[7] S. Elaydi. *An introduction to difference equations.* Undergraduate Texts in Mathematics. Springer, New York, third edition, 2005.

[8] G. Garcia Lorenzana and A. Altieri. Well-mixed Lotka-Volterra model with random strongly competitive interactions. *Phys. Rev. E*, 105(2):Paper No. 024307, 15, 2022.

[9] W. Gautschi. *Numerical analysis.* Birkhäuser Boston, Inc., Boston, MA, 2012.

[10] A. Kamimura, G. F. Burani, and H. M. França. The economic system seen as a living system: A Lotka–Volterra framework. *Emergence: Complexity and Organization*, 13(3):80–93, 2011.

[11] M. Lemos-Silva, and D. F. M. Torres. A note on a prey-predator model with constant-effort harvesting. In *Dynamic control and optimization*, pages 201–209. Springer Nature Switzerland, AG, 2022.

[12] P. Liu and S. N. Elaydi. Discrete competitive and cooperative models of Lotka-Volterra type. *J. Comput. Anal. Appl.*, 3(1):53–73, 2001.

[13] R. E. Mickens. *Nonstandard finite difference models of differential equations.* World Scientific Publishing Co., Inc., River Edge, NJ, 1994.

[14] R. E. Mickens. Nonstandard finite difference schemes for differential equations. *J. Difference Equ. Appl.*, 8(9):823–847, 2002.

[15] R. E. Mickens. Dynamic consistency: a fundamental principle for constructing non-standard finite difference schemes for differential equations. *J. Difference Equ. Appl.*, 11(7):645–653, 2005.

[16] R. E. Mickens. Calculation of denominator functions for nonstandard finite difference schemes for differential equations satisfying a positivity condition. *Numer. Meth. Part. Diff. Equ.*, 23(3):672–691, 2007.

[17] R. E. Mickens and T. M. Washington. NSFD discretizations of interacting population models satisfying conservation laws. *Comput. Math. Appl.*, 66(11):2307–2316, 2013.

[18] J. D. Murray. *Mathematical biology. I*, volume 17 of *Interdisciplinary Applied Mathematics.* Springer-Verlag, New York, third edition, 2002.

[19] Z. M. Odibat and S. Momani. An algorithm for the numerical solution of differential equations of fractional order. *J. Appl. Math Inf.*, 26(1-2):15–27, 2008.

[20] B. Ross. The development of fractional calculus 1695–1900. *Historia Math.*, 4:75–89, 1977.

[21] S. G. Samko, A. A. Kilbas, and O. I. Marichev. *Fractional integrals and derivatives.* Gordon and Breach Science Publishers, Yverdon, 1993.

[22] P. Shi and L. Dong. Dynamical behaviors of a discrete HIV-1 virus model with bilinear infective rate. *Math. Methods Appl. Sci.*, 37(15):2271–2280, 2014.

[23] M. R. Sidi Ammi, M. Tahiri, and D. F. M. Torres. Global stability of a Caputo fractional SIRS model with general incidence rate. *Math. Comput. Sci.*, 15(1):91–105, 2021.

[24] N. Sumarti, R. Nurfitriyana, and W. Nurwenda. A dynamical system of deposit and loan volumes based on the Lotka–Volterra model. In *AIP Conference Proceedings*, volume 1587, pages 92–94, 2014.

[25] D. Valério, J. Tenreiro Machado, and V. Kiryakova. Some pioneers of the applications of fractional calculus. *Fract. Calc. Appl. Anal.*, 17(2):552–578, 2014.

[26] J. Véron. Alfred J. Lotka and the mathematics of population. *J. Électron. Hist. Probab. Stat.*, 4(1):10, 2008.

[27] S. Yan and Z. Du. Hopf bifurcation in a Lotka-Volterra competition-diffusion-advection model with time delay. *J. Diff. Equ.*, 344:74–101, 2023.

CHAPTER 13

Existence and Uniqueness of Solutions to Proper Fractional Riemann-Liouville Initial Value Problems on Time Scales

Nedjoua Zine and Benaoumeur Bayour

University of Mascara

Delfim F. M. Torres

University of Aveiro

CONTENTS

13.1 INTRODUCTION

Let $\mathbb{T}$ be a time scale, that is, a nonempty closed subset of $\mathbb{R}$. In [2], Benkhettou, Hammoudi and Torres introduced a concept of fractional integral,

$$ {}_a^{\mathbb{T}}I_t^{\alpha}h(t) = \frac{1}{\mu(\alpha)}\left(\int_a^t (t-s)^{\alpha-1}h(s)\Delta s\right)^{\Delta}, \tag{13.1} $$

and the concept of fractional derivative

$$ {}_a^{\mathbb{T}}D_t^{\alpha}h(t) = \frac{1}{\mu(1-\alpha)}\left(\int_a^t (t-s)^{-\alpha}h(s)\Delta s\right)^{\Delta} \tag{13.2} $$

DOI: 10.1201/9781003388678-13

of Riemann–Liouville on time scales. In [6], Torres gives more suitable definitions of fractional integral (13.1) and fractional derivative (13.2) of Riemann–Liouville on time scales, introducing the forward jump σ operator of time scales in their definition:

$$ {}_a^{\mathbb{T}}I_t^{\alpha}h(t) = \frac{1}{\mu(\alpha)}\left(\int_a^t (t-\sigma(s))^{\alpha-1}h(s)\Delta s\right)^{\Delta} \tag{13.3}$$

and

$$ {}_a^{\mathbb{T}}D_t^{\alpha}h(t) = \frac{1}{\mu(1-\alpha)}\left(\int_a^t (t-\sigma(s))^{-\alpha}h(s)\Delta s\right)^{\Delta}. \tag{13.4}$$

Here we focus on definitions (13.3) and (13.4), but changing the operator σ into the backward jump operator ρ. As we shall prove, the new definitions with ρ provide proper notions with respect to the existence and uniqueness of solution to the following initial value problem:

$$ ({}_{t_0}^{\alpha}Dy)(t) = f(t, y(t)), \quad t \in [t_0, t_0 + d] = \mathcal{J} \subseteq \mathbb{T}, \tag{13.5}$$

$$ ({}_{t_0}^{1-\alpha}Iy)(t_0) = 0, \tag{13.6}$$

where $\mathbb{T}$ is a given time scale, $0 < \alpha < 1$, $d > 0$, ${}_{t_0}^{\alpha}D$ is the proper (left) Riemann–Liouville fractional derivative operator or order α defined on $\mathbb{T}$ with ρ, ${}_{t_0}^{1-\alpha}I$ is the proper (left) Riemann–Liouville fractional integral operator of order $1-\alpha$ defined on $\mathbb{T}$ with ρ, and function $f : \mathcal{J} \times \mathbb{T} \to \mathbb{R}$ is a right dense continuous function. Our main results give sufficient conditions for the existence (Theorem 13.2) and uniqueness (Theorem 13.3) of solution to problem (13.5)–(13.6).

13.2 PRELIMINARIES

A time scale $\mathbb{T}$ is an arbitrary nonempty closed subset of the real numbers $\mathbb{R}$. For $t \in \mathbb{T}$, we define the forward jump operator $\sigma : \mathbb{T} \to \mathbb{T}$ by

$$\sigma(t) = \inf\{s \in \mathbb{T} : s > t\}$$

and the backward jump operator $\rho : \mathbb{T} \to \mathbb{T}$ by

$$\rho(t) := \sup\{s \in \mathbb{T} : s < t\}.$$

Then, one defines the graininess function $\mu : \mathbb{T} \to [0, +\infty[$ by $\mu(t) = \sigma(t) - t$. If $\sigma(t) > t$, then we say that t is right-scattered; if $\rho(t) < t$, then t is left-scattered. Moreover, if $t < \sup \mathbb{T}$ and $\sigma(t) = t$, then t is called right-dense; if $t > \inf \mathbb{T}$ and $\rho(t) = t$, then t is called left-dense. If $\mathbb{T}$ has a left-scattered maximum m, then we define $\mathbb{T}^{\kappa} = \mathbb{T} \setminus \{m\}$; otherwise $\mathbb{T}^{\kappa} = \mathbb{T}$. If $f : \mathbb{T} \to \mathbb{R}$, then we define $f^{\sigma} : \mathbb{T} \to \mathbb{R}$ by $f^{\sigma}(t) = f(\sigma(t))$ for all $t \in \mathbb{T}$.

Definition 13.1 (See [4]) *Let $f : \mathbb{T} \to \mathbb{R}$ and $t \in \mathbb{T}$. We define $f^{\Delta}(t)$ to be the number, provided it exists, with the property that given any $\epsilon > 0$ there is a neighborhood U of t (i.e., $U = (t - \delta, t + \delta) \cap \mathbb{T}$ for some $\delta > 0$) such that*

$$\left|[f^{\sigma}(t) - f(s)] - f^{\Delta}(t)[\sigma(t) - s]\right| \le \epsilon|\sigma(t) - s|$$

for all $s \in U$. We call $f^{\Delta}(t)$ the Hilger (or the time-scale) derivative of f at t.

For more on the calculus on time scales, we refer the reader to the books [3, 4].

Definition 13.2 (See [3]) *Let $[a, b]$ denote a closed bounded interval in $\mathbb{T}$. A function $F : [a, b] \to \mathbb{R}$ is called a delta anti-derivative of function $f : \mathbb{T} \to \mathbb{R}$ provided F is continuous on $[a, b]$, delta differentiable on $[a, b)$, and $F^{\Delta}(t) = f(t)$ for all $t \in [a, b)$. Then, we define the Δ-integral of f from a to b by*

$$\int_a^b f(t)\Delta t := F(b) - F(a).$$

Definition 13.3 (See [4]) *A function $f : \mathbb{T} \to \mathbb{R}$ is called rd-continuous provided it is continuous at right-dense points in $\mathbb{T}$ and its left-sided limits exist (finite) at left-dense points in $\mathbb{T}$. The set of rd-continuous functions $f : \mathbb{T} \to \mathbb{R}$ is denoted by $C_{rd}(\mathbb{T}, \mathbb{R})$.*

Proposition 13.1 (See [1]) *Suppose $\mathbb{T}$ is a time scale and f is an increasing continuous function on the time-scale interval $[a, b]$ (i.e., $\mathbb{T} \subseteq [a, b]$). If F is the extension of f to the real interval $[a, b]$ given by*

$$F(s) := \begin{cases} f(s) & if s \in \mathbb{T}, \\ f(t) & if s \in (t, \sigma(t)) \notin \mathbb{T}, \end{cases}$$

then

$$\int_a^b f(t)\Delta t \leq \int_a^b F(t)dt.$$

We now recall the celebrated gamma function.

Definition 13.4 (Gamma function) *For complex numbers with a positive real part, the gamma function $\Gamma(t)$ is defined by the following convergent improper integral:*

$$\Gamma(t) := \int_0^\infty s^{t-1}e^{-s}ds.$$

Remark 13.1 *The gamma function satisfies the following useful property:*

$$\Gamma(t+1) = t\Gamma(t).$$

Now we introduce new notions of fractional operators, analogous to the Riemann–Liouville fractional operators on time scales proposed in [6].

Definition 13.5 (Fractional integral on time scales) *Suppose $\mathbb{T}$ is a time scale, $[a, b]$ is an interval of $\mathbb{T}$, and f is an integrable function on $[a, b]$. Let $\alpha > 0$ and $t \in [a, b]$. Then the (left) fractional integral of order α of f is defined by*

$$({}_a^{\alpha}If)(t) := \int_a^t \frac{(t-\rho(s))^{\alpha-1}}{\Gamma(\alpha)} f(s)\Delta s, \tag{13.7}$$

where Γ is the gamma function.

Definition 13.6 (Fractional derivative on time scales) *Suppose $\mathbb{T}$ is a time scale, $[a,b]$ is an interval of $\mathbb{T}$, and f is an integrable function on $[a,b]$. Let $0<\alpha<1$, $t\in[a,b]$. The (left) Riemann–Liouville fractional derivative of order α of f is defined by*

$$({}_a^\alpha Df)(t) := \frac{1}{\Gamma(1-\alpha)}\left(\int_a^t (t-\rho(s))^{-\alpha} f(s)\Delta s\right)^{\Delta}. \tag{13.8}$$

Fractional operators of negative order are defined as follows.

Definition 13.7 *If $-1<\alpha<0$, then the (Riemann–Liouville) fractional derivative of order α is defined as the fractional integral of order $-\alpha$. Moreover, the fractional integral of order α is defined as the (Riemann–Liouville) fractional derivative of order $-\alpha$:*

$$({}_a^\alpha Df)(t) := ({}_a^{-\alpha} If)(t), \qquad ({}_a^\alpha If)(t) := ({}_a^{-\alpha} Df)(t).$$

Remark 13.2 *Along the work, we consider the order α of the fractional derivatives in the real interval $(0,1)$. We can, however, easily generalize our definitions to any positive real α. Indeed, let $\alpha\in\mathbb{R}^+\setminus\mathbb{N}$. Then there exists $\beta\in(0,1)$ such that $\alpha=[\alpha]+\beta$, where $[\alpha]$ is the integer part of α, and we can set*

$$({}_a^\alpha Df)(t) := {}_a^\beta D\left(f^{\Delta^{[\alpha]}}\right)(t).$$

13.3 MAIN RESULTS

We begin by proving some fundamental properties of the fractional operators on time scales (Section 13.3.1). After that, we prove the existence of a solution to the fractional order initial value problem (13.5)–(13.6) defined on a time scale $\mathbb{T}$ (Section 13.3.2).

13.3.1 Properties of the Time-Scale Fractional Operators

Proposition 13.2 *Let $\mathbb{T}$ be a time scale with derivative Δ; $0<\alpha<1$. Then,*

$$({}_a^\alpha Dg)(t) = \left(\Delta\circ{}_a^{1-\alpha} Ig\right)(t).$$

Proof. Let $g:\mathbb{T}\to\mathbb{R}$. From (13.8) we have

$$\begin{aligned} ({}_a^\alpha Dg)(t) &= \frac{1}{\Gamma(1-\alpha)}\left(\int_a^t (t-\rho(s))^{-\alpha} g(s)\Delta s\right)^{\Delta} \\ &= \left({}_a^{1-\alpha} Ig)(t)\right)^{\Delta} \\ &= (\Delta\circ({}_a^{1-\alpha} Ig))(t). \end{aligned}$$

The proof is complete. □

Proposition 13.3 *For any integrable function g on $[a,b]\cap\mathbb{T}$, the Riemann–Liouville Δ-fractional integral satisfies*

$${}_a^\alpha I\circ{}_a^\beta I(g) = {}_a^{\alpha+\beta} I(g), \qquad \text{for } \alpha>0 \text{ and } \beta>0.$$

Proof. Similar to the proof of Proposition 16 of [2]. □

Proposition 13.4 *For any integrable function g on $[a,b] \cap \mathbb{T}$ one has*

$$ {}_a^{\alpha}D \circ {}_a^{\alpha}Ig = g, \qquad 0 < \alpha < 1. $$

Proof. By Propositions 13.2 and 13.3, we have

$$ {}_a^{\alpha}D \circ {}_a^{\alpha}Ig = \left[\left({}_a^{1-\alpha}I\right)\left({}_a^{\alpha}Ig\right)(t)\right]^{\Delta} = \left[\left({}_a^{1}Ig\right)(t)\right]^{\Delta} = g(t). $$

The proof is complete. □

Corollary 13.1 *For $0 < \alpha < 1$, we have*

$$ \left({}_a^{\alpha}D\right) \circ \left({}_a^{-\alpha}D\right) = Id $$

and

$$ {}_a^{-\alpha}I \circ {}_a^{\alpha}I = Id, $$

where Id denotes the identity operator.

Proof. From Definition 13.7 and Proposition 13.4, we have that

$$ {}_a^{\alpha}D \circ {}_a^{-\alpha}D = {}_a^{\alpha}D \circ {}_a^{\alpha}I = Id $$

and

$$ {}_a^{-\alpha}I \circ {}_a^{\alpha}I = {}_a^{\alpha}D \circ {}_a^{\alpha}I = Id. $$

The proof is complete. □

Definition 13.8 *For $\alpha > 0$, we denote by ${}_a^{\alpha}I([a,b])$ the space of functions that can be represented by the Riemann–Liouville Δ-integral of order α of some $C_{rd}([a,b])$ function.*

Theorem 13.1 *Let $f \in C_{rd}([a,b])$ and $\alpha > 0$. In order that $f \in {}_a^{\alpha}I([a,b])$, it is necessary and sufficient that*

$$ \left({}_a^{1-\alpha}If\right) \in C_{rd}^{1}([a,b]) \tag{13.9} $$

and

$$ \left(\left({}_a^{1-\alpha}If\right)(t)\right)\big|_{t=a} = 0. \tag{13.10} $$

Proof. Assume $f \in {}_a^{\alpha}I([a,b])$, $f(t) = \left({}_a^{\alpha}Ih\right)(t)$ for some $h \in C_{rd}([a,b])$, and

$$ \left({}_a^{1-\alpha}If\right)(t) = \left({}_a^{1-\alpha}I\right)\left({}_a^{\alpha}Ih\right)(t). $$

From Proposition 13.3, we have

$$ \left({}_a^{1-\alpha}If\right)(t) = \left({}_a^{1}Ih\right)(t) = \int_a^t h(s)\Delta s. $$

Therefore,

$$({}_a^{1-\alpha}If) \in C_{rd}^1([a,b])$$

and

$$({}_a^{1-\alpha}If)(t))|_{t=a} = \int_a^a h(s)\Delta s = 0.$$

Conversely, assume that $f \in C_{rd}([a,b])$ satisfies (13.9) and (13.10). From Taylor's formula applied to function $I_a^{1-\alpha}f$, one has

$$({}_a^{1-\alpha}If)(t) = \int_a^t \frac{\Delta}{\Delta s}({}_a^{1-\alpha}If)(s)\Delta s, \quad \forall t \in [a,b].$$

Let $\varphi(t) := \frac{\Delta}{\Delta t}({}_a^{1-\alpha}If)(t)$. Note that, by (13.9), $\varphi \in C_{rd}([a,b])$. From Proposition 13.3, we have

$$({}_a^{1-\alpha}If)(t) = ({}_a^1I\varphi)(t) = ({}_a^{1-\alpha}I)({}_a^{\alpha}I\varphi)(t)$$

and thus

$$({}_a^{1-\alpha}If)(t) - ({}_a^{1-\alpha}I)({}_a^{\alpha}I\varphi)(t) \equiv 0.$$

Then,

$$[{}_a^{1-\alpha}I(f - ({}_a^{\alpha}I\varphi)](t) \equiv 0.$$

This implies that

$$f - ({}_a^{\alpha}I\varphi) \equiv 0.$$

We conclude that $f = {}_a^{\alpha}I\varphi$ and $f \in {}_a^{\alpha}I([a,b])$. □

Corollary 13.2 *Let $0 < \alpha < 1$ and $f \in C_{rd}([a,b])$ satisfy the condition in Theorem 13.1. Then,*

$$\left({}_a^{\alpha}I \circ {}_a^{\alpha}D\right)(f) = f.$$

13.3.2 Existence of Solutions to Fractional IVPs on Time Scales

Let $\mathbb{T}$ be a time scale and $\mathcal{J} = [t_0, t_0 + d] \subset \mathbb{T}$. A function $y \in C_{rd}(\mathcal{J}, \mathbb{R})$ is a solution to problem (13.5)–(13.6) if

$$({}_{t_0}^{\alpha}Dy)(t) = f(t,y) \text{ on } \mathcal{J}, \quad 0 < \alpha < 1,$$
$$\left({}_{t_0}^{1-\alpha}Iy\right)(t_0) = 0.$$

To establish the existence of such solution, first we recall the definition of compact map.

Definition 13.9 (See p. 112 of [5]) *Let X and Y be topological spaces. A map $f : X \to Y$ is called compact if $f(X)$ is contained in a compact subset of Y.*

Let us define the operator

$$T : \mathcal{C}_{rd}(\mathcal{J}, \mathbb{R}) \to \mathcal{C}_{rd}(\mathcal{J}, \mathbb{R})$$

by

$$T(y)(t) = \frac{1}{\Gamma(\alpha)} \int_{t_0}^{t} (t - \rho(s))^{\alpha-1} f(s, y(s)) \Delta s.$$

Lemma 13.1 *Let $0 < \alpha < 1$, $\mathcal{J} \subseteq \mathbb{T}$, and $f : \mathcal{J} \times \mathbb{R} \to \mathbb{R}$. A function y is a solution to problem (13.5)–(13.6) if, and only if, this function is a solution to the integral equation*

$$y(t) = \frac{1}{\Gamma(\alpha)} \int_{t_0}^{t} (t - \rho(s))^{\alpha-1} f(s, y(s)) \Delta s,$$

that is, y is a fixed point of operator T: $T(y)(t) = y(t)$.

Proof. By Corollary 13.2, $({}_{t_0}^{\alpha} I) \circ ({}_{t_0}^{\alpha} D y)(t) = y(t)$. From (13.8) we have

$$y(t) = \frac{1}{\Gamma(\alpha)} \int_{t_0}^{t} (t - \rho(s))^{\alpha-1} f(s, y(s)) \Delta s$$

and the proof is complete. □

Theorem 13.2 (Existence of solution) *Suppose $f : \mathcal{J} \times \mathbb{R} \to \mathbb{R}$ is a rd-continuous bounded function such that there exists $M > 0$ with $|f(t, y(t))| < M$ for all $t \in \mathcal{J}$, $y \in \mathbb{R}$. Then problem (13.5)–(13.6) has a solution on $\mathcal{J}$.*

Proof. The proof is given in three steps.

Step 1: T is continuous. Let y_n be a sequence such that $y_n \to y$ in $\mathcal{C}(\mathcal{J}, \mathbb{R})$. Then, for each $t \in \mathcal{J}$,

$$\begin{aligned}
&|T(y_n)(t) - T(y)(t)| \\
&\quad \leq \frac{1}{\Gamma(\alpha)} \int_{t_0}^{t} (t - \sigma(s))^{\alpha-1} \left| f(s, y_n(s)) - f(s, y(s)) \right| \Delta s \\
&\quad \leq \frac{1}{\Gamma(\alpha)} \int_{t_0}^{t} (t - \rho(s))^{\alpha-1} \sup_{s \in \mathcal{J}} \left| f(s, y_n(s)) - f(s, y(s)) \right| \Delta s \qquad (13.11) \\
&\quad \leq \frac{\|f(\cdot, y_n(\cdot)) - f(\cdot, y(\cdot))\|_\infty}{\Gamma(\alpha)} \int_{t_0}^{t} (t - \rho(s))^{\alpha-1} \Delta s \\
&\quad \leq \frac{\|f(\cdot, y_n(\cdot)) - f(\cdot, y(\cdot))\|_\infty}{\Gamma(\alpha)} \int_{t_0}^{t} (t - \rho(s))^{\alpha-1} ds.
\end{aligned}$$

For $0 < \alpha < 1$ we have

$$(t - \rho(s))^{\alpha-1} < (t - s)^{\alpha-1}$$

and from (13.11) it follows that

$$\begin{aligned} |T(y_n)(t) - T(y)(t)| &\leq \frac{\|f(\cdot, y_n(\cdot)) - f(\cdot, y(\cdot))\|_\infty}{\Gamma(\alpha)} \frac{a^\alpha}{\alpha} \\ &\leq \frac{a^\alpha \|f(\cdot, y_n(\cdot)) - f(\cdot, y(\cdot))\|_\infty}{\Gamma(\alpha+1)}. \end{aligned}$$

Since f is a continuous function, one has

$$|T(y_n)(t) - T(y)(t)|_\infty \leq \frac{a^\alpha}{\Gamma(\alpha+1)} \|f(\cdot, y_n(\cdot)) - f(\cdot, y(\cdot))\|_\infty \to 0$$

as $n \to \infty$.

Step 2: For the second part of the proof, we have to show that the set $T(\mathcal{C}(\mathcal{J}, \mathbb{R}))$ is relatively compact. Let $T(y) \in T(\mathcal{C}(\mathcal{J}, \mathbb{R}))$. Then, $\|T(y)\|_\infty \leq l$. By hypothesis, for each $t \in \mathcal{J}$, we have

$$\begin{aligned} |T(y)(t)| &\leq \frac{1}{\Gamma(\alpha)} \int_{t_0}^{t} (t - \rho(s))^{\alpha-1} |f(s, y(s))| \Delta s \\ &\leq \frac{M}{\Gamma(\alpha)} \int_{t_0}^{t} (t - \rho(s))^{\alpha-1} \Delta s \\ &\leq \frac{M}{\Gamma(\alpha)} \int_{t_0}^{t} (t - \rho(s))^{\alpha-1} ds. \end{aligned} \tag{13.12}$$

For $0 < \alpha < 1$, we know that

$$(t - \rho(s))^{\alpha-1} \leq (t - s)^{\alpha-1}$$

and from (13.12) and Proposition 13.1 we can write that

$$|T(y)(t)| \leq \frac{Ma^\alpha}{\alpha\Gamma(\alpha)} = \frac{Ma^\alpha}{\Gamma(\alpha+1)} = l.$$

Therefore, $T(\mathcal{C}(\mathcal{J}, \mathbb{R}))$ is uniformly bounded. This set is also equicontinuous since for every t_1, $t_2 \in \mathcal{J}$, $t_1 < t_2$. Let $A = |T(y)(t_1) - T(y)(t_2)|$. Then we can write that

$$A \leq \frac{1}{\Gamma(\alpha)} \left| \int_{t_0}^{t_1} (t_1 - \rho(s))^{\alpha-1} f(s, y(s)) \Delta s - \int_{t_0}^{t_2} (t_2 - \rho(s))^{\alpha-1} f(s, y(s)) \Delta s \right|,$$

that is,

$$\begin{aligned} A \leq \frac{M}{\Gamma(\alpha)} \Bigg(&\int_{t_0}^{t_1} ((t_1 - \rho(s))^{\alpha-1} - (t_2 - \rho(s))^{\alpha-1}) \Delta s \\ &+ \int_{t_1}^{t_2} (t_2 - \rho(s))^{\alpha-1} \Delta s \Bigg). \end{aligned} \tag{13.13}$$

For $0 < \alpha < 1$

$$(t - \rho(s))^{\alpha-1} < (t - s)^{\alpha-1}$$

and it follows that

$$\begin{aligned}
&|T(y)(t_1) - T(y)(t_2)| \\
&\quad \leq \frac{M}{\Gamma(\alpha)}\left(\int_{t_0}^{t_1}((t_1 - s)^{\alpha-1} - (t_2 - s)^{\alpha-1})ds + \int_{t_1}^{t_2}(t_2 - s)^{\alpha-1}\Delta s\right) \\
&\quad \leq \frac{M}{\alpha\Gamma(\alpha+1)}[(t_2 - t_1)^{\alpha} + (t_1 - t_0)^{\alpha} - (t_2 - t_0)^{\alpha} + (t_2 - t_1)^{\alpha}] \\
&\quad = \frac{2M}{\alpha\Gamma(\alpha+1)}(t_2 - t_1)^{\alpha} + \frac{M}{\alpha\Gamma(\alpha+1)}[(t_1 - t_0)^{\alpha} - (t_2 - t_0)^{\alpha}].
\end{aligned}$$

As $t_1 \to t_2$, the right-hand side of the above inequality tends to zero. From the Arzela–Ascoli theorem, adapted to our context, it follows that $T(\mathcal{C}(\mathcal{J}, \mathbb{R}))$ is relatively compact.

Step 3: conclusion. As a consequence of Schauder's fixed point theorem, we conclude that T has a fixed point, which is solution of problem (13.5)–(13.6). □

Theorem 13.3 (Existence and uniqueness of solution) *Let $\mathcal{J} = [t_0, t_0 + d] \subseteq \mathbb{T}$. The initial value problem (13.5)–(13.6) has a unique solution on $\mathcal{J}$ if function $f(t, y)$ is a right-dense continuous bounded function such that there exists $M > 0$ for which $|f(t, y(t))| < M$ on $\mathcal{J}$ and the Lipshitz condition*

$$\|f(t, x) - f(t, y)\| \leq L\,\|x - y\|$$

holds for some $L > 0$, for all $t \in \mathcal{J}$ and all x, $y \in \mathbb{R}$.

Proof. Let $\mathcal{S}$ be the set of rd-continuous functions on $\mathcal{J} \subseteq \mathbb{T}$. For $y \in \mathcal{S}$, define $\|y\| = \sup_{t\in\mathcal{J}} \|y(t)\|$. It is easy to see that S is a Banach space with this norm. The subset of $\mathcal{S}(R)$ and the operator T are defined by

$$\mathcal{S}(R) = \{X \in \mathcal{S} : \|X_s\| \leq R\}$$

and

$$T(y) = \frac{1}{\Gamma(\alpha)}\int_{t_0}^{t}(t - \rho(s))^{\alpha-1}f(s, y(s))\Delta s.$$

Then,

$$\begin{aligned}
|T(y(t))| &\leq \frac{1}{\Gamma(\alpha)}\int_{t_0}^{t}(t - \rho(s))^{\alpha-1}\,|f(t, y(t))|\,\Delta s \\
&\leq \frac{M}{\Gamma(\alpha)}\int_{t_0}^{t}(t - \rho(s))^{\alpha-1}\Delta s.
\end{aligned}$$

Since $(t-\rho(s))^{\alpha-1}$ is an increasing monotone function, by using Proposition 13.1 we can write that

$$\int_{t_0}^{t}(t-\rho(s))^{\alpha-1}\Delta s \leq \int_{t_0}^{t}(t-\rho(s))^{\alpha-1}ds.$$

Consequently,

$$|T(y(t))| \leq \frac{M}{\Gamma(\alpha)}\int_{t_0}^{t}(t-\rho(s))^{\alpha-1}ds.$$

For $0<\alpha<1$ we have

$$(t-\rho(s))^{\alpha-1} < (t-s)^{\alpha-1}$$

and from (13.14) it follows that

$$|T(y(t))| \leq \frac{M}{\Gamma(\alpha)}\frac{a^\alpha}{\alpha} =: \bar{R}.$$

With $\bar{R}=\frac{Ma^\alpha}{\Gamma(\alpha+1)}$, we conclude that T is an operator from $\mathcal{S}(R)$ to $\mathcal{S}(\bar{R})$. Moreover,

$$\begin{aligned}
\|T(x)-T(y)\| &\leq \frac{1}{\Gamma(\alpha)}\int_{t_0}^{t}(t-\rho(s))^{\alpha-1} \mid f(t,x(s))-f(t,y(s)) \mid \Delta s \\
&\leq \frac{L\,\|x-y\|_\infty}{\Gamma(\alpha)}\int_{t_0}^{t}(t-\rho(s))^{\alpha-1}\Delta s \\
&\leq \frac{L\,\|x-y\|_\infty}{\Gamma(\alpha)}\int_{t_0}^{t}(t-\rho(s))^{\alpha-1}ds. \qquad (13.14)
\end{aligned}$$

It follows from (13.14) that

$$\begin{aligned}
\|T(x)-T(y)\| &\leq \frac{L\,\|x-y\|_\infty}{\Gamma(\alpha)}\frac{a^\alpha}{\alpha} \\
&= \frac{La^\alpha}{\Gamma(\alpha+1)}\|x-y\|_\infty
\end{aligned}$$

for x, $y \in \mathcal{S}(\rho)$. If $\frac{La^\alpha}{\Gamma(\alpha+1)} \leq 1$, then one has a contraction map. This implies the uniqueness of solution to problem (13.5)–(13.6). □

13.4 ACKNOWLEDGMENTS

This work is part of Zine's PhD project. Torres was supported by FCT (Fundação para a Ciência e a Tecnologia) through the R&D Unit CIDMA and project UIDB/04106/2020.

BIBLIOGRAPHY

[1] A. Ahmadkhanlu and M. Jahanshahi. On the existence and uniqueness of solution of initial value problem for fractional order differential equations on time scales. *Bull. Iranian Math. Soc.*, 38(1):241–252, 2012.

[2] N. Benkhettou, A. Hammoudi, and D. F. M. Torres. Existence and uniqueness of solution for a fractional Riemann–Liouville initial value problem on time scales. *J. King Saud Univ. Sci.*, 28(1):87–92, 2016.

[3] M. Bohner and A. Peterson. *Dynamic equations on time scales.* Birkhäuser Boston, Inc., Boston, MA, 2001.

[4] M. Bohner and A. Peterson. *Advances in dynamic equations on time scales.* Birkhäuser Boston, Inc., Boston, MA, 2003.

[5] A. Granas and J. Dugundji. *Fixed point theory.* Springer Monographs in Mathematics. Springer-Verlag, New York, 2003.

[6] D. F. M. Torres. Cauchy's formula on nonempty closed sets and a new notion of Riemann-Liouville fractional integral on time scales. *Appl. Math. Lett.*, 121:Paper No. 107407, 6, 2021.

CHAPTER 14

Ostrowski Type Inequalities for Conformable Fractional Calculus via a Parameter

Miguel Vivas-Cortez
Pontifical Catholic University of Ecuador

Seth Kermausuor
Alabama State University

Juan E. Nápoles Valdés
Notheast National University

CONTENTS

14.1 INTRODUCTION

The classical Ostrowski inequality was established by Alexander M. Ostrowski in 1938 [17], through the next theorem:

Theorem 14.1 *Let $f : [\kappa, \mu] \to \mathbb{R}$ be continuous on $[\kappa, \mu]$ and differentiable in (κ, μ). If $|f'(\nu)| \leq M$ then the following inequality holds for all $\xi \in [\kappa, \mu]$*

$$\left| f(\xi) - \frac{1}{\mu-\kappa}\int_{\kappa}^{\mu} f(\nu)d\nu \right| \leq \left(\frac{1}{4} + \frac{\left(\xi - \frac{\kappa+\mu}{2}\right)^2}{(\mu-\kappa)^2} \right)(\mu-\kappa)M, \tag{14.1}$$

where $M := \sup_{\xi\in(\kappa,\mu)} |f'(\xi)| < \infty$. The inequality is sharp in the sense that the constant $\frac{1}{4}$ cannot be replaced by a smaller one.

DOI: 10.1201/9781003388678-14

Inequality (14.1) has found numerous fields of application, such as numerical integration, probability calculus, and approximation theory, among others; this is one of the fundamental reasons that have motivated many researchers to study and generalize this inequality in various directions [8, 3, 7, 5, 6, 9, 10, 12, 14]. For instance, S. S. Dragomir and his collaborators gave a new generalization of Ostrowski's integral inequality for mappings whose derivatives and applications in numerical integration and special means using a paramater [8]:

Theorem 14.2 *Let $f : [\kappa, \mu] \to \mathbb{R}$ be continuous on $[\kappa, \mu]$, differentiable on (a, b) and whose derivative $f' : (\kappa, \mu) \to \mathbb{R}$ is bounded on $(\kappa.\mu)$. Denote $M := \sup_{\xi\in(\kappa,\mu)} |f'(\xi)| < \infty$. Then*

$$\left|(1-\lambda)f(\xi) + \lambda\frac{f(\kappa)+f(\mu)}{2} - \frac{1}{\mu-\kappa}\int_\kappa^\mu f(\nu)d\nu\right| \leq \frac{M}{\mu-\kappa}\left[\frac{(\mu-\kappa)^2}{4}\left[\lambda^2+(\lambda-1)^2\right] + \left(\xi-\frac{\kappa+\mu}{2}\right)^2\right]$$

for all $\lambda \in [0, 1]$ and $a + \lambda\frac{\mu-\kappa}{2} \leq \xi \leq b - \lambda\frac{\mu-\kappa}{2}$.

Numerous applications in the applied sciences, engineering, economics, finance, and many more over the last few decades have been the reason for a vertiginous development and evolution in the theory of fractional calculus. New tools and extending existing classical results in fractional calculus have been some of the major areas of research in recent years [1, 2, 4, 11, 16, 15, 18, 19, 20, 21].

Given the scope of the cited research, in this chapter, we focus on the relationship between conformable fractional calculus and inequality (14.1). This work is divided as follows: the first section contains some preliminary concepts that serve as a foundation for the new proposed results; the second section contains Ostrowski's integral inequality with a parameter version in conformable fractional calculus and some other related results; the third section contains other results regarding the boundedness of the inequality (14.1); and a conclusion section.

14.2 PRELIMINARY NOTES

R. Khalil et. al., in 2014, introduced a new fractional derivative of order $\alpha \in (0, 1]$ (see [15]), as follows:

Definition 14.1 (Conformable fractional derivative) *For any function $f : [0, \infty) \to \mathbb{R}$, the conformable fractional order derivative of f of order α is defined by*

$$D_\alpha f(\nu) = \lim_{\epsilon\to 0} \frac{f(\nu+\epsilon\nu^{1-\alpha}) - f(\nu)}{\epsilon}$$

for all $\nu > 0$ and $\alpha \in (0, 1]$. We say the f is α-differentiable, if the conformable fractional derivative of order α exists. If f is α-differentiable on $(0, a)$ and $\lim_{t\to 0^+} D_\alpha f(\nu)$

exists, then we define

$$D_\alpha f(0) = \lim_{t\to 0+} D_\alpha f(\nu).$$

Using the Definition 14.1, the author of the aforementioned article proves the following result:

Theorem 14.3 *Let $\alpha \in (0,1]$ and f, g be α-differentiable at $\nu > 0$. Then the following hold:*

(a) $D_\alpha(af(\nu) + bg(\nu)) = aD_\alpha f(\nu) + bD_\alpha g(\nu)$, for all $a, b \in \mathbb{R}$.

(b) $D_\alpha(fg)(\nu) = f(\nu)D_\alpha g(\nu) + g(\nu)D_\alpha f(\nu)$.

(c) $D_\alpha(f/g)(\nu) = \dfrac{g(\nu)D_\alpha f(\nu) - f(\nu)D_\alpha g(\nu)}{g^2(\nu)}$, where $g(\nu) \neq 0$.

(d) $D_\alpha(f \circ g)(\nu) = f'(g(\nu))D_\alpha g(\nu)$, if f is differentiable at $g(\nu)$.

(e) If f is differentiable, then $D_\alpha f(\nu) = \nu^{1-\alpha} f'(\nu)$.

(f) $D_\alpha(c) = 0$ for any constant function $f(\nu) = c$.

(g) $D_\alpha(\nu^\beta) = \beta\nu^{\beta-\alpha}$ for all $\beta \in \mathbb{R}$. In particular, $D_\alpha(\nu^\alpha) = \alpha$.

Also Khalil et al. define a conformable fractional integral operator, and its applications were verified by D.R. Anderson in [4].

Definition 14.2 (Conformable fractional integral) *Let $\alpha \in (0,1]$ and $0 \leq \kappa < \mu$. A function $f : [\kappa, \mu] \to \mathbb{R}$ is α-fractional integrable on $[\kappa, \mu]$, if the integral*

$$\int_\kappa^\mu f(\nu) d_\alpha \nu := \int_\kappa^\mu \nu^{\alpha-1} f(\nu) d\nu$$

exist and is finite. We denote the set of all α-fractional integrable functions on $[\kappa, \mu]$ is denoted by $L^1_\alpha([\kappa, \mu])$. Furthermore, we say that $f \in L^p_\alpha([\kappa, \mu])$ for $p > 1$, if $|f|^p \in L^1_\alpha([\kappa, \mu])$ and we write

$${}^\alpha\|f\|_p := \left(\int_\kappa^\mu |f(\nu)|^p d_\alpha \nu\right)^{1/p}.$$

Theorem 14.4 (Hölder's inequality for conformable fractional integrals: See [20]) *If $f, g : [\kappa, \mu] \to \mathbb{R}$ and $p, q > 1$ with $\dfrac{1}{p} + \dfrac{1}{q} = 1$, then*

$$\int_\kappa^\mu |f(\nu)g(\nu)| d_\alpha \nu \leq \left(\int_\kappa^\mu |f(\nu)|^p d_\alpha \nu\right)^{1/p} \left(\int_\kappa^\mu |g(\nu)|^q d_\alpha \nu\right)^{1/q}.$$

Theorem 14.5 (Integration by parts: See [1]) *Let $f, g : [\kappa, \mu] \to \mathbb{R}$ be two α-differentiable functions such that fg is α-differentiable. Then*

$$\int_\kappa^\mu f(\nu) D_\alpha g(\nu) d_\alpha \nu = fg\Big|_\kappa^\mu - \int_\kappa^\mu g(\nu) D_\alpha f(\nu) d_\alpha \nu.$$

With these preliminary concepts, we can establish the main results of our work.

14.3 THE CONFORMABLE OSTROWSKI'S INTEGRAL INEQUALITY WITH A PARAMETER

First, we establish the parametric version of the Montgomery identity.

Lemma 14.1 *Let $\kappa, \mu, \xi, \lambda \in \mathbb{R}$ with $0 \leq \kappa < \mu$ and $f : [\kappa, \mu] \to \mathbb{R}$ be a α-differentiable for $0 < \alpha \leq 1$. Then the following equality holds:*

$$(1-\lambda)f(\xi) + \lambda\frac{f(\kappa)+f(\mu)}{2} - \frac{\alpha}{\mu^\alpha - \kappa^\alpha}\int_\kappa^\mu f(\nu)d_\alpha\nu \tag{14.1}$$
$$= \frac{\alpha}{\mu^\alpha - \kappa^\alpha}\int_\kappa^\mu K(\xi,\nu)D_\alpha f(\nu)d_\alpha\nu$$

where $K(\xi,\nu) = \begin{cases} \dfrac{\nu^\alpha - \kappa^\alpha}{\alpha} - \lambda\dfrac{\mu^\alpha - \kappa^\alpha}{2\alpha} & \text{if } \nu \in [\kappa, \xi] \\ \dfrac{\nu^\alpha - \mu^\alpha}{\alpha} + \lambda\dfrac{\mu^\alpha - \kappa^\alpha}{2\alpha} & \text{if } \nu \in (\xi, \mu] \end{cases}.$

Proof 14.1 *First, we observe that*

$$\int_\kappa^\mu K(\xi,\nu)D_\alpha f(\nu)d_\alpha\nu = I_1 + I_2 \tag{14.2}$$

where

$$I_1 = \int_\kappa^\xi \left[\frac{\nu^\alpha - \kappa^\alpha}{\alpha} - \lambda\frac{\mu^\alpha - \kappa^\alpha}{2\alpha}\right] D_\alpha f(\nu)d_\alpha\nu$$

and

$$I_2 = \int_\xi^\mu \left[\frac{\nu^\alpha - \mu^\alpha}{\alpha} + \lambda\frac{\mu^\alpha - \kappa^\alpha}{2\alpha}\right] D_\alpha f(\nu)d_\alpha\nu.$$

By using the integration by parts formula for the conformable fractional integral, we have

$$I_1 = \left[\frac{\xi^\alpha - \kappa^\alpha}{\alpha} - \lambda\frac{\mu^\alpha - \kappa^\alpha}{2\alpha}\right] f(\xi) + \lambda\frac{\mu^\alpha - \kappa^\alpha}{2\alpha} f(\kappa)$$
$$- \int_\kappa^\xi f(\nu)d_\alpha\nu \tag{14.3}$$

and

$$I_2 = \lambda\frac{\mu^\alpha - \kappa^\alpha}{2\alpha} f(\mu) - \left[\frac{\mu^\alpha - \xi^\alpha}{\alpha} + \lambda\frac{\mu^\alpha - \kappa^\alpha}{2\alpha}\right] f(\xi)$$
$$- \int_\xi^\mu f(\nu)d_\alpha\nu. \tag{14.4}$$

Adding (14.3) *and* (14.4) *yields*

$$I_1 + I_2 = \frac{\mu^\alpha - \kappa^\alpha}{\alpha}\left[(1-\lambda)f(\xi) + \lambda\frac{f(\kappa)+f(\mu)}{2}\right]$$

$$-\int_{\kappa}^{\mu} f(\nu)d_{\alpha}\nu. \tag{14.5}$$

The desired identity in (14.1) *follows from* (14.2) *and* (14.5).

Lemma 14.2 *For any* $\alpha \in (0,1]$,

$$\int_{s}^{r} |\nu^{\alpha}-\rho|^{\varsigma} d_{\alpha}\nu = \frac{1}{\alpha(\varsigma+1)}\left[(\rho-s^{\alpha})^{\varsigma+1}+(r^{\alpha}-\rho)^{\varsigma+1}\right]$$

for all $s,\rho,r>0$ *such that* $s^{\alpha}\le\rho\le r^{\alpha}$.

Proof 14.2 *We observe that*

$$\begin{aligned}\int_{s}^{r} |\nu^{\alpha}-\rho|^{\varsigma} d_{\alpha}\nu &= \int_{s}^{r} |\nu^{\alpha}-\rho|^{\varsigma}\nu^{\alpha-1}d\nu\\ &=\frac{1}{\alpha}\int_{s^{\alpha}}^{r^{\alpha}}|u-\rho|^{\varsigma}du\\ &=\frac{1}{\alpha}\left[\int_{s^{\alpha}}^{\rho}(\rho-u)^{\varsigma}du+\int_{\rho}^{r^{\alpha}}(u-\rho)^{\varsigma}du\right]\\ &=\frac{1}{\alpha(\varsigma+1)}\left[(\rho-s^{\alpha})^{\varsigma+1}+(r^{\alpha}-\rho)^{\varsigma+1}\right].\end{aligned}$$

Now we establish our main results.

Theorem 14.6 *Under the conditions of Lemma 14.1, if* $\lambda\in[0,1]$ *and* $M=\sup_{\nu\in[\kappa,\mu]}|D_{\alpha}f(\nu)|<\infty$, *then the following inequality holds:*

$$\begin{aligned}&\left|(1-\lambda)f(\xi)+\lambda\frac{f(\kappa)+f(\mu)}{2}-\frac{\alpha}{\mu^{\alpha}-\kappa^{\alpha}}\int_{\kappa}^{\mu}f(\nu)d_{\alpha}\nu\right|\\ &\le\frac{M}{\alpha(\mu^{\alpha}-\kappa^{\alpha})}\left[\frac{(\mu^{\alpha}-\kappa^{\alpha})^2}{4}\left[\lambda^2+(\lambda-1)^2\right]+\left(\xi^{\alpha}-\frac{\kappa^{\alpha}+\mu^{\alpha}}{2}\right)^2\right]\end{aligned}$$

provided $\kappa^{\alpha}+\lambda\frac{\mu^{\alpha}-\kappa\alpha}{2}\le\xi^{\alpha}\le\mu^{\alpha}-\lambda\frac{\mu^{\alpha}-\kappa^{\alpha}}{2}$.

Proof 14.3 *By using the definition of* $K(\xi,\nu)$ *in Lemma 14.1, we observe that*

$$\begin{aligned}\left|\int_{\kappa}^{\mu}K(\xi,\nu)D_{\alpha}f(\nu)d_{\alpha}\nu\right| &\le\frac{1}{\alpha}\Bigg[\int_{\kappa}^{\xi}\left|\nu^{\alpha}-\left(\kappa^{\alpha}+\lambda\frac{\mu^{\alpha}-\kappa^{\alpha}}{2}\right)\right||D_{\alpha}f(\nu)|d_{\alpha}\nu\\ &\quad+\int_{\xi}^{\mu}\left|\nu^{\alpha}-\left(\mu^{\alpha}-\lambda\frac{\mu^{\alpha}-\kappa^{\alpha}}{2}\right)\right||D_{\alpha}f(\nu)|d_{\alpha}\nu\Bigg].\end{aligned} \tag{14.6}$$

Using the condition that $M=\sup_{\xi\in[\kappa,\mu]}|D_{\alpha}f(\xi)|<\infty$ *in* (14.6) *yields*

$$\left|\int_{\kappa}^{\mu}K(\xi,\nu)D_{\alpha}f(\nu)d_{\alpha}\nu\right| \le \frac{M}{\alpha}\Bigg[\int_{\kappa}^{\xi}\left|\nu^{\alpha}-\left(\kappa^{\alpha}+\lambda\frac{\mu^{\alpha}-\kappa^{\alpha}}{2}\right)\right|d_{\alpha}\nu$$

$$+\int_{\xi}^{\mu}\left|\nu^{\alpha}-\left(\mu^{\alpha}-\lambda\frac{\mu^{\alpha}-\kappa^{\alpha}}{2}\right)\right|d_{\alpha}\nu\Bigg]. \tag{14.7}$$

By using Lemma 14.2 when $\varsigma = 1$ and the identity

$$(A-B)^2+(A-C)^2=\frac{1}{2}(B-C)^2+2\left(A-\frac{B+C}{2}\right)^2, \tag{14.8}$$

we obtain

$$\begin{aligned}
T_1 &= \int_{\kappa}^{\xi}\left|\nu^{\alpha}-\left(\kappa^{\alpha}+\lambda\frac{\mu^{\alpha}-\kappa^{\alpha}}{2}\right)\right|d_{\alpha}\nu \\
&= \frac{1}{\alpha}\left[\frac{1}{4}(\xi^{\alpha}-\kappa^{\alpha})^2+\left(\kappa^{\alpha}+\lambda\frac{\mu^{\alpha}-\kappa^{\alpha}}{2}-\frac{\xi^{\alpha}+\kappa^{\alpha}}{2}\right)^2\right] \\
&= \frac{1}{\alpha}\left[\frac{1}{4}(\xi^{\alpha}-\kappa^{\alpha})^2+\left(\lambda\frac{\mu^{\alpha}-\kappa^{\alpha}}{2}-\frac{\xi^{\alpha}-\kappa^{\alpha}}{2}\right)^2\right]
\end{aligned}$$

and

$$\begin{aligned}
T_2 &= \int_{\xi}^{\mu}\left|\nu^{\alpha}-\left(\mu^{\alpha}-\lambda\frac{\mu^{\alpha}-\kappa^{\alpha}}{2}\right)\right|d_{\alpha}\nu \\
&= \frac{1}{\alpha}\left[\frac{1}{4}(\mu^{\alpha}-\xi^{\alpha})^2+\left(\mu^{\alpha}-\lambda\frac{\mu^{\alpha}-\kappa^{\alpha}}{2}-\frac{\mu^{\alpha}+\xi^{\alpha}}{2}\right)^2\right] \\
&= \frac{1}{\alpha}\left[\frac{1}{4}(\mu^{\alpha}-\xi^{\alpha})^2+\left(\frac{\mu^{\alpha}-\xi^{\alpha}}{2}-\lambda\frac{\mu^{\alpha}-\kappa^{\alpha}}{2}\right)^2\right]
\end{aligned}$$

Adding the aforementioned equations and using (14.8), *we get*

$$\begin{aligned}
T_1+T_2 &= \frac{1}{\alpha}\Bigg[\frac{1}{4}\left[(\xi^{\alpha}-\kappa^{\alpha})^2+(\mu^{\alpha}-\xi^{\alpha})^2\right] \\
&\quad +\left(\lambda\frac{\mu^{\alpha}-\kappa^{\alpha}}{2}-\frac{\xi^{\alpha}-\kappa^{\alpha}}{2}\right)^2+\left(\frac{\mu^{\alpha}-\xi^{\alpha}}{2}-\lambda\frac{\mu^{\alpha}-\kappa^{\alpha}}{2}\right)^2\Bigg] \\
&= \frac{1}{\alpha}\Bigg[\frac{1}{4}\left[\frac{1}{2}(\mu^{\alpha}-\kappa^{\alpha})^2+2\left(\xi^{\alpha}-\frac{\kappa^{\alpha}+\mu^{\alpha}}{2}\right)^2\right] \\
&\quad +\frac{1}{2}\left(\xi^{\alpha}-\frac{\kappa^{\alpha}+\mu^{\alpha}}{2}\right)^2+2\left(\lambda\frac{\mu^{\alpha}-\kappa^{\alpha}}{2}-\frac{\mu^{\alpha}-\kappa^{\alpha}}{4}\right)^2\Bigg] \\
&= \frac{1}{\alpha}\Bigg[\frac{1}{8}(\mu^{\alpha}-\kappa^{\alpha})^2+\frac{1}{2}\left(\xi^{\alpha}-\frac{\kappa^{\alpha}+\mu^{\alpha}}{2}\right)^2 \\
&\quad +\frac{1}{2}\left(\xi^{\alpha}-\frac{\kappa^{\alpha}+\mu^{\alpha}}{2}\right)^2+2\left(\frac{\mu^{\alpha}-\kappa^{\alpha}}{2}\right)^2\left(\lambda-\frac{1}{2}\right)^2\Bigg]
\end{aligned}$$

$$= \frac{1}{\alpha}\left[\frac{1}{8}(\mu^\alpha - \kappa^\alpha)^2 + \left(\xi^\alpha - \frac{\kappa^\alpha + \mu^\alpha}{2}\right)^2 + 2\left(\frac{\mu^\alpha - \kappa^\alpha}{2}\right)^2\left(\lambda - \frac{1}{2}\right)^2\right]$$
$$= \frac{1}{\alpha}\left[\left(\frac{\mu^\alpha - \kappa^\alpha}{2}\right)^2\left[\frac{1}{2} + 2\left(\lambda - \frac{1}{2}\right)^2\right] + \left(\xi^\alpha - \frac{\kappa^\alpha + \mu^\alpha}{2}\right)^2\right]$$
$$= \frac{1}{\alpha}\left[\left(\frac{\mu^\alpha - \kappa^\alpha}{2}\right)^2\left[\lambda^2 + (\lambda - 1)^2\right] + \left(\xi^\alpha - \frac{\kappa^\alpha + \mu^\alpha}{2}\right)^2\right]. \tag{14.9}$$

Substituting (14.9) *in* (14.7), *we have*

$$\left|\int_\kappa^\mu K(\xi, \nu))D_\alpha f(\nu)d_\alpha \nu\right|$$
$$\leq \frac{M}{\alpha^2}\left[\left(\frac{\mu^\alpha - \kappa^\alpha}{2}\right)^2\left[\lambda^2 + (\lambda - 1)^2\right] + \left(\xi^\alpha - \frac{\mu^\alpha + \kappa^\alpha}{2}\right)^2\right] \tag{14.10}$$

The desired inequality follows from (14.1) *and* (14.10).

Remark 14.1

(a) *If* $\alpha = 1$ *in Theorem 14.6, then we recover the result in [8, Theorem 2].*

(b) *If* $\lambda = 0$ *in Theorem 14.6, then we recover the result in [4, Theorem 6.2].*

(c) *If* $\lambda = 1$ *and* $\xi^\alpha = (\kappa^\alpha + \mu^\alpha)/2$ *in Theorem 14.6, then we can establish an inequality of the trapezoidal type:*

$$\left|\frac{f(\kappa) + f(\mu)}{2} - \frac{\alpha}{\mu^\alpha - \kappa^\alpha}\int_\kappa^\mu f(\nu)d_\alpha\nu\right| \leq \frac{(\mu^\alpha - \kappa^\alpha)M}{4\alpha}.$$

(d) *If* $\lambda = 1/3$ *in Theorem 14.6, then it is possible to establish an inequality of the Simpson type:*

$$\left|\frac{1}{6}\Big[f(\kappa) + 4f(\xi) + f(\mu)\Big] - \frac{\alpha}{\mu^\alpha - \kappa^\alpha}\int_\kappa^\mu f(\nu)d_\alpha\nu\right|$$
$$\leq \frac{M}{\alpha(\mu^\alpha - \kappa^\alpha)}\left[\frac{5(\mu^\alpha - \kappa^\alpha)^2}{36} + \left(\xi^\alpha - \frac{\kappa^\alpha + \mu^\alpha}{2}\right)^2\right]$$

provided $\dfrac{5\kappa^\alpha + \mu^\alpha}{6} \leq \xi^\alpha \leq \dfrac{5\mu^\alpha + \kappa^\alpha}{6}$.

(e) *If* $\lambda = 1/2$ *in Theorem 14.6, then we have the following inequality:*

$$\left|\frac{1}{2}\left[f(\xi) + \frac{f(\kappa) + f(\mu)}{2}\right] - \frac{\alpha}{\mu^\alpha - \kappa^\alpha}\int_\kappa^\mu f(\nu)d_\alpha\nu\right|$$
$$\leq \frac{M}{\alpha(\mu^\alpha - \kappa^\alpha)}\left[\frac{(\mu^\alpha - \kappa^\alpha)^2}{8} + \left(\xi^\alpha - \frac{\kappa^\alpha + \mu^\alpha}{2}\right)^2\right]$$

provided $\dfrac{3\kappa^\alpha + \mu^\alpha}{4} \leq \xi^\alpha \leq \dfrac{3\mu^\alpha + \kappa^\alpha}{4}$.

Theorem 14.7 *Under the conditions of Lemma 14.1, if $\lambda \in [0,1]$ and $D_\alpha f \in L^p_\alpha([\kappa,\mu])$ for $p>1$, then the following inequality holds:*

$$\left|(1-\lambda)f(\xi)+\lambda\frac{f(\kappa)+f(\mu)}{2}-\frac{\alpha}{\mu^\alpha-\kappa^\alpha}\int_\kappa^\mu f(\nu)d_\alpha\nu\right|$$
$$\leq \frac{\alpha\|D_\alpha f\|_p}{\mu^\alpha-\kappa^\alpha}\left(\frac{1}{\alpha(q+1)}\right)^{1/q}\times$$
$$\left[\left(\left(\xi^\alpha-\kappa^\alpha-\lambda\frac{\mu^\alpha-\kappa^\alpha}{2}\right)^{q+1}+\left(\lambda\frac{\mu^\alpha-\kappa^\alpha}{2}\right)^{q+1}\right)^{1/q}\right.$$
$$\left.+\left(\left(\mu^\alpha-\xi^\alpha+\lambda\frac{\mu^\alpha-\kappa^\alpha}{2}\right)^{q+1}+\left(\lambda\frac{\mu^\alpha-\kappa^\alpha}{2}\right)^{q+1}\right)^{1/q}\right]$$

provided $\kappa^\alpha+\lambda\frac{\mu^\alpha-\kappa^\alpha}{2}\leq\xi^\alpha\leq\mu^\alpha-\lambda\frac{\mu^\alpha-\kappa^\alpha}{2}$, *and $q>1$ such that $1/p+1/q=1$.*

Proof 14.4 *Applying the Hölder's inequality for the conformable α-integral on* (14.6), *we have*

$$\left|\int_\kappa^\mu K(\xi,\nu)D_\alpha f(\nu)d_\alpha\nu\right|\leq\frac{\alpha\|D_\alpha f\|_p}{\alpha}\left[\left(\int_\kappa^\xi\left|\nu^\alpha-\left(\kappa^\alpha+\lambda\frac{\mu^\alpha-\kappa^\alpha}{2}\right)\right|^q d_\alpha\nu\right)^{1/q}\right.$$
$$\left.+\left(\int_\xi^\mu\left|\nu^\alpha-\left(\mu^\alpha-\lambda\frac{\mu^\alpha-\kappa^\alpha}{2}\right)\right|^q d_\alpha\nu\right)^{1/q}\right]. \tag{14.11}$$

By using Lemma 14.2, we deduce that

$$\int_\kappa^\xi\left|\nu^\alpha-\left(\kappa^\alpha+\lambda\frac{\mu^\alpha-\kappa^\alpha}{2}\right)\right|^q d_\alpha\nu$$
$$=\frac{1}{\alpha(q+1)}\left[\left(\xi^\alpha-\kappa^\alpha-\lambda\frac{\mu^\alpha-\kappa^\alpha}{2}\right)^{q+1}+\left(\lambda\frac{\mu^\alpha-\kappa^\alpha}{2}\right)^{q+1}\right] \tag{14.12}$$

and

$$\int_\xi^\mu\left|\nu^\alpha-\left(\mu^\alpha-\lambda\frac{\mu^\alpha-\kappa^\alpha}{2}\right)\right|^q d_\alpha\nu$$
$$=\frac{1}{\alpha(q+1)}\left[\left(\mu^\alpha-\xi^\alpha+\lambda\frac{\mu^\alpha-\kappa^\alpha}{2}\right)^{q+1}+\left(\lambda\frac{\mu^\alpha-\kappa^\alpha}{2}\right)^{q+1}\right]. \tag{14.13}$$

Substituting (14.12) *and* (14.13) *in* (14.11), *we have*

$$\left|\int_\kappa^\mu K(\xi,\nu)D_\alpha f(\nu)d_\alpha\nu\right|\leq\frac{\alpha\|D_\alpha f\|_p}{\alpha}\times$$
$$\left[\left(\frac{1}{\alpha(q+1)}\left[\left(\xi^\alpha-\kappa^\alpha-\lambda\frac{\mu^\alpha-\kappa^\alpha}{2}\right)^{q+1}+\left(\lambda\frac{\mu^\alpha-\kappa^\alpha}{2}\right)^{q+1}\right]\right)^{1/q}\right. \tag{14.14}$$
$$\left.+\left(\frac{1}{\alpha(q+1)}\left[\left(\mu^\alpha-\xi^\alpha+\lambda\frac{\mu^\alpha-\kappa^\alpha}{2}\right)^{q+1}+\left(\lambda\frac{\mu^\alpha-\kappa^\alpha}{2}\right)^{q+1}\right)^{1/q}\right].$$

The desired inequality follows from (14.1) *and* (14.14).

Remark 14.2 *(a) If we take $\lambda = 0$ in Theorem 14.7, then we have the following Ostrowski type inequality:*

$$\left| f(\xi) - \frac{\alpha}{\mu^\alpha - \kappa^\alpha} \int_\kappa^\mu f(\nu) d_\alpha \nu \right| \\ \leq \frac{\alpha \| D_\alpha f \|_p}{\mu^\alpha - \kappa^\alpha} \left(\frac{1}{\alpha(q+1)} \right)^{1/q} \left[(\xi^\alpha - \kappa^\alpha)^{1+\frac{1}{q}} + (\mu^\alpha - \xi^\alpha)^{1+\frac{1}{q}} \right] \tag{14.15}$$

for all $\xi \in [\kappa, \mu]$.

(b) If we take $\alpha = 1$ in (14.15), then we obtain the result in [10].

Theorem 14.8 *Under the conditions of Lemma 14.1, if $\lambda \in [0,1]$ and $D_\alpha f \in L_\alpha^1([\kappa, \mu])$, then the following inequality holds:*

$$\left| (1-\lambda) f(\xi) + \lambda \frac{f(\kappa) + f(\mu)}{2} - \frac{\alpha}{\mu^\alpha - \kappa^\alpha} \int_\kappa^\mu f(\nu) d_\alpha \nu \right| \\ \leq \frac{\alpha \| D_\alpha f \|_1}{\mu^\alpha - \kappa^\alpha} \left[\max \left\{ \lambda \frac{\mu^\alpha - \kappa^\alpha}{2}, \left| \xi^\alpha - \kappa^\alpha - \lambda \frac{\mu^\alpha - \kappa^\alpha}{2} \right|, \right. \right. \\ \left. \left. \left| \xi^\alpha - \mu^\alpha + \lambda \frac{\mu^\alpha - \kappa^\alpha}{2} \right| \right\} \right]$$

for all $\xi \in [\kappa, \mu]$. Furthermore, if $\kappa^\alpha + \lambda \frac{\mu^\alpha - \kappa^\alpha}{2} \leq \xi^\alpha \leq \mu^\alpha - \lambda \frac{\mu^\alpha - \kappa^\alpha}{2}$, then we have

$$\left| (1-\lambda) f(\xi) + \lambda \frac{f(\kappa) + f(\mu)}{2} - \frac{\alpha}{\mu^\alpha - \kappa^\alpha} \int_\kappa^\mu f(\nu) d_\alpha \nu \right| \\ \leq \frac{\alpha \| D_\alpha f \|_1}{\mu^\alpha - \kappa^\alpha} \left[\max \left\{ \lambda \frac{\mu^\alpha - \kappa^\alpha}{2}, \xi^\alpha - \kappa^\alpha - \lambda \frac{\mu^\alpha - \kappa^\alpha}{2}, \right. \right. \\ \left. \left. \mu^\alpha - \xi^\alpha - \lambda \frac{\mu^\alpha - \kappa^\alpha}{2} \right\} \right].$$

Proof 14.5 *From (14.6), we have*

$$\left| \int_\kappa^\mu K(\xi, \nu) D_\alpha f(\nu) d_\alpha \nu \right| \\ \leq \frac{1}{\alpha} \left[\max_{\nu \in [\kappa, \xi]} \left| \nu^\alpha - \left(\kappa^\alpha + \lambda \frac{\mu^\alpha - \kappa^\alpha}{2} \right) \right| \int_\kappa^\xi |D_\alpha f(\nu)| d_\alpha \nu \right. \\ \left. + \max_{\nu \in [\xi, \mu]} \left| \nu^\alpha - \left(\mu^\alpha - \lambda \frac{\mu^\alpha - \kappa^\alpha}{2} \right) \right| \int_\xi^\mu |D_\alpha f(\nu)| d_\alpha \nu \right] \\ \leq \frac{\alpha \| D_\alpha f \|_1}{\alpha} \left[\max \left\{ \lambda \frac{\mu^\alpha - \kappa^\alpha}{2}, \left| \xi^\alpha - \left(\kappa^\alpha + \lambda \frac{\mu^\alpha - \kappa^\alpha}{2} \right) \right| \right\} \right. \\ \left. + \max \left\{ \lambda \frac{\mu^\alpha - \kappa^\alpha}{2}, \left| \xi^\alpha - \left(\mu^\alpha - \lambda \frac{\mu^\alpha - \kappa^\alpha}{2} \right) \right| \right\} \right]$$

$$\leq \frac{{}^{\alpha}\|D_{\alpha}f\|_1}{\alpha}\left[\max\left\{\lambda\frac{\mu^{\alpha}-\kappa^{\alpha}}{2}, \left|\xi^{\alpha}-\kappa^{\alpha}-\lambda\frac{\mu^{\alpha}-\kappa^{\alpha}}{2}\right|,\right.\right.$$
$$\left.\left.\left|\xi^{\alpha}-\mu^{\alpha}+\lambda\frac{\mu^{\alpha}-\kappa^{\alpha}}{2}\right)\right|\right\}\right]. \tag{14.16}$$

The desired inequalities follow from (14.1) *and* (14.16).

Remark 14.3

(a) If we take $\lambda = 0$ *in Theorem 14.8, then we have the following Ostrowski type inequality:*

$$\begin{aligned}&\left|f(\xi)-\frac{\alpha}{\mu^{\alpha}-\kappa^{\alpha}}\int_{\kappa}^{\mu}f(\nu)d_{\alpha}\nu\right|\\&\leq\frac{{}^{\alpha}\|D_{\alpha}f\|_1}{\mu^{\alpha}-\kappa^{\alpha}}\max\left\{\xi^{\alpha}-\kappa^{\alpha},\mu^{\alpha}-\xi^{\alpha}\right\}\\&=\frac{{}^{\alpha}\|D_{\alpha}f\|_1}{\mu^{\alpha}-\kappa^{\alpha}}\left[\frac{\mu^{\alpha}-\kappa^{\alpha}}{2}+\left|\xi^{\alpha}-\frac{\kappa^{\alpha}+\kappa^{\alpha}}{2}\right|\right]\end{aligned} \tag{14.17}$$

for all $\xi \in [\kappa, \mu]$.

(b) If we take $\alpha = 1$ *in* (14.17), *then we obtain the result in [9, Theorem 2.1].*

14.4 OTHER BOUNDS FOR THE CONFORMABLE FRACTIONAL OSTROWSKI TYPE INEQUALITIES VIA A PARAMETER

In this section, we obtain new estimates for the left-hand side of the inequalities established above by dropping the condition that

$$\kappa^{\alpha}+\lambda\frac{\mu^{\alpha}-\kappa^{\alpha}}{2}\leq\xi^{\alpha}\leq\mu^{\alpha}-\lambda\frac{\mu^{\alpha}-\kappa^{\alpha}}{2}.$$

Lemma 14.3 *Under the conditions of Lemma 14.1, we have*

$$\begin{aligned}&\left|(1-\lambda)f(\xi)+\lambda\frac{f(\kappa)+f(\mu)}{2}-\frac{\alpha}{\mu^{\alpha}-\kappa^{\alpha}}\int_{\kappa}^{\mu}f(\nu)d_{\alpha}\nu\right|\\&\leq\frac{1}{\mu^{\alpha}-\kappa^{\alpha}}\left[\int_{\kappa}^{\xi}(\nu^{\alpha}-\kappa^{\alpha})|D_{\alpha}f(\nu)|d_{\alpha}\nu+\int_{\xi}^{\mu}(\mu^{\alpha}-\nu^{\alpha})|D_{\alpha}f(\nu)|d_{\alpha}\nu\right.\\&\qquad\left.+\lambda\frac{\mu^{\alpha}-\kappa^{\alpha}}{2}\int_{\kappa}^{\mu}|D_{\alpha}f(\nu)|d_{\alpha}\nu\right].\end{aligned}$$

Proof 14.6 *The result follows directly from* (14.1) *by using the definition of* $K(\xi,\nu)$ *and the property of the absolute value.*

Theorem 14.9 *Under the conditions of Lemma 14.1, if* $\lambda \in [0,1]$ *and* $M = \sup_{t\in[\kappa,\mu]} |D_\alpha f(\nu)| < \infty$, *then the following inequality holds:*

$$\left|(1-\lambda)f(\xi)+\lambda\frac{f(\kappa)+f(\mu)}{2}-\frac{\alpha}{\mu^\alpha-\kappa^\alpha}\int_\kappa^\mu f(\nu)d_\alpha\nu\right|$$
$$\leq \frac{M}{2\alpha(\mu^\alpha-\kappa^\alpha)}\left[(\xi^\alpha-\kappa^\alpha)^2+(\mu^\alpha-\xi^\alpha)^2+\lambda(\mu^\alpha-\kappa^\alpha)^2\right]$$
$$=\frac{M}{\alpha(\mu^\alpha-\kappa^\alpha)}\left[\frac{(\mu^\alpha-\kappa^\alpha)^2}{4}[1+2\lambda]+\left(\xi^\alpha-\frac{\kappa^\alpha+\mu^\alpha}{2}\right)^2\right]$$

for all $\xi \in [\kappa,\mu]$.

Proof 14.7 *The result follows directly from Lemma 14.3 and using the identities in Lemma 14.2 when* $\gamma = 1$ *and using the identity*

$$(\xi^\alpha-\kappa^\alpha)^2+(\mu^\alpha-\xi^\alpha)^2=\frac{1}{2}(\mu^\alpha-\kappa^\alpha)+2\left(\xi^\alpha-\frac{\kappa^\alpha+\mu^\alpha}{2}\right)^2.$$

Theorem 14.10 *Under the conditions of Lemma 14.1, if* $\lambda \in [0,1]$ *and* $D_\alpha f \in L^p_\alpha([\kappa,\mu])$ *for* $p > 1$, *then the following inequality holds:*

$$\left|(1-\lambda)f(\xi)+\lambda\frac{f(\kappa)+f(\mu)}{2}-\frac{\alpha}{\mu^\alpha-\kappa^\alpha}\int_\kappa^\mu f(\nu)d_\alpha\nu\right|$$
$$\leq \frac{{}^\alpha\|D_\alpha f\|_p}{\mu^\alpha-\kappa^\alpha}\left(\frac{1}{\alpha(q+1)}\right)^{1/q}\left[(\xi^\alpha-\kappa^\alpha)^{\frac{q+1}{q}}+(\mu^\alpha-\xi^\alpha)^{\frac{q+1}{q}}\right.$$
$$\left.+\lambda\frac{(\mu^\alpha-\kappa^\alpha)^{\frac{q+1}{q}}}{2}\right]$$

for all $\xi \in [\kappa,\mu]$ *and* $q > 1$ *for which* $\frac{1}{p}+\frac{1}{q}=1$.

Proof 14.8 *Using Lemma 14.3 and the Hölder's inequality for the conformable* α*-integral, we have*

$$\left|(1-\lambda)f(\xi)+\lambda\frac{f(\kappa)+f(\mu)}{2}-\frac{\alpha}{\mu^\alpha-\kappa^\alpha}\int_\kappa^\mu f(\nu)d_\alpha\nu\right|$$
$$\leq\frac{1}{\mu^\alpha-\kappa^\alpha}\left[\left(\int_\kappa^\xi(\nu^\alpha-\kappa^\alpha)^q d_\alpha\nu\right)^{1/q}\left(\int_\kappa^\xi|D_\alpha f(\nu)|^p d_\alpha\nu\right)^{1/p}\right.$$
$$+\left(\int_\xi^\mu(\mu^\alpha-\nu^\alpha)^q d_\alpha\nu\right)^{1/q}\left(\int_\xi^\mu|D_\alpha f(\nu)|^p d_\alpha\nu\right)^{1/p}$$
$$\left.+\lambda\frac{\mu^\alpha-\kappa^\alpha}{2}\left(\int_\kappa^\mu 1^q\, d_\alpha\nu\right)^{1/q}\left(\int_\kappa^\mu|D_\alpha f(\nu)|^p d_\alpha\nu\right)^{1/p}\right]$$

$$\leq \frac{1}{\mu^\alpha-\kappa^\alpha}\left[\left(\int_\kappa^\xi (\nu^\alpha-\kappa^\alpha)^q d_\alpha\nu\right)^{1/q}+\left(\int_\xi^\mu (\mu^\alpha-\xi^\alpha)^q d_\alpha\nu\right)^{1/q}\right.$$
$$\left.+\lambda\frac{\mu^\alpha-\kappa^\alpha}{2}\left(\int_\kappa^\mu 1^q\, d_\alpha\nu\right)^{1/q}\right]\left(\int_\kappa^\mu |D_\alpha f(\nu)|^p d_\alpha\nu\right)^{1/p} \tag{14.1}$$

The desired result follows from (14.1) *and Lemma 14.2.*

Theorem 14.11 *Under the conditions of Lemma 14.1, if* $\lambda \in [0,1]$ *and* $D_\alpha f \in L^1_\alpha([\kappa,\mu])$*, then the following inequality holds:*

$$\left|(1-\lambda)f(\xi)+\lambda\frac{f(\kappa)+f(\mu)}{2}-\frac{\alpha}{\mu^\alpha-\kappa^\alpha}\int_\kappa^\mu f(\nu)d_\alpha\nu\right|$$
$$\leq \frac{\alpha\|D_\alpha f\|_1}{\mu^\alpha-\kappa^\alpha}\left[\frac{(1+\lambda)(\mu^\alpha-\kappa^\alpha)}{2}+\left|\xi^\alpha-\frac{\kappa^\alpha+\mu^\alpha}{2}\right|\right]$$

for all $\xi \in [\kappa,\mu]$.

Proof 14.9 *From Lemma 14.3, we have*

$$\left|(1-\lambda)f(\xi)+\lambda\frac{f(\kappa)+f(\mu)}{2}-\frac{\alpha}{\mu^\alpha-\kappa^\alpha}\int_\kappa^\mu f(\nu)d_\alpha\nu\right|$$
$$\leq \frac{1}{\mu^\alpha-\kappa^\alpha}\left[(\xi^\alpha-\kappa^\alpha)\int_\kappa^\xi |D_\alpha f(\nu)|d_\alpha\nu+(\mu^\alpha-\xi^\alpha)\int_\xi^\mu |D_\alpha f(\nu)|d_\alpha\nu\right.$$
$$\left.+\lambda\frac{\mu^\alpha-\kappa^\alpha}{2}\int_\kappa^\mu |D_\alpha f(\nu)|d_\alpha\nu\right]$$
$$\leq \frac{1}{\mu^\alpha-\kappa^\alpha}\left[\max\{\xi^\alpha-\kappa^\alpha,\mu^\alpha-\xi^\alpha\}\int_\kappa^\mu |D_\alpha f(\nu)|d_\alpha\nu\right.$$
$$\left.+\lambda\frac{\mu^\alpha-\kappa^\alpha}{2}\int_\kappa^\mu |D_\alpha f(\nu)|d_\alpha\nu\right]$$
$$=\frac{\alpha\|D_\alpha f\|_1}{\mu^\alpha-\kappa^\alpha}\left[\max\{\xi^\alpha-\kappa^\alpha,\mu^\alpha-\xi^\alpha\}+\lambda\frac{\mu^\alpha-\kappa^\alpha}{2}\right]. \tag{14.2}$$

We note that for any real numbers A and B,

$$\max\{A,B\}=\frac{A+B}{2}+\frac{|A-B|}{2}.$$

So, it follows that

$$\max\{\xi^\alpha-\kappa^\alpha,\mu^\alpha-\xi^\alpha\}=\frac{(\xi^\alpha-\kappa^\alpha)+(\mu^\alpha-\xi^\alpha)}{2}$$
$$+\frac{|(\xi^\alpha-\kappa^\alpha)-(\mu^\alpha-\xi^\alpha)|}{2}$$
$$=\frac{\mu^\alpha-\kappa^\alpha}{2}+\frac{|2\xi^\alpha-(\kappa^\alpha+\mu^\alpha)|}{2}$$

$$= \frac{\mu^\alpha - \kappa^\alpha}{2} + \left|\xi^\alpha - \frac{\kappa^\alpha + \mu^\alpha}{2}\right|. \tag{14.3}$$

The desired result follows from (14.2) *and* (14.3).

14.5 CONCLUSION

We established some new Ostrowski type inequalities in conformable fractional calculus by introducing a parameter $\lambda \in [0, 1]$. Our results generalize several results in the literature – some of which are pointed out in Remarks 20.2, 14.2 and 14.3. Furthermore, by considering some specific values of the variables involved in the main results, we will obtain some inequalities of the Midpoint type, Trapezoidal type and Simpson's type in conformable fractional calculus. These results will be useful in the study of numerical quadrature rules, special means and many more in the theory of conformable fractional calculus. Another open problem is to extend the results obtained for the generalized integral operators defined in [13].

BIBLIOGRAPHY

[1] T. Abdeljawad, On conformable fractional calculus, *J. Comput. Appl. Math.*, 279 (2015) 57–66.

[2] G. AINemer, M. Kenawy, M. Zakarya, C. Cesarano, H.M. Rezk, Generalizations of Hardy's type inequalities via conformable Calculus, *Symmetry* 13(2) (2021) 242.

[3] M. Alomari, M. Darus, S. S. Dragomir, P. Cerone, Ostrowski type inequalities for the functions whose derivative are s-convex in second sense, *Appl. Math. Lett.*, 23(9) (2010) 1071–1076.

[4] D. R. Anderson, Taylor's formula and integral inequalities for conformable fractional derivatives, in Panos M. Pardalos, Themistocles M. Rassias (Eds), *Contributions in Mathematics and Engineering*, pp. 25-43, Springer, Switzerland, 2016.

[5] S. S. Dragomir, A generalization of Ostrowski integral inequality for mappings whose derivatives belong to $L_1[a, b]$ and applications in numerical integration, *J. Comput. Anal. Appl.* 3(4) (2001) 343–360.

[6] S. S. Dragomir, A generalization of the Ostrowski integral inequality for mappings whose derivatives belong to $L_p[a, b]$ and applications in numerical integration, *J. Math. Anal. Appl.* 255 (2001) 605–626.

[7] S. S. Dragomir, An Ostrowski type inequality for convex functions, *Uni. Beograd. Publ. Elektrotehn. Fak. Ser. Mat.* 16 (2005), 12–25.

[8] S. S. Dragomir, P. Cerone, J. Roumeliotis, A new generalization of Ostrowski's Integral inequality for mappings whose derivatives are bounded and applications in numerical integration and for special means, *Appl. Math. Lett.* 13 (2000), 19–25.

[9] S. S. Dragomir, S. Wang, A new inequality of Ostrowski's type in L_1-norm and applications to some special means and to some numerical quadrature rules, *Tamkang J. Math.* 28 (1997), 239–244.

[10] S. S. Dragomir, S. Wang, A new inequality of Ostrowski's type in L_p-norm, *Indian J. Math.* 40(3) (1998) 299–304.

[11] S. Erden, M. Z. Sarikaya, Pompeiu type inequalities using conformable fractional calculus and its applications, *Jordan J. Math. Stat.*, 14(3) (2021) 527–544.

[12] G. Farid, M. Usman, Ostrowski type k-fractional integral inequalities for MT-convex and h-convex functions, *Nonlinear Funct. Anal. Appl.* 22(3) (2017) 627–639.

[13] P. M. Guzmán, L. M. Lugo, J.E. Nápoles Valdés , M. Vivas-Cortez, On a new generalized integral operator and certain operating properties, *Axioms* 8 (2020) 69.

[14] S. Kermausuor, Ostrowski type inequalities for functions whose derivatives are strongly (α, m)-convex via k-fractional integrals, *Studia Universitatis Babeş-Bolyai Mathematica* 64(1) (2019) 25–34.

[15] R. Khalil, M. Al Horani, A. Yousef, M. Sababheh, A new definition of fractional derivative, *J. Comput. Appl. Math.* 264 (2014) 65–70.

[16] M. A. Khan, Y.-M. Chu, A. Kashuri, R. Liko, G. Ali, Conformable fractional integrals versions of Hermite–Hadamard inequalities and their generalizations, *J. Funct. Spaces* 2018 (2018) Article ID 6928130, 1–9.

[17] A. M. Ostrowski, Über die Absolutabweichung einer differentiebaren Funktion von ihrem Integralmitelwert, *Comment. Math. Helv.* 10 (1938) 226–227.

[18] S. Sitho, S. K. Ntouyas, P. Agarwal, J. Tariboon, Noninstantaneous impulsive inequalities via conformable fractional calculus, *J. Inequal Appl.* 2018 (2018) 261–270.

[19] T. Tunč, H. Budak, M. Z. Sarikaya, On functional generalization of Ostrowski inequality for conformable fractional integrals, TWMS, *J. App. Eng. Math.* 8(2) (2018) 495–508.

[20] F. Usta, H. Budak, T. Tunc,, M. Z. Sarikaya, New bounds for the Ostrowski-type inequalities via conformable fractional calculus, *Arab. J. Math.* 7 (2018) 317–328.

[21] F. Usta, M. Z. Sarikaya, Explicit bounds on certain integral inequalities via conformable fractional calculus, *Cogent Math.* 4(1) (2017) Article ID 1277505, 1–15 .

CHAPTER 15

The Regional Observability Problem for a Class of Semilinear Time-Fractional Systems With Riemann-Liouville Derivative

Zguaid Khalid and El Alaoui Fatima-Zahrae

Moulay Ismail University

CONTENTS

15.1 INTRODUCTION

Fractional Calculus (FC) is the field of mathematics that generalizes integrals and derivatives to a non-integer (arbitrary) order. Some researchers think that this rapidly growing field, both in theory and in applications, is new, but it is, in fact, as old as classical calculus. The first discussion about non-integer differentiation and integration traces back to 300 years ago, to the days of Leibniz, but it was ignored and excluded since there was, and still is, no acceptable geometrical interpretation for these kinds of operators. For more information about the subject, see [11, 11, 15].

Many real-world processes that expand in space and evolve with time are naturally heterogeneous, for instance, crowded systems and anomalous diffusion in porous media. It has been shown that mathematical models built with classical calculus fail

DOI: 10.1201/9781003388678-15

to represent such unpredictable processes. In contrast, the ones based on FC proved their capability to explain these complex processes better [5, 12]. One of the reasons that make fractional operators (differentiation and integration) powerful in modeling is that they have non-local and hereditary properties, depending on the past (left derivative or integral) or on the future (right derivative or integral). This makes these properties useful when working with processes that possess long-term memory.

One of the domains being rediscovered for fractional systems is control theory, which links applied fields such as engineering, physics, and mathematics. This domain splits into several branches, from which we mention controllability, observability, stability, etc. Our primary interest, in this chapter, is focused on observability. This concept investigates the possibility of knowing and recovering the value of the initial state for the studied system; it was first introduced by Rudolf Kalman in the 60s [10]. From then until now, this subject has gained a high level of maturity, mostly for classical (integer order) systems. It should be mentioned that not all states can be observed and recovered in the whole evolution domain [1]. Hence, we introduce the notion of regional observability. In other terms, it is observability but in a wanted subset of the spatial domain. To learn more about observability or general control theory for classical and fractional systems, we refer the reader to [2, 3, 6, 7, 8, 17, 21, 18, 19]. The main objective of this chapter is to reconstruct the initial state of the considered system in a desired subregion of the evolution domain, and in order to achieve this goal, we use a prolongation of the HUM technique to the fractional framework involving the Caputo derivative. This method was first brought up in [13] and then developed to deal with other types of systems [1, 4].

The remaining of this chapter is arranged as follows: Section 15.2 is dedicated to giving the considered system and some preliminary facts and results for a better understanding of the chapter. In Section 15.3, we illustrate the steps of the HUM approach, which lead to our main theorem (Theorem (15.1)); we also give an algorithm for the regional reconstruction. Section 4 provides a numerical example based on the proposed algorithm to support our theoretical result. Finally, in Section 15.5, a brief conclusion is given.

15.2 AN OVERVIEW OF THE CONSIDERED SYSTEM

In this section, we give the general form of the considered system and some of its properties and characteristics. We also present a few ingredients regarding the concept of regional observability. We consider a system that evolves in a spatial domain Ω, which is a subset of $\mathbb{R}^n$ $(n \geq 1)$, over a period of time $[0, r]$. The set Ω is called the evolution domain, and it is assumed to have a Lipschitz continuous boundary $\partial\Omega$. Let us consider $\alpha \in]\frac{1}{2}, 1]$ to be the non-integer differentiation order. The general form of the considered semilinear time-fractional system is written as follows:

$$\begin{cases} {}^{RL}\mathcal{D}^{\alpha}_{0+} z(\rho,t) = \mathcal{G}z(\rho,t) + \mathcal{T}z(\rho,t) & in\ \Delta_r, \\ z(\nu,t) = 0 & on\ \chi_r, \\ \lim\limits_{t\mapsto 0+} \mathcal{I}^{1-\alpha}_{0+} z(\rho,t) = z_0(\rho) & in\ \Omega, \end{cases} \tag{15.1}$$

with $\Delta_r := \Omega \times [0,r]$ and $\chi_r := \partial\Omega \times [0,r]$.

The system (15.1) is augmented with the following output equation,

$$m(\tau) = Cz(.,\tau), \qquad \tau \in [0,r], \tag{15.2}$$

where :
The operator $\mathcal{G}$ is a linear and unbounded operator that generates a strongly continuous semigroup $\{Q(t)\}_{t\geq 0}$ on the state space $S = L^2(\Omega)$. The operator C, called the observation operator, is a linear and bounded operator defined from S to $\mathcal{U}$ (the observation space). The nonlinear operator $\mathcal{T}$ is locally Lipschitz and satisfies the following condition:

$$(H_0) \quad \exists c > 0, \quad \|\mathcal{T}z(.,s)\|_S \leq c\|z(.,s)\|^2_S, \quad \forall s \in [0,r],\ \forall z \in L^2(0,r;S). \tag{15.3}$$

The quantity ${}^{RL}\mathcal{D}^{\alpha}_{0+} z(\rho,\tau) := \dfrac{1}{\Gamma(1-\alpha)} \dfrac{\partial}{\partial t} \displaystyle\int_0^{\tau} (\tau - s)^{-\alpha} z(\rho,s)ds$, is the α-th order left-sided Riemann-Liouville derivative with respect to t and with lower terminal 0. We will also need in what follows the right-sided Caputo derivative of order α with upper terminal r, which is defined as: ${}^{C}\mathcal{D}^{\alpha}_{r-} z(\rho,\tau) := \dfrac{1}{\Gamma(1-\alpha)} \displaystyle\int_{\tau}^{r} (s-\tau)^{-\alpha} \dfrac{\partial}{\partial s} z(\rho,s)ds$. Without any loss of generality, let us consider from now on $z(t) = z(.,t)$.

A function z in $C(0,r;S)$ is said to be a mild solution of (15.1) if it verifies the upcoming formula [22]:

$$z(t) = t^{\alpha-1}\mathcal{H}_\alpha(t)z_0 + \int_0^t (t-s)^{\alpha-1}\mathcal{H}_\alpha(t-s)\mathcal{T}z(s)ds, \quad \forall t \in]0,r]. \tag{15.4}$$

where

$$\mathcal{H}_\alpha(t) = \alpha \int_0^{\infty} \theta \mathcal{M}_\alpha(\theta) Q(t^\alpha \theta) d\theta.$$

The function $\mathcal{M}_\alpha$ is defined by:

$$\mathcal{M}_\alpha(\theta) = \sum_{n=1}^{\infty} \frac{(-\theta)^{n-1}}{\Gamma(n)\Gamma(1-\alpha n)}, \qquad \theta \geq 0, \tag{15.5}$$

and it is referred to as "The Mainardi function".

We present this upcoming proposition regarding the operator $\mathcal{H}_\alpha$.

Proposition 15.1 *The operator $\mathcal{H}_\alpha$ is strongly continuous and bounded. More precisely, $\exists M > 0$, such that for all $t \geq 0$, we have:*

$$\|\mathcal{H}_\alpha(t)\|_{\mathcal{L}(S,S)} \leq \frac{M}{\Gamma(\alpha)}. \tag{15.6}$$

Let us choose ω to be the desired non-empty subregion, which is included in the evolution domain Ω. The restriction operator in ω is given by:

$$\begin{array}{rcccl} \kappa_\omega & : & S & \longrightarrow & L^2(\omega) \\ & & v & \longmapsto & v_{|\omega} \end{array},$$

and its adjoint by

$$\begin{array}{rcccl} \kappa_\omega^* & : & L^2(\omega) & \longrightarrow & S \\ & & h & \longmapsto & \begin{cases} h \text{ in } \omega \\ 0 \text{ in } \Omega \setminus \omega \end{cases} \end{array}.$$

We call the linear part associated to the semilinear system (15.1), the below system,

$$\begin{cases} {}^{RL}\mathcal{D}^{\alpha}_{0+} z(\rho,t) = \mathcal{G}z(\rho,t) & \text{in } \Delta_r, \\ z(\nu,t) = 0 & \text{on } \chi_r, \\ \lim\limits_{t\mapsto 0+} \mathcal{I}^{1-\alpha}_{0+} z(\rho,t) = z_0(\rho) & \text{in } \Omega, \end{cases} \tag{15.7}$$

which is said to be approximately ω-observable, if

$$Ker(Cf_\alpha(.)\kappa_\omega^*) = \{0\},$$

where

$$f_\alpha(t) = t^{\alpha-1}\mathcal{H}_\alpha(t). \tag{15.8}$$

For the rest of this chapter, we adopt a very important hypothesis, which is indispensable in our framework:

Hypothesis (H_1) : The linear system (15.7) is approximately ω-observable.

Problem: Giving any system (15.1), is it possible to recover the initial state z_0, using the output equation (15.2).

15.3 THE RECONSTRUCTION APPROACH

We start by defining the following set,

$$W = \left\{ v \in S \mid v_{|\Omega\setminus\omega} = 0 \right\},$$

on which we consider the semi-norm,

$$\|v\|_W = \sqrt{\int_0^r \|Cf_\alpha(t)v\|_{\mathcal{U}}^2 dt},$$

and since the linear part (15.7) is approximately ω-observable, then $\|.\|_W$ becomes a norm on W[9], and we denote again by W its completion by the norm $\|.\|_W$.

Let φ_0 be an element of W, we consider the following semilinear fractional system:

$$\begin{cases} {}^{RL}\mathcal{D}^{\alpha}_{0+}\varphi(\rho,t) = \mathcal{G}\varphi(\rho,t) + \mathcal{T}\varphi(\rho,t) & in\ \Delta_r, \\ \varphi(\nu,t) = 0 & on\ \chi_r, \\ \lim\limits_{t\mapsto 0+} \mathcal{I}^{1-\alpha}_{0+}\varphi(\rho,t) = \varphi_0(\rho) & in\ \Omega. \end{cases} \tag{15.9}$$

This system (15.9) has a unique mild solution giving by:

$$\varphi(t) = f_\alpha(t)\varphi_0 + \int_0^t f_\alpha(t-\tau)\mathcal{T}\varphi(\tau)d\tau, \quad \forall t \in [0,r]. \tag{15.10}$$

We first decompose the function φ into two parts $\varphi = \phi_0 + \phi_1$, where ϕ_0 and ϕ_1 are mild solutions, respectively, of the following systems,

$$\begin{cases} {}^{RL}\mathcal{D}^{\alpha}_{0+}\phi_0(\rho,t) = \mathcal{G}\phi_0(\rho,t) & in\ \Delta_r, \\ \phi_0(\nu,t) = 0 & on\ \chi_r, \\ \lim\limits_{t\mapsto 0+} \mathcal{I}^{1-\alpha}_{0+}\phi_0(\rho,t) = \varphi_0(\rho) & in\ \Omega, \end{cases} \tag{15.11}$$

and

$$\begin{cases} {}^{RL}\mathcal{D}^{\alpha}_{0+}\phi_1(\rho,t) = \mathcal{G}\phi_1(\rho,t) + \mathcal{T}\left[\phi_0(\rho,t) + \phi_1(\rho,t)\right] & in\ \Delta_r, \\ \phi_1(\nu,t) = 0 & on\ \chi_r, \\ \lim\limits_{t\mapsto 0+} \mathcal{I}^{1-\alpha}_{0+}\phi_1(\rho,t) = 0 & in\ \Omega. \end{cases} \tag{15.12}$$

Hence,

$$\phi_0(t) = f_\alpha(t)\varphi_0, \quad \forall t \in [0,r], \tag{15.13}$$

and

$$\phi_1(t) = \int_0^t f_\alpha(t-\tau)\mathcal{T}[\phi_0(\tau) + \phi_1(\tau)]d\tau, \quad \forall t \in [0,r]. \tag{15.14}$$

The next step is to give the auxiliary system,

$$\begin{cases} {}^{C}D^{\alpha}_{r-}\theta(\rho,t) = \mathcal{G}^*\theta(\rho,t) - \mathcal{T}\theta(\rho,t) - C^*C\phi_0(\rho,t) & \text{in } \Delta_r, \\ \theta(\nu,t) = 0 & \text{on } \chi_r, \\ \theta(x,r) = 0 & \text{in } \Omega, \end{cases} \tag{15.15}$$

whose mild solution is written:

$$\theta(t) = \int_t^r f^*_\alpha(\tau-t)\left[\mathcal{T}\theta(\tau) - C^*C\phi_0(\tau)\right]d\tau, \quad \forall t \in [0,r], \tag{15.16}$$

by choosing φ_0 properly in W (i.e. $C\phi_0(t) = m(t)$), the previous system is the adjoint system of (15.9). Let us decompose the solution of system (15.15) into two parts $\theta = \Psi_0 + \Psi_1$, which are mild solutions of the following systems,

$$\begin{cases} {}^C D^{\alpha}_{r-}\Psi_0(\rho,t) = \mathcal{G}^*\Psi_0(\rho,t) - C^*C\phi_0(\rho,t) & \text{in } \Delta_r, \\ \Psi_0(\nu,t) = 0 & \text{on } \chi_r, \\ \Psi_0(x,r) = 0 & \text{in } \Omega, \end{cases} \tag{15.17}$$

and

$$\begin{cases} {}^C D^{\alpha}_{r-}\Psi_1(\rho,t) = \mathcal{G}^*\Psi_1(\rho,t) - \mathcal{T}\left[\Psi_1(\rho,t) + \Psi_0(\rho,t)\right] & \text{in } \Delta_r, \\ \Psi_1(\nu,t) = 0 & \text{on } \chi_r, \\ \Psi_1(x,r) = 0 & \text{in } \Omega, \end{cases} \tag{15.18}$$

therefore Ψ_0 and Ψ_1 are, respectively, given by,

$$\Psi_0(t) = -\int_t^r f^*_{\alpha}(\tau - t)\left[C^*C\phi_0(\tau)\right] d\tau, \quad \forall t \in [0,r], \tag{15.19}$$

and

$$\Psi_1(t) = -\int_t^r f^*_{\alpha}(\tau - t)\left[\mathcal{T}\left(\Psi_0(\tau) + \Psi_1(\tau)\right)\right] d\tau, \quad \forall t \in [0,r]. \tag{15.20}$$

We define $P_\omega := \kappa^*_\omega \kappa_\omega$, which is a projection operator on W. One can see that:

$$\begin{aligned} P_\omega(\theta(0)) &= P_\omega(\Psi_0(0)) + P_\omega(\Psi_1(0)) \\ &= \Lambda(\varphi_0) + \mathcal{K}(\varphi_0), \end{aligned}$$

where

$$\begin{aligned} \mathcal{K} \,:\, W &\longrightarrow W \\ \varphi_0 &\longmapsto P_\omega(\Psi_1(0)), \end{aligned} \qquad \begin{aligned} \Lambda \,:\, W &\longrightarrow W \\ \varphi_0 &\longmapsto P_\omega(\Psi_0(0)), \end{aligned}$$

hence,

$$\Lambda(\varphi_0) = P_\omega(\theta(0)) - \mathcal{K}(\varphi_0).$$

It has been shown in [20] that Λ is an isomorphism if (15.7) approximately observable in ω, which gives that:

$$\varphi_0 = \Lambda^{-1} P_\omega(\theta(0)) - \Lambda^{-1}\mathcal{K}(\varphi_0) := \mathcal{F}\varphi_0.$$

This means that the reconstruction problem can be reduced to a fixed point problem of the operator $\mathcal{F}$.

Theorem 15.1 *Let us assume that H_0 and H_1 are verified. Then, the nonlinear operator $\mathcal{F}$ admits one, and only one, fixed point, which coincides with z_0 in ω.*

Proof 15.1 *Let us show that the operator $\mathcal{F}$ is compact and that for some $p > 0$, $\mathcal{F}$ maps the closed ball of center 0 and radius p, $B(0,p)$, into itself. Note that it is sufficient to show that $\mathcal{K}$ is compact in order for $\mathcal{F}$ to be compact.*
Let's consider $p > 0$, and show that $\mathcal{K}(B(0,p))$ is relatively compact, since

$$\mathcal{K}(B(0,p)) = \left\{\mathcal{K}\varphi_0 = P_\omega(\Psi_1(0)) \mid \varphi_0 \in W\right\} \subset \tilde{B}_p = \left\{P_\omega(\Psi_1(.)) \mid \varphi_0 \in W\right\},$$

we only have to show that $\tilde{B}_p$ is relatively compact.

Step 1: *$\tilde{B}_p$ is uniformly bounded (i.e. $\underset{t\in[0,r]}{Sup} \|P_\omega(\Psi_1(t))\|_W < \infty$).*
We have that P_ω is a projection operator, hence continuous, then

$$\exists c_\omega > 0 \quad \text{such that} \quad \|P_\omega(\Psi_1(t))\|_W \leq c_\omega \|\Psi_1(t)\|_S. \tag{15.21}$$

From (15.8) and (15.20), we get

$$\|\Psi_1(t)\|_S \leq \int_t^r (\tau - t)^{\alpha-1} \|\mathcal{H}_\alpha(\tau - t)\|_{\mathcal{L}(S,S)} \|\mathcal{T}(\Psi_0(\tau) + \Psi_1(\tau))\|_S d\tau,$$

also from (15.6) and (15.3), we obtain

$$\|\Psi_1(t)\|_S \leq \frac{Mc}{\Gamma(\alpha)} \int_t^r (\tau - t)^{\alpha-1} \|\Psi_0(\tau) + \Psi_1(\tau)\|_S^2 d\tau,$$

which gives that,

$$\begin{aligned} \|\Psi_1(t)\|_S \quad \leq \quad & \frac{2Mc}{\Gamma(\alpha)} \int_t^r (s - t)^{\alpha-1} \|\Psi_1(s)\|_S^2 ds \\ & + \frac{2Mc}{\Gamma(\alpha)} \int_t^r (s - t)^{\alpha-1} \|\Psi_0(s)\|_S^2 ds. \end{aligned} \tag{15.22}$$

By using (15.6) and (15.19), we have

$$\|\Psi_0(s)\|_S \leq \frac{M\|C\|_{\mathcal{L}(S,\mathcal{U})}}{\Gamma(\alpha)} \int_s^r (\tau - s)^{\alpha-1} \|C\phi_0(\tau)\|_{\mathcal{U}} d\tau,$$

and by Cauhcy-Shwartz, we obtain

$$\|\Psi_0(s)\|_S \leq \frac{M\|C\|_{\mathcal{L}(S,\mathcal{U})}}{\Gamma(\alpha)} \left[\int_s^r (\tau - s)^{2(\alpha-1)} d\tau\right]^{\frac{1}{2}} \left[\int_0^r \|C\phi_0(\tau)\|_{\mathcal{U}}^2 d\tau\right]^{\frac{1}{2}},$$

thus

$$\|\Psi_0(s)\|_S \leq \frac{M\|C\|_{\mathcal{L}(S,\mathcal{U})}}{\Gamma(\alpha)} \cdot \frac{r^{\alpha-\frac{1}{2}}}{\sqrt{2\alpha - 1}} \cdot \|\varphi_0\|_W. \tag{15.23}$$

By substituting the inequality (15.23) in (15.22), we obtain

$$\|\Psi_1(t)\|_S \leq \int_t^r h(s) \|\Psi_1(s)\|_S^2 ds + \mathcal{R},$$

where $h(s) = \dfrac{2Mc}{\Gamma(\alpha)}(s-t)^{\alpha-1}$ *and* $\mathcal{R} = \dfrac{2M^3c}{\Gamma(\alpha)^3}\|C\|^2_{\mathcal{L}(S,\mathcal{U})}\|\varphi_0\|^2_g\left[\dfrac{r^{\alpha-\frac{1}{2}}}{\sqrt{2\alpha-1}}\right]^2$,

and by assuming that,

$$(H_3) \;-\; \left(\sqrt{\alpha}(2\alpha-1)\Gamma(\alpha)^2 - 2M^2c\|C\|_{\mathcal{L}(S,\mathcal{U})}\|\varphi_0\|_W r^{\frac{3\alpha-1}{2}}\right) > 0,$$

we can apply a generalization of Gronwall's lemma in [14], and get

$$\|\Psi_1(t)\|_S \leq \frac{\mathcal{R}}{1-\mathcal{R}\left[\frac{r\alpha}{\alpha}\right]}, \tag{15.24}$$

hence,

$$\underset{t\in[0,r]}{Sup}\|P_\omega(\Psi_1(t))\|_W \leq \frac{c_\omega\mathcal{R}}{1-\mathcal{R}\left[\frac{r\alpha}{\alpha}\right]}.$$

therefore $\tilde{B_p}$ is uniformly bounded.

Step 2: *$\tilde{B_p}$ is equicontinuous.*

Let's consider $\varepsilon > 0$, $\varphi_0 \in W$ and $t_1 \leq t_2 \leq r$, we have

$$\begin{aligned}
\Psi_1(t_1) - \Psi_1(t_2) &= \int_{t_1}^{r}(\tau-t_1)^{\alpha-1}\mathcal{H}^*_\alpha(\tau-t_1)\mathcal{T}(\Psi_1(\tau)+\Psi_0(\tau))\,d\tau \\
&\quad - \int_{t_2}^{r}(\tau-t_2)^{\alpha-1}\mathcal{H}^*_\alpha(\tau-t_2)\mathcal{T}(\Psi_1(\tau)+\Psi_0(\tau))\,d\tau \\
&= \int_{t_2}^{r}(\tau-t_1)^{\alpha-1}\left[\mathcal{H}^*_\alpha(\tau-t_1)-\mathcal{H}^*_\alpha(\tau-t_2)\right] \\
&\quad \mathcal{T}(\Psi_1(\tau)+\Psi_0(\tau))\,d\tau \\
&\quad + \int_{t_1}^{t_2}(\tau-t_1)^{\alpha-1}\mathcal{H}^*_\alpha(\tau-t_1) \\
&\quad \mathcal{T}(\Psi_1(\tau)+\Psi_0(\tau))\,d\tau \\
&\quad + \int_{t_2}^{r}\left[(\tau-t_1)^{\alpha-1}-(\tau-t_2)^{\alpha-1}\right]\mathcal{H}^*_\alpha(\tau-t_2) \\
&\quad \mathcal{T}(\Psi_1(\tau)+\Psi_0(\tau))\,d\tau \\
&= A_1 + A_2 + A_3,
\end{aligned} \tag{15.25}$$

where :

$$A_1 = \int_{t_2}^{r}(\tau-t_1)^{\alpha-1}\left[\mathcal{H}^*_\alpha(\tau-t_1)-\mathcal{H}^*_\alpha(\tau-t_2)\right]\mathcal{T}(\Psi_1(\tau)+\Psi_0(\tau))\,d\tau,$$

$$A_2 = \int_{t_1}^{t_2}(\tau-t_1)^{\alpha-1}\mathcal{H}^*_\alpha(\tau-t_1)\mathcal{T}(\Psi_1(\tau)+\Psi_0(\tau))\,d\tau,$$

$$A_3 = \int_{t_2}^{r}\left[(\tau-t_1)^{\alpha-1}-(\tau-t_2)^{\alpha-1}\right]\mathcal{H}^*_\alpha(\tau-t_2)\mathcal{T}(\Psi_1(\tau)+\Psi_0(\tau))\,d\tau\cdot$$

We have, by using (15.23), (15.24) *and* (15.3), *that*

$$
\begin{aligned}
&\|\mathcal{T}\left[\Psi_0(t)+\Psi_1(t)\right]\|_S \\
&\leq c\left(\frac{\mathcal{R}}{1-\mathcal{R}\left[\frac{r^\alpha}{\alpha}\right]}+\frac{M\|C\|_{\mathcal{L}(S,\mathcal{U})}}{\Gamma(\alpha)}.\frac{r^{\alpha-\frac{1}{2}}}{\sqrt{2\alpha-1}}.\|\varphi_0\|_W\right)^2 \\
&:= c\mathcal{J}.
\end{aligned}
\tag{15.26}
$$

hence from (15.26) *and the fact that* $\mathcal{H}_\alpha^*$ *is strongly continuous,* $\forall\varepsilon_1>0, \exists\sigma>0,$ *such that* $|t_1-t_2|<\sigma$ *implies that*

$$
\|A_1\|_S \leq c\mathcal{J}\varepsilon_1\int_{t_1}^{r}(\tau-t_1)^{\alpha-1}d\tau \leq \frac{c\mathcal{J}r^\alpha}{\alpha}\varepsilon_1, \tag{15.27}
$$

and by using (15.26) *and* (15.6), *we obtain*

$$
\|A_2\|_S \leq \frac{cM\mathcal{J}}{\Gamma(\alpha)}\int_{t_1}^{t_2}(\tau-t_1)^{\alpha-1}d\tau \leq \frac{cM\mathcal{J}}{\Gamma(\alpha+1)}(t_2-t_1)^\alpha. \tag{15.28}
$$

Again by (15.26) *and* (15.6), *we have*

$$
\begin{aligned}
\|A_3\|_S &\leq \frac{cM\mathcal{J}}{\Gamma(\alpha)}\int_{t_2}^{r}\left((\tau-t_1)^{\alpha-1}-(\tau-t_2)^{\alpha-1}\right)d\tau, \\
&\leq \frac{cM\mathcal{J}}{\Gamma(\alpha+1)}\left[(r-t_1)^\alpha-(r-t_2)^\alpha-(t_2-t_1)^\alpha\right],
\end{aligned}
\tag{15.29}
$$

therefore $\|A_3\|_S \longrightarrow 0$ *whenever* $|t_1-t_2|\longrightarrow 0$, *which mean that* $\exists\nu>0$ *such that*

$$
|t_1-t_2|<\nu \implies c_\omega\|A_3\|_S \leq \frac{\varepsilon}{3}.
$$

We now have, by using (15.21) *and* (15.25),

$$
\|P_\omega\Psi_1(t_1)-P_\omega\Psi_1(t_2)\|_W \leq c_\omega\|A_1\|_S + c_\omega\|A_2\|_S + c_\omega\|A_3\|_S
$$

Finally, by taking $\varepsilon_1 < \dfrac{\alpha\varepsilon}{3c_\omega c\mathcal{J}r^\alpha}$, $\exists\delta=\min\left\{\sigma,\nu,\left(\dfrac{\varepsilon\Gamma(\alpha+1)}{3c_\omega cM\mathcal{J}}\right)^{\frac{1}{\alpha}}\right\}$, *such that*

$$
|t_1-t_2|<\delta \implies \|P_\omega\Psi_1(t_1)-P_\omega\Psi_1(t_2)\|_W \leq \frac{\varepsilon}{3}+\frac{\varepsilon}{3}+\frac{\varepsilon}{3} \leq \varepsilon.
$$

Thus $\tilde{B_p}$ *is equicontinuous.*

Step 3: $\mathcal{F}$ *maps* $B(0,p)$ *into itself for some* $p>0$.
We have

$$
\|\mathcal{F}\varphi_0\|_S \leq c_\omega\|\Lambda^{-1}\|\left(\|\theta(0)\|_S+\|\Psi_1(0)\|_S\right).
$$

Since θ *and* Ψ_1 *are in* $C(0,r;S)$, *they are also in* $L^\infty(0,r;S)$. *Thus* $\exists M_1, M_2>0$, *such that*

$$
\|\theta(0)\|_S \leq M_1 \text{ and } \|\Psi_1(0)\|_S \leq M_2.
$$

Hence

$$\|\mathcal{F}\varphi_0\|_S \leq c_\omega \|\Lambda^{-1}\| (M_1 + M_2).$$

By taking $p > c_\omega \|\Lambda^{-1}\| (M_1 + M_2)$, *we obtain* $\mathcal{F}(B(0,p)) \subset B(0,p)$.
Therefore, by the Schauder fixed point theorem, $\mathcal{F}$ *has a fixed point.*

Step 4: *The uniqueness of the fixed point.*
Let $\tilde{\varphi}_0$ *and* $\overline{\varphi}_0$ *be two fixed points of* $\mathcal{F}$. *Using the discussion in the paragraph before equation* (15.17), *we get* $z = Cf_\alpha\tilde{\varphi}_0 = Cf_\alpha\overline{\varphi}_0$, *which implies that* $Cf_\alpha(\tilde{\varphi}_0 - \overline{\varphi}_0) = 0$. *With the help of the fact that,* $\kappa_\omega^* \kappa_\omega g = g$, *for all* g *in* W,*we obtain* $Cf_\alpha \kappa_\omega^* \kappa_\omega (\tilde{\varphi}_0 - \overline{\varphi}_0) = 0$, *and since* (15.7) *is approximately observable in* ω, *we have* $\kappa_\omega(\tilde{\varphi}_0 - \overline{\varphi}_0) = 0$, *thus* $\tilde{\varphi}_0 = \overline{\varphi}_0$ *in* ω. *Taking into consideration that* $\tilde{\varphi}_0$ *and* $\overline{\varphi}_0$ *are in* W, *we conclude that* $\tilde{\varphi}_0 = \overline{\varphi}_0$.
Therefore $\mathcal{F}$, *has a unique fixed point in* W.

Algorithm

1 - Initialization of :

- α : the order of differentiation.
- ω : the desired subregion.
- ε : the highest margin of error
- φ_0 : the starting value of the initial state

2 - Solve (15.13) and get ϕ_0.

3 - Solve (15.15) and get θ.

4 - Solve (15.17) and get Ψ_0.

5 - Solve (15.18) and get Ψ_1.

6 - Calculate or approximate Λ^{-1}.

7 - If $\|\varphi_0 - \mathcal{F}\varphi_0\| > \varepsilon$, then

- $\varphi_0 = \mathcal{F}\varphi_0$.
- go back to step 2.

else

- Stop.

The initial sate in ω is $\kappa_\omega \varphi_0$.

15.4 EXAMPLE

We give now an example to support the previously established results. Let us take the evolution domain $\Omega = [0,1]$ and the final time $T = 2$, we consider the following fractional system:

$$\begin{cases} {}^{RL}\mathcal{D}^{0.8}_{0+} z(\rho,t) = \dfrac{\partial^2}{\partial \rho^2} z(\rho,t) + \displaystyle\sum_{i=1}^{\infty} \langle z(t),\varphi_i\rangle_S \left| \langle z(t),\varphi_i\rangle_S \right| \varphi_i(\rho) & in\ \Delta_2, \\ z(\nu,t) = 0 & on\ \chi_2, \\ \lim\limits_{t\mapsto 0+} \mathcal{I}^{0.2}_{0+} z(\rho,t) = z_0(\rho) & in\ [0,1], \end{cases} \quad (15.30)$$

where $\varphi_i(\rho) = \sqrt{2}\sin(i\pi\rho)$. Note that $(\varphi_i)_{i\geq 1}$ is family of eigenfunctions of the operator $\dfrac{\partial^2}{\partial \rho^2}$ which is also an orthonormal basis of S. We take the system (15.30) to be observed by a zonal sensor (B,y), that is :
$B \subset \Omega$ is the geometrical domain of the sensor and $y \in L^2(B)$ is the spatial distribution of the sensors. The observation space $\mathcal{U}$ in this case is $\mathbb{R}$ and the output function takes the form

$$m(t) = \langle z(t), y\rangle_{L^2(B)}, \quad t \in [0,2].$$

For this simulation, we consider $B = [0.2, 0.4]$, $y \equiv 1$, $\omega = [0.3, 0.6]$, and $z_0(\rho) = (\rho - 1)(\rho - 0.5)(e^{\rho} - 1)$. After the application of the proposed algorithm and in 23 iterations, we obtain the following figure: We can see the two curves, real and reconstructed initial state, are almost identical, with an error of

$$\|z_0 - \varphi_0\|_{L^2(\omega)} = 8.40 \times 10^{-4}.$$

Table 15.1 shows that the reconstruction error varies if we change the sensors' geometrical support.

TABLE 15.1 Some Values of the Reconstruction error in Function of the Domain of the Sensor

The Spatial Support B	The Reconstruction Error
[0.0, 0.2]	1.43×10^{-2}
[0.2, 0.4]	8.40×10^{-4}
[0.4, 0.6]	4.25×10^{13}
[0.6, 0.8]	5.80×10^{-3}
[0.8, 1.0]	1.39×10^{-1}

It is obvious that if the sensor is placed in the region $[0.4, 0.6]$, then the error explodes, which mean that the sensor is not strategic (i.e. the system is not observable in $[0.4, 0.6]$). This means that the placement of the sensor affects the regional observability of the system.

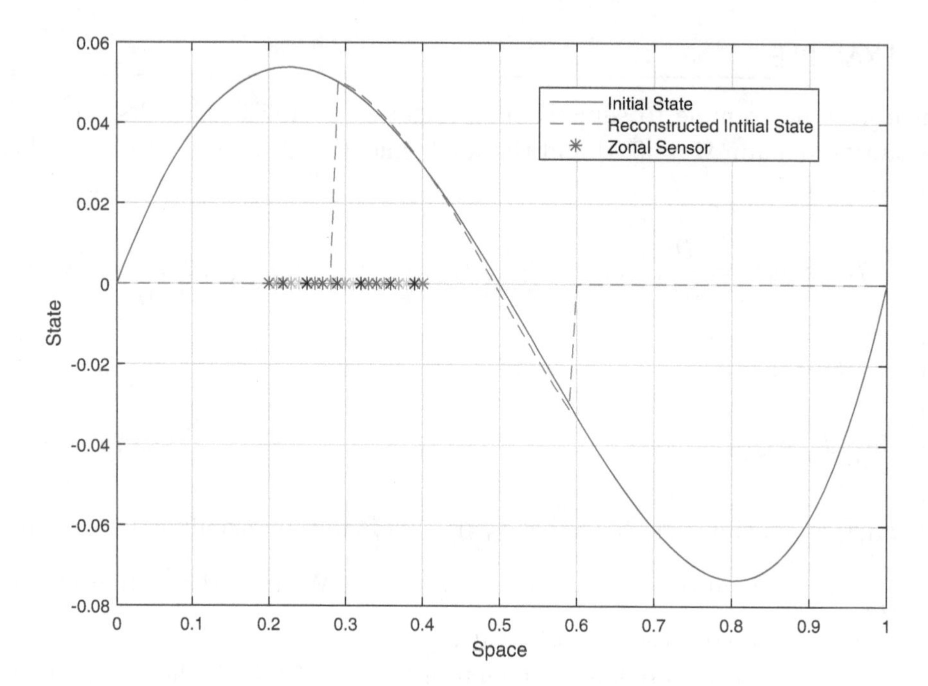

Figure 15.1 The initial state and the reconstructed one in $\omega = [0.3, 0.6]$.

15.5 CONCLUSION

We discussed, in this chapter, the possibility of recovering the initial state of a class of semilinear time-fractional systems with Caputo derivative. The used method is HUM approach, which transforms the problem in hand into a solvability one. The two Hypotheses (H_0) and (H_1) seem, until the time this chapter is being constructed, to be only sufficient, so it is intriguing to treat this problem without these conditions and see what it would give. We also gave an algorithm for the regional reconstruction, which can be optimized for more efficiency. Finally, we show a numerical simulation to back up the theoretical result of this work. As a future direction, we are heading toward the study of regional boundary observability for various kinds of fractional systems. Furthermore, we are eager to see what happens if we take α in $]0, 2[$.

BIBLIOGRAPHY

[1] M. Amouroux, A. El Jai, and E. Zerrik, Regional observability of distributed systems, *Int. J. Syst. Sci.* 25 (2)(1994) 301–313.

[2] A. Boutoulout, H. Bourray, and F. Z. El Alaoui, Regional boundary observability for semi-linear systems approach and simulation, *Int. J. Math. Anal.* 4 (2010) 1153–1173.

[3] A. Boutoulout, H. Bourray, and F. Z. El Alaoui, Regional gradient observability for distributed semilinear parabolic systems, *J. Dyn. Control Syst.* 18 (2012) 159–179.

[4] A. Boutoulout, H. Bourray, F. Z. El Alaoui, and S. Benhadid, Regional observability for distributed semi-linear hyperbolic systems, *Int. J. Control,* 87 (5)(2014) 898–910.

[5] K. Cao, Y. Chen, and D. Stuart, A fractional micro-macro model for crowds of pedestrians based on fractional mean field games, *IEEE/CAA J. Autom. Sin.* 3 (3)(2016) 261–270.

[6] F. Z. El Alaoui, *Regional observability of semilinear systems*, Ph.D thesis, Faculty of Sciences, Moulay Ismail University, Meknes, 2011.

[7] F. Z. El Alaoui, A. Boutoulout, and K. Zguaid, Regional reconstruction of semilinear Caputo type time-fractional systems using the analytical approach, *Adv. Theory Nonlinear Anal. Appl.* 5 (4)(2021) 580–599.

[8] A. El Jai, *Eléments d'analyse et de contrôle des systèmes*, Perpignan: Presses Universitaires de Perpignan, 2005.

[9] F. Ge, Y. C. Quan, and C. Kou, *Regional analysis of time-fractional diffusion processes*, Singapore: Springer International Publishing, 2018.

[10] R. E. Kalman, On the general theory of control systems, *IFAC Proc.* 1 (1)(1960) 491–502.

[11] A. A. Kilbas, H. M. Srivastava, and J. J. Trujillo, *Theory and applications of fractional differential equations*, Amsterdam: Elsevier, 2006.

[12] D. L. Koch and J. F. Brady, Anomalous diffusion in heterogeneous porous media, *Phys. Fluids*, 31 (5)(1998) 965.

[13] J. L. Lions, *Contrôlabilité exacte, perturbations et stabilisation de systèmes distribués, tome 1. Contrôlabilité exacte*, Paris: Dunod, 1997.

[14] Y. Louartassi, E. H. El Mazoudi, and N. Elalami, A new generalization of lemma Gronwall-Bellman, *Appl. Math. Nonlinear Sci.* 6 (13)(2012) 621–628.

[15] K. B. Oldham and J. Spanier, *Fractional calculus*, 1st edition. New York: Elsevier Science, 1974.

[16] I. Podlubny, *Fractional differential equations: an introduction to fractional derivatives, fractional differential equations, to methods of their solution and some of their applications*, 1st edition. Cambridge, MA: Academic Press, 1998.

[17] A. J. Pritchard and A. Wirth, Unbounded control and observation systems and their duality, *SIAM J Control Optim.* 16 (4)(1978) 535–545.

[18] K. Zguaid and F.Z. El Alaoui, Regional boundary observability for Riemann–Liouville linear fractional evolution systems, *Math. Comput. Simul.* 199 (2022) 272–286.

[19] K. Zguaid and F. Z. El Alaoui, Regional boundary observability for linear time-fractional systems, *Partial. Differ. Equ. Appl. Math.* 6 (2022) 100432.

[20] K. Zguaid, F. Z. El Alaoui, and A. Boutoulout, Regional observability of linear fractional systems, involving Riemann-Liouville fractional derivative, series: lecture notes in networks and systems, 168, In: *Nonlinear analysis: problems, applications and computational methods*, Editors: Hammouch Z. Dutta H. Melliani S. and Ruzhansky M. Singapore: Springer International Publishing, 2021, 164–178.

[21] K. Zguaid, F. Z. El Alaoui, and A. Boutoulout, Regional observability for linear time fractional systems, *Math. Comput. Simul.* 185 (2021) 77–87.

[22] Y. Zhou, L. Zhang, and X. H. Shen, Existence of mild solutions for fractional evolution equations, *J. Integral Equ. Appl.* 25 (4)(2013) 557–586.

CHAPTER 16

Construction of Fractional Extended Nabla Operator and Strong Convergence Analysis

Leila Khitri-Kazi-Tani and Hacen Dib

Aboubekr Belkaid University

CONTENTS

16.1 INTRODUCTION

It is well known that fractional calculus is a developing field both from the theoretical and applied point of view. The fractional differential equations turned out to be the best tool for modeling memory-dependent processes [7]. We refer to the monograph [29], which contains almost complete qualitative fractional differential equation theory, and to the monograph [10] for an application-oriented exposition.

DOI: 10.1201/9781003388678-16

Furthermore, the most recent progress achieved in connection with the fixed point theory is presented in Ref. [8].

Besides this rapid development, the notion of difference operators has been extended to fractional calculus in different ways [14, 20, 22]. Discrete calculus provides a natural setting to define such operators. However, in literature, there is no single definition of fractional difference operators, and this situation can be confusing (see for example [1, 2, 19]).

Another way to define these operators is to consider the fractional power of positive discrete operators. See Refs. [4, 16].

Effectively, functional calculus is a consistent way to define operators of the form A^α for a given linear operator A in a Banach space. The fundamental aspects of the theory of fractional powers of non-negative operators are given in Ref. [18]. Sectorial operators satisfy a resolvent condition that leads to define the fractional power of such operators. The functional calculus for sectorial operators has been developed by Haase in the book [13].

The Riemann–Liouville fractional difference derivative presented by Ashyralyev [4] is nothing, but the Grünwald–Letnikov approximation of the Riemann–Liouville fractional derivative as established in Ref. [17]. The pointwise error of the Grünwald-Letnikov approximation was analyzed in Ref. [5] for a fractional initial value problem and in Ref. [6] for an initial-boundary value problem. The order of consistency of this approximation in a bounded domain has been the subject of the work of Sousa in Ref. [24].

In this chapter,[1] we define the fractional difference as fractional power of the nabla operator in a Hölder space which offers an interesting setting in the analysis of fractional integrals and derivatives. This framework was developed by Samko et al. for fractional operators in the sense of Marchaud [22, 23]. In this functional framework, we study the strong convergence of the extended backward differences to the derivative. To our knowledge, there is no result which establish convergence of discrete fractional derivative to continuous fractional derivative with respect to some continuous norm. Only discrete convergence was studied [6, 24, 25]. This is the main contribution of the present work. Our goal was to provide users some Euler-like formula, which helps to do numerical analysis for fractional differential equations. We construct the fractional operators associated and the strong convergence result is proved. Finally, some examples are provided to show the effectiveness of the approach.

This chapter is organized as follows: Section 16.2 is devoted to preliminaries and some Hölderian tools. In Section 16.3, we give the basic definitions and results for the fractional power of a sectorial operator, and we construct the different fractional operators as fractional powers of sectorial operators in Hölder spaces setting. In Section 16.4, we discuss in this context the strong convergence of the operators involved. Some examples are given in Section 16.5.

16.2 PRELIMINARIES ON OPERATORS IN HÖLDER SPACES

Without loss of generality, we assume that the functions are defined on the interval $[0,1]$. Let H^β be the Banach space of Hölderian functions on $[0,1]$ with exponent β, where $0<\beta<1$ and such that $f(0)=0$, endowed with the norm $\|f\|_\beta=\omega_\beta(f,1)$, where

$$\omega_\beta(f,\delta)=\sup_{\substack{s,t\in[0,1]\\0<|s-t|\leq\delta}}\frac{|f(t)-f(s)|}{|t-s|^\beta}.$$

Let H_0^β be the subspace defined by

$$H_0^\beta=\left\{f\in H^\beta,\lim_{\delta\to0}\omega_\beta(f,\delta)=0\right\}.$$

Remark 16.1 *If $f\in H_0^\beta$ then $|f(t)-f(t-h)|=o(h^\beta)$ uniformly in t, for $t=h$ we get $|f(h)|=o(h^\beta)$.*

Remark 16.2 *If $f\in H^\beta$ then $\|f\|_\infty\leq\|f\|_\beta$. Indeed, for every $x\in\,]0,1]$, $|f(x)|=\dfrac{|f(x)-f(0)|}{x^\beta}x^\beta\leq\|f\|_\beta$.*

Remark 16.3 *If $f'\in H^\beta$ then $\omega_\beta(f,h)$ tends to 0 as h tends to 0.*
Indeed, for every x , $y\in[0,1]$, $x\neq y$ there exists some $\xi\in\,]x,y[$ such that

$$\frac{|f(x)-f(y)|}{|x-y|^\beta}=|x-y|^{1-\beta}|f'(\xi)|\leq|x-y|^{1-\beta}\|f'\|_\beta,$$

which leads to

$$\omega_\beta(f,h)\leq h^{1-\beta}\|f'\|_\beta. \tag{16.1}$$

The Hölder norm of the piecewise linear interpolation is given by the next lemma. As far as we know, this result was first proved by H. E. White, Jr, in a general setting see [28, 3.2 Corollary, p. 106], but we follow [21] in the presentation.

Lemma 16.1 (see lemma 3.1 [21]) *Let $t_0=0<t_1<\cdots<t_n=1$ be a partition of $[0,1]$ and f be a real valued polygonal line function on $[0,1]$ with vertices at t_i's, i.e. f is continuous on $[0,1]$ and its restriction to each interval $[t_i,t_{i+1}]$ is an affine function. Then for any $0\leq\beta<1$,*

$$\sup_{0\leq s<t\leq1}\frac{|f(t)-f(s)|}{(t-s)^\beta}=\max_{0\leq i<j\leq1}\frac{|f(t_j)-f(t_i)|}{(t_j-t_i)^\beta}.$$

Definition 16.1 *For $0 < h < 1$ fixed, let Δ_h be the subdivision of $[0,1]$ in n subintervals with $n = [1/h]$ and $t_k = kh$, for each $k = 0, 1, \ldots, n$ where $[a]$ means the integer part of a. We denote by $\mathcal{I}_h \in \mathcal{L}(H^\beta)$ the piecewise linear interpolation operator defined by*

$$(\mathcal{I}_h f)(x) := \sum_{k=1}^{n} \left(\frac{x - t_{k-1}}{h} f(t_k) + \frac{t_k - x}{h} f(t_{k-1}) \right) \mathbf{1}_{[\mathbf{t_{k-1}}, \mathbf{t_k}]}(\mathbf{x}).$$

In the following lemma, the remainder of piecewise linear interpolation is expressed in Hölder norm.

Lemma 16.2 *Let $(r_h f)(x) = (I - \mathcal{I}_h) f(x)$ then*

$$\|(r_h f)\|_\beta \leq 4\omega_\beta(f, h) \ .$$

Proof 16.1 *First let us suppose that $x, y \in \,]t_{k-1}, t_k]$ then*

$$(r_h f)(x) - (r_h f)(y) = f(x) - f(y) - \frac{x - y}{h} \left(f(t_k) - f(t_{k-1}) \right).$$

It follows, from $|x - y| < h$ that

$$\begin{aligned} \frac{|(r_h f)(x) - (r_h f)(y)|}{|x - y|^\beta} &\leq \frac{|f(x) - f(y)|}{|x - y|^\beta} + \frac{|f(t_k) - f(t_{k-1})|}{h^\beta} \\ &\leq 2\omega_\beta(f, h). \end{aligned}$$

Second, suppose that $x \in [t_{k-1}, t_k]$ and $y \in [t_k, t_{k+1}]$ then from the first case

$$\begin{aligned} \frac{|(r_h f)(x) - (r_h f)(y)|}{|x - y|^\beta} &\leq \frac{|(r_h f)(x) - (r_h f)(t_k)|}{|x - t_k|^\beta} + \frac{|(r_h f)(t_k) - (r_h f)(y)|}{|t_k - y|^\beta} \\ &\leq 4\omega_\beta(f, h). \end{aligned}$$

Third, suppose that $x \in [t_{k-1}, t_k]$ and $y \in [t_{m-1}, t_m]$ with $|x - y| > h$ then

$$\begin{aligned} \frac{|(r_h f)(x) - (r_h f)(y)|}{|x - y|^\beta} \leq & \frac{|(r_h f)(x) - (r_h f)(t_k)|}{|x - y|^\beta} \\ & + \frac{|(r_h f)(t_k) - (r_h f)(t_{m-1})|}{|x - y|^\beta} \\ & + \frac{|(r_h f)(t_{m-1}) - (r_h f)(y)|}{|x - y|^\beta}. \end{aligned}$$

Knowing that $(r_h f)(t_k) = (r_h f)(t_{m-1}) = 0$, then

$$\frac{|(r_h f)(x) - (r_h f)(y)|}{|x - y|^\beta} \leq 4\omega_\beta(f, h).$$

We denote by $A = \frac{d}{dx}$ the differential operator acting on H^{β} with domain,

$$D(A) = \{f \in H^{\beta}, f' \in H^{\beta}\},$$

and for $0 < h < 1$, let $\nabla_h \in \mathcal{L}(H^{\beta})$ the nabla operator defined by

$$(\nabla_h f)(x) := \frac{f(x) - f(x-h)}{h}, \qquad \text{for } x \in [h, 1]$$

We set $(\nabla_h f)(x) = \frac{f(x)}{h}$ for all $0 < x < h$.

Finally, we introduce the extended nabla operator as the polygonal line with vertices $(t_k, \nabla f(t_k))$, $k = 0, 1, \ldots, n$, in the following definition.

Definition 16.2 *We define the operator A_h for any $f \in H^{\beta}$ by*

$$A_h f(x) = (\mathcal{I}_h \nabla_h f)(x)$$

Obviously A_h is a linear bounded operator with $\|A_h\|_{\beta} \leq \frac{2}{h}$.

In the next proposition, it can be pointed out that the sequence $(A_h)_h$ has no uniform limit as h tends to 0.

Proposition 16.1 *The sequence $(A_h)_h$ of extended nabla operator is not a convergent sequence in $\mathcal{L}(H^{\beta})$ as h tends to 0.*

Proof 16.2 *We need only to prove that $(A_h)_h$ is not a Cauchy sequence.*

Let Δ_h , $\Delta_{h/2}$ be two subdivisions of $[0, 1]$ and $f(x) = x^{\beta}$. We then have for $t = h/2$,

$$\begin{aligned}
\left|(A_h - A_{h/2}) f(t)\right| &= \left|\frac{h/2}{h}\frac{f(h)}{h} - \frac{f(h/2)}{h/2}\right| \\
&= \left|\frac{h^{\beta-1}}{2} - \left(\frac{h}{2}\right)^{\beta-1}\right| \geq \left|\frac{1}{2} - \left(\frac{1}{2}\right)^{\beta-1}\right|.
\end{aligned}$$

Since

$$\|(A_h - A_{2h}) f\|_{\beta} \geq |(A_h - A_{2h}) f(h/2)|,$$

then

$$\|A_h - A_{2h}\|_{\beta} \|f\|_{\beta} \geq \|(A_h - A_{2h}) f\|_{\beta} \geq \left(\frac{1}{2}\right)^{\beta-1} - \frac{1}{2}.$$

We have thus seen that $(A_h)_h$ is not a convergent sequence in $\mathcal{L}(H^{\beta})$.

In the next proposition, the strong convergence of extended nabla operator to the derivative operator is proved.

Proposition 16.2 *For every $f \in D(A)$, such that $f' \in H_0^\beta$ the sequence $(A_h)_h$ converges strongly to A as h tends to 0.*

Proof 16.3 *Note that*

$$\begin{aligned}(A - A_h) f(x) &= (I - \mathcal{I}_h) Af(x) + \mathcal{I}_h (A - \nabla_h) f(x) \\ &= (r_h f')(x) + \mathcal{I}_h (A - \nabla_h) f(x).\end{aligned}$$

Then

$$\frac{|(A - A_h) f(x) - (A - A_h) f(y)|}{|x-y|^\beta} \leq \|(r_h f')\|_\beta + \frac{|\mathcal{I}_h (A - \nabla_h) f(x) - \mathcal{I}_h (A - \nabla_h) f(y)|}{|x-y|^\beta}.$$

From Lemma 16.2

$$\|(r_h f')\|_\beta \leq 4\omega_\beta (f', h).$$

For some $i, j, \xi_i \in [t_{i-1}, t_i]$ and $\xi_j \in [t_{j-1}, t_j]$ by Lemma 16.1 we have

$$\begin{aligned}\frac{|\mathcal{I}_h (A - \nabla_h) f(x) - \mathcal{I}_h (A - \nabla_h) f(y)|}{|x-y|^\beta} &\leq \frac{|f'(t_i) - \nabla_h f(t_i)|}{|t_i - t_j|^\beta} \\ &\quad + \frac{|f'(t_j) - \nabla_h f(t_j)|}{|t_i - t_j|^\beta} \\ &\leq \frac{|f'(t_i) - f'(\xi_i)|}{|t_i - t_j|^\beta} \\ &\quad + \frac{|f'(t_j) - f'(\xi_j)|}{|t_i - t_j|^\beta} \\ &\leq 2\omega_\beta (f', h).\end{aligned}$$

Hence

$$\frac{|(A - A_h) f(x) - (A - A_h) f(y)|}{|x-y|^\beta} \leq 6\omega_\beta (f', h). \tag{16.2}$$

Therefore, $\lim_{h \to 0} \|Af - A_h f\|_\beta = 0$.

16.3 FRACTIONAL POWER OF SECTORIAL OPERATOR

16.3.1 Sectorial Property

We first recall the Haase concept of sectorial operators [13, Section 2.1 , p. 19]. In the following, $R(\lambda, B) = (\lambda I - B)^{-1}, \rho(B)$ and $\sigma(B) = \mathbb{C} \backslash \rho(B)$ denote, respectively, the resolvent, the resolvent set and the spectrum of a linear operator B on a Banach space Z. Let S_ω denote the open sector,

$$\{z \in \mathbb{C}, z \neq 0 \text{ and } |\arg z| < \omega\}, 0 < \omega \leq \pi.$$

Definition 16.3 *The operator B is sectorial of angle $\omega < \pi$ (in short: $B \in Sect(\omega)$) if:*

1) $\sigma(B) \subset \overline{S_\omega}$ and

2) $M(B,\omega') := \sup\left\{\|\lambda R(\lambda,B)\|, \lambda \notin \overline{S_{\omega'}}\right\} < \infty$ for all $\omega < \omega' < \pi$.

A family of operators $(B_\iota)_\iota$ is uniformly sectorial of angle ω if $B_\iota \in Sect(\omega)$ for each ι and $\sup_\iota M(B_\iota,\omega') < \infty$ for all $\omega < \omega' < \pi$.

Remark 16.4 *Sectorial operator in Haase's definition doesn't have to be densely defined. See [15, Definition 3.8, p. 97].*

We are now able to define the fractional power of a sectorial operator with the help of the Balakrishnan representation (see [13, Proposition 3.1.12]).

Proposition 16.3 *Let B an operator with domain $\mathcal{D}(B)$, $B \in Sect(\omega)$, and let $0 < \alpha < 1$. Then for all $v \in \mathcal{D}(B)$*

$$B^\alpha v = -\frac{\sin \alpha\pi}{\pi}\int_0^\infty \lambda^{\alpha-1}R(-\lambda,B)Bv\,d\lambda. \tag{16.3}$$

16.3.2 Fractional Power of the Derivative

We define the fractional derivative as fractional power of a sectorial operator in Hölder space. To do so, we examine the sectoriality of A.

Proposition 16.4 *The operator A on H^β is sectorial of angle $\frac{\pi}{2}$.*

Proof 16.4 *For all $\lambda \in \mathbb{C}$, the resolvent of the operator A on H^β is given by*

$$R(\lambda,A)f(x) = -\int_0^x e^{\lambda(x-t)}f(t)dt.$$

Let's take $\lambda \in \mathbb{C}$ with $\mathrm{Re}(\lambda) < 0$ and let x,h be such that $0 \le x-h < x \le 1$. We have

$$\begin{aligned}&|R(\lambda,A)f(x) - R(\lambda,A)f(x-h)|\\ &= \left|\int_0^{x-h} -e^{\lambda t}\left[f(x-t) - f(x-h-t)\right]dt + \int_{x-h}^{x} -e^{\lambda t}f(x-t)\,dt\right|.\end{aligned}$$

If p and q are two conjugates positive reals i.e, $1/p + 1/q = 1$, then the Hölder inequality implies:

$$\begin{aligned}&|\lambda|\,|R(\lambda,A)f(x) - R(\lambda,A)f(x-h)|\\ &\le |\lambda|\left(\int_0^{x-h} e^{p\,\mathrm{Re}(\lambda)t}dt\right)^{\frac{1}{p}}\left(\int_0^{x-h} |f(x-t) - f(x-h-t)|^q\,dt\right)^{\frac{1}{q}}\end{aligned}$$

$$+|\lambda|\left(\int_{x-h}^{x} e^{p\operatorname{Re}(\lambda)t}dt\right)^{\frac{1}{p}}\left(\int_{x-h}^{x}|f(x-t)|^{q}\right)^{\frac{1}{q}}dt$$

$$\leq \sup_{0\leq t\leq x-h}|f(t)-f(t-h)|\,(x-h)^{\frac{1}{q}}\,|\lambda|\left(\frac{e^{p\operatorname{Re}(\lambda)(x-h)}-1}{p\operatorname{Re}(\lambda)}\right)^{\frac{1}{p}}$$

$$+\sup_{x-h\leq t\leq x}|f(x-t)|\left|(h)^{\frac{1}{q}}\lambda\right|\left(\frac{e^{p\operatorname{Re}(\lambda)x}-e^{p\operatorname{Re}(\lambda)(x-h)}}{p\operatorname{Re}(\lambda)}\right)^{\frac{1}{p}}$$

$$\leq \omega_{\beta}(f,h)\,h^{\beta}2\,|\lambda|\left(\frac{-1}{p\operatorname{Re}(\lambda)}\right)^{\frac{1}{p}}.$$

It is not hard to check that for every parameter $\operatorname{Re}(\lambda)<0$, *the infimum, on* $[1,\infty]$, *of the function (of p)*

$$\left(\frac{-1}{p\operatorname{Re}(\lambda)}\right)^{\frac{1}{p}},$$

is given by

$$\inf_{p>1}\left(\frac{-1}{p\operatorname{Re}(\lambda)}\right)^{1/p}=\begin{cases}-\frac{1}{\operatorname{Re}(\lambda)}, & \text{if } -\frac{e}{\operatorname{Re}(\lambda)}\leq 1,\\ e^{\frac{1}{e}\operatorname{Re}(\lambda)}, & \text{if } -\frac{e}{\operatorname{Re}(\lambda)}>1.\end{cases}$$

In addition, by $\operatorname{Re}(\lambda)=|\lambda|\cos\omega$, *we have*

$$|\lambda|\left(-\frac{1}{\operatorname{Re}(\lambda)}\right)=|\lambda|\left(-\frac{1}{|\lambda|\cos\omega}\right)=-\frac{1}{\cos\omega},$$

and

$$|\lambda|\,e^{\frac{1}{e}\operatorname{Re}(\lambda)}=|\lambda|\,e^{\frac{1}{e}|\lambda|\cos\omega}.$$

Now, the function defined on $[0,\infty]$ *by* $xe^{x\frac{1}{e}\cos\omega}$ *admits for maximum value* $-\dfrac{1}{\cos\omega}$, *which gives the estimate*

$$|\lambda|\inf_{p>1}\left(\frac{-1}{p\operatorname{Re}(\lambda)}\right)^{1/p}\leq -\frac{1}{\cos\omega}.$$

According to the previous arguments, we get

$$|\lambda|\,|R(\lambda,A)f(x)-R(\lambda,A)f(x-h)|\leq -\frac{2}{\cos\omega}\omega_{\beta}(f,h)\,h^{\beta},$$

which implies

$$|\lambda|\,\omega_{\beta}(R(\lambda,A)f,h)\leq -\frac{2}{\cos\omega}\omega_{\beta}(f,h)\,h^{\beta}$$

and

$$|\lambda|\,\|R(\lambda,A)f\|_{\beta}\leq -\frac{2}{\cos\omega}\,\|f\|_{\beta}\,.$$

Therefore, for every $\lambda \in \mathbb{C}\backslash\overline{S_\omega}$, $\frac{\pi}{2} < \omega \leq \pi$,

$$\|\lambda R(\lambda, A)\| \leq -\frac{2}{\cos \omega}.$$

Corollary 16.1 *Let* $0 < \alpha < 1$ *and* $f \in D(A)$. *Then*

$$A^\alpha f(x) = \frac{1}{\Gamma(1-\alpha)} \int_0^x (x-t)^{-\alpha} f'(t)dt.$$

Proof 16.5 *Using the Balakrishnan representation of fractional power of sectorial operator (equation 16.3), the previous representation follows.*

In the next subsection, the fractional power of operator ∇_h and A_h is constructed.

16.3.3 Fractional Nabla Operators

Before studying the sectoriality of ∇_h and A_h, we begin by somehow a surprising and useful result. Elementary calculations show that the operator $\mathcal{I}_h$ commutes with ∇_h. This property has an interesting consequence for the resolvent operator given in the next lemma and stated in general framework.

Lemma 16.3 *Let* $\mathcal{X}$ *be a Banach space,* $B, T \in \mathcal{L}(\mathcal{X})$ *such that* T *is idempotent and* T *commute with* B *then, for every* $\lambda \in \rho(B)$, $\lambda \neq 0$,

$$R(\lambda, TB) = TR(\lambda, B) + \frac{1}{\lambda}(I - T).$$

Proof 16.6 *To obtain the resolvent operator for* TB *we consider the equation, for* $f, g \in \mathcal{X}$,

$$f = (\lambda I - TB)\, g,$$

then by idempotence of the operator T *and commutative property we get*

$$Tf = (\lambda I - B)\, Tg,$$

combining the above two equations we have

$$f - Tf = \lambda\,(I - T)\, g.$$

Using the fact that

$$Tg = (\lambda I - B)^{-1}\, Tf$$

then

$$g = R(\lambda, B)Tf + \frac{1}{\lambda}(f - Tf).$$

Proposition 16.5 *The family* $(\nabla_h)_h$ *is uniformly sectorial of angle* $\frac{\pi}{2}$ *on* H^β.

Proof 16.7 *It can be easily proved using Laplace and inverse Laplace transforms that*

$$R(\lambda,\nabla_h)f(x) = -h\sum_{j=0}^{n}\frac{1}{(1-\lambda h)^{j+1}}f(x-t_j).$$

First, we check the boundedness of $R(\lambda,\nabla_h)$ *in* $\mathcal{L}\left(H^\beta\right)$.
For every $0\le x<y\le 1$,

$$\begin{aligned}
&\frac{|R(\lambda,\nabla_h)f(x)-R(\lambda,\nabla_h)f(y)|}{|x-y|^\beta}\\
&\le \frac{1}{|x-y|^\beta}\left|-h\sum_{j=0}^{[x/h]}\frac{1}{(1-\lambda h)^{j+1}}\left[f(x-t_j)-f(y-t_j)\right]\right|\\
&+\frac{1}{|x-y|^\beta}\left|-h\sum_{[x/h]+1}^{[y/h]}\frac{1}{(1-\lambda h)^{j+1}}f(x-t_j)\right|\\
&\le \|f\|_\beta\sum_{j=1}^{n}\frac{h}{|1-\lambda h|^j}.
\end{aligned}$$

Using the sum of a geometric series, we have

$$\frac{|R(\lambda,\nabla_h)f(x)-R(\lambda,\nabla_h)f(y)|}{|x-y|^\beta}\le\frac{h}{|1-\lambda h|-1}\|f\|_\beta\,. \tag{16.4}$$

Now, observe that for any $\lambda\in\mathbb{C}\backslash\overline{S}_\omega\,,\frac{\pi}{2}<\omega<\pi$, *we have* $|\lambda h-1|>1$ *and*

$$\frac{|\lambda|\,|R(\lambda,\nabla_h)f(x)-R(\lambda,\nabla_h)f(y)|}{|x-y|^\beta}\le\frac{|\lambda|\,h}{|1-\lambda h|-1}\|f\|_\beta\,.$$

Knowing that

$$|1-\lambda h|^2 = h^2\,|\lambda|^2+1-2\,|\lambda|\cos(\arg\lambda)\ge h^2\,|\lambda|^2+1-2h\,|\lambda|\cos\omega.$$

Then

$$\frac{|\lambda|\,h}{|1-\lambda h|-1}\le\frac{|\lambda|\,h}{\sqrt{h^2\,|\lambda|^2+1-2h\,|\lambda|\cos\omega}-1}.$$

Put $\chi(z)=\dfrac{z}{\sqrt{z^2+1-2z\cos\omega}-1}$ *for* $z>0$. *It is easy to see that* $\chi(+\infty)=1$, $\chi(0^+)=\dfrac{-1}{\cos\omega}$ *and the derivative satisfies*

$$\chi'(z) = \frac{\left(-1+\cos^2\omega\right)z^2}{\left(\sqrt{z^2+1-2z\cos\omega}-1\right)^2\sqrt{z^2+1-2z\cos\omega}\left(1-z\cos\omega+\sqrt{z^2+1-2z\cos\omega}\right)} < 0.$$

Then, for every $z \in [0,+\infty]$ $1 < \chi(z) \leq \dfrac{-1}{\cos\omega}$, *which implies that*

$$\frac{|\lambda|\,|R(\lambda,\nabla_h)f(x) - R(\lambda,\nabla_h)f(y)|}{|x-y|^\beta} \leq \frac{-1}{\cos\omega}\,\|f\|_\beta\,. \tag{16.5}$$

We conclude that the family $(\nabla_h)_h$ *is uniformly sectorial of angle* $\dfrac{\pi}{2}$.

Consequently, the extended nabla operator A_h is also sectorial as shown in the next corollary.

Corollary 16.2 *The family* $(A_h)_h$ *is uniformly sectorial of angle* $\dfrac{\pi}{2}$ *on* H^β.

Proof 16.8 *From Lemma 16.3 we have*

$$R(\lambda,A_h)f(x) = \mathcal{I}_h\left(R(\lambda,\nabla_h)f\right)(x) + \frac{1}{\lambda}\left(f-\mathcal{I}_hf\right)(x).$$

Then

$$\begin{aligned}&\frac{|\lambda|\,|R(\lambda,A_h)f(x) - R(\lambda,A_h)f(y)|}{|x-y|^\beta}\\ &\leq \frac{|\lambda|\,|\mathcal{I}_hR(\lambda,A_h)f(x) - \mathcal{I}_hR(\lambda,A_h)f(y)|}{|x-y|^\beta}\\ &+\frac{|\left(I-\mathcal{I}_h\right)f(x) - \left(I-\mathcal{I}_h\right)f(y)|}{|x-y|^\beta},\end{aligned}$$

From Lemmas 16.1 and 16.2, there exist $0 \leq m,l \leq n$ *such that,*

$$\frac{|\lambda|\,|R(\lambda,A_h)f(x) - R(\lambda,A_h)f(y)|}{|x-y|^\beta} \leq \frac{|\lambda|\,|R(\lambda,A_h)f(t_m) - R(\lambda,A_h)f(t_l)|}{|t_m-t_l|^\beta} + 4\omega_\beta(f,h).$$

From Proposition 16.5, we get the estimate:

$$\frac{|\lambda|\,|R(\lambda,A_h)f(x) - R(\lambda,A_h)f(y)|}{|x-y|^\beta} \leq \left(\frac{-1}{\cos\omega}+4\right)\|f\|_\beta\,.$$

As a result, we are able to define the fractional power of ∇_h and A_h. This is the purpose of the following theorem.

Theorem 16.1 *Let $0 < \alpha < 1$, then fractional nabla operator is given by*

$$\nabla_h^\alpha f(x) = \frac{h^{1-\alpha}}{\Gamma(1-\alpha)} \sum_{j=1}^{[x/h]} \frac{\Gamma(j+1-\alpha)}{\Gamma(j+1)} \nabla_h f(x-t_j)$$

and the fractional operator A_h^α is:

$$A_h^\alpha f(x) = \sum_{k=0}^{n} \left(\frac{x-t_{k-1}}{h} \sum_{j=1}^{k} \frac{\Gamma(j+1-\alpha)}{\Gamma(j+1)} \nabla_h f(t_k - t_j) + \right.$$

$$\left. + \frac{t_k - x}{h} \sum_{j=1}^{k-1} \frac{\Gamma(j+1-\alpha)}{\Gamma(j+1)} \nabla_h f(t_{k-1} - t_j) \right) \mathbf{1}_{[\mathbf{t_{k-1}},\mathbf{t_k}]}(\mathbf{x}).$$

We call A_h^α the fractional extended nabla operator.

Proof 16.9 *Using Balakrishnan representation of fractional power of sectorial operator (equation 16.3) , we get when $0 < \alpha < 1$,*

$$\begin{aligned} \nabla_h^\alpha f(x) &= \frac{\sin \alpha\pi}{\pi} \int_0^{+\infty} \lambda^{\alpha-1} (\lambda + \nabla_h)^{-1} \nabla_h f(x) d\lambda \\ &= -\frac{\sin \alpha\pi}{\pi} \int_0^{+\infty} \lambda^{\alpha-1} R(-\lambda, \nabla_h) \nabla_h f(x) d\lambda. \end{aligned}$$

Then,

$$\begin{aligned} &\nabla_h^\alpha f(x) \\ &= \frac{\sin \alpha\pi}{\pi} \int_0^{+\infty} \lambda^{\alpha-1} h \sum_{j=0}^{[x/h]} \frac{1}{(1+\lambda h)^{j+1}} \nabla_h f(x-t_j) d\lambda \\ &= h \sum_{j=0}^{[x/h]} \left(\frac{\sin \alpha\pi}{\pi} \int_0^{+\infty} \lambda^{\alpha-1} \frac{1}{(1+\lambda h)^{j+1}} d\lambda \right) \nabla_h f(x-t_j). \end{aligned}$$

Similar calculations to those in [4, Theorem 3.1] gives

$$\int_0^{+\infty} \lambda^{\alpha-1} \frac{1}{(1+\lambda h)^{j+1}} d\lambda = h^{-\alpha} \frac{\Gamma(j+1-\alpha)\Gamma(\alpha)}{\Gamma(j+1)}.$$

Therefore,

$$\nabla_h^\alpha f(x) = \frac{h^{1-\alpha}}{\Gamma(1-\alpha)} \sum_{j=0}^{[x/h]} \frac{\Gamma(j+1-\alpha)}{\Gamma(j+1)} \nabla_h f(x-t_j).$$

We now turn to the evaluation of $A_h^\alpha f$. From Lemma 16.3 we get

$$\begin{aligned} A_h^\alpha f(x) &= -\frac{\sin\alpha\pi}{\pi}\int_0^{+\infty}\lambda^{\alpha-1}R\left(-\lambda, A_h\right)A_h f(x)d\lambda \\ &= -\frac{\sin\alpha\pi}{\pi}\int_0^{+\infty}\lambda^{\alpha-1}\mathcal{I}_h R\left(-\lambda, \nabla_h\right)\mathcal{I}_h\nabla_h f(x)d\lambda \\ &= -\frac{\sin\alpha\pi}{\pi}\int_0^{+\infty}\lambda^{\alpha-1}\mathcal{I}_h R\left(-\lambda, \nabla_h\right)\nabla_h f(x)d\lambda \end{aligned}$$

The required evaluation of $A_h^\alpha f$ then follows.

The remaining problem is to study if the strong convergence $(A_h f)_h$ to Af can give rise to the convergence of power operators; this is the aim of the next section.

16.4 HÖLDERIAN CONVERGENCE OF FRACTIONAL EXTENDED NABLA OPERATOR TO FRACTIONAL DERIVATIVE

The following result, known elsewhere, is given in a suitable form for later uses.

Lemma 16.4 *There exists a function Φ_α, such that*

$$\forall m \geq 1 \quad \frac{\Gamma(m+\alpha)}{\Gamma(m+1)} = m^{\alpha-1} + \frac{1}{\Gamma(1-\alpha)}\Phi_\alpha(m),$$
$$\text{with } |\Phi_\alpha(m)| \leq \frac{\Gamma(2-\alpha)}{2}m^{\alpha-2}.$$

Proof 16.10 *From the definition of the beta function, we have*

$$\frac{\Gamma(m+\alpha)}{\Gamma(m+1)} = \frac{1}{\Gamma(1-\alpha)}\int_0^1 t^{(m+\alpha-1)}(1-t)^{-\alpha}dt. \tag{16.6}$$

Set $t = e^{-u}$, then the equality (equation 16.6) becomes

$$\begin{aligned} \frac{\Gamma(m+\alpha)}{\Gamma(m+1)} &= \frac{1}{\Gamma(1-\alpha)}\int_0^{+\infty} e^{-mu}(e^u-1)^{-\alpha}du \\ &= \frac{1}{\Gamma(1-\alpha)}\int_0^{+\infty} e^{-mu}u^{-\alpha}(\frac{u}{e^u-1})^{\alpha}du. \end{aligned}$$

Using the generating function of the Bernoulli numbers,

$$G(u) = \frac{u}{e^u-1} = \sum_{k=0}^{+\infty} B_k\frac{u^k}{k!} = 1+\theta(u) > 0,$$

where $\theta : [0, +\infty] \to [-1, 0]$ is a continuous function. We have

$$\frac{\Gamma(m+\alpha)}{\Gamma(m+1)} = \frac{1}{\Gamma(1-\alpha)}\int_0^{+\infty} e^{-mu}u^{-\alpha}(1+\theta(u))^{\alpha}du.$$

Now, Taylor's formula with integral remainder applied to the function $(1+\theta(u))^\alpha$ *gives*

$$(1+\theta(u))^\alpha = 1 + \alpha\theta(u)\int_0^1 (1+\xi\theta(u))^{\alpha-1}d\xi.$$

Therefore,

$$\frac{\Gamma(m+\alpha)}{\Gamma(m+1)} = \frac{1}{\Gamma(1-\alpha)}\int_0^{+\infty} e^{-mu}u^{-\alpha}du+$$
$$+\frac{\alpha}{\Gamma(1-\alpha)}\int_0^{+\infty} e^{-mu}u^{-\alpha}\theta(u)\int_0^1 (1+\xi\theta(u))^{\alpha-1}d\xi du,$$

and then,

$$\frac{\Gamma(m+\alpha)}{\Gamma(m+1)} = m^{\alpha-1} + \frac{1}{\Gamma(1-\alpha)}\Phi_\alpha(m),$$

where

$$\Phi_\alpha(m) = \alpha\int_0^{+\infty} e^{-mu}u^{-\alpha}\theta(u)\int_0^1 (1+\xi\theta(u))^{\alpha-1}d\xi du.$$

From the identity $1+\xi\theta(u) = 1-\xi+\xi(1+\theta(u))$ *and the fact that* $1+\theta(u) > 0$ *for every* $u \geq 0$, *we have*

$$1+\xi\theta(u) \geq 1-\xi \qquad \textit{and} \qquad (1+\xi\theta(u))^{\alpha-1} \leq (1-\xi)^{\alpha-1}.$$

Consequently,

$$\int_0^1 (1+\xi\theta(u))^{\alpha-1}d\xi \leq \int_0^1 (1-\xi)^{\alpha-1}d\xi \leq \frac{1}{\alpha}.$$

Hence,

$$\alpha\left|\int_0^{+\infty} e^{-mu}u^{-\alpha}\theta(u)\int_0^1 (1+\xi\theta(u))^{\alpha-1}d\xi du\right| \leq \int_0^{+\infty} e^{-mu}u^{-\alpha}\left|\theta(u)\right| du. \quad (16.7)$$

The function $\frac{\theta(u)}{u}$ *is strictly increasing on* $[0,+\infty]$, $\lim_{u\to 0+}\frac{\theta(u)}{u} = -\frac{1}{2}$ *and* $\lim_{u\to\infty}\frac{\theta(u)}{u} = 0$. *So,* $\left|\frac{\theta(u)}{u}\right| \leq \frac{1}{2}$, *and the inequality (equation 16.7) becomes*

$$|\Phi_\alpha(m)| \leq \frac{1}{2}\int_0^{+\infty} e^{-mu}u^{1-\alpha}du,$$

and

$$|\Phi_\alpha(m)| \leq \frac{\Gamma(2-\alpha)m^{\alpha-2}}{2}.$$

Before stating the convergence theorem, we define for all f in $D(A)$, the function φ by

$$\varphi(x) := \frac{1}{\Gamma(1-\alpha)}\int_0^x (x-t)^{-\alpha}A_h f(t)\,dt.$$

For the construction of the convergence result proof, we need the following lemmas

Lemma 16.5 *For all* $0 < \beta < 1$ *such that* $1 - \alpha - \beta > 0$ *we have*

$$\omega_\beta(\varphi, h) \leq \frac{8}{\Gamma(2-\alpha)} \|f'\|_\beta h^{1-\alpha-\beta}$$

Proof 16.11 *Remark that the function* φ *satisfies the following estimation*

$$\frac{|\varphi(x) - \varphi(y)|}{|x-y|^\beta} \leq \frac{1}{\Gamma(2-\alpha)} \|A_h f\|_\beta (\max(x,y))^{1-\alpha} \leq \frac{1}{\Gamma(2-\alpha)} \|A_h f\|_\beta,$$

then $\varphi \in H^\beta[0,1]$.
Let us now estimate $\omega_\beta(\varphi, h)$, *we distinguish two cases,*
First if, $t_{k-1} < x < y \leq t_k$:
Notice that

$$\Gamma(1-\alpha)\varphi(x) = \sum_{i=1}^{k-1} \int_{t_{i-1}}^{t_i} (x-t)^{-\alpha} \left(\frac{t - t_{i-1}}{h} \nabla_h f(t_i) + \frac{t_i - t}{h} \nabla_h f(t_{i-1}) \right) dt$$
$$+ \int_{t_{k-1}}^{x} (x-t)^{-\alpha} \left(\frac{t - t_{k-1}}{h} \nabla_h f(t_k) + \frac{t_k - t}{h} \nabla_h f(t_{k-1}) \right) dt.$$

Hence,

$$\Gamma(1-\alpha)(\varphi(x) - \varphi(y))$$
$$= \sum_{i=1}^{k-1} \int_{t_{i-1}}^{t_i} ((x-t)^{-\alpha} - (y-t)^{-\alpha}) \left(\frac{t - t_{i-1}}{h} \nabla_h f(t_i) + \frac{t_i - t}{h} \nabla_h f(t_{i-1}) \right) dt$$
$$+ \int_{t_{k-1}}^{x} ((x-t)^{-\alpha} - (y-t)^{-\alpha}) \left(\frac{t - t_{k-1}}{h} \nabla_h f(t_k) + \frac{t_k - t}{h} \nabla_h f(t_{k-1}) \right) dt$$
$$+ \int_{x}^{y} (y-t)^{-\alpha} \left(\frac{t - t_{k-1}}{h} \nabla_h f(t_k) + \frac{t_k - t}{h} \nabla_h f(t_{k-1}) \right) dt,$$

and

$$\Gamma(1-\alpha)|\varphi(x) - \varphi(y)|$$
$$\leq 2\|f'\|_\beta \left\{ \sum_{i=1}^{k-1} \int_{t_{i-1}}^{t_i} ((x-t)^{-\alpha} - (y-t)^{-\alpha}) dt \right.$$
$$\left. + \int_{t_{k-1}}^{x} ((x-t)^{-\alpha} - (y-t)^{-\alpha}) dt + \int_x^y (y-t)^{-\alpha} \right\}.$$

which leads to

$$
\begin{aligned}
&|\varphi(x) - \varphi(y)| \\
&\leq \frac{2}{\Gamma(2-\alpha)} \|f'\|_\beta \sum_{i=1}^{k-1} \left((x - t_{i-1})^{1-\alpha} - (y - t_{i-1})^{1-\alpha} - (x - t_i)^{1-\alpha}\right. \\
&\left. +(y - t_i)^{1-\alpha}\right) + \frac{2}{\Gamma(2-\alpha)} \|f'\|_\beta \left[(x - t_{k-1})^{1-\alpha}\right. \\
&\left. -(y - t_{k-1})^{1-\alpha} + 2(y - x)^{1-\alpha}\right].
\end{aligned}
$$

Finally,

$$
\begin{aligned}
\frac{|\varphi(x) - \varphi(y)|}{|x - y|^\beta} &\leq \frac{4}{\Gamma(2-\alpha)} \|f'\|_\beta |y - x|^{1-\alpha-\beta} \\
&\leq \frac{4}{\Gamma(2-\alpha)} \|f'\|_\beta h^{1-\alpha-\beta}.
\end{aligned}
$$

Second if, $t_{k-1} < x \leq t_k < y \leq t_{k+1}$, *then*

$$
\begin{aligned}
\frac{|\varphi(x) - \varphi(y)|}{|x - y|^\beta} &\leq \frac{|\varphi(x) - \varphi(t_k)|}{|x - t_k|^\beta} + \frac{|\varphi(t_k) - \varphi(y)|}{|t_k - y|^\beta} \\
&\leq \frac{8}{\Gamma(2-\alpha)} \|f'\|_\beta h^{1-\alpha-\beta}.
\end{aligned}
$$

Therefore

$$
\omega_\beta(\varphi, h) \leq \frac{8}{\Gamma(2-\alpha)} \|f'\|_\beta h^{1-\alpha-\beta}.
$$

Lemma 16.6 *There exists* $C > 0$ *such that for every* $0 \leq k \leq n$,

$$
\frac{|\varphi(t_k) - \nabla_h^\alpha f(t_k)|}{h^\beta} \leq \left(1 + \frac{C}{\Gamma(1-\alpha)}\right) \|f'\|_\beta h^{1-\alpha-\beta} + \frac{2^\beta}{\Gamma(2-\alpha)} \omega_\beta(f', 2h).
$$

Proof 16.12 *Obviously if* $k = 0$ *the lemma holds for every* $C > 0$. *Assume now that* $k > 0$, *then*

$$
\varphi(t_k) = \frac{1}{\Gamma(1-\alpha)} \sum_{i=1}^{k} \int_{t_{j-1}}^{t_j} (t_k - t)^{-\alpha} \left[\frac{t - t_{j-1}}{h} \nabla_h f(t_j) + \frac{t_j - t}{h} \nabla_h f(t_{j-1})\right] dt.
$$

A simple integration leads to

$$
\varphi(t_k) = \frac{1}{\Gamma(2-\alpha)} \sum_{i=1}^{k} (t_k - t_{j-1})^{1-\alpha} \nabla_h f(t_{j-1}) - (t_k - t_j)^{1-\alpha} \nabla_h f(t_j)
$$

$$+\frac{1}{\Gamma(2-\alpha)}\sum_{i=1}^{k}\frac{(t_k-t_{j-1})^{2-\alpha}-(t_k-t_j)^{2-\alpha}}{(2-\alpha)h}\left[\nabla_h f(t_j)-\nabla_h f(t_{j-1})\right],$$

which can be arranged as follows:

$$\varphi(t_k)=\frac{1}{\Gamma(2-\alpha)}\sum_{i=1}^{k-1}\left[(t_k-t_j)^{1-\alpha}-(t_k-t_{j+1})^{1-\alpha}\right]\nabla_h f(t_j)$$
$$+\frac{1}{\Gamma(2-\alpha)}\sum_{i=1}^{k}\left(\frac{(t_k-t_{j-1})^{2-\alpha}-(t_k-t_j)^{2-\alpha}}{(2-\alpha)h}-(t_k-t_j)^{1-\alpha}\right)$$
$$\cdot\left(\nabla_h f(t_j)-\nabla_h f(t_{j-1})\right).$$

Then,

$$\varphi(t_k)-\nabla_h^{\alpha} f(t_k)=S_1+S_2-h^{1-\alpha}\nabla_h f(t_k),$$

where

$$S_1=\frac{h^{1-\alpha}}{\Gamma(1-\alpha)}\sum_{i=1}^{k-1}\left[\frac{j^{1-\alpha}-(j-1)^{1-\alpha}}{1-\alpha}-\frac{\Gamma(j+1-\alpha)}{\Gamma(j+1)}\right]\nabla_h f(t_k-t_j)$$

and

$$S_2=\frac{1}{\Gamma(2-\alpha)}\sum_{i=1}^{k}\left(\frac{(t_k-t_{j-1})^{2-\alpha}-(t_k-t_j)^{2-\alpha}}{(2-\alpha)h}-(t_k-t_j)^{1-\alpha}\right)$$
$$\cdot\left(\nabla_h f(t_j)-\nabla_h f(t_{j-1})\right).$$

By using the fact that,

$$\begin{aligned}0&\le\frac{(t_k-t_{j-1})^{2-\alpha}-(t_k-t_j)^{2-\alpha}}{(2-\alpha)h}-(t_k-t_j)^{1-\alpha}\\&=\frac{1}{h}\int_{t_{j-1}}^{t_j}\left((t_k-t)^{1-\alpha}-(t_k-t_j)^{1-\alpha}\right)dt\\&\le(t_k-t_{j-1})^{1-\alpha}-(t_k-t_j)^{1-\alpha},\end{aligned}$$

and

$$\left|\nabla_h f(t_j)-\nabla_h f(t_{j-1})\right|\le(2h)^{\beta}\,\omega_\beta(f',2h),$$

$|S_2|$ *can be estimated by*

$$|S_2| \leq \frac{(2h)^{\beta}\, \omega_{\beta}(f', 2h)}{\Gamma(2-\alpha)}.$$

It remain to estimate $|S_1|$. To do so, we use the Lemma 16.4,

$$\frac{\Gamma(j+1-\alpha)}{\Gamma(j+1)} = j^{-\alpha} + \frac{1}{\Gamma(\alpha)}\Phi_{1-\alpha}(j),$$

with

$$|\Phi_{1-\alpha}(j)| \leq \frac{\Gamma(1+\alpha)}{2} j^{-\alpha-1}.$$

Therefore,

$$\frac{j^{1-\alpha} - (j-1)^{1-\alpha}}{1-\alpha} - \frac{\Gamma(j+1-\alpha)}{\Gamma(j+1)} = \int_{t_{j-1}}^{t_j} \left(s^{-\alpha} - j^{-\alpha}\right) ds - \frac{1}{\Gamma(\alpha)}\Phi_{1-\alpha}(j)$$

and for every $j \geq 2$

$$\left|\frac{j^{1-\alpha} - (j-1)^{1-\alpha}}{1-\alpha} - \frac{\Gamma(j+1-\alpha)}{\Gamma(j+1)}\right| \leq (j-1)^{-\alpha} - j^{-\alpha} + \frac{1}{\Gamma(\alpha)}|\Phi_{1-\alpha}(j)|.$$

This leads to

$$\begin{aligned}
&\sum_{j=1}^{k-1}\left|\frac{j^{1-\alpha} - (j-1)^{1-\alpha}}{1-\alpha} - \frac{\Gamma(j+1-\alpha)}{\Gamma(j+1)}\right| \\
&\leq \frac{1}{1-\alpha} - \Gamma(2-\alpha) + 1 - (k-1)^{-\alpha} + \frac{\alpha}{2}\sum_{j=1}^{k-1} j^{-\alpha-1} \\
&\leq C,
\end{aligned}$$

with

$$C = \frac{1}{1-\alpha} - \Gamma(2-\alpha) + 1 + \frac{\alpha}{2}\zeta(1+\alpha) > 0,$$

where $\zeta(\cdot)$ is the Riemann zeta function.
Finally,

$$|S_1| \leq \frac{h^{1-\alpha}}{\Gamma(1-\alpha)} C \, \|f'\|_{\beta}.$$

Now we can put the pieces together to get

$$\frac{|\varphi(t_k) - \nabla_h^{\alpha} f(t_k)|}{h^{\beta}} \leq \left(1 + \frac{C}{\Gamma(1-\alpha)}\right) \|f'\|_{\beta}\, h^{1-\alpha-\beta} + \frac{2^{\beta}}{\Gamma(2-\alpha)}\omega_{\beta}(f', 2h).$$

The following theorem shows that the sequence $(A_h^\alpha)_h$ converges strongly to A^α.

Theorem 16.2 *Let X_β be the space $X_\beta = \left\{ f \in H^\beta \text{ such that } f' \in H_0^\beta \right\}$. Then for all β such that $1-\alpha-\beta>0$ the sequence $(A_h^\alpha)_h$ converges strongly to the fractional derivative A^α on X_β as h tends to 0.*

Proof 16.13 *For every $0 \leq x < y \leq 1$,*

$$\begin{aligned} &(A^\alpha - A_h^\alpha)(f)(x) - (A^\alpha - A_h^\alpha)(f)(y) \\ &= -\frac{\sin \pi\alpha}{\pi} \int_0^{+\infty} \lambda^{\alpha-1} \left(R(-\lambda, A)(Af) - R(-\lambda, A_h)(A_h f) \right)(x)\, d\lambda \\ &+ \frac{\sin \pi\alpha}{\pi} \int_0^{+\infty} \lambda^{\alpha-1} \left(R(-\lambda, A)(Af) - R(-\lambda, A_h)(A_h f) \right)(y)\, d\lambda, \end{aligned}$$

by introducing a mixed term we get

$$\begin{aligned} &(A^\alpha - A_h^\alpha)(f)(x) - (A^\alpha - A_h^\alpha)(f)(y) \\ &= -\frac{\sin \pi\alpha}{\pi} \int_0^{+\infty} \lambda^{\alpha-1} R(-\lambda, A) \left((Af - A_h f)(x) - (Af - A_h f)(y) \right) d\lambda \\ &+ \frac{\sin \pi\alpha}{\pi} \int_0^{+\infty} \lambda^{\alpha-1} \left(R(-\lambda, A) - R(-\lambda, A_h) \right) \left(A_h f(x) - (A_h f)(y) \right) d\lambda. \end{aligned}$$

Denote by $I_1(x,y)$ and $I_2(x,y)$ respectively the first and the second integral in the equality above.

We begin by estimating the first integral

$$\begin{aligned} &|I_1(x,y)| \\ &\leq \int_0^{+\infty} \lambda^{\alpha-1} \left| R(-\lambda, A) \left((Af - A_h f)(x) - (Af - A_h f)(y) \right) \right| d\lambda \\ &\leq \int_0^{+\infty} \lambda^{\alpha-1} \int_0^x e^{-\lambda t} \left| (Af - A_h f)(y-t) - (Af - A_h f)(x-t) \right| dt d\lambda \\ &+ \int_0^{+\infty} \lambda^{\alpha-1} \int_x^y e^{-\lambda t} \left| (Af - A_h f)(y-t) \right| dt d\lambda, \end{aligned}$$

which leads to

$$\frac{I_1(x,y)}{|x-y|^\beta} \leq \|Af - A_h f\|_\beta \int_0^{+\infty} \lambda^{\alpha-1} \left(\int_0^y e^{-\lambda t} dt \right) d\lambda.$$

The estimate

$$\frac{I_1(x,y)}{|x-y|^\beta} \leq \frac{6\Gamma(\alpha)}{1-\alpha} \omega_\beta(f', h),$$

follows from Fubini's theorem and inequality (equation 16.2).

Consider now the second integral.
First notice that

$$\begin{aligned}
&\frac{\sin \pi\alpha}{\pi}\int_0^{+\infty} \lambda^{\alpha-1} R(-\lambda, A)\, A_h f(x) d\lambda \\
&= \frac{\sin \pi\alpha}{\pi}\int_0^{+\infty} \lambda^{\alpha-1}\left(\int_0^x e^{-\lambda(x-t)} A_h f(t)\, dt\right) d\lambda \\
&= \frac{\sin \pi\alpha}{\pi}\int_0^{x}\left(\int_0^{+\infty} \lambda^{\alpha-1} e^{-\lambda(x-t)} d\lambda\right) A_h f(t)\, dt \\
&= \frac{1}{\Gamma(1-\alpha)}\int_0^x (x-t)^{-\alpha} A_h f(t)\, dt.
\end{aligned}$$

Then,

$$\begin{aligned}
&\frac{\sin \pi\alpha}{\pi}\int_0^{+\infty} \lambda^{\alpha-1}\left(R(-\lambda, A) - R(-\lambda, A_h)\right) A_h f(x)\, d\lambda \\
&= \frac{1}{\Gamma(1-\alpha)}\int_0^x (x-t)^{-\alpha} A_h f(t)\, dt - \mathcal{I}_h \nabla_h^{\alpha} f(x) \\
&= (r_h\varphi)(x) + \mathcal{I}_h\left(\varphi(x) - \nabla_h^{\alpha} f(x)\right).
\end{aligned}$$

From Lemmas 16.2 and 16.1, we have for some k and m

$$\frac{\sin \pi\alpha}{\pi}\frac{|I_2(x,y)|}{|x-y|^{\beta}} \leq 4\omega_{\beta}(\varphi, h) + \frac{|\varphi(t_k) - \nabla_h^{\alpha} f(t_k) - \varphi(t_m) + \nabla_h^{\alpha} f(t_m)|}{|t_k - t_m|^{\beta}}.$$

From Lemmas 16.5 and 16.6, we deduce

$$\begin{aligned}
\frac{\sin \pi\alpha}{\pi}\frac{|I_2(x,y)|}{|x-y|^{\beta}} \leq\ & 2\left(\frac{16}{\Gamma(2-\alpha)} + \frac{C}{\Gamma(1-\alpha)} + 1\right)\|f'\|_{\beta}\, h^{1-\alpha-\beta} \\
& + \frac{2^{\beta+1}}{\Gamma(2-\alpha)}\omega_{\beta}(f', 2h).
\end{aligned}$$

Hence,

$$\begin{aligned}
&\|(A^{\alpha} - \nabla_h^{\alpha})(f)\|_{\beta} \\
&\leq 2\left(\frac{16}{\Gamma(2-\alpha)} + \frac{C}{\Gamma(1-\alpha)} + 1\right)\|f'\|_{\beta}\, h^{1-\alpha-\beta} + \frac{2^{\beta+1}+6}{\Gamma(2-\alpha)}\omega_{\beta}(f', 2h)
\end{aligned}$$

and the conclusion of the theorem holds.

16.5 NUMERICAL EXAMPLES

In this section, two examples are discussed.

Example 16.1

Consider the fractional derivative of $f(x) = x^{\mu} \ln x$. The analytical expression of the fractional derivative of f is

$$A^{\alpha} f(x) = \frac{\Gamma(\mu+1)}{\Gamma(\mu+1-\alpha)} x^{\mu-\alpha} \left[\ln x + \psi(\mu+1) - \psi(\mu+1-\alpha)\right],$$

where $\psi(\cdot)$ denote the digamma function see [26, Formula (103)].
In the next tables, error at the step size h is the Hölderian error defined by

$$\max_{0 \leq i < j \leq 1/h} \frac{\left|\left(A^{\alpha} f - \nabla_h^{\alpha} f\right)(t_i) - \left(A^{\alpha} f - \nabla_h^{\alpha} f\right)(t_j)\right|}{|t_i - t_j|^{\beta}}. \tag{16.8}$$

According to our theoretical consideration, the convergence is ensured by $h^{1-\alpha-\beta}$ and $\omega_{\beta}(f', 2h)$.
Let us establish an estimation of $\omega_{\beta}(f', h)$.
For every $0 \leq x < y \leq 1$,

$$f'(y) - f'(x) = \mu(y^{\mu-1} \ln y - x^{\mu-1} \ln x) + y^{\mu-1} - x^{\mu-1}.$$

Using the fact that

$$\begin{aligned} y^{\mu-1} \ln y - x^{\mu-1} \ln x &= \int_x^y \frac{d}{dt}(t^{\mu-1} \ln t) dt \\ &= (\mu-1) \int_x^y t^{\mu-2} \ln t dt + \frac{1}{\mu-1}\left(y^{\mu-1} - x^{\mu-1}\right), \end{aligned}$$

then for every $1 + \beta < \beta' < \mu$, we have

$$\int_x^y t^{\mu-2} \ln t dt = \int_x^y t^{\mu-\beta'+\beta'-2} \ln t dt.$$

Setting $M = \max_{t \in]0,1]} \left|t^{\mu-\beta'} \ln t\right|$,
we have

$$\left|\int_x^y t^{\mu-2} \ln t dt\right| \leq M \int_x^y t^{\beta'-2} dt = \frac{M}{\beta'-1}\left(y^{\beta'-1} - x^{\beta'-1}\right).$$

Therefore,

$$|f'(y) - f'(x)| \leq \frac{M\mu(\mu-1)}{\beta'-1}\left(y^{\beta'-1} - x^{\beta'-1}\right) + \left(\frac{\mu}{\mu-1} + 1\right)\left(y^{\mu-1} - x^{\mu-1}\right).$$

It follows that

$$\frac{|f'(y) - f'(x)|}{|y-x|^{\beta}} \leq \frac{M\mu(\mu-1)}{\beta'-1} |y-x|^{\beta'-1-\beta} + \left(\frac{\mu}{\mu-1} + 1\right) |y-x|^{\mu-1-\beta}$$

$$\leq \left(\frac{M\mu(\mu-1)}{\beta'-1} + \left(\frac{\mu}{\mu-1}+1\right)|y-x|^{\mu-\beta'}\right)|y-x|^{\beta'-1-\beta}$$

If $|y-x| \leq h$ then

$$\frac{|f'(y)-f'(x)|}{|y-x|^{\beta}} \leq \left(\frac{M\mu(\mu-1)}{\beta'-1} + \left(\frac{\mu}{\mu-1}+1\right)h^{\mu-\beta'}\right)h^{\beta'-1-\beta},$$

and

$$\omega_{\beta}(f',h) \leq \left(\frac{M\mu(\mu-1)}{\beta'-1} + \left(\frac{\mu}{\mu-1}+1\right)h^{\mu-\beta'}\right)h^{\mu-1-\beta}.$$

In case $\mu = 3/2, \alpha = 0.3, \beta = 0.1$, the convergence is ensured since

$$\mu - 1 - \beta = 0.4 > 0.$$

The results concerning errors are presented in Table 21.1 for $\mu = 3/2$, $\alpha = 0.3$ *and* $\beta = 0.1$.

TABLE 16.1 Error Defined by Equation (16.8) at Different Values h, When $f(x) = x^{\mu}\ln x$ for $\mu = 3/2$, $\alpha = 0.3$ and $\beta = 0.1$

h	**Error**
2^{-6}	0.0079082
2^{-7}	0.0040833
2^{-8}	0.0021392
2^{-9}	0.0011478
2^{-10}	0.0006054
2^{-11}	0.0003150
2^{-12}	0.0001622

In Figure 16.1, on the left, the graphs of $A^{\alpha}f$ and $A^{\alpha}_{h}f$ are shown. On the right, we give the Hölderian errors.

Example 16.2

For the second example, we consider the fractional differential equation presented in Ref. [9], for $t \in [0,1]$.

$$\begin{aligned} D^{\alpha}y(t) &= \frac{40320}{\Gamma(9-\alpha)}t^{8-\alpha} - 3\frac{\Gamma(5+\alpha/2)}{\Gamma(5-\alpha/2)}t^{4-\frac{\alpha}{2}} + \frac{9}{4}\Gamma(\alpha+1) \\ &+ \left(\frac{3}{2}t^{\frac{\alpha}{2}} - t^{4}\right)^{3} - [y(t)]^{\frac{3}{2}}. \end{aligned} \tag{16.9}$$

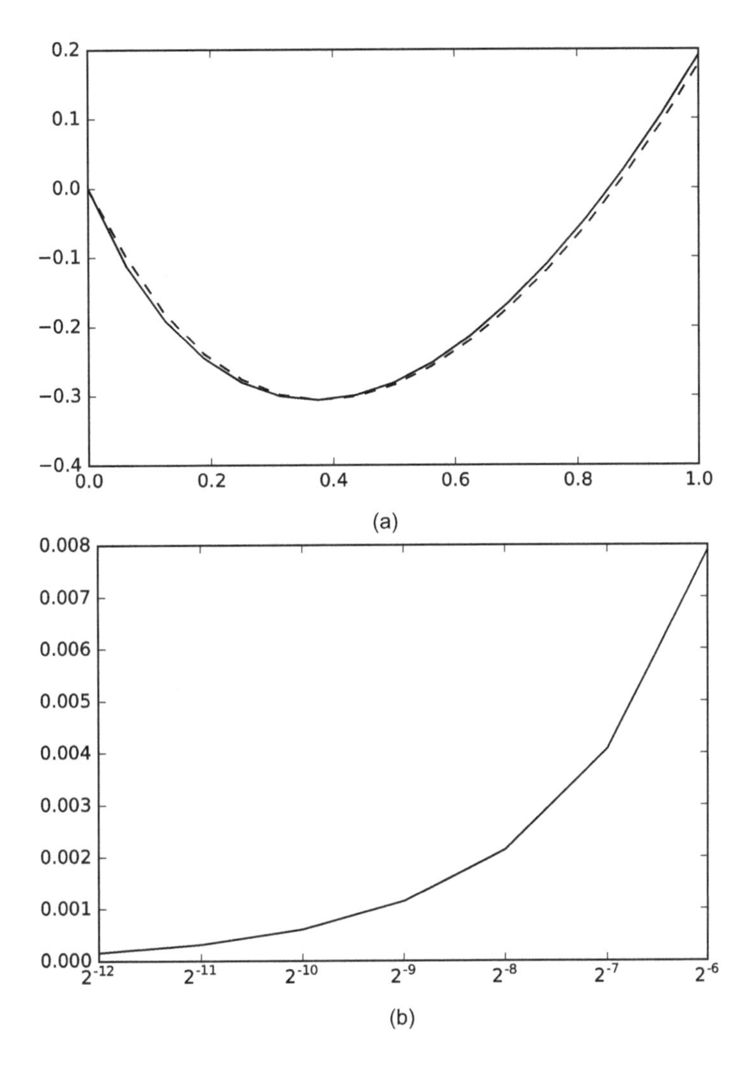

Figure 16.1 Comparison between $A^{\alpha}f$ and $A_h^{\alpha}f$ when $f(x) = x^{\mu}\ln x$ for $\mu = 3/2, \alpha = 0.3$ and $\beta = 0.1$. (a) $A^{\alpha}f$ in the continuous line, $A_h^{\alpha}f$ in dotted line for $h = 2^{-4}$. Hölderian error with respect to step size h.

The initial condition is $y(0) = 0$. The exact solution of this problem is

$$y(t) = t^8 - 3t^{4+\frac{\alpha}{2}} + \frac{9}{4}t^{\alpha}.$$

For $\alpha = 0.5$, we display the results in Table 21.2 for $\beta = 0.1$ and $\beta = 0.01$ respectively. Apparently, we need to use small values for β to increase the accuracy.

16.6 CONCLUSION

Throughout this chapter, we have defined a fractional operator as a fractional power of a piecewise linear interpolation of a backward difference on a Hölder space. We

TABLE 16.2 Hölderian Errors for Problem (16.9) with $\alpha = 0.5$

h	**Errors for** $\beta = 0.1$	**Errors for** $\beta = 0.01$
2^{-7}	0.0347581	0.0224598
2^{-8}	0.0269360	0.0163528
2^{-9}	0.0206910	0.0118018
2^{-10}	0.0158085	0.0084716
2^{-11}	0.0120388	0.0060613
2^{-12}	0.0091502	0.0043283
2^{-13}	0.0069465	0.0030872

have proved the strong convergence of this operator to fractional derivative, and we have supported our results with examples.

We think that we have now a kind of process to define Euler-like formulas which contribute to solve numerically fractional differential equations in Hölder spaces. In our approach, there is in addition a freedom degree in the choice of the space X_β , where β have to fulfill only the condition $0 < \beta < 1 - \alpha$. However, several questions can be the subject of further works. In particular, the analysis of the order of approximation, and what results can be expected if one replaces the linear spline $\mathcal{I}_h$ by a spline of higher degree, or if one replaces the operator ∇_h by another more accurate approximation of the derivative.

Note

[1]This work has been deposited in hal-archives-ouvertes as a preprint entitled *Hölderian convergence of fractional extended nabla operator to fractional derivative*, see the following link https://hal.archives-ouvertes.fr/hal-01824977/document.

BIBLIOGRAPHY

[1] T. Abdeljawad, On Riemann and Caputo fractional differences, *Comput. Math. Appl.* 62 (2011), 1602–1611. https://doi.org/10.1016/j.camwa.2011.03.036.

[2] T. Abdeljawad and F.M. Atici, On the definitions of nabla fractional operators, *Abstr. Appl. Anal.* 2012 (2012), 1–13. http://dx.doi.org/10.1155/2012/406757.

[3] M. Abramowitz and I.A. Stegun, *Handbook of Mathematical Functions with Formulas, Graphs and Mathematical Tables*, National Bureau of Standards: Gaithersburg, MD (1964).

[4] A. Ashyralyev, A note on fractional derivatives and fractional powers of operators, *J. Math. Anal. Appl.* 357 (2009), 232–236. https://doi.org/10.1016/j.jmaa.2009.04.012.

[5] H. Chen, F. Holland, and M. Stynes, An analysis of the Grünwald–Letnikov scheme for initial-value problems with weakly singular solutions, *Appl. Numer. Math.* 139 (2019), 52–61. https://doi.org/10.1016/j.apnum.2019.01.004.

[6] D. Cao and H. Chen, Sharp error estimate of Grünwald-Letnikov scheme for a multi-term time fractional diffusion equation, *Adv. Comput. Math.* 48 (2022), 82. https://doi.org/10.1007/s10444-022-09999-3.

[7] S. Das, *Functional Fractional Calculus*, Springer: Berlin/Heidelberg, Germany (2011).

[8] P. Debnath, H.M. Srivastava, P. Kumam, and B. Hazarika, *Fixed Point Theory and Fractional Calculus: Recent Advances and Applications*, Springer: Singapore (2022).

[9] K. Diethelm, N.J. Ford, and A.D. Freed, Detailed error analysis for fractional Adams method, *Numer. Algorithms* 36 (2004), 31–52.

[10] K. Diethelm, *The Analysis of Fractional Differential Equations, An Application-Oriented Exposition Using Differential Operators of Caputo Type*, Lecture Notes in Mathematics, Springer: Berlin/Heidelberg, Germany (2010).

[11] N. Elezović, L. Lin, and L. Vukšić, Inequalities and asymptotic expansions of the Wallis sequence and the sum of the Wallis ratio. *J. Math. Inequal.* 7(4) (2013), 679–695.

[12] R. Garrappa, Numerical solution of fractional differential equations: A survey and a software tutorial, *Mathematics* 6(2) (2018), 16. https://doi.org/10.3390/math6020016.

[13] M. Haase, *The Functional Calculus for Sectorial Operators, Operator Theory: Advances and Applications*, Birkhäuser: Basel (2006).

[14] R. Hilfer, *Applications of Fractional Calculus in Physics*, World Scientific Publishing Co: Singapore (2000).

[15] K. Ito and F. Kappel, *Evolutions Equations and Approximations*, World Scientific Publishing Co: Singapore (2002).

[16] L. Khitri-Kazi-Tani and H. Dib, A new h-discrete fractional operator, fractional power and finite summation of hypergeometric polynomials, *Mem. Differ. Equations Math. Phys.* 86 (2022), 85–96.

[17] L. Liu, Z. Fan, G. Li, and S. Piskarev, Discrete almost maximal regularity and stability for fractional differential equations in $L^p([0,1],\Omega)$, *Appl. Math. Comput.* 389 (2021). https://doi.org/10.1016/j.amc.2020.125574.

[18] C. Martinez and M. Sanz, *The Theory of Fractional Powers of Operators*, vol. 187, North Holland: Amsterdam, Netherlands (2001).

[19] D. Mozyrska and E. Girejko, Overview of fractional h-difference operators, In: A.Y. Karlovich, L. Castro, and M. Amelia Bastos (Eds.), *Operator Theory: Advances and Applications*, vol. 229, pp. 253–267, Birkhäuser (2013). https://doi.org/10.1007/978-3-0348-0516-2_14.

[20] M.D. Ortigueira, *Fractional Calculus for Scientists and Engineers*, Lecture Notes in Electrical Engineering, vol. 84, Springer: Berlin/Heidelberg, Germany (2011).

[21] A. Račkauskas and C. Suquet, Functional laws of large numbers in Hölder spaces, *ALEA, Lat. Am. J. Probab. Math. Stat.* 10(2) (2013), 609–624

[22] S.G. Samko, A.A. Kilbas, and O.I. Marichev, *Fractional Integrals and Derivatives: Theory and Applications*, Gordon and Breach Science Publishers: Philadelphia, PA (1993).

[23] S.G. Samko and Z.U. Mussalaeva, Fractional type operators in weighted generalized Hölder spaces, *Georgian Math. J.* 1 (1994), 537–559.

[24] E. Sousa, Consistency analysis of the Grünwald-Letnikov approximation in a bounded domain, *IMA J. Numer. Anal.* 42 (2022), 2771–2793. https://doi.org/10.1093/imanum/drab051.

[25] E. Sousa, The convergence rate for difference approximations to fractional boundary value problems, *J. Comput. Appl. Math.* 415 (2022). https://doi.org/10.1016/j.cam.2022.114486.

[26] D. Valério, J.J. Trujillo, M. Rivero, J.A.T. Machado, and D. Baleanu, Fractional calculus: A survey of useful formulas, *Eur. Phys. J. Special Topics* 222 (2013),1827–1846. https://doi.org/10.1140/epjst/e2013-01967-y.

[27] J.G. Wendel, Note on the gamma function. *Am. Math. Mon.* 55 (1948), 563–564.

[28] H.E. White, Functions with a concave modulus of continuity, *Proc. Am. Math. Soc.* 42(1) (1974), 104–112.

[29] Y. Zhou, *Basic Theory of Fractional Differential Equations*, World Scientific Publishing Co. Pte. Ltd.: Singapore (2014).

CHAPTER 17

Stability Analysis of Fractional Nonlinear Dynamical Systems

Priyadharsini Sivaraj

Sri Krishna Arts and Science College

CONTENTS

17.1 INTRODUCTION

A number of academic fields and engineering applications use fractional differential equations (FDE). It has been found that the fractional order theory is a powerful tool for describing the behaviors of various physical systems. In actuality, systems with fractional order are typically or most often used in real-world processes. For instance, from an aircraft, we may view city streets and track the flow of traffic. The car appears to be traveling straight ahead. As a result, as observers, we construct the velocity curve using a straightforward first-order integer displacement derivative and discover that it maps to a straight line [1].

DOI: 10.1201/9781003388678-17

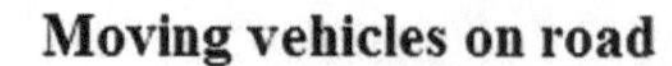

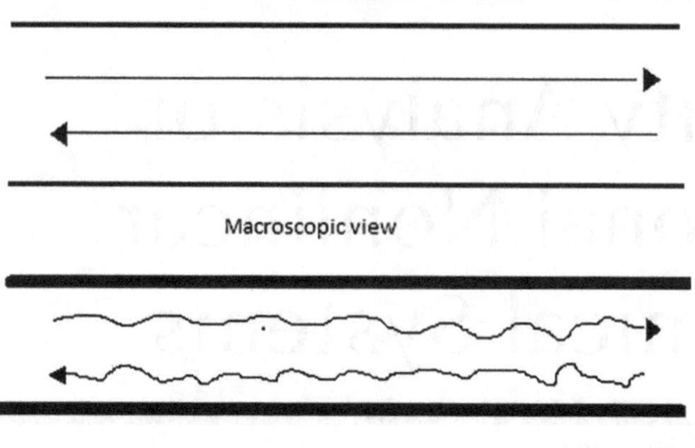

The pair of straight lines in the previous illustration show the velocity trajectory of the upstream and downstream vehicles, as seen at a macro-scale. When the same vehicle is seen in greater detail, it moves in a zigzag pattern. The speeds of upstream and downstream vehicles follow a continuous, non-differentiable curve rather than a pair of straight lines. Here the quantity $d^{1+\alpha}y/dx^{1+\alpha}$, gives the pattern of real meander. The quantity dy/dx is velocity, and d^2y/dx^2 is the acceleration; however, $d^{1.23}y/dx^{1.23}$ is difficult to picture. The nature of meander is referred to as a fractal curve; however, it is actually a continuous function that is not differentiable everywhere. Fractional calculus can explain discontinuity and singularity formation. One can say that nature itself uses fractional derivatives.

17.1.1 Birth of Fractional Calculus

The concept of fractional calculus is widely thought to have originated from a query asked by L'Hopital to Leibeniz [5]. What is the value of d^ny/dx^n when $n = 1/2$?. Although fractional derivatives have a lengthy mathematical history, physics did not utilize them for a very long time.

There exists a long gap because of numerous non-equivalent definitions of fractional derivatives. Another reason for its unpopularity is that non-local nature, fractional derivatives failed to give a clear geometrical interpretation. However, over the past ten years, mathematicians and physicists have begun to pay much greater attention to fractional calculus. The primary impediment to solving fractional differential equations was the use of integer-order models.

17.1.2 Motivation

This field has been a matter of study, because of its applications in various fields, including fluid mechanics, viscoelasticity, electric conductance of biological systems, fractional order model of neurons, electroanalytical chemistry, electrode- electrolyte interface, physics, engineering, etc. [11]. One may notice that the differential operator of integer order is a local one. While the fractional differential operator is non-local one, that is, a system's next state depends on all of its previous states as well as its current state. The FDE's non-local aspect is their main benefit. The distinctive characteristics of FDE have drawn researchers to this field of study. It is worth mentioning that many methods [2, 7, 12, 13, 14, 15] are broadly available, which enrich as the concept of asymptotic stability.

This chapter is outlined as shown. In Section 17.2, some basic definitions along with preliminaries are stated. Various nonlinear systems and its stability analysis are developed in Section 17.3, whereas the examples appear in Section 17.4. This chapter ends with conclusion.

17.2 PRELIMINARIES

This section contains basic definitions and standard results [3, 6].

Definition 17.1 Riemann - Liouville Fractional Integral [6]

$$I^{\alpha}y(x) = \frac{1}{\Gamma(\alpha)}\int_0^x (x-s)^{\alpha-1}y(s)\mathrm{d}s, \tag{17.1}$$

where $y \in L^1(\mathcal{R}^+), \alpha > 0.$

Definition 17.2 Riemann - Liouville Fractional Derivative [6]

$$\begin{aligned} D^{\alpha}y(x) &= D^n I^{n-\alpha}y(x) \\ &= \frac{1}{\Gamma(n-\alpha)}\left(\frac{d}{dx}\right)^n \int_0^x (x-s)^{n-\alpha-1}y(s)\mathrm{d}s, \end{aligned} \tag{17.2}$$

where $\alpha > 0, n-1 < \alpha < n, n \in N, y^{n-1}(x) \in \mathbb{AB}(\mathcal{R}^+).$

Definition 17.3 Caputo Fractional Derivative [6]

$${}^{C}D^{\alpha}y(x) = I^{n-\alpha}y^n(x) = \frac{1}{\Gamma(n-\alpha)}\int_0^x (x-s)^{n-\alpha-1}y^n(s)\mathrm{d}s, \tag{17.3}$$

where $\alpha > 0, n-1 < \alpha < n, n \in N,\ y^{n-1}(x) \in \mathbb{AB}(\mathcal{R}^+).$

$${}^{C}D^{\alpha}y(x) = I^{1-\alpha}y'(x) = \frac{1}{\Gamma(1-\alpha)}\int_0^x (x-s)^{-\alpha}y'(s)\mathrm{d}s, \tag{17.4}$$

Property 17.1 Relation between Caputo and Riemann-Liouville

$$\begin{aligned} D^{\alpha}y(x) &= {}^{C}D^{\alpha}y(x) + \sum_{k=0}^{n-1}\frac{y^{(k)}(0)}{\Gamma(k-\alpha+1)}x^{k-\alpha}, \qquad (n-1 < \alpha \le n), \\ D^{\alpha}y(x) &= {}^{C}D^{\alpha}y(x) + \frac{y(0)}{\Gamma(1-\alpha)}x^{-\alpha}, \qquad (0 < \alpha < 1). \end{aligned}$$

Property 17.2 Linearity Property

$$I^{\alpha}(ay_1(x)+by_2(x)) = aI^{\alpha}y_1(x)+bI^{\alpha}y_2(x),$$
$${}^{C}D^{\alpha}(ay_1(x)+by_1(x)) = a^{C}D^{\alpha}y_1(x)+b^{C}D^{\alpha}y_2(x), \textit{ where } a \textit{ and } b \textit{ are constants.}$$

Property 17.3 Semi Group Property

$$\begin{aligned} I^{\alpha}I^{\beta}y(x) &= I^{\beta}I^{\alpha}y(x) = I^{\alpha+\beta}y(x), \\ {}^{C}D^{\alpha C}D^{n}y(x) &= {}^{C}D^{nC}D^{\alpha}y(x) = {}^{C}D^{\alpha+n}y(x), \quad n \in \mathrm{N}, \\ {}^{C}D^{\alpha C}D^{\beta}y(x) &\neq {}^{C}D^{\beta C}D^{\alpha}y(x) \neq {}^{C}D^{\alpha+\beta}y(x), \textit{ where } \alpha>0, \beta>0. \end{aligned}$$

Property 17.4 Fractional Leibniz Formula

$$\begin{aligned} {}^{C}D^{\alpha}I^{\alpha}y(x) &= y(x), \quad \textit{where} \quad 0<\alpha\leq 1, \\ I^{\alpha C}D^{\alpha}y(x) &= y(x)-y(0), \quad \textit{where} \quad 0<\alpha\leq 1, \\ I^{\alpha C}D^{\beta}y(x) &= {}^{C}D^{\beta}I^{\alpha}y(x) = {}^{C}D^{\beta-\alpha}y(x) \quad \textit{with } \alpha,\beta \geq 0, \\ I^{\alpha C}D^{\beta}y(x) &= y(x)-\sum_{k=0}^{m-1} y^{(k)}(0)\frac{x^k}{k!}, \quad x>0,\ m-1<\alpha\leq m. \end{aligned}$$

Property 17.5 Laplace Transform

$$\begin{aligned} \mathrm{L}\{I^{\alpha}y(x)\} &= s^{-\alpha}Y(s) \quad \textit{where } \mathrm{R}(s), \mathrm{R}(\alpha)>0, \\ \mathrm{L}\{{}^{C}D^{\alpha}y(x)\} &= s^{\alpha}Y(s)-\sum_{k=0}^{n-1} s^{\alpha-k-1}y^{(k)}(0), \qquad (n-1<\alpha\leq n). \end{aligned}$$

Definition 17.4 The Queen Function of Fractional Calculus [6] *The Mittag-Leffler functions in two parameter is defined by*

$$E_{\alpha,\beta}(\theta) = \sum_{k=0}^{\infty}\frac{\theta^k}{\Gamma(\alpha k+\beta)}, (\alpha,\beta>0, \theta\in\mathbb{C}). \tag{17.5}$$

17.2.1 Laplace Transform

$$\mathcal{L}\{E_{\alpha}(-\lambda\theta^{\alpha})\} = \frac{s^{\alpha-1}}{s^{\alpha}+\lambda}, \quad \mathcal{L}\{\theta^{\beta-1}E_{\alpha,\beta}(-\lambda\theta^{\alpha})\} = \frac{s^{\alpha-\beta}}{s^{\alpha}+\lambda},$$

$$\mathcal{L}\{\theta^{\alpha-1}E_{\alpha,\alpha}(-\lambda\theta^{\alpha})\} = \frac{s^{\alpha-\alpha}}{s^{\alpha}+\lambda},$$

$$\mathcal{L}\{\theta^{\alpha k+\beta-1}E^{(k)}_{\alpha,\beta}(-\lambda\theta^{\alpha})\} = \frac{k!s^{\alpha-\beta}}{(s^{\alpha}+\lambda)^{k+1}},$$

where $\theta \geq 0, \lambda \in \mathbb{R}, \ (\mathcal{R}(s) > |\lambda|^{\frac{1}{\alpha}}),$

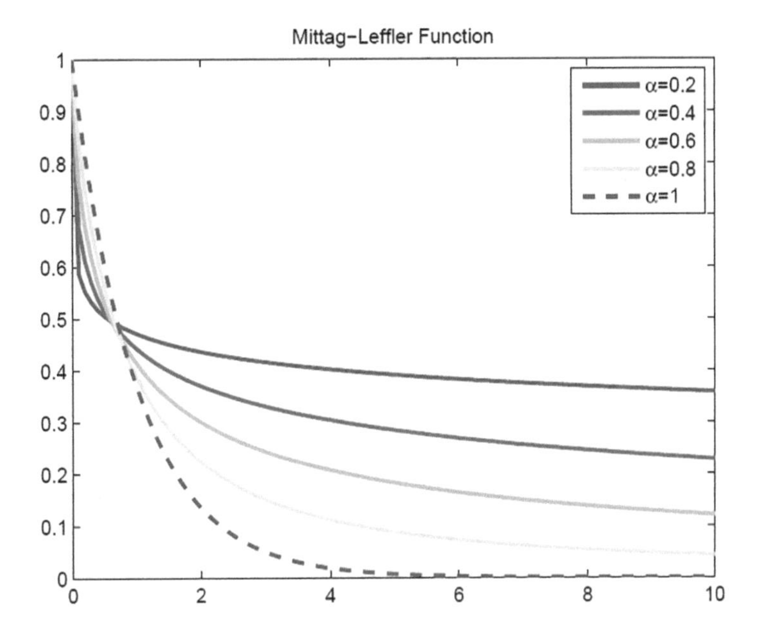

17.3 STABILITY ANALYSIS OF SOME SPECIAL NONLINEAR FRACTIONAL DIFFERENTIAL SYSTEMS

Stability of some special nonlinear fractional dynamical systems was analyzed in this section. It could be observed that it depends on the spectrum of linear system. So, Take the fractional equation into consideration [10]

$$^{C}D^{\alpha}y(x) = Ay(x), \tag{17.6}$$

with initial value $y(0) = y_0$ where $y \in \mathbb{R}^n, 0 < \alpha < 1$ and $A \in \mathbb{R}^{n\times n}$ whose solution is given by

$$y(x) = E_{\alpha}(Ax^{\alpha})y_0.$$

Definition 17.5 *The autonomous linear system (equation 22.1) is said to be*

(i) stable iff for any x_0, there exists $\epsilon > 0$ such that $||y(x)|| \leq \epsilon$ for $x \geq 0$,

(ii) asymptotically stable iff $\lim_{x\to\infty} ||y(x)|| = 0$.

Theorem 17.1 *The autonomous system (equation 22.1) is asymptotically stable iff*

$$|\arg(spc(A))| > \frac{\alpha\pi}{2}.$$

17.3.1 Stability of Fractional Nonlinear System

Take the nonlinear system into consideration [8]

$$^{C}D^{\alpha}y(x) = Ay(x) + f(x, y(x)), \quad 0 < \alpha < 1, \tag{17.7}$$

where $f(x,y) \in C(\mathbb{R} \times \mathbb{R}^n, \mathbb{R}^n), f(x,0) = 0$, and the initial condition is given by $y(0) = y_0$, where $y = (y_1, \ldots, y_n)^T$, and $A \in \mathbb{R}^{n \times n}$.
To prove the main theorem we need the following lemmas.

Lemma 17.1 *Whenever the condition is satisfied*

$$\frac{\pi\alpha}{2} < \mu < \min\{\pi, \pi\alpha\}, \mu \leq |\arg(z)| \leq \pi,$$

we have

$$E_{\alpha,\beta}(z) = -\sum_{k=1}^{p} \frac{z^{-k}}{\Gamma(\beta - \alpha k)} + O(|z|^{-1-p}), \tag{17.8}$$

for $p \geq 1, 0 < \alpha < 2, \beta$ *and* $|z| \to \infty$.

Lemma 17.2 Gronwall Inequality
Suppose that $f(x)$ *and* $g(x)$ *are continuous in* $[x_0, x_1], f(x) \geq 0, \lambda \geq 0$ *and* $r \geq 0$ *are two constants. If*

$$g(x) \leq \lambda + \int_{x_0}^{x} [f(t)g(t) + r]\mathrm{d}t, \tag{17.9}$$

then

$$g(x) \leq (\lambda + r(x_1 - x_0)) \exp\left(\int_{x_0}^{x} f(t))\mathrm{d}t\right), x_0 \leq x \leq x_1. \tag{17.10}$$

Lemma 17.3 Spectrum Condition*[12]*
If every eigenvalue of A meet the requirement

$$|arg(spc(A))| > \frac{\alpha\pi}{2}, \tag{17.11}$$

then there is a constant $K > 0$ *satisfies,*

$$\int_0^t ||\theta^{\alpha-1} E_{\alpha,\alpha}(A\theta^\alpha)||\mathrm{d}\theta \leq K. \tag{17.12}$$

Theorem 17.2 *Suppose* $||f(x, y(x))|| \leq M||y||$ *and every eigenvalue of A meet the requirement (equation 17.11). Then, there exists a asymptotically stable solution of equation* (17.7).

Proof
The solution representation of the system (equation 17.7) is given by

$$y(x) = E_\alpha(Ax^\alpha)y_0 + \int_0^x (x-\theta)^{\alpha-1} E_{\alpha,\alpha}(A(x-\theta)^\alpha) f(\theta, y(\theta))\mathrm{d}\theta. \tag{17.13}$$

From which it follows that,

$$||y(x)|| \leq ||E_\alpha(Ax^\alpha)y_0||$$

$$
\begin{aligned}
&+\int_0^x ||(x-\theta)^{\alpha-1}E_{\alpha,\alpha}(A(x-\theta)^\alpha)||\ ||f(\theta,y(\theta))||\mathrm{d}\theta \\
\leq\ & ||E_\alpha(Ax^\alpha)y_0|| \\
&+\int_0^x ||(\theta)^{\alpha-1}E_{\alpha,\alpha}(A(\theta)^\alpha)||\ ||f(x-\theta,y(x-\theta))||\mathrm{d}\theta \\
\leq\ & ||E_\alpha(Ax^\alpha)y_0|| + M\int_0^x ||(\theta)^{\alpha-1}E_{\alpha,\alpha}(A(\theta)^\alpha)||\ ||y||\mathrm{d}\theta.
\end{aligned}
$$

From Gronwall's inequality, we have

$$||y(x)|| \leq ||E_\alpha(Ax^\alpha)y_0|| \exp\left\{M\int_0^x ||(\theta)^{\alpha-1}E_{\alpha,\alpha}(A(\theta)^\alpha)||\mathrm{d}\theta\right\}.$$

By using Lemma 17.3, we have

$$\exp\left\{M\int_0^x ||(\theta)^{\alpha-1}E_{\alpha,\alpha}(A(\theta)^\alpha)||\mathrm{d}\theta\right\} \text{ is bounded.} \tag{17.14}$$

Further, by using Lemma 17.1, we have $||E_\alpha(Ax^\alpha)y_0|| \to 0$ as $x \to \infty$. Thereby, $\lim_{x\to\infty} y(x) \to 0$. That is, there exists an asymptotically stable solution of equation 17.7.

17.3.2 Stability of Fractional Neutral Differential Equations

Take the nonlinear neutral differential system into consideration [9]

$${}^CD^\alpha[y(x) - g(x,y(x))] = Ay(x) + f(x,y(x)), \quad 0<\alpha<1, \tag{17.15}$$

where $f(x,y)$, $g(x,y) \in \mathbb{C}(J\times\mathbb{R}^n,\mathbb{R}^n)$, $y(0) = y_0$, $f(x,0) = 0$, $g(0,y_0) \neq y_0$, $y = (y_1,\ldots,y_n)^T$, and $A \in \mathbb{R}^{n\times n}$.

Theorem 17.3 *Let $f(x,y(x))$ and $g(x,y(x))$ satisfies the condition*

$$
\begin{aligned}
||f(x,y(x))|| &\leq M_1||y||, \\
||g(x,y(x))|| &\leq M_2||y||,
\end{aligned} \tag{17.16}
$$

with $M_2 \neq 1$ and every eigenvalue of A meets the requirement of spectral condition (17.11). Then, there exists an asymptotically stable solution for the given system (equation 17.15).

Proof

The standard form of given equation (17.15) is

$${}^CD^\alpha y(x) = Ay(x) + f(x,y(x)) + {}^CD^\alpha g(x,y(x)).$$

The solution representation of the given system is,

$$y(x) \ = \ E_\alpha(A(x)^\alpha)x_0 + \int_0^x (x-s)^{\alpha-1}E_{\alpha,\alpha}(A(x-s)^\alpha)$$

$$\begin{aligned}
& \{f(s,y(s)) +^C D^\alpha g(s,y(s))\}\mathrm{d}s \\
= \; & E_\alpha(Ax^\alpha)y_0 + \int_0^x (x-s)^{\alpha-1}E_{\alpha,\alpha}(A(x-s)^\alpha)f(s,y(s))\mathrm{d}s \\
& + \int_0^x (x-s)^{\alpha-1}E_{\alpha,\alpha}(A(x-s)^\alpha)^C D^\alpha g(s,y(s))\mathrm{d}s \\
= \; & T_1 + T_2.
\end{aligned}$$

Let evaluate T_2,

$$\begin{aligned}
T_2 \; = \; & \int_0^x (x-s)^{\alpha-1}E_{\alpha,\alpha}(A(x-s)^\alpha)^C D^\alpha g(s,y(s))\mathrm{d}s \\
= \; & \frac{1}{\Gamma(1-\alpha)}\int_0^x\int_0^s (x-s)^{\alpha-1}(s-\tau)^{-\alpha} \\
& \cdot E_{\alpha,\alpha}(A(x-s)^\alpha)g'(\tau,y(\tau))\mathrm{d}\tau\mathrm{d}s \\
= \; & \frac{1}{\Gamma(1-\alpha)}\int_0^x g'(\tau,y(\tau))\sum_{k=0}^{\infty}\frac{A^k}{\Gamma(\alpha k+\alpha)} \\
& \cdot\int_\tau^x (x-s)^{\alpha k+\alpha-1}(s-\tau)^{-\alpha}\mathrm{d}s\mathrm{d}\tau \\
= \; & \frac{1}{\Gamma(1-\alpha)}\int_0^x E_\alpha(A(x-\tau)^\alpha)g'(\tau,y(\tau))\mathrm{d}\tau, \\
= \; & g(x,y(x)) - E_\alpha(Ax^\alpha)g(0,y_0) \\
& + A\int_0^x (x-\tau)^{\alpha-1}E_{\alpha,\alpha}(A(x-\tau)^\alpha)g(\tau,y(\tau))\mathrm{d}\tau.
\end{aligned}$$

Hence,

$$\begin{aligned}
y(x) \; = \; & E_\alpha(As^\alpha)\{y_0 - g(0,y_0)\} + g(x,y(x))) \\
& + \int_0^x (x-s)^{\alpha-1}E_{\alpha,\alpha}(A(x-s)^\alpha)\{f(s,y(s)) + Ag(s,y(s))\}\mathrm{d}\tau.
\end{aligned}$$

From this,

$$\begin{aligned}
||y(x)|| \; \leq \; & ||E_\alpha(Ax^\alpha)(y_0 - g(0,y_0))|| + ||g(x,y(x))|| \\
& + \int_0^x ||(x-s)^{\alpha-1}E_{\alpha,\alpha}(A(x-s)^\alpha)|| \; ||f(s,y(s)) \\
& + Ag(s,y(s))||\mathrm{d}\tau \\
\leq \; & ||E_\alpha(Ax^\alpha)(y_0 - g(0,y_0))|| + M_2||y|| \\
& + \{M_1 + ||A||M_2\}\int_0^x ||(x-s)^{\alpha-1}E_{\alpha,\alpha}(A(x-s)^\alpha)|| \; ||y||\mathrm{d}\tau \\
\leq \; & ||E_\alpha(Ax^\alpha)\,\frac{(y_0 - g(0,y_0))}{1-M_2}|| \\
& + \left\{\frac{M_1 + ||A||M_2}{1-M_2}\right\}\int_0^x ||(x-s)^{\alpha-1}E_{\alpha,\alpha}(A(x-s)^\alpha)|| \; ||y||\mathrm{d}\tau.
\end{aligned}$$

By using Lemma 17.2,

$$||y(x)|| \leq ||E_\alpha(Ax^\alpha)C_1|| \exp\{C_2 \int_0^x ||(x-s)^{\alpha-1}E_{\alpha,\alpha}(A(x-s)^\alpha)||d\tau\},$$

where $C_1 = \dfrac{(y_0 - g(0, y_0))}{1 - M_2}$, $C_2 = \dfrac{M_1 + ||A||M_2}{1 - M_2}$.

We can prove that $\lim\limits_{x\to\infty} y(x) \to 0$. Then, there exists an asymptotically stable solution for the given system (equation 17.15).

17.3.3 Stability of Fractional Langevin Differential Equations

Take the nonlinear fractional Langevin system [4] into consideration

$$\begin{aligned} &{}^C D^\beta ({}^C D^\alpha - A) y(x) = f(x, y(x)), \quad 0 < \alpha, \beta \leq 1, \quad x \in [0, \infty), \\ &y(0) = y_0, \quad {}^C D^\alpha y(x)|_{x=0} = y_1, \end{aligned} \tag{17.17}$$

Theorem 17.4 *If $||f(x, y(x))|| \leq M||y||$, for some $M > 0$ and every eigenvalue of A meet the requirement of spectral condition $|\arg(spc(A))| > \dfrac{\alpha\pi}{2}$, then there exists an asymptotically stable solution of system (equation 17.17).*

Proof

The solution representation of equation (17.17) is

$$\begin{aligned} y(x) &= E_\alpha(A(x)^\alpha)y_0 + x^\alpha E_{\alpha,\alpha+1}(Ax^\alpha)(Iy_1 - Ay_0) \\ &\quad + \int_0^x (x-s)^{\alpha+\beta-1} E_{\alpha,\alpha+\beta}(A(x-s)^\alpha) f(s, y(s)) ds. \end{aligned}$$

Thus

$$\begin{aligned} ||y(x)|| &\leq ||E_\alpha(A(x)^\alpha)|| ||y_0|| + ||x^\alpha E_{\alpha,\alpha+1}(Ax^\alpha)|| ||(Iy_1 - Ay_0)|| \\ &\quad + M \int_0^x ||(x-s)^{\alpha+\beta-1} E_{\alpha,\alpha+\beta}(A(x-s)^\alpha)|| ||y|| ds. \end{aligned}$$

By using Gronwall inequality,

$$\begin{aligned} ||y(x)|| &\leq \{||E_\alpha(A(x)^\alpha)|| ||y_0|| + ||t^\alpha E_{\alpha,\alpha+1}(Ax^\alpha)|| ||(Iy_1 - Ay_0)||\} \\ &\quad \exp\left\{ M \int_0^x ||(x-s)^{\alpha+\beta-1} E_{\alpha,\alpha+\beta}(A(x-s)^\alpha)|| ds \right\}. \end{aligned}$$

By using Lemma 17.3 we may conclude that

$$\exp\left\{ M \int_0^x |x^{\alpha+\beta-1} E_{\alpha,\alpha+\beta}(Ax^\alpha)| dx \right\} \leq K.$$

is bounded.

$$||E_\alpha(A(x)^\alpha)|| ||y_0|| + ||x^\alpha E_{\alpha,\alpha+1}(Ax^\alpha)|| ||Iy_1 - Ay_0|| \to 0 \text{ as } x \to \infty.$$

$$\lim_{x \to +\infty} y(x) \to 0.$$

Therefore there exists an asymptotically stable solution of the system (equation 17.17). Hence the theorem is proved.

17.3.4 Stability of a Fractional Delay Differential System

Take the time-delayed nonlinear fractional differential equation into consideration

$$^{C}D^{\alpha}y(x) = Ay(x) + By(x-\tau)y(x) + f(x, y(x)), \quad 0 < \alpha < 1, \tag{17.18}$$

where $y(0) = \phi(x), x \in [-1, 0]$ and $f(x, y) \in \mathbb{C}[0, T], T > 0$, and $A, B \in \mathbb{R}^{n \times n}$.

Lemma 17.4 *Let* $y_0 = \phi(x)$

$$g_k(x) = \left\{ \begin{array}{c} y_0(x-\tau), 0 < x \leq \tau \\ y_1(x-\tau), \tau < x \leq 2\tau \\ \vdots \\ y_{(k-1)}(x-\tau), (k-1)\tau < x \leq k\tau \end{array} \right\} \tag{17.19}$$

is continuous, where k be the greatest positive integer such that the function. Then the IVP (equation 17.18) has a unique solution on $[0, k\tau]$. *By using method of steps it can be represented by* $y(x) = y_i(x), (i-1)\tau \leq x \leq i\tau, c_i$ *is a constant,* $i = 1, \ldots, k$, *here*

$$\begin{aligned} y_i(x) &= E_{\alpha}(Ax^{\alpha})c_i + \int_0^x (x-\theta)^{\alpha-1} E_{\alpha,\alpha}(A(x-\theta)^{\alpha}) \\ &\quad [b\phi(\theta-1)(g_i(\theta)y(\theta) + f(\theta, y(\theta)))]d\theta. \end{aligned} \tag{17.20}$$

Theorem 17.5 *Suppose* $||g_k(x)|| \leq M, ||f(x, y(x))|| \leq M'||y||$ *for some* $M, M' > 0$ *and every eigenvalue of A meet the requirement spectrum condition (17.11). Then, there exists an asymptotically stable solution of equation* (17.18).

Proof

The solution representation of system (equation 17.18) is

$$\begin{aligned} y(x) &= E_{\alpha}(Ax^{\alpha})c_i + \int_0^x (x-\theta)^{\alpha-1} E_{\alpha,\alpha}(A(x-\theta)^{\alpha}) \\ &\quad [B\phi(\theta-1)(g_i(\theta)y(\theta) + f(\theta, y(\theta)))]d\theta. \end{aligned}$$

From which it follows that,

$$\begin{aligned} ||y(x)|| &\leq ||E_{\alpha}(Ax^{\alpha})c_i|| + \int_0^x ||(x-\theta)^{\alpha-1} E_{\alpha,\alpha}(A(x-\theta)^{\alpha})|| \\ &\quad ||[B\phi(\theta-1)(g_i(\theta)y(\theta) + f(\theta, y(\theta)))]||d\theta \\ &\leq ||E_{\alpha}(Ax^{\alpha})c_i|| + N\int_0^x ||(\theta)^{\alpha-1} E_{\alpha,\alpha}(A(\theta)^{\alpha})|| \; ||y||d\theta. \end{aligned}$$

Here $N = (bCM + M')$, $||\phi|| = max_{-1\leq x\leq 0}\phi(x) = C$, $||B|| = b$. From Gronwall's inequality, we have

$$||y(x)|| \leq ||E_\alpha(Ax^\alpha)c_i|| \exp\left\{N\int_0^x ||(\theta)^{\alpha-1}E_{\alpha,\alpha}(A(\theta)^\alpha)||d\theta\right\}.$$

By using Lemma 17.3, we have

$$\exp\left\{N\int_0^x ||(\theta)^{\alpha-1}E_{\alpha,\alpha}(A(\theta)^\alpha)||d\theta\right\} \text{ is bounded.} \tag{17.21}$$

Further, $||E_\alpha(Ax^\alpha)c_i|| \to 0$ as $x \to \infty$.

$$\lim_{x\to+\infty} y(x) \to 0.$$

Therefore, there exists an asymptotically stable solution of equation 17.18. .

17.4 NUMERICAL EXAMPLES

To demonstrate the usefulness of suggested spectral criteria, some examples are provided. The problems are resolved numerically, utilizing the fractional Euler's technique as a forecast and the modified trapezoidal rule as a corrective measure to arrive at the final value [7, 8].

Example 17.1 Linear Homogeneous System
Take the linear fractional differential equation into consideration

$$({}^cD^{1/2}y)(x) = -2y(x), \tag{17.22}$$

with $y(0) = 1$.

Solution: This equation is of the standard form with $A = -2$, and $\alpha = 1/2$, $0 < \alpha \leq 1$. Here one may notice that the eigenvalue of A satisfies the spectrum condition, thereby the given system is asymptotically stable. We may cross-verify with the solution of the system. Consider the corresponding homogeneous equation

$$({}^cD^{1/2}y)(x) + 2y(x) = 0.$$

Applying Laplace transforms, we get

$$\mathcal{L}[({}^cD^{1/2}y)(x)] + 2\mathcal{L}[y(x)] = 0,$$

using the initial condition $y(0) = 1$,

$$s^{1/2}(\mathcal{L}y)(s) - s^{-1/2} + 2(\mathcal{L}y)(s) = 0, \implies (\mathcal{L}y)(s) = \frac{s^{1/2}}{s^{\frac{1}{2}} + 2}.$$

Applying inverse Laplace transforms, we get

$$y(x) = \mathcal{L}^{-1}\frac{s^{1/2}}{s^{1/2}+2},$$
$$y(x) = E_{1/2,1}(-2\sqrt{x}).$$

The solution is asymptotically stable, since $\lim_{x\to+\infty} y(x) \to 0$.

Example 17.2 Linear Non-homogeneous System
Consider the equation

$$({}^{c}D^{1/2}y)(x) = y(x) + x, \tag{17.23}$$

with $y(0) = 1$.

Solution: Given equation can be written in the standard form with $A = 1$, $f(x) = x$ and $\alpha = 1/2, 0 < \alpha \leq 1$. One may easily notice that the eigen value does not satisfy the spectrum condition, thereby the given system is not asymptotically stable. We may cross-verify the results by finding its solution. Consider the corresponding homogeneous equation $({}^{c}D^{1/2}y)(x) - y(x) = 0$. Applying Laplace transforms, we get

$$\mathcal{L}[({}^{c}D^{1/2}y)(x)] - \mathcal{L}[y(x)] = 0,$$

using the initial condition $y(0) = 1$

$$s^{1/2}(\mathcal{L}y)(s) - s^{-1/2} - (\mathcal{L}y)(s) = 0, \implies (\mathcal{L}y)(s) = \frac{s^{1/2}}{s^{\frac{1}{2}}-1}.$$

Applying inverse Laplace transforms, we get

$$y(x) = \mathcal{L}^{-1}\frac{s^{1/2}}{s^{1/2}-1}, \implies y(x) = E_{1/2,1}(\sqrt{x}).$$

Particular solution is given by

$$\begin{aligned} y(x) &= \int_0^x (x-t)^{\alpha-1}E_{\alpha,\alpha}[\lambda(x-t)^{\alpha}]f(t)dt, \\ &= \int_0^x (x-t)^{-1/2}E_{1/2,1/2}[(x-t)^{1/2}]tdt. \\ \textit{Put } z = x-t & \\ &= \int_0^x (z)^{-1/2}E_{1/2,1/2}(\sqrt{z})(x-z)dt. \end{aligned}$$

By using the Mittag-Leffler formula with $\alpha = -1/2, \beta = -1/2, \nu = 2$,

$$y(x) = x^{3/2}E_{1/2,5/2}(\sqrt{x}).$$

The solution of given system is

$$y(x) = E_{1/2,1}(\sqrt{x}) + x^{3/2}E_{1/2,5/2}(\sqrt{x}).$$

The solution is not asymptotically stable, $\lim_{x\to+\infty} y(x) \not\longrightarrow 0$.

Example 17.3 Linear Non-Homogeneous System
Consider the fractional non-homogeneous system

$$^{C}D^{\alpha}y(x) + y(x) = f_0(x), \tag{17.24}$$

where $\alpha = 3/2$, *and* $f_0(x) = e^{-x}\sin(0.2x)$ *with* $y(0) = y'(0) = 0$.

The given equation (17.24) can be rewritten by using the substitution

$$\begin{aligned} ^{C}D^{\frac{1}{2}}y_1(x) &= y_2(x), \\ ^{C}D^{\frac{1}{2}}y_2(x) &= y_3(x), \\ ^{C}D^{\frac{1}{2}}y_3(x) &= -y_1(x) + f_0(x), \text{ where } \quad y_1(x) = y(x). \end{aligned} \tag{17.25}$$

The standard form of given system is $^{C}D^{1/2}Y(X) = AY(X) + F(X)$, and $F(X) = (0, 0, e^{-x}\sin(0.2x))^T$ with initial condition $y_1(x) = y_2(x) = y_3(x) = 0$, here $Y(X) = (y_1(x), y_2(x), y_3(x))$, where

$$A = \begin{bmatrix} 0 & 1 & 0 \\ 0 & 0 & 1 \\ -1 & 0 & 0 \end{bmatrix}.$$

The eigenvalues of A satisfies the spectrum condition. Hence, there exists an asymptotically stable solution.

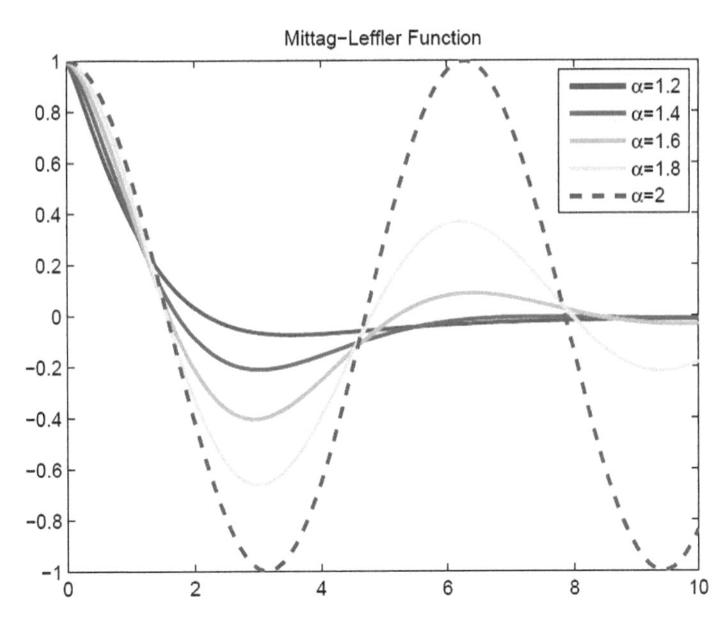

Example 17.4 Duffing Equation
Take the nonlinear fractional equations into consideration,

$$^{C}D^{\alpha}y(x) = -y(x) - y(x)^3, \quad 0 < \alpha < 2.$$

Case I: If $\alpha = 1/2$

$$\begin{aligned} {}^{C}D^{\frac{1}{2}}y(x) &= -y(x) - y(x)^3, \\ x(0) &= 1. \end{aligned} \tag{17.26}$$

Eigenvalue meet the requirement condition of spectrum $|\arg(-1)| > \frac{\pi}{4}$. Also $f(x,y) = -y^3$, which satisfies $f(x,0) = 0$. Since the necessary conditions are satisfied. Hence the given system (equation 17.26) is asymptotically stable.

Case II: If $\alpha = 3/2$, The given system can be written as follows

$$\begin{aligned} {}^{C}D^{\frac{1}{2}}y_1(x) &= y_2(x), \\ {}^{C}D^{\frac{1}{2}}y_2(x) &= y_3(x), \\ {}^{C}D^{\frac{1}{2}}y_3(x) &= -y_1(x) - y_1^3(x), \end{aligned} \tag{17.27}$$

where $y_1(x) = y(x)$, with initial condition $y_1(0) = 0, y_2(0) = 1, y_3(0) = 0$. The standard form is given by, ${}^{C}D^{1/2}y(x) = Ay(x) + f(x,y)$, and $f(x,y) = (0,0,y_1^3(x))^T$ where

$$A = \begin{bmatrix} 0 & 1 & 0 \\ 0 & 0 & 1 \\ -1 & 0 & 0 \end{bmatrix}.$$

Every eigenvalues of A meet spectrum condition $|\arg(\text{spc}(A))| > \frac{\pi}{4}$. Also, the nonlinear term $f(x,y)$ satisfies the necessary condition $f(x,0) = 0$. Hence there exists an asymptotically stable solution of the given system.

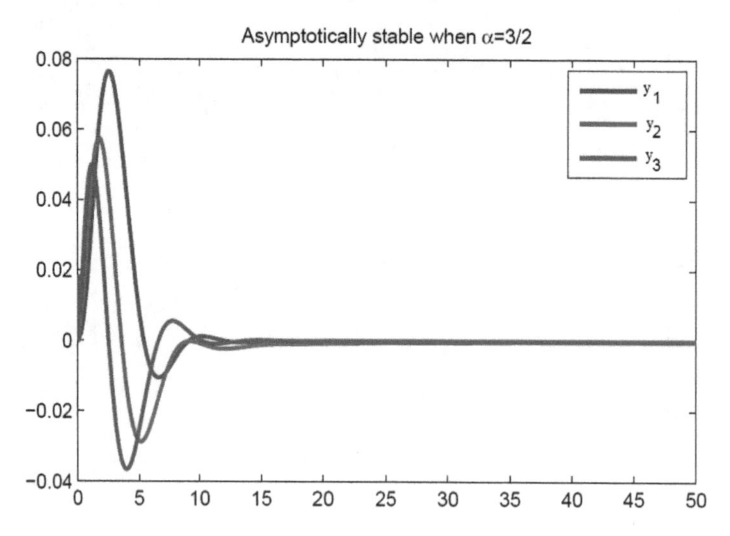

Example 17.5 Nonlinear System

Now we shall discuss the stability of nonlinear system

$${}^{C}D^{\alpha}y(x) + Ay(x) = f(x,y), \quad 0 < \alpha \leq 1, \tag{17.28}$$

where $A = \begin{bmatrix} -3 & 5 \\ -2 & 3 \end{bmatrix}$ *and* $f(x,y) = (0, -\sin(y(x)))^T$.

The eigenvalues of matrix A is given by $\pm i$. Let us consider the two cases.
Case 1: When $\alpha = 1$. Since the eigenvalues of A do not satisfy meet the requirement of spectrum condition. There does not exist a asymptotically stable solution.
Case 2: When $\alpha = 1/2$. The eigenvalues of A satisfies the spectrum condition $|\arg(\text{spc}(A))| = \frac{\pi}{2} > \frac{\pi}{4}$. Also the nonlinear term satisfies the necessary condition. Hence there exists an asymptotically stable solution.

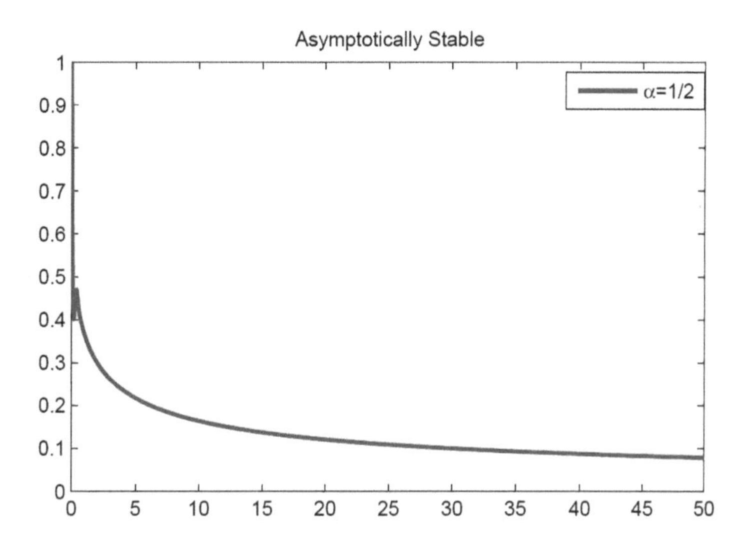

Example 17.6 Fractional Neutral System
Take the linear fractional neutral differential system into consideration

$$^{C}D^{\alpha}[^{C}D^{\alpha}y(x)] + 3^{C}D^{\alpha}y(x) = -y(x), \tag{17.29}$$

with $\alpha = 1/2$, *and* $y(0) = 2$.

The standard equation having corresponding values of $A = -3$, $\alpha = \beta = 1/2$, and $f(x,y) = -y(x)$. The given equation (17.29) meets the requirement of necessary conditions. Hence there exist an asymptotically stable solution.

Example 17.7 Fractional Langevin System
Consider the fractional nonlinear system

$$^{C}D^{3/4}(^{C}D^{3/4} - A)y(x) = f(x, y(x)), \tag{17.30}$$

Where $A = \begin{bmatrix} -2 & 0 \\ 0 & -4 \end{bmatrix}$, $f(x,y(x)) = \begin{bmatrix} y^2(x) \\ \sin y(x) \end{bmatrix}$ *and with the initial conditions* $y(0) = \begin{bmatrix} 0.21 \\ 0.21 \end{bmatrix}$ *and* $^{C}D^{\alpha}y(x)|_{x=0} = \begin{bmatrix} 0.31 \\ 0.31 \end{bmatrix}$.

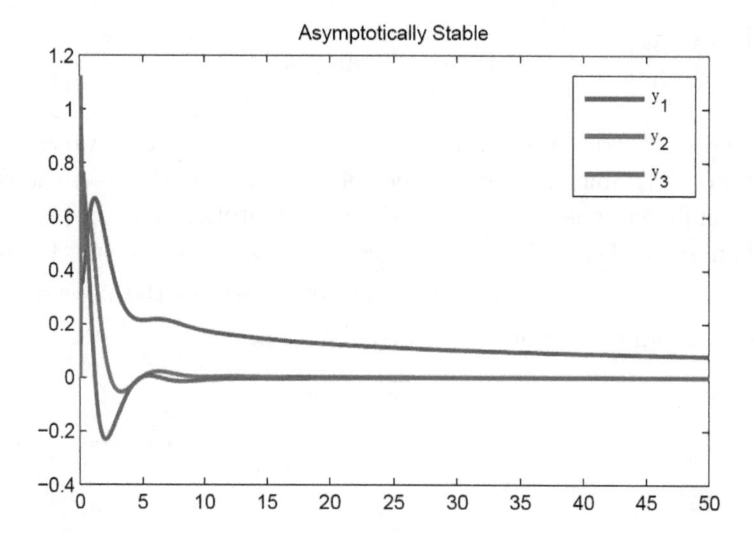

The eigenvalues of matrix satisfies spectrum condition

$$|\arg(-2)| = |\arg(-4)| = \pi > \pi/8,$$

Also, the nonlinear function $f(x, y(x)) = (y^2(x), \sin y(x))^T$ satisfies the necessary conditions. Hence there exists an asymptotically stable solution of the system (equation 17.31).

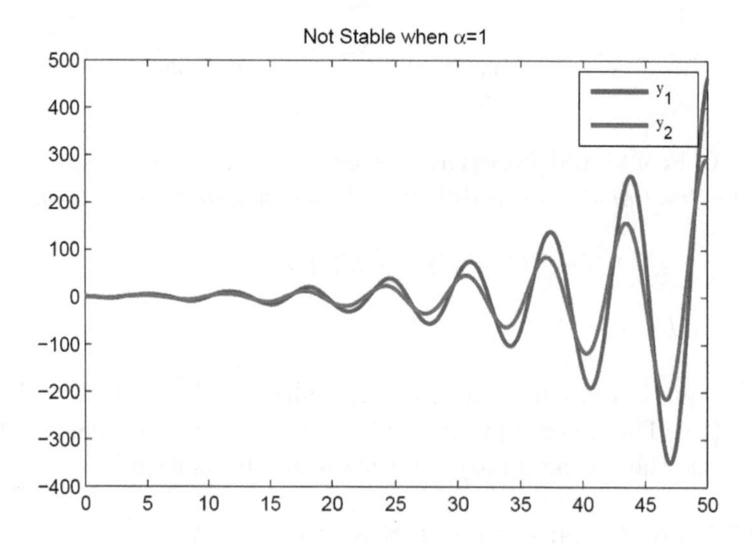

Example 17.8 Fractional Logistic delayed System
Consider the fractional delayed nonlinear logistic system

$$^{C}D^{1/2}y(x) = ay(x) + by(x-\tau)(1-y(x)) + f(x, y(x)), \tag{17.31}$$

$$y(x) = 0.8, -\tau < x < 0 \tag{17.32}$$

Where $\alpha = 1/2, \tau = 1, f(x, y(x)) = -siny(x)$

When we take $a = -1 < 0$ and $b = -2 < 0$. Then

$$|arg(-1)| = \pi > \pi/8, |arg(-2)| = \pi > \pi/8.$$

Also, the nonlinear function $f(x, y(x)) = -\sin y(x)$ satisfies the necessary conditions. Hence the system (equation 17.31) is asymptotically stable.
When we take $a = 1 > 0$ and $b = -2 > 0$. Then

$$|\arg(1)| = \pi < \pi/8, |\arg(-2)| = \pi > \pi/8.$$

Though the nonlinear term $f(x, y(x)) = -\sin y(x)$ satisfies the necessary conditions, one of the eigenvalues fails to satisfy the spectrum condition. Hence there does not exist an asymptotically stable solution of the system (equation 17.31).

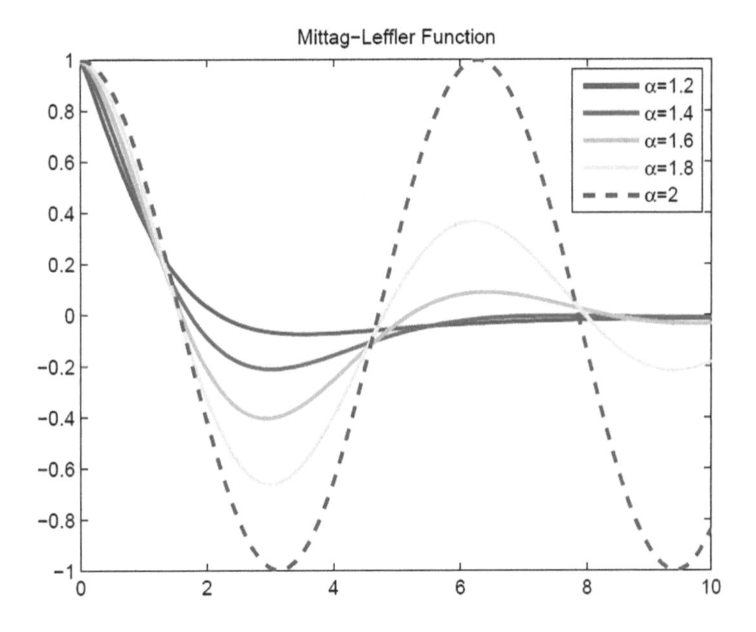

17.5 CONCLUSION

Every system in nature exists as a fractional order. This leads us to the logical conclusion that, in contrast to integer order, fractional order system offers favorable features. In this article, it was possible to develop some straightforward spectral condition adequate conditions on nonlinear term that ensure the stability of FDEs. The stability conditions for the delayed, Langevin, and fractional neutral systems were also addressed. Numerous examples were provided to demonstrate how the acquired conditions.

BIBLIOGRAPHY

[1] S. Das, *Functional Fractional Calculus for System Identification and Controls*, Springer Publications, New York (2008).

[2] P. Debnath, N. Konwar, S. Radenovic, *Metric Fixed Point Theory: Applications in Science, Engineering and Behavioural Sciences*, Springer Verlag, Singapore (2021).

[3] K. Diethelm, N.J. Ford, *Analysis of Fractional Differential Equations*, Springer, New York (2004).

[4] V. Govindaraj, S. Priyadharsini, P. Suresh Kumar, K. Balachandan, Asymptotic stability of fractional Langevin systems, *J. Appl. Nonlinear Dyn.* 11(3) (2022), 635–650.

[5] A.A. Kilbas, H.M. Srivastava, J.J. Trujillo, *Theory and Applications of Fractional Differential Equation*, Elsevier, Amsterdam (2006).

[6] K.S. Miller, B. Ross, *An Introduction to the Fractional Calculus and Fractional Differential Equation*, Wiley, New York (1993).

[7] M. Odibat, S. Momani, An algorithm for the numerical solutions of differential equations of fractional order, *J. Appl. Math. Inf.* 26 (2008), 15–27.

[8] S. Priyadharsini, Stability of fractional neutral and integrodifferential systems, *J. Fract. Calc. & Appl.* 76 (2016), 87–102.

[9] S. Priyadharsini, Stability analysis of fractional differential systems with constant delay, *J. Indian Math. Soc.* 83 (2016), 337–350.

[10] S. Priyadharsini, Some numerical examples on the stability of fractional linear dynamical systems, *Mapana J. Sci.* 17(3) (2018), 51–66.

[11] I. Podlubny, *Fractional Differential Equation*, Academic Press, New York (1999).

[12] D. Qian, C. Li, R.P. Agarwal, P.J.Y. Wong, Stability analysis of fractional differential system with Riemann-Liouville derivative, *Math. Comput. Model.* 52(5–6) (2010), 86–874.

[13] P. Debnath, H.M. Srivastava, P. Kumam, B. Hazarika, *Fixed Point Theory and Fractional Calculus: Recent Advances and Applications*, Springer Verlag, Singapore (2022).

[14] P. Debnath, H.M. Srivastava, K. Chakrabory, P. Kumam, *Advances in Number Theory and Applied Analysis*, World Scientific, Singapore (2023).

[15] M. Zurigat, S. Momani, Z. Odibat, A. Alawneh, The homotopy analysis method for handling systems of fractional differential equations, *Appl. Math. Model.* 34 (2010), 24–35.

CHAPTER 18

On Periodic Dirichlet Series and Special Functions

Jay Mehta
Sardar Patel University

Imre Kátai
Eötvös Loránd University

Shigeru Kanemitsu
KSCSTE-Kerala School of Mathematics

CONTENTS

18.1 INTRODUCTION

Our research is based partly on some portions of Ref. [2] and we quote basic results freely from it. For special functions, we refer to [4, 9, 10, 16], etc.

Let $h(d) = |I/P|$ denote the class number of the quadratic field $K := \mathbb{Q}(\sqrt{D})$ of discriminant d. The *Dirichlet class number formula*

$$h(d) = \frac{w\sqrt{|d|}}{2\pi} L(1, \chi_d), \quad d < 0 \tag{18.1}$$

and

$$h(d) = \frac{2\log \varepsilon}{\sqrt{d}} L(1.\chi_d) \quad d > 0, \tag{18.2}$$

where w is the number of roots of unity in K, ε the fundamental unit and χ_d the Kronecker character associated with K. It is a special Dirichlet character introduced in Section 18.2.

DOI: 10.1201/9781003388678-18

We remark that the Dirichlet class number formula is merging of algebraic number theory and analytic number theory. From the former side, it gives a closed form for the class number, and from the latter side, it assures the non-vanishing of the value $L(1,\chi)$, which implies infinitude of primes in an arithmetic progression with common difference $|d|$. The requirement from algebraic number theory that a finite expression is needed for the infinite series $L(1,\chi)$ for the class number which is finite drove the study of finite form for $L(1,\chi)$. One of the culminations is achieved by Ref. [8], cf. Corollary 18.2.

18.1.1 Notation and Terminology

We assemble here notation and symbols which are used throughout.

The boundary Lerch zeta-function (or the polylogarithm with complex exponential argument) $\ell_s(x)$ is defined by

$$\ell_s(x) = \sum_{n=1}^{\infty} e^{2\pi i x n} n^{-s}, \quad \operatorname{Re} s = \sigma > 1 \quad \text{or} \quad \sigma > 0, x \notin \mathbb{Z}, \tag{18.3}$$

which has its counterpart, the Hurwitz zeta-function defined by

$$\zeta(s,x) = \sum_{n=0}^{\infty} \frac{1}{(n+x)^s}, \quad \sigma > 1. \tag{18.4}$$

It is continued meromorphically over the whole plane with a simple pole at $s=1$ with the Laurent expansion (equation 18.7). Both of them reduce to the Riemann zeta-function

$$\zeta(s,1) = \ell_s(1) = \zeta(s).$$

Equations (18.3) and (18.4) are connected by the *Hurwitz formula* (i.e. the functional equation for the Hurwitz zeta-function): for $\sigma > 1, 0 < x \le 1$,

$$\zeta(1-s,x) = \frac{\Gamma(s)}{(2\pi)^s}\left(e^{-\frac{\pi i s}{2}} \ell_s(x) + e^{\frac{\pi i s}{2}} \ell_s(1-x)\right), \tag{18.5}$$

while its reciprocal is

$$\ell_{1-s}(x) = \frac{\Gamma(s)}{(2\pi)^s}\left(e^{\frac{\pi i s}{2}} \zeta(s,x) + e^{-\frac{\pi i s}{2}} \zeta(s,1-x)\right), \quad 0 < x < 1. \tag{18.6}$$

By equation (18.6), the Lerch zeta-function $\ell_{1-s}(x)$ is continued meromorphically over the whole plane with a possible pole at $s=1$. We assume the Laurent expansion, cf. Section 18.4.

$$\zeta(s,x) = \frac{1}{s-1} - \psi(x) + O(s-1), \quad \text{as} \quad s \to 1, \tag{18.7}$$

where

$$\psi(x) = \frac{\Gamma'}{\Gamma}(x) \tag{18.8}$$

is the *Euler digamma function* and γ indicates the *Euler constant*

$$\gamma = -\psi(1) = 0.5772156649\cdots. \tag{18.9}$$

Hence equation (18.7) reduces to

$$\zeta(s) = \frac{1}{s-1} + \gamma + O(s-1), \quad \text{as} \quad s \to 1. \tag{18.10}$$

$s = 0$ is a removable singularity of $\ell_s(x)$ if $x \in \mathbb{R} - \mathbb{Z}$ in view of equation (18.7). Indeed, comparison of both sides of equation (18.6) as $s \to 1$ leads to the relation

$$\psi(x) - \psi(1-x) = 2\pi i \ell_0(x) + \pi i, \tag{18.11}$$

which together with

$$\ell_0(x) = \frac{e^{2\pi i x}}{1 - e^{2\pi i x}} = \frac{1}{2}(-1 + i \cot \pi x) \tag{18.12}$$

for $x \in \mathbb{R} \setminus \mathbb{Z}$ entails the odd part formula for the digamma function

$$\psi(x) - \psi(1-x) = -\pi \cot \pi x \tag{18.13}$$

valid for non-integral values of x.

We assemble the identities for $\ell_1(x)$, cf. e.g. [17, 19].

$$\begin{aligned} &\sum_{n=1}^{\infty} \frac{\cos(2\pi n x)}{n} + i \sum_{n=1}^{\infty} \frac{\sin(2\pi n x)}{n} = \ell_1(x) \\ &= -\log\left(1 - e^{2\pi i x}\right) = \sum_{n=1}^{\infty} \frac{e^{2\pi i n x}}{n} = A_1(x) - \pi i \overline{B}_1(x), \end{aligned} \tag{18.14}$$

$0 < x < 1$, where

$$A_1(x) = -\log 2|\sin \pi x| = \sum_{n=1}^{\infty} \frac{\cos(2\pi n x)}{n}, \tag{18.15}$$

is the real part (even part) of $\ell_1(x)$, called the first Clausen function (or the logsine function) and the imaginary part (odd part) is the first periodic Bernoulli polynomial

$$x - [x] - \frac{1}{2} = \bar{B}_1(x) = -\frac{1}{\pi} \sum_{n=1}^{\infty} \frac{\sin(2\pi n x)}{n} \tag{18.16}$$

where the second equality holds for $x \notin \mathbb{Z}$, and where $[x]$ is the greatest integer not exceeding x.

18.2 DIRICHLET SERIES WITH PERIODIC COEFFICIENTS

The theory of discrete Fourier transforms (DFT) (for arithmetic functions) has been developed in Refs. [6, 19, 5], and [12, Chapter 3] etc.. Also DFT in the case of a finite group is exposed in Refs. [7], [13, pp. 109–114] etc.

Let $M \geq 3$ be a fixed integer used as the modulus. Let $C(M)$ be the vector space of all periodic arithmetic functions f with period M:

$$C(M) = \{f : \mathbb{Z} \to \mathbb{C} | f(n+M) = f(n)\}. \tag{18.17}$$

The inner product of $f_1, f_2 \in C(M)$ is defined by

$$(f_1, f_2) = \sum_{a \bmod M} f_1(a)\overline{f_2(a)}, \tag{18.18}$$

where the bar $\bar{\cdot}$ means the complex conjugation of $\cdot$. $C(M)$ becomes an inner product space.

Let

$$\varepsilon_j(a) = e^{2\pi i j a/M}, \quad 1 \leq j \leq M, \tag{18.19}$$

where a runs through a complete set of representatives mod M and we mostly choose $1 \leq a \leq M$. Then

$$(\varepsilon_j, \varepsilon_k) = \delta_{jk} M, \quad 1 \leq j, k \leq M, \tag{18.20}$$

so that $\mathcal{E} := \{\frac{1}{\sqrt{M}}\varepsilon_j | 1 \leq j \leq M\}$ is an orthonormal system (ONS).

The *DFT* (discrete Fourier transform) $\hat{f}$ (or the bth Fourier coefficient) of $f \in C(M)$ is defined by

$$\hat{f}(b) = \left(f, \frac{1}{\sqrt{M}}\varepsilon_b\right) = \frac{1}{\sqrt{M}}\sum_{a=1}^{M} \varepsilon_b(-a) f(a). \tag{18.21}$$

Then by orthogonality we have

$$\hat{\hat{f}}(-a) = \frac{1}{\sqrt{M}}\sum_{b=1}^{M} \hat{f}(b)\varepsilon_b(a) = f(a), \tag{18.22}$$

the *Fourier inversion formula* or the Fourier expansion . This also shows that $C(M)$ is generated by $\mathcal{E}$, which therefore forms a basis (i.e. ONB) of $C(M)$ and $\dim C(M) = M$

By equations (18.20) and (18.21) reads for $f = \varepsilon_j$

$$\hat{\varepsilon}_j(b) = \sqrt{M}\delta_{bj} = \sqrt{M}\chi_j(b), \tag{18.23}$$

where χ_j is the characteristic function (18.27).

In what follows we use the notation

$$f^e(x) = \frac{1}{2}(f(x) + f(-x)), \quad f^o(x) = \frac{1}{2}(f(x) - f(-x)) \tag{18.24}$$

$$\text{or}\quad f^e(x) = \frac{1}{2}(f(x) + f(1-x)), \quad f^o(x) = \frac{1}{2}(f(x) - f(1-x))$$

as the case may be, so that

$$f = f^e + f^o. \tag{18.25}$$

Hence f is even $\iff f = f^e$ resp. f is odd $\iff f = f^o$.

As in equation (18.24) we let

$$\begin{aligned} f^e &= \frac{1}{2}\left(f(n \bmod M) + f(-n \bmod M)\right), \\ f^o &= \frac{1}{2}\left(f(n \bmod M) - f(-n \bmod M)\right) \end{aligned} \tag{18.26}$$

be the even resp. odd part of f. Then the parity inherits to the DFT and we have

$$\begin{aligned} \hat{f^e}(b) &= \frac{1}{2}(\hat{f}(b) + \hat{f}(-b)) = \frac{1}{\sqrt{M}}\sum_{a=1}^{M} \cos\left(2\pi\frac{b}{M}a\right) f(a), \\ \hat{f^o}(b) &= \frac{1}{2}(\hat{f}(b) - \hat{f}(-b)) = i\frac{1}{\sqrt{M}}\sum_{a=1}^{M-1} \sin\left(2\pi\frac{b}{M}a\right) f(a). \end{aligned}$$

To find another natural basis, let χ_a be the characteristic function a mod M.

$$\chi_a(n) = \begin{cases} 1 & n \equiv a \bmod M \\ 0 & n \not\equiv a \bmod M \end{cases}. \tag{18.27}$$

Then $\mathcal{X} := \{\chi_a | 1 \le a \le M\}$ is a basis of $C(M)$ and

$$\sqrt{M}\hat{\chi}_a(n) = \sum_{j=1}^{M} \varepsilon_n(-j)\chi_a(j) = \varepsilon_a(-n). \tag{18.28}$$

For $f \in C(M)$ let

$$D(s,f) = \sum_{n=1}^{\infty} \frac{f(n)}{n^s}. \tag{18.29}$$

Since

$$\sum_{n=1}^{\infty} \frac{|f(n)|}{n^\sigma} \ll \zeta(\sigma),$$

the series in equation (18.29) is absolutely convergent for $\sigma > 1$. Let

$$D(M) = \{D(s,f) | f \in C(M)\} \tag{18.30}$$

the set of all Dirichlet series of the form equation (18.29) for $\sigma > 1$ in the first instance. Then it forms a vector space of dimension M canonically isomorphic to $C(M)$. The canonical basis of $D(M)$ is $\{\ell_s\left(\frac{a}{M}\right) | 1 \le a \le M\}$ corresponding to $\mathcal{E}$, where $\ell_s(x)$ is the Lerch zeta-function (LZF). Hence the first equality of equation (18.22) implies that

$$D(s,f) = \frac{1}{\sqrt{M}}\sum_{a=1}^{M}\hat{f}(a)\ell_s\left(\frac{a}{M}\right) = \frac{1}{\sqrt{M}}\sum_{a=1}^{M-1}\hat{f}(a)\ell_s\left(\frac{a}{M}\right) + \frac{\hat{f}(M)}{\sqrt{M}}\zeta(s). \quad (18.31)$$

It follows that $D(s,f)$ can be continued meromorphically over the whole plane with a possible simple pole at $s=1$ with residue $\frac{\hat{f}(M)}{\sqrt{M}}$. It is an entire function if and only if $\hat{f}(M)=0$ and the value $D(1,f)$ is meaningful. The entireness condition is

$$\hat{f}(M) = \frac{1}{\sqrt{M}}\sum_{a=1}^{M} f(a) = 0, \quad (18.32)$$

vanishing of the whole sum. This holds if f is odd. The additive characters (equation 18.19) also satisfy the condition for $1 \le j \le M-1$.

Another basis of $D(M)$ corresponding to $\mathcal{X}$ is $\{D(s,\chi_a) | 1 \le a \le M\}$, where

$$D(s,\chi_a) = \sum_{n=1}^{\infty}\frac{\chi_a(n)}{n^s} = \sum_{\substack{n=1 \\ n\equiv a \bmod M}}^{\infty}\frac{1}{n^s} = M^{-s}\zeta\left(s, \frac{a}{M}\right). \quad (18.33)$$

Note that it is $\zeta(s, 1-\{\frac{a}{M}\})$ that belongs to $D(M)$ rather than $\zeta(s,\{\frac{a}{M}\})$.Hence in contrast to equation (18.31), we have another expression

$$D(s,f) = \frac{1}{M^s}\sum_{a=1}^{M} f(a)\zeta\left(s, \frac{a}{M}\right). \quad (18.34)$$

We let

$$D(s,f) = \frac{\frac{\hat{f}(M)}{\sqrt{M}}}{s-1} + \gamma_0(f) + O(s-1), \quad s \to 1 \quad (18.35)$$

be the Laurent expansion of $D(s,f)$, whence

$$\gamma_0(f) = \lim_{s\to 1}\left(D(s,f) - \frac{\hat{f}(M)}{\sqrt{M}}\frac{1}{s-1}\right).$$

The most important example of periodic Dirichlet series is the *Dirichlet L-functions*, which are the Dirichlet series (equation 18.29) with Dirichlet character coefficients defined by

$$L(s,\chi) = D(s,\chi) = \sum_{n=1}^{\infty}\frac{\chi(n)}{n^s} = \prod_p\left(1 - \frac{\chi(p)}{p^s}\right)^{-1} \quad (18.36)$$

for $\sigma > 1$ in the first instance. It is continued meromorphically over the whole plane with a possible simple pole at $s=1$.

From each reduced residue class character $\chi \in (\widehat{\mathbb{Z}/M\mathbb{Z}})^\times$ there arises its 0-extension denoted by the same symbol:

$$\chi(n) = \begin{cases} \chi(n+M\mathbb{Z}) & (n,M)=1 \\ 0 & (n,M)>1, \end{cases} \tag{18.37}$$

where (n,M) is the greatest common divisor of n and M. Then $\chi \in C(M)$ and is called a Dirichlet character mod M. The 0-extension of the trivial character (with the constant value 1) is called the principal Dirichlet character, denoted χ_0. Cf. [1, 3] for Dirichlet L-functions.

For non-principal Dirichlet character χ, the vanishingness condition (18.32) holds and the Laurent expansion (18.35) amounts to the Taylor expansion

$$L(s,\chi) = L(1,\chi) + O(s-1), \quad s \to 1 \tag{18.38}$$

It is to $L(s,\chi_0)$ that equation (18.35) is applied and creates a complication.

Lemma 18.1 *The space $D(M)$ in equation* (18.30) *is an inner product space of dimension M isomorphic to $C(M)$. $C(M)$ resp. $D(M)$ has ONBs equations (18.19) and (18.27).*

$D(s,\hat{f}^o)$ is an entire function and so is $L(s,\chi)$ for non-principal Dirichlet character χ.

The following is a slightly enhanced version of Ref. [17] and is stated as Ref. [2, Proposition 1.3].

Lemma 18.2 (base change formula) *For $f \in C(M)$*

$$\frac{1}{M^s}\sum_{a=1}^{M} f(a)\zeta\left(s,\frac{a}{M}\right) = D(s,f) \tag{18.39}$$

$$= \frac{1}{\sqrt{M}}\sum_{a=1}^{M} \hat{f}(a)\ell_s\left(\frac{a}{M}\right) = \frac{1}{\sqrt{M}}\sum_{a=1}^{M-1} \hat{f}(a)\ell_s\left(\frac{a}{M}\right) + \frac{\hat{f}(M)}{\sqrt{M}}\zeta(s).$$

This entails the formula for the Laurent constant, cf. equation (18.35)

$$\begin{aligned} \gamma_0(f) &= -\frac{\hat{f}(M)}{\sqrt{M}}\log M - \frac{1}{M}\sum_{a=1}^{M} f(a)\psi\left(\frac{a}{M}\right) \\ &= \frac{1}{\sqrt{M}}\sum_{a=1}^{M-1} \hat{f}(a)\ell_1\left(\frac{a}{M}\right) + \frac{\hat{f}(M)}{\sqrt{M}}\gamma. \end{aligned} \tag{18.40}$$

Corollary 18.1 ([6, Theorem 5])

$$\gamma_0(f) = \lim_{s\to 1}\left(D(s,f) - \frac{\frac{\hat{f}(M)}{\sqrt{M}}}{s-1}\right) = \frac{1}{\sqrt{M}}\sum_{a=1}^{M-1} \hat{f}(a)\ell_1\left(\frac{a}{M}\right) + \frac{\hat{f}(M)}{\sqrt{M}}\gamma, \tag{18.41}$$

which amounts to

$$= \frac{1}{\sqrt{M}} \sum_{a=1}^{M-1} \hat{f}(a) A_1 \left(\frac{a}{M}\right) - \frac{\pi i}{\sqrt{M}} \sum_{a=1}^{M-1} \hat{f}(a) \bar{B}_1 \left(\frac{a}{M}\right) + \frac{\hat{f}(M)}{\sqrt{M}} \gamma \tag{18.42}$$

in view of equation (18.14). *We have further*

$$\gamma_0(f) = -\frac{1}{\sqrt{M}} \sum_{a=1}^{M-1} \hat{f}(a) \log 2 \sin \frac{\pi a}{M} + \frac{\pi}{2M} \sum_{a=1}^{M-1} f(a) \cot \frac{\pi a}{M} + \frac{\hat{f}(M)}{\sqrt{M}} \gamma, \tag{18.43}$$

where the expression for the second term follows from Eisenstein formula Corollary 18.3 below.

Equation (18.40) with $\hat{f}(M) = 0$ reads

$$-\frac{1}{M} \sum_{a=1}^{M} f(a) \psi \left(\frac{a}{M}\right) = D(1, f) = \frac{1}{\sqrt{M}} \sum_{a=1}^{M-1} \hat{f}(a) \ell_1 \left(\frac{a}{M}\right). \tag{18.44}$$

Lemma 18.3 ([2, Theorem 1.5]) *Suppose* $a \not\equiv 0 \bmod M$ *by which we understand either we extend* ψ *to a periodic function or* $1 \le a \le M-1$. *Then Gauss' first formula for the digamma function at rational argument*

$$\begin{aligned} \psi\left(\frac{a}{M}\right) &= -\gamma - \log M + \pi i \sum_{b=1}^{M-1} \varepsilon_a(-b) \bar{B}_1 \left(\frac{b}{M}\right) - \sum_{b=1}^{M-1} \varepsilon_a(-b) A_1 \left(\frac{b}{M}\right) \quad (18.45) \\ &= -\gamma - \log M + \pi i \sum_{b=1}^{M-1} \varepsilon_a(-b) \bar{B}_1 \left(\frac{b}{M}\right) - \sum_{j=1}^{M-1} \cos \frac{2\pi a j}{M} A_1 \left(\frac{j}{M}\right) \\ &= -\gamma - \log M - \frac{\pi}{2} \cot \frac{a}{M} \pi + \sum_{j=1}^{M-1} \cos \frac{2\pi a j}{M} \log 2 \sin \frac{\pi j}{M}, \end{aligned}$$

resp. Gauss' second formula, cf. [16, p. 19, (49)]

$$\sum_{a=1}^{M} e^{2\pi i \frac{n}{M} a} \psi\left(\frac{a}{M}\right) = \ell_1 \left(\frac{n}{M}\right) = -\log\left(1 - e^{2\pi i \frac{n}{M}}\right) \tag{18.46}$$

is a consequence of equation (18.40) *with* $f = \chi_a$ *resp.* $f = \varepsilon_a$.

Proof 18.1 *We rewrite equation* (18.40) *as*

$$-\frac{1}{M} \sum_{a=1}^{M} f(k) \psi\left(\frac{a}{M}\right) = \frac{1}{\sqrt{M}} \sum_{a=1}^{M-1} \hat{f}(a) \ell_1 \left(\frac{a}{M}\right) + \frac{\hat{f}(M)}{\sqrt{M}} (\log M + \gamma). \tag{18.47}$$

Now apply this with $f = \chi_a$ *resp.* $f = \varepsilon_a$*, for which we have*

$$\hat{\varepsilon}_a(b) = \sqrt{M}\chi_a(b) \tag{18.48}$$

resp.

$$\hat{\chi}_a(b) = \frac{1}{\sqrt{M}}\varepsilon_a(-b) \tag{18.49}$$

where (equation 18.48) resp. (equation 18.49) is a restatement of equations (18.23) *resp.* (18.28).

We are in a position to state our main theorem. It is restated with parity consideration as Theorem 18.2 below.

Theorem 18.1 *The two expressions* (18.40) *for the Laurent constant* $\gamma_0(f)$

$$-\frac{\hat{f}(M)}{\sqrt{M}}\log M - \frac{1}{M}\sum_{a=1}^{M} f(a)\psi\left(\frac{a}{M}\right) = \gamma_0(f) \tag{18.50}$$

$$= \frac{1}{\sqrt{M}}\sum_{a=1}^{M-1} \hat{f}(a)\ell_1\left(\frac{a}{M}\right) + \frac{\hat{f}(M)}{\sqrt{M}}\gamma.$$

are equivalent to Gauss' first formula (18.45) *in the form* $(1 \le a \le M-1)$

$$\psi\left(\frac{a}{M}\right) + \log M = -\sqrt{M}\hat{\ell}_1\left(\frac{a}{M}\right) = \pi i\sqrt{M}\hat{B}\left(\frac{a}{M}\right) - \sqrt{M}\hat{A}\left(\frac{a}{M}\right), \tag{18.51}$$

where

$$\hat{\ell}_1\left(\frac{a}{M}\right) = \frac{1}{\sqrt{M}}\sum_{b=1}^{M}\varepsilon_a(-b)\ell_1\left(\frac{b}{M}\right), \ell_1(1) = \gamma, \tag{18.52}$$

$$\hat{A}\left(\frac{a}{M}\right) = \frac{1}{\sqrt{M}}\sum_{b=1}^{M}\varepsilon_a(-b)A_1\left(\frac{b}{M}\right), A_1(1) = \gamma + \frac{\pi i}{2} \tag{18.53}$$

and

$$\hat{B}\left(\frac{a}{M}\right) = \frac{1}{\sqrt{M}}\sum_{b=1}^{M}\varepsilon_a(-b)B_1\left(\frac{b}{M}\right), B_1(1) = \frac{1}{2} \tag{18.54}$$

Proof 18.2 *Equation* (18.50) $\implies$ (18.51) *is proved in Lemma 18.3. The reverse implication follows by substitution.*

Remark 18.1 *In view of the Kubert identity with* $x = 1$

$$M^{-1}\sum_{r=0}^{M-1}\psi\left(\frac{x+r}{M}\right) = \psi(x) - \log M, \tag{$*_1$}$$

the left-hand side of equation (18.50) *seems natural. The definition of the value* $A_1(1)$ *is arbitrary since it is not the function value at* $x = 1$ *but can be any constant.*

18.3 PARITY AND RESTATEMENT OF THE MAIN THEOREM

Here we give a restatement of Theorem 18.1 in terms of parity and state results on the odd part.

Lemma 18.4 *$A_1(x)$ is an even function $0 < x < 1$ and $\bar{B}_1(x)$ is an odd function, where its value at integers is 0 as in the second equality of equation (18.16). If f is odd, then $\hat{f}(M) = 0$.*

Lemma 18.5 *For ease of notation, we set $f^e = f^+$ and $f^o = f^-$. Then for any function g, we have*

$$\sum_{a=1}^{M} f^{\pm}(a)g(a) = \sum_{a=1}^{M} f(a)g^{\pm}(a). \tag{18.55}$$

and in particular

$$\sum_{a=1}^{M} f^{\pm}(a)g^{\mp}(a) = \sum_{a=1}^{M} f(a)g^{\pm\mp}(a) = 0, \tag{18.56}$$

meaning that both the odd part of an even function and the even part of an odd function is 0.

$A_1(s)$ $(0 < x < 1)$ is even and $\bar{B}_1(x)$ is odd, which entails

$$\psi^e\left(\frac{a}{M}\right) = -\gamma - \log M - \sum_{b=1}^{M-1} \varepsilon_a(-b)A_1\left(\frac{b}{M}\right) \tag{18.57}$$

$$\psi^o\left(\frac{a}{M}\right) = \pi i \sum_{b=1}^{M-1} \varepsilon_a(-b)\bar{B}_1\left(\frac{b}{M}\right).$$

Proof 18.3 *Equation (18.55) follows from*

$$\begin{aligned}\sum_{a=1}^{M} f^{\pm}(a)g(a) &= \frac{1}{2}\left(\sum_{a=1}^{M} f(a) \pm \sum_{a=1}^{M} f(a)g(M-a)\right)\\ &= \sum_{a=1}^{M} f(a)\frac{1}{2}(g(a) \pm g(M-a)).\end{aligned}$$

Since

$$\psi\left(1-\frac{a}{M}\right) = -\gamma - \log M + \pi i \sum_{b=1}^{M-1} \varepsilon_a(-b)\bar{B}_1\left(1-\frac{b}{M}\right) - \sum_{b=1}^{M-1} \varepsilon_a(-b)A_1\left(\frac{b}{M}\right),$$

Equation (18.56) follows from

$$\sum_{a=1}^{M} f(a)g^{\pm\mp}(a) = \sum_{a=1}^{M} f(a)\frac{1}{2}(g^{\mp}(a) \pm g^{\mp}(-a)) = 0.$$

Equation (18.57) *follows on applying equation* (18.56) *and parity of* $A, \bar{B}$ *to equation* (18.45).

E.g. assuming that $f(M) = 0$, we have

$$\sum_{a=1}^{M} f^o(a)\zeta\left(s, \frac{a}{M}\right) = \sum_{a=1}^{M-1} f(a)\zeta^o\left(s, \frac{a}{M}\right), \tag{18.58}$$

which in the limit as $s \to 1$ leads to

$$\lim_{s\to1} \sum_{a=1}^{M} f^o(a)\zeta\left(s, \frac{a}{M}\right) = \lim_{s\to1} \sum_{a=1}^{M} f(a)\zeta^o\left(s, \frac{a}{M}\right) = -\pi \sum_{a=1}^{M-1} f(a) \cot \frac{a}{M}\pi. \tag{18.59}$$

This entails (equation 18.62) since the left-hand side is $L(1, f^o)$.

Theorem 18.2 *The two expressions* (18.50) *for the Laurent constant* $\gamma_0(f)$ *rephrased as*

$$-\frac{\hat{f}^e(M)}{\sqrt{M}} \log M - \frac{1}{M} \sum_{a=1}^{M} f^e(a)\psi\left(\frac{a}{M}\right) = \gamma_0(f^e) \tag{18.60}$$

$$= \frac{1}{\sqrt{M}} \sum_{a=1}^{M-1} \hat{f}(a) A_1\left(\frac{a}{M}\right) + \frac{\hat{f}^e(M)}{\sqrt{M}}\gamma$$

and

$$-\frac{1}{M} \sum_{a=1}^{M} f^o(a)\psi\left(\frac{a}{M}\right) = \gamma_0(f^o) = L(1, f^o) = \frac{\pi}{2M} \sum_{a=1}^{M-1} f(a) \cot \frac{\pi a}{M} \tag{18.61}$$

are equivalent to Gauss' first formula (18.51) $(1 \le a \le M-1)$.

Proof 18.4 *It suffices to deduce the right-hand side expression for* $\gamma_0(f^\pm)$ *from Gauss' first formula. Equation* (18.61) *follows from the second equality of equation* (18.57).

On the other hand,

$$\gamma_0(f^e) = -\frac{\hat{f}(M)}{\sqrt{M}} \log M - \frac{1}{M} \sum_{a=1}^{M} f(a)\psi^e\left(\frac{a}{M}\right)$$

since $\widehat{f^o}(M)=0$.

Substituting equation (18.57) *in the right-hand side, we obtain*

$$\gamma_0(f^e) = -\frac{\hat{f}(M)}{\sqrt{M}} \log M - \frac{1}{M} \sum_{a=1}^{M} f(a) \left(-\gamma - \log M - \sum_{b=1}^{M-1} \varepsilon_a(-b) A_1\left(\frac{b}{M}\right)\right)$$

$$= \frac{\hat{f}(M)}{\sqrt{M}}\gamma + \frac{1}{M} \sum_{b=1}^{M-1} A_1\left(\frac{b}{M}\right) \sum_{a=1}^{M} \varepsilon_a(-b) f(a),$$

which leads to the second equality of equation (18.60), *completing the proof.*

Corollary 18.2 ([8]) *Suppose $a \not\equiv 0 \bmod M$ (similarly as in Theorem 18.3). Then Gauss' first formula* (18.45) *for the digamma function at rational argument, which we restate as*

$$\psi\left(\frac{a}{M}\right) = -\gamma - \log M + \pi i \sum_{b=1}^{M-1} \varepsilon_a(-b)\bar{B}_1\left(\frac{b}{M}\right) - \sum_{j=1}^{M-1} \cos\frac{2\pi aj}{M} A_1\left(\frac{j}{M}\right)$$

$$= -\gamma - \log M - \frac{\pi}{2}\cot\frac{a}{M}\pi + \sum_{j=1}^{M-1} \cos\frac{2\pi aj}{M}\log 2\sin\frac{\pi j}{M}$$

is equivalent to finite expressions for $L(1,\chi)$ ($\chi \bmod M$)

$$L(1,\chi) = \frac{\pi}{2M}\sum_{a=1}^{M-1} \chi(a)\cot\frac{a}{M}\pi \tag{18.62}$$

for χ odd and

$$L(1,\chi) = \frac{1}{\sqrt{M}}\sum_{a=1}^{M-1} \widehat{\chi}(a)A_1\left(\frac{a}{M}\right) = -\frac{1}{\sqrt{M}}\sum_{a=1}^{M-1} \widehat{\chi}(a)\log\left(2\sin\frac{a}{M}\pi\right) \tag{18.63}$$

for $\chi \neq \chi_0$ even, where $\widehat{\chi}$ is the DFT of χ, cf. equation (18.21)*:*

$$\widehat{\chi}(a) = \frac{1}{\sqrt{M}}\sum_{k \bmod M} \chi(k)\, e^{-2\pi i \frac{k}{M}a}$$

also known as the general Gauss sum.

The two expressions

$$\psi^o\left(\frac{a}{M}\right) = \pi i \sum_{b=1}^{M-1} \varepsilon_a(-b)\bar{B}_1\left(\frac{b}{M}\right) = -\frac{\pi}{2}\cot\frac{a}{M}\pi. \tag{18.64}$$

arising from equation (18.39) has been known as the general Eisenstein formula:

Corollary 18.3 *The pair of the Eisenstein formula*

$$\sqrt{M}\widehat{\cot}\left(\frac{a}{M}\right) = \sum_{b=1}^{M-1} \varepsilon_a(-b)\cot\left(\frac{b}{M}\pi\right) = -\frac{1}{2i}\bar{B}_1\left(\frac{a}{M}\right) \tag{18.65}$$

and the (inverse) Eisenstein formula

$$\sqrt{M}\hat{\bar{B}}_1\left(\frac{a}{M}\right) = \sum_{b=1}^{M-1} \varepsilon_a(-b)\bar{B}_1\left(\frac{b}{M}\right) = -\frac{1}{2i}\cot\frac{\pi a}{M} \tag{18.66}$$

is a consequence of

$$\frac{\pi}{M}\sum_{b=1}^{M-1} f(b)\cot\left(\frac{\pi b}{M}\right) = \frac{-2\pi i}{\sqrt{M}}\sum_{b=1}^{M-1} \hat{f}(b)\bar{B}_1\left(\frac{b}{M}\right) \tag{18.67}$$

with $f = \varepsilon_a$ resp. $f = \chi_a$ for which equation (18.48) *resp. equation* (18.49) *holds.*

Equation (18.64) gives

$$L(1,\chi^o) = -\frac{\pi i}{\sqrt{M}} \sum_{a=1}^{M-1} \widehat{\chi}(a) B_1\left(\frac{a}{M}\right) \tag{18.68}$$

corresponding to equation (18.63). This reduces to a simpler form only when the character is primitive:

Suppose χ is a primitive character with conductor f. Then we have

$$L(1,\chi) = \pi i \frac{\tau(\chi)}{f} B_{1,\bar{\chi}} \tag{18.69}$$

where

$$B_{1,\bar{\chi}} = \sum_{a=1}^{f-1} \bar{\chi}(a) \bar{B}_1\left(\frac{a}{f}\right) \tag{18.70}$$

is the first generalized Bernoulli number. It is this form that has been developed in the aspect of p-adic interpolation, cf. e.g. [18].

18.4 ALGEBRAIC ELUCIDATION OF ANALYTIC EXPRESSIONS

In this section we shall give algebraic elucidation of analytic expression proved above. 1. *The equivalence of the finite expression for $L(1,\chi)$ and Gauss' first formula*: There are two known proofs [8] as expounded in Refs. [9, Theorem 8.2, pp. 174–175] and [2, Chapter 4]. The latter is another manifestation of algebraic elucidation of analytic expressions and in both of them the proof of deducing Gauss' first formula from the finite expressions is naturally more involved than that of the reverse implication. For the reverse implication needs inly substitution of equation (18.51).

In our setting, however, the reverse implication looks rather simple as shown in Lemma 18.3 by just choosing $f = \chi_a$ but it looks an ad hoc treatment. But indeed, there is a hidden algebraic structure as indicated in Ref. [2, Chapter 4] which is exhibited in

Lemma 18.6 ([2, Theorem 4.4]) *Let G be a finite Abelian group of order M written additively and $\hat{G}$ its character group. Consider a vector space of dimension M*

$$V = \{f : G \to \mathbb{C}\}. \tag{18.71}$$

which is isomorphic to $\mathbb{C}[\hat{G}]$.

Let χ_a be the characteristic function

$$\chi_a(b) = \delta_{ab} = \begin{cases} 1 & a = b \\ 0 & a \neq b \end{cases}, \tag{18.72}$$

which means orthogonality

$$\chi_a(b) = \delta_{ab} = \frac{1}{|G|} \sum_{\varepsilon \in \hat{G}} \bar{\varepsilon}(a) \varepsilon(b). \tag{18.73}$$

Then for any function $f \in V$ we have the base change formula

$$\sum_{a \in G} f(a)\chi_a = f = \frac{1}{M} \sum_{\varepsilon \in \hat{G}} \sum_{a \in G} f(a)\varepsilon(-a)\varepsilon \tag{18.74}$$

or

$$\sum_{a=1}^{M} f(a)\chi_a(n) = f(n) = \frac{1}{\sqrt{M}} \sum_{a=1}^{M} \hat{f}(a)\varepsilon_a(n) = \hat{\hat{f}}(-n), \tag{18.75}$$

2. *The case* $\hat{f}(M) \neq 0$: The following assertions about the principal character χ_0 appear both of two proofs.

$$\eta_{\chi_0} := \sum_{\substack{a=1 \\ (a,M)=1}}^{M-1} \psi\left(\frac{a}{M}\right) = -\varphi(M)(\log M + \gamma) - \log N_M, \tag{18.76}$$

where

$$\log N_M = \sum_{d|M} \varphi(d)\Lambda\left(\frac{M}{d}\right) = \varphi(M) \sum_{p|M} \frac{\log p}{p-1} \tag{18.77}$$

by Refs. [11, p. 136] and [9, Theorem 8.3, p. 176]. And $\eta_{\chi_0}^e$ is found in Ref. [9, (8.52), p. 182]

$$\eta_{\chi_0}^e = \sum_{b=1}^{M-1} \chi_0(b) \sum_{a=1}^{M-1} \varepsilon_b(-a) \log\left(2 \sin \frac{a}{M}\pi\right) = -\log N_M. \tag{18.78}$$

Hence

$$\eta_{\chi_0} = \eta_{\chi_0^e} - \varphi(M)(\log M + \gamma). \tag{18.79}$$

But there was no explanation given as to the origin of equation (18.76). The following theorem reveals the underlying structure.

Let χ_0 be the principal character mod M and

$$L(s, \chi_0) = \zeta(s) \prod_{p|M} (1 - p^{-s}) = \frac{\prod_{p|M} (1 - p^{-1})}{s-1} + c)\chi_0 + O(s-1)$$

Theorem 18.3 *The Laurent expansion holds:*

$$L(s, \chi_0) = \zeta(s) \prod_{p|M} (1 - p^{-s}) = \frac{\prod_{p|M} (1 - p^{-1})}{s-1} + c(\chi_0) + O(s-1) \tag{18.80}$$

around $s = 1$. *For* $f^e = \chi_0$ *Equation* (18.60) *reads*

$$-\frac{\varphi(M)}{\sqrt{M}} - \eta_{\chi_0} = \gamma_0(\chi_0)$$

$$= \frac{1}{\sqrt{M}} \sum_{a=1}^{M-1} \hat{\chi}_0(a) A_1 \left(\frac{a}{M}\right) + \frac{\varphi(M)}{\sqrt{M}} \gamma$$

$$= \eta_{\chi_0}^e + \frac{\varphi(M)}{\sqrt{M}} \gamma, \tag{18.81}$$

which leads to equation (18.79).

3. *The Laurent expansion* (18.7): Gauss' second formula involves the logarithm $\log(1-z)$, $|z| \le 1$, $z \neq 1$ which is used by Lehmer [11] and Funakura et al. [6]. They treated only this Lerch zeta-aspect, so that as Corollary 18.1 gives, only the right-hand side expression of equation (18.60). The superseding ingredient is the use of the Euler digamma function for expression the Laurent constant $\gamma_0(f)$. Equation (18.7) looks rather foreign but it follows from the functional equation for the Riemann zeta-function as is proved in Ref. [2, Chapter 3]. There equation (18.7) is proved by

Corollary 18.4 (Hermite's formula [16, p. 91, (12)]). *For* $\operatorname{Re} a > 0$ *we have*

$$\zeta(s,a) = \frac{1}{2} a^{-s} + \frac{a^{1-s}}{s-1} + 2 \int_0^\infty \left(a^2 + x^2\right)^{-\frac{s}{2}} \sin\left(s \arctan \frac{x}{a}\right) \frac{dx}{e^{2\pi x} - 1}. \tag{18.82}$$

The corollary in turn is deduced from the general Plana summation formula, which is proved by the functional equation for the Riemann zeta-function. As is known, the functional equation is equivalent to the Bochner modular relation which reflects the group structure of the positive reals: $x \leftrightarrow x^{-1}$ $(x > 0)$.

As is shown in Ref. [2, Chapter 1], in the Dirichlet class number formula in the form

$$L(1, \chi_d) = \frac{2\pi h(d)}{w\sqrt{|d|}} \quad d < 0; \quad L(1, \chi_d) = \frac{\sqrt{d} h(d)}{2 \log \varepsilon}, \quad d > 0, \tag{18.83}$$

the expression for $L(1, \chi_d)$ are due to the Fourier-Bessel expansion which is equivalent to the functional equation. The equalities in equation (18.83) depend on the prime decomposition in quadratic fields.

4. It is interesting to characterize the Dirichlet characters among periodic functions. As we have seen, periodic functions in general have both even and odd parts while Dirichler characters are strictly divided into even and odd. Non-principal Dirichlet characters satisfy the vanishing condition and the treatment is easier. Some results toward this direction are stated in Refs. [14, 15].

ACKNOWLEDGMENT

The authors would like to thank the referee for suggesting typos and general remarks.

BIBLIOGRAPHY

[1] Apostol, T. M. (1976). *Introduction to Analytic Number Theory*, Springer Verlag, Berlin.

[2] Chakraborty, K., Kanemitsu, S., and Kuzumaki, T. (2023). Modular relations and parity in number theory: Unification and generalization vom etwas anderen Standpunkte aus

[3] Davenport, H. (1980). *Multiplicative Number Theory*, 1st ed. Markham, Chicago 1967, 2nd ed. Springer Verlag, New York.

[4] Erdélyi, A, Magnus, W., Oberhettinger, F., and Tricomi, F. G. (1953). *Higher Transcendental Functions*, I–III McGraw-Hill, New York, Toronto and London.

[5] Frazier, M. W. (1999). *An Introduction to Wavelettes through Linear Algebra*, Springer, New York.

[6] Funakura, T. (1990). On Kronecker's limit formula for Dirichlet series with periodic coefficients, *Acta Arith.* **55**, 59–73.

[7] Gal, I. (1961). *Lectures on Number Theory*, Jones Letter Service, Minneapolis.

[8] Hashimoto, M., Kanemitsu, S., and Toda, M. (2008). On Gauss' formula for ψ and finite expressions for the L-series at 1, *J. Math. Soc. Japan* **60**, 219–236.

[9] Kanemitsu, S. and Tsukada, H. (2007). *Vistas of Special Functions*, World Scientific, Singapore.

[10] Kanemitsu, S. and Tsukada, H. (2014). *Contributions to the Theory of Zeta-Functions: The Modular Relation Supremacy*, World Scientific, Singapore.

[11] Lehmer, D. H. (1975). Euler constants for arithmetic progressions, *Acta Arith.* **27**, 125–142; Selected Papers of D. H. Lehmer, Vol. II, 591-6oe, Charles Babbage Research Center, Manitoba 1981.

[12] Li, F. -H., Wang, N. -L., and Kanemitsu, S. (2018). *Number Theory and Its Applications*, World Scienctific, London-Singapore-New Jersey.

[13] Li, H. -L., Li, F. -H, Wang, N. -L, and Kanemitsu, S. (2017). *Number Theory and Its Applications II*, World Scienctific, London-Singapore-New Jersey.

[14] Prachar, K. (1957). *Primzahlverteilung*, Springer Verlag, Berlin-Göttingen-Heidelberg (Second edition, Springer Verlag 1978).

[15] Spira, R. (1970). Residue class characters, *Duke Math. J.* **37**, 633–637.

[16] Srivastava, H. M. and Choi, J. -S. (2001). *Series Associated with the Zeta and Related Functions*, Kluwer Academic Publication, Dordrecht.

[17] Wang, N -L., Agarwal, K., and Kanemitsu, S. (2020). Limiting values and functional and difference equations, *Math. Differ. Differ. Equations* **8**, 407. doi:10.3390/math8030407.

[18] Washington, L. (1983). *Introduction to Cyclotomic Fields*, Spring Verlay, New York.

[19] Yamamoto, Y. (1977). Dirichlet series with periodic coefficients, *Proceedings of the International Symposium on Algebraic Number Theory*, Kyoto 1976, pp. 275–289. JSPS, Tokyo.

CHAPTER 19

The Lotka-Volterra Dynamical System and Its Discretization

Márcia Lemos-Silva and Delfim F. M. Torres

University of Aveiro

CONTENTS

19.1 INTRODUCTION TO THE LOTKA–VOLTERRA MODEL

Prey-predator equations intend to describe the dynamics of an ecological system where two species interact with each other. Alfred J. Lotka (1880–1949) introduced such equations in 1925 [4]; and Vito Volterra (1860-1940) studied them, independently [11]. For this reason, these equations are known as Lotka–Volterra equations. There have been great developments regarding this topic ever since [1].

Here we consider the classical model of Lotka–Volterra, which is composed of two autonomous and nonlinear differential equations given by

$$\begin{cases} \dot{x} = \alpha x - \beta xy, \\ \dot{y} = -\delta y + \gamma xy, \end{cases} \tag{19.1}$$

where $x(t)$ and $y(t)$ represent the size at time t of prey and predator populations, respectively. Moreover, all the parameters α, β, γ, and δ are assumed to be positive.

If the density of both species reaches the zero value at any moment t, then they will remain there indefinitely, which represents the natural extinction of both species. The absence of prey leads to the extinction of predators since in that case $y(t)$ converges

DOI: 10.1201/9781003388678-19

to 0 when $t \to +\infty$. On the other hand, the absence of predators leads to exponential growth of prey, since $x(t) \to +\infty$ when $t \to +\infty$.

From an ecological point of view, population densities must always be nonnegative, restricting the system trajectories to $\mathbb{R}^2_+$. From the equations of system (19.1), we have

$$\begin{cases} \dot{x}\big|_{x=0} = 0, \\ \dot{y}\big|_{y=0} = 0, \end{cases}$$

from which, according to Lemma 2 of [13], we can conclude that the solution of the system is nonnegative, meaning that $\mathbb{R}^2_+$ is the invariant domain of the system.

This system has equilibria at two different points: $p_1 = (0,0)$ and $p_2 = \left(\frac{\delta}{\gamma}, \frac{\alpha}{\beta}\right)$. To observe the approximate behavior of the solutions over time near these equilibrium points, we start by computing the Jacobian matrix of the system, which is given by

$$J(x,y) = \begin{pmatrix} \alpha - \beta y & -\beta x \\ \gamma y & \gamma x - \delta \end{pmatrix}. \tag{19.2}$$

The Jacobian matrix (19.2), evaluated at the equilibrium $(0,0)$, is given by

$$J(0,0) = \begin{pmatrix} \alpha & 0 \\ 0 & -\delta \end{pmatrix}.$$

The corresponding eigenvalues are $\lambda_1 = \alpha$ and $\lambda_2 = -\delta$ and, as $\alpha, \delta > 0$, it turns out that $(0,0)$ is a saddle point. In contrast, the Jacobian matrix (19.2) evaluated at the coexistence equilibrium point p_2 is

$$J\left(\frac{\delta}{\gamma}, \frac{\alpha}{\beta}\right) = \begin{pmatrix} 0 & -\frac{\beta\delta}{\gamma} \\ \frac{\alpha\gamma}{\beta} & 0 \end{pmatrix},$$

for which eigenvalues are pure imaginary: $\lambda = \pm i\sqrt{\alpha\delta}$. This means that p_2 is a stable center in the linearized system. However, with this analysis, nothing can be concluded regarding the stability for the nonlinear system at this equilibrium.

To investigate the phase portrait of system (19.1), we start by drawing the two lines

$$x = \frac{\delta}{\gamma}, \quad y = \frac{\alpha}{\beta}.$$

By doing so, the first quadrant of the xy-plane is divided into four different regions, as shown in Figure 19.1.

In each region, the signs of $\dot{x}$ and $\dot{y}$ determine the behavior of the solution of the system. By analyzing the equations of system (19.1), the following result holds.

Proposition 19.1 *The trajectory x of the system will*

- *decrease in regions I and II, since $\dot{x} < 0$;*

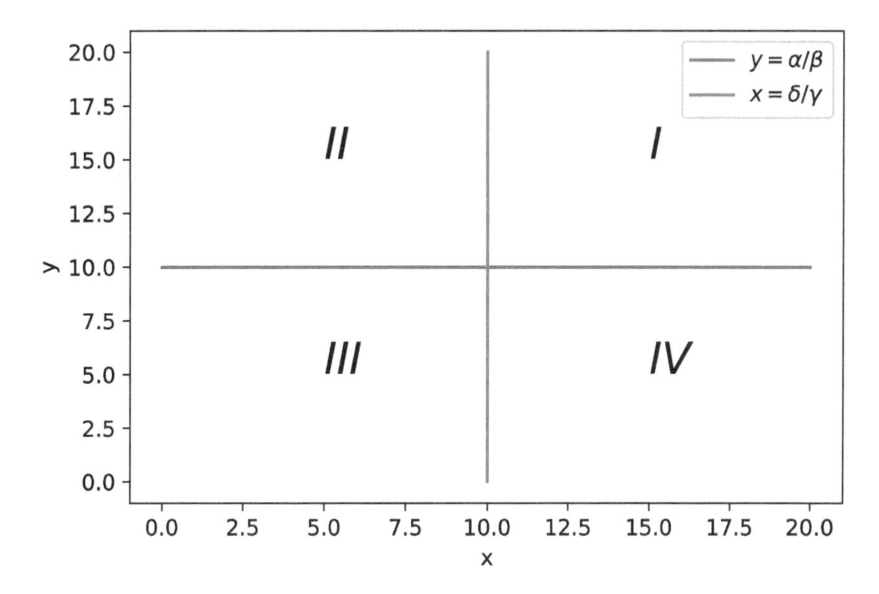

Figure 19.1 Regions defined by the lines $x = \frac{\delta}{\gamma}$ and $y = \frac{\alpha}{\beta}$, with $\alpha = 1$, $\beta = 0.1$, $\gamma = 0.075$, $\delta = 0.75$.

- *increase in regions III and IV, since* $\dot{x} > 0$.

Regarding the trajectory y *of the system, it will*

- *decrease in regions II and III, since* $\dot{y} < 0$*;*
- *increase in regions I and IV, since* $\dot{y} > 0$.

Proposition 19.1 suggests that the curve of the system in the phase plane will be counterclockwise around the equilibrium point p_2, but that it is not enough to conclude whether the trajectory spiral toward p_2; spiral out, toward infinity; or it is a closed curve. Despite this, it has already been explained that this equilibrium is, in fact, a center in the nonlinear system, meaning the trajectories will be closed curves. This allows to write the following result.

Proposition 19.2 *Except for those beginning at the equilibrium* p_2 *or at coordinate axes, every trajectory of the system is a closed orbit that turns counterclockwise around the equilibrium point* p_2.

The behavior described by Proposition 19.2 can be seen in Figure 19.2, for several different initial conditions.

From Proposition 19.2, it comes directly that the densities of predators and prey will oscillate periodically, as can be seen in Figure 19.3, with the amplitude and frequency of oscillations depending only on the considered initial conditions.

All the results described so far are well known. In particular, both Propositions 19.1 and 19.2 can be found, e.g., in [2]. In Sections 19.2 and 19.3 we provide new insights.

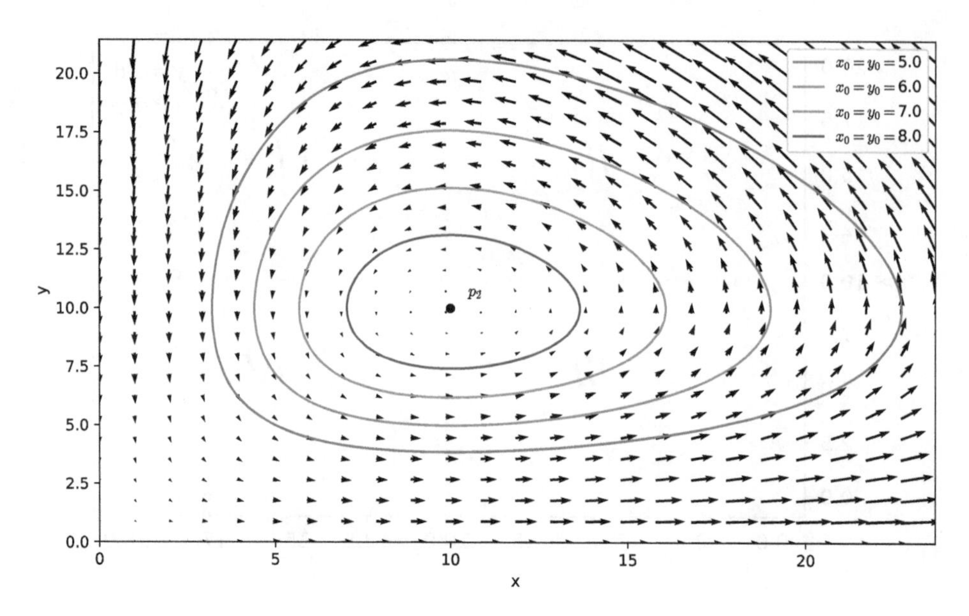

Figure 19.2 Phase portrait of system (19.1) with $\alpha = 1$, $\beta = 0.1$, $\gamma = 0.075$, and $\delta = 0.75$.

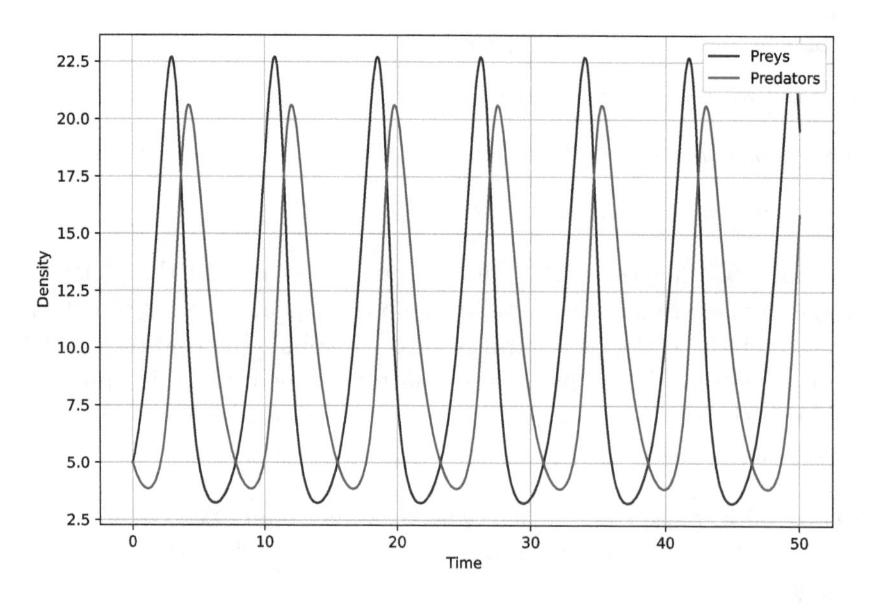

Figure 19.3 Oscillations of prey and predator densities for system (19.1) with $\alpha = 1$, $\beta = 0.1$, $\gamma = 0.075$, and $\delta = 0.75$.

19.2 DISCRETIZATION BY EULER'S METHOD

There are several methods for converting continuous systems into discrete counterparts. The most conventional way to do so is to implement a standard difference scheme, the most classical one being the progressive Euler's method. However, it is known that this method can raise several problems such as lack of dynamical consistency, even when applied to the simplest systems [10]. A discrete-time model is said to be dynamically consistent with its continuous analog if they both exhibit the same dynamical behavior, namely the stability behavior of fixed points, bifurcation, and chaos. In [5], Mickens points out that the fundamental reason for the existence of numerical instabilities is that discrete models have a larger parameter space than the corresponding differential equations: one has the step size h as an additional parameter. Nevertheless, this step h is, obviously, inherent to any discretization and any discrete dynamical system. Therefore, it is crucial to consider a numerical method that is able to overcome this setback during discretization.

Here we prove that Euler's method applied to the Lotka–Volterra model brings a discrete system that is not dynamically consistent with its continuous counterpart.

Applying the progressive Euler's method to both equations of system (19.1), we obtain that

$$\begin{cases} x_{i+1} = x_i + h(\alpha x_i - \beta x_i y_i), \\ y_{i+1} = y_i + h(\gamma x_i y_i - \delta y_i), \end{cases} \tag{19.3}$$

where h denotes the step size and x_i and y_i define the density of the prey and predators' populations, respectively, at time i.

The fixed points of system (19.3) are $p_1 = (0,0)$ and $p_2 = \left(\frac{\delta}{\gamma}, \frac{\alpha}{\beta}\right)$. To determine the nature of the fixed points, one must compute the Jacobian matrix $J(x,y)$ of system (19.3). This matrix is given by

$$J(x,y) = \begin{pmatrix} -\beta hy + \alpha h + 1 & -\beta hx \\ \gamma hy & \gamma hx - \delta h + 1 \end{pmatrix}. \tag{19.4}$$

Follows our first result.

Theorem 19.1 *The fixed point* $(0,0)$ *of system (19.3) is*

- *a saddle point if* $h \in \left]0, \frac{2}{\delta}\right[$;
- *a source if* $h \in \left]\frac{2}{\delta}, +\infty\right[$.

Proof. The Jacobian matrix (19.4) evaluated at the fixed point $(0,0)$ is

$$J(0,0) = \begin{pmatrix} \alpha h + 1 & 0 \\ 0 & -\delta h + 1 \end{pmatrix},$$

whose eigenvalues are $\lambda_1 = -\delta h + 1$ and $\lambda_2 = \alpha h + 1$. As all the parameters are positive, one can easily conclude that $|\lambda_2| > 1$. On the other hand, $|\lambda_1|$ can either be greater or less than one. In particular,

$$\begin{aligned} |\lambda_1| < 1 &\Leftrightarrow -\delta h + 1 < 1 \wedge -\delta h + 1 > -1 \\ &\Leftrightarrow h > 0 \wedge h < \frac{2}{\delta}, \end{aligned}$$

and

$$|\lambda_1| > 1 \Leftrightarrow -\delta h + 1 > 1 \vee -\delta h + 1 < -1$$
$$\Leftrightarrow h < 0 \vee h > \frac{2}{\delta}.$$

Therefore, the fixed point $(0,0)$ is a saddle point if $h \in \left]0, \frac{2}{\delta}\right[$ or a source if $h \in \left]\frac{2}{\delta}, +\infty\right[$. As h is strictly positive, the condition $h < 0$ is not considered. □

For both possibilities of Theorem 19.1, the point $p_1 = (0,0)$ is unstable, which brings no major changes to what is obtained in the continuous case, as described in Section 19.1. We now study what happens with the second fixed point p_2.

Theorem 19.2 *The fixed point* $\left(\frac{\delta}{\gamma}, \frac{\alpha}{\beta}\right)$ *is an unstable focus.*

Proof. The Jacobian matrix (19.4) evaluated at the fixed point $\left(\frac{\delta}{\gamma}, \frac{\alpha}{\beta}\right)$ is given by

$$J\left(\frac{\delta}{\gamma}, \frac{\alpha}{\beta}\right) = \begin{pmatrix} 1 & -\frac{\beta\delta h}{\gamma} \\ \frac{\alpha\gamma h}{\beta} & 1 \end{pmatrix},$$

whose eigenvalues are the complex conjugates $\lambda = 1 \pm \sqrt{\alpha\delta h}$. As α, δ, and h are strictly positive, it is clear that $|\lambda| > 1$, meaning that the fixed point $\left(\frac{\delta}{\gamma}, \frac{\alpha}{\beta}\right)$ is an unstable focus. □

Theorem 19.2 asserts that the orbits of system (19.3) near the fixed point p_2 will not be closed, but spirals that spiral out toward infinity.

Through a simple analysis of the system equations, taking into consideration the four regions defined in Figure 19.1, it is possible to understand the direction of the solution in those regions.

Theorem 19.3 *The trajectory x of system (19.3) will*

- *decrease in regions I and II, i.e., $x_{i+1} < x_i$;*
- *increase in regions III and IV, i.e., $x_{i+1} > x_i$.*

Regarding the trajectory y of system (19.3), it will

- *decrease in regions II and III, i.e., $y_{i+1} < y_i$;*
- *increase in regions I and IV, i.e., $y_{i+1} > y_i$.*

Proof. We start by analyzing the trajectory of x by looking to the first equation of system (19.3). In regions I and II, $y_i > \frac{\alpha}{\beta}$. This means that

$$h(\alpha x_i - \beta x_i y_i) < 0,$$

and

$$x_i + h(\alpha x_i - \beta x_i y_i) < x_i \Rightarrow x_{i+1} < x_i.$$

On the other hand, in regions III and IV, we have $y_i < \frac{\alpha}{\beta}$. In this case,

$$h(\alpha x_i - \beta x_i y_i) > 0,$$

from which we can conclude that

$$x_i + h(\alpha x_i - \beta x_i y_i) > x_i \Rightarrow x_{i+1} > x_i.$$

Through the second equation of the system, by an analogous reasoning, the intended conclusions are obtained for y. □

Theorem 19.3 implies a counterclockwise displacement of the system. Precisely, it follows directly from Theorem 19.3 that a trajectory of equation (19.3) near the fixed point p_2 will spiral in a counterclockwise direction, as can be seen in Figure 19.4.

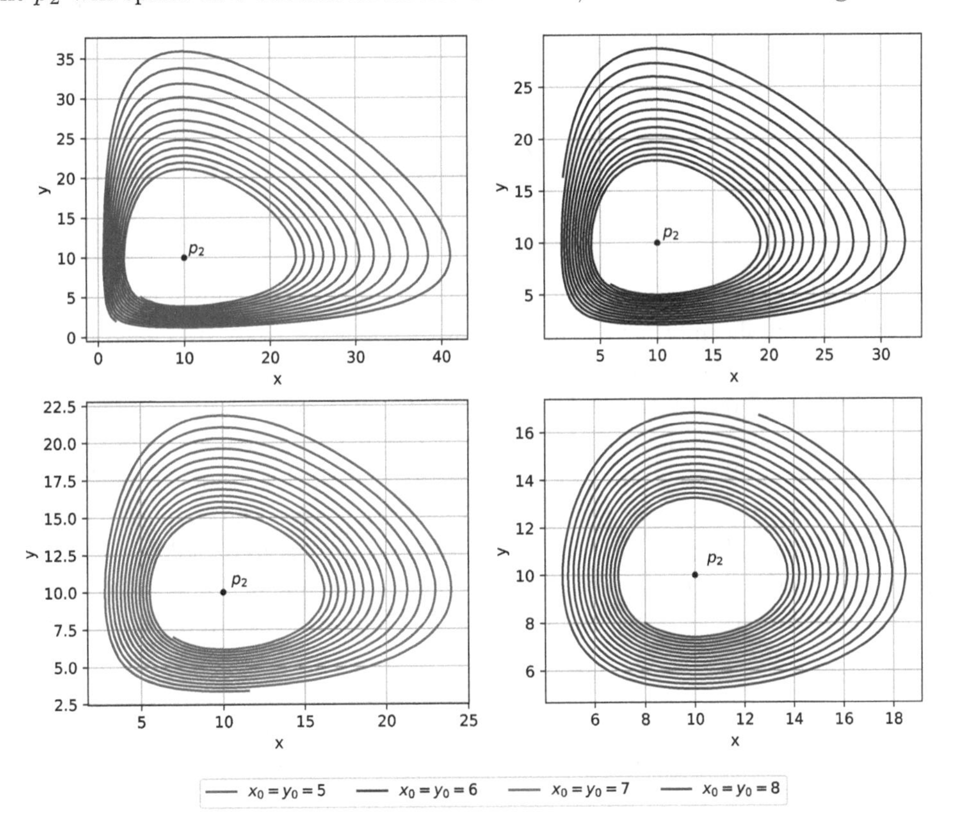

Figure 19.4 Trajectory of the system near the fixed point p_2 with $\alpha = 1$, $\beta = 0.1$, $\gamma = 0.075$, $\delta = 0.75$, and $h = 0.02$.

Since the trajectories of system (19.3) are not closed curves, they cease to be periodic orbits. Here, the trajectories are expansive, not converging to a particular fixed point, and the amplitude of the curves does not remain constant. In this case, the amplitude increases over time: see Figure 19.5.

In addition to the aforementioned dynamic inconsistency, caused by the progressive Euler method, the considered discrete system also makes it possible to predict

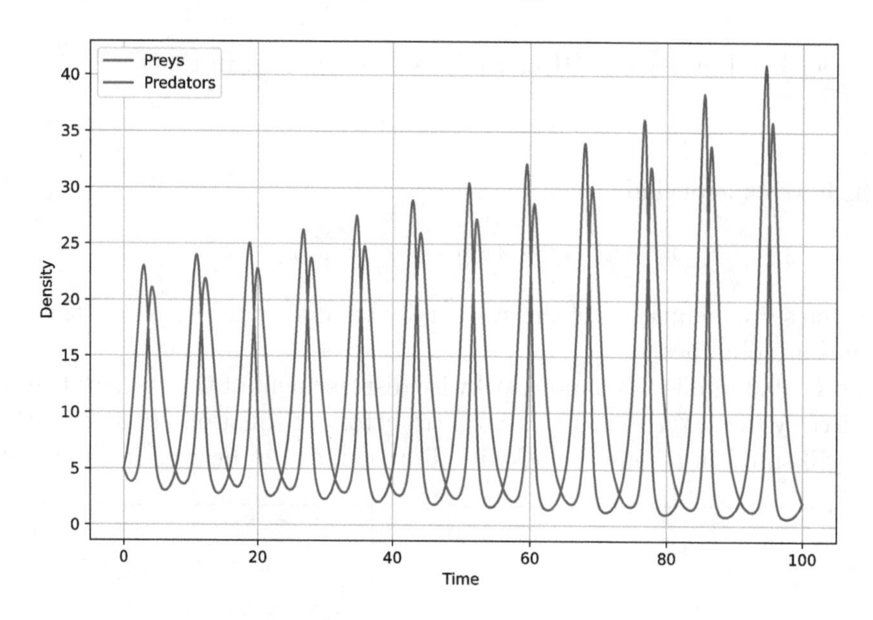

Figure 19.5 Oscillations of prey and predator densities for system (19.3) with $\alpha = 1$, $\beta = 0.1$, $\gamma = 0.075$, $\delta = 0.75$, $h = 0.02$, and $x_0 = y_0 = 5$.

negative population densities, even when all parameters and initial conditions are taken positive. Furthermore, it is also possible to prove that, under some circumstances, negative solutions can return to positive values. Although mathematically possible, these two possibilities do not make any sense, neither in the context of the problem (the problem is defined only in $\mathbb{R}^2_+$) nor from an ecological point of view. However, as we shall prove analytically, and geometrically, under Euler's method, both mentioned situations are indeed possible. This shows the inconsistency of the discrete-time system (19.3). Next we study such situations in detail.

According to the orientation of the solution pointed out in Theorem 19.3, system (19.3) can only predict negative solutions in two different cases and, in each of them, for only one of the variables.

1. Let (x_i, y_i) be a point in region II of Figure 19.1. Here we have $y_i > \frac{\alpha}{\beta}$, $h(\alpha x_i - \beta x_i y_i) < 0$, and $x_{i+1} < x_i$. From the first equation of system (19.3), x_{i+1} can assume a negative value if

$$x_i < -h(\alpha x_i - \beta x_i y_i).$$

Now, two situations can also occur. When the trajectory crosses the positive semi-axis yy, predicting a $x_{i+1} < 0$, this intersection can happen in such a way that y_i remains greater than $\frac{\alpha}{\beta}$ or y_i becomes less than that same value. We now note that, with $x_i < 0$, the first equation of the system can be rewritten as

$$x_{i+1} = -x_i + h(-\alpha x_i + \beta x_i y_i) \quad \text{with} \quad x_i, y_i > 0.$$

- If $y_i > \frac{\alpha}{\beta}$, then we have $h(-\alpha x_i + \beta x_i y_i) > 0$. Thus, x_{i+1} can assume a positive value, as long as $h(-\alpha x_i + \beta x_i y_i) > x_i$. If this happens, the system enters region III of Figure 19.1. Otherwise, the system goes outside the four mentioned regions, resulting in negative values for prey density.
- If $y_i < \frac{\alpha}{\beta}$, then $h(\alpha x_i - \beta x_i y_i) < 0$, which leads to

$$-x_i + h(-\alpha x_i + \beta x_i y_i) < 0 \Rightarrow x_{i+1} < 0,$$

meaning that the system will go outside the four admissible regions.

2. Let (x_i, y_i) be a point in region III of the Figure 19.1. Here $x_i < \frac{\delta}{\gamma}$, $h(\gamma x_i y_i - \delta y_i) < 0$, and $y_{i+1} < y_i$. According to the second equation of system (19.3), y_{i+1} can assume a negative value if

$$y_i < -h(\gamma x_i y_i - \delta y_i).$$

When the trajectory of the system crosses the positive semi-axis xx, obtaining $y_{i+1} < 0$, one can continue to have $x_i < \frac{\delta}{\gamma}$ or there can be a change in its value such that $x_i > \frac{\delta}{\gamma}$. Rewriting the second equation of system (19.3), knowing that now $y_i < 0$, we obtain

$$y_{i+1} = -y_i + h(-\gamma x_i y_i + \delta y_i) \quad \text{with} \quad x_i, y_i > 0.$$

- If $x_i < \frac{\delta}{\gamma}$, then $h(\gamma x_i y_i - \delta y_i) > 0$. Consequently, y_{i+1} may be positive as long as $h(\gamma x_i y_i - \delta y_i) > y_i$. In this case, the trajectory of the system will enter in region IV. Otherwise, it will remain outside the four regions under study, with negative values for the density of predators.
- If, on the other hand, $x_i > \frac{\delta}{\gamma}$, then $h(\gamma x_i y_i - \delta y_i) < 0$. Thus,

$$-y_i + h(-\gamma x_i y_i + \delta y_i) < 0 \Rightarrow y_{i+1} < 0,$$

which means that the system obtained by Euler's method gives negative values for y_{i+1}, with values outside the four admissible regions.

By way of example, changing the value of h from 0.02 to 0.03, it is possible to observe negative values for the variable x_i, as seen in Figure 19.6. In addition, it is verified that, after some time, the solutions that were previously negative return to positive values.

19.3 DISCRETIZATION BY MICKENS' METHOD

As seen in Section 19.2, the progressive Euler method, when applied to the classical Lotka–Volterra system (19.1), has the particularity of losing the periodic solutions, which correspond to closed curves in the phase space. In [7], Mickens points out that

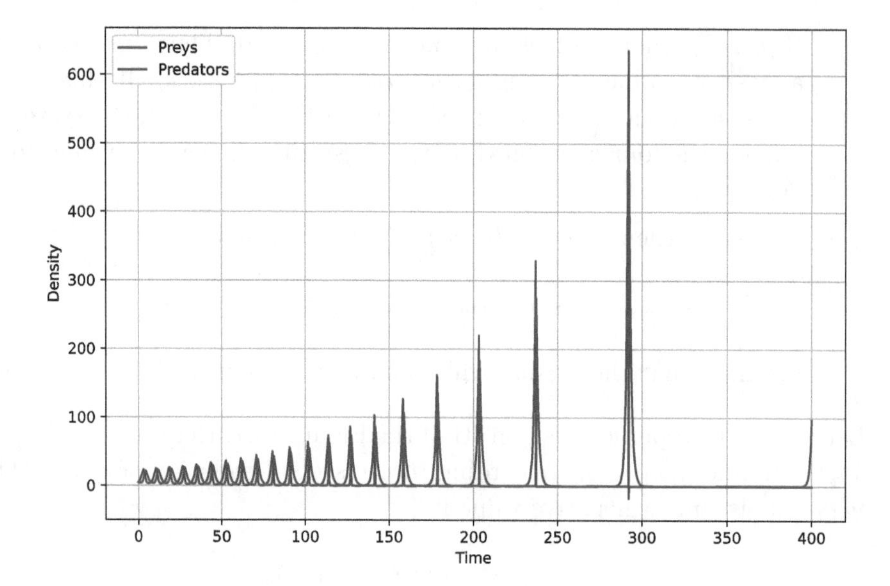

Figure 19.6 Oscillations of prey and predator densities for system (19.3) with $\alpha = 1$, $\beta = 0.1$, $\gamma = 0.075$, $\delta = 0.75$, $h = 0.03$, and $x_0 = y_0 = 5$.

the most likely reason for the loss of periodic solutions is the fact that the Lotka–Volterra system is not structurally stable, i.e., a small perturbation in the equations of the system may change its topological properties. In particular, it can change the closed curves into ones that can spiral into or out of the fixed point. It is known that the application of most classical numerical methods to a system with periodic solutions, transforms the original system into a very close one whose trajectories are not closed [9].

Here we intend to demonstrate that a nonstandard finite difference scheme, as generated according to the rules suggested by Mickens [6], can be applied consistently to a structurally unstable dynamical system such as the one of Lotka–Volterra. In addition to proving that this scheme preserves the periodic solutions, it is also ensured that the positivity of the system is kept unchanged.

In [7], Mickens suggests a discretization of the Lotka–Volterra model where, for simplicity, it is considered that all parameters – α, β, δ, and γ – are equal to one. Here, the same strategy suggested by Mickens is followed, with the difference that the parameters are general, assuming any value in $\mathbb{R}_+$.

Following the rules stated by Mickens, the first-order derivatives are approximated by

$$\dot{x} \to \frac{x_{i+1} - x_i}{\phi}$$

and

$$\dot{y} \to \frac{y_{i+1} - y_i}{\phi},$$

where in both cases ϕ is such that $\phi(h) = h + \mathcal{O}(h^2)$.

Starting with the first equation of system (19.1), the linear and nonlinear terms are all substituted by nonlocal forms given by

$$\alpha x = 2\alpha x - x \to 2\alpha x_i - \alpha x_{i+1},$$
$$-\beta xy \to -\beta x_{i+1} y_i.$$

Thus, through the above substitutions, the first equation of system (19.1) can be rewritten as

$$\frac{x_{i+1} - x_i}{\phi} = 2\alpha x_i - \alpha x_{i+1} - \beta x_{i+1} y_i,$$

which is equivalent to

$$x_{i+1} = \frac{x_i(2\alpha\phi + 1)}{1 + \alpha\phi + \beta\phi y_i}. \tag{19.5}$$

Regarding the second equation of system (19.1), the following substitutions are proposed:

$$\gamma xy = 2\gamma xy - \gamma xy \to 2\gamma x_{i+1} y_i - \gamma x_{i+1} y_{i+1},$$
$$-\delta y \to -\delta y_{i+1}.$$

Applying the two substitutions above, the second equation of the system is defined as

$$\frac{y_{i+1} - y_i}{\phi} = 2\gamma x_{i+1} y_i - \gamma x_{i+1} y_{i+1} - \delta y_{i+1},$$

which is equivalent to

$$y_{i+1} = \frac{y_i(2\gamma\phi x_{i+1} + 1)}{1 + \gamma\phi x_{i+1} + \delta\phi}. \tag{19.6}$$

Substituting (19.5) into (19.6), and joining both equations, we obtain the Lotka–Volterra model discretized by the Mickens method as

$$\begin{cases} x_{i+1} = \dfrac{x_i(2\alpha\phi + 1)}{1 + \alpha\phi + \beta\phi y_i}, \\ y_{i+1} = \dfrac{2\gamma\phi x_i y_i(2\alpha\phi + 1) + y_i(1 + \alpha\phi + \beta\phi y_i)}{(1 + \delta\phi)(1 + \alpha\phi + \beta\phi y_i) + \gamma\phi x_i(2\alpha\phi + 1)}, \end{cases} \tag{19.7}$$

which, as we shall show next, recovers the periodic solutions and ensure that the positivity property of the Lotka–Volterra system is maintained. In concrete, through a simple analysis of the equations of system (19.7), it is clear that the Mickens method guarantees that the positivity property is maintained. Indeed, by choosing $(x_0, y_0) \in \mathbb{R}^2_+$, and as a consequence of all the parameters being positive, it is impossible to have negative values for any of the variables, since both equations will be quotients of strictly positive quantities.

The fixed points of system (19.7) coincide with the ones of Sections 19.1 and 19.2: $p_1 = (0, 0)$ and $p_2 = \left(\frac{\delta}{\gamma}, \frac{\alpha}{\beta}\right)$.

Given the complexity of the system (19.7) under study, we make use of the free open-source mathematics software system SageMath [14] to analyze the nature of each one of the fixed points. For this purpose, we start by computing the Jacobian matrix of the system (19.7) in an arbitrary point (x, y). This matrix is given by

$$Jf_{(x,y)} = \begin{pmatrix} a & b \\ c & d \end{pmatrix}, \tag{19.8}$$

where

$$\begin{aligned}
a =& \frac{2\alpha\phi + 1}{\beta\phi y + \alpha\phi + 1}, \\
b =& -\frac{(2\alpha\phi + 1)\beta\phi x}{(\beta\phi y + \alpha\phi + 1)^2}, \\
c =& \frac{2(2\alpha\phi + 1)\gamma\phi y}{(2\alpha\phi + 1)\gamma\phi x + (\beta\phi y + \alpha\phi + 1)(\delta\phi + 1)} \\
& - \frac{(2(2\alpha\phi + 1)\gamma\phi xy + (\beta\phi y + \alpha\phi + 1)y)(2\alpha\phi + 1)\gamma\phi}{((2\alpha\phi + 1)\gamma\phi x + (\beta\phi y + \alpha\phi + 1)(\delta\phi + 1))^2}, \\
d =& -\frac{(2(2\alpha\phi + 1)\gamma\phi xy + (\beta\phi y + \alpha\phi + 1)y)(\delta\phi + 1)\beta\phi}{((2\alpha\phi + 1)\gamma\phi x + (\beta\phi y + \alpha\phi + 1)(\delta\phi + 1))^2} \\
& + \frac{2(2\alpha\phi + 1)\gamma\phi x + 2\beta\phi y + \alpha\phi + 1}{(2\alpha\phi + 1)\gamma\phi x + (\beta\phi y + \alpha\phi + 1)(\delta\phi + 1)}.
\end{aligned}$$

Theorem 19.4 *The fixed point* $(0, 0)$ *of system (19.7) is a saddle point.*

Proof. The Jacobian matrix (19.8) evaluated at the fixed point $(0, 0)$ is

$$Jf_{(0,0)} = \begin{pmatrix} \dfrac{2\alpha\phi + 1}{\alpha\phi + 1} & 0 \\ 0 & \dfrac{1}{\delta\phi + 1} \end{pmatrix},$$

whose eigenvalues are $\lambda_1 = \dfrac{1}{\delta\phi + 1}$ and $\lambda_2 = \dfrac{2\alpha\phi + 1}{\alpha\phi + 1}$. From these results, it is possible to draw the following conclusions:

- Since $\delta, \phi > 0$, it follows that $\delta\phi + 1 > 1$. Thus, λ_1 is always less than one, regardless of the values of δ and ϕ. Moreover, by the positivity of the parameters, it is clear that λ_1 is always greater than zero. Thereby, $|\lambda_1| < 1$.

- On the other hand, since $\alpha, \phi > 0$, then $\alpha\phi + 1$ is always less than $2\alpha\phi + 1$. For this reason, $\lambda_2 > 1$, which leads to $|\lambda_2| > 1$.

Thus, p_1 is a saddle point and, therefore, unstable. □

In contrast, the Jacobian matrix (19.2) evaluated at the coexistence equilibrium point p_2 is

$$Jf_{\left(\frac{\delta}{\gamma},\frac{\alpha}{\beta}\right)} = \begin{pmatrix} 1 & -\dfrac{\beta\delta\phi}{(2\alpha\phi+1)\gamma} \\ \dfrac{\alpha\gamma\phi}{2\beta\delta\phi+\beta} & \dfrac{3\alpha\delta\phi^2+2(\alpha+\delta)\phi+1}{4\alpha\delta\phi^2+2(\alpha+\delta)\phi+1} \end{pmatrix},$$

whose eigenvalues are complex conjugates

$$\lambda = \frac{7\alpha\delta\phi^2+4(\alpha+\delta)\phi+2\pm i\phi\sqrt{15\alpha^2\delta^2\phi^2+4\alpha\delta+8\left(\alpha^2\delta+\alpha\delta^2\right)\phi}}{2(4\alpha\delta\phi^2+2(\alpha+\delta)\phi+1)}.$$

With the help of **SageMath**, it is easily verified that $|\lambda| = 1$, which means that the point p_2 is a center point in the linearized system, while nothing can be concluded regarding the stability for the nonlinear system at this equilibrium. However, it is possible to verify numerically that the orbits are periodic, corresponding to closed curves in the phase space, meaning that, at least for the indicated parameter values, p_2 is, in fact, a center. This effect can be seen in Figures 19.7 and 19.8, which simultaneously show the results obtained here and those obtained in the continuous case.

Despite the fact that one of Mickens' rules mention that a more complex expression should be used for the step function $\phi(h)$, it appears that all the results achieved are valid regardless of the expression used for $\phi(h)$. Accordingly, in our simulations, we chose the simplest function given by $\phi(h) = h$. In particular, a step size given by $h = 0.01$ was considered. In both Figures 19.7 and 19.8, it is observed that the

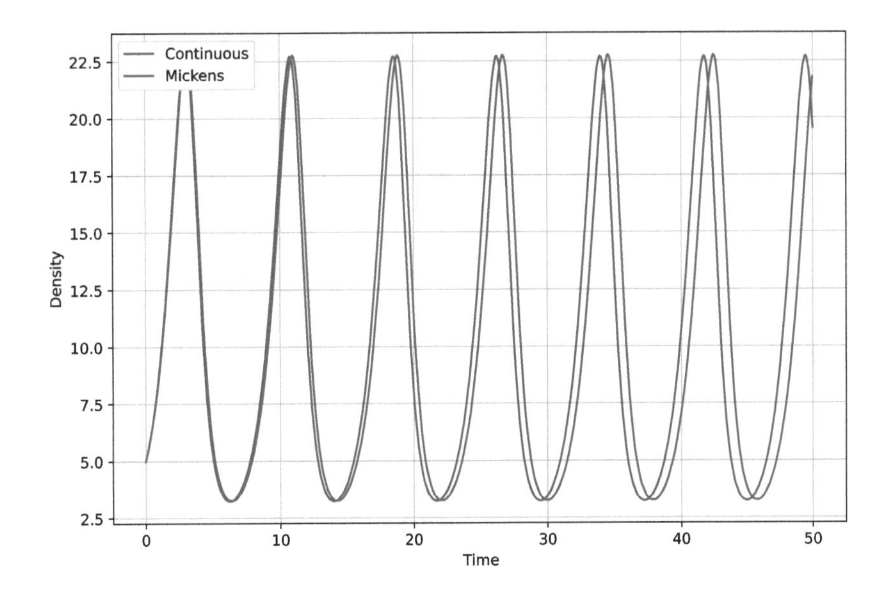

Figure 19.7 Oscillations of preys for the system (19.1) versus system (19.7) with $\alpha = 1$, $\beta = 0.1$, $\gamma = 0.075$, and $\delta = 0.75$.

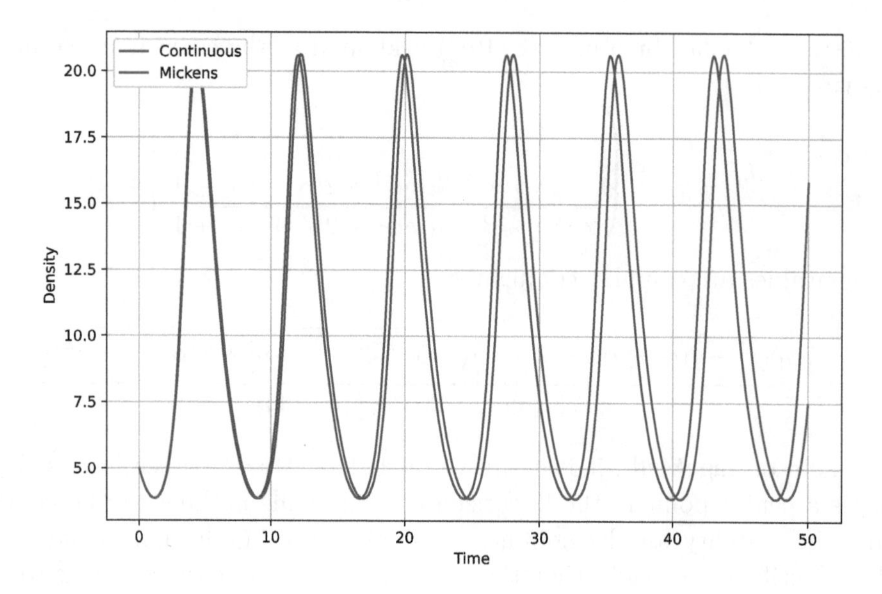

Figure 19.8 Oscillations of predators for the system (19.1) versus system (19.7) with $\alpha = 1$, $\beta = 0.1$, $\gamma = 0.075$, and $\delta = 0.75$.

periodic oscillations of the discrete system practically overlap those of the original continuous system. Logically, the smaller the value of the chosen step h, the more superimposed the curves for each of the systems will be.

To complete the analysis of system (19.7), we end by proving that Theorem 19.3 remains valid in this case, from which one can conclude that the direction of the trajectories of the Mickens' discrete system will continue to be counterclockwise.

Consider the first equation of system (19.7). Equivalently, one can write that

$$\frac{x_{i+1}}{x_i} = \frac{2\alpha\phi + 1}{1 + \alpha\phi + \beta\phi y_i}.$$

- Let y_i be a value that either belongs to regions I or II considered in Figure 19.1. Then, we have $y_i > \frac{\alpha}{\beta}$. For this reason,

$$\alpha\phi < \beta\phi y_i \Rightarrow 2\alpha\phi < \alpha\phi + \beta\phi y_i.$$

  Thus,

$$2\alpha\phi + 1 < 1 + \alpha\phi + \beta\phi y_i \Rightarrow \frac{x_{i+1}}{x_i} < 1 \Leftrightarrow x_{i+1} < x_i.$$

- Now, let y_i be a value that either belongs to regions III or IV. In both cases, we have $y_i < \frac{\alpha}{\beta}$, which leads to

$$\alpha\phi > \beta\phi y_i \Rightarrow 2\alpha\phi > \alpha\phi + \beta\phi y_i.$$

  Finally,

$$2\alpha\phi + 1 > 1 + \alpha\phi + \beta\phi y_i \Rightarrow \frac{x_{i+1}}{x_i} > 1 \Leftrightarrow x_{i+1} > x_i.$$

We now consider equation (19.6) that is equivalent to

$$\frac{y_{i+1}}{y_i} = \frac{2\gamma\phi x_{i+1} + 1}{1 + \gamma\phi x_{i+1} + \delta\phi}.$$

- Let x_{i+1} be a value that either belongs to regions II or III of Figure 19.1. In there we have $x_{i+1} < \frac{\delta}{\gamma}$, which is equivalent to $\delta > \gamma x_{i+1}$. In this way,

$$\delta\phi > \gamma\phi x_{i+1} \Rightarrow \gamma\phi x_{i+1} + \delta\phi > 2\gamma\phi x_{i+1}.$$

  Therefore,

$$2\gamma\phi x_{i+1} + 1 < 1 + \gamma\phi x_{i+1} + \delta\phi \Rightarrow \frac{y_{i+1}}{y_i} < 1 \Leftrightarrow y_{i+1} < y_i.$$

- On the other hand, let x_{i+1} be a value that either belongs to regions I or IV. In this case, $x_{i+1} > \frac{\delta}{\gamma}$, which means that $\delta < \gamma x_{i+1}$, and we obtain

$$\delta\phi < \gamma\phi x_{i+1} \Rightarrow \gamma\phi x_{i+1} + \delta\phi < 2\gamma\phi x_{i+1}.$$

  Finally,

$$2\gamma\phi x_{i+1} + 1 > 1 + \gamma\phi x_{i+1} + \delta\phi \Rightarrow \frac{y_{i+1}}{y_i} > 1 \Leftrightarrow y_{i+1} > y_i.$$

We conclude that Theorem 19.3 also holds for system (19.7).

19.4 CONCLUSION

In this work, our goal was to show that the choice of the numerical method for the discretization of a continuous dynamical system is crucial in order to obtain consistent results. It was proved that the progressive Euler method, although appealing for its simplicity, is not able to deal with structurally unstable systems, making the solutions of the classical Lotka–Volterra model, that should be closed curves in phase space, become spirals. Furthermore, Euler's discretization does not take into account special fundamental properties of the systems, such as positivity. On the other hand, Mickens' method, despite generating an apparently more complex system, manages to guarantee that the qualitative behavior of the system, in a neighborhood of the fixed points, is identical to the one found in its continuous counterpart. Additionally, this method takes into account basic rules so that positivity is never compromised.

Our conclusions open the possibility of applying Mickens' method to other structurally unstable dynamical systems of particular interest, recovering properties that may have been lost through different standard discretizations. We also concluded that the Computer Algebra System SageMath is a strong tool that allows to do computations in a reliable way, serving as a good support when the systems under study are complex. In addition, it produces numerical simulations of good quality and in a very simple way. All the figures were generated with SageMath.

ACKNOWLEDGMENTS

Torres was partially supported by the Portuguese Foundation for Science and Technology (FCT) through the Center for Research and Development in Mathematics and Applications (CIDMA), projects UIDB/04106/2020 and UIDP/04106/2020.

BIBLIOGRAPHY

[1] É. Diz-Pita and M. V. Otero-Espinar. Predator–prey models: A review of some recent advances. *Mathematics*, 9(15):1783, 2021.

[2] M. W. Hirsch and S. Smale. *Differential Equations, Dynamical Systems, and Linear Algebra.* Pure and Applied Mathematics, vol. 60. Academic Press [Harcourt Brace Jovanovich, Publishers], New York-London, 1974.

[3] S. Li, S. Yuan, and H. Wang. Disease transmission dynamics of an epidemiological predator-prey system in open advective environments. *Discrete Contin. Dyn. Syst. Ser. B*, 28(2):1480–1502, 2023.

[4] A. J. Lotka. *Elements of Physical Biology.* Williams and Wilkins, Baltimore, 1925.

[5] R. E. Mickens. *Nonstandard Finite Difference Models of Differential Equations.* World Scientific Publishing Co., Inc., River Edge, NJ, 1994.

[6] R. E. Mickens. Nonstandard finite difference schemes for differential equations. *J. Differ. Equations Appl.*, 8(9):823–847, 2002.

[7] R. E. Mickens. A nonstandard finite-difference scheme for the Lotka-Volterra system. *Appl. Numer. Math.*, 45(2–3):309–314, 2003.

[8] B. T. Mulugeta, L. Yu, Q. Yuan, and J. Ren. Bifurcation analysis of a predator-prey model with strong Allee effect and Beddington-DeAngelis functional response. *Discrete Contin. Dyn. Syst. Ser. B*, 28(3):1938–1963, 2023.

[9] J. M. Sanz-Serna. An unconventional symplectic integrator of W. Kahan. *Appl. Numer. Math.*, 16(1–2):245–250, 1994.

[10] M. S. Shabbir, Q. Din, M. Safeer, M. A. Khan, and K. Ahmad. A dynamically consistent nonstandard finite difference scheme for a predator-prey model. *Adv. Differ. Equations*, 2019:381, 2019.

[11] V. Volterra. Variations and fluctuations of the number of individuals in animal species living together. *ICES J. Mar. Sci.*, 3(1):3–51, 1928.

[12] D. Yan, Y. Yuan, and X. Fu. Asymptotic analysis of an age-structured predator-prey model with ratio-dependent Holling III functional response and delays. *Evol. Equ. Control Theory*, 12(1):391–414, 2023.

[13] X. Yang, L. Chen, and J. Chen. Permanence and positive periodic solution for the single-species nonautonomous delay diffusive models. *Comput. Math. Appl.*, 32(4):109–116, 1996.

[14] P. Zimmermann, A. Casamayou, N. Cohen, G. Connan, T. Dumont, L. Fousse, F. Maltey, M. Meulien, M. Mezzarobba, C. Pernet, et al. *Computational Mathematics with SageMath.* SIAM, Philadelphia, PA, 2018.

CHAPTER 20

A New Inertial Projection Algorithm for Solving Pseudomonotone Equilibrium Problems

Tran Van Thang
Electric Power University

Le Dung Muu
Thang Long University

CONTENTS

20.1 INTRODUCTION

Let C be a nonempty closed convex subset in finite dimensional Euclidean space $\mathbb{R}^n$ and be defined by

$$C := \{x \in \mathbb{R}^n : \ g_i(x) \le 0, \ i = 1, 2, \ldots, r\}, \tag{20.1}$$

where $g_i : \mathbb{R}^n \to \mathbb{R}$ is a lower semicontinuous convex function on $\mathbb{R}^n$ for every $i = 1, 2, \ldots, r$. Let $f : C \times C \to \mathbb{R}$ be a bifunction such that $f(x, x) = 0$ for every $x \in C$. The equilibrium problem is formulated as follows:

$$\text{Find } \ p \in C : f(p, x) \ge 0, \forall x \in C. \tag{EP}$$

DOI: 10.1201/9781003388678-20

In the last few decades, this problem attracted the attention of many authors because this problem contains some other ones such as optimization, variational inequality, the Kakutani fixed point problems and the Nash, Walras equilibrium models as special cases, see e.g. the comprehensive monograph [5] and the paper [6, 12, 13]. A basic method for solving this problem is the projection one under the assumption that the function $f(x,.)$ is convex subdifferentiable for each $x \in C$. Then the projection algorithms use a subgradient as line search direction at each iteration. However, since the subgradient is a convex subset rather than a singleton, the efficiency of the projection algorithms depends crucially upon the choice of the line search direction, see e.g. [9, 13, 14, 17]. In this chapter, we propose an inertial smoothing projection algorithm for solving the pseudomonotone equilibrium problem (EP), where the bifunction f is convex, subdifferentiable, but may not be diffrentiable, in its second variable. In this algorithm, we take the advantage that the convex function $f(x,.)$ is differentiable everywhere to use a diagonal gradient rather than a diagonal subgradient as a linesearch direction. Moreover, the stepsize of the algorithm is updated by using the currently obtained iterate at each iteration. We also provide a convergence rate for the proposed algorithm. To our best knowledge, up to now there is no such a smoothing algorithm for pseudomonotone equilibrium problems of the type (EP).

This chapter is organized as follows. The next section contains some preliminaries on the monotonicity of a bifunction, the metric projection, the subdifferential of a convex function as well as lemmas that will be used in the third section, where we describe the algorithm and study its convergence as well as convergence rate. This chapter closes with some reports for computational results obtained by the algorithm for a Nash-Cournot equilibrium model.

20.2 PRELIMINARIES

In this section, we review some concepts and results needed to prove the main results presented in this chapter.

Definition 20.1 *([6])* Let C be a nonempty closed convex subset of a real space $\mathbb{R}^n$. A bifunction $f : C \times C \to \mathbb{R}$ is said to be

(a) *τ-strongly monotone on C*, if $f(x,y) + f(y,x) \leq -\tau\|x-y\|^2 \;\; \forall x, y \in C$;

(b) *monotone on C*, if $f(x,y) + f(y,x) \leq 0 \;\; \forall x, y \in C$;

(c) *pseudomonotone on C*, if $f(x,y) \geq 0 \Rightarrow f(y,x) \leq 0 \; \forall x, y \in C$.

Definition 20.2 *([4, Chapter 4])*Let C be a nonempty closed convex subset in $\mathbb{R}^n$. The metric projection from $\mathbb{R}^n$ onto C is defined by Π_C and

$$\Pi_C(x) = \operatorname{argmin}_{y \in C}\{\|x-y\|\}, \; \forall x \in \mathbb{R}^n.$$

It is well known [4, Chapter 4] that Π_C has the following properties.

Lemma 20.1 (a) *For any* $x \in \mathbb{R}^n, u = \Pi_C(x)$ *if and only if* $\langle u - x, y - z\rangle \leq 0, \quad \forall y \in C$;

(b) $\|\Pi_C(x) - \Pi_C(y)\| \leq \|x - y\|, \quad \forall x, y \in \mathbb{R}^n$;

(c) $\|\Pi_C(x) - z\|^2 + \|\Pi_C(x) - x\|^2 \leq \|x - z\|^2, \quad \forall x \in \mathbb{R}^n, \ u \in C.$

Lemma 20.2 *It holds that*

(a) $\|\lambda x + (1 - \lambda)y\|^2 = \lambda\|x\|^2 + (1 - \lambda)\|y\|^2 - \lambda(1 - \lambda)\|x - y\|^2$;

(b) $\|\alpha x+\beta y+\gamma z\|^2 = \alpha\|x\|^2+\beta\|y\|^2+\gamma\|z\|^2-\alpha\beta\|x-y\|^2-\beta\gamma\|y-z\|^2-\alpha\gamma\|x-z\|^2,$

for every $x, \ y, \ z \in \mathbb{R}^n, \ \lambda \in \mathbb{R}$ *and* $\alpha, \ \beta, \ \gamma \in [0, 1], \ \alpha + \beta + \gamma = 1.$

Lemma 20.3 *([4, Chapter 2]) For every* $a, \ b \in \mathbb{R}^n$*, one has.*

(a) $\|a + b\|^2 = \|a\|^2 + 2\langle a, b\rangle + \|b\|^2$;

(b) $\|a + b\|^2 \leq \|a\|^2 + 2\langle b, a + b\rangle$.

The subdifferential of a convex function $h : \mathbb{R}^n \to \mathbb{R} \cup \{+\infty\}$ is defined by

$$\partial h(\bar{x}) = \{u \in \mathbb{R}^n : \langle u, x - \bar{x}\rangle \leq h(x) - h(\bar{x}) \ \ \forall x \in \mathbb{R}^n\}.$$

From the convex programming, we have the following result.

Lemma 20.4 *([15, Theorem 27.4]) Let* $h : \mathbb{R}^n \to \mathbb{R} \cup \{+\infty\}$ *and* C *be a convex subset* $\mathbb{R}^n$ *and be subdifferentiable on* C*. Assume either that* h *is continuous at some point of* C*, or that there is an interior point of* C *where* h *is finite. Then,* $\bar{x}$ *is a solution to the convex program*

$$\min\{h(x) : x \in C\}$$

if and only if $0 \in \partial h(\bar{x}) + N_C(\bar{x})$*, where* $N_C(\bar{x})$ *is the (outer) normal cone of* C *at* $\bar{x} \in C$*, that is,* $N_C(\bar{x}) = \{u \in \mathbb{R}^n : \langle u, y - \bar{x}\rangle \leq 0 \ \ \forall y \in C\}$.

Lemma 20.5 *([15, Theorem 25.5]) Let* $h : \mathbb{R}^n \to \mathbb{R}$ *be a proper convex function, and let* S *be the set of points where* h *is differentiable. Then* S *is a dense subset of* $\mathbb{R}^n$*, and its complement in* $\mathbb{R}^n$ *is a set of measure zero.*

Definition 20.3 *([15]) Let* $h : \mathbb{R}^n \to \mathbb{R}$ *be a proper convex function, and let* S *be the set of points where* g *is differentiable. The D-subdifferential of a function* h *is defined by*

$$Dh(x) = \{u \in \mathbb{R}^n : \ x^k \in S, \ x^k \to x, \ \nabla h(x^k) \to u\}.$$

The D- subdifferential with respect to a set K *is defined by*

$$D^K h(x) := \{u \in \mathbb{R}^n : \ x^k \in S \cap K, \ x^k \to x, \ \nabla h(x^k) \to u\}.$$

Lemma 20.6 *([15, Theorem 25.6]) Let $g : \mathbb{R}^n \to \mathbb{R}$ be a convex function. Then, for any $x \in \mathbb{R}^n$ we have*

$$\partial g(x) = \text{co}Dg(x).$$

Lemma 20.7 *([1]) Let $\{\sigma_k\}, \{\eta_k\}, \{\zeta_k\}$ be sequences of nonnegative real numbers satisfying two following conditions:*

(i) $\sigma_{k+1} \leq \sigma_k + \zeta_k(\sigma_k - \sigma_{k-1}) + \eta_k, \quad \forall k \geq 1, \sum_{k=1}^{+\infty} \eta_k < \infty;$

(ii) *there exists a real number c such that $0 \leq \zeta_k \leq p < 1$ for every $k \geq 1$.*

Then,

(a) $\sum_{k=1}^{\infty} [\sigma_k - \sigma_{k-1}]_+ < \infty$, *where* $[\sigma_k - \sigma_{k-1}]_+ := \max\{\sigma_k - \sigma_{k-1}, 0\}$;

(b) *there exists $\sigma \in [0, +\infty)$ such that $\lim_{k\to\infty} \sigma_k = \sigma$.*

Lemma 20.8 *([4, Lemma 2.39]) Let C be a nonempty subset in real space $\mathbb{R}^n$ and $\{x^k\} \subset \mathbb{R}^n$ satisfies the following conditions:*

(a) *for every $x \in C$, $\lim_{k\to+\infty} \|x^k - x\|$ exists;*

(b) *every sequentially cluster point of $\{x^k\}$ is in C.*

Then, the sequence $\{x^k\}$ converges to a point in C.

20.3 THE ALGORITHM AND CONVERGENCE ANALYSIS

In this section, we introduce a new inertial algorithm for finding the solution of the (EP) and show its convergence. This algorithm combines the subgradient projection method with the inertial iteration technique.

To obtain the convergence result of the proposed algorithm, we assume that the bifunction $f : C \times C \to \mathbb{R}$ satisfies the following assumptions:

M_1. f is pseudomonotone on C and, for each $y \in C$, $f(\cdot, y)$ is sequentially upper semicontinuous on C;

M_2. for every $x \in C$ and $\epsilon > 0$ there exists $y \in B(x, \epsilon) \cap C$ such that $f(x, \cdot)$ is differentiable at y;

M_3. $Sol(EP)$ is nonempty;

M_4. for each $x \in C$, $f(x, \cdot)$ is convex subdifferentiable on C;

M_5. for every x^1, $x^2 \in C$ if $f(x^1, \cdot)$ and $f(x^2, \cdot)$ are differentiable at y^1 and y^2 respectively then

$$\|\nabla_2 f(x^1, y^1) - \nabla_2 f(x^2, y^2)\| \leq L_1\|x^2 - x^1\| + L_2\|y^2 - y^1\|.$$

Remark 20.1 *(a) If the set C has a non-empty interior, then we get from the condition M_4 and Lemma 20.5 that f satisfies the assumption M_2.*
(b) Let F be a Lipschitz continuous and sequentially upper semicontinuous function on a convex subset C. In the case ỏ variational inequality, when $f(x,y) := \langle Fx, y-x\rangle$, it is not hard to prove that f satisfies conditions $M_1 - M_2$, M_4 and M_5.
(c) Let $f(x,y) := h(y) - h(x)$ and $C = \{x \in \mathbb{R}^n : \|x\|^2 - 2\langle a, x\rangle - 1 \leq 0\}$, where $h(x) = \max\{2\|x\|^2, \|x\|^2 + 2\langle a, x\rangle + 1\}$. We have that h is convex subdifferentiable on $\mathbb{R}^n$ with

$$\partial h(x) = \begin{cases} \{4x\} & \text{if } \|x\|^2 - 2\langle a, x\rangle - 1 > 0 \\ [4x, 2x + 2a] & \text{if } \|x\|^2 - 2\langle a, x\rangle - 1 = 0 \\ \{2x + 2a\} & \text{if } \|x\|^2 - 2\langle a, x\rangle - 1 < 0, \end{cases} \tag{20.2}$$

where $[4x, 2x + 2a] = \{t4x + (1-t)(2x + 2a) : t \in [0,1]\}$. It is easy to check that $f(x, \cdot)$ is not differentiable on C and f satisfies conditions $M_1 - M_2$, M_4 and M_5.

Let C be given by equation (20.1). We can use the following procedure to find a point in C.

Procedure A: ([10]) Data: A point $x \in \mathbb{R}^n$. Output: A point $R(x) \in C$.

Step a. If $x \in C$, set $R(x) = x$, else, set $z^0 = x, k = 0$.

Step b. Take $g^k \in \partial g(z^k)$, if $g^k = 0$, terminate: $z^k \in C$. If $g^k \neq 0$, calculate

$$z^{k+1} = z^k - 2g(z^k)\frac{g^k}{\|g^k\|^2},$$

where $g(x) = \max\{g_i(x) : i = 1, 2, \ldots, r\}, \forall x \in \mathbb{R}^n$.

Step c. If $z^{k+1} \in C$, then set $R(x) := z^{k+1}$ and stop, else, let $k \leftarrow k+1$ and go to Step b.

In the case $g^k = 0$ then we deduce from $g^k \in \partial g(z^k)$ that

$$g(x) - g(z^k) \geq \langle g^k, x - z^k\rangle = 0, \ \forall x \in C,$$

which implies that $g(z^k) \leq g(x) \leq 0$ for every $x \in C$, and so $z^k \in C$.

Lemma 20.9 *([10]) Production A is well defined and for all $x \in \mathbb{R}^n$, $y \in C$ we have*

$$\|R(x) - y\| \leq \|x - y\|, \ x \in \mathbb{R}^n, \ \ \forall y \in C.$$

Now, we are ready to describe the algorithm. The following relevant monographs in references may be updated where similar projection algorithms are studied [7, 8].

Algorithm 1

Take starting points $x^0,\ x^1 \in \mathbb{R}^n$, the parameter numbers $\alpha_0 > 0,\ \rho,\ \nu \in (0,1),\ \varrho \in (0,2)$ and the positive sequences $\{\theta_k\},\ \{\mu_k\},\ \{\kappa_k\}$ satisfying

$$\sum_{k=0}^{+\infty} \theta_k < +\infty,\ \mu_k \in (0,1),\ \sum_{k=0}^{+\infty} \mu_k < +\infty,\ \lim_{k\to+\infty} \kappa_k = 0. \tag{20.3}$$

Step 1. Take ρ_k such that $0 \leq \rho_k \leq \bar{\rho}_k$ with

$$\bar{\rho}_k := \begin{cases} \min\{\frac{\mu_k}{\|x^k - x^{k-1}\|^2}, \rho\}, & \textit{if } x^k \neq x^{k-1}, \\ \rho & \text{otherwise.} \end{cases} \tag{20.4}$$

Step 2. Calculate $w^k = x^k + \rho_k(x^k - x^{k-1})$ and $\bar{w}^k = R(w^k)$.

Step 3. Take $U_{\bar{w}^k} \in D_2^C f(\bar{w}^k, \bar{w}^k)$. If $U_{\bar{w}^k} = 0$ then Stop. Otherwise, find $y^k \in C$ such that

$$y^k = \Pi_C(\bar{w}^k - \alpha_k U_{\bar{w}^k}), \tag{20.5}$$

If $y^k = \bar{w}^k$ then Stop. Otherwise, go to the next step.

Step 4. Take $U_{y^k} \in D_2^C f(y^k, y^k)$. Set $d^k := \bar{w}^k - y^k - \alpha_k(U_{\bar{w}^k} - U_{y^k})$ and calculate

$$x^{k+1} := \Pi_C(\bar{w}^k - \tau_k \alpha_k U_{y^k}), \tag{20.6}$$

where τ_k is defined by taking

$$\tau_k := \begin{cases} (\varrho + \kappa_k)\frac{\langle \bar{w}^k - y^k, d^k \rangle}{\|d^k\|^2}, & \textit{if } \|d^k\| \neq 0, \\ 0 & \text{otherwise.} \end{cases} \tag{20.7}$$

Calculate

$$\alpha_{k+1} = \begin{cases} \min\left\{\frac{\nu\|x^k - y^k\|}{\|U_{\bar{w}^k} - U_{y^k}\|}, \alpha_k + \theta_k\right\}, & \textit{if } U_{\bar{w}^k} - U_{y^k} \neq 0, \\ \alpha_k + \theta_k & \text{otherwise,} \end{cases} \tag{20.8}$$

Step 5. Increase k by one and return to Step 1.

Remark 20.2 *(a) If $U_{\bar{w}^k} = 0$ then we have from Lemma 20.6 that $U_{\bar{w}^k} \in \partial_2 f(\bar{w}^k, \bar{w}^k)$, and so*

$$f(\bar{w}^k, x) = f(\bar{w}^k, y) - f(\bar{w}^k, \bar{w}^k) \geq \langle U_{\bar{w}^k}, x - \bar{w}^k \rangle = 0,\ \forall x \in C.$$

Hence, $\bar{w}^k$ is a solution of Problem (EP) and the algorithm terminates at iteration k.
(b) If $y^k = \bar{w}^k$, it follows from equation (20.5) that $\bar{w}^k = \Pi_C(\bar{w}^k - \alpha_k U_{\bar{w}^k})$. By Lemma 20.1 (a), we obtain that

$$0 \leq \langle U_{\bar{w}^k}, x - \bar{w}^k \rangle \leq f(\bar{w}^k, x),\ \forall x \in C.$$

So, $\bar{w}^k$ is a solution of Problem (EP).
(c) *We have*

$$\|U_{\bar{w}^k} - U_{y^k}\| \le (L_1 + L_2)\|\bar{w}^k - y^k\|.$$

Indeed, by $U_{y^k} \in D_2^C f(y^k, y^k)$, $U_{\bar{w}^k} \in D_2^C f(\bar{w}^k, \bar{w}^k)$ and the assumption M_2, there exist the sequences $\{\bar{w}^j\}$ and $\{y^j\}$ such that $\bar{w}^j \to \bar{w}^k$, $\nabla_2 f(\bar{w}^k, \bar{w}^j) \to U_{\bar{w}^k}$ and $y^j \to y^k$, $\nabla_2 f(y^k, y^j) \to U_{y^k}$. From the assumption M_5, it follows that

$$\|\nabla_2 f(\bar{w}^k, \bar{w}^j) - \nabla_2 f(y^k, y^j)\| \le L_1\|\bar{w}^k - y^k\| + L_2\|\bar{w}^j - y^j\|.$$

Taking the limit as $j \to \infty$ on both sides of the last inequality, we have

$$\|U_{\bar{w}^k} - U_{y^k}\| \le (L_1 + L_2)\|\bar{w}^k - y^k\|.$$

Lemma 20.10 *Under the assumptions $(M_1) - (M_5)$, for every $x^s \in Sol(EP)$ and the sequences $\{x^k\}, \{w^k\}$, $\{\bar{w}^k\}$, $\{y^k\}$ generated by Algorithm 1, we have*

(a) $\lim_{k\to\infty} \|w^k - x^k\| = \lim_{k\to\infty} \|\bar{w}^k - x^k\| = \lim_{k\to\infty} \rho_k\|x^k - x^{k-1}\| = 0$;

(b) $\alpha_k \in [\min\{\frac{\nu}{(L_1+L_2)}, \alpha_0\}, \ \alpha_0 + \Theta]$, $\forall k \ge 0$ *and* $\lim_{k\to\infty} \alpha_k = \alpha$, *where* $\sum_{k=0}^{+\infty} \theta_k = \Theta$;

(c) $\|x^{k+1} - x^s\|^2 \le \|w^k - x^s\|^2 - (\varrho + \kappa_k)(2 - \varrho - \kappa_k)\frac{(\alpha_{k+1} - \nu\alpha_k)^2}{(\alpha_{k+1} + \nu\alpha_k)^2}\|w^k - y^k\|^2$.

Proof It follows from equation (20.4) that $\rho_k\|x^k - x^{k-1}\|^2 \le \mu_k$, which together with Condition (20.3) implies that

$$\sum_{k=1}^{\infty} \rho_k\|x^k - x^{k-1}\|^2 < \infty.$$

By this and the definition of w^k, we get $\lim_{k\to\infty} \|w^k - x^k\| = \lim_{k\to\infty} \rho_k\|x^k - x^{k-1}\| = 0$, which together with Lemma 20.9 (b) implies that $\lim_{k\to\infty} \|\bar{w}^k - x^k\| = 0$.

To prove (b), we observe that if $U_{\bar{w}^k} - U_{y^k} \ne 0$, then by Remark 20.2 (c) we have

$$\frac{\nu\|\bar{w}^k - y^k\|}{\|U_{\bar{w}^k} - U_{y^k}\|} \ge \frac{\nu\|\bar{w}^k - y^k\|}{(L_1 + L_2)\|\bar{w}^k - y^k\|} = \frac{\nu}{(L_1 + L_2)}. \tag{20.9}$$

Using the mathematical induction proof method and equation (20.8), it is not hard to prove that $\{\alpha_k\}$ belongs to $[\min\{\frac{\nu}{(L_1+L_2)}, \alpha_0\}, \ \alpha_0 + \Theta]$, $\forall k \ge 0$. Set $(\alpha_{k+1} - \alpha_k)^+ = \max\{0, \alpha_{k+1} - \alpha_k\}$ and $(\alpha_{k+1} - \alpha_k)^- = \max\{0, -(\alpha_{k+1} - \alpha_k)\}$. It is easy to see from equation (20.8) that

$$\sum_{k=0}^{+\infty}(\alpha_{k+1} - \alpha_k)^+ \le \sum_{k=0}^{+\infty}\theta_k < +\infty. \tag{20.10}$$

Suppose that $\sum_{k=0}^{+\infty}(\alpha_{k+1}-\alpha_k)^- = +\infty$. Again, using the mathematical induction, from the quality

$$\alpha_{k+1}-\alpha_k = (\alpha_{k+1}-\alpha_k)^+ - (\alpha_{k+1}-\alpha_k)^-,$$

it follows that

$$\begin{aligned}\alpha_{k+1}-\alpha_0 &= \sum_{k=0}^{+\infty}(\alpha_{k+1}-\alpha_k)\\ &= \sum_{k=0}^{+\infty}(\alpha_{k+1}-\alpha_k)^+ - \sum_{k=0}^{+\infty}(\alpha_{k+1}-\alpha_k)^-. \end{aligned} \tag{20.11}$$

By equation (20.10), we get in the limit of the above inequality that $\alpha_k \to -\infty$, which is a contradiction. Hence, $\sum_{k=0}^{+\infty}(\alpha_{k+1}-\alpha_k)^- < +\infty$. From equation (20.10), it follows that $\lim_{k\to\infty}\alpha_k = \alpha \in [\min\{\frac{\nu}{L}, \alpha_0\},\ \alpha_0+\Theta]$.

Next, we prove (c). Using $x^s \in Sol(EP)$, $f(y^k,y^k)=0$, $U_{y^k} \in \partial_2 f(y^k,y^k)$ and the pseudomonotonicity of f, we obtain

$$\tau_k\alpha_k\langle U_{y^k}, y^k - x^s\rangle \geq \tau_k\alpha_k[f(y^k,y^k) - f(y^k,x^s)] \geq 0. \tag{20.12}$$

From equation (20.6) and Lemma 20.1 (c), we get

$$\begin{aligned}\|x^{k+1}-x^s\|^2 &= \|\Pi_C(\bar{w}^k - \tau_k\alpha_k U_{y^k}) - x^s\|\\ &\leq \|\bar{w}^k - \tau_k\alpha_k U_{y^k} - x^s\|^2 - \|\bar{w}^k - \tau_k\alpha_n U_{y^k} - x^{k+1}\|^2\\ &= \|\bar{w}^k - x^s\|^2 - 2\tau_k\alpha_k\langle \bar{w}^k - x^s, U_{y^k}\rangle\\ &\quad +2\tau_k\alpha_k\langle \bar{w}^k - x^{k+1}, U_{y^k}\rangle - \|x^{k+1}-\bar{w}^k\|^2\\ &= \|\bar{w}^k - x^s\|^2 - \|x^{k+1}-\bar{w}^k\|^2 - 2\tau_k\alpha_k\langle y^k - x^s, U_{y^k}\rangle\\ &\quad +2\tau_k\alpha_k\langle y^k - x^{k+1}, U_{y^k}\rangle.\end{aligned}$$

Combining this and equation (20.12), we obtain

$$\begin{aligned}\|x^{k+1}-x^s\| &\leq \|\bar{w}^k - x^s\|^2 - \|x^{k+1}-\bar{w}^k\|^2\\ &\quad +2\tau_k\alpha_k\langle y^k - x^{k+1}, U_{y^k}\rangle.\end{aligned} \tag{20.13}$$

We deduce from Step 3 that

$$0 \leq 2\tau_k\langle \bar{w}^k - y^k - \alpha_k U_{\bar{w}^k}, y^k - x\rangle,\ \forall x \in C,$$

it follows that

$$0 \leq 2\tau_k\langle \bar{w}^k - y^k - \alpha_k U_{\bar{w}^k}, y^k - x^{k+1}\rangle. \tag{20.14}$$

Combining equations (21.12) and (21.13) to get

$$\|x^{k+1}-x^s\|^2 \leq \|\bar{w}^k - x^s\|^2 - \|x^{k+1}-\bar{w}^k\|^2 + 2\tau_k\alpha_k\langle y^k - x^{k+1}, U_{y^k}\rangle$$

$$\begin{aligned}
& +2\tau_k\langle \bar{w}^k - y^k - \alpha_k U_{\bar{w}^k}, y^k - x^{k+1}\rangle \\
= \;& \|\bar{w}^k - x^s\|^2 - \|\bar{w}^k - x^{k+1} - \tau_k d^k\|^2 \\
& +(\tau_k\|d^k\|)^2 - 2\tau_k\langle \bar{w}^k - x^{k+1}, d^k\rangle + 2\tau_k\langle y^k - x^{k+1}, d^k\rangle \\
= \;& \|\bar{w}^k - x^s\|^2 - \|\bar{w}^k - x^{k+1} - \tau_k d^k\|^2 + (\tau_k\|d^k\|)^2 \\
& +2\tau_k\langle y^k - \bar{w}^k, d^k\rangle \\
\leq \;& \|\bar{w}^k - x^s\|^2 + (\tau_k\|d^k\|)^2 + 2\tau_k\langle y^k - \bar{w}^k, d^k\rangle.
\end{aligned}$$

We deduce, from the above inequality and equation (20.7), that

$$\begin{aligned}
\|x^{k+1} - x^s\|^2 \;\leq\; & \|\bar{w}^k - x^s\|^2 + (\varrho + \kappa_k)^2\left(\frac{\langle \bar{w}^k - y^k, d^k\rangle}{\|d^k\|^2}\right)^2\|d^k\|^2 \\
& +2(\varrho + \kappa_k)\frac{\langle \bar{w}^k - y^k, d^k\rangle}{\|d^k\|^2}\langle y^k - \bar{w}^k, d^k\rangle \\
\leq\; & \|\bar{w}^k - x^s\|^2 \\
& -(\varrho + \kappa_k)(2 - \varrho - \kappa_k)\frac{(\langle \bar{w}^k - y^k, d^k\rangle)^2}{\|d^k\|^2}. \qquad (20.15)
\end{aligned}$$

On the other hand, by the definition of d^k and equation (20.8), we have

$$\begin{aligned}
\langle \bar{w}^k - y^k, d^k\rangle \;=\; & \langle \bar{w}^k - y^k, \bar{w}^k - y^k - \alpha_k(U_{\bar{w}^k} - U_{y^k})\rangle \\
=\; & \|\bar{w}^k - y^k\|^2 - \alpha_k\langle \bar{w}^k - y^k, U_{\bar{w}^k} - U_{y^k}\rangle \\
\geq\; & \|\bar{w}^k - y^k\|^2 - \alpha_k\|\bar{w}^k - y^k\|\|U_{\bar{w}^k} - U_{y^k}\| \\
\geq\; & (1 - \frac{\nu\alpha_k}{\alpha_{k+1}})\|\bar{w}^k - y^k\|^2. \qquad (20.16)
\end{aligned}$$

Thank to Step 3, we obtain

$$\begin{aligned}
\|d^k\| \;=\; & \|\bar{w}^k - y^k - \alpha_k(U_{\bar{w}^k} - U_{y^k})\| \\
\leq\; & \|\bar{w}^k - y^k\| + \alpha_k\|U_{\bar{w}^k} - U_{y^k}\| \\
\leq\; & (1 + \frac{\nu\alpha_k}{\alpha_{k+1}})\|\bar{w}^k - y^k\|.
\end{aligned}$$

Thus, by equation (20.16), we can write

$$\begin{aligned}
\langle \bar{w}^k - y^k, d^k\rangle \;\geq\; & (1 - \frac{\nu\alpha_k}{\alpha_{k+1}})\|\bar{w}^k - y^k\|^2 \\
\geq\; & \alpha_{k+1}\frac{\alpha_{k+1} - \nu\alpha_k}{(\alpha_{k+1} + \nu\alpha_k)^2}\|d^k\|^2. \qquad (20.17)
\end{aligned}$$

We have from equations (20.15), (20.16), (20.17) and Lemma 20.9 (b) that

$$\begin{aligned}
\|x^{k+1} - x^s\|^2 \;\leq\; & \|\bar{w}^k - x^s\|^2 \\
& -(\varrho + \kappa_k)(2 - \varrho - \kappa_k)\alpha_{k+1}\frac{\alpha_{k+1} - \nu\alpha_k}{(\alpha_{k+1} + \nu\alpha_k)^2}\langle \bar{w}^k - y^k, d^k\rangle
\end{aligned}$$

$$\leq \|w^k - x^s\|^2$$
$$-(\varrho + \kappa_k)(2 - \varrho - \kappa_k)\frac{(\alpha_{k+1} - \nu\alpha_k)^2}{(\alpha_{k+1} + \nu\alpha_k)^2}\|\bar{w}^k - y^k\|^2.$$

□

Lemma 20.11 *Under the assumptions $(M_1) - (M_5)$, for any $x^s \in Sol(EP)$, we have*

(a) *the sequence $\{\|x^k - x^s\|\}$ has limit as $k \to \infty$;*

(b) *the $\{x^k\}$, $\{w^k\}$ and $\{\bar{w}^k\}$ are bounded sequences;*

(c) $\lim_{k\to\infty} \left[\|x^k - x^s\|^2 - \|x^{k-1} - x^s\|^2\right]_+ = 0$, *where $[\xi]_+ := \max\{\xi, 0\}$ for each $\xi \in \mathbb{R}$.*

Proof. Let $x^s \in Sol(EP)$. Thanks to Lemma 20.2 (a) and the definition of w^k, we obtain

$$\begin{aligned}
\|w^k - x^s\|^2 &= \|x^k - \rho_k(x^k - x^{k-1}) - x^s\|^2 \\
&= \|(1+\rho_k)(x^k - x^s) + (-\rho_k)(x^{k-1} - x^s)\|^2 \\
&= (1+\rho_k)\|x^k - x^s\|^2 + (-\rho_k)\|x^{k-1} - x^s\|^2 \\
&\quad +\rho_k(1+\rho_k)\|x^k - x^{k-1}\|^2. \qquad (20.18)
\end{aligned}$$

We have from Condition (20.3) and Lemma 20.10 (b) that

$$\lim_{k\to\infty}(\varrho + \kappa_k)(2 - \varrho - \kappa_k)\frac{(\alpha_{k+1} - \nu\alpha_k)^2}{(\alpha_{k+1} + \nu\alpha_k)^2} = \varrho(2-\varrho)\frac{(1-\nu)^2}{(1+\nu)^2} > 0,$$

which means that there exits a positive integer number K_0 such that

$$(\varrho + \kappa_k)(2 - \varrho - \kappa_k)\frac{(\alpha_{k+1} - \nu\alpha_k)^2}{(\alpha_{k+1} + \nu\alpha_k)^2} > 0, \ \forall k \geq K_0.$$

It follows from Lemma 20.10 (c) that

$$\|x^{k+1} - x^s\|^2 \leq \|w^k - x^s\|^2, \ \forall k \geq K_0,$$

which together with equation (21.25) implies that, for every $k \geq K_0$,

$$\begin{aligned}
\|x^{k+1} - x^s\|^2 &\leq (1+\rho_k)\|x^k - x^s\|^2 + (-\rho_k)\|x^{k-1} - x^s\|^2 \\
&\quad +\rho_k(1+\rho_k)\|x^k - x^{k-1}\|^2 \\
&= \|x^k - x^s\|^2 + \rho_k(\|x^k - x^s\|^2 - \|x^{k-1} - x^s\|^2) \\
&\quad +\rho_k(1+\rho)\|x^k - x^{k-1}\|^2.
\end{aligned}$$

Setting $\sigma_k := \|x^k - x^s\|^2$, $\eta_k = \rho_k(1+\rho)\|x^k - x^{k-1}\|^2$ and $\zeta_k := \rho_k$, then by Lemma 20.10 (a), we obtain

$$\sum_{k=1}^{\infty} \eta_k = \sum_{k=1}^{\infty} \rho_k(1+\rho)\|x^k - x^{k-1}\|^2 < +\infty. \tag{20.19}$$

Thanks to Lemma 20.7, we can conclude that $\lim_{k\to\infty} \|x^k - x^s\|^2$ exists and

$$\sum_{k=1}^{\infty} \left[\|x^k - x^s\|^2 - \|x^{k-1} - x^s\|^2\right]_+ < +\infty,$$

which implies that

$$\lim_{k\to\infty} [\|x^k - x^s\|^2 - \|x^{k-1} - x^s\|^2]_+ = 0 \tag{20.20}$$

and the sequence $\{x^k\}$ is bounded, which together with Lemma 20.10 (a) implies that both $\{w^k\}$ and $\{\bar{w}^k\}$ are bounded. □

Now we can state the following convergence.

Theorem 20.1 *Under the assumptions (M_1) - (M_5), the sequence $\{x^k\}$ generated by Algorithm 1 converges to a point $x^s \in Sol(EP)$.*

Proof For $x^s \in Sol(EP)$. Thank to Lemma 20.10 (b), $\rho_k \leq \rho$ and equation (21.25), we get

$$\begin{aligned}
\|x^{k+1} - x^s\|^2 \leq\; & (1+\rho_k)\|x^k - x^s\|^2 - \rho_k\|x^{k-1} - x^s\|^2 \\
& + \rho_k(1+\rho_k)\|x^k - x^{k-1}\|^2 \\
& - (\varrho + \kappa_k)(2 - \varrho - \kappa_k)\frac{(\alpha_{k+1} - \nu\alpha_k)^2}{(\alpha_{k+1} + \nu\alpha_k)^2}\|\bar{w}^k - y^k\|^2 \\
\leq\; & \|x^k - x^s\|^2 + \rho_k(\|x^k - x^s\|^2 - \|x^{k-1} - x^s\|^2) \\
& + \rho_k(1+\rho)\|x^k - x^{k-1}\|^2 \\
& - (\varrho + \kappa_k)(2 - \varrho - \kappa_k)\frac{(\alpha_{k+1} - \nu\alpha_k)^2}{(\alpha_{k+1} + \nu\alpha_k)^2}\|\bar{w}^k - y^k\|^2 \\
\leq\; & \|x^k - x^s\|^2 + \rho_k[\|x^k - x^s\|^2 - \|x^{k-1} - x^s\|^2]_+ \\
& + \rho_k(1+\rho)\|x^k - x^{k-1}\|^2 - (\varrho + \kappa_k)(2 - \varrho \\
& - \kappa_k)\frac{(\alpha_{k+1} - \nu\alpha_k)^2}{(\alpha_{k+1} + \nu\alpha_k)^2}\|\bar{w}^k - y^k\|^2, \quad \forall k \geq 1.
\end{aligned} \tag{20.21}$$

Consequently

$$\begin{aligned}
& (\varrho + \kappa_k)(2 - \varrho - \kappa_k)\frac{(\alpha_{k+1} - \nu\alpha_k)^2}{(\alpha_{k+1} + \nu\alpha_k)^2}\|\bar{w}^k - y^k\|^2 \\
& \leq \|x^k - x^s\|^2 - \|x^{k+1} - x^s\|^2 + \rho_k[\|x^k - x^s\|^2 - \|x^{k-1} - x^s\|^2]_+
\end{aligned}$$

$$+ \rho_k(1+\rho)\|x^k - x^{k-1}\|^2, \quad \forall k \geq 1. \tag{20.22}$$

Letting $k \to \infty$ in the last inequality, by using conditions (20.3), (21.27), Lemma 20.10 (a) and Lemma 20.11, we obtain

$$\lim_{k\to\infty} \|\bar{w}^k - y^k\| = 0.$$

Let $\Xi(x^k)$ be the set of all cluster points of the sequence $\{x^k\}$. Then, we can prove that $\Xi(x^k) \subset Sol(EP)$. In fact, assume that $\hat{x}$ is arbitrary point in $\Xi(x^k)$. It follows that there exists a subsequence $\{x^{k_i}\}$ of $\{x^k\}$ converging to $\hat{x}$. Thank to Lemma 20.10 (a), the sequence $\{\bar{w}^{k_i}\}$ also converges to $\hat{x}$. By equation (20.5), one has

$$\langle \bar{w}^{k_i} - y^{k_i} - \alpha_{k_i} U_{\bar{w}^{k_i}}, y^{k_i} - x\rangle \geq 0, \ \forall x \in C,$$

which together with $U_{y^{k_i}} \in \partial_2 f(y^{k_i}, y^{k_i})$ implies that

$$\begin{aligned}\langle \bar{w}^{k_i} - y^{k_i}, x - y^{k_i}\rangle \leq & \alpha_{k_i}\langle U_{\bar{w}^{k_i}}, x - y^{k_i}\rangle \\ \leq & \alpha_{k_i}(\langle U_{\bar{w}^{k_i}}, x - \bar{w}^{k_i}\rangle + \langle U_{\bar{w}^{k_i}}, \bar{w}^{k_i} - y^{k_i}\rangle.\end{aligned}$$

Since $\|\bar{w}^k - y^k\| \to 0$ as $k \to \infty$ and $\{\bar{w}^{k_i}\}$ is bounded, $\{y^{k_i}\}$ is bounded too. For each $x \in C$, take the limit as $i \to \infty$, using $\lim_{i\to\infty} \|\bar{w}^{k_i} - y^{k_i}\| = 0$, Condition (20.3) and the sequentially upper semicontinuity of bifunction $f(\cdot, y)$, one has

$$f(\hat{x}, x) \geq 0 \ \ \forall x \in C.$$

Therefore, $\hat{x} \in Sol(EP)$. Thanks to Lemma 20.8, we can conclude that the sequence $\{x^k\}$ converges to a solution $x^s \in Sol(EP)$. The theorem is proven □

20.4 RATE OF CONVERGENCE

The non-asymptotic convergence rate results of algorithms is shown in [16, 17] when solving variational inequality and equilibrium problems. In this section, we develop the non-asymptotic convergence rate of the proposed algorithm for (EP). We have from Remark 20.2 (b) that y^k is a solution of the (EP) whenever $y^k = \bar{w}^k$, which together with $\lim_{k\to\infty} \|y^k - \bar{w}^k\| = 0$ implies that the y^k is s an ϵ -approximate solution of (EP), that meas $\|y^k - \bar{w}^k\|^2 < \epsilon$.

Theorem 20.2 *Assume that the conditions (M_1)–(M_5) are satisfied. Let $\{x^k\}$ be the sequence of iterates generated by Algorithm 1 and x^s be an arbitrary solution in $Sol(EP)$. Then, there exist positive numbers $\bar{\varrho}$, $\bar{\nu}$ and $\bar{\mu}$ such that*

$$\min_{K\leq i\leq k} \|y^i - \bar{w}^i\|^2 \leq \frac{\|x^1 - x^s\|^2 + \frac{\rho}{1-\rho}[\|x^1 - x^s\|^2 - \|x^0 - x^s\|^2]_+ + \frac{1+\rho}{1-\rho}\bar{\mu}}{\bar{\varrho}(2-\bar{\varrho})\left[\frac{(1-\nu)^2}{(1+\nu)^2} - \bar{\nu}\right](k - K + 1)}.$$

Proof. Let $\bar{\varrho}$ and $\bar{\nu}$ be positive real numbers satisfying $\varrho < \bar{\varrho} < 2$ and $\frac{(1-\nu)^2}{(1+\nu)^2} - \bar{\nu} > 0$. Since $\lim_{k\to\infty} \kappa_k = 0$, we have $\lim_{k\to\infty}(\bar{\varrho} - \varrho - \kappa_k) = \bar{\varrho} - \varrho > 0$, which means that there is a integer number K_1 satisfying $\bar{\varrho} - \varrho - \kappa_k > 0,\ \forall k \geq K_1$, and so

$$2 - \varrho - \kappa_k > 2 - \bar{\varrho},\ \forall k \geq K_1 \tag{20.23}$$

From Condition (20.3) and Lemma 20.10 (b), it follows that

$$\lim_{k\to\infty} \frac{(\alpha_{k+1} - \nu\alpha_k)^2}{(\alpha_{k+1} + \nu\alpha_k)^2} = \frac{(1-\nu)^2}{(1+\nu)^2} > 0.$$

Hence, there is a integer number K_2 satisfying

$$\frac{(\alpha_{k+1} - \nu\alpha_k)^2}{(\alpha_{k+1} + \nu\alpha_k)^2} > \frac{(1-\nu)^2}{(1+\nu)^2} - \bar{\nu},\ \forall k \geq K_2.$$

Set $K = \max\{K_1, K_2\}$ and $M = \varrho(2 - \bar{\varrho})\left[\frac{(1-\nu)^2}{(1+\nu)^2} - \bar{\nu}\right]$. From equation (20.23) and the last inequality, we obtain

$$(\varrho + \kappa_k)(2 - \varrho - \kappa_k)\frac{(\alpha_{k+1} - \nu\alpha_k)^2}{(\alpha_{k+1} + \nu\alpha_k)^2} \geq M > 0$$

for every $k \geq K$. We have from Lemma 20.10 (b) that

$$\|x^{k+1} - x^s\|^2 \leq \|w^k - x^s\|^2 - M\|\bar{w}^k - y^k\|^2,$$

which together with equation (21.25) implies that

$$\begin{aligned}\|x^{k+1} - x^s\|^2 \leq{}& (1 + \rho_k)\|x^k - x^s\|^2 + (-\rho_k)\|x^{k-1} - x^s\|^2\\ &+ \rho_k(1 + \rho_k)\|x^k - x^{k-1}\|^2 - M\|\bar{w}^k - y^k\|^2,\end{aligned}$$

for every $k \geq K$. That is equivalent to

$$\begin{aligned}\|x^{k+1} - x^s\|^2 - \|x^k - x^s\|^2 \leq{}& \rho_k(\|x^k - x^s\|^2 - \|x^{k-1} - x^s\|^2)\\ &+ \rho_k(1 + \rho_k)\|x^k - x^{k-1}\|^2\\ &- M\|y^k - \bar{w}^k\|^2\end{aligned} \tag{20.24}$$

for every $k \geq K$. Since $\rho_k \leq \rho$ for every k, we have

$$\begin{aligned}M\|y^k - \bar{w}^k\|^2 \leq{}& \|x^k - x^s\|^2 - \|x^{k+1} - x^s\|^2\\ &+\rho(\|x^k - x^s\|^2 - \|x^{k-1} - x^s\|^2)\\ &+(1 + \rho)\rho_k\|x^k - x^{k-1}\|^2\end{aligned}$$

for every $k \geq K$. Setting $\sigma_k = \|x^k - x^s\|^2$, $\Lambda_k = \sigma_k - \sigma_{k-1}$ and $\eta_k = (1 + \rho)\rho_k\|x^k - x^{k-1}\|^2$, we get

$$M\|y^k - \bar{w}^k\|^2 \leq \sigma_k - \sigma_{k+1} + \rho\Lambda_k + \eta_k \leq \sigma_k - \sigma_{k+1} + \rho[\Lambda_k]_+ + \eta_k, \tag{20.25}$$

for every $k \geq K$. By equation (20.19), there exists a constant $\bar{\mu} > 0$ such that

$$\sum_{k=1}^{\infty} \eta_k = (1+\rho) \sum_{k=1}^{\infty} \rho_k \|x^k - x^{k-1}\|^2 \leq (1+\rho) \sum_{k=1}^{\infty} \mu_k \leq (1+\rho)\bar{\mu}. \tag{20.26}$$

From equation (21.29), it follows that

$$\Lambda_{k+1} \leq \rho_k \Lambda_k + \eta_k \leq \rho[\Lambda_k]_+ + \eta_k, \tag{20.27}$$

for every $k \geq K$ and hence

$$[\Lambda_{k+1}]_+ \leq \rho[\Lambda_k]_+ + \eta_k \leq \rho^{k-K+1}[\Lambda_K]_+ + \sum_{i=1}^{k-K+1} \rho^{i-1}\eta_{k+1-i}.$$

Combining the last inequality and equation (21.32), we obtain

$$\begin{aligned}\sum_{k=K}^{\infty} [\Lambda_{k+1}]_+ &\leq [\Lambda_K]_+ \sum_{k=1}^{\infty} \rho^k + \sum_{k=1}^{\infty} \rho^{k-1} \sum_{k=N}^{\infty} \eta_k \\ &\leq \frac{\rho}{1-\rho}[\Lambda_K]_+ + \frac{1}{1-\rho}(1+\rho)\bar{\mu},\end{aligned}$$

which together with equations (21.31) and (21.32) implies that

$$\begin{aligned}M \sum_{i=K}^{k} \|y^i - \bar{w}^i\|^2 &\leq \sigma_K - \sigma_{k+1} + \rho \sum_{i=K}^{k} [\Lambda_i]_+ + \sum_{i=K}^{k} \eta_i \\ &\leq \sigma_K + \rho \left([\Lambda_K]_+ + \sum_{i=K}^{k-1} [\Lambda_{i+1}]_+ \right) + \sum_{i=K}^{k} \eta_i \\ &= \sigma_K + \rho[\Lambda_K]_+ + \rho \left(\frac{\rho}{1-\rho}[\Lambda_K]_+ + \frac{1+\rho}{1-\rho}\bar{\mu} \right) + (1+\rho)\bar{\mu} \\ &\leq \sigma_K + \frac{\rho}{1-\rho}[\Lambda_K]_+ + \frac{1+\rho}{1-\rho}\bar{\mu},\end{aligned}$$

equivalently,

$$\begin{aligned}&\sum_{i=K}^{k} \|y^i - \bar{w}^i\|^2 \\ &\leq \frac{1}{M} \left(\|x^1 - x^s\|^2 + \frac{\rho}{1-\rho} \left[\|x^1 - x^s\|^2 - \|x^0 - x^s\|^2 \right]_+ + \frac{1+\rho}{1-\rho}\bar{\mu} \right).\end{aligned}$$

Hence

$$\min_{K \leq i \leq k} \|y^i - \bar{w}^i\|^2 \leq \frac{\|x^1 - x^s\|^2 + \frac{\rho}{1-\rho}[\|x^1 - x^s\|^2 - \|x^0 - x^s\|^2]_+ + \frac{1+\rho}{1-\rho}\bar{\mu}}{\varrho(2-\bar{\varrho}) \left[\frac{(1-\nu)^2}{(1+\nu)^2} - \bar{\nu} \right] (k - K + 1)}.$$

The proof is complete. □

The result of theorem 4 shows that to get an ϵ-optimal solution of (EP), our algorithm needs to perform at least

$$\left\lfloor \frac{\|x^1 - x^s\|^2 + \frac{\rho}{1-\rho}[\|x^1 - x^s\|^2 - \|x^0 - x^s\|^2]_+ + \frac{1+\rho}{1-\rho}\bar{\mu}}{\varrho(2-\bar{\varrho})\left[\frac{(1-\nu)^2}{(1+\nu)^2} - \bar{\nu}\right]\epsilon} \right\rfloor + K$$

iterations, where $\lfloor t \rfloor$ is a integer number such that $t \leq \lfloor t \rfloor \leq t+1$.

20.5 COMPUTATIONAL EXPERIMENTS

In the last section, we make some numerical examples to illustrate our algorithm. We used Matlab R2016a to code all computer programs on a PC Intel(R) Core(TM) i5-2430M CPU @ 2.40 GHz 4GB Ram. The Optimization Toolbox (fmincon) is used to solve strongly convex subproblems.

Example 20.1 *Let us consider a Cournot-Nash equilibrium model in* $\mathbb{R}^n$ *with n-firms, each firm i has a profit function defined by* $f_i(x_1, x_2, \ldots, x_n) = x_i r_i(\pi_x) - h_i(x_i)$, *where* $h_i(x_i)$, $r_i(\pi_x)$ *denote the total cost, the price function of firm i respectively, and* $\pi_x := \sum_{i=1}^n x_i$, *see e. g. [2, 3, 11]). Let* $C_i \subset \mathbb{R}$ *be the nonempty bounded strategy set of firm i including its possible production levels. Then the strategy set of the model is* $C = \{x = (x_1, \ldots, x_n) \in \mathbb{R}^n : \ x_i \in C_i, \ i = 1, \ldots, n\}$.

Each firm seeks to maximize its profit by choosing the level of production corresponding to the assumption that the other firms' production is the parameter input. In this context, the Nash equilibrium is a model of production in which no firm can increase profits by varying the variables controlled. As well known e.g. [13, 14] the Nash equilibrium point of this model is a point $x^s := (x_1^s, \ldots, x_n^s)^T \in C$ *such that*

$$f_i(x^s) \geq f_i(x^s[x_i]) \ \ \forall x_i \in C_i, \ \forall i,$$

where the notation $x^s([x_i])$ *is the vector obtained from* x^s *by replacing* x_i^s *by* $x_i \in C_i$. *The problem of finding a Nash-Cournot oligopolistic market equilibrium problem can be expressed as follows:*

$$f(p, x) \geq 0, \quad \forall x \in C,$$

where

$$f(x, y) = \langle Fx, y - x\rangle + h(y) - h(x),$$

$h(x) = \sum_{i=1}^n h_i(x_i)$, $Fx = (F_1x, F_2x, \ldots, F_nx)$, $F_ix = -r_i(\pi_x) - x_i r_i'(\pi_x)$. *Obviously, if* $F : \mathbb{R}^n \to \mathbb{R}^n$ *is pseudomonotone,* L_1*-Lipschitz continuous and* $g : \mathbb{R}^n \to \mathbb{R}$ *is convex, subdifferentiable with subgradient operator being* L_2-*Lipschitz Hausdorff continuous, then the bifunction f satisfies conditions* M_1, M_2, M_4 *and* M_5.

As in the classical case,, when the price function $r_i(\pi_x)$ is affine $(i = 1, \ldots, n)$ and given by

$$r_i(\pi_x) = \rho_i - q_i\pi_x, \quad q_i \geq 0, \rho_i \geq 0, \quad \forall i = 1, \ldots, n.$$

Then

$$F_i x = -r_i(\pi_x) - x_i r_i'(\pi_x) = q_i \pi_x - \rho_i + q_i x_i = 2q_i x_i + q_i \sum_{j=1, j\neq i}^{n} x_j - \rho_i,$$

and so, $Fx = Qx - \rho$, where

$$Q = \begin{pmatrix} 2q_1 & q_1 & \dots & q_1 \\ q_2 & 2q_2 & \dots & q_2 \\ \dots \\ q_n & q_n & \dots & 2q_n \end{pmatrix}, \rho = (\rho_1, \rho_2, \dots, \rho_n)^T.$$

We show that F is monotone on $\mathbb{R}^n$. In fact, by induction, we can show that $|\Delta_k| > 0$ for every $k = 1, 2, \dots, n$, where

$$\Delta_k = \begin{pmatrix} 2q_1 & q_1 & \dots & q_1 \\ q_2 & 2q_2 & \dots & q_2 \\ \dots \\ q_k & q_k & \dots & 2q_k \end{pmatrix}.$$

Then, by using a triangular transformation of the unknowns, the quadratic form $\langle Fx - Fy, x - y\rangle = \langle Q(x - y), x - y\rangle$ can be reduced to the canonical form

$$\langle Fx - Fy, x - y\rangle = \sum_{i=1}^{n} \frac{|\Delta_{i-1}|}{|\Delta_i|} X_i^2,$$

where $|\Delta_0| = 1$, which implies that F monotone on $\mathbb{R}^n$. It is easy to check that F is $\|Q\|$-Lipschitz continuous.

20.5.1 Test 1

Consider $n = 5$. We do some numerical tests to solve Example 20.1 when total cost function $g(x)$ is determined by $g(x) = \sum_{i=1}^{5} |x_i - i|$; $q_i = \frac{i}{5}$, $\forall i = 1, 2, \dots, 5$, $\rho = (10, 10, 10, 10, 10,)^T$ and

$$C := \left\{ (x_1, x_2, \dots, x_5)^T \in \mathbb{R}^5 : i \le x_i \le 15 + \frac{i}{3i - 2}, \quad \forall i = 1, 2, \dots, 5 \right\}.$$

It is easy to see that $f(x, \cdot) = \langle Fx, \cdot - x\rangle + g(\cdot) - g(x)$ is not differentiable on C. Figure 20.1 shows convergent results of $\|x^k - x^{k-1}\|$ with $x^0 = (-1, 2, -3, 4, -5)^T$ and the stopping criteria being $Err = \|x^k - x^{k-1}\| \le 10^{-3}$.

20.5.2 Test 2

Let $g(x) := \max\{2\|x\|^2, \|x\|^2 + 2\langle a, x\rangle + 1\}$ and $C := \{x \in \mathbb{R}^n : \|x\|^2 - 2\langle a, x\rangle - 1 \le 0\}$. We have from Remark 20.1 that $f(x, \cdot) = \langle Fx, \cdot - x\rangle + g(\cdot) - g(x)$ is not differentiable on C. In this test, we use Algorithm 3.1 for solving Example 20.1 with

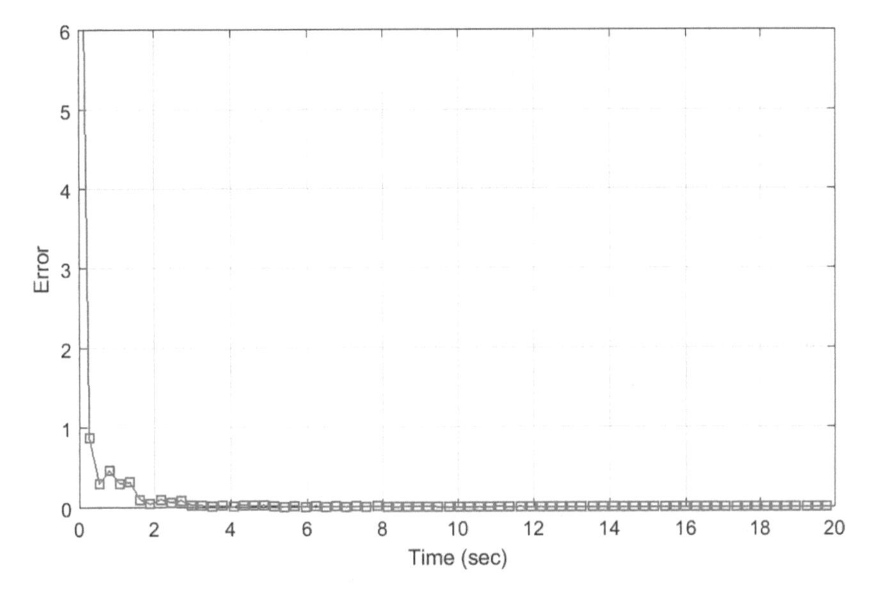

Figure 20.1 Convergence of Algorithm 1 with the tolerance $\epsilon = 10^{-3}$.

the data: $q_i := \frac{i}{5}$, $\forall i = 1, 2, \ldots, n$, $\rho := (0, 0, \cdots, 0)^T$ and $a := (0, 0, \ldots, 0)^T$. The computed results are shown in **Table 20.1** with different parameters θ_k and κ_k. From this table, we see that the efficiency of our algorithm is less affected by the parameters θ_k and κ_k. It shows that the parameters α_k and τ_k are mostly updated based on previous iteration points. **Table 20.2** shows computed results with $x^0 := (1, 1, \ldots, 1)$ and different given tolerance parameters ϵ and dimensions n.

TABLE 20.1 The Computed Results for Different Parameters

	Parameters		**Algorithm 1**	
Init. Point x^0	θ_k	κ_k	**Iterations.**	**CPU Times**
$(-4, 0, 4, 2, 0)^\top$	$\frac{1}{k^2+1}$	$\frac{1}{k+1}$	21	0.1716
$(-4, 0, 4, 2, 0)^\top$	$\frac{1}{k^2+10}$	$\frac{1}{k+1}$	19	0.1560
$(-4, 0, 4, 2, 0)^\top$	$\frac{1}{k^2+100}$	$\frac{1}{k+1}$	19	0.1560
$(-4, 0, 4, 2, 0)^\top$	$\frac{1}{k^2+100}$	$\frac{1}{k+1}$	18	0.1560
$(-4, 0, 4, 2, 0)^\top$	$\frac{1}{k^2+1000}$	$\frac{1}{k+1}$	19	0.1716
$(-4, 0, 4, 2, 0)^\top$	$\frac{1}{5k^2+1}$	$\frac{1}{k+1}$	21	0.2028
$(-4, 0, 4, 2, 0)^\top$	$\frac{1}{10k^2+1}$	$\frac{1}{k+1}$	21	0.2312
$(-4, 0, 4, 2, 0)^\top$	$\frac{1}{20k^2+1}$	$\frac{1}{k+1}$	21	0.2034
$(-4, 0, 4, 2, 0)^\top$	$\frac{1}{100k^2+1}$	$\frac{1}{k+1}$	21	0.2475

TABLE 20.2 The Computed Results with Different Tolerance Parameter ϵ, the Dimension n and the Stopping Criterion Being $\|x^{k+1} - x^k\| \leq \epsilon$

	Algorithm 1					
	$\epsilon = 10^{-3}$		$\epsilon = 10^{-4}$		$\epsilon = 10^{-5}$	
n	Iter.	**CPU-Times**	Iter.	**CPU-Times**	Iter.	**CPU-Times**
5	15	0.1560	21	0.1716	21	0.1892
10	21	0.1614	30	0.1788	40	0.1978
20	29	0.2092	43	0.2116	58	0.2496
30	36	0.2172	54	0.2384	74	0.2596
50	47	0.2716	75	0.2988	103	0.3208
70	58	0.3120	93	0.3489	131	0.4616
100	71	0.3338	120	0.5864	171	0.7332

20.6 CONCLUSION

We have proposed an inertial projection algorithm for solving pseudomonotone equilibrium problems, where the bifunctions may not be differentiable in its second variable. In the proposed algorithm, unlike the existing ones, we have used a diagonal gradient rather than a diagonal subgradient as a linesearch direction. This smoothing technique avoids the difficulty arising from the fact that the efficiency of a projection algorithm crucially depends upon the choice of the subgradient. Some computed results have been reported that show the behaviour and efficiency of the algorithm for a Cournot-Nash model with no differentiable convex cost function.

BIBLIOGRAPHY

[1] F. Alvarez, H. Attouch, An inertial proximal method for maximal monotone operators via discretization of a nonlinear oscillator with damping, *Set-Valued Anal.* 9 (2001) 3–11.

[2] P. N. Anh, T. V. Thang, H. T. C. Thach, Halpern projection methods for solving pseudomonotone multivalued variational inequalities in Hilbert spaces. *Num. Alg.* 87 (2021) 335–363.

[3] P. N. Anh, T. V. Thang, H. T. C. Thach, A subgradient proximal method for solving a class of monotone multivalued variational inequality problems. *Num. Alg.* 89 (2022) 409–430.

[4] H. H. Bauschke, P. L. Combettes, *Convex Analysis and Monotone Operator Theory in Hilbert Spaces*, Springer, New York (2011).

[5] G. Bigi, M. Castellani, M. Pappalardo, M. Passacantando, *Nonlinear Programming Techniques for Equilibria*, Springer, New York (2019).

[6] E. Blum, W. Oettli, From optimization and variational inequalities to equilibrium problems. *Math. Stud.* 63 (1994) 123–145.

[7] P. Debnath, H. M. Srivastava, P. Kumam, B. Hazarika, *Fixed Point Theory and Fractional Calculus: Recent Advances and Applications*, Springer, Singapore (2022).

[8] P. Debnath, H. M. Srivastava, K. Chakraborty, P. Kumam, *Advances in Number Theory and Applied Analysis*, World Scientific, Singapore (2023).

[9] N. M. Khoa, T. V. Thang, Approximate projection algorithms for solving equilibrium and multivalued variational inequality problems in Hilbert space. *Bull. Korean Math. Soc.* 59 (2022) 1019–1044.

[10] I. V. Konnov, *Combined Relaxation Methods for Variational Inequalities*, Springer-Verlag, Berlin (2000).

[11] G. M. Korpelevich, Extragradient method for finding saddle points and other problems. *Matecon* 12 (1976) 747–756.

[12] L. D. Muu, W. Oettli, Convergence of an adaptive penalty scheme for finding constrained equilibria. *Nonlinear Anal.* 18 (1992) 1159–1166.

[13] L. D Muu, T. Q. Quoc, Regularization algorithms for solving monotone Ky Fan inequalities with application to a Nash-Cournot equilibrium model. *J. Optim. Theory Appl.* 142 (2009) 185–204.

[14] T. Q. Quoc, L. D Muu, V. H. Nguyen, Extragradient algorithms extended to equilibrium problems. *Optimization* 57(6) (2008) 749–776.

[15] R. T. Rockafellar, *Convex Analysis.* Princeton University Press, Princeton, NJ (1970).

[16] Y. Shehu, O. S. Iyiola, X. H. Li, et al., Convergence analysis of projection method for variational inequalities. *Comput. Appl. Math.* 38 (2019) 161.

[17] T. V. Thang, Inertial subgradient projection algorithms extended to equilibrium problems. *Bull. Iranian Math. Soc.* 48 (2022) 2349–2370.

CHAPTER 21

Convergence Analysis of a Relaxed Inertial Alternating Minimization Algorithm with Applications

Yuchao Tang

Guangzhou University

Yang Yang

Sun Yat-sen University
Guangdong Key Laboratory of Information Security Technology

Jigen Peng

Guangzhou University

CONTENTS

21.1 INTRODUCTION

Many problems in signal and image processing can be modeled as convex minimization problems, whose objective functions may be two-block separable with linear equality constraints. The alternating direction method of multipliers (ADMM) dated

DOI: 10.1201/9781003388678-21

back to the work of Glowinski et al. [1] and Gabay et al. [2] is a widely used method for solving two-block separable convex minimization problems with linear equality constraints. The ADMM was received much attention in recent years due to its simplicity in solving various inverse problems arising in image restoration and medical image reconstruction. See for example [3, 4, 5]. We refer interested readers to [6, 7, 8, 9] for theoretical results on ADMM with two-block including convergence analysis and convergence rates analysis.

Since the popularity of the two-block ADMM, it is natural to consider how to generalize it to solve a three-block separable convex minimization problem. There exist many problems that are suitable for representing in the formulation of three-block other than two-block. For instance, the stable principal component pursuit (SPCP) [10], the latent variable Gaussian graphical model selection [11], the robust principal component analysis model with noisy and incomplete data [12], and so on. The three-block separable convex minimization problem is modeled as follows:

$$\begin{aligned} \min_{x_1,x_2,x_3} \ & f_1(x_1) + f_2(x_2) + f_3(x_3) \\ \text{s.t.}\ & L_1x_1 + L_2x_2 + L_3x_3 = b, \end{aligned} \tag{21.1}$$

where $f_i : H_i \to (-\infty, +\infty]$ with $i = 1, 2, 3$ are proper, lower semi-continuous convex functions (not necessarily smooth); $L_i : H_i \to H$ with $i = 1, 2, 3$ are bounded linear operators; and $b \in H$ is a given vector; H and H_i with $i = 1, 2, 3$ are real Hilbert spaces. Throughout this chapter, we assume that the solution set of problem (21.1) exists. For solving the convex minimization problem (21.1), the direct extension of the three-block ADMM iterative scheme is as follows:

$$\begin{cases} x_1^{k+1} = \arg\min_{x_1}\{f_1(x_1) + \frac{\gamma}{2}\|L_1x_1 + L_2x_2^k + L_3x_3^k - b - \frac{1}{\gamma}w^k\|^2\}, \\ x_2^{k+1} = \arg\min_{x_2}\{f_2(x_2) + \frac{\gamma}{2}\|L_1x_1^{k+1} + L_2x_2 + L_3x_3^k - b - \frac{1}{\gamma}w^k\|^2\}, \\ x_3^{k+1} = \arg\min_{x_3}\{f_3(x_3) + \frac{\gamma}{2}\|L_1x_1^{k+1} + L_2x_2^{k+1} + L_3x_3 - b - \frac{1}{\gamma}w^k\|^2\}, \\ w^{k+1} = w^k - \gamma(L_1x_1^{k+1} + L_2x_2^{k+1} + L_3x_3^{k+1} - b), \end{cases} \tag{21.2}$$

where w is the Lagrange multiplier and $\gamma > 0$ is the penalty parameter. However, Chen et al. [13] showed that the direct extension of the two-block ADMM to three-block ADMM is divergent if no further condition is imposed. Therefore, many efforts have been made to overcome this shortage. We can roughly divide them into two categories. The first is to make some minor changes to the direct extension of the three-block ADMM. For example, in [14, 15], He et al. generated a new iteration point by correcting the output of each step to guarantee the convergence of three-block ADMM. He et al. [14] proposed an alternating direction method for prediction correction (ADM-G), which guarantees convergence by adding a Gaussian back substitution correction step. ADM-G specific iteration format is read as:

$$\begin{cases} \tilde{x_1}^k = \arg\min_{x_1}\{f_1(x_1) + \frac{\gamma}{2}\|L_1x_1 + L_2x_2^k + L_3x_3^k - b - \frac{1}{\gamma}w^k\|^2\}, \\ \tilde{x_2}^k = \arg\min_{x_2}\{f_2(x_2) + \frac{\gamma}{2}\|L_1\tilde{x_1}^k + L_2x_2 + L_3x_3^k - b - \frac{1}{\gamma}w^k\|^2\}, \\ \tilde{x_3}^k = \arg\min_{x_3}\{f_3(x_3) + \frac{\gamma}{2}\|L_1\tilde{x_1}^k + L_2\tilde{x_2}^k + L_3x_3 - b - \frac{1}{\gamma}w^k\|^2\}, \\ \tilde{w}^k = w^k - \gamma(L_1\tilde{x_1}^k + L_2\tilde{x_2}^k + L_3\tilde{x_3}^k - b), \\ x_1^{k+1} = \tilde{x_1}^k, \\ v^{k+1} = v^k - \theta G^{-1}(v^k - \tilde{v}^k), \end{cases} \tag{21.3}$$

where $\theta \in (0,1)$ and

$$v^k = \begin{pmatrix} x_2^k \\ x_3^k \\ w^k \end{pmatrix}, \tilde{v}^k = \begin{pmatrix} \tilde{x_2}^k \\ \tilde{x_3}^k \\ \tilde{w}^k \end{pmatrix}, G = \begin{pmatrix} I_2 & (L_2^TL_2)^{-1}L_2^TL_3 & 0 \\ 0 & I_3 & 0 \\ 0 & 0 & I \end{pmatrix}. \tag{21.4}$$

We can see that G is an upper triangular matrix, so step 6 in (21.3) is easy to perform. In [16], Hong and Luo added a contraction factor to the Lagrange multiplier update step and established its global linear convergence under some assumptions. Deng et al. [17] and Sun et al. [18] each proposed a variant of three-block ADMM and proved the convergence. Their variants not only add the contraction factor in the Lagrange multiplier update step but also employ an appropriate proximal term in the subproblem of ADMM. The three-block ADMM variant proposed by Sun et al. [18] uses the Gauss-Seidel cycle to update variables. The three-block ADMM variant proposed by Deng et al. [17] uses the Jacobi cycle to update variables and considers the general m-block case for any m larger than or equal to 3. The algorithms in [14, 15] belong to the algorithmic framework of prediction-correction methods. We refer interested readers to [19, 20, 21, 22, 23] for other types of prediction-correction three-block ADMM. The second way is to add more conditions to the objective function or/and linear equality constraints to ensure the convergence of three-block ADMM. For instance, Han and Yuan in [24] have proved the convergence of three-block ADMM by assuming that the objective functions are strongly convex and the penalty parameter has a small upper bound. In a few years, this condition has been relaxed. The authors of [25, 26] proved the convergence of a three-block ADMM iteration scheme if only two of the objective functions are strongly convex and the penalty parameter is limited to a small range. Furthermore, Lin et al. [27] proved the globally linear convergence rate of the method under some additional conditions. This chapter mainly studies the three-block convex optimization problem with a strongly convex function in the objective function. Cai et al. [28] proved the convergence of (21.2) when f_3 is strongly convex with a constant $\mu_3 > 0$, and L_1, L_2 are full column rank. Li et al. [29] proposed a semi-proximal alternating direction method of multipliers (sPADMM) by hiring appropriate proximal terms on the subproblem of (21.2) and proved its global convergence. sPADMM iteration details are presented below.

$$\begin{cases} x_1^{k+1} = \arg\min_{x_1}\{f_1(x_1) + \frac{\gamma}{2}\|L_1x_1 + L_2x_2^k + L_3x_3^k - b - \frac{1}{\gamma}w^k\|^2 \\ \quad + \frac{1}{2}\|x_1 - x_1^k\|_{T_1}^2\}, \\ x_2^{k+1} = \arg\min_{x_2}\{f_2(x_2) + \frac{\gamma}{2}\|L_1x_1^{k+1} + L_2x_2 + L_3x_3^k - b - \frac{1}{\gamma}w^k\|^2 \\ \quad + \frac{1}{2}\|x_2 - x_2^k\|_{T_2}^2\}, \\ x_3^{k+1} = \arg\min_{x_3}\{f_3(x_3) + \frac{\gamma}{2}\|L_1x_1^{k+1} + L_2x_2^{k+1} + L_3x_3 - b - \frac{1}{\gamma}w^k\|^2 \\ \quad + \frac{1}{2}\|x_3 - x_3^k\|_{T_3}^2\}, \\ w^{k+1} = w^k - \tau\gamma(L_1x_1^{k+1} + L_2x_2^{k+1} + L_3x_3^{k+1} - b), \end{cases} \tag{21.5}$$

where $\tau \in (0, \frac{1+\sqrt{5}}{2})$, $\gamma \in (0, +\infty)$ and T_i with $i = 1, 2, 3$ are self adjoint and positive semi-definite operators. And function f_2 is strongly convex with constant $\mu_2 > 0$. The operator T_i with $i = 1, 2, 3$ may be 0 if γ is smaller than a threshold. This makes sPADMM (21.5) return to directly extended three-block ADMM (21.2) with $\tau \in (0, \frac{1+\sqrt{5}}{2})$. The aforementioned three-block ADMM convergence guarantee with strong convexity requirements requires that the penalty parameters be relatively small. Recently, Lin et al. [30] proved that the convergence of the three-block ADMM has only one strong convexity and smoothness in the objective function and the penalty parameter is larger than zero.

Besides the ADMM and its variants, the alternating minimization algorithm (AMA) proposed by Tseng [31] is an important algorithm for solving a two-block separable convex minimization problem with linear equality constraints, where one of the convex functions is assumed to be strongly convex. It is worth noting that the AMA algorithm is equivalent to the forward-backward splitting algorithm applied to the corresponding dual problem. Recently, Davis and Yin [32] proposed a so-called three-block ADMM for solving a three-block separable convex minimization problem, where one of them is strongly convex. The three-block ADMM is derived from the three-operator splitting algorithm applied to the dual problem. They pointed out the three-block ADMM included the Tseng's AMA algorithm, the classical ADMM, and the augmented Lagrangian method. In comparison with the three-block extension of ADMM, the first step of Davis and Yin's three-block ADMM [32] does not involve a quadratic penalty term, which is the same as the AMA algorithm. Therefore, we think it is better to name the three-block ADMM proposed by Davis and Yin [32] as a three-block AMA algorithm. The iteration scheme of the three-block AMA algorithm is as follows:

$$\begin{cases} x_1^{k+1} = \arg\min\limits_{x_1}\{f_1(x_1) - \langle w^k, L_1x_1\rangle\}, \\ x_2^{k+1} = \arg\min\limits_{x_2}\{f_2(x_2) - \langle w^k, L_2x_2\rangle + \frac{\gamma}{2}\|L_1x_1^{k+1} + L_2x_2 \\ \qquad + L_3x_3^k - b\|^2\}, \\ x_3^{k+1} = \arg\min\limits_{x_3}\{f_3(x_3) - \langle w^k, L_3x_3\rangle + \frac{\gamma}{2}\|L_1x_1^{k+1} + L_2x_2^{k+1} \\ \qquad + L_3x_3 - b\|^2\}, \\ w^{k+1} = w^k - \gamma(L_1x_1^{k+1} + L_2x_2^{k+1} + L_3x_3^{k+1} - b), \end{cases} \tag{21.6}$$

where f_1 is a strongly convex function with constant $\mu_1 > 0$. As an important method, the alternating minimization algorithm (AMA) has received extensive attention from scholars.

As Goldstein et al. [33] pointed out that the ADMM and the AMA are preferred ways to solve two-block separable convex programming because of their simplicity, they often perform poorly in situations where the problem is poorly conditioned or when high precision is required. Eckstein and Bertsekas [34] first proposed a relaxed ADMM (RADMM), which included the classical ADMM as a special case. Numerical experiments have confirmed that the RADMM can accelerate the classical ADMM when the relaxation parameter belongs to $(1, 2)$. Further, Xu et al. [35] proposed an adaptive relaxed ADMM that automatically tuned the algorithm parameters. Goldstein et al. [33] proposed two accelerated variants of the ADMM and the AMA, which are based on Nesterov's accelerated gradient method. Kadkhodaie et al. [36] proposed a so-called accelerated alternating direction method of multipliers (A2DM2) and proved that the algorithm achieved $O(1/k^2)$ convergence rate, where k is the iteration number. They weakened the assumptions required in [33]. We would like to point out that there also exist some other approaches for accelerating the ADMM, such as accelerated ADMM [37] based on the Douglas-Rachford envelope (DRE), adaptive accelerated ADMM [38] and accelerated ADMM based on accelerated proximal point algorithm [39]. Similar to the idea of Nesterov's accelerated gradient method, the inertial method becomes popular in recent years. It provides a general way to select the inertia parameters. Chen et al. [40] proposed an inertial proximal ADMM, which was derived from the inertial proximal point algorithm. On the other hand, Boţ and Csetnek [41] proposed an inertial ADMM, which is based on the inertial Douglas-Rachford splitting algorithm [42]. To the best of our knowledge, we have not seen any generalization work on the three-block AMA algorithm to the relaxation or the inertia.

The purpose of this chapter is to introduce a relaxed inertial three-block AMA algorithm for solving the three-block separable convex minimization problem. The idea is to employ the inertial three-operator splitting algorithm [43] to the dual problem. As a by-product, we obtain a relaxed three-block AMA algorithm, which generalizes the three-block AMA algorithm of Davis and Yin [32]. Under mild conditions, we prove the convergence of the proposed algorithms. To verify the efficiency and effectiveness of the proposed algorithms, we apply them to solve the SPCP [10] problem. We also report numerical results compared with other algorithms for solving the SPCP.

We highlight the contributions of this chapter: (i) We propose a generalization of the three-block AMA with relaxation and inertia. The obtained algorithm includes several algorithms as its special cases; (ii) We study the convergence of the proposed algorithm under different conditions on the parameters in infinite-dimensional Hilbert spaces. Compared with other existing three-block ADMM and its variants, we obtain weak and strong convergence of the iteration schemes; (iii) We conduct extensive numerical experiments on SPCP to verify the impact of the introduced relaxation and inertia parameters.

The rest of this chapter is organized as follows. In Section 21.2, we present some preliminaries on the maximally monotone operators and convex functions. In particular, we review several results on the inertial three-operator splitting algorithm. In Section 21.3, we present the main algorithm and prove the convergence of it. In Section 21.4, we conduct numerical experiments on the SPCP to demonstrate the efficiency and effectiveness of the proposed algorithms. Finally, we give some conclusions. We also present two open questions for further study.

21.2 PRELIMINARIES

In this section, we review some basic definitions in convex analysis and monotone operator theory. Most of the following definitions can be found in [44]. Let H is a real Hilbert space, which endowed with an inner product $\langle \cdot, \cdot \rangle$ and associated norm $\|\cdot\| = \sqrt{\langle \cdot, \cdot \rangle}$. The symbols $\rightharpoonup$ and $\rightarrow$ denote weak and strong convergence, respectively.

Let $A : H \rightarrow 2^H$ be a set-valued operator. We denote by $zerA = \{x \in H : 0 \in Ax\}$ its set of zeros, by $graA = \{(x, u) \in H \times H : u \in Ax\}$ its graph and by $ranA = \{u \in H : \exists x \in H, u \in Ax\}$ its range. The resolvent of an operator $A : H \rightarrow 2^H$ is denoted by $J_A = (I + A)^{-1}$.

Definition 21.1 *([44]) Let $A : H \rightarrow 2^H$ be a set-valued operator. Then*
(i) A is monotone if

$$\langle x - y, u - v \rangle \geq 0, \forall (x, u), (y, v) \in graA.$$

Further, A is maximally monotone if there exists no monotone operator $A' : H \rightarrow 2^H$ such that $graA'$ properly contains $graA$.

(ii) A is uniformly monotone if there exists an increasing function $\phi : [0, +\infty) \rightarrow [0, +\infty]$ that vanishes only at 0 such that

$$\langle x - y, u - v \rangle \geq \phi(\|x - y\|), \forall (x, u), (y, v) \in graA.$$

Definition 21.2 *([44]) Let $B : H \rightarrow H$ is a single valued operator. $B : H \rightarrow H$ is said to be $\beta -$ cocoercive, for some $\beta > 0$, if*

$$\langle x - y, Bx - By \rangle \geq \beta \|Bx - By\|^2, \forall x, y \in H.$$

Let a function $f : H \to (-\infty, +\infty]$. We denote by $\Gamma_0(H)$ be the class of proper, lower semi-continuous convex functions $f : H \to (-\infty, +\infty]$. Let $f \in \Gamma_0(H)$, the conjugate of f is $f^* \in \Gamma_0(H)$ defined by $f^* : u \mapsto sup_{x\in H}(\langle x, u\rangle - f(x))$, and the subdifferential of f is the maximally monotone operator

$$\partial f : H \to 2^H : x \mapsto \{u \in H | f(y) \geq f(x) + \langle u, y - x\rangle, \forall y \in H\}.$$

f is uniformly convex if there exists an increasing function $\phi : [0, +\infty) \to [0, +\infty]$ that vanishes only at 0 such that

$$\alpha f(x) + (1 - \alpha)f(y) \geq f(\alpha x + (1 - \alpha)y) + \alpha(1 - \alpha)\phi(\|x - y\|),$$

$\forall x, y \in H, \alpha \in (0, 1)$.

f is $\sigma -$ *strongly* convex for some $\sigma > 0$ if $f - \frac{\sigma}{2}\|\cdot\|^2$ is convex.

The proximity operator $prox_{\lambda f} : x \mapsto \arg\min_y\{f(y) + \frac{1}{2\lambda}\|x - y\|^2\}$, where $\lambda > 0$. Let $f \in \Gamma_0(H)$, then we have $J_{\lambda\partial f} = (I + \lambda\partial f)^{-1} = prox_{\lambda f}$.

To analyze the convergence of the algorithm proposed in this chapter, we recall the main results of the inertial three-operator splitting algorithm in [43].

Theorem 21.1 *([43]) Let H be real Hilbert space. Let $A, B : H \to 2^H$ be two maximally monotone operators. Let $C : H \to H$ be a $\beta -$ cocoercive operator, for some $\beta > 0$. Let $z^0, z^1 \in H$, and set*

$$\begin{cases} y^k = z^k + \alpha_k(z^k - z^{k-1}), \\ w^k = J_{\gamma B} y^k, \\ u^k = J_{\gamma A}(2w^k - y^k - \gamma C w^k), \\ z^{k+1} = y^k + \lambda_k(u^k - w^k), \end{cases} \tag{21.7}$$

where the parameters γ, $\{\alpha_k\}$ and $\{\lambda_k\}$ satisfy the following conditions:

(c1) $\gamma \in (0, 2\beta\bar{\varepsilon})$, where $\bar{\varepsilon} \in (0, 1)$;

(c2) $\{\alpha_k\}$ is nondecreasing with $k \geq 1$, $\alpha_1 = 0$ and $0 \leq \alpha_k \leq \alpha < 1$;

(c3) for every $k \geq 1$, and $\lambda, \sigma, \delta > 0$ such that

$$\delta > \frac{\alpha^2(1 + \alpha) + \alpha\sigma}{1 - \alpha^2} \text{ and } 0 < \lambda \leq \lambda_k \leq \frac{\delta - \alpha[\alpha(1 + \alpha) + \alpha\delta + \sigma]}{\bar{\alpha}\delta[1 + \alpha(1 + \alpha) + \alpha\delta + \sigma]}, \tag{21.8}$$

where $\bar{\alpha} = \frac{1}{2-\bar{\varepsilon}}$. Then the following hold:

(i) $\{z^k\}$ converges weakly to z^*;

(ii) $\{w^k\}$ converges weakly to $J_{\gamma B} z^* \in zer(A + B + C)$;

(iii) $\{u^k\}$ converges weakly to $J_{\gamma B} z^* = J_{\gamma A}(2J_{\gamma B} z^* - z^* - \gamma C J_{\gamma B} z^*) \in zer(A + B + C)$;

(iv) $\{z^k - z^{k-1}\}$ converges strongly to 0;

(v) $\{w^k - u^k\}$ converges strongly to 0;

(vi) $\{Cw^k\}$ converges strongly to $CJ_{\gamma B} z^*$;

(vii) Suppose that one of the following conditions holds:

(a) A be uniformly monotone on every nonempty bounded subset of $domA$;
(b) B be uniformly monotone on every nonempty bounded subset of $domB$;
(c) C be demiregular at every point $x \in zer(A+B+C)$.
Then $\{w^k\}$ and $\{u^k\}$ converge strongly to $J_{\gamma B}z^* \in zer(A+B+C)$.

Proof. (i), (ii), (iii) and (vii) are directly obtained from Theorem 3.1 of [43]. (iv), (v) and (vi) can be easily obtained from Theorem 3.1 of [43]. Here we omit the proof. □

Theorem 21.2 *([43]) Let H be a real Hilbert space. Let $A, B : H \to 2^H$ be two maximally monotone operators. Let $C : H \to H$ be a β-cocoercive operator, for some $\beta > 0$. Let the iterative sequences $\{z^k\}$, $\{w^k\}$ and $\{u^k\}$ are generated by (21.7). Assume that the parameters γ, $\{\alpha_k\}$ and $\{\lambda_k\}$ satisfy the following conditions:*

(c1) $\gamma \in (0, 2\beta\bar{\varepsilon})$, *where* $\bar{\varepsilon} \in (0,1)$;

(c2) $0 \le \alpha_k \le \alpha < 1$ *and* $0 < \underline{\lambda} \le \lambda_k\bar{\alpha} \le \overline{\lambda} < 1$, *where* $\bar{\alpha} = \frac{1}{2-\bar{\varepsilon}}$;

(c3) $\sum_{k=0}^{+\infty} \alpha_k \|z^k - z^{k-1}\|^2 < +\infty$.

Then the following hold:

(i) $\{z^k\}$ *converges weakly to* z^*;

(ii) $\{w^k\}$ *converges weakly to* $J_{\gamma B}z^* \in zer(A+B+C)$;

(iii) $\{u^k\}$ *converges weakly to* $J_{\gamma B}z^* = J_{\gamma A}(2J_{\gamma B}z^* - z^* - \gamma CJ_{\gamma B}z^*) \in zer(A+B+C)$;

(iv) $\{\alpha_k(z^k - z^{k-1})\}$ *converges strongly to* 0;

(v) $\{w^k - u^k\}$ *converges strongly to* 0;

(vi) $\{Cw^k\}$ *converges strongly to* $CJ_{\gamma B}z^*$;

(vii) Suppose that one of the following conditions holds:

(a) A *be uniformly monotone on every nonempty bounded subset of* $domA$;

(b) B *be uniformly monotone on every nonempty bounded subset of* $domB$;

(c) C *be demiregular at every point* $x \in zer(A+B+C)$.

Then $\{w^k\}$ *and* $\{u^k\}$ *converge strongly to* $J_{\gamma B}z^* \in zer(A+B+C)$.

Proof. The proof of Theorem 21.2 is similar to Theorem 21.1, so we omit it here. □

Remark 21.1 *The condition* $\alpha_1 = 0$ *in the Theorem 21.1 can be replaced by the assumption* $z^0 = z^1$.

21.3 RELAXED INERTIAL THREE-BLOCK AMA FOR SOLVING THREE-BLOCK SEPARABLE CONVEX MINIMIZATION PROBLEM

In this section, we present the main results of this chapter including our proposed algorithm and its convergence theorem. The Lagrange function of problem (21.1) is defined as follows:

$$L(x_1, x_2, x_3, w) = f_1(x_1) + f_2(x_2) + f_3(x_3) - \langle L_1x_1 + L_2x_2 + L_3x_3 - b, w\rangle, \quad (21.9)$$

where w is a Lagrange multiplier. Through to Lagrange function (21.9), the dual problem of problem (21.1) is

$$\min_{w \in H} f_1^*(L_1^*w) + f_2^*(L_2^*w) + f_3^*(L_3^*w) - \langle b, w\rangle, \quad (21.10)$$

where f_i^* are the Fenchel-conjugate functions of f_i, respectively. According to the first-order optimality condition of problem (21.1), the solution of problem (21.1) is equivalent to finding $x_i^* \in H_i$ and $w^* \in H$ satisfying the following formula:

$$\begin{aligned} &0 \in \partial f_1(x_1^*) - L_1^*w^*, \\ &0 \in \partial f_2(x_2^*) - L_2^*w^*, \\ &0 \in \partial f_3(x_3^*) - L_3^*w^*, \\ &L_1x_1^* + L_2x_2^* + L_3x_3^* - b = 0, \end{aligned} \quad (21.11)$$

this is what we usually call the KKT condition.

Next, we present the main algorithm of this chapter and prove its convergence.

Algorithm 2 Relaxed Inertial Three-Block AMA

Require: For arbitrary $w^1 \in H$, $p^1 = 0$ and $x_3^1 \in H_3$. Choose γ, α_k and λ_k.

For $k = 1, 2, 3, \ldots$, compute

1: $x_1^{k+1} = \arg\min_{x_1}\{f_1(x_1) - \langle w^k, L_1x_1\rangle\}$,

2: $x_2^{k+1} = \arg\min_{x_2}\{f_2(x_2) - \langle w^k, L_2x_2\rangle + \frac{\gamma}{2}\|L_1x_1^{k+1} + L_2x_2 + L_3x_3^k - b\|^2\}$,

3: $x_3^{k+1} = \arg\min_{x_3}\{f_3(x_3) - \langle w^k + \alpha_{k+1}p^k, L_3x_3\rangle + \frac{\gamma}{2}\|L_3(x_3 - x_3^k) + (1 + \alpha_{k+1})\lambda_k(L_1x_1^{k+1} + L_2x_2^{k+1} + L_3x_3^k - b)\|^2\}$,

4: $w^{k+1} = w^k + \alpha_{k+1}p^k - \gamma(L_3(x_3^{k+1} - x_3^k) + (1 + \alpha_{k+1})\lambda_k(L_1x_1^{k+1} + L_2x_2^{k+1} + L_3x_3^k - b))$,

5: $p^{k+1} = \alpha_{k+1}(p^k - \gamma\lambda_k(L_1x_1^{k+1} + L_2x_2^{k+1} + L_3x_3^k - b))$.

Stop when a given stopping criterion is met.

Ensure: $x_1^{k+1}, x_2^{k+1}, x_3^{k+1}$ and w^{k+1}.

To study the convergence analysis of Algorithm 2, we make the following assumptions:

(A1). Assume that f_1 is μ-strongly convex, for some $\mu > 0$.

(A2). The optimal solution of problem (21.1) is nonempty, and the exists $x' = (x'_1, x'_2, x'_3) \in ri(dom(f_1) \times dom(f_2) \times dom(f_3)) \cap C$, where $C = \{(x_1, x_2, x_3) \in H_1 \times H_2 \times H_3 | L_1x_1 + L_2x_2 + L_3x_3 = b\}$.

(A3). For each $i = 1, 2, 3$, let the bounded linear operator L_i satisfies that $\|L_ix_i\| \geq \theta_i\|x_i\|$, for some $\theta_i > 0$ and $\forall x_i \in H_i$.

Under the assumption (A2), we know that the dual solution of problem (21.1) is nonempty, and the strong duality holds, i.e., $v(P) = v(D)$.

Next, we will prove the convergence theorem of Algorithm 2 under two different conditions. For the convergence proof of Algorithm 2, we roughly divide it into two steps. First, we prove that Algorithm 2 is equivalent to (21.7), and Algorithm 2 is derived from (21.7) through variable substitution. Second, after proving the equivalence of Algorithm 2 and (21.7), we can prove the convergence conclusions of the Algorithm 2 by using Theorem 21.1 and Theorem 21.2, respectively.

Theorem 21.3 *Suppose that the assumptions (A1)–(A3) are valid. Let $\{(x_1^k, x_2^k, x_3^k, w^k)\}$ be the sequence generated by Algorithm 2. Let $\gamma \in (0, 2\beta\bar{\varepsilon})$, where $\bar{\varepsilon} \in (0,1)$ and $\beta = \mu/\|L_1\|^2$. Let $\{\alpha_k\}$ is nondecreasing with $\alpha_1 = 0$ and $0 \leq \alpha_k \leq \alpha < 1$. Let $\lambda > 0, \sigma > 0, \delta > 0$ and $\{\lambda_k\}$ such that*

$$\delta > \frac{\alpha^2(1+\alpha) + \alpha\sigma}{1-\alpha^2} \quad \text{and } 0 < \lambda \leq \lambda_k \leq \frac{\delta - \alpha[\alpha(1+\alpha) + \alpha\delta + \sigma]}{\bar{\alpha}\delta[1 + \alpha(1+\alpha) + \alpha\delta + \sigma]}$$

$\forall k \geq 1$, where $\bar{\alpha} = \frac{1}{2-\bar{\varepsilon}}$. Then there exists a point pair $(x_1^*, x_2^*, x_3^*, w^*)$, which is the saddle point of the Lagrange function (21.9) such that the following hold:

(i) $\{(x_1^{k+1}, x_2^{k+1}, x_3^{k+1})\}_{k\geq 1}$ converges weakly to (x_1^*, x_2^*, x_3^*). In particular, $\{x_1^{k+1}\}_{k\geq 1}$ converges strongly to x_1^*;

(ii) $\{w^{k+1}\}_{k\geq 1}$ converges weakly to w^*;

(iii) $\{L_1x_1^{k+1} + L_2x_2^{k+1} + L_3x_3^k\}_{k\geq 2}$ converges strongly to b;

(iv) Suppose that one of the following conditions holds:

(a) f_1^* is uniformly convex on every nonempty bounded subset of $dom f_1^*$;
(b) f_2^* is uniformly convex on every nonempty bounded subset of $dom f_2^*$;
(c) f_3^* is uniformly convex on every nonempty bounded subset of $dom f_3^*$;

then $\{w^{k+1}\}_{k\geq 1}$ converges strongly to the unique optimal solution of (D);

(v) $\lim_{k\to+\infty}(f_1(x_1^{k+1}) + f_2(x_2^{k+1}) + f_3(x_3^k)) = v(P) = v(D) = \lim_{k\to+\infty}(-f_1^*(L_1^*w^k) - f_2^*(L_2^*u^k) - f_3^*(L_3^*w^k) + \langle w^k, b\rangle)$, where u^k is defined as follows

$$u^k = w^k - \gamma(L_1x_1^{k+1} + L_2x_2^{k+1} + L_3x_3^k - b). \tag{21.12}$$

Proof. Let $A = \partial(f_2^* \circ (L_2^* \cdot))$, $B = \partial(f_3^* \circ (L_3^* \cdot) - \langle b, \cdot\rangle)$, and $C = \nabla(f_1^* \circ (L_1^* \cdot))$. Since f_1 is μ-strongly convex, $\nabla(f_1^* \circ (L_1^* \cdot))$ is $\mu/\|L_1\|^2$-cocoercive, and A, B are maximally monotone. Then, we obtain the following inertial three-operator splitting algorithm from [43] to solve the dual problem (21.10).

$$\begin{aligned} y^k &= z^k + \alpha_k(z^k - z^{k-1}), \\ w^k &= J_{\gamma B} y^k, \\ u^k &= J_{\gamma A}(2w^k - y^k - \gamma C w^k), \\ z^{k+1} &= y^k + \lambda_k(u^k - w^k). \end{aligned} \tag{21.13}$$

Next, we prove that the iterative sequence $\{(x_1^k, x_2^k, x_3^k, w^k)\}$ generated by Algorithm 2 is equivalent to the inertial three-operator splitting algorithm (21.13).

From $w^k = J_{\gamma B} y^k$ and $B = \partial(f_3^* \circ (L_3^* \cdot) - \langle b, \cdot \rangle)$, we have

$$y^k - w^k \in \gamma(L_3 \partial f_3^*(L_3^* w^k) - b). \tag{21.14}$$

Let $x_3^k \in \partial f_3^*(L_3^* w^k)$, we get

$$L_3^* w^k \in \partial f_3(x_3^k), \text{ and } z^k + \alpha_k(z^k - z^{k-1}) - w^k = \gamma L_3 x_3^k - \gamma b. \tag{21.15}$$

It follows from the definition of C that $Cw^k = L_1 \nabla f_1^*(L_1^* w^k)$. Let $x_1^{k+1} = \nabla f_1^*(L_1^* w^k)$, then we obtain

$$L_1^* w^k \in \partial f_1(x_1^{k+1}), \text{ and } Cw^k = L_1 x_1^{k+1}. \tag{21.16}$$

From the relation $L_1^* w^k \in \partial f_1(x_1^{k+1})$ yields,

$$x_1^{k+1} = \arg\min_{x_1}\{f_1(x_1) - \langle w^k, L_1 x_1 \rangle\}, \tag{21.17}$$

which is the first step of Algorithm 2.

Notice that $u^k = J_{\gamma A}(2w^k - y^k - \gamma C w^k)$ and (21.16), we have

$$2w^k - y^k - \gamma L_1 x_1^{k+1} - u^k \in \gamma L_2 \partial f_2^*(L_2^* u^k). \tag{21.18}$$

Let $x_2^{k+1} \in \partial f_2^*(L_2^* u^k)$, we obtain that

$$L_2^* u^k \in \partial f_2(x_2^{k+1}) \text{ and } 2w^k - y^k - \gamma L_1 x_1^{k+1} - u^k = \gamma L_2 x_2^{k+1}. \tag{21.19}$$

By combining (21.15) and (21.19), we get

$$u^k = w^k - \gamma(L_1 x_1^{k+1} + L_2 x_2^{k+1} + L_3 x_3^k - b). \tag{21.20}$$

Consequently, we obtain

$$\begin{aligned} 0 \in & \partial f_2(x_2^{k+1}) - L_2^*(w^k - \gamma(L_1 x_1^{k+1} + L_2 x_2^{k+1} + L_3 x_3^k - b)), \\ \Leftrightarrow x_2^{k+1} &= \arg\min_{x_2}\{f_2(x_2) - \langle w^k, L_2 x_2 \rangle + \frac{\gamma}{2}\|L_1 x_1^{k+1} + L_2 x_2 \\ &\quad + L_3 x_3^k - b\|^2\}, \end{aligned} \tag{21.21}$$

which is the second step of Algorithm 2.

Set $p^k = \alpha_k(z^k - z^{k-1})$. Then it follows from (21.15) that

$$p^k = w^k + \gamma L_3 x_3^k - \gamma b - z^k. \tag{21.22}$$

Further, we have

$$\begin{aligned} p^{k+1} &= \alpha_{k+1}(z^{k+1} - z^k) \\ &= \alpha_{k+1}(w^{k+1} + \gamma L_3 x_3^{k+1} - \gamma b - p^{k+1} - (w^k + \gamma L_3 x_3^k - \gamma b - p^k)), \end{aligned} \tag{21.23}$$

which implies that

$$p^{k+1} = \frac{\alpha_{k+1}}{1+\alpha_{k+1}}(p^k + (w^{k+1} - w^k) + \gamma L_3(x_3^{k+1} - x_3^k)). \tag{21.24}$$

By (21.15) and $z^{k+1} = y^k + \lambda_k(u^k - w^k)$, we have

$$\begin{aligned} z^k + \alpha_k(z^k - z^{k-1}) + \lambda_k(u^k - w^k) &= z^{k+1} \\ &= w^{k+1} + \gamma L_3 x_3^{k+1} - \gamma b \\ &\quad - \alpha_{k+1}(z^{k+1} - z^k). \end{aligned} \tag{21.25}$$

From (21.15), (21.20), (21.24) and (21.25), we get

$$\begin{aligned} w^{k+1} &= z^k + \alpha_k(z^k - z^{k-1}) + \lambda_k(u^k - w^k) + \alpha_{k+1}(z^{k+1} - z^k) \\ &\quad - \gamma L_3 x_3^{k+1} + \gamma b \\ &= w^k + \gamma L_3 x_3^k - \gamma b - \gamma\lambda_k(L_1 x_1^{k+1} + L_2 x_2^{k+1} + L_3 x_3^k - b) \\ &\quad + p^{k+1} - \gamma L_3 x_3^{k+1} + \gamma b. \end{aligned} \tag{21.26}$$

Consequently, we obtain

$$\begin{aligned} w^{k+1} &= w^k + \alpha_{k+1}p^k - \gamma(L_3(x_3^{k+1} - x_3^k) + \lambda_k(1+\alpha_{k+1})(L_1 x_1^{k+1} \\ &\quad + L_2 x_2^{k+1} + L_3 x_3^k - b)), \end{aligned} \tag{21.27}$$

which is the fourth step of Algorithm 2.

Combining (21.24) with (21.27), we get

$$p^{k+1} = \alpha_{k+1}(p^k - \gamma\lambda_k(L_1 x_1^{k+1} + L_2 x_2^{k+1} + L_3 x_3^k - b)), \tag{21.28}$$

which is the fifth step of Algorithm 2.

According to $L_3^* w^{k+1} \in \partial f_3(x_3^{k+1})$, we have

$$\begin{aligned} 0 \in \partial f_3(x_3^{k+1}) - L_3^*\{&w^k + \alpha_{k+1}p^k \\ &- \gamma[L_3(x_3^{k+1} - x_3^k) + \lambda_k(1+\alpha_{k+1})(L_1 x_1^{k+1} + L_2 x_2^{k+1} \\ &+ L_3 x_3^k - b)]\}, \end{aligned} \tag{21.29}$$

which is equivalent to

$$\begin{aligned} x_3^{k+1} = \arg\min_{x_3}\{ & f_3(x_3) - \langle w^k + \alpha_{k+1}p^k, L_3x_3\rangle \\ & + \frac{\gamma}{2}\|L_3(x_3 - x_3^k) + \lambda_k(1+\alpha_{k+1})(L_1x_1^{k+1} + L_2x_2^{k+1} \\ & + L_3x_3^k - b)\|^2\}, \end{aligned} \tag{21.30}$$

and is the third step of Algorithm 2. Therefore, we can conclude from the above that Algorithm 2 is equivalent to (21.13).

It follows from Theorem 21.1 that there exists $z^* \in H$ such that

$$z^k \rightharpoonup z^* \quad as \quad k \to +\infty, \tag{21.31}$$

$$w^k \rightharpoonup J_{\gamma B}z^* \quad as \quad k \to +\infty, \tag{21.32}$$

$$u^k \rightharpoonup J_{\gamma B}z^* = J_{\gamma A}(2J_{\gamma B}z^* - z^* - \gamma CJ_{\gamma B}z^*) \quad as \quad k \to +\infty, \tag{21.33}$$

$$z^k - z^{k-1} \to 0 \quad as \quad k \to +\infty, \tag{21.34}$$

$$w^k - u^k \to 0 \quad as \quad k \to +\infty, \tag{21.35}$$

$$Cw^k \to CJ_{\gamma B}z^* \quad as \quad k \to +\infty. \tag{21.36}$$

(i) From (21.15), we have $\gamma L_3x_3^k = z^k + \alpha_k(z^k - z^{k-1}) - w^k + \gamma b$. By using (21.31), (21.32) and (21.34), we get

$$L_3x_3^k \rightharpoonup b + \frac{1}{\gamma}(z^* - J_{\gamma B}z^*). \tag{21.37}$$

According to (21.16), we have $L_1x_1^{k+1} = Cw^k$. Together with (21.36), we obtain

$$L_1x_1^{k+1} \to CJ_{\gamma B}z^*. \tag{21.38}$$

From (21.19), we know that $\gamma L_2x_2^{k+1} = 2w^k - z^k - \alpha_k(z^k - z^{k-1}) - \gamma L_1x_1^{k+1} - u^k$. By using (21.31), (21.32), (21.34), (21.35) and (21.38), we get

$$L_2x_2^{k+1} \rightharpoonup \frac{1}{\gamma}(J_{\gamma B}z^* - z^*) - CJ_{\gamma B}z^*. \tag{21.39}$$

Since for any $i \in \{1,2,3\}$, $\|L_ix_i\| \geq \theta_i\|x_i\|$, $\forall x_i \in H_i$. Then there exist x_1^*, x_2^*, and x_3^* such that

$$L_1x_1^* = CJ_{\gamma B}z^*, \; L_2x_2^* = \frac{1}{\gamma}(J_{\gamma B}z^* - z^*) - CJ_{\gamma B}z^*, \quad \text{and } L_3x_3^* = b + \frac{1}{\gamma}(z^* - J_{\gamma B}z^*). \tag{21.40}$$

According to (21.37), (21.38) and (21.39), we get

$$x_1^{k+1} \to x_1^*, \; x_2^{k+1} \rightharpoonup x_2^*, \text{ and } x_3^k \rightharpoonup x_3^*, \tag{21.41}$$

and

$$L_1x_1^* + L_2x_2^* + L_3x_3^* = b. \tag{21.42}$$

Let $w^* = J_{\gamma B}z^*$, then

$$z^* - w^* \in \gamma Bw^*. \tag{21.43}$$

Thus $L_3x_3^* = b + \frac{1}{\gamma}(z^* - J_{\gamma B}z^*) \in b + Bw^*$, which is equivalent to

$$0 \in \partial f_3(x_3^*) - L_3^*w^*. \tag{21.44}$$

According to $L_1x_1^* = CJ_{\gamma B}z^* = Cw^* = L_1\nabla f_1^*(L_1^*w^*)$ which is

$$0 \in \partial f_1(x_1^*) - L_1^*w^*. \tag{21.45}$$

Again from (21.33), $w^* = J_{\gamma A}(2w^* - z^* - \gamma Cw^*)$, we have $\frac{1}{\gamma}(w^* - z^*) - Cw^* \in Aw^*$. Since $L_2x_2^* = \frac{1}{\gamma}(J_{\gamma B}z^* - z^*) - CJ_{\gamma B}z^* = \frac{1}{\gamma}(w^* - z^*) - Cw^*$, we get

$$0 \in \partial f_2(x_2^*) - L_2^*w^*. \tag{21.46}$$

According to (21.42), (21.44), (21.45) and (21.46), we prove that point pair $(x_1^*, x_2^*, x_3^*, w^*)$ satisfies optimality condition (21.11), that is, point pair $(x_1^*, x_2^*, x_3^*, w^*)$ is saddle point of Lagrangian function (21.9).

(ii) We can get it directly from $w^k \rightharpoonup J_{\gamma B}z^*$ and $w^* = J_{\gamma B}z^*$.

(iii) From (21.20) we have $L_1x_1^{k+1} + L_2x_2^{k+1} + L_3x_3^k = \frac{1}{\gamma}(w^k - u^k) + b$, and then use (21.35).

(iv) Suppose that one of the following conditions holds:

(a) f_1^* is uniformly convex on every nonempty bounded subset of $dom f_1^*$;

(b) f_2^* is uniformly convex on every nonempty bounded subset of $dom f_2^*$;

(c) f_3^* is uniformly convex on every nonempty bounded subset of $dom f_3^*$.

Assume that for any $i \in \{1, 2, 3\}$, $\|L_i^*x\| \geq \beta_i\|x\|, \forall x \in H$, for some $\beta_i > 0$.

(a) Suppose that f_1^* is uniformly convex. Let $x_1 \in H, x_2 \in H$, then there exists an nondecreasing function $\phi_{f_1^*} : [0, +\infty) \to [0, +\infty)$ that vanishes only at 0 such that

$$\begin{aligned} \langle x_1 - x_2, L_1\nabla f_1^*(L_1^*x_1) - L_1\nabla f_1^*(L_1^*x_2)\rangle &= \langle L_1^*x_1 - L_1^*x_2, \nabla f_1^*(L_1^*x_1) \\ &\quad - \nabla f_1^*(L_1^*x_2)\rangle \\ &\geq \phi_{f_1^*}(\|L_1^*x_1 - L_1^*x_2\|) \\ &\geq \phi_{f_1^*}(\beta_1\|x_1 - x_2\|), \end{aligned} \tag{21.47}$$

which implies that $C = L_1 \circ \nabla f_1^* \circ L_1^*$ is uniformly monotone.

Similarly, we can prove that $A = L_2 \circ \partial f_2^* \circ L_2^*$ and $B = L_3 \circ \partial f_3^* \circ L_3^*$ are also uniformly monotone under the conditions of (b) or (c).

(v) We know that $f_i (i = 1, 2, 3)$ are lower semi-continuous, therefore we have

$$\begin{aligned} \liminf_{k\to+\infty}(f_1(x_1^{k+1}) + f_2(x_2^{k+1}) + f_3(x_3^k)) &\geq \liminf_{k\to+\infty} f_1(x_1^{k+1}) \\ &\quad + \liminf_{k\to+\infty} f_2(x_2^{k+1}) + \liminf_{k\to+\infty} f_3(x_3^k), \end{aligned}$$

$$\geq f_1(x_1^*) + f_2(x_2^*) + f_3(x_3^*)$$
$$= v(P). \tag{21.48}$$

On the other hand, since $L_1^* w^k \in \partial f_1(x_1^{k+1})$, $L_2^* u^k \in \partial f_2(x_2^{k+1})$ and $L_3^* w^k \in \partial f_3(x_3^k)$, we get

$$f_1(x_1^*) \geq f_1(x_1^{k+1}) + \langle x_1^* - x_1^{k+1}, L_1^* w^k \rangle, \tag{21.49}$$
$$f_2(x_2^*) \geq f_2(x_2^{k+1}) + \langle x_2^* - x_2^{k+1}, L_2^* u^k \rangle, \tag{21.50}$$
$$f_3(x_3^*) \geq f_3(x_3^k) + \langle x_3^* - x_3^k, L_3^* w^k \rangle. \tag{21.51}$$

Adding (21.49)–(21.51), we obtain

$$\begin{aligned} v(P) \geq\ & f_1(x_1^{k+1}) + f_2(x_2^{k+1}) + f_3(x_3^k) \\ & + \langle b - L_1 x_1^{k+1} - L_2 x_2^{k+1} - L_3 x_3^k, w^k \rangle \\ & + \langle L_2 x_2^* - L_2 x_2^{k+1}, u^k - w^k \rangle. \end{aligned} \tag{21.52}$$

Again from (i), (ii), (iii) and (21.35), we have

$$\limsup_{k \to +\infty} (f_1(x_1^{k+1}) + f_2(x_2^{k+1}) + f_3(x_3^k)) \leq v(P). \tag{21.53}$$

Combined with (21.48) and (21.53), we complete the first part of the Theorem 21.3 (v).

From $L_1^* w^k \in \partial f_1(x_1^{k+1})$, $L_2^* u^k \in \partial f_2(x_2^{k+1})$ and $L_3^* w^k \in \partial f_3(x_3^k)$, then

$$f_1(x_1^{k+1}) + f_1^*(L_1^* w^k) = \langle x_1^{k+1}, L_1^* w^k \rangle, \tag{21.54}$$
$$f_2(x_2^{k+1}) + f_2^*(L_2^* u^k) = \langle x_2^{k+1}, L_2^* u^k \rangle, \tag{21.55}$$
$$f_3(x_3^k) + f_3^*(L_3^* w^k) = \langle x_3^k, L_3^* w^k \rangle. \tag{21.56}$$

Adding (21.54)–(21.56), we obtain

$$\begin{aligned} f_1(x_1^{k+1}) + f_2(x_2^{k+1}) + f_3(x_3^k) =& - f_1^*(L_1^* w^k) - f_2^*(L_2^* u^k) - f_3^*(L_3^* w^k) \\ & + \langle L_1 x_1^{k+1} + L_2 x_2^{k+1} + L_3 x_3^k, w^k \rangle \\ & + \langle L_2 x_2^{k+1}, u^k - w^k \rangle. \end{aligned} \tag{21.57}$$

Finally, taking into account (i), (iii), (21.35) and the first part of Theorem 21.3 (v), we get

$$\lim_{k \to +\infty} (-f_1^*(L_1^* w^k) - f_2^*(L_2^* u^k) - f_3^*(L_3^* w^k) + \langle w^k, b \rangle) = v(D) = v(P). \tag{21.58}$$

This completes the proof. □

Theorem 21.4 *Suppose that the assumptions (A1)–(A3) are valid. Let $\{(x_1^k, x_2^k, x_3^k, w^k)\}$ be the sequence generated by Algorithm 2. Let $\gamma \in (0, 2\beta\bar{\varepsilon})$, where $\bar{\varepsilon} \in (0,1)$ and $\beta = \mu/\|L_1\|^2$. Let $0 \leq \alpha_k \leq \alpha < 1$ and $0 < \underline{\lambda} \leq \lambda_k \bar{\alpha} \leq \overline{\lambda} < 1$, where $\bar{\alpha} = \frac{1}{2-\bar{\varepsilon}}$. Let $\sum_{k=1}^{+\infty} \alpha_{k+1} \|p^k - \gamma\lambda_k(L_1 x_1^{k+1} + L_2 x_2^{k+1} + L_3 x_3^k - b)\|^2 < +\infty$. Then there exists a point pair $(x_1^*, x_2^*, x_3^*, w^*)$, which is the saddle point of the Lagrange function (21.9) such that the following hold:*

(i) $\{(x_1^{k+1}, x_2^{k+1}, x_3^{k+1})\}_{k\geq 1}$ *converges weakly to* (x_1^*, x_2^*, x_3^*). *In particular,* $\{x_1^{k+1}\}_{k\geq 1}$ *converges strongly to* x_1^**;*

(ii) $\{w^{k+1}\}_{k\geq 1}$ *converges weakly to* w^**;*

(iii) $\{L_1 x_1^{k+1} + L_2 x_2^{k+1} + L_3 x_3^k\}_{k\geq 2}$ *converges strongly to* b*;*

(iv) Suppose that one of the following conditions holds:

(a) f_1^* *is uniformly convex on every nonempty bounded subset of* $\operatorname{dom} f_1^*$*;*

(b) f_2^* *is uniformly convex on every nonempty bounded subset of* $\operatorname{dom} f_2^*$*;*

(c) f_3^* *is uniformly convex on every nonempty bounded subset of* $\operatorname{dom} f_3^*$*;*

then $\{w^{k+1}\}_{k\geq 1}$ *converges strongly to the unique optimal solution of* (D)*;*

(v) $\lim_{k\to+\infty}(f_1(x_1^{k+1}) + f_2(x_2^{k+1}) + f_3(x_3^k)) = v(P) = v(D) = \lim_{k\to+\infty}(-f_1^*(L_1^* w^k) - f_2^*(L_2^* u^k) - f_3^*(L_3^* w^k) + \langle w^k, b\rangle)$, *where* u^k *is defined as (21.12).*

Proof. The proof of Theorem 21.4 is similar to Theorem 21.3, so we omit it here. □

Remark 21.2 *Notice that, in finite-dimensional case, the assumption on* L_i $(i = 1, 2, 3)$ *in Theorem 21.3 and Theorem 21.4 means that* L_i *are matrices with full column rank and all weak convergences in Theorem 21.3 and Theorem 21.4 are strong convergences.*

Remark 21.3 *In comparison with the other three-block ADMM, such as (21.3) and (21.5). We prove the weak and strong convergence of the iteration sequences generated by Algorithm 2. However, the strong convergence of three-block ADMM (21.3) and (21.5) are only proved in finite-dimensional Hilbert spaces. It's not clear whether they still have strong convergence in infinite-dimensional Hilbert spaces. It is well known that weak and strong convergence is not equivalent to each other in infinite-dimensional Hilbert spaces.*

In the following, we present several particular cases of the proposed Algorithm 2.

Let $\alpha_k = 0$ in Algorithm 2, then we get the relaxed three-block alternating minimization algorithm (R-AMA)

$$\begin{cases} x_1^{k+1} = \arg\min\limits_{x_1}\{f_1(x_1) - \langle w^k, L_1 x_1\rangle\}, \\ x_2^{k+1} = \arg\min\limits_{x_2}\{f_2(x_2) - \langle w^k, L_2 x_2\rangle + \frac{\gamma}{2}\|L_1 x_1^{k+1} + L_2 x_2 \\ \qquad + L_3 x_3^k - b\|^2\}, \\ x_3^{k+1} = \arg\min\limits_{x_3}\{f_3(x_3) - \langle w^k, L_3 x_3\rangle + \frac{\gamma}{2}\|L_3(x_3 - x_3^k) \\ \qquad + \lambda_k(L_1 x_1^{k+1} + L_2 x_2^{k+1} + L_3 x_3^k - b)\|^2\}, \\ w^{k+1} = w^k - \gamma(L_3(x_3^{k+1} - x_3^k) + \lambda_k(L_1 x_1^{k+1} + L_2 x_2^{k+1} + L_3 x_3^k - b)). \end{cases}$$

Further, let $\lambda_k = 1$ in (21.59), we recover the three-block AMA proposed by Davis and Yin [32].

In Algorithm 2, when f_1 and x_1 vanish, the iterative sequences of Algorithm 2 becomes for every $k \geq 1$,

$$\begin{cases} x_2^{k+1} = \arg\min_{x_2}\{f_2(x_2) - \langle w^k, L_2x_2\rangle + \frac{\gamma}{2}\|L_2x_2 + L_3x_3^k - b\|^2\}, \\ x_3^{k+1} = \arg\min_{x_3}\{f_3(x_3) - \langle w^k + \alpha_{k+1}p^k, L_3x_3\rangle + \frac{\gamma}{2}\|L_3(x_3 - x_3^k) \\ \quad + (1+\alpha_{k+1})\lambda_k(L_2x_2^{k+1} + L_3x_3^k - b)\|^2\}, \\ w^{k+1} = w^k + \alpha_{k+1}p^k - \gamma(L_3(x_3^{k+1} - x_3^k) + (1+\alpha_{k+1})\lambda_k(L_2x_2^{k+1} \\ \quad + L_3x_3^k - b)), \\ p^{k+1} = \alpha_{k+1}(p^k - \gamma\lambda_k(L_2x_2^{k+1} + L_3x_3^k - b)), \end{cases}$$

which is the two-block inertial ADMM proposed in [45]. Moreover, when f_2, x_2 and f_3, x_3 vanish respectively, and $\alpha_k = 0$, then the Algorithm 2 reduces to the following two different relaxed alternating minimization algorithms

$$\begin{cases} x_1^{k+1} = \arg\min_{x_1}\{f_1(x_1) - \langle w^k, L_1x_1\rangle\}, \\ x_3^{k+1} = \arg\min_{x_3}\{f_3(x_3) - \langle w^k, L_3x_3\rangle + \frac{\gamma}{2}\|L_3(x_3 - x_3^k) \\ + \lambda_k(L_1x_1^{k+1} + L_3x_3^k - b)\|^2\}, \\ w^{k+1} = w^k - \gamma(L_3(x_3^{k+1} - x_3^k) + \lambda_k(L_1x_1^{k+1} + L_3x_3^k - b)), \end{cases}$$

and

$$\begin{cases} x_1^{k+1} = \arg\min_{x_1}\{f_1(x_1) - \langle w^k, L_1x_1\rangle\}, \\ x_2^{k+1} = \arg\min_{x_2}\{f_2(x_2) - \langle w^k, L_2x_2\rangle + \frac{\gamma}{2}\|L_1x_1^{k+1} + L_2x_2 - b\|^2\}, \\ w^{k+1} = w^k - \gamma\lambda_k(L_1x_1^{k+1} + L_2x_2^{k+1} - b). \end{cases} \tag{21.59}$$

Let $\lambda_k = 1$, then (21.59) and (21.59) are reduced to the alternating minimization algorithm (AMA) proposed by Tseng [31].

21.4 NUMERICAL EXPERIMENTS

In this section, we carry out simulation experiments and compare the proposed algorithm (Algorithm 2) and its by-product relaxed alternative minimization algorithm (R-AMA (21.59)) with other state-of-the-art algorithms include the three-block ADMM (21.2) [28], the ADM-G (21.3) [14], the sPADMM (21.5) [29] and three-block alternative minimization algorithm (AMA (21.6)) proposed by Davis and Yin [32] on the SPCP. All the experiments are conducted on a 64-bit Windows 10 operating system with an Intel(R) Core(TM) i5-7200U CPU and 8GB memory. All the codes are tested in MATLAB R2016a.

21.4.1 Stable Principal Component Pursuit

The purpose of the SPCP [10] is to recover the low-rank matrix from the high dimensional data matrix with sparse error and small noise. This problem is a special case of (21.1), which can be formulated as:

$$\begin{aligned} \min_{L,S,Z} & \ \beta_1\|L\|_* + \beta_2\|S\|_1 + \frac{1}{2}\|Z\|_F^2 \\ \text{s.t.} & \ L + S + Z = b, \end{aligned} \tag{21.60}$$

where $\|L\|_*$ is defined as the sum of all singular values of the matrix L, $\|S\|_1$ is the l_1 norm of the matrix S and $\|Z\|_F$ is the Frobenius norm of the matrix Z; b is a given damaged data matrix, L, S and Z are a low rank, sparse and noise components of b, respectively. We conduct numerical experiments with the generated simulation data to show the effectiveness of the proposed algorithm. The generation of simulation data is similar to [46]. The observed damaged data matrix b is generated as follows. The low rank matrix L^* is generated by $L^* = L_1 L_2^T$, where $L_1 = randn(m, r)$ and $L_2 = randn(m, r)$ are two independently generated random matrices of $m \times r$ ($r < m$ is rank of matrix L^*) scale. The S^* is a sparse matrix with nonzero elements uniformly distributed and values uniformly distributed between $[-500{,}500]$. The Z^* is a matrix with Gaussian noise whose mean value is 0 and standard deviation is 10^{-5}. Finally, we set $b = L^* + S^* + Z^*$.

We put the actual problem (21.60) into Algorithm 2. Let $x_1 := Z, x_2 := L$ and $x_3 := S$, it is obvious that problem (21.60) is a special case of model (21.1). Accordingly, $f_1(x_1) := \frac{1}{2}\|Z\|_F^2$, $f_2(x_2) := \beta_1\|L\|_*$ and $f_3(x_3) := \beta_2\|S\|_1$, coefficient matrixes $L_1 = L_2 = L_3 := I$, where I is the identity operator.

The following is the detailed calculation process of the problem (21.60) executing Algorithm 2.

1. Z-subproblem in Algorithm 2:

$$\begin{aligned} Z^{k+1} &= \arg\min_Z\{\frac{1}{2}\|Z\|^2 - \langle w^k, Z\rangle\} \\ &= w^k. \end{aligned}$$

2. L-subproblem in Algorithm 2:

$$\begin{aligned} L^{k+1} &= \arg\min_L\{\beta_1\|L\|_* - \langle w^k, L\rangle + \frac{\gamma}{2}\|Z^{k+1} + L + S^k - b\|^2\} \\ &= \arg\min_L\{\beta_1\|L\|_* + \frac{\gamma}{2}\|Z^{k+1} + L + S^k - b - \frac{1}{\gamma}w^k\|^2\} \\ &= prox_{\frac{\beta_1}{\gamma}\|\cdot\|_*}(b + \frac{1}{\gamma}w^k - Z^{k+1} - S^k), \end{aligned}$$

where $prox_{c\|\cdot\|_*}(\cdot)$ is the proximal function [47] of the function $c\|\cdot\|_*$ with a constant $c > 0$. For any matrix $L \in R^{m\times n}$ with $rank(L) = r$, let its singular value decomposition be $L = Udiag(\{\sigma_i\}_{1\le i\le r})V^T$, where $U \in R^{m\times r}$, $V \in R^{n\times r}$, then $prox_{c\|\cdot\|_*}(L) = Udiag(max\{\{\sigma_i\}_{1\le i\le r} - c, 0\})V^T$.

3. S-subproblem in Algorithm 2:

$$\begin{aligned} S^{k+1} &= \arg\min_S\{\beta_2\|S\|_1 - \langle w^k + \alpha_{k+1}p^k, S\rangle + \frac{\gamma}{2}\|(S - S^k) \\ &+ (1+\alpha_{k+1})\lambda_k(Z^{k+1} + L^{k+1} + S^k - b)\|^2\} \\ &= \arg\min_S\{\beta_2\|S\|_1 + \frac{\gamma}{2}\|(S - S^k) + (1+\alpha_{k+1})\lambda_k(Z^{k+1} + L^{k+1} \\ &+ S^k - b) - \frac{1}{\gamma}(w^k + \alpha_{k+1}p^k)\|^2\} \\ &= prox_{\frac{\beta_2}{\gamma}\|\cdot\|_1}(S^k + \frac{1}{\gamma}(w^k + \alpha_{k+1}p^k) - (1+\alpha_{k+1})\lambda_k(Z^{k+1} + L^{k+1} \\ &+ S^k - b)). \end{aligned}$$

where $prox_{c\|\cdot\|_1}(S) = sign(S). * max(abs(S) - c, 0)$.

4. Update of Lagrange multiplier w in Algorithm 2:

$$\begin{aligned} w^{k+1} &= w^k + \alpha_{k+1}p^k - \gamma(S^{k+1} - S^k + (1+\alpha_{k+1})\lambda_k(Z^{k+1} + L^{k+1} \\ &+ S^k - b)). \end{aligned}$$

5. Update of variable p in Algorithm 2:

$$p^{k+1} = \alpha_{k+1}(p^k - \gamma\lambda_k(Z^{k+1} + L^{k+1} + S^k - b)).$$

21.4.2 Parameters Setting

The specific setting of each parameter in the algorithm is given in this subsection. Let $\beta_1 = 0.05$ and $\beta_2 = \beta_1/\sqrt{m}$. Let the relative error of L and S be the stopping criterion, i.e.,

$$rel\,L := \frac{\|L^{k+1} - L^k\|_F}{\|L^k\|_F}, \quad rel\,S := \frac{\|S^{k+1} - S^k\|_F}{\|S^k\|_F},$$
$$\max(rel\,L, rel\,S) \le \varepsilon,$$

where ε is a small constant. We first conduct a numerical experiment to illustrate the relationship between the value of the penalty parameter γ and the experimental results such as the number of iteration steps in the three-block AMA (21.6) algorithm. In this experiment, we set $m = 200$, $rank(L^*) = 0.05m$, $\|S^*\|_0 = 0.05m^2$, $\varepsilon = 10^{-5}$, and the initial variables $(S^1, w^1) = (0, 0)$.

From Table 21.1, we can see that when the value of the penalty parameter γ is large, the iteration step k of the three-block AMA (21.6) algorithm is large. When $\gamma = 0.0005$, the three-block AMA (21.6) algorithm has the fastest convergence speed. In the following experiments, we fix the $\gamma = 0.0005$ and compare the effects of different relaxation parameters λ_k on the numerical experimental results of the three-block R-AMA (21.59) algorithm. The above experimental data is still used, and the λ_k takes ten different values as 0.5, 0.8, 1, 1.1, 1.2, 1.3, 1.5, 1.6, 1.7 and 1.8, respectively.

TABLE 21.1 Numerical Experimental Results of Three-Block AMA (21.6) Algorithm under Different Penalty Parameters γ($rel\ L^*$ and $rel\ S^*$ Are defined as $\frac{\|L^k-L^*\|_F}{\|L^*\|_F}$ and $\frac{\|S^k-S^*\|_F}{\|S^*\|_F}$, respectively)

Methods	$m=200$ $rank(L^*)=0.05m$ $\|S^*\|_0=0.05m^2$ $\varepsilon=10^{-5}$				
	γ	k	$rank(L^k)$	$rel\ L^*$	$rel\ S^*$
AMA (21.6)	0.0005	37	10	3.2242e-4	2.3868e-5
	0.005	224	10	2.8690e-4	1.4136e-5
	0.05	2160	10	2.8738e-4	1.4145e-5
	0.1	4311	10	2.8814e-4	1.4161e-5
	0.5	21534	10	2.8719e-4	1.4140e-5
	1	43066	10	2.8759e-4	1.4150e-5
	1.2	51679	10	2.8765e-4	1.4151e-5
	1.5	64598	10	2.8778e-4	1.4153e-5
	1.8	77516	10	2.8853e-4	1.4168e-5

From Table 21.2, we can see that the relaxation parameter λ_k can effectively improve the convergence speed of the AMA algorithm. When $\lambda_k \in [1.1, 1.7]$, the relaxation parameter λ_k can accelerate the three-block AMA (21.6) algorithm, and the optimal acceleration effect is $\lambda_k = 1.5$. In the following experiments, we fix the relaxation parameter $\lambda_k = 1.5$ of the relaxed three-block AMA (21.59) algorithm. Subsequently, we compare the three-block ADMM (21.2), ADM-G (21.3), sPADMM (21.5), AMA (21.6), R-AMA (21.59) and Algorithm 2 with different conditions. When $\gamma = 0.0005$, it satisfies the three-block ADMM (21.2), ADM-G (21.3) and sPADMM (21.5) restrictions on penalty parameters. Make the parameter $\theta = 0.99999$ in ADM-G (21.3) and $\tau = 1.2$ in sPADMM (21.5). We know that $\mu = 1$, $L_1 = I$, that is, $\beta = \mu/\|L_1\|^2 = 1$. And $\gamma \in (0, 2\beta\bar{\varepsilon})$, so we make $\bar{\varepsilon} = 0.00026$ and $\bar{\alpha} = \frac{1}{2-\bar{\varepsilon}} \approx 0.5001$. We define their parameters in Table 21.3.

TABLE 21.2 Numerical Experimental Results of Three-Block R-AMA (21.59) Algorithm under different Relaxation Parameters λ_k($rel\ L^*$ and $rel\ S^*$ Are defined as $\frac{\|L^k-L^*\|_F}{\|L^*\|_F}$ and $\frac{\|S^k-S^*\|_F}{\|S^*\|_F}$, respectively)

Methods	$m=200$ $rank(L^*)=0.05m$ $\|S^*\|_0=0.05m^2$ $\varepsilon=10^{-5}$					
	γ	λ_k	k	$rank(L^k)$	$rel\ L^*$	$rel\ S^*$
R-AMA (21.59)	0.0005	0.5	69	10	3.2301e-4	2.3880e-5
		0.8	46	10	3.2245e-4	2.3867e-5
		1	37	10	3.2242e-4	2.3868e-5
		1.1	34	10	3.2209e-4	2.3866e-5
		1.2	31	10	3.2202e-4	2.3865e-5
		1.3	29	10	3.2210e-4	2.3866e-5
		1.5	27	10	3.2210e-4	2.3865e-5
		1.6	33	10	2.8743e-4	1.4149e-5
		1.7	36	10	2.8743e-4	1.4157e-5
		1.8	43	10	2.8753e-4	1.4203e-5

TABLE 21.3 Parameters Selection of the Compared Iterative Algorithms

Methods	γ	λ_k	**Inertial parameter** α_k
Three-block ADMM (21.2)		None	None
ADM-G (21.3)		None	None
sPADMM (21.5)		None	None
AMA (21.6)	0.0005	1	None
R-AMA (21.59)		1.5	None
Algorithm 2-1		1.25	0.15
Algorithm 2-2		1.5	$\min\{\frac{1}{k^2\|p^k-\gamma\lambda_k(L_1x_1^{k+1}+L_2x_2^{k+1}+L_3x_3^k-b)\|^2}, 0.005\}$

21.4.3 Results and Discussion

To make the experimental results more convincing, we conduct a number of numerical experiments. Let the order m of the matrix be 200, 400 and 500, respectively. The rank of low rank matrix L^* and the sparsity of sparse matrix S^* are also divided into two combinations: $rank(L^*) = 0.05m$ and $\|S^*\|_0 = 0.05m^2$, $rank(L^*) = 0.1m$ and $\|S^*\|_0 = 0.1m^2$.

We test the performance of the studied iterative algorithms including three-block ADMM (21.2), ADM-G (21.3), sPADMM (21.5), three-block AMA (21.6), three-block R-AMA (21.59), Algorithm 2-1 and Algorithm 2-2 with parameters selection in Table 21.3. The results of numerical experiments are reported in Table 21.4. Several indicators are listed here including the number of iteration steps, error accuracy, and running CPU time. From Table 21.4, we can find that both the three-block R-AMA (21.59) algorithm and the two relaxed inertial three-block AMA (Algorithm 2) algorithms with different conditions can accelerate the convergence speed of the three-block AMA (21.6) algorithm, and their accuracy is higher. Table 21.4 also conveys a message: Inertia technology does not seem to be able to effectively accelerate the three-block AMA (21.6) algorithm. The numerical performance of the two relaxed inertial three-block AMA (Algorithm 2) algorithms is almost the same or slightly worse than the three-block R-AMA (21.59) algorithm. However, their performance is not as good as the three-block ADMM (21.2), ADM-G (21.3) and sPADMM (21.5). The iteration speed of the three-block ADMM (21.2) and sPADMM (21.5) is almost the same, and they are faster than ADM-G (21.3), which further proves that the direct promotion of the three-block ADMM (21.2) numerical experiment is better than other variants of ADMM.

21.5 CONCLUSIONS

The ADMM and the alternating minimization algorithm (AMA) are two common splitting methods for solving separable convex programming with linear equality constraints. Recently, Davis and Yin [32] have generalized the AMA to the case of three-block AMA (21.6). In this chapter, we proposed a relaxed inertial three-block AMA (Algorithm 2), which is derived from the inertial three-operator splitting

TABLE 21.4 Comparison of Numerical Experimental Results of Three-Block ADMM, ADM-G, sPADMM, AMA, R-AMA and Algorithm 2 (*rel* L^* and *rel* S^* are defined as $\frac{\|L^k-L^*\|_F}{\|L^*\|_F}$ and $\frac{\|S^k-S^*\|_F}{\|S^*\|_F}$, respectively)

	m	Methods	k	$rank(L^k)$	$rel\ L^*$	$rel\ S^*$	CPU
$rank(L^*) = 0.05m$ $\|S^*\|_0 = 0.05m^2$ $\varepsilon = 10^{-5}$	200	Three-block ADMM (21.2)	20	10	3.2221e-4	2.3867e-5	0.1946
		ADM-G (21.3)	21	10	3.2216e-4	2.3858e-5	0.1934
		sPADMM (21.5)	17	10	3.2219e-4	2.3868e-5	0.1740
		AMA (21.6)	37	10	3.2242e-4	2.3868e-5	0.3450
		R-AMA (21.59)	27	10	3.2210e-4	2.3865e-5	0.2379
		Algorithm 2-1	28	10	3.2216e-4	2.3865e-5	0.2709
		Algorithm 2-2	27	10	3.2214e-4	2.3866e-5	0.2507
	400	Three-block ADMM (21.2)	15	20	1.9399e-4	3.0007e-5	0.7093
		ADM-G (21.3)	23	20	1.5296e-4	1.5123e-5	1.1701
		sPADMM (21.5)	17	20	1.6064e-4	1.8351e-5	0.8746
		AMA (21.6)	38	20	1.8482e-4	2.6547e-5	2.3205
		R-AMA (21.59)	33	20	1.6071e-4	1.8348e-5	1.6953
		Algorithm 2-1	34	20	1.6064e-4	1.8348e-5	1.7018
		Algorithm 2-2	33	20	1.6070e-4	1.8349e-5	1.7227
	500	Three-block ADMM (21.2)	17	25	1.3543e-4	2.0612e-5	1.5268
		ADM-G (21.3)	23	25	1.1591e-4	1.0546e-5	2.0847
		sPADMM (21.5)	18	25	1.1576e-4	1.0526e-5	1.5371
		AMA (21.6)	42	25	1.2165e-4	1.3823e-5	3.5990
		R-AMA (21.59)	35	25	1.1694e-4	1.1138e-5	2.9207
		Algorithm 2-1	35	25	1.1697e-4	1.1137e-5	3.0514
		Algorithm 2-2	35	25	1.1695e-4	1.1138e-5	2.9553
$rank(L^*) = 0.1m$ $\|S^*\|_0 = 0.1m^2$ $\varepsilon = 10^{-5}$	200	Three-block ADMM (21.2)	19	20	4.1054e-4	3.0653e-5	0.1733
		ADM-G (21.3)	27	20	4.0819e-4	3.0568e-5	0.2612
		sPADMM (21.5)	18	20	4.0954e-4	3.0639e-5	0.2018
		AMA (21.6)	48	20	3.3991e-4	1.8898e-5	0.4429
		R-AMA (21.59)	36	20	3.3905e-4	1.8876e-5	0.3082
		Algorithm 2-1	37	20	3.3883e-4	1.8872e-5	0.3320
		Algorithm 2-2	36	20	3.3898e-4	1.88774e-5	0.3014
	400	Three-block ADMM (21.2)	28	40	2.1397e-4	2.2150e-5	1.4032
		ADM-G (21.3)	32	40	2.1420e-4	2.1907e-5	1.8402
		sPADMM (21.5)	34	40	1.7984e-4	1.5447e-5	1.6659
		AMA (21.6)	60	40	2.2231e-4	2.3951e-5	3.3423
		R-AMA (21.59)	52	40	1.7343e-4	1.3852e-5	2.6572
		Algorithm 2-1	53	40	1.7347e-4	1.3853e-5	2.8001
		Algorithm 2-2	52	40	1.7336e-4	1.3850e-5	2.8206
	500	Three-block ADMM (21.2)	32	50	1.3999e-4	1.2946e-5	2.6338
		ADM-G (21.3)	41	50	1.3394e-4	1.1220e-5	3.4820
		sPADMM (21.5)	28	50	1.4008e-4	1.2967e-5	2.5707
		AMA (21.6)	74	50	1.4346e-4	1.3376e-5	7.9132
		R-AMA (21.59)	52	50	1.4302e-4	1.0980e-5	5.3290
		Algorithm 2-1	53	50	1.4303e-4	1.0981e-5	4.5382
		Algorithm 2-2	52	50	1.4201e-4	1.1018e-5	5.9416

algorithm [43]. The obtained algorithm generalized and recovered some existing algorithms. In particular, we obtain a relaxed three-block AMA (21.59). We analyze the convergence of the proposed algorithm in infinite-dimensional Hilbert spaces. Compared with other three-block ADMM, our convergence conclusions have not only weak convergence but also strong convergence. To demonstrate the efficiency and effectiveness of the proposed algorithm, we conduct numerical experiments on the SPCP [10].

Numerical results showed that the relaxed three-block AMA (21.59) performs better than the three-block AMA (21.6) when the relaxation parameter belongs to $[1.1, 1.7]$. We also observed that the performance of the relaxed inertial three-block AMA is similar to the relaxed three-block AMA. Our numerical results also confirmed the limitations of the inertial accelerated ADMM pointed out by Poon and Liang [38].

Recently, Bitterlich et al. [48] proposed a proximal AMA, which added proximal terms to the subproblem of the original AMA. Therefore, we would like to present the first open question:

Question 21.1. *Can we study the convergence of the following proximal three-block AMA (21.61)?*

$$\begin{cases} x_1^{k+1} = \arg\min_{x_1}\{f_1(x_1) - \langle w^k, L_1x_1\rangle + \frac{1}{2}\|x_1 - x_1^k\|_{M_1^k}\}, \\ x_2^{k+1} = \arg\min_{x_2}\{f_2(x_2) - \langle w^k, L_2x_2\rangle + \frac{\gamma}{2}\|L_1x_1^{k+1} + L_2x_2 + L_3x_3^k - b\|^2 \\ \quad + \frac{1}{2}\|x_2 - x_2^k\|_{M_2^k}\}, \\ x_3^{k+1} = \arg\min_{x_3}\{f_3(x_3) - \langle w^k, L_3x_3\rangle + \frac{\gamma}{2}\|L_1x_1^{k+1} + L_2x_2^{k+1} \\ \quad + L_3x_3 - b\|^2 + \frac{1}{2}\|x_3 - x_3^k\|_{M_3^k}\}, \\ w^{k+1} = w^k - \gamma(L_1x_1^{k+1} + L_2x_2^{k+1} + L_3x_3^{k+1} - b), \end{cases}$$

where $\{M_1^k\}$, $\{M_2^k\}$ and $\{M_3^k\}$ are self-adjoint positive semi-definite operators.

As we know, the AMA is equivalent to the forward-backward splitting algorithm applied to the corresponding dual problem. In 2013, Raguet et al. [49] proposed a generalized forward-backward splitting algorithm for finding a zero of the sum of a cocoercive operator B (See Definition 21.2) and a finite sum of maximally monotone operators $\{A\}_{i=1}^m$, that is, find $x \in H$, such that $0 \in Bx + \sum_{i=1}^m A_ix$. It is natural to employ the generalized forward-backward splitting algorithm to solve the dual of the following multi-block convex separable optimization problem.

$$\begin{aligned} \min_{x_1,\cdots,x_m} \; & \sum_{i=1}^{m} f_i(x_i) \\ s.t. L_1x_1 + & L_2x_2 + \cdots + L_mx_m = b, \end{aligned} \tag{21.61}$$

where $\{f_i\}_{i=1}^m : H_i \to (-\infty, +\infty]$ are proper, lower semi-continuous convex functions, $\{L_i\}_{i=1}^m : H_i \to H$ are nonzero bounded linear operators, and f_1 is a strongly convex function. Then, we raise the second open question:

Question 21.2. *Can we obtain a primal-dual iteration scheme for solving (21.61) from the generalized forward-backward splitting algorithm?*

BIBLIOGRAPHY

[1] R. Glowinski and A. Marroco. Sur l'approximation, par elements finis d'ordre un, et la resolution, par penalisation-dualite, d'une classe de problemes de dirichlet non lineares. *Revue Francaise d'Automatique, Informatique et Recherche Operationelle*, 9:41–76, 1975.

[2] D. Gabay and B. Mercier. A dual algorithm for the solution of nonlinear variational problems via finite element approximation. *Comput. Math. Appl.*, 2:17–40, 1976.

[3] Y.L. Wang, J.F. Yang, W.T. Yin, and Y. Zhang. A new alternating minimization algorithm for total variation image reconstruction. *SIAM J. Imaging Sci.*, 1(3):248–272, 2008.

[4] J.F. Yang, Y. Zhang, and W.T. YIn. An efficient tvl1 algorithm for deblurring multichannel images corrupted by impulsive noise. *SIAM J. Sci. Comput.*, 31(4):2842–2865, 2009.

[5] C.L. Wu and X.C. Tai. Augmented lagrangian method, dual methods and split bregman iteration for rof, vectorial tv and high order models. *SIAM J. Imaging Sci.*, 3(3):300–339, 2010.

[6] B. He and X.M. Yuan. On the o(1/n) convergence rate of the Douglas-Rachford alternating direction method. *SIAM J. Numer. Anal.*, 50(2):700–709, 2012.

[7] R.D. Monteiro and B.F. Svaiter. Iteration-complexity of block-decomposition algorithms and the alternating direction method of multipliers. *SIAM J. Optim.*, 23(1):475–507, 2013.

[8] E.X. Fang, B.S. He, H. Liu, and X.M. Yuan. Generalized alternating direction method of multipliers: new theoretical insights and applications. *Math. Program. Comput.*, 7(2):149–187, 2015.

[9] B.S. He and X.M. Yuan. On non-ergodic convergence rate of Douglas-Rachford alternating direction method of multipliers. *Numer. Math.*, 130:567–577, 2015.

[10] Z. Zhou, X. Li, J. Wright, E. Candès, and Y. Ma. Stable principal component pursuit. In *2010 IEEE International Symposium on Information Theory*, Austin, TX, USA, pages 1518–1522, 2010.

[11] V. Chandrasekaran, P.A. Parrilo, and A.S. Willsky. Latent variable graphical model selection via convex optimization. *Ann. Stat.*, 40(4):1610–1613, 2012.

[12] M. Tao and X.M. Yuan. Recovering low-rank and sparse components of matrices from incomplete and noisy observations. *SIAM J. Optim.*, 21(1):57–81, 2011.

[13] C.H. Chen, B.S. He, Y.Y. Ye, and X.M. Yuan. The direct extension of ADMM for multi-block convex minimization problems is not necessarily convergent. *Math. Program.*, 155:57–79, 2016.

[14] B.S. He, M. Tao, and X.M. Yuan. Alternating direction method with Gaussian back substitution for separable convex programming. *SIAM J. Optim.*, 22(2):313–340, 2012.

[15] B.S. He, M. Tao, and X.M. Yuan. Convergence rate and iteration complexity on the alternating direction method of multipliers with a substitution procedure for separable convex programming. *Preprint*, 2012.

[16] M.Y. Hong and Z.Q. Luo. On the linear convergence of the alternating direction method of multipliers. *Math. Programming*, 162:165–199, 2017.

[17] W. Deng, M.J. Lai, Z.M. Peng, and W.T. Yin. Parallel multi-block ADMM with o(1/k) convergence. *J. Sci. Comput.*, 71:712–736, 2017.

[18] D.F. Sun, K.-C. Toh, and L.Q. Yang. A convergent 3-block semiproximal alternating direction method of multipliers for conic programming with 4-type constraints. *SIAM J. Optim.*, 25(2):882–915, 2015.

[19] D.R. Han, W.W. Kong, and W.X. Zhang. A partial splitting augmented lagrangian method for low patch-rank image decomposition. *J. Math. Imaging Vis.*, 51(1):145–160, 2015.

[20] K. Wang, J. Desai, and H.J. He. A proximal partially-parallel splitting method for separable convex programs. *Optim. Methods Softw.*, 32(1):39–68, 2017.

[21] X.K. Chang, S.Y. Liu, P.J. Li, and X. Li. Convergent prediction-correction-based ADMM for multi-block separable convex programming. *J. Comput. Appl. Math.*, 335:270–288, 2018.

[22] M. Sun and Y.J. Wang. Modified hybrid decomposition of the augmented lagrangian method with larger step size for three-block separable convex programming. *J. Inequal. Appl.*, 2018:269, 2018.

[23] Y. Shen, X.Y. Zhang, and X.Y. Zhang. A partial ppa block-wise ADMM for multi-block linearly constrained separable convex optimization. *Optimization*, 2020. URL: https://optimization-online.org/2020/03/7707/

[24] D. Han and X.M. Yuan. A note on the alternating direction method of multipliers. *J. Optim. Theory Appl.*, 155:227–238, 2012.

[25] C.H. Chen, S.Yuan, and Y.F. You. On the convergence analysis of the alternating direction method of multipliers with three blocks. *Abstr. Appl. Anal.*, 2013:Article ID 183961, 2013.

[26] T.Y. Lin, S.Q.Ma, and S.Z. Zhang. On the sublinear convergence rate of multi-block ADMM. *J. Oper. Res. Soc. China*, 3:251–274, 2015.

[27] T.Y. Lin, S.Q. Ma, and S.Z. Zhang. On the global linear convergence of the ADMM with multi-block variables. *SIAM J. Optim.*, 25(3), 2014.

[28] X.J. Cai, D. Han, and X.M. Yuan. On the convergence of the direct extension of ADMM for three-block separable convex minimization models with one strongly convex function. *Comput. Optim. Appl.*, 66(1):39–73, 2017.

[29] M. Li, D.F. Sun, and K.C. Toh. A convergent 3-block semi-proximal ADMM for convex minimization problems with one strongly convex block. *Asia-Pacific J. Oper. Res.*, 32(4):1550024, 2015.

[30] T.Y. Lin, S.Q. Ma, and S.Z. Zhang. Global convergence of unmodified 3-block ADMM for a class of convex minimization problems. *J. Sci. Comput.*, 76:69–88, 2017.

[31] P. Tseng. Applications of splitting algorithm to decomposition in convex programming and variational inequalities. *SIAM J. Control Optim.*, 29:119–138, 1991.

[32] D. Davis and W.T. Yin. A three-operator splitting scheme and its optimization applications. *Set-Valued Var. Anal.*, 25(4):829–858, 2017.

[33] T. Goldstein, B. O'Donoghue, S. Setzer, and R. Baraniuk. Fast alternating direction optimization methods. *SIAM J. Imaging Sci.*, 7(3):1588–1623, 2014.

[34] J. Eckstein and D. Bertsekas. On the Douglas-Rachford splitting method and the proximal point algorithm for maximal monotone operators. *Math. Program.*, 55(1):293–318, 1992.

[35] Z. Xu, M.A.T. Figueiredo, X.M. Yuan, C. Studer, and T. Goldstein. Adaptive relaxed ADMM: convergence theory and practical implementation. In *IEEE Conference on Computer Vision and Pattern Recognition (CVPR)*, Honolulu, HI, USA, pages 7234–7243, 2017. doi: 10.1109/CVPR.2017.765.

[36] M. Kadkhodaie, K. Christakopoulou, M. Sanjabi, and A. Banerjee. Accelerated alternating direction method of multipliers. In *Proceedings of the 21th ACM SIGKDD International Conference on Knowledge Discovery and Data Mining*, Sydney, Australia pages 497–506, 2015. doi: 10.1145/2783258.2783400.

[37] I. Pejcic and C. Jones. Accelerated ADMM based on accelerated Douglas-Rachford splitting. *2016 European Control Conference (Ecc)*, Aalborg, Denmark, pages 1952–1957, 2016. doi: 10.1109/ECC.2016.7810577.

[38] C. Poon and J.W. Liang. Trajectory of alternating direction method of multipliers and adaptive acceleration. In H. Wallach, H. Larochelle, A. Beygelzimer, F. dtextquotesingle Alché-Buc, E. Fox, and R. Garnett, editors, *Advances in Neural Information Processing Systems*, pages 7355–7363, USA United States, December 2019. Curran Associates, Inc.

[39] D. Kim. Accelerated proximal point method for maximally monotone operators. *arXiv eprint*, arXiv:1905.05149, 2019.

[40] C.H. Chen, R.H. Chan, S.Q. Ma, and J.F. Yang. Inertial proximal ADMM for linearly constrained separable convex optimization. *SIAM J. Imaging Sci.*, 8(4):2239–2267, 2015.

[41] R.I. Boţ and E.R. Csetnek. An inertial alternating direction method of multipliers. *Minimax Theory Appl.*, 1:29–49, 2016.

[42] R.I. Boţ, E.R. Csetnek, and C. Hendrich. Inertial Douglas-Rachford splitting for monotone inclusion problems. *Appl. Math. Comput.*, 256:472–487, 2015.

[43] F.Y. Cui, Y.C. Tang, and Y. Yang. An inertial three-operator splitting algorithm with applications to image inpainting. *Appl. Set-Valued Anal. Optim.*, 1(2):113–134, 2019.

[44] H.H. Bauschke and P.L. Combettes. *Convex Analysis and Monotone Operator Theory in Hilbert Spaces.* Springer, London, second edition, 2017.

[45] Y. Yang and Y.C. Tang. An inertial alternating direction method of multipliers for solving a two-block separable convex minimization problem. *arXiv eprint*, arXiv:2002.12670, 2020.

[46] E.J. Candès, X. Li, Y. Ma, and J. Wright. Robust principal component analysis? *J. ACM*, 58(1):1–37, 2009.

[47] J.F. Cai, E.J. Candes, and Z. Shen. A singular value thresholding algorithm for matrix completion. *SIAM J. Optim.*, 20:1956–1982, 2010.

[48] S. Bitterlich, R.I. Boţ, E.R. Csetnek, and G. Wanka. The proximal alternating minimization algorithm for two-block separable convex optimization problems with linear constraints. *J. Optim. Theory Appl.*, 182:110–132, 2019.

[49] H. Raguet, J. Fadili, and G. Peyré. A generalized forward-backward splitting. *SIAM J. Imaging Sci.*, 6(3):1199–1226, 2013.

CHAPTER 22

Ball Convergence of Iterative Methods without Derivatives with or without Memory Relying on the Weight Operator Technique

Ioannis K. Argyros
Cameron University

Santhosh George
National Institute of Technology Karnataka

Christopher Argyros
Cameron University

CONTENTS

22.1 INTRODUCTION

Let $F : D \subset X \longrightarrow Y$ be differentiable in the Fréchet sense with D being nonempty, convex, open set and X, Y be Banach spaces.

A plethora of problems are modeled using an equation

$$F(x) = 0. \tag{22.1}$$

DOI: 10.1201/9781003388678-22

Then, to determine a solution x_* we rely mostly on iterative methods. This is the case, since solutions in closed form are found in special cases.

Methods without memory use the current iteration, whereas those with memory rely on the current iteration and the preceding ones. The idea of using the latter methods is to elevate the convergence order without additional operator evaluations. These types of methods are important, since they are derivative-free.

That is these methods are limited to solving equations with operators which are at least differentiable five times. However, these methods can converge (see also the numerical section). These problems appear in other methods where Taylor expansions are used to show convergence [1, 2, 3, 4, 5, 6, 7, 8, 9, 10, 12, 13, 14, 15, 16, 17, 18, 19, 20, 21, 22].

We develop the ball convergence analysis of, one without memory and another method with memory. The methods are defined, respectively as:

$$\begin{aligned} z_n &= x_n + \alpha F(x_n) \\ y_n &= x_n - F^{-1}_{[x_n,z_n]}F(x_n) \\ x_{n+1} &= y_n - A_n F^{-1}_{[y_n,z_n]}F(z_n), \end{aligned} \tag{22.2}$$

where $F_{[x,y]} := [x,y;F]$ $\alpha \in \mathbb{R}$ or $\alpha \in \mathbb{C}$, $A_n : D \times D \longrightarrow L(X,Y)$, and

$$\begin{aligned} z_n &= x_n - F^{-1}_{[x_n,x_{n-1}]}F(x_n) \\ y_n &= x_n - F^{-1}_{[x_n,z_n]}F(x_n) \\ x_{n+1} &= y_n - A_n F^{-1}_{[y_n,z_n]}F(y_n), \end{aligned} \tag{22.3}$$

These methods are extensions of Traub's work on Steffensen like methods [22]. Methods (22.2) and (22.3) are studied in [11], when $X = Y = \mathbb{R}$. They are of order four and $2+\sqrt{6}$, respectively provided that $A(0) = A'(0) = 1$ and $A''(0) < \infty$ [11] in the scalar case. The convergence order was shown based on derivatives up to order five (not on these methods)) limiting their applicability. As a motivational example:

Let $\Omega = [-\frac{1}{2}, \frac{3}{2}]$. Define function f on the interval Ω by

$$f(t) = \begin{cases} t^5 - t^4 + t^3 \log t^2 & if\ t \neq 0 \\ 0 & if\ t = 0. \end{cases}$$

Obviously, in this example, the exact solution is $t_* = 1$. Moreover, we get

$$f'''(t) = 6\log t^2 + 60t^2 - 24t + 22.$$

Hence, the convergence is not assured by the work in [11], since the function $f'''(t)$ is discontinuous on Ω.

Other concerns are: the lack of upper error distances on e_n or on the location of x_*. These concerns constitute our motivation for writing this chapter. In particular, we find computable convergence radius and error estimates relying only on the derivative appearing on these methods and generalized conditions on F'. That is how we extend the utilization in a Banach space. Our results are shown using only conditions on operators appearing in these methods. Notice that local convergence results on iterative methods are significant, since they reveal how difficult it is to pick starting points x_0. Our idea can be used analogously on other methods and for the same reasons because it is so general.

The local analysis is developed in Section 22.2, whereas the examples appear in Section 22.3. Some remarks conclude this chapter.

22.2 BALL CONVERGENCE

The ball convergence is developed first for the method (22.2) depending on real parameters and functions. Let $M = [0, \infty)$ and $a \geq 0$.

Suppose function(s):

(i)

$$\xi_0(t, at) - 1$$

has a minimal zero $R_0 \in M - \{0\}$, for function $\xi_0 : M \times M \longrightarrow M$ nondecreasing and continuous. Let $M_0 = [0, R_0)$.

(ii)

$$\zeta_1(t) - 1$$

has a minimal zero $d_1 \in M_0 - \{0\}$, where $\xi : M_0 \times M_0 \longrightarrow M$, is nondecreasing and continuous and $\zeta_1 : M_0 \longrightarrow M$ is given by

$$\zeta_1(t) = \frac{\xi(t, at)}{1 - \xi_0(t, at)}.$$

(iii)

$$\xi_0(\zeta_1(t)t, 0) - 1, \; \xi_0(\zeta_1(t)t, at) - 1$$

have minimal zeros $R_1, R_2 \in M_0 - \{0\}$, respectively. Let $R = \min\{R_1, R_2\}$ and $M_1 = [0, R)$.

(iv)

$$\zeta_2(t) - 1$$

has a minimal zero $d_2 \in M_1 - \{0\}$, or some functions $\xi_1 : M_1 \longrightarrow M, \xi_2 : M_1 \times M_1 \longrightarrow M$ nondecreasing and continuous and $\zeta_2 : M_1 \longrightarrow M$ defined by

$$\begin{aligned}\zeta_2(t) \quad = \quad & \left[\frac{\xi(\zeta_1(t)t, at)\xi_1(\zeta_1(t)t)}{(1 - \xi_0(\zeta_1(t)t, 0))(1 - \xi_0(\zeta_1(t)t, at))}\right. \\ & \left. + \frac{\xi_2(t, \zeta_1(t)t)\xi_1(\zeta_1(t)t)}{1 - \xi_0(\zeta_1(t)t, at)}\right] \zeta_1(t).\end{aligned}$$

The parameter

$$d = \min\{d_i\}, \ i = 1, 2 \tag{22.4}$$

is proven to be a radius of convergence of method (22.2). Set $M_2 = [0, d)$.

The definition of d implies

$$0 \leq \xi_0(t, at) < 1 \tag{22.5}$$

$$0 \leq \xi_0(\zeta_1(t)t, 0) < 1 \tag{22.6}$$

$$0 \leq \xi_0(\zeta_1(t)t, at) < 1 \tag{22.7}$$

and

$$0 \leq \zeta_i(t) < 1 \tag{22.8}$$

hold for all $t \in M_2$.

The ball $\bar{U}(x_*, \lambda)$ is the closure of the open ball $U(x_*, \lambda)$ of center $x_* \in X$ and with radius $\lambda > 0$. Throughout this chapter, we use the notation $e_n =: \|x_n - x_*\|$.

The following conditions (H) are employed for the ξ functions defined before and x_* denoting a simple solution of F.

Suppose:

(h1)

$$\|F'(x_*)^{-1}(F_{[x,y]} - F'(x_*))\| \leq \xi_0(\|x - x_*\|, \|y - x_*\|)$$

and

$$\|I + \alpha F_{[x,x_*]}\| \leq a$$

for each $x, y \in D$. Set $D_0 = U(x_*, R_0) \cap D$.

(h2)

$$\|F'(x_*)^{-1}(F_{[x,z]} - F_{[x,x_*]})\| \leq \xi(\|x - x_*\|, \|z - x_*\|),$$

$$\|F'(x_*)^{-1}F'(x)\| \leq \xi_1(\|x - x_*\|)$$

and

$$\|I - A(x, y)\| \leq \xi_2(\|x - x_*\|, \|y - x_*\|)$$

provided that $x, y, z \in D_0$.

(h3) $\bar{U}(x_*, \tilde{d}_*) \subset D$ for $\tilde{d}_* = \max\{a\tilde{d}, \tilde{d}\}$ and $\tilde{d}$ to be given later and

(h4) There exists $d_* \geq \tilde{d}_*$ satisfying $\xi_0(0, d_*) < 1$ or $\xi_0(d_*, 0) < 1$. Let $D_1 = \bar{U}(x_*, d_*) \cap D$.

The ball convergence analysis of method (22.2) uses conditions (H).

Theorem 22.1 *Under conditions (H) for $\tilde{d} = d$ pick $x_0 \in U(x_*, d) - \{x_*\}$. Then,* $\lim_{n \longrightarrow +\infty} x_n = x_*$. *Moreover, the limit point x_* uniquely solves the equation (22.1) in the domain D_1 provided in the condition (h4).*

Proof. The following items shall be established by induction on n

$$\|y_n - x_*\| \leq \zeta_1(e_n)e_n \leq e_n < d \tag{22.9}$$

and

$$e_{n+1} \leq \zeta_2(e_n)e_n \leq e_n, \tag{22.10}$$

with radius d as defined in (22.4) and functions ζ_i as given previously. We have

$$\begin{aligned}\|z_0 - x_*\| &= \|x_0 - x_* + \alpha F(x_0)\| \\ &= \|x_0 - x_* + \alpha F_{[x_0,x_*]}(x_0 - x_*)\| \\ &= \|(I + \alpha F_{[x_0,x_*]})(x_0 - x_*)\| \\ &\leq a e_0 < \tilde{d}_*\end{aligned}$$

It follows using (22.4), (22.5), (h1) and (h3) that

$$\|F'(x_*)^{-1}(F_{[x_0,z_0]} - F'(x_*))\| \leq \xi_0(e_0, \|z_0 - x_*\|) \leq \xi_0(d, ad) < 1. \tag{22.11}$$

But a lemma on inverses of linear operators attributed to Banach [18] give $F_{[x_0,z_0]}$ is invertible with

$$\|F_{[x_0,z_0]}^{-1} F'(x_*)\| \leq \frac{1}{1 - \xi_0(\|x - x_*\|, \|z_0 - x_*\|)}. \tag{22.12}$$

Notice that the iterate y_0 is well defined in view of the first sub-step of the method (22.2) from which we also have

$$\begin{aligned}y_0 - x_* &= x_0 - x_* - F_{[x_0,z_0]}^{-1} F(x_0) \\ &= F_{[x_0,z_0]}^{-1} \\ &\quad \times (F_{[x_0,z_0]} - F_{[x_0,x_*]})(x_0 - x_*).\end{aligned} \tag{22.13}$$

By (22.4), (22.8) (for $i = 1$), (h2), (h3), (22.12) and (22.13), we get

$$\begin{aligned}\|y_0 - x_*\| &\leq \frac{\xi(e_0, \|z_0 - x_*\|)e_0}{1 - \xi_0(e_0, \|z_0 - x_*\|)} \\ &\leq \zeta_1(e_0)e_0 \leq e_0 < d,\end{aligned} \tag{22.14}$$

showing (22.9) for $n = 0$ and $y_0 \in U(x_*, d)$. Notice also that x_1 is well defined in view of the second sub-step of method (22.2) by which

$$\begin{aligned}x_1 - x_* &= y_0 - x_* - F_{[y_0,x_0]}^{-1} F(y_0) \\ &\quad + (F_{[y_0,x_*]}^{-1} - F_{[y_0,z_0]}^{-1}) F(y_0) \\ &\quad + (I - A_0) F_{[y_0,z_0]}^{-1} F(y_0) \\ &= F_{[y_0,x_*]}^{-1} (F_{[y_0,z_0]} - F_{[y_0,x_*]}) F_{[y_0,z_0]}^{-1} F(y_0)\end{aligned}$$

$$+(I - A_0)F^{-1}_{[y_0,z_0]}F(y_0). \tag{22.15}$$

Then, utilizing (22.4), (22.8) (for $i = 1$), (22.12), (22.14) and (22.15), we obtain

$$\begin{aligned} e_1 &\leq \left[\frac{\xi(\|y_0 - x_*\|, \|z_0 - x_*\|)\xi_1(\|y_0 - x_*\|)}{(1 - \xi_0(\|y_0 - x_*\|, 0))(1 - \xi_0(\|y_0 - x_*\|, \|z_0 - x_*\|))} \right. \\ &\quad \left. + \frac{\xi_2(e_0, \|y_0 - x_*\|)\xi_1(\|y_0 - x_*\|)}{1 - \xi_0(\|y_0 - x_*\|, e_0)}\right] \|y_0 - x_*\| \\ &\leq \zeta_2(e_0)e_0 \leq e_0, \end{aligned} \tag{22.16}$$

showing (22.10) for $m = 0$ and $x_1 \in U(x_*, d)$. Then, switch z_0, x_0, y_0, x_1 by z_m, x_m, y_m, x_{m+1} in the previous calculations to complete the induction for (22.9) and (22.10). Then, from the estimation

$$e_{m+1} \leq \gamma e_m < d, \tag{22.17}$$

for $\gamma = \zeta_2(e_0) \in [0, 1)$, we conclude $\lim_{m \longrightarrow +\infty} x_m = x_*$ and the iterate $x_{m+1} \in U(x_*, \rho)$. Let $Q = F_{[x_*,q]}$ for some $q \in D_1$ with $F(q) = 0$. By (h1) and (h4), we get

$$\|F'(x_*)^{-1}(Q - F'(x_*))\| \leq \xi_0(0, \|q - x_*\|) \leq \xi_0(0, d_*) < 1.$$

Hence, $q = x_*$ is defined by the existence of Q^{-1} and the estimate $0 = F(q) - F(x_*) = Q(q - x_*)$.

□

Remark 22.1 *The choice $A(t) = 1 + t + \beta t^2$, $t = F'(x)^{-1}F(y)$ verifies $A(0) = A'(0) = 1$ and $A''(0) < \infty$ required to show the fourth convergence order of method (22.20. Next, we show how to choose function ξ_2 in this case. Notice that we have*

$$\begin{aligned} &\|(F'(x_*)(x - x_*))^{-1}(F(x) - F(x_*) - F'(x_*)(x - x_*))\| \\ \leq\ & \frac{1}{\|x - x_*\|}\|F'(x_*)^{-1}(F_{[x,x_*]} - F'(x_*))\|\|x - x_*\| \text{ for } x \neq x_* \\ \leq\ & \xi_0(\|x - x_*\|, 0), \end{aligned}$$

so

$$\xi_2(s, t) = \xi_0(s, 0).$$

Next, the study of method (22.3) is provided analogously. But this time the "ζ" functions are defined as

$$\begin{aligned} \bar{\zeta}_1(t) &= \frac{\xi(t, t)}{1 - \xi_0(t, t)}, \\ \bar{\zeta}_2(t) &= \frac{\xi(t, \bar{\zeta}_1(t)t)}{1 - \xi_0(t, \bar{\zeta}_1(t)t)} \end{aligned}$$

and

$$\begin{aligned}
\bar{\zeta}_3(t) &= \left[\frac{\xi_0(\bar{\zeta}_2(t)t, \bar{\zeta}_1(t)t)\xi_1(\bar{\zeta}_2(t)t)}{(1-\xi_0(\bar{\zeta}_2(t)t, 0))(1-\xi_0(\bar{\zeta}_2(t)t, \bar{\zeta}_1(t)t))}\right. \\
&\quad \left. + \frac{\xi_2(t, \bar{\zeta}_2(t)t)\xi_1(\bar{\zeta}_2(t)t)}{1-\xi_0(\bar{\zeta}_2(t)t), \xi_0(\bar{\zeta}_1(t)t))}\right] \bar{\zeta}_2(t).
\end{aligned}$$

and

$$\bar{d} = \min\{\bar{d}_j\},\ j = 1, 2, 3,\ \tilde{d} = \bar{d},$$

and the least zeros of $\bar{\zeta}_i, i = 1, 2, 3$ functions in $M_0 - \{0\}$, denoted by $\bar{d}_1, \bar{d}_2, \bar{d}_3$, exist, respectively exist.

The motivation for the introduction of functions $\bar{\zeta}_i$ comes from the estimations

$$\begin{aligned}
\|z_n - x_*\| &= \|x_n - x_* - F^{-1}_{[x_n, x_{n-1}]}F(x_n)\| \\
&= \|F^{-1}_{[x_n, x_{n-1}]}(F_{[x_n, x_{n-1}]} - F_{[x_n, x_*]})(x_n - x_*)\| \\
&\leq \frac{\xi(e_n, \|x_{n-1} - x_*\|)e_n}{1 - \xi_0(e_n, \|x_{n-1} - x_*\|)} \\
&\leq \bar{\zeta}_1(e_0)e_0 \leq e_0 < \bar{d},
\end{aligned}$$

$$\begin{aligned}
\|y_n - x_*\| &= \|x_n - x_* - F^{-1}_{[x_n, z_n]}F(x_n)\| \\
&= \|F^{-1}_{[x_n, z_n]}(F_{[x_n, z_n]} - F_{[x_n, x_*]})(x_n - x_*)\| \\
&\leq \frac{\xi(e_n, \|z_n - x_*\|)e_n}{1 - \xi_0(e_n, \|z_n - x_*\|)} \\
&\leq \bar{\zeta}_2(e_n)e_n \leq e_n
\end{aligned}$$

and as in (22.15)

$$\begin{aligned}
e_{n+1} &\leq \left[\frac{\xi_0(\bar{\zeta}_2(e_n)e_n, \bar{\zeta}_1(e_n)e_n)\xi_1(\bar{\zeta}_2 e_n)e_n)}{(1-\xi_0(\bar{\zeta}_2(e_n)e_n, 0))(1-\xi_0(\bar{\zeta}_2(e_n)e_n, \bar{\zeta}_1(e_n)e_n))}\right. \\
&\quad \left. + \frac{\xi_2(e_n, \bar{\zeta}_2(e_n)e_n)\xi_1(\bar{\zeta}_2 e_n)e_n)}{1-\xi_0(\bar{\zeta}_2(e_n)e_n, \bar{\zeta}_1(e_n)e_n))}\right] \bar{\zeta}_2(e_n)e_n \\
&\leq \bar{\zeta}_3(e_n)e_n \leq e_n.
\end{aligned}$$

Therefore, the ball convergence analysis t for method (22.3) is

Theorem 22.2 *Under the conditions (H) with $\tilde{d} = \bar{d}$, the conclusions of Theorem 22.1 hold for method (22.3) with d, ζ replaced by $\bar{d}, \bar{\zeta}$ respectively.*

22.3 NUMERICAL EXAMPLES

We use for simplicity the choices

$$F_{[x,y]} = \int_0^1 F'(y+\theta(x-y))d\theta,\ \alpha = -1,$$

and ξ_2 as in Remark 2.2. But according to conditions (H) only differentiability of F at $x = x_*$ is required. Hence, the results apply to solve nondifferentiable operator equations too.

Example 3.1 We solve the motion system

$$F_1'(x) = e^x,\ F_2'(y) = (e-1)y+1,\ F_3'(z) = 1$$

with $F_1(0) = F_2(0) = F_3(0) = 0$. Let $F = (F_1, F_2, F_3)$. Let $X = Y = \mathbb{R}^3, D = \bar{U}(0,1), x_* = (0,0,0)^T$. Define function F on D for $w = (x,y,z)^T$ by

$$F(w) = (e^x - 1, \frac{e-1}{2}y^2 + y, z)^T.$$

The Jacobian is

$$F'(w) = \begin{bmatrix} e^x & 0 & 0 \\ 0 & (e-1)y+1 & 0 \\ 0 & 0 & 1 \end{bmatrix},$$

so $\xi_0(s,t) = \frac{1}{2}(e-1)(s+t), \xi(t) = \frac{1}{2}e^{\frac{1}{e-1}}(s+t), \xi_1(t) = e^{\frac{1}{e-1}}$, and $\xi_2(s,t) = \frac{1}{2}(e-1)s$, and $a = \frac{e}{2}$. Then, the radii are:

$$d_1 = 0.241677,\ d_2 = 0.192518,\ \bar{d}_1 = 0.285075,$$

$$\bar{d}_2 = 0.285075,\ \bar{d}_3 = 0.251558.$$

The iterates are given in Table 22.1.

Example 3.2 Let $D = \overline{U}(0,1) \subset Y = X = C[0,1]$ and the operator $F : D \longrightarrow \mathcal{B}_2$ is given by

$$F(\psi)(x) = \varphi(x) - 5\int_0^1 x\theta\psi(\theta)^3 d\theta. \tag{22.18}$$

Then, the derivative is

$$F'(\psi(\xi))(x) = \xi(x) - 15\int_0^1 x\theta\psi(\theta)^2\xi(\theta)d\theta,\ \text{for each } \xi \in D.$$

Notice that $x_* = 0$. Thus, $\xi_0(s,t) = \frac{15}{4}(s+t), \xi(s,t) = \frac{15}{2}(s+t), \xi_1(t) = 2, \xi_2(s,t) = \frac{15}{4}t$ and $a = 7$. Then, the radii are:

$$d_1 = 0.011111,\ d_2 = 0.00865678,\ \bar{d}_1 = 0.044444,$$

TABLE 22.1 Iterates of Method (22.2) and Method (22.3)

n	x_n **by (22.2)**	x_n **by 22.3)**
−1	(0.2000,0.2000, 0.2000)	(0.2000, 0.2000, 0.2000)
0	(0.1000, 0.1000, 0.1000)	(0.1000, 0.1000, 0.1000)
1	(0.0044, 0.0526, 0)	(0.0000, 0.0457, 0)
2	(0.0000, 0.0325, 0)	(0.0000, 0.0276, 0)
3	(0.0000, 0.0215, 0)	(−0.0000, 0.0181, 0)
4	(0.0000, 0.0147, 0)	(−0.0000, 0.0124, 0)
5	(0.0000, 0.0103, 0)	(−0.0000, 0.0087, 0)
6	(0.0000, 0.0074, 0)	(−0.0000, 0.0062, 0)
7	(0.0000, 0.0053, 0)	(−0.0000, 0.0045, 0)
8	(0.0000, 0.0038, 0)	(−0.0000, 0.0032, 0)
9	(0.0000, 0.0028, 0)	(−0.0000, 0.0024, 0)
10	(0.0000, 0.0020, 0)	(−0.0000, 0.0017, 0)
11	(0.0000, 0.0015, 0)	(−0.0000, 0.0013, 0)
12	(0.0000, 0.0011, 0)	(−0.0000, 0.0009, 0)
13	(0.0000, 0.0008, 0)	(−0.0000, 0.0007, 0)
14	(0.0000, 0.0006, 0)	(−0.0000, 0.0005, 0)
15	(0.0000, 0.0004, 0)	(−0.0000, 0.0004, 0)
16	(0.0000, 0.0003, 0)	(−0.0000, 0.0003, 0)
17	(0.0000, 0.0002, 0)	(−0.0000, 0.0002, 0)
18	(0.0000, 0.0002, 0)	(−0.0000, 0.0001, 0)
19	(0.0000, 0.0001, 0)	(−0.0000, 0.0001, 0)
20	(0.0000, 0.0001, 0)	(−0.0000, 0.0001, 0)
21	(0.0000, 0.0001, 0)	(−0.0000, 0.0001, 0)
22	(0.0000, 0.0001, 0)	(0,0,0)

$$\bar{d}_2 = 0.044444,\ \bar{d}_3 = 0.0413453.$$

Example 3.3 In view of the motivational example, the conditions are verified for $\xi_0(s,t) = \xi(s,t) = \frac{96.6629073}{2}(s+t), \xi_1(t) = 2$, $\xi_2(s,t) = \frac{96.6629073}{2}s$ and $a = 5$. Then, the radii are:

$$d_1 = 0.0017242,\ d_2 = 0.00138313,\ \bar{d}_1 = 0.00517261,$$

$$\bar{d}_2 = 0.005172613,\ \bar{d}_3 = 0.0044859.$$

22.4 CONCLUSION

There are some drawbacks when Taylor series expansions are used to find the convergence order for iterative methods. Some of these are: (a) high-order derivatives not on the methods must exist; (b) computable estimates on e_n or (c) uniqueness of the solution x_* results are not given. These drawbacks create problems like: we do not know how to pick the initial points or the number of iterations required to achieve a predecided error tolerance. We developed a technique in this chapter. In particular, we addressed problems (a)–(c) using generalized conditions only on the first derivative and divided differences of order one. Notice that only divided differences of order one appear in these methods. Hence, we extended the usage of these methods and in a more general Banach space. Similar work can be used on other methods due to the generality of the technique [1, 2, 3, 4, 5, 6, 7, 8, 9, 10, 12, 13, 14, 15, 16, 17, 18, 19, 20, 21, 22]. Experiments where the convergence criteria are tested complete this chapter.

BIBLIOGRAPHY

[1] S. Amat, S. Busquier, S. Plaza, Chaotic dynamics of a third-order Newton-type method, *J. Math. Anal. Appl.* 366 (2010) 24–32.

[2] I. K. Argyros, A unifying local-semilocal convergence analysis and applications for two-point Newton-like methods in Banach spaces, *J. Math. Anal. Appl.* 298 (2004) 374–397.

[3] I. K. Argyros, *Convergence and Applications of Newton-Type Iterations*, Springer-Verlag, New York, 2008.

[4] I. K. Argyros, *Computational Theory of Iterative Methods, Series: Studies in Computational Mathematics*, 15, Editors: Chui C.K. and Wuytack L. Elsevier Publication Company, New York, 2007.

[5] I. K. Argyros, Unified convergence criteria for iterative Banach space valued methods with applications, *Mathematics* 9(16) (2021) 1942; https://doi.org/10.3390/math9161942.

[6] I. K. Argyros, A. A. Magreñán, *Iterative Method and Their Dynamics with Applications*, CRC Press, New York, 2017.

[7] I. K. Argyros, S. George, A. A. Magreñán, Local convergence for multi-point-parametric Chebyshev-Halley-type method of higher convergence order. *J. Comput. Appl. Math.* 282 (2015) 215–224.

[8] I. K. Argyros, A. A., Magreñán, A study on the local convergence and the dynamics of Chebyshev-Halley-type methods free from second derivative. *Numer. Algorithms* 71 (2015) 1–23.

[9] I. K. Argyros, S. George, *Mathematical Modeling for the Solution of Equations and Systems of Equations with Applications*, Volume-IV, Nova Publisher, New York, 2020.

[10] I. K. Argyros, *The Theory and Application of Iterative Methods*, Second Edition, CRC Press, Taylor and Francis, Boca Raton, FL, 2022.

[11] F. I. Chicharro, A. Cordero, J. R. Torregrosa, Drawing dynamical and parameters planes of iterative families and methods, *Sci. World J.* 2013(780513) (2013) 1–11.

[12] A. Cordero, J. R. Torregrosa, Variants of Newton's method using fifth-order quadrature formulas, *Appl. Math. Comput.* 190 (2007) 686–698.

[13] J. Dzunic, M. S. Petkovic, On generalized biparametric multi point root finding methods with memory, *J. Comput. Appl. Math.* 255 (2014) 362–375.

[14] H. T. Kung, J. F. Traub, Optimal order of one point and multi point iteration, *J. Assoc. Comput. Mach.* 21(4) (1974) 643–651.

[15] A. A. Magrenan, A new tool to study real dynamics: The convergence plane, *Appl. Math. Comput.* 248 (2014) 215–224.

[16] B. Neta, A new family of high order methods for solving equations, *Int. J. Comput. Math.* 14 (1983) 191–195.

[17] J. M. Ortega, W. G. Rheinboldt, *Iterative Solutions of Nonlinear Equations in Several Variables*, SIAM, Philadelphia, 1970.

[18] M. S. Petkovic, J. Dzunic, L. D. Petkovid, A family of two point methods with memory for solving non linear equations, *Appl. Anal. Discrete Math.* 5 (2011) 298–317.

[19] J. R. Sharma, P. Gupta, On some highly efficient derivative free methods with and without memory for solving nonlinear equations, *Int. J. Comput. Methods*, 12 (2015) 1–28.

[20] J. F. Steffensen, Remarks on iteration, *Scand. Actuar. J.* 16 (1993) 64–72.

[21] J. F. Traub, *Iterative Methods for the Solution of Equations*, Prentice Hall, Hoboken, NJ, 1964.

[22] F. I. Chicharro, A. Cordero, J. R. Torregrosa, Dynamics of iterative families with memory based on weight functions procedure, *J. Comput. Appl. Math.* 354 (2019) 286–298.

CHAPTER 23

Inner Product Generalized Trapezoid Type Inequalities in Hilbert Spaces

Silvestru Sever Dragomir

Victoria University
University of the Witwatersrand

CONTENTS

23.1 INTRODUCTION

In 2001, Dragomir et al. [8] obtained the following generalized trapezoid inequality:

Theorem 23.1 *If $\psi : [a,b] \to \mathbb{R}$ is Riemann integrable on $[a,b]$ and $\varphi : [a,b] \to \mathbb{R}$ is of bounded variation on $[a,b]$, then*

$$\left| \int_a^b \varphi(t)\psi(t)\,dt - \varphi(b)\int_u^b \psi(t)\,dt - \varphi(a)\int_a^u \psi(t)\,dt \right| \tag{23.1}$$
$$\leq \left[\frac{1}{2}(b-a) + \left| u - \frac{a+b}{2} \right| \right] \sup_{t\in[a,b]} |\psi(t)| \bigvee_a^b (\varphi)$$

for all $u \in [a,b]$, where $\bigvee_a^b(\varphi)$ is the total variation of φ on $[a,b]$.

In particular, we have the mid-point trapezoid inequality

$$\left| \int_a^b \varphi(t)\psi(t)\,dt - \varphi(b)\int_{\frac{a+b}{2}}^b \psi(t)\,dt - \varphi(a)\int_a^{\frac{a+b}{2}} \psi(t)\,dt \right| \tag{23.2}$$

DOI: 10.1201/9781003388678-23

$$\leq \frac{1}{2}(b-a)\sup_{t\in[a,b]}|\psi(t)|\bigvee_a^b(\varphi).$$

The constant $1/2$ *is sharp in the sense that it cannot be replaced by a smaller quantity.*

For some recent results related to the trapezoid type inequalities, see [1] and [9]–[13].

Let U be a self-adjoint operator on the complex Hilbert space $(H, \langle .,. \rangle)$ with the spectrum $\mathrm{Sp}(U)$ included in the interval $[m, M]$ for some real numbers $m < M$ and let $\{E_\lambda\}_\lambda$ be its *spectral family.* Then for any continuous function $\varphi : [m, M] \to \mathbb{C}$, it is well known that we have the following *spectral representation in terms of the Riemann-Stieltjes integral*:

$$\langle \varphi(U)x, y\rangle = \int_{m-0}^{M} \varphi(\lambda)\, d(\langle E_\lambda x, y\rangle), \tag{23.3}$$

for any x, $y \in H$. The function $\psi_{x,y}(\lambda) := \langle E_\lambda x, y\rangle$ is of *bounded variation* on the interval $[m, M]$ and

$$\psi_{x,y}(m-0) = 0 \text{ and } \psi_{x,y}(M) = \langle x, y\rangle$$

for any x, $y \in H$. It is also well known that $\psi_x(\lambda) := \langle E_\lambda x, x\rangle$ is *monotonic nondecreasing* and *right continuous* on $[m, M]$.

With the notations introduced above, we have considered in the recent paper [3] the problem of bounding the error

$$\frac{\varphi(M)+\varphi(m)}{2}\langle x, y\rangle - \langle \varphi(A)x, y\rangle$$

in approximating $\langle \varphi(A)x, y\rangle$ by the trapezoidal type formula $\frac{\varphi(M)+\varphi(m)}{2}\langle x, y\rangle$, where x, y are vectors in the Hilbert space H and φ is a continuous functions of the self-adjoint operator A with the spectrum in the compact interval of real numbers $[m, M]$.

We recall here only two such results. The first result deals with the case of continuous functions of bounded variation and is incorporated in the following theorem [3]:

Theorem 23.2 *Let A be a self-adjoint operator in the Hilbert space H with the spectrum $\mathrm{Sp}(A) \subseteq [m, M]$ for some real numbers $m < M$ and let $\{E_\lambda\}_\lambda$ be its spectral family. If $\varphi : [m, M] \to \mathbb{C}$ is a continuous function of bounded variation on $[m, M]$, then we have the inequality*

$$\left|\frac{\varphi(M)+\varphi(m)}{2}\langle x, y\rangle - \langle \varphi(A)x, y\rangle\right| \tag{23.4}$$

$$\leq \frac{1}{2}\max_{\lambda\in[m,M]}\left[\langle E_\lambda x, x\rangle^{1/2}\langle E_\lambda y, y\rangle^{1/2}\right.$$

$$+\left\langle \left(1_H-E_\lambda\right)x,x\right\rangle^{1/2}\left\langle \left(1_H-E_\lambda\right)y,y\right\rangle^{1/2}\Big]\bigvee_m^M(\varphi)$$
$$\le \frac{1}{2}\|x\|\,\|y\|\bigvee_m^M(\varphi)$$

for any x, $y\in H$.

The case of Lipschitzian functions is as follows [3]:

Theorem 23.3 *Let A be a self-adjoint operator in the Hilbert space H with the spectrum $\mathrm{Sp}(A)\subseteq[m,M]$ for some real numbers $m<M$ and let $\{E_\lambda\}_\lambda$ be its spectral family. If $\varphi:[m,M]\to\mathbb{C}$ is Lipschitzian with the constant $L>0$ on $[m,M]$, then we have the inequality*

$$\left|\frac{\varphi(M)+\varphi(m)}{2}\langle x,y\rangle-\langle\varphi(A)x,y\rangle\right| \tag{23.5}$$
$$\le\frac{1}{2}L\int_{m-0}^{M}\Big[\left\langle E_\lambda x,x\right\rangle^{1/2}\left\langle E_\lambda y,y\right\rangle^{1/2}$$
$$+\left\langle \left(1_H-E_\lambda\right)x,x\right\rangle^{1/2}\left\langle \left(1_H-E_\lambda\right)y,y\right\rangle^{1/2}\Big]d\lambda$$
$$\le\frac{1}{2}(M-m)L\|x\|\,\|y\|$$

for any x, $y\in H$.

For some trapezoid operator inequalities, see [5], [6], [4] and [7].

In this chapter, we show among others that, if $\varphi:[a,b]\to H$ is continuous and $\psi:[a,b]\to H$ is strongly differentiable on (a,b), then

$$\left|\left\langle\int_u^b\varphi(s)\,ds,\psi(b)\right\rangle+\left\langle\int_a^u\varphi(s)\,ds,\psi(a)\right\rangle-\int_a^b\langle\varphi(t),\psi(t)\rangle\,dt\right|$$
$$\le\sup_{t\in[a,b]}\|\psi'(t)\|$$
$$\times\begin{cases}\left[(b-u)\int_u^b\|\varphi(t)\|\,dt+(u-a)\int_a^u\|\varphi(t)\|\,dt\right],\\[2ex] \frac{1}{(q+1)^{1/q}}\left[(b-u)^{1+1/q}\left(\int_u^b\|\varphi(t)\|^p\,dt\right)^{1/p}\right.\\ \left.+(u-a)^{1+1/q}\left(\int_a^u\|\varphi(t)\|^p\,dt\right)^{1/p}\right],\\[2ex] \frac{1}{2}\left[(b-u)^2\sup_{t\in[u,b]}\|\varphi(t)\|+(u-a)^2\sup_{t\in[a,u]}\|\varphi(t)\|\right]\end{cases}$$

for all $u\in[a,b]$, where p, $q>1$ with $\frac{1}{p}+\frac{1}{q}=1$. Applications for operator monotone functions with examples for power and logarithmic functions are also given.

23.2 MAIN RESULTS

We have the following weighted version of generalized trapezoid inequality for two functions with values in Hilbert spaces:

Theorem 23.4 *Assume that φ, $\psi : [a,b] \to H$ are continuous and ψ is strongly differentiable on (a,b), then for all $u \in [a,b]$*

$$\left| \left\langle \int_u^b \varphi(s)\,ds, \psi(b) \right\rangle + \left\langle \int_a^u \varphi(s)\,ds, \psi(a) \right\rangle - \int_a^b \langle \varphi(t), \psi(t) \rangle\,dt \right| \leq C(\varphi, \psi, u), \tag{23.6}$$

where

$$C(\varphi,\psi,u) := \int_u^b \left(\int_u^t \|\varphi(s)\|\,ds \right) \|\psi'(t)\|\,dt + \int_a^u \left(\int_t^u \|\varphi(s)\|\,ds \right) \|\psi'(t)\|\,dt.$$

We also have the bounds

$$C(\varphi,\psi,u) \leq \begin{cases} \int_u^b \|\varphi(s)\|\,ds \int_u^b \|\psi'(t)\|\,dt, \\ \left[\int_u^b \left(\int_u^t \|\varphi(s)\|\,ds \right)^p dt \right]^{1/p} \left(\int_u^b \|\psi'(t)\|^q\,dt \right)^{1/q}, \\ \int_u^b \left(\int_u^t \|\varphi(s)\|\,ds \right) dt \sup_{t\in[u,b]} \|\psi'(t)\|, \end{cases} + \begin{cases} \left(\int_a^u \|\varphi(s)\|\,ds \right) \int_a^u \|\psi'(t)\|\,dt, \\ \left[\int_a^u \left(\int_t^u \|\varphi(s)\|\,ds \right)^p dt \right]^{1/p} \left(\int_a^u \|\psi'(t)\|^q\,dt \right)^{1/q}, \\ \int_a^u \left(\int_t^u \|\varphi(s)\|\,ds \right) dt \sup_{t\in[a,u]} \|\psi'(t)\|, \end{cases} \tag{23.7}$$

where p, $q > 1$ with $\frac{1}{p} + \frac{1}{q} = 1$.

Proof 23.1 *Let $u \in [a,b]$. Using the integration by parts formula for inner products*

$$\int_a^b \langle h(t), l'(t) \rangle\,dt = \langle h(b), l(b) \rangle - \langle h(a), l(a) \rangle - \int_a^b \langle h'(t), l(t) \rangle\,dt,$$

where h, l are strongly differentiable on (a,b), we have

$$\int_a^b \left\langle \left(\int_a^t \varphi(s)\,ds - \int_a^u \varphi(s)\,ds \right), \psi'(t) \right\rangle dt \tag{23.8}$$

$$= \left\langle \left(\int_a^t \varphi(s)\,ds - \int_a^u \varphi(s)\,ds \right), \psi(t) \right\rangle \Big|_a^b$$
$$- \int_a^b \left\langle \left(\int_a^t \varphi(s)\,ds - \int_a^u \varphi(s)\,ds \right)', \psi(t) \right\rangle dt$$

$$= \left\langle \left(\int_a^b \varphi(s)\,ds - \int_a^u \varphi(s)\,ds \right), \psi(b) \right\rangle$$
$$- \left\langle \left(\int_a^a \varphi(s)\,ds - \int_a^u \varphi(s)\,ds \right), \psi(a) \right\rangle$$
$$- \int_a^b \left\langle \left(\int_a^t \varphi(s)\,ds - \int_a^u \varphi(s)\,ds \right)', \psi(t) \right\rangle dt$$
$$= \left\langle \left(\int_u^b \varphi(s)\,ds \right), \psi(b) \right\rangle + \left\langle \left(\int_a^u \varphi(s)\,ds \right), \psi(a) \right\rangle \tag{23.9}$$
$$- \int_a^b \langle \varphi(t), \psi(t) \rangle\, dt.$$

Also,

$$\int_a^b \left\langle \left(\int_a^t \varphi(s)\,ds - \int_a^u \varphi(s)\,ds \right), \psi'(t) \right\rangle dt \tag{23.10}$$
$$= \int_a^u \left\langle \left(\int_a^t \varphi(s)\,ds - \int_a^u \varphi(s)\,ds \right), \psi'(t) \right\rangle dt$$
$$+ \int_u^b \left\langle \left(\int_a^t \varphi(s)\,ds - \int_a^u \varphi(s)\,ds \right), \psi'(t) \right\rangle dt$$
$$= - \int_a^u \left\langle \left(\int_t^u \varphi(s)\,ds \right), \psi'(t) \right\rangle dt + \int_u^b \left\langle \left(\int_u^t \varphi(s)\,ds \right), \psi'(t) \right\rangle dt.$$

By utilizing (23.8) and (23.10), we derive the following identity of interest

$$\left\langle \int_u^b \varphi(s)\,ds, \psi(b) \right\rangle + \left\langle \int_a^u \varphi(s)\,ds, \psi(a) \right\rangle - \int_a^b \langle \varphi(t), \psi(t) \rangle\, dt \tag{23.11}$$
$$= \int_u^b \left\langle \int_u^t \varphi(s)\,ds, \psi'(t) \right\rangle dt - \int_a^u \left\langle \int_t^u \varphi(s)\,ds, \psi'(t) \right\rangle dt$$

for all $u \in [a, b]$.

Taking the norm in (23.11) and using the properties of the integral and Schwarz's inequality, we get

$$\left| \left\langle \int_u^b \varphi(s)\,ds, \psi(b) \right\rangle + \left\langle \int_a^u \varphi(s)\,ds, \psi(a) \right\rangle - \int_a^b \langle \varphi(t), \psi(t) \rangle\, dt \right| \tag{23.12}$$

$$\le \left| \int_u^b \left\langle \int_u^t \varphi(s)\, ds, \psi'(t) \right\rangle dt \right| + \left| \int_a^u \left\langle \int_t^u \varphi(s)\, ds, \psi'(t) \right\rangle dt \right|$$
$$\le \int_u^b \left| \left\langle \int_u^t \varphi(s)\, ds, \psi'(t) \right\rangle \right| dt + \int_a^u \left| \left\langle \int_t^u \varphi(s)\, ds, \psi'(t) \right\rangle \right| dt$$
$$\le \int_u^b \left\| \int_u^t \varphi(s)\, ds \right\| \|\psi'(t)\|\, dt + \int_a^u \left\| \int_t^u \varphi(s)\, ds \right\| \|\psi'(t)\|\, dt$$
$$\le \int_u^b \left(\int_u^t \|\varphi(s)\|\, ds \right) \|\psi'(t)\|\, dt + \int_a^u \left(\int_t^u \|\varphi(s)\|\, ds \right) \|\psi'(t)\|\, dt$$
$$= C(\varphi, \psi, u),$$

which proves (23.6).

Using Hölder's inequality, we get for $p,\ q > 1,\ \frac{1}{p} + \frac{1}{q} = 1$, that

$$\int_u^b \left(\int_u^t \|\varphi(s)\|\, ds \right) \|\psi'(t)\|\, dt$$
$$\le \begin{cases} \sup_{t\in[u,b]} \left(\int_u^t \|\varphi(s)\|\, ds \right) \int_u^b \|\psi'(t)\|\, dt, \\ \left[\int_u^b \left(\int_u^t \|\varphi(s)\|\, ds \right)^p dt \right]^{1/p} \left(\int_u^b \|\psi'(t)\|^q\, dt \right)^{1/q}, \\ \int_u^b \left(\int_u^t \|\varphi(s)\|\, ds \right) dt \sup_{t\in[u,b]} \|\psi'(t)\|, \end{cases}$$
$$= \begin{cases} \left(\int_u^b \|\varphi(s)\|\, ds \right) \int_u^b \|\psi'(t)\|\, dt, \\ \left[\int_u^b \left(\int_u^t \|\varphi(s)\|\, ds \right)^p dt \right]^{1/p} \left(\int_u^b \|\psi'(t)\|^q\, dt \right)^{1/q}, \\ \int_u^b \left(\int_u^t \|\varphi(s)\|\, ds \right) dt \sup_{t\in[u,b]} \|\psi'(t)\|, \end{cases}$$

and

$$\int_a^u \left(\int_t^u \|\varphi(s)\|\, ds \right) \|\psi'(t)\|\, dt$$
$$\le \begin{cases} \sup_{t\in[a,u]} \left(\int_t^u \|\varphi(s)\|\, ds \right) \int_a^u \|\psi'(t)\|\, dt \\ \left[\int_a^u \left(\int_t^u \|\varphi(s)\|\, ds \right)^p dt \right]^{1/p} \left(\int_a^u \|\psi'(t)\|^q\, dt \right)^{1/q} \\ \int_a^u \left(\int_t^u \|\varphi(s)\|\, ds \right) dt \sup_{t\in[a,u]} \|\psi'(t)\|, \end{cases}$$

$$= \begin{cases} \left(\int_a^u \|\varphi(s)\| ds\right) \int_a^u \|\psi'(t)\| dt, \\ \left[\int_a^u \left(\int_t^u \|\varphi(s)\| ds\right)^p dt\right]^{1/p} \left(\int_a^u \|\psi'(t)\|^q dt\right)^{1/q}, \\ \int_a^u \left(\int_t^u \|\varphi(s)\| ds\right) dt \sup_{t\in[a,u]} \|\psi'(t)\|. \end{cases}$$

By making use of (23.12), we deduce (23.7).

Corollary 23.1 *With the assumptions of Theorem 23.4, we have*

$$\left|\left\langle \int_u^b \varphi(s)\, ds, \psi(b)\right\rangle + \left\langle \int_a^u \varphi(s)\, ds, \psi(a)\right\rangle - \int_a^b \langle \varphi(t), \psi(t)\rangle\, dt\right| \tag{23.13}$$

$$\leq \int_u^b \|\varphi(s)\| ds \int_u^b \|\psi'(t)\| dt + \int_a^u \|\varphi(s)\| ds \int_a^u \|\psi'(t)\| dt$$

$$\leq \begin{cases} \max\left\{\int_u^b \|\varphi(s)\| ds, \int_a^u \|\varphi(s)\| ds\right\} \int_a^b \|\psi'(t)\| dt \\ \int_a^b \|\varphi(s)\| ds \max\left\{\int_u^b \|\psi'(t)\| dt, \int_a^u \|\psi'(t)\| dt\right\} \end{cases}$$

$$\leq \int_a^b \|\varphi(s)\| ds \int_a^b \|\psi'(t)\| dt,$$

for all $u \in [a, b]$.

The proof follows by the first branch in the bounds (23.7).

Remark 23.1 *If $m \in (a, b)$ is such that*

$$\int_a^u \|\varphi(s)\| ds = \int_u^b \|\varphi(s)\| ds = \frac{1}{2}\int_a^b \|\varphi(s)\| ds, \tag{23.14}$$

then by (23.12) we get

$$\left|\left\langle \int_m^b \varphi(s)\, ds, \psi(b)\right\rangle + \left\langle \int_a^m \varphi(s)\, ds, \psi(a)\right\rangle - \int_a^b \langle \varphi(t), \psi(t)\rangle\, dt\right| \tag{23.15}$$

$$\leq \frac{1}{2}\int_a^b \|\varphi(s)\| ds \int_a^b \|\psi'(t)\| dt.$$

Corollary 23.2 *With the assumptions of Theorem 23.4, we have*

$$\left|\left\langle \int_u^b \varphi(s)\, ds, \psi(b)\right\rangle + \left\langle \int_a^u \varphi(s)\, ds, \psi(a)\right\rangle - \int_a^b \langle \varphi(t), \psi(t)\rangle\, dt\right| \tag{23.16}$$

$$\leq \left[\int_u^b (b-t)\|\varphi(t)\| dt + \int_a^u (t-a)\|\varphi(t)\| dt\right] \sup_{t\in[a,b]} \|\psi'(t)\|$$

for all $u \in [a, b]$.

Proof 23.2 *From the third branch in the bounds in (23.7), we have*

$$\begin{aligned}
&\left|\left\langle \int_u^b \varphi(s)\,ds, \psi(b)\right\rangle + \left\langle \int_a^u \varphi(s)\,ds, \psi(a)\right\rangle - \int_a^b \langle \varphi(t), \psi(t)\rangle\, dt\right| \qquad (23.17)\\
&\le \int_u^b \left(\int_u^t \|\varphi(s)\|\,ds\right) dt \sup_{t\in[u,b]} \|\psi'(t)\| \\
&+ \int_a^u \left(\int_t^u \|\varphi(s)\|\,ds\right) dt \sup_{t\in[a,u]} \|\psi'(t)\| \\
&\le \sup_{t\in[a,b]} \|\psi'(t)\| \left[\int_u^b \left(\int_u^t \|\varphi(s)\|\,ds\right) dt + \int_a^u \left(\int_t^u \|\varphi(s)\|\,ds\right) dt\right].
\end{aligned}$$

Using integration by parts, we have for $u \in [a,b]$ that

$$\begin{aligned}
\int_u^b \left(\int_u^t \|\varphi(s)\|\,ds\right) dt &= \left(\int_u^t \|\varphi(s)\|\,ds\right) t\Big|_u^b - \int_u^b t\,\|\varphi(t)\|\,dt \\
&= \left(\int_u^b \|\varphi(s)\|\,ds\right) b - \int_u^b t\,\|\varphi(t)\|\,dt \\
&= \int_u^b (b-t)\,\|\varphi(t)\|\,dt
\end{aligned}$$

and

$$\begin{aligned}
\int_a^u \left(\int_t^u \|\varphi(s)\|\,ds\right) dt &= \left(\int_t^u \|\varphi(s)\|\,ds\right) t\Big|_a^u + \int_a^u t\,\|\varphi(t)\|\,dt \\
&= -\left(\int_a^u \|\varphi(s)\|\,ds\right) a + \int_a^u t\,\|\varphi(t)\|\,dt \\
&= \int_a^u (t-a)\,\|\varphi(t)\|\,dt,
\end{aligned}$$

which, by (23.17) provides the desired result (23.16).

Remark 23.2 *By making use of Hölder's integral inequality, we have for $p,\ q > 1$ with $\frac{1}{p} + \frac{1}{q} = 1$ that*

$$\int_u^b (b-t)\,\|\varphi(t)\|\,dt \le \begin{cases} \sup_{t\in[u,b]} (b-t) \int_u^b \|\varphi(t)\|\,dt, \\[2ex] \left(\int_u^b (b-t)^q\,dt\right)^{1/q} \left(\int_u^b \|\varphi(t)\|^p\,dt\right)^{1/p}, \\[2ex] \int_u^b (b-t)\,dt \sup_{t\in[u,b]} \|\varphi(t)\|, \end{cases}$$

$$= \begin{cases} (b-u)\int_u^b \|\varphi(t)\| \, dt, \\ \frac{(b-u)^{1+1/q}}{(q+1)^{1/q}} \left(\int_u^b \|\varphi(t)\|^p \, dt\right)^{1/p}, \\ \frac{1}{2}(b-u)^2 \sup_{t\in[u,b]} \|\varphi(t)\| \end{cases}$$

and

$$\int_a^u (t-a)\|\varphi(t)\| \, dt \le \begin{cases} \sup_{t\in[a,u]}(t-a)\int_a^u \|\varphi(t)\| \, dt, \\ \left(\int_a^u (t-a)^q \, dt\right)^{1/q} \left(\int_a^u \|\varphi(t)\|^p \, dt\right)^{1/p}, \\ \int_a^u (t-a) \, dt \sup_{t\in[a,u]} \|\varphi(t)\|, \end{cases}$$

$$= \begin{cases} (u-a)\int_a^u \|\varphi(t)\| \, dt, \\ \frac{(u-a)^{1+1/q}}{(q+1)^{1/q}} \left(\int_a^u \|\varphi(t)\|^p \, dt\right)^{1/p}, \\ \frac{1}{2}(u-a)^2 \sup_{t\in[a,u]} \|\varphi(t)\|. \end{cases}$$

By (23.16) we then get

$$\left|\left\langle \int_u^b \varphi(s) \, ds, \psi(b)\right\rangle + \left\langle \int_a^u \varphi(s) \, ds, \psi(a)\right\rangle - \int_a^b \langle \varphi(t), \psi(t)\rangle \, dt\right| \tag{23.18}$$
$$\le \sup_{t\in[a,b]} \|\psi'(t)\|$$
$$\times \begin{cases} \left[(b-u)\int_u^b \|\varphi(t)\| \, dt + (u-a)\int_a^u \|\varphi(t)\| \, dt\right], \\ \frac{1}{(q+1)^{1/q}} \left[(b-u)^{1+1/q} \left(\int_u^b \|\varphi(t)\|^p \, dt\right)^{1/p} \right. \\ \left. + (u-a)^{1+1/q} \left(\int_a^u \|\varphi(t)\|^p \, dt\right)^{1/p}\right], \\ \frac{1}{2}\left[(b-u)^2 \sup_{t\in[u,b]} \|\varphi(t)\| + (u-a)^2 \sup_{t\in[a,u]} \|\varphi(t)\|\right] \end{cases}$$

for all $u \in [a,b]$.

Observe that

$$(b-u)\int_u^b \|\varphi(t)\| \, dt + (u-a)\int_a^u \|\varphi(t)\| \, dt$$
$$\le \max\{b-u, u-a\}\left[\int_u^b \|\varphi(t)\| \, dt + \int_a^u \|\varphi(t)\| \, dt\right]$$
$$= \left[\frac{1}{2}(b-a) + \left|u - \frac{a+b}{2}\right|\right] \int_a^b \|\varphi(t)\| \, dt.$$

By using the elementary inequality for a, b, c, $d \geq 0$ *and* p, $q > 1$ *with* $\frac{1}{p} + \frac{1}{q} = 1$,

$$(ab + cd) \leq (a^p + c^p)^{1/p} (b^q + d^q)^{1/q}$$

we get

$$\begin{aligned}
&(b-u)^{1+1/q} \left(\int_u^b \|\varphi(t)\|^p dt \right)^{1/p} + (u-a)^{1+1/q} \left(\int_a^u \|\varphi(t)\|^p dt \right)^{1/p} \\
&\leq \left[\left((b-u)^{1+1/q} \right)^q + \left((u-a)^{1+1/q} \right)^q \right]^{1/q} \\
&\times \left[\left(\left(\int_u^b \|\varphi(t)\|^p dt \right)^{1/p} \right)^p + \left(\left(\int_a^u \|\varphi(t)\|^p dt \right)^{1/p} \right)^p \right]^{1/p} \\
&= \left[(b-u)^{q+1} + (u-a)^{q+1} \right]^{1/q} \left[\int_u^b \|\varphi(t)\|^p dt + \int_a^u \|\varphi(t)\|^p dt \right]^{1/p} \\
&= \left[(b-u)^{q+1} + (u-a)^{q+1} \right]^{1/q} \left(\int_a^b \|\varphi(t)\|^p dt \right)^{1/p}.
\end{aligned}$$

Also,

$$\begin{aligned}
&\frac{1}{2} \left[(b-u)^2 \sup_{t\in[u,b]} \|\varphi(t)\| + (u-a)^2 \sup_{t\in[a,u]} \|\varphi(t)\| \right] \\
&\leq \frac{1}{2} \left[(b-u)^2 + (u-a)^2 \right] \sup_{t\in[a,b]} \|\varphi(t)\| \\
&= \left[\frac{1}{4}(b-a) + \left(u - \frac{a+b}{2} \right)^2 \right] \sup_{t\in[a,b]} \|\varphi(t)\|.
\end{aligned}$$

Then by (23.18) we get for p, $q > 1$ *with* $\frac{1}{p} + \frac{1}{q} = 1$ *that*

$$\begin{aligned}
&\left| \left\langle \int_u^b \varphi(s)\, ds, \psi(b) \right\rangle + \left\langle \int_a^u \varphi(s)\, ds, \psi(a) \right\rangle - \int_a^b \langle \varphi(t), \psi(t) \rangle dt \right| \qquad (23.19) \\
&\leq \sup_{t\in[a,b]} \|\psi'(t)\| \\
&\times \begin{cases} \left[\frac{1}{2}(b-a) + \left| u - \frac{a+b}{2} \right| \right] \int_a^b \|\varphi(t)\| dt, \\ \frac{1}{(q+1)^{1/q}} \left[(b-u)^{q+1} + (u-a)^{q+1} \right]^{1/q} \left(\int_a^b \|\varphi(t)\|^p dt \right)^{1/p}, \\ \left[\frac{1}{4}(b-a) + \left(u - \frac{a+b}{2} \right)^2 \right] \sup_{t\in[a,b]} \|\varphi(t)\| \end{cases}
\end{aligned}$$

for all $u \in [a, b]$.

We also have:

Corollary 23.3 *With the assumptions of Theorem 23.4, we have for all* $u \in [a,b]$,

$$\left|\left\langle \int_u^b \varphi(s)\,ds, \psi(b) \right\rangle + \left\langle \int_a^u \varphi(s)\,ds, \psi(a) \right\rangle - \int_a^b \langle \varphi(t), \psi(t) \rangle\,dt\right| \tag{23.20}$$
$$\leq \left[\left(\int_u^b \|\varphi(s)\|\,ds\right)^p (b-u) + \left(\int_a^u \|\varphi(s)\|\,ds\right)^p (u-a)\right]^{1/p}$$
$$\times \left(\int_a^b \|\psi'(t)\|^q\,dt\right)^{1/q}$$
$$\leq (b-a)^{1/p} \left[\left(\int_u^b \|\varphi(s)\|\,ds\right)^p + \left(\int_a^u \|\varphi(s)\|\,ds\right)^p\right]^{1/p}$$
$$\times \left(\int_a^b \|\psi'(t)\|^q\,dt\right)^{1/q}$$

for $p, q > 1$ *with* $\frac{1}{p} + \frac{1}{q} = 1$

Proof 23.3 *Observe that, by the elementary inequality for* $a, b, c, d \geq 0$ *and* $p, q > 1$ *with* $\frac{1}{p} + \frac{1}{q} = 1$,

$$(ab + cd) \leq (a^p + c^p)^{1/p} (b^q + d^q)^{1/q},$$

we have

$$\left[\int_u^b \left(\int_u^t \|\varphi(s)\|\,ds\right)^p dt\right]^{1/p} \left(\int_u^b \|\psi'(t)\|^q\,dt\right)^{1/q}$$
$$+ \left[\int_a^u \left(\int_t^u \|\varphi(s)\|\,ds\right)^p dt\right]^{1/p} \left(\int_a^u \|\psi'(t)\|^q\,dt\right)^{1/q}$$
$$\leq \left(\int_u^b \left(\int_u^t \|\varphi(s)\|\,ds\right)^p dt + \int_a^u \left(\int_t^u \|\varphi(s)\|\,ds\right)^p dt\right)^{1/p}$$
$$\times \left(\int_u^b \|\psi'(t)\|^q\,dt + \int_a^u \|\psi'(t)\|^q\,dt\right)^{1/q}$$
$$= \left(\int_u^b \left(\int_u^t \|\varphi(s)\|\,ds\right)^p dt + \int_a^u \left(\int_t^u \|\varphi(s)\|\,ds\right)^p dt\right)^{1/p}$$
$$\times \left(\int_a^b \|\psi'(t)\|^q\,dt\right)^{1/q}$$

$$\leq \left(\left(\int_u^b \|\varphi(s)\| \, ds \right)^p \int_u^b dt + \left(\int_a^u \|\varphi(s)\| \, ds \right)^p \int_a^u dt \right)^{1/p}$$
$$\times \left(\int_a^b \|\psi'(t)\|^q \, dt \right)^{1/q}$$
$$= \left(\left(\int_u^b \|\varphi(s)\| \, ds \right)^p (b-u) + \left(\int_a^u \|\varphi(s)\| \, ds \right)^p (u-a) \right)^{1/p}$$
$$\times \left(\int_a^b \|\psi'(t)\|^q \, dt \right)^{1/q}$$
$$\leq (b-a)^{1/p} \left(\left(\int_u^b \|\varphi(s)\| \, ds \right)^p + \left(\int_a^u \|\varphi(s)\| \, ds \right)^p \right)^{1/p}$$
$$\times \left(\int_a^b \|\psi'(t)\|^q \, dt \right)^{1/q},$$

which proves (23.20).

Remark 23.3 *If $m \in (a, b)$ is such that (23.14) is valid, then by (23.20) we get*

$$\left| \left\langle \int_m^b \varphi(s) \, ds, \psi(b) \right\rangle + \left\langle \int_a^m \varphi(s) \, ds, \psi(a) \right\rangle - \int_a^b \langle \varphi(t), \psi(t) \rangle \, dt \right| \tag{23.21}$$
$$\leq \frac{1}{2} (b-a)^{1/p} \int_a^b \|\varphi(s)\| \, ds \left(\int_a^b \|\psi'(t)\|^q \, dt \right)^{1/q}.$$

Assume that $\varphi, \psi : [a, b] \to H$ are continuous and ψ is strongly differentiable on (a, b), then

$$\left| \left\langle \int_{\frac{a+b}{2}}^b \varphi(s) \, ds, \psi(b) \right\rangle + \left\langle \int_a^{\frac{a+b}{2}} \varphi(s) \, ds, \psi(a) \right\rangle - \int_a^b \langle \varphi(t), \psi(t) \rangle \, dt \right| \tag{23.22}$$
$$\leq C(\varphi, \psi),$$

where

$$C(\varphi, \psi) := \int_{\frac{a+b}{2}}^b \left(\int_{\frac{a+b}{2}}^t \|\varphi(s)\| \, ds \right) \|\psi'(t)\| \, dt$$
$$+ \int_a^{\frac{a+b}{2}} \left(\int_t^{\frac{a+b}{2}} \|\varphi(s)\| \, ds \right) \|\psi'(t)\| \, dt.$$

We also have the bounds

$$C(\varphi, \psi) \tag{23.23}$$

$$\leq \begin{cases} \int_{\frac{a+b}{2}}^{b} \|\varphi(s)\| \, ds \int_{\frac{a+b}{2}}^{b} \|\psi'(t)\| \, dt, \\ \left[\int_{u}^{b} \left(\int_{\frac{a+b}{2}}^{t} \|\varphi(s)\| \, ds\right)^{p} dt\right]^{1/p} \left(\int_{\frac{a+b}{2}}^{b} \|\psi'(t)\|^{q} \, dt\right)^{1/q}, \\ \int_{\frac{a+b}{2}}^{b} \left(\int_{\frac{a+b}{2}}^{t} \|\varphi(s)\| \, ds\right) dt \sup_{t\in\left[\frac{a+b}{2}, b\right]} \|\psi'(t)\|, \end{cases}$$

$$+ \begin{cases} \left(\int_{a}^{\frac{a+b}{2}} \|\varphi(s)\| \, ds\right) \int_{a}^{\frac{a+b}{2}} \|\psi'(t)\| \, dt, \\ \left[\int_{a}^{\frac{a+b}{2}} \left(\int_{t}^{\frac{a+b}{2}} \|\varphi(s)\| \, ds\right)^{p} dt\right]^{1/p} \left(\int_{a}^{\frac{a+b}{2}} \|\psi'(t)\|^{q} \, dt\right)^{1/q}, \\ \int_{a}^{\frac{a+b}{2}} \left(\int_{t}^{\frac{a+b}{2}} \|\varphi(s)\| \, ds\right) dt \sup_{t\in\left[a, \frac{a+b}{2}\right]} \|\psi'(t)\|, \end{cases}$$

where p, $q > 1$ with $\frac{1}{p} + \frac{1}{q} = 1$.

From (23.13) we get

$$\left| \left\langle \int_{\frac{a+b}{2}}^{b} \varphi(s) \, ds, \psi(b) \right\rangle + \left\langle \int_{a}^{\frac{a+b}{2}} \varphi(s) \, ds, \psi(a) \right\rangle - \int_{a}^{b} \langle \varphi(t), \psi(t) \rangle \, dt \right| \tag{23.24}$$

$$\leq \int_{\frac{a+b}{2}}^{b} \|\varphi(s)\| \, ds \int_{\frac{a+b}{2}}^{b} \|\psi'(t)\| \, dt + \int_{a}^{\frac{a+b}{2}} \|\varphi(s)\| \, ds \int_{a}^{\frac{a+b}{2}} \|\psi'(t)\| \, dt$$

$$\leq \begin{cases} \max\left\{\int_{\frac{a+b}{2}}^{b} \|\varphi(s)\| \, ds, \int_{a}^{\frac{a+b}{2}} \|\varphi(s)\| \, ds\right\} \int_{a}^{b} \|\psi'(t)\| \, dt \\ \int_{a}^{b} \|\varphi(s)\| \, ds \max\left\{\int_{\frac{a+b}{2}}^{b} \|\psi'(t)\| \, dt, \int_{a}^{\frac{a+b}{2}} \|\psi'(t)\| \, dt\right\} \end{cases}$$

$$\leq \int_{a}^{b} \|\varphi(s)\| \, ds \int_{a}^{b} \|\psi'(t)\| \, dt,$$

while from (23.19) we get

$$\left| \left\langle \int_{\frac{a+b}{2}}^{b} \varphi(s) \, ds, \psi(b) \right\rangle + \left\langle \int_{a}^{\frac{a+b}{2}} \varphi(s) \, ds, \psi(a) \right\rangle - \int_{a}^{b} \langle \varphi(t), \psi(t) \rangle \, dt \right| \tag{23.25}$$

$$\leq \sup_{t\in[a,b]} \|\psi'(t)\| \times \begin{cases} \frac{1}{2}(b-a) \int_{a}^{b} \|\varphi(t)\| \, dt, \\ \frac{1}{2(q+1)^{1/q}} (b-a)^{1+1/q} \left(\int_{a}^{b} \|\varphi(t)\|^{p} \, dt\right)^{1/p}, \\ \frac{1}{4}(b-a) \sup_{t\in[a,b]} \|\varphi(t)\|. \end{cases}$$

From (23.20) we also get

$$\left| \left\langle \int_{\frac{a+b}{2}}^{b} \varphi(s) \, ds, \psi(b) \right\rangle + \left\langle \int_{a}^{\frac{a+b}{2}} \varphi(s) \, ds, \psi(a) \right\rangle - \int_{a}^{b} \langle \varphi(t), \psi(t) \rangle \, dt \right| \tag{23.26}$$

$$\leq \frac{(b-a)^{1/p}}{2^{1/p}}\left[\left(\int_{\frac{a+b}{2}}^{b}\|\varphi(s)\|\,ds\right)^{p}+\left(\int_{a}^{\frac{a+b}{2}}\|\varphi(s)\|\,ds\right)^{p}\right]^{1/p}$$
$$\times\left(\int_{a}^{b}\|\psi'(t)\|^{q}\,dt\right)^{1/q}.$$

23.3 INEQUALITIES FOR OPERATOR MONOTONE FUNCTIONS

A real valued continuous function h on $[0,\infty)$ is said to be operator monotone if $h(A)\geq h(B)$ holds for any $A\geq B\geq 0$.

We have the following representation of operator monotone functions, see for instance [2, p. 144-145]:

Theorem 23.5 *A function $h:[0,\infty)\to\mathbb{R}$ is operator monotone in $[0,\infty)$ if and only if it has the representation*

$$h(t)=h(0)+bt+\int_{0}^{\infty}\frac{t\lambda}{t+\lambda}d\mu(\lambda), \tag{23.27}$$

where $b\geq 0$ and a positive measure μ on $[0,\infty)$ such that

$$\int_{0}^{\infty}\frac{\lambda}{1+\lambda}d\mu(\lambda)<\infty. \tag{23.28}$$

Lemma 23.1 *Let $h:[0,\infty)\to\mathbb{R}$ be operator monotone in $[0,\infty)$. Assume that $U\geq 0$, then for all self-adjoint operators V we have*

$$Dh(U)(V)=bV+\int_{0}^{\infty}\lambda^{2}\left[(\lambda+U)^{-1}V(\lambda+U)^{-1}\right]d\mu(\lambda). \tag{23.29}$$

Proof 23.4 *From (23.27) we get*

$$h(t)=h(0)+bt+\int_{0}^{\infty}\left(\lambda-\frac{\lambda^{2}}{t+\lambda}\right)d\mu(\lambda).$$

Assume that $U\geq 0$, then for all self-adjoint operator V we have, by the representation of h and for t in a small open interval around 0, that

$$\begin{aligned}
&h(U+tV)-h(U)\\
&=btV+\int_{0}^{\infty}\left(\lambda-\lambda^{2}(U+tV+\lambda)^{-1}\right)d\mu(\lambda)-\int_{0}^{\infty}\left(\lambda-\lambda^{2}(U+\lambda)^{-1}\right)d\mu(\lambda)\\
&=btV+\int_{0}^{\infty}\lambda^{2}\left[(\lambda+U)^{-1}-(\lambda+U+tV)^{-1}\right]d\mu(\lambda)\\
&=btV+\int_{0}^{\infty}\lambda^{2}\left[(\lambda+U)^{-1}(\lambda+U+tV-\lambda-U)(\lambda+U+tV)^{-1}\right]d\mu(\lambda)
\end{aligned}$$

$$= btV + t\int_0^\infty \lambda^2 \left[(\lambda + U)^{-1} V (\lambda + U + tV)^{-1}\right] d\mu(\lambda).$$

Dividing by $t \neq 0$, we get

$$\frac{h(U + tV) - h(U)}{t} = bV + \int_0^\infty \lambda^2 \left[(\lambda + U)^{-1} V (\lambda + U + tV)^{-1}\right] d\mu(\lambda)$$

and by taking the limit over $t \to 0$, we get

$$Dh(U)(V) = bV + \int_0^\infty \lambda^2 \left[(\lambda + U)^{-1} V (\lambda + U)^{-1}\right] d\mu(\lambda)$$

for all self-adjoint operator V we have (23.29).

Theorem 23.6 *Let $h : [0, \infty) \to \mathbb{R}$ be operator monotone in $[0, \infty)$. Assume that $U \geq u > 0$, then for all self-adjoint operators V we have*

$$\|Dh(U)(V)\| \leq h'(u)\|V\|. \tag{23.30}$$

Proof 23.5 *From (23.29) we get*

$$\|Dh(U)(V) - bV\| \leq \int_0^\infty \lambda^2 \left\|(\lambda + U)^{-1} V (\lambda + U)^{-1}\right\| d\mu(\lambda) \tag{23.31}$$

$$\leq \|V\| \int_0^\infty \lambda^2 \left\|(\lambda + U)^{-1}\right\|^2 d\mu(\lambda).$$

Observe that $\lambda + U \geq \lambda + u > 0$ for $\lambda \in [0, \infty)$. Then $0 < (\lambda + U)^{-1} \leq (\lambda + u)^{-1}$, which implies that $\left\|(\lambda + U)^{-1}\right\| \leq (\lambda + u)^{-1}$, namely $\left\|(\lambda + U)^{-1}\right\|^2 \leq (\lambda + u)^{-2}$ for $\lambda \in [0, \infty)$.

Therefore by (23.31) we get

$$\|Dh(U)(V) - bV\| \leq \|V\| \int_0^\infty \lambda^2 (\lambda + u)^{-2} d\mu(\lambda). \tag{23.32}$$

If we take the derivative over t in (23.27), then we have

$$h'(t) = b + \int_0^\infty \frac{\lambda(t + \lambda) - \lambda t}{(t + \lambda)^2} d\mu(\lambda) = b + \int_0^\infty \frac{\lambda^2}{(t + \lambda)^2} d\mu(\lambda) \tag{23.33}$$

for $t > 0$.

From (23.33) we get

$$\int_0^\infty \lambda^2 (\lambda + u)^{-2} d\mu(\lambda) = h'(u) - b,$$

and by (23.32) we derive

$$\|Dh(U)(V) - bV\| \leq \|V\| h'(u) - b\|V\|.$$

Finally, by the triangle inequality and by the fact that $b \geq 0$, we obtain that

$$\|Dh(U)(V)\| - b\|V\| \leq \|Dh(U)(V) - bV\|,$$

which proves the desired result (23.30).

For a continuous function h on $(0,\infty)$ and $A,\ B>0$, we consider the auxiliary function $h_{A,B}:[0,1]\to\mathbb{R}$ defined by

$$h_{A,B}(t):=h((1-t)A+tB),\ t\in[0,1].$$

We have the following representations of the derivatives:

Lemma 23.2 *Assume that the operator function generated by h is Fréchet differentiable in each $A\geq 0$, then for $B\geq 0$ we have that $h_{A,B}$ is differentiable on $[0,1]$ and*

$$h'_{A,B}(t)=D(h)((1-t)A+tB)(B-A) \tag{23.34}$$

for $t\in[0,1]$, where in 0 and 1 the derivatives are the right and left derivatives.

Proof 23.6 *We prove only for the interior points $t\in(0,1)$. Let h be in a small interval around 0 such that $t+h\in(0,1)$. Then for $h\neq 0$,*

$$\begin{aligned}&\frac{h_{A,B}(t+h)-h(t)}{h}\\&=\frac{h((1-(t+h))A+(t+h)B)-h((1-t)A+tB)}{h}\\&=\frac{h((1-t)A+tB+h(B-A))-h((1-t)A+tB)}{h}\end{aligned}$$

and by taking the limit over $h\to 0$, we get

$$\begin{aligned}h'_{A,B}(t)&=\lim_{h\to 0}\frac{h_{A,B}(t+h)-h(t)}{h}\\&=\lim_{h\to 0}\left[\frac{h((1-t)A+tB+h(B-A))-h((1-t)A+tB)}{h}\right]\\&=D(h)((1-t)A+tB)(B-A),\end{aligned}$$

which proves (23.34).

Corollary 23.4 *Let $h:[0,\infty)\to\mathbb{R}$ be operator monotone in $[0,\infty)$. Then for all $A\geq a>0,\ B\geq b>0$, we have*

$$\begin{aligned}\left\|h'_{A,B}(t)\right\|&=\|D(h)((1-t)A+tB)(B-A)\|\\&\leq h'((1-t)a+tb)\|B-A\|\end{aligned} \tag{23.35}$$

for all $t\in[0,1]$.

The proof follows by Theorem 23.6 and Lemma 23.2.

One can observe that the inequality (23.35) remains valid for operator monotone functions on $(0,\infty)$. This follows by considering the function $h_\varepsilon(t):=h(t+\varepsilon)$ for $\varepsilon>0$, which is operator monotone on $[0,\infty)$ and then by letting $\varepsilon\to 0+$ and using the continuity of h and h'.

We define the generalized trapezoid functional

$$T(\varphi, \psi, A, B, x, y; u) := \left\langle \int_u^1 \varphi((1-s)A + sB)\, x ds, \psi(b)\, y \right\rangle + \left\langle \int_0^u \varphi((1-s)A + sB)\, x ds, \psi(a)\, y \right\rangle - \int_0^1 \langle \varphi((1-t)A + tB)\, x, \psi((1-t)A + tB)\, y \rangle\, dt, \tag{23.36}$$

where φ and ψ are continuous on $[0, \infty)$, A, $B \geq 0$ and x, $y \in H$.

We have the following result:

Theorem 23.7 *Let φ be continuous on $[0, \infty)$ and ψ be operator monotone in $[0, \infty)$. Then for all $A \geq a > 0$, $B \geq b > 0$, we have for all x, $y \in H$ that*

$$\begin{aligned} &|T(\varphi, \psi, A, B, x, y; u)| \\ &\leq \|B - A\| \|x\| \|y\| \left[\int_u^1 \|\varphi((1-t)A + tB)\|\, dt \int_u^1 \psi'((1-t)a + tb)\, dt \right. \\ &\left. + \int_0^u \|\varphi((1-t)A + tB)\|\, dt \int_0^u \psi'((1-t)a + tb)\, dt \right] \\ &\leq \|B - A\| \|x\| \|y\| \\ &\times \max\left\{ \int_u^1 \|\varphi((1-t)A + tB)\|\, dt, \int_0^u \|\varphi((1-t)A + tB)\|\, dt \right\} \\ &\times \begin{cases} \frac{\psi(b) - \psi(a)}{b-a} & \textit{if } b \neq a, \\ \psi'(a) & \textit{if } b = a, \end{cases} \end{aligned} \tag{23.37}$$

$$\begin{aligned} &|T(\varphi, \psi, A, B, x, y; u)| \\ &\leq \|B - A\| \|x\| \|y\| \sup_{t \in [a,b]} \psi'((1-t)a + tb) \\ &\times \left[\int_u^1 (1-t) \|\varphi((1-t)A + tB)\|\, dt + \int_0^u t \|\varphi((1-t)A + tB)\|\, dt \right] \end{aligned} \tag{23.38}$$

and

$$\begin{aligned} &|T(\varphi, \psi, A, B, x, y; u)| \\ &\qquad \leq \|B - A\| \|x\| \|y\| \left(\int_0^1 [\psi'((1-t)a + tb)]^q\, dt \right)^{1/q} \\ &\times \left[(1-u) \left(\int_u^1 \|\varphi((1-t)A + tB)\|\, dt \right)^p + u \left(\int_0^u \|\varphi((1-t)A + tB)\|\, dt \right)^p \right]^{1/p} \end{aligned}$$

$$\leq \|B-A\|\|x\|\|y\|\left(\int_0^1 \left[\psi'\left((1-t)a+tb\right)\right]^q dt\right)^{1/q}$$

$$\times\left[\left(\int_u^1 \|\varphi\left((1-t)A+tB\right)\| dt\right)^p+\left(\int_0^u \|\varphi\left((1-t)A+tB\right)\| dt\right)^p\right]^{1/p} \quad (23.39)$$

for $p,\ q>1$ *with* $\frac{1}{p}+\frac{1}{q}=1$

The proof follows by Corollaries 23.1–23.3 for $\varphi_{A,B}x$ and $\psi_{A,B}y$. The details are omitted.

From (23.19) we also have

$$\begin{aligned}
&\left|T\left(\varphi,\psi,A,B,x,y;u\right)\right| \qquad (23.40)\\
&\leq \|B-A\|\|x\|\|y\| \sup_{t\in[a,b]} \psi'\left((1-t)a+tb\right)\\
&\times\begin{cases}
\left[\frac{1}{2}+\left|u-\frac{1}{2}\right|\right]\int_0^1 \|\varphi\left((1-t)A+tB\right)\| dt,\\[2mm]
\frac{1}{(q+1)^{1/q}}\left[(1-u)^{q+1}+u^{q+1}\right]^{1/q}\left(\int_0^1 \|\varphi\left((1-t)A+tB\right)\|^p dt\right)^{1/p},\\[2mm]
\left[\frac{1}{4}+\left(u-\frac{1}{2}\right)^2\right]\sup_{t\in[0,1]}\|\varphi\left((1-t)A+tB\right)\|
\end{cases}
\end{aligned}$$

provided that φ is continuous on $[0,\infty)$, ψ is operator monotone in $[0,\infty)$ and $A\geq a>0$, $B\geq b>0$, while $x,\ y\in H$.

In particular, we have

$$\begin{aligned}
\left|M\left(\varphi,\psi,A,B,x,y\right)\right| &\leq \frac{1}{2}\|B-A\|\|x\|\|y\| \sup_{t\in[a,b]} \psi'\left((1-t)a+tb\right) \qquad (23.41)\\
&\times\begin{cases}
\int_0^1 \|\varphi\left((1-t)A+tB\right)\| dt,\\[2mm]
\frac{1}{(q+1)^{1/q}}\left(\int_0^1 \|\varphi\left((1-t)A+tB\right)\|^p dt\right)^{1/p},\\[2mm]
\frac{1}{2}\sup_{t\in[0,1]}\|\varphi\left((1-t)A+tB\right)\|,
\end{cases}
\end{aligned}$$

where

$$\begin{aligned}
M\left(\varphi,\psi,A,B,x,y\right) :=& \left\langle \int_{1/2}^1 \varphi\left((1-s)A+sB\right)xds,\psi(b)y\right\rangle\\
&+\left\langle \int_0^{1/2} \varphi\left((1-s)A+sB\right)xds,\psi(a)y\right\rangle\\
&-\int_a^b \left\langle \varphi\left((1-t)A+tB\right)x,\psi\left((1-t)A+tB\right)y\right\rangle dt.
\end{aligned}$$

23.4 SOME EXAMPLES

We consider the function $\varphi(t) = \ell^r(t) = t^r$ for $r \in (0,1)$. Then for $A,\ B \geq 0$ and $t \in [0,1]$ we have

$$\|((1-t)A + tB)^r\| \leq \|(1-t)A + tB\|^r \leq [(1-t)\|A\| + t\|B\|]^r.$$

Therefore

$$\begin{aligned}\int_0^1 \|((1-t)A + tB)^r\| dt &\leq \int_0^1 [(1-t)\|A\| + t\|B\|]^r \\ &= \begin{cases} \frac{\|B\|^{r+1} - \|A\|^{r+1}}{(r+1)(\|B\| - \|A\|)} \text{ if } \|B\| \neq \|A\| \\ \\ \|A\|^r \text{ if } \|B\| = \|A\|.\end{cases}\end{aligned}$$

Also

$$\begin{aligned}\int_0^1 \|((1-t)A + tB)^r\|^p dt &\leq \int_0^1 [(1-t)\|A\| + t\|B\|]^{pr} \\ &= \begin{cases} \frac{\|B\|^{pr+1} - \|A\|^{pr+1}}{(pr+1)(\|B\| - \|A\|)} \text{ if } \|B\| \neq \|A\|, \\ \\ \|A\|^{pr} \text{ if } \|B\| = \|A\|\end{cases}\end{aligned}$$

and

$$\sup_{t\in[0,1]} \|((1-t)A + tB)^r\| \leq \sup_{t\in[0,1]} [(1-t)\|A\| + t\|B\|]^r = \max\{\|A\|^r, \|B\|^r\}.$$

From the inequality (23.40) we obtain

$$\begin{aligned}&|T(\ell^r, \psi, A, B, x, y; u)| \qquad (23.42)\\ &\leq \|B - A\| \|x\| \|y\| \sup_{t\in[a,b]} \psi'((1-t)a + tb) \\ &\times \begin{cases} \left[\frac{1}{2} + \left|u - \frac{1}{2}\right|\right] \times \begin{cases} \frac{\|B\|^{r+1} - \|A\|^{r+1}}{(r+1)(\|B\| - \|A\|)} \text{ if } \|B\| \neq \|A\|, \\ \\ \|A\|^r \text{ if } \|B\| = \|A\|, \end{cases} \\ \\ \frac{1}{(q+1)^{1/q}} \left[(1-u)^{q+1} + u^{q+1}\right]^{1/q} \\ \times \begin{cases} \left[\frac{\|B\|^{pr+1} - \|A\|^{pr+1}}{(pr+1)(\|B\| - \|A\|)}\right]^{1/p} \text{ if } \|B\| \neq \|A\|, \\ \\ \|A\|^r \text{ if } \|B\| = \|A\|, \end{cases} \\ \\ \left[\frac{1}{4} + \left(u - \frac{1}{2}\right)^2\right] \max\{\|A\|^r, \|B\|^r\}. \end{cases}\end{aligned}$$

if ψ is operator monotone in $[0,\infty)$ and $A \geq a > 0$, $B \geq b > 0$, while $x,\ y \in H$.

In particular, we have

$$\begin{aligned}
&\left|M\left(\ell^{r},\psi,A,B,x,y\right)\right| \qquad (23.43)\\
&\leq\frac{1}{2}\left\|B-A\right\|\left\|x\right\|\left\|y\right\|\sup_{t\in[a,b]}\psi'\left((1-t)\,a+tb\right)\\
&\times\begin{cases}\begin{cases}\frac{\|B\|^{r+1}-\|A\|^{r+1}}{(r+1)(\|B\|-\|A\|)} \text{ if } \|B\|\neq\|A\|,\\ \|A\|^{r} \text{ if } \|B\|=\|A\|,\end{cases}\\ \frac{1}{(q+1)^{1/q}}\times\begin{cases}\left[\frac{\|B\|^{pr+1}-\|A\|^{pr+1}}{(pr+1)(\|B\|-\|A\|)}\right]^{1/p} \text{ if } \|B\|\neq\|A\|,\\ \|A\|^{r} \text{ if } \|B\|=\|A\|,\end{cases}\\ \frac{1}{2}\max\left\{\|A\|^{r},\|B\|^{r}\right\}.\end{cases},
\end{aligned}$$

If in (23.42) we take $\psi(t)=\ln t$, then for $A\geq a>0$, $B\geq b>0$ and $x,\ y\in H$ we derive

$$\begin{aligned}
&\left|T\left(\ell^{r},\ln,A,B,x,y;u\right)\right| \qquad (23.44)\\
&\leq\frac{1}{\min\{a,b\}}\left\|B-A\right\|\left\|x\right\|\left\|y\right\|\\
&\times\begin{cases}\left[\frac{1}{2}+\left|u-\frac{1}{2}\right|\right]\times\begin{cases}\frac{\|B\|^{r+1}-\|A\|^{r+1}}{(r+1)(\|B\|-\|A\|)} \text{ if } \|B\|\neq\|A\|\\ \|A\|^{r} \text{ if } \|B\|=\|A\|.\end{cases}\\ \frac{1}{(q+1)^{1/q}}\left[(1-u)^{q+1}+u^{q+1}\right]^{1/q}\\ \quad\times\begin{cases}\left[\frac{\|B\|^{pr+1}-\|A\|^{pr+1}}{(pr+1)(\|B\|-\|A\|)}\right]^{1/p} \text{ if } \|B\|\neq\|A\|\\ \|A\|^{r} \text{ if } \|B\|=\|A\|\end{cases}\\ \left[\frac{1}{4}+\left(u-\frac{1}{2}\right)^{2}\right]\max\left\{\|A\|^{r},\|B\|^{r}\right\}.\end{cases},
\end{aligned}$$

and

$$\begin{aligned}
&\left|M\left(\ell^{r},\ln,A,B,x,y\right)\right| \qquad (23.45)\\
&\leq\frac{1}{2\min\{a,b\}}\left\|B-A\right\|\left\|x\right\|\left\|y\right\|
\end{aligned}$$

$$\times\begin{cases}\begin{cases}\frac{\|B\|^{r+1}-\|A\|^{r+1}}{(r+1)(\|B\|-\|A\|)} \text{ if } \|B\|\neq\|A\|,\\[2ex] \|A\|^{r} \text{ if } \|B\|=\|A\|,\end{cases}\\[6ex] \frac{1}{(q+1)^{1/q}}\times\begin{cases}\left[\frac{\|B\|^{pr+1}-\|A\|^{pr+1}}{(pr+1)(\|B\|-\|A\|)}\right]^{1/p} \text{ if } \|B\|\neq\|A\|,\\[2ex] \|A\|^{r} \text{ if } \|B\|=\|A\|,\end{cases}\\[6ex] \frac{1}{2}\max\left\{\|A\|^{r},\|B\|^{r}\right\}.\end{cases},$$

BIBLIOGRAPHY

[1] M. W. Alomari, New upper and lower bounds for the trapezoid inequality of absolutely continuous functions and applications. *Konuralp J. Math.* **7** (2019), no. 2, 319–323.

[2] R. Bhatia, Matrix analysis. *Graduate Texts in Mathematics*, 169. Springer-Verlag, New York, 1997. xii+347 pp. ISBN: 0-387-94846-5.

[3] S. S. Dragomir, Some trapezoidal vector inequalities for continuous functions of self-adjoint operators in Hilbert spaces. *Abstr. Appl. Anal.* (2011), Art. ID 941286, 13 pp.

[4] S. S. Dragomir, *Operator Inequalities of Ostrowski and Trapezoidal Type.* SpringerBriefs in Mathematics. Springer, New York, 2012. x+112 pp. ISBN: 978-1-4614-1778-1.

[5] S. S. Dragomir, Generalised trapezoid-type inequalities for complex functions defined on unit circle with applications for unitary operators in Hilbert spaces. *Mediterr. J. Math.* **12** (2015), no. 3, 573–591.

[6] S. S. Dragomir, Trapezoid type inequalities for complex functions defined on the unit circle with applications for unitary operators in Hilbert spaces. *Georgian Math. J.* **23** (2016), no. 2, 199–210.

[7] S. S. Dragomir, *Riemann–Stieltjes Integral Inequalities for Complex Functions Defined on Unit Circle with Applications to Unitary Operators in Hilbert Spaces*, 160 Pages, CRC Press, Boca Raton, 2019, ISBN 9780367337100.

[8] S. S. Dragomir, C. Buşe, M. V. Boldea and L. Braescu, A generalization of the trapezoidal rule for the Riemann-Stieltjes integral and applications. *Nonlinear Anal. Forum* **6** (2001), no. 2, 337–351.

[9] A. Kashuri and R. Liko, Generalized trapezoidal type integral inequalities and their applications. *J. Anal.* **28** (2020), no. 4, 1023–1043.

[10] W. Liu and J. Park, Some perturbed versions of the generalized trapezoid inequality for functions of bounded variation. *J. Comput. Anal. Appl.* **22** (2017), no. 1, 11–18.

[11] W. Liu and H. Zhang, Refinements of the weighted generalized trapezoid inequality in terms of cumulative variation and applications. *Georgian Math. J.* **25** (2018), no. 1, 47–64.

[12] K. L. Tseng and S. R. Hwang, Some extended trapezoid-type inequalities and applications. *Hacet. J. Math. Stat.* **45** (2016), no. 3, 827–850.

[13] W. Yang, A companion for the generalized Ostrowski and the generalized trapezoid type inequalities. *Tamsui Oxf. J. Inf. Math. Sci.* **29** (2013), no. 2, 113–127.

CHAPTER 24

A Note on Degenerate Gamma Random Variables

Taekyun Kim

Xi'an Technological University
Kwangwoon University

Dae San Kim

Sogang University

Jongkyum Kwon

Gyeongsang National University

Hyunseok Lee

Kwangwoon University

CONTENTS

24.1 INTRODUCTION

For any $0 \neq \lambda \in \mathbb{R}$, the degenerate exponentials are defined by

$$e_{\lambda}^{x}(t) = \sum_{n=0}^{\infty}(x)_{n,\lambda}\frac{t^{n}}{n!}, \quad e_{\lambda}(t) = e_{\lambda}^{1}(t) = \sum_{n=0}^{\infty}(1)_{n,\lambda}\frac{t^{n}}{n!}, \tag{24.1}$$

where $(x)_{0,\lambda} = 1$, $(x)_{n,\lambda} = x(x-\lambda)\cdots(x-(n-1)\lambda)$, $(n \geq 1)$, (see [4,6,7,8,9]). Note that $\lim_{\lambda\to 0} e_{\lambda}^{x}(t) = e^{xt}$.

Observe also that $\log_{\lambda}(e_{\lambda}(t)) = e_{\lambda}(\log_{\lambda}(t)) = t$, for $\log_{\lambda}(t) = \frac{1}{\lambda}(t^{\lambda} - 1)$.

As is known, the λ-binomial coefficients are defined as

$$\binom{x}{n}_{\lambda} = \frac{(x)_{n,\lambda}}{n!} = \frac{x(x-\lambda)\cdots(x-(n-1)\lambda)}{n!}, \ (n \geq 1),$$

DOI: 10.1201/9781003388678-24

$$\binom{x}{0}_\lambda = 1, \text{ (see [7,10]).} \tag{24.2}$$

The degenerate Stirling numbers of the first kind are defined by Kim-Kim as

$$\begin{aligned}(x)_n &= \sum_{l=0}^{n} S_{1,\lambda}(n,l)(x)_{l,\lambda}, \frac{1}{n!}\Big(\log_\lambda(1+t)\Big)^n \\ &= \sum_{k=n}^{\infty} S_{1,\lambda}(k,n)\frac{t^k}{k!}, \ (n \geq 0), \text{ (see [4]),}\end{aligned} \tag{24.3}$$

where $(x)_0 = 1$, $(x)_n = x(x-1)\cdots(x-n+1)$, $(n \geq 1)$.
As an inversion formula of (24.3), the degenerate Stirling numbers of the second kind are given by

$$\begin{aligned}(x)_{n,\lambda} &= \sum_{l=0}^{n} S_{2,\lambda}(n,l)(x)_l, \ \frac{1}{n!}\Big(e_\lambda(t)-1\Big)^n \\ &= \sum_{k=n}^{\infty} S_{2,\lambda}(k,n)\frac{t^k}{k!}, \ (n \geq 0), \text{ (see [5]).}\end{aligned} \tag{24.4}$$

Recently, Kim-Kim introduced the degenerate gamma functions which are given by

$$\Gamma_\lambda(s) = \int_0^\infty e_\lambda^{-1}(t)t^{s-1}dt, \quad \big(\lambda \in (0,1)\big), \quad \text{(see [6, 9]),} \tag{24.5}$$

where $s \in \mathbb{C}$ with $0 < \mathrm{Re}(s) < \frac{1}{\lambda}$.
From (24.5), we note that

$$\Gamma_\lambda(k) = \frac{\Gamma(k)}{(1)_{k+1,\lambda}}, \quad \left(k \in \mathbb{N}, \ \lambda \neq 1, \frac{1}{2}, \frac{1}{3}, \ldots, \frac{1}{k}\right), \quad \text{(see [9]),} \tag{24.6}$$

where Γ is the usual gamma function.
Note that $\lim_{\lambda \to 0+} \Gamma_\lambda(s) = \Gamma(s) = \int_0^\infty e^{-t}t^{s-1}dt, \quad (\mathrm{Re}(s) > 0)$.
The random variables are real valued functions defined on sample spaces. We say that X is a continuous random variable if there exists a non-negative function f, defined for all real $x \in (-\infty, \infty)$, having the property that for any set B of real numbers

$$P\{X \in B\} = \int_B f(x)dx, \quad \text{(see [14]).} \tag{24.7}$$

The function f is called the probability density function of X. If X is a continuous random variable having the probability density function of f, then the expectation of X is defined by

$$E[X] = \int_{-\infty}^{\infty} xf(x)dx. \tag{24.8}$$

Let X be a continuous random variable with the probability density function f. Then, for any real valued function g, we have

$$E[g(X)] = \int_{-\infty}^{\infty} g(x)f(x)dx. \tag{24.9}$$

The expected value of the random variable X, $E[X]$, is also referred to as the mean or the first moment of X. The quantity $E[X^n]$, $n \geq 1$, is said to be the n-th moment of X. That is,

$$E[X^n] = \int_{-\infty}^{\infty} x^n f(x)dx, \quad (n \geq 1), \tag{24.10}$$

Another quantity of interest is the variance of random variable X which is defined by

$$\text{Var}(X) = E\Big[\big(X - E[X^2]\big)^2\Big] = E[X^2] - \big(E[X]\big)^2. \tag{24.11}$$

We say that the random variables X and Y are jointly continuous random variable if there exists a function $f(x, y)$, defined for all x and y, having the probability that for all sets A and B of real numbers

$$P\{X \in A,\ Y \in B\} = \int_B \int_A f(x, y)dxdy. \tag{24.12}$$

The function $f(x, y)$ is called the joint probability density function of X and Y. The random variable X and Y are said to be independent if, for all a, b,

$$P\{X \leq a,\ Y \leq b\} = P\{X \leq a\} \cdot P\{Y \leq b\},$$

that is to say if

$$E[XY] = E[X]E[Y].$$

Let X, Y be independent random variables. For any real valued functions h and g, we have

$$E[g(X)h(Y)] = E[g(X)]E[h(Y)].$$

A continuous random variable X is said to have a gamma distribution with parameters $\alpha > 0$ and $\beta > 0$ if its probability density function has the form

$$f(x) = \begin{cases} \frac{1}{\Gamma(\alpha)}\beta e^{-\beta x}(\beta x)^{\alpha-1}, & \text{if } x > 0, \\ 0, & \text{otherwise.} \end{cases}$$

In this case, we shall say X is the gamma random variable with parameters α and β , for which we indicate by $X \sim \Gamma(\alpha, \beta)$. The gamma random variables have long been studied by many researchers (see [1–3, 11–13,15,16]). They are widely used in science, engineering and business and occur naturally in the processes in which the waiting times between Poisson distributed events are relevant to each other. Note that

$$P\{X \in (-\infty, \infty)\} = \int_{-\infty}^{\infty} f(x)dx$$

$$
\begin{aligned}
&= \frac{\beta}{\Gamma(\alpha)} \int_0^\infty (\beta x)^{\alpha-1} e^{-\beta x} dx \\
&= \frac{1}{\Gamma(\alpha)} \int_0^\infty y^{\alpha-1} e^{-y} dy \\
&= 1.
\end{aligned}
$$

In this chapter, we study the degenerate gamma random variables with parameters α and β arising from the degenerate gamma function and deduce the expectation and variance of these random variables. They may be viewed as a 'degenerate version' of the gamma random variables.

24.2 DEGENERATE GAMMA RANDOM VARIABLES

For any $\lambda \in (0,1)$, a continuous random variable $X = X_\lambda$ is said to have a degenerate gamma distribution with parameter $\alpha > 0$ and $\beta > 0$, $(\frac{1}{\lambda} > \alpha > 0, \beta > 0)$ if its probability density function has the form

$$
f_\lambda(x) = \begin{cases} \frac{1}{\Gamma_\lambda(\alpha)} \beta e_\lambda^{-1}(\beta x)(\beta x)^{\alpha-1}, & \text{if } x > 0, \\ 0, & \text{otherwise.} \end{cases} \tag{24.13}
$$

In this case, we shall say that X is the degenerate gamma random variable with parameters α and β , for which we write as $X \sim \Gamma_\lambda(\alpha,\beta)$.
Not that

$$
\begin{aligned}
P\{X \in (-\infty,\infty)\} &= \int_{-\infty}^\infty f_\lambda(x) dx \\
&= \frac{\beta}{\Gamma_\lambda(\alpha)} \int_0^\infty e_\lambda^{-1}(\beta x)(\beta x)^{\alpha-1} dx \\
&= \frac{1}{\Gamma_\lambda(\alpha)} \int_0^\infty e_\lambda(y)^{-1} y^{\alpha-1} dy \\
&= \frac{1}{\Gamma_\lambda(\alpha)} \Gamma_\lambda(\alpha) \\
&= 1.
\end{aligned}
$$

From (24.5), we note that

$$
\Gamma_\lambda(s+1) = \frac{s}{1-\lambda(s+1)} \Gamma_\lambda(s). \tag{24.14}
$$

Let $X \sim \Gamma_\lambda(\alpha,\beta)$. For $n \in \mathbb{N}$, the n-th moment of X is given by

$$
\begin{aligned}
E[X^n] &= \int_{-\infty}^\infty x^n f_\lambda(x) dx \\
&= \frac{\beta}{\Gamma_\lambda(\alpha)} \int_0^\infty e_\lambda^{-1}(\beta x)(\beta x)^{\alpha-1} x^n dx
\end{aligned}
$$

$$
\begin{aligned}
&= \frac{1}{\Gamma_\lambda(\alpha)} \frac{1}{\beta^n} \int_0^\infty e_\lambda^{-1}(y) y^{n+\alpha-1} dy \\
&= \frac{1}{\beta^n} \frac{\Gamma_\lambda(n+\alpha)}{\Gamma_\lambda(\alpha)}. \qquad (24.15)
\end{aligned}
$$

From (24.14), we can derive the following equation.

$$
\begin{aligned}
\Gamma_\lambda(n+\alpha) &= \frac{(n+\alpha-1)\Gamma_\lambda(n+\alpha-1)}{1-\lambda(n+\alpha)} \qquad (24.16) \\
&= \frac{(n+\alpha-1)(n+\alpha-2)}{\big(1-\lambda(n+\alpha)\big)\big(1-\lambda(n+\alpha-1)\big)} \Gamma_\lambda(n+\alpha-2) \qquad (24.17) \\
&= \cdots \\
&= \frac{(n+\alpha-1)(n+\alpha-2)\cdots\alpha\Gamma_\lambda(\alpha)}{\big(1-\lambda(n+\alpha)\big)\big(1-\lambda(n+\alpha-1)\big)\cdots\big(1-\lambda(\alpha+1)\big)}.
\end{aligned}
$$

By (24.15) and (24.16), we get

$$
\begin{aligned}
E[X^n] &= \frac{1}{\beta^n} \frac{(n+\alpha-1)_n}{\big(1-\lambda(\alpha+1)\big)_{n,\lambda}} \qquad (24.18) \\
&= \frac{1}{\beta^n} \frac{\binom{n+\alpha-1}{n}}{\binom{1-\lambda(\alpha+1)}{n}_\lambda}.
\end{aligned}
$$

Therefore, by (24.18), we obtain the following theorem.

Theorem 24.1 *Let $X \sim \Gamma_\lambda(\alpha, \beta)$. For $n \in \mathbb{N}$, we have*

$$
E[X^n] = \frac{1}{\beta^n} \frac{\binom{n+\alpha-1}{n}}{\binom{1-\lambda(\alpha+1)}{n}_\lambda}.
$$

From Theorem 24.1, we note that

$$
\begin{aligned}
E[X] &= \frac{\alpha}{\beta\big(1-\lambda(\alpha+1)\big)}, \\
E[X^2] &= \frac{1}{\beta^2} \frac{\binom{\alpha+1}{2}}{\binom{1-\lambda(\alpha+1)}{2}_\lambda} = \frac{1}{\beta^2}\left(\frac{\alpha(\alpha+1)}{\big(1-\lambda(\alpha+1)\big)\big(1-\lambda(\alpha+2)\big)}\right).
\end{aligned}
$$

Thus, the variance of X is given by

$$
\begin{aligned}
\mathrm{Var}(X) &= E[X^2] - \big(E[X]\big)^2 \\
&= \frac{1}{\beta^2}\left(\frac{\alpha(\alpha+1)}{\big(1-\lambda(\alpha+1)\big)\big(1-\lambda(\alpha+2)\big)}\right) - \frac{\alpha^2}{\beta^2}\left(\frac{1}{1-\lambda(\alpha+1)}\right)^2 \qquad (24.19) \\
&= \frac{\alpha}{\beta^2}\left(\frac{(\alpha+1)\big(1-\lambda(\alpha+1)\big) - \alpha\big(1-\lambda(\alpha+2)\big)}{\big(1-\lambda(\alpha+1)\big)^2\big(1-\lambda(\alpha+2)\big)}\right)
\end{aligned}
$$

$$= \frac{\alpha}{\beta^2}\left(\frac{1-\lambda}{\big(1-\lambda(\alpha+1)\big)^2\big(1-\lambda(\alpha+2)\big)}\right).$$

Therefore, by (24.19), we obtain the following theorem

Theorem 24.2 *Let $X \sim \Gamma_\lambda(\alpha,\beta)$. Then we have*

$$\mathrm{Var}(X) = \frac{\alpha}{\beta^2}\left(\frac{1-\lambda}{\big(1-\lambda(\alpha+1)\big)^2\big(1-\lambda(\alpha+2)\big)}\right).$$

24.3 FURTHER REMARK

Let $X \sim \Gamma_\lambda(1,1)$. Then we have

$$\begin{aligned} E[e_\lambda(Xt)] &= (1-\lambda)\int_0^\infty e_\lambda(xt)e_\lambda^{-1}(x)dx \\ &= (1-\lambda)\sum_{n=0}^\infty (1)_{n,\lambda}\int_0^\infty x^n e_\lambda^{-1}(x)dx\frac{t^n}{n!} \\ &= (1-\lambda)\sum_{n=0}^\infty (1)_{n,\lambda}\Gamma_\lambda(n+1)\frac{t^n}{n!} \\ &= (1-\lambda)\sum_{n=0}^\infty (1)_{n,\lambda}\frac{\Gamma(n+1)}{(1)_{n+2,\lambda}}\frac{t^n}{n!} \\ &= \sum_{n=0}^\infty \frac{1-\lambda}{(1-n\lambda)(1-(n+1)\lambda)}t^n. \end{aligned} \tag{24.20}$$

By (24.20), we get

$$(1)_{n,\lambda}\frac{E[X^n]}{n!} = \frac{1-\lambda}{(1-n\lambda)(1-(n+1)\lambda)}, \quad (n \in \mathbb{N}),$$

which agrees with the result in (24.18).
Let $X_1 \sim \Gamma_\lambda(\alpha_1,\beta_1)$, $X_2 \sim \Gamma_\lambda(\alpha_2,\beta_2)$, $\ldots$, $X_r \sim \Gamma_\lambda(\alpha_r,\beta_r)$. If $X_1, X_2, \ldots, X_r$ are identically independent, then we have

$$E[(X_1+\cdots+X_r)^n] = \frac{1}{\beta^n}\sum_{l_1+\cdots+l_r=n}\binom{n}{l_1,\ldots,l_r}\prod_{i=1}^r \frac{\binom{l_i+\alpha_i-1}{l_i}}{\binom{1-\lambda(\alpha_i+1)}{l_i}_\lambda},$$

where n is a positive integer.
Let $X \sim \Gamma_\lambda(\alpha,1)$. Then we have

$$E[e_\lambda^X(t)] = \sum_{n=0}^\infty \frac{t^n}{n!}E[(x)_{n,\lambda}] \tag{24.21}$$

$$= \sum_{n=0}^{\infty} \sum_{l=0}^{n} S_{2,\lambda}(n,l) E[(X)_l] \frac{t^n}{n!}$$

$$= \sum_{n=0}^{\infty} \left(\sum_{l=0}^{n} \sum_{m=0}^{l} S_{2,\lambda}(n,l) S_1(l,m) E[X^m] \right) \frac{t^n}{n!}$$

$$= \sum_{n=0}^{\infty} \left(\sum_{l=0}^{n} \sum_{m=0}^{l} S_{2,\lambda}(n,l) S_1(l,m) \frac{\binom{m+\alpha-1}{m}}{\binom{1-(\alpha+1)\lambda}{m}_{\lambda}} \right) \frac{t^n}{n!}.$$

Thus, by (24.21), we get

$$\begin{aligned} E[(1+t)^X] &= E[e_{\lambda}^X (\log_{\lambda}(1+t))] \qquad (24.22) \\ &= \sum_{n=0}^{\infty} \left(\sum_{l=0}^{n} \sum_{m=0}^{l} S_{2,\lambda}(m,l) S_1(l,m) \frac{\binom{m+\alpha-1}{m}}{\binom{1-(\alpha+1)\lambda}{m}_{\lambda}} \right) \frac{1}{n!} \Big(\log_{\lambda}(1+t) \Big)^n \\ &= \sum_{n=0}^{\infty} \left(\sum_{l=0}^{n} \sum_{m=0}^{l} S_{2,\lambda}(m,l) S_1(l,m) \frac{\binom{m+\alpha-1}{m}}{\binom{1-(\alpha+1)\lambda}{m}_{\lambda}} \right) \sum_{k=n}^{\infty} S_{1,\lambda}(k,n) \frac{t^k}{k!} \\ &= \sum_{k=0}^{\infty} \left(\sum_{n=0}^{k} \sum_{l=0}^{n} \sum_{m=0}^{l} S_{2,\lambda}(n,l) S_1(l,m) S_{1,\lambda}(k,n) \frac{\binom{m+\alpha-1}{m}}{\binom{1-(\alpha+1)\lambda}{m}_{\lambda}} \right) \frac{t^k}{k!}. \end{aligned}$$

On the other hand,

$$\begin{aligned} E[(1+t)^X] &= \sum_{n=0}^{\infty} E[X^n] \frac{1}{n!} \big(\log(1+t) \big)^n \qquad (24.23) \\ &= \sum_{n=0}^{\infty} E[X^n] \sum_{k=n}^{\infty} S_1(k,n) \frac{t^k}{k!} \\ &= \sum_{k=0}^{\infty} \left(\sum_{n=0}^{k} S_1(k,n) E[X^n] \right) \frac{t^k}{k!} \\ &= \sum_{k=0}^{\infty} \left(\sum_{n=0}^{k} S_1(k,n) \frac{\binom{n+\alpha-1}{n}}{\binom{1-(\alpha+1)\lambda}{n}_{\lambda}} \right) \frac{t^k}{k!}. \end{aligned}$$

Therefore, by (24.22) and (24.23), we obtain the following theorem.

Theorem 24.3 *Let $X \sim \Gamma_{\lambda}(\alpha, 1)$. Then we have*

$$\sum_{n=0}^{k} S_1(k,n) \frac{\binom{n+\alpha-1}{n}}{\binom{1-(\alpha+1)\lambda}{n}_{\lambda}} = \sum_{n=0}^{k} \sum_{l=0}^{n} \sum_{m=0}^{l} S_{2,\lambda}(n,l) S_1(l,m) S_{1,\lambda}(k,n) \frac{\binom{m+\alpha-1}{m}}{\binom{1-(\alpha+1)\lambda}{m}_{\lambda}}.$$

BIBLIOGRAPHY

[1] J. Aitchison, Inverse distributions and independent gamma-distributed products of random variables. *Biometrika* **50** (1963), 505–508.

[2] D. Barbu, A new fast method for computer generation of gamma and beta random variables by transformations of uniform variables. *Statistics* **18** (1987), no. 3, 453–464.

[3] R. J. Henery, Permutation probabilities for gamma random variables. *J. Appl. Probab.* **20** (1983), no. 4, 822–834.

[4] D. S. Kim, T. Kim, A note on a new type degenerate Bernoulli numbers. *Russ. J. Math. Phys.* **27** (2020), no. 2, 227–235.

[5] T. Kim, A note on degenerate Stirling polynomials of the second kind. *Proc. Jangjeon Math. Soc.* **20** (2017), no. 3, 319–331.

[6] T. Kim, D. S. Kim, Degenerate Laplace transform and degenerate gamma function. *Russ. J. Math. Phys.* **24** (2017), no. 2, 241–248.

[7] T. Kim, D. S. Kim, Degenerate polyexponential functions and degenerate Bell polynomials. *J. Math. Anal. Appl.* **487** (2020), no. 2, 124017.

[8] T. Kim, D. S. Kim, Some identities of extended degenerate r-central Bell polynomials arising from umbral calculus. *Rev. R. Acad. Cienc. Exactas Fis. Nat. Ser. A Mat. RACSAM* **114** (2020), no. 1, Art. 1, 19 pp.

[9] T. Kim, D. S. Kim, Note on the degenerate gamma function. *Russ. J. Math. Phys.* **27** (2020), no. 3, 352–358.

[10] T. Kim, D. S. Kim, H. Lee, J. Kwon, Degenerate binomial coefficients and degenerate hypergeometric functions. *Adv. Difference Equ.* 2020, Paper No. 115.

[11] H. Podolski, The distribution of a product of n-independent random variables with generalized gamma distribution. *Demonstratio Math.* **4** (1972), 119–123.

[12] S. B. Provost, On the distribution of the ratio of powers of sums of gamma random variables. *Pakistan J. Statist.* **5** (1989), no. 2, 157–174.

[13] C. D. Roberts, On the distribution of random variables whose m-th absolute power is gamma. *Sankhyā Ser. A* **33** (1971), 229–232.

[14] S. M. Ross, *Introduction to probability models. Eleventh edition.* Elsevier/Academic Press, Amsterdam, 2014. xvi+767 pp. ISBN: 978-0-12-407948-9 60-01.

[15] I. Vaduva, On computer generation of gamma random variables by rejection and composition procedures. *Math. Operationsforsch. Statist. Ser. Statist.* **8** (1977), no. 4, 545–576.

[16] J. Whittaker, Generation gamma and beta random variables with non-integral shape parameters. *J. Roy. Statist. Soc. Ser. C* **23** (1974), 210–214.

CHAPTER 25

Dynamical Systems on Free Random Variables Followed by the Semicircular Law

Ilwoo Cho

St. Ambrose University

CONTENTS

DOI: 10.1201/9781003388678-25

25.1 INTRODUCTION

For a topological $*$-algebra (e.g., a C^*-algebra, or a von Neumann algebra, or a Banach $*$-algebra) B over the complex field $\mathbb{C}$, if ψ is a bounded linear functional on B, then the pair (B, ψ) forms a topological (noncommutative free) $*$-probability space (e.g., a C^*-probability space, respectively, a W^*-probability space, respectively, a Banach $*$-probability space, etc.). A topological $*$-probability space (B, ψ) is said to be unital, if B contains its unity $1_B \in B$, satisfying $\psi(1_B) = 1$. If an element $T \in B$ is regarded as an element of (B, ψ), then we call T, a free random variable.

The free distribution of a free random variable $T \in (B, \psi)$ is characterized by the joint free moments of $\{T, T^*\}$,

$$\psi\left(\prod_{l=1}^{n} T^{r_l}\right) = \psi\left(T^{r_1} T^{r_2} ... T^{r_n}\right),$$

or, the joint free cumulants of $\{T, T^*\}$,

$$k_n^{\psi}\left(T^{r_1}, ..., T^{r_n}\right) = \sum_{\pi \in NC(n)} \left(\prod_{V \in \pi} \psi\left(\prod_{l \in V} T^{r_l}\right)\right) \mu\left(\pi, 1_n\right),$$

for all $(r_1, ..., r_n) \in \{1, *\}^n$, for all $n \in \mathbb{N}$, where $k_\bullet^{\psi}(.)$ is the free cumulant on B in terms of the linear functional ψ, by the Möbius inversion (e.g., [17], [22–25]). Here, $NC(n)$ is the lattice of consisting of all "noncrossing" partitions over $\{1, ..., n\}$, with its maximal element,

$$1_n = \{(1, ..., n)\}, \text{ the single-block partition,}$$

where $(1, ..., n)$ means a block of a partition 1_n, for all $n \in \mathbb{N}$; and μ is the Möbius functional satisfying

$$\mu\left(0_n, 1_n\right) = (-1)^{n+1} c_n, \; \forall n \in \mathbb{N},$$

and

$$\sum_{\theta \in NC(n)} \mu\left(\theta, 1_n\right) = 0,$$

where

$$c_n = \frac{1}{n+1}\begin{pmatrix} 2n \\ n \end{pmatrix} = \frac{(2n)!}{n!(n+1)!},$$

is the n-th Catalan number, and

$$0_n = \{(1), (2), ..., (n)\}, \text{ the } n\text{-block partition,}$$

the minimal element of $NC(n)$, and (θ_1, θ_2) means the interval in $NC(n)$ under the partial ordering,

$$\theta_1 \leq \theta_2, \Longleftrightarrow \forall U \in \theta_1, \; \exists V \in \theta_2, \text{ s.t., } U \subseteq V,$$

where "$U \in \theta_1$" means "U is a block of θ_1."

For example, the free distribution of a self-adjoint free random variable $S \in (B, \psi)$ is characterized by the free moment sequence,

$$(\psi(S^n))_{n=1}^{\infty},$$

or, by the free cumulant sequence,

$$\left(k_n^{\psi}\left(\underbrace{S, S, ..., S}_{n\text{-times}}\right)\right)_{n=1}^{\infty},$$

since $S = S^*$ in B.

In this chapter, we study "non-self-adjoint" free random variables $T \in (B, \psi)$, having their free distributions satisfying

$$\psi\left(\prod_{l=1}^{n} T^{r_l}\right) = \omega_n c_{\frac{n}{2}},$$

for all $(r_1, ..., r_n) \in \{1, *\}^n$, where

$$\omega_n = \begin{cases} 1 & \text{if } n \text{ is even} \\ 0 & \text{if } n \text{ is odd,} \end{cases}$$

for all $n \in \mathbb{N}$, and c_k are the k-th Catalan numbers for all $k \in \mathbb{N}$. In particular, we study a unital C^*-probability space generated by countable-infinitely many such free random variables, and a dynamical system of the infinite cyclic abelian discrete group $(\mathbb{Z}, +)$ acting on this C^*-probability space. It is characterized how this dynamical system affects the original free distributions.

25.1.1 Motivations

In classical and free probability theory, and related applied fields including quantum statistical physics, *the semicircular law* is a major topic (e.g., [1, 2, 4–7, 8, 10–12, 20, 21, 29, 30]). Semicircular elements, free random variables whose free distributions are the semicircular law, have been studied in free probability theory as one of the most important objects. They are well characterized both analytically and combinatorially (e.g., [1, 17, 18, 21, 28, 29, 30]). They play a key role in free probability because their free distributions, the semicircular law is a noncommutative counterpart of the Gaussian (or, the normal) distribution of commutative function theory by the *(free) central limit theorem(s)* (e.g., see [2, 17, 19, 28, 29, 30]). Interestingly, the semicircular law is fully characterized by the Catalan numbers $\{c_k\}_{k=1}^{\infty}$ (e.g., [9, 17, 22–24]).

Recently, we constructed semicircular elements from p-adic numbers $\mathbb{Q}_p$ over primes p in [5, 12], illustrating relations among number theory, operator algebra and quantum statistical physics (e.g., [26, 27]) under free probability. By applying the constructions of [5, 12], semicircular elements are generated in [13, 14] from $|\mathbb{Z}|$-many *orthogonal projections* in a C^*-algebra, and the operator-theoretic, and free-probabilistic properties of these semicircular elements have been studied in [6–8, 10, 11].

25.1.2 Overview

In Sections 25.2 and 25.3, we consider free-probabilistic concepts and tools used in the text. In particular, we construct and study a unital C^*-probability space generated by mutually free, countable-infinitely many semicircular elements. It is shown that such C^*-probability spaces are free-isomorphic (i.e., $*$-isomorphic preserving free probability) to a unital C^*-probability space $\mathfrak{X}_{\varphi}$ generated by mutually free, $|\mathbb{Z}|$-many semicircular elements $X = \{x_j\}_{j \in \mathbb{Z}}$. The free-distributional data on $\mathfrak{X}_{\varphi}$ are characterized, and certain free-isomorphisms ($*$-isomorphisms preserving free probability) on $\mathfrak{X}_{\varphi}$ are considered naturally.

In Section 25.4, free random variables followed by the semicircular law are constructed-and-studied from the semicircular elements of Section 25.4, and the free-isomorphisms of Section 25.5. In particular, the free probability on a unital C^*-probability space $\mathscr{X}_{\tau}$, generated by $|\mathbb{Z}|$-many free random variables $\mathcal{X} = \{u_{k,j}\}_{k,j \in \mathbb{Z}}$ followed by semicircular law, is considered. And we show that free-distributional data on $\mathscr{X}_{\tau}$ are determined by those on $\mathfrak{X}_{\varphi}$ in a certain sense.

In Section 25.5, we study the dynamics of the infinite cyclic abelian group $\mathbb{Z}$, acting on $\mathscr{X}_{\tau}$. A corresponding group-dynamical system, $\Gamma = (\mathbb{Z}, \mathscr{X}_{\tau}, \alpha)$, is established. This dynamical system Γ induces the crossed product C^*-algebra $\mathscr{X}[\Gamma]$, and a suitable free-probabilistic structure. The free probability on this C^*-probability space $\mathscr{X}[\Gamma]$ is investigated under dynamics. In the long run, we show that the free probability on $\mathscr{X}_{\tau}$ is deformed by our Γ-dynamics (in $\mathscr{X}[\Gamma]$); however, such deformations are manageable or characterizable.

In Section 25.6, as applications of Sections 25.6 and 25.7, certain free random variables of $\mathscr{X}_{\tau}$, and those of $\mathscr{X}[\Gamma]$, whose free distributions are followed by the circular law, and, by some free Poisson distributions, in certain manners.

25.2 PRELIMINARIES

For basic free probability theory, e.g., see [3, 15, 16, 17, 28, 29, 30]. Throughout this section, we let (A, φ) be an arbitrary *unital topological $*$-probability space.*

Definition 25.1 *A free random variable $x \in (A, \varphi)$ is said to be semicircular, if it is self-adjoint in A, and*

$$\varphi(x^n) = \omega_n c_{\frac{n}{2}}, \; \forall n \in \mathbb{N}, \tag{25.1}$$

where $\omega_n = 1$, if n is even, while $\omega_n = 0$, if n is odd, and c_k are the k-th Catalan numbers for all $k \in \mathbb{N}$.

By the Möbius inversion, a free random variable x is semicircular in (A, φ), if and only if if is self-adjoint in A, and

$$k_n(x, \ldots, x) = \delta_{n,2} \tag{25.2}$$

for all $n \in \mathbb{N}$, where δ is the Kronecker delta.

By (25.1) and (25.2), *the semicircular law* is characterized by the free moment sequence,

$$(0,\ c_1,\ 0,\ c_2,\ 0,\ c_3,\ 0,\ c_4,\ ...), \tag{25.3}$$

equivalently, by the free-cumulant sequence,

$$(0,\ 1,\ 0,\ 0,\ 0,\ 0,\ ...). \tag{25.4}$$

So, the free distributions of "all" semicircular elements are said to be "the" semicircular law, by the universality (25.3), or (25.4), of the semicircular law.

Suppose $x_1,\ ...,\ x_N$ are N-many mutually distinct semicircular elements of $(A,\ \varphi)$, for $N \in \mathbb{N}$, and assume that they are mutually free in $(A,\ \varphi)$, in the sense that all mixed free cumulants of them vanish (e.g., [17]). By the self-adjointness of $x_1,\ ...,\ x_N$ in A, the (joint) free distribution of $x_1, ..., x_N$ are characterized by the *joint free moments*,

$$\overset{\infty}{\underset{n=1}{\cup}} \left(\underset{(i_1,...,i_n)\in\{1,...,N\}^n}{\cup} \{\varphi\left(x_{i_1} x_{i_2} ... x_{i_n}\right)\} \right). \tag{25.5}$$

Fix $s \in \mathbb{N}$, and an s-tuple I_s,

$$I_s \overset{denote}{=} (i_1,\ ...,\ i_s) \in \{1,\ ...,\ N\}^s, \tag{25.6}$$

in $\{1, ..., N\}$. From the sequence I_s of (25.6), define a set,

$$[I_s] = \{i_1,\ i_2,\ ...,\ i_s\}, \tag{25.7}$$

ignoring repetition of entries. That is, even though $i_{j_1} = i_{j_2}$ as entries of the sequence I_s of (3.3) for $j_1 \neq j_2 \in \{1, ..., s\}$, regard them as distinct elements of the set $[I_s]$ of (24.7).

Then, from the set $[I_s]$ of (24.7), define a "noncrossing" partition $\pi\,(i_1)$ in the noncrossing-partition lattice $NC\,([I_s])$ as follows; (i) starting from the first entry i_1 of I_s, construct the maximal block U_1 satisfying

$$U_1 = \left(i_1 = i_{j_1}, i_{j_2}, ..., i_{j_{|U_1|}}\right) \in \pi_{(I_s)},$$

with the rule: (25.8)

$$i_1 = i_{j_1} = i_{j_2} = \ ...\ = i_{j_{|U_1|}}\ ,$$

in I_s, (ii) and then, by fixing the very next entry of

$$[I_s] \setminus U_1,$$

construct the second maximal block U_2 of $\pi\,(i_1)$ containing the entry, as in (25.8), and do these processes until end to have the noncrossing partition $\pi\,(i_1)$, and (iii) such a resulted partition $\pi\,(i_1)$ must be "maximal" in $NC([I_s])$, satisfying both (i) and (ii). For example, if

$$I_{10} = (1, 1, 2, 2, 1, 1, 1, 2, 1, 2)$$

and

$$[I_{10}] = \{i_1, i_2, ..., i_{10}\}\,,$$

with

$$i_1 = i_2 = i_5 = i_6 = i_7 = i_9 = 1 \text{ and } i_3 = i_4 = i_8 = i_{10} = 2,$$

then there exists a noncrossing partition,

$$\begin{aligned} \pi(i_1) &= \{(i_1, i_2, i_5, i_6, i_7, i_9),\ (i_3, i_4),\ (i_8),\ (i_{10})\} \\ &= \{(1,1,1,1,1,1),\ (2,2),\ (2),\ (2)\}, \end{aligned}$$

in $NC([I_8])$, satisfying the conditions (i), (ii) and (iii). Remark here that, even though $i_3 = i_4 = i_8 = i_{10} = 2$, one cannot have the block (i_3, i_4, i_8, i_{10}) in $\pi(i_1)$, because it has two crossings with the first block $(i_1, i_2, i_5, i_6, i_7, i_9)$.

Now, similar to the noncrossing partition $\pi(i_1)$ for the first entry i_1 of I_s, construct noncrossing partitions,

$$\pi(i_2), ..., \pi(i_s) \text{ in } NC([I_s]),$$

similarly satisfying the aforementioned conditions (i), (ii) and (iii) by replacing i_1 to i_l, for all $l = 2, ..., s$. By collecting all such partitions, define the subset $\Pi([I_s])$ of $NC([I_s])$ by

$$\Pi([I_s]) = \{\pi(i_l) : l = 1, ..., s\}.$$

Also, define a subset $\Pi_e([I_s])$ of $\Pi([I_s])$ by

$$\Pi_e([I_s]) = \{\theta \in \Pi([I_s]) : |V| \in 2\mathbb{N},\ \forall V \in \theta\}, \tag{25.9}$$

where $2\mathbb{N} = \{2n : n \in \mathbb{N}\}$.

For a given s-tuple I_s of (25.6), if x_{i_1}, ..., x_{i_s} are the corresponding semicircular elements of (A, φ) in $\{x_1, ..., x_N\}$, without considering repetition, then define a free random variable $X[I_s]$ by

$$X[I_s] \overset{def}{=} \prod_{l=1}^{s} x_{i_l} \in (A, \varphi). \tag{25.10}$$

Theorem 25.1 *For an s-tuple I_s of (25.6), if $X[I_s] = \prod_{l=1}^{s} x_{i_l}$ is the free random variable (25.10) of (A, φ), then*

$$\varphi(X[I_s]) = \sum_{\theta \in \Pi_e([I_s])} \varphi_\theta(x_{i_1}, ..., x_{i_s}),$$

with (25.11)

$$\varphi_\theta(x_{i_1}, ..., x_{i_s}) = \prod_{V \in \theta} c_{\frac{|V|}{2}},$$

where c_k are the k-th Catalan numbers.

Proof 25.1 *The formula (25.11) is proven in [9].*

25.3 ON $|\mathbb{N}|$-MANY SEMICIRCULAR ELEMENTS

In this section, we construct and study the C^*-subalgebra $\mathfrak{X}$ generated by mutually free, $|\mathbb{N}|$-many semicircular elements $\{s_n\}_{n=1}^{\infty}$ in a C^*-probability space (A, φ). To do that, let $X = \{s_n\}_{n=1}^{\infty}$ be a family of mutually free, countable-infinitely many semicircular elements s_n's in (A, φ). For convenience, re-index the free semicircular family X to be

$$\{x_n\}_{n=0}^{\infty} = \{x_0, x_1, x_2, ...\}.$$

25.3.1 A C^*-Algebra $\mathfrak{X}$ Generated by X

A unital C^*-probability space (A_1, φ_1) is free-homomorphic to a unital C^*-probability space (A_2, φ_2), if there is a $*$-homomorphism $\Omega : A_1 \to A_2$, such that,

$$\varphi_2(\Omega(a)) = \varphi_1(a), \text{ for all } a \in (A_1, \varphi_1).$$

Such a $*$-homomorphism Ω is called a *free-homomorphism.* We denote this free-homomorphic relation by

$$(A_1, \varphi_1) \overset{\text{free-homo}}{\longrightarrow} (A_2, \varphi_2). \tag{25.12}$$

Definition 25.2 *Assume that* $(A_1, \varphi_1) \overset{\textit{free-homo}}{\longrightarrow} (A_2, \varphi_2)$ *in the sense of (25.12), by a free-homomorphism* Ω. *If* Ω *is a* $*$*-isomorphism, then it is called a free-isomorphism, and* (A_1, φ_1) *is said to be free-isomorphic to* (A_2, φ_2). *We denote this relation is denoted by*

$$(A_1, \varphi_1) \overset{\textit{free-iso}}{=} (A_2, \varphi_2). \tag{25.13}$$

Let (A, φ) be a unital C^*-probability space containing a family $X = \{x_n\}_{n=0}^{\infty}$ of mutually free semicircular elements. Define the C^*-subalgebra $\mathfrak{X}$ of A by the C^*-algebra $C_A^*(X)$ generated by X in A, where $C_A^*(Y)$ is the C^*-subalgebra of A generated by $\{y, y^* : y \in Y\}$. Then one can construct a C^*-probabilistic sub-structure,

$$\mathfrak{X}_\varphi \overset{denote}{=} (\mathfrak{X}, \varphi = \varphi\mid_{\mathfrak{X}}) \tag{25.14}$$

in (A, φ).

Independently, let (B, ψ) be a unital C^*-probability space, containing a family $S = \{y_n\}_{n\in\mathbb{Z}}$ of mutually free, $|\mathbb{Z}|$-many semicircular elements, and let

$$\mathfrak{S}_\psi \overset{denote}{=} (C_B^*(S), \psi = \psi\mid_{\mathfrak{S}}) \tag{25.15}$$

be the C^*-probabilistic sub-structure of (B, ψ), as in (25.14). Note that such a C^*-probability space (25.14) does exist canonically, or artificially (e.g., [5, 6–8, 10, 20, 21]).

Proposition 25.1 *If* $\mathfrak{X} = C^*(X)$ *be a* C^**-subalgebra (25.14), then*

$$\mathfrak{X} \overset{*\text{-}iso}{=} \underset{n=0}{\overset{\infty}{\star}} (C_A^*(\{x_n\})) \overset{*\text{-}iso}{=} C_A^*\left(\underset{n=0}{\overset{\infty}{\star}}\{x_n\}\right), \tag{25.16}$$

in (A, φ), where $(\star)$ in the first $*$-isomorphic relation of (25.16) is the free-probabilistic free product, and the $(\star)$ in the second $*$-isomorphic relation of (25.16) is the pure-algebraic free product inducing the noncommutative free words in X.

Proof 25.2 *The freeness on the family X guarantees the relation (25.16), by (25.14).*

By (25.14) and (25.16), one has

$$\mathfrak{X}_{\varphi} = \left(\underset{n \in \mathbb{N}_0}{\star} C_A^* (\{x_n\}), \ \underset{n \in \mathbb{N}_0}{\star} \varphi \mid_{C_A^*(\{x_n\})} \right). \tag{25.17}$$

In a similar manner, the C^*-probability space $\mathfrak{S}_{\psi}$ of (25.15) satisfies

$$\mathfrak{S}_{\psi} = \left(\underset{j \in \mathbb{Z}}{\star} C_B^* (\{s_j\}), \ \underset{j \in \mathbb{Z}}{\star} \psi \mid_{C_B^*(\{s_j\})} \right). \tag{25.18}$$

Note here that

$$C_A^* (\{x_n\}) \overset{*\text{-iso}}{=} \overline{\mathbb{C}[\{x_n\}]}^A \overset{*\text{-iso}}{=} \overline{\mathbb{C}[\{y_j\}]}^B \overset{*\text{-iso}}{=} C_B^* (\{y_j\}),$$

for all $n \in \mathbb{N}$, and $j \in \mathbb{Z}$, where $\overline{Y}^A$ and $\overline{Z}^B$ are C^*-topology closures of subsets $Y \subseteq A$ and $Z \subseteq B$, respectively.

Now, let

$$\mathbb{N}_0 = \{0\} \sqcup (2\mathbb{N}) \sqcup (2\mathbb{N} - 1),$$

and (25.19)

$$\mathbb{Z} = (-\mathbb{N}) \sqcup \{0\} \sqcup \mathbb{N},$$

where $\sqcup$ is the disjoint union, and

$$2\mathbb{N} = \{2n : n \in \mathbb{N}\}, \ 2\mathbb{N} - 1 = \{2n - 1 : n \in \mathbb{N}\},$$

and

$$-\mathbb{N} = \{-n : n \in \mathbb{N}\}.$$

Then, one can define a bijection $g : \mathbb{N}_0 \to \mathbb{Z}$ by

$$g(n) = \begin{cases} 0 & \text{if } n = 0 \\ \frac{n+1}{2} & \text{if } n \in 2\mathbb{N} - 1 \\ -\frac{n}{2} & \text{if } n \in 2\mathbb{N}, \end{cases} \tag{25.20}$$

in $\mathbb{Z}$, for all $n \in \mathbb{N}_0$, by (25.19). This bijection g of (25.20) induces a bijection,

$$G : X \to S,$$

by (25.21)

$$G(x_n) = y_{g(n)}, \text{ for all } n \in \mathbb{N}_0.$$

where X is the generator set (25.14) of $\mathfrak{X}_\varphi$, S is the generator set (25.15) of $\mathfrak{S}_\psi$. Therefore, one can define the "multiplicative" linear transformation,

$$\Psi : \mathfrak{X} \to \mathfrak{S}$$

satisfying (25.22)

$$\Psi(x_n) = G(x_n) = s_{g(n)} \in S, \forall x_n \in X,$$

in $\mathfrak{S}$, where G is the bijection (25.21). More precisely, for any alternating N-tuple $(n_1, ..., n_N) \in \mathbb{N}_0^N$, with

$$n_1 \neq n_2, n_2 \neq n_3, ..., n_{N-1} \neq n_N \text{ in } \mathbb{N}_0,$$

if $T = \prod_{l=1}^{N} x_{n_l}^{k_l} \in \mathfrak{X}_\varphi$, where $x_{n_1}, ..., x_{n_N} \in X$, for $k_1, ..., k_N \in \mathbb{N}$, for $N \in \mathbb{N}$, then

$$\Psi(T) = \Psi\left(\prod_{l=1}^{N} x_{n_l}^{k_l}\right) = \prod_{l=1}^{N} \Psi\left(x_{n_l}^{k_l}\right)$$

by the multiplicativity of Ψ

$$= \prod_{l=1}^{N} \left(\Psi(x_{n_l})\right)^{k_l}$$

by the multiplicativity of Ψ

$$= \prod_{l=1}^{N} s_{g(n_l)}^{k_l},$$

in $\mathfrak{S}_\psi$. Since $(n_1, ..., n_N) \in \mathbb{N}_0^N$ is assumed to be alternating in $\mathbb{N}_0$, the N-tuple,

$$(g(n_1), \; ..., \; g(n_N)) \in \mathbb{Z}^N,$$

is an alternating N-tuple in $\mathbb{Z}$, too, by (25.20) and (25.21). It means that this bijective morphism Ψ assigns free reduced words of $\mathfrak{X}$ to those of $\mathfrak{X}$, preserving their lengths, by (25.22).

Theorem 25.2 *If $\mathfrak{X}_\varphi$ and $\mathfrak{S}_\psi$ are the unital C^*-probability spaces of (25.14), respectively, (25.15), then*

$$\mathfrak{X}_\varphi \overset{\text{free-iso}}{=} \mathfrak{S}_\psi. \tag{25.23}$$

Proof 25.3 *By the bijectivity, the multiplicative linear transformation Ψ of (25.22) is a $*$-isomorphism from $\mathfrak{X}$ onto $\mathfrak{S}$. Remark that, if $x_n \in X \subset \mathfrak{X}_\varphi$, then*

$$\psi\left((\Psi(x_n))^k\right) = \psi\left(s_{g(n)}^k\right) = \omega_k c_{\frac{k}{2}} = \varphi\left(x_n^k\right),$$

for all $k \in \mathbb{N}$, by (25.2) and (25.3). It shows that Ψ preserves the free probability on $\mathfrak{X}_\varphi$ to that on $\mathfrak{S}_\psi$ by (25.11), equivalently, it is a free-isomorphism. Therefore, the free-isomorphic relation (25.23) holds.

From below, we identify $\mathfrak{X}_\varphi$ and $\mathfrak{S}_\psi$ as the same unital C^*-probability space, denote by $\mathfrak{X}_\varphi$, by (25.23). i.e., from below, we let $\mathfrak{X}_\varphi = (\mathfrak{X}, \varphi)$ be an independent unital C^*-probability space generated by the free semicircular family $X = \{x_j\}_{j \in \mathbb{Z}}$. Note that free-probabilistic information on $\mathfrak{X}_\varphi$ are characterized by the free-distributional data (25.11).

25.3.2 Certain Free-Isomorphisms on $\mathfrak{X}_\varphi$

By (25.16), (25.17), (25.18) and (25.23), if $(\mathfrak{A}, \tau)$ be a unital C^*-probability space generated by mutually free, $|\mathbb{N}|$-many semicircular elements, then it is free-isomorphic to the C^*-probability space $\mathfrak{X}_\varphi$ in the sense of (25.18). Therefore, the C^*-probability space $\mathfrak{X}_\varphi$ becomes a representative of all unital C^*-probability spaces generated by mutually free, $|\mathbb{N}|$-many, or $|\mathbb{Z}|$-many semicircular elements. So, as we emphasized at the end of Section 25.3.1, we let $\mathfrak{X}_\varphi$ be the C^*-probability space (25.18) generated by the free semicircular family $X = \{x_j\}_{j \in \mathbb{Z}}$.

Let h be a bijection on the set $\mathbb{Z}$ of all integers by

$$h(j) = j + 1,$$

with its inverse function, (25.24)

$$h^{-1}(j) = j - 1,$$

for all $j \in \mathbb{Z}$. From the bijection h of (25.24), construct the bijections $h^{(n)}$ on $\mathbb{Z}$, by

$$h^{(n)} \overset{def}{=} \begin{cases} id_{\mathbb{Z}}, \text{ the identity function on } \mathbb{Z} & \text{if } n = 0 \\ \underbrace{h \circ h \circ h \circ \ldots\ldots\ldots\ldots \circ h}_{n\text{-times}} & \text{if } n > 0 \\ \underbrace{h^{-1} \circ h^{-1} \circ \ldots \circ h^{-1}}_{|n|\text{-times}} & \text{if } n < 0, \end{cases} \tag{25.25}$$

for all $n \in \mathbb{Z}$, where $(\circ)$ is the functional composition. By (25.25),

$$h^{(n)}(j) = j + n,$$

satisfying $\left(h^{(n)}\right)^{-1} = h^{(-n)}$, for all $j, n \in \mathbb{Z}$.

Definition 25.3 *We call the bijections $h^{(n)}$ of (25.25), the n-th shifts on $\mathbb{Z}$, for all $n \in \mathbb{Z}$.*

For $k \in \mathbb{Z}$, define a "multiplicative" linear transformation λ^k acting on $\mathfrak{X}_\varphi$ by a map satisfying

$$\lambda^k(T) \overset{\text{def}}{=} \prod_{l=1}^{N} \lambda^k (x_{j_l})^{n_l} = \prod_{l=1}^{N} x_{j_l+k}^{n_l}, \tag{25.26}$$

for any free reduced word $T = \prod_{l=1}^{N} x_{j_l}^{n_l} \in \mathfrak{X}_\varphi$ with its length-N in the generator set X, for all alternating N-tuple $(j_1, \ldots, j_N) \in \mathbb{Z}^N$, and $n_1, \ldots, n_N \in \mathbb{N}$.

Note that, for any alternating N-tuple $(j_1, ..., j_N) \in \mathbb{Z}^N$, a new N-tuple,

$$(j_1 + k, \ ..., \ j_N + k) \in \mathbb{Z}^N,$$

is alternating in $\mathbb{Z}$, too. So, the definition (25.26) illustrates that the morphism λ^k maps free reduced words to free reduced words in $\mathfrak{X}_\varphi$, preserving length. Clearly, by (25.25) and (25.26), one has

$$\lambda^k = \begin{cases} 1_{\mathfrak{X}_\varphi}, \text{ the identity map on } \mathfrak{X}_\varphi & \text{if } k = 0 \\ \underbrace{\lambda \cdot \lambda \cdot \lambda \cdot \ldots\ldots \cdot \lambda}_{k\text{-times}} & \text{if } k > 0 \\ \underbrace{\lambda^{-1} \cdot \lambda^{-1} \cdot \ldots\ldots \cdot \lambda^{-1}}_{|k|\text{-times}} & \text{if } k < 0, \end{cases}$$

for all $k \in \mathbb{Z}$, where $(\cdot)$ is the multiplication of linear transformations. So, for any $t \in \mathbb{C}$, and $x_j \in X \subset \mathfrak{X}$,

$$\lambda^k \left((tx_j)^* \right) = \bar{t} \ x_{j+k}^* = \left(\lambda^k (tx_j) \right)^*,$$

implying that

$$\lambda^k \left(T^* \right) = \left(\lambda^k (T) \right)^*, \text{ for all } T \in \mathfrak{X}_\varphi, \tag{25.27}$$

in $\mathfrak{X}_\varphi$.

Theorem 25.3 *A morphism λ^k of (25.26) is a free-isomorphism on $\mathfrak{X}_\varphi$, for all $k \in \mathbb{Z}$.*

Proof 25.4 *By the definition (25.26) and the adjoint-preserving property (25.27), each morphism λ^k is a $*$-homomorphism on $\mathfrak{X}_\varphi$, for all $k \in \mathbb{Z}$. By the bijectivity of $h^{(k)}$ on $\mathbb{Z}$, the restriction $\lambda^{(k)} \mid_X$ is a bijection on the generating free family X of $\mathfrak{X}_\varphi$ by (25.24), i.e., is a $*$-isomorphism on $\mathfrak{X}_\varphi$, for all $k \in \mathbb{N}$. Observe that*

$$\varphi \left(\left(\lambda^k (x_j) \right)^n \right) = \varphi \left(x_{j+k}^n \right) = \omega_n c_{\frac{n}{2}} = \varphi \left(x_j^n \right),$$

for all $n \in \mathbb{N}$, for all $x_j \in X$, implying that

$$\varphi \left(X[I_s] \right) = \varphi \left(\lambda^k \left(X[I_s] \right) \right) \ in \ \mathfrak{X}_\varphi,$$

for all s-tuples $I_s \in \mathbb{Z}^s$, for all $s \in \mathbb{N}$, where $X[I_s]$ are in the sense of (25.10). Therefore,

$$\varphi (T) = \varphi \left(\lambda^k (T) \right), \ for \ all \ T \in \mathfrak{X}_\varphi,$$

in $\mathfrak{X}_\varphi$, for all $k \in \mathbb{N}$. Therefore, $\left\{ \lambda^k \right\}_{k \in \mathbb{Z}}$ are free-isomorphisms on $\mathfrak{X}_\varphi$.

Every C^*-algebra B induces its automorphism group $Aut\,(B)$ consisting of all $*$-isomorphisms on B (or, $*$-automorphisms) equipped with the product (or the composition) on the $*$-isomorphisms. Define a subset λ of $Aut\,(\mathfrak{X}_\varphi)$ by

$$\lambda = \{ \lambda^k : k \in \mathbb{Z} \}, \tag{25.28}$$

where λ^k are the free-isomorphisms (25.26) on $\mathfrak{X}_\varphi$.

Theorem 25.4 *The family λ of (25.28) forms an commutative subgroup of $Aut(\mathfrak{X}_\varphi)$, and*

$$\lambda \overset{\text{Group}}{=} (\mathbb{Z}, +), \text{ the infinite abelian cyclic group,} \tag{25.29}$$

where "$\overset{\text{Group}}{=}$" means "being group-isomorphic."

Proof 25.5 *By the definitions (25.26) and (25.28), the $*$-isomorphism product $(\cdot)$, inherited from that on $Aut(\mathfrak{X}_\varphi)$ is not only closed on the set λ but also associative and commutative, since*

$$\lambda^{k_1}\lambda^{k_2} = \lambda^{k_1+k_2}, \text{ in } \lambda,$$

for all $k_1, k_2 \in \mathbb{Z}$. Moreover, the identity $\lambda^0 = id_{\mathfrak{X}}$ exists in λ. So, the family λ of (25.28) forms an abelian group. Furthermore, one can define a bijection,

$$\lambda^k \in \lambda \longmapsto k \in \mathbb{Z},$$

which becomes a group-isomorphism from $(\lambda, \cdot)$ to $(\mathbb{Z}, +)$, implying the group-isomorphic relation (25.29).

Definition 25.4 *We call the group λ of (25.29), the integer-shift group on $\mathfrak{X}_\varphi$.*

By (25.26) and (25.29), there does exist a group-action θ of λ acting on $\mathfrak{X}_\varphi$,

$$\theta\left(\lambda^k\right)(T) = \lambda^k(T), \text{ for all } T \in \mathfrak{X}_\varphi, \tag{25.30}$$

for all $k \in \mathbb{Z}$. By construction, this group-action θ of (25.30) preserves the free probability on $\mathfrak{X}_\varphi$.

Theorem 25.5 *If θ is the group-action (25.30) of the integer-shift group λ acting on $\mathfrak{X}_\varphi$, then*

$$\varphi\left(\theta\left(\lambda^k\right)(T)\right) = \varphi(T), \text{ for all } T \in \mathfrak{X}_\varphi, \tag{25.31}$$

for all $\lambda^k \in \lambda$.

Proof 25.6 *By (25.30), for any $T \in \mathfrak{X}_\varphi$,*

$$\theta\left(\lambda^k\right)(T) = \lambda^k(T), \text{ in } \mathfrak{X}_\varphi, \forall \lambda^k \in \lambda.$$

Since all elements of the integer-shift group λ are free-isomorphisms, the relation (25.31) holds.

If there are no confusions, then the images $\theta\left(\lambda^k\right) \in Aut(\mathfrak{X}_\varphi)$ are simply denoted by λ^k, for $k \in \mathbb{Z}$, from below.

25.4 FREE RANDOM VARIABLES FOLLOWED BY THE SEMICIRCULAR LAW

Let $\lambda \subset Aut(\mathfrak{X}_\varphi)$ be the integer-shift group (25.28) acting on the C^*-probability space $\mathfrak{X}_\varphi$ generated by the free semicircular family $\{x_j\}_{j \in \mathbb{Z}}$.

Definition 25.5 *Let (B, ψ) be a topological $*$-probability space. A free random variable $y \in (B, \psi)$ is followed by the semicircular law, if*

$$\psi\left(\prod_{l=1}^{n} y^{r_l}\right) = \omega_n c_{\frac{n}{2}},$$

for all $(r_1, ..., r_n) \in \{1, *\}^n$, for all $n \in \mathbb{N}$, where $\omega_n = 1$, if n is even, meanwhile, $\omega_n = 0$, if n is odd, and where c_k is the k-th Catalan number, for all $k \in \mathbb{N}_0$.

By the definition, all semicircular elements are free random variables followed by the semicircular law, but not all free random variables followed by the semicircular law are semicircular. In particular, if such an element is not self-adjoint, then it is not semicircular.

25.4.1 The C^*-Algebra Λ Generated by the Integer-Shift Group λ

For a discrete group Γ, define the group-Hilbert space H by

$$H = l^2(\Gamma), \text{ the } l^2\text{-space},$$

with its orthonormal basis, (25.32)

$$\mathcal{B} = \{\xi_g : g \in \Gamma \setminus \{e\}\},$$

where $e \in \Gamma$ is the group-identity, inducing $\xi_e = 1_H$, the identity vector of H, satisfying the orthonormality,

$$\langle \xi_{g_1}, \xi_{g_2} \rangle_2 = \delta_{g_1, g_2},$$

and (25.33)

$$\|\xi_g\|_2 = \sqrt{\langle \xi_g, \xi_g \rangle_2} = 1,$$

for all $g_1, g_2, g \in \Gamma$, where $\langle , \rangle_2$ is the usual l^2-inner product inducing the l^2-norm $\|.\|_2$ on H. By (25.32), every vector $\xi \in H$ has the form,

$$\xi = \sum_{g \in \Gamma} t_g \xi_g, \text{ for } t_g \in \mathbb{C},$$

where $\sum$ is the infinite sum under l^2-norm topology of (25.33). Note that

$$\xi_{g_1} \xi_{g_2} = \xi_{g_1 g_2} \text{ in } H, \ \forall g_1, g_2 \in \Gamma. \tag{25.34}$$

By (25.34), each group-element $g \in \Gamma$ is regarded as a multiplication operator $m_g \in B(H)$,

$$m_g\left(\sum_{u \in \Gamma} t_u \xi_u\right) = \sum_{u \in \Gamma} t_u \xi_g \xi_u = \sum_{u \in \Gamma} t_u \xi_{gu}, \tag{25.35}$$

satisfying $m_g^* = m_{g^{-1}}$ in $B(H)$.

For a subset $\mathcal{M}$ of $B(H)$,

$$\mathcal{M} \overset{\text{def}}{=} \{m_g \in B(H) : m_g \text{ is in the sense of (25.35)}\},$$

define the group (C^*-)algebra $\mathscr{M}$ of Γ by

$$\mathscr{M} \overset{\text{def}}{=} C^*_{B(H)}(\mathcal{M}) \text{ of } B(H). \tag{25.36}$$

By (25.36), for our integer-shift group λ, one can obtain the corresponding group algebra Λ.

Definition 25.6 *We call the group algebra Λ of the integer-shift group λ, the integer-shift algebra, and all elements of Λ are called (integer-)shift operators.*

Every shift operator T is expressed by

$$T = \sum_{k\in\mathbb{Z}} t_k m_{\lambda^k}, \text{ with } t_{\lambda^k} = t_k, \text{ in } \Lambda.$$

By the aforementioned definition and (25.30), the integer-shift algebra Λ naturally acts on the C^*-probability space $\mathfrak{X}_\varphi$, via an action,

$$\Theta : \Lambda \to B(\mathfrak{X}_\varphi),$$

defined by (25.37)

$$\Theta\left(\sum_{k\in\mathbb{Z}} t_k m_{\lambda^k}\right)(S) = \sum_{k\in\mathbb{Z}} t_k \lambda^k(S),$$

for all $S \in \mathfrak{X}_\varphi$, where $B(\mathfrak{X}_\varphi)$ is the operator space (e.g., [13]) of all bounded linear transformations on $\mathfrak{X}_\varphi$, by understanding $\mathfrak{X}_\varphi$ as a Banach space with its C^*-norm. Note that the morphism Θ of (25.37) is a $*$-algebra-action satisfying

$$\Theta(S_1 S_2) = \Theta(S_1)\Theta(S_2), \text{ and } \Theta(X^*) = \Theta(S)^*,$$

by (25.30) and (25.37), for all $S_1, S_2, S \in \mathfrak{X}_\varphi$.

Proposition 25.2 *If $T = \sum_{k\in\mathbb{Z}} t_k m_{\lambda^k} \in \Lambda$ is a shift operator, then, for any generating element $x_j \in X$ of $\mathfrak{X}_\varphi$,*

$$\varphi\left(\Theta(T)\left(x_j^n\right)\right) = \left(\omega_n c_{\frac{n}{2}}\right)\left(\sum_{k\in\mathbb{Z}} t_k\right), \forall n \in \mathbb{N}. \tag{25.38}$$

Proof 25.7 *For any $n \in \mathbb{N}$, and $x_j \in X \subset \mathfrak{X}_\varphi$,*

$$\begin{aligned}\varphi\left(\Theta\left(\sum_{k\in\mathbb{Z}} t_k m_{\lambda^k}\right)\left(x_j^n\right)\right) &= \varphi\left(\sum_{k\in\mathbb{Z}} t_k \lambda^k\left(x_j^n\right)\right) \\ &= \sum_{k\in\mathbb{Z}} t_k \varphi\left(x_{j+k}^n\right) = \sum_{k\in\mathbb{Z}} t_k \varphi\left(x_j^n\right) \\ &= \left(\omega_n c_{\frac{n}{2}}\right)\left(\sum_{k\in\mathbb{Z}} t_k\right).\end{aligned}$$

Therefore, the free-distributional data (25.38) holds.

25.4.2 On the Tensor Product $\Lambda \otimes \mathfrak{X}$

Let $\mathscr{X}$ be the tensor product C^*-algebra,

$$\mathscr{X} \overset{\text{def}}{=} \Lambda \otimes \mathfrak{X}, \tag{25.39}$$

of the integer-shift algebra Λ, and the C^*-algebra $\mathfrak{X}$ generated by the free semicircular family $X = \{x_j\}_{j \in \mathbb{Z}}$, where $\otimes$ is the tensor product of C^*-algebras. Define a linear functional τ on $\mathscr{X}$ by a linear map satisfying

$$\tau(S \otimes T) \overset{\text{def}}{=} \varphi(S(T)), \tag{25.40}$$

for all $S \otimes T \in \mathscr{X}$, with $S \in \Lambda$ and $T \in \mathfrak{X}$.

Definition 25.7 *The C^*-probability space $\mathscr{X}_\tau \overset{\text{denote}}{=} (\mathscr{X}, \tau)$ is called the (integer-)shift-semicircular C^*-probability space, where $\mathscr{X}$ and τ are in the sense of (25.39) and (25.40), respectively.*

Observe that the unity $I = m_{\lambda^0} \otimes 1_{\mathfrak{X}}$ of $\mathscr{X}_\tau$ satisfies

$$\tau(I) = \varphi\left(\lambda^0 (1_{\mathfrak{X}})\right) = \varphi(1_{\mathfrak{X}}) = 1,$$

and hence, the shift-semicircular C^*-probability space is unital. By (25.39), $\mathscr{X}_\tau$ is generated by the elements formed by

$$u_{k,j} \overset{\text{denote}}{=} \lambda^k \otimes x_j \in \mathscr{X}_\tau, \text{ for } k, j \in \mathbb{Z}. \tag{25.41}$$

i.e., the subset of all elements of (25.41),

$$\mathcal{X} = \{u_{k,j} \in \mathscr{X}_\tau : k, j \in \mathbb{Z}\},$$

generates $\mathscr{X}_\tau$. Consider that

$$\begin{aligned} \tau\left((u_{k,j})^n\right) &= \tau\left(\left(\lambda^k \otimes x_j\right)^n\right) = \tau\left(\lambda^{kn} \otimes x_j^n\right) \\ &= \varphi\left(\lambda^{kn}\left(x_j^n\right)\right) = \varphi\left(x_{j+kn}^n\right) = \varphi\left(x_j^n\right) = \omega_n c_{\frac{n}{2}}, \end{aligned} \tag{25.42}$$

on $\mathfrak{X}_\varphi$, by (25.38) and (25.40). Also, consider that, for a generating element $u_{k,j} \in \mathcal{X}$ of $\mathscr{X}_\tau$,

$$(u_{k,j})^* = \left(\lambda^k\right)^* \otimes x_j^* = \lambda^{-k} \otimes x_j = u_{-k,j}, \text{ in } \mathscr{X}_\tau. \tag{25.43}$$

It illustrates that if $k \neq 0$, then the generating operators $u_{k,j}$ are not self-adjoint in $\mathscr{X}_\tau$, for all $j \in \mathbb{Z}$. However, one has that

$$\tau\left(\left(u_{k,j}^*\right)^n\right) = \tau\left(u_{-k,j}^n\right) = \omega_n c_{\frac{n}{2}} = \varphi\left(x_j^n\right), \tag{25.44}$$

by (25.43), for all $k, j \in \mathbb{Z}$, and $n \in \mathbb{N}$.

Now, let $k, j \in \mathbb{Z}$, and $u_{k,j} \in \mathcal{X}$, a generating element of $\mathscr{X}_\tau$. If $(r_1, ..., r_n) \in \{1, *\}^n$ is a n-tuple of $\{1, *\}$, for $n \in \mathbb{N}_{>1} = \mathbb{N} \setminus \{1\}$, for some $m \in \{1, ..., n\}$, then

$$\prod_{l=1}^{n} (u_{k,j})^{r_l} = \lambda^{\#(1)k - \#(*)k} \otimes x_j^n,$$

where (25.45)

$$\#(1) = \text{the number of 1's in } (r_1, ..., r_n),$$

and

$$\#(*) = \text{the number of } * \text{'s in} (r_1, ..., r_n),$$

in $\mathscr{X}$. So, by (25.40) and (25.45),

$$\tau\left(\lambda^{(\#(1)-\#(*))k} \otimes x_j^n\right) = \varphi\left(\lambda^{(\#(1)-\#(*))k}\left(x_j^n\right)\right). \tag{25.46}$$

Theorem 25.6 *Every generating free random variable* $u_{k,j} \in \mathcal{X}$ *of the shift-semicircular* C^**-probability space* $\mathscr{X}_\tau$ *is followed by the semicircular law. i.e.,*

$$\tau\left(\prod_{l=1}^{n} (u_{k,j})^{r_l}\right) = \omega_n c_{\frac{n}{2}} = \varphi\left(x_j^n\right), \tag{25.47}$$

for all $(r_1, ..., r_n) \in \{1, *\}^n$*, for all* $n \in \mathbb{N}$*.*

Proof 25.8 *By (25.42), (25.44) and (25.46), we have*

$$\tau\left(\prod_{l=1}^{n} (u_{k,j})^{r_l}\right) = \varphi\left(\lambda^{(\#(1)-\#(*))k}\left(x_j^n\right)\right)$$

$$= \varphi\left(x_{j+(\#(1)-\#(*))k}^n\right) = \varphi\left(x_j^n\right) = \omega_n c_{\frac{n}{2}},$$

for all $(r_1, ..., r_n) \in \{1, *\}^n$*, for all* $n \in \mathbb{N}$*.*

The aforementioned theorem guarantees the existence of free random variables followed by the semicircular law.

Theorem 25.7 *If* (B, ψ) *is a unital* C^**-probability space generated by mutually free, semicircular elements* $\{y_1, ..., y_N\}$*, for* $N \in \mathbb{N}^\infty = \mathbb{N} \cup \{\infty\}$*, then there is a* C^**-probability space* $(\mathscr{B}, \tau)$ *and a free random variable* $y \in (\mathscr{B}, \tau)$*, such that* y *is followed by the semicircular law.*

Proof 25.9 *If a unital* C^**-probability space* (B, ψ) *is generated by mutually free* $|\mathbb{N}|$*-many semicircular elements,*

$$\{y_1, y_2, y_3, ...\},$$

in $\mathbb{N}^\infty$*, then it is free-isomorphic to our* C^**-probability space* $\mathfrak{X}_\varphi = (\mathfrak{X}, \varphi)$*. So, the shift-semicircular* C^**-probability space* $\mathscr{X}_\tau = (\mathscr{X}, \tau)$ *is well-constructed by (25.39) and (25.40), containing its generating free random variables* $u_{k,j}$ *of (25.41) followed by the semicircular law by (25.47).*

Now, if a unital C^**-probability space* (B, ψ) *is generated by mutually free, N-many semicircular elements,*

$$\{y_1, ..., y_N\}, \text{ for } N < \infty,$$

then a unital C^**-probability space* $(\mathcal{B}, \nu)$ *of*

$$\mathcal{B} = \mathop{\star}_{i=1}^{\infty} B_i, \text{ with } B_i = B, \ \forall i \in \mathbb{N},$$

and

$$\nu = \psi^{\star\infty}, \text{ on } \mathcal{B},$$

where $(\star)$ *is the free product of* C^**-algebras (e.g., [1, 30]), generated by the free semicircular family,*

$$Y = \underset{i=1}{\overset{\infty}{\sqcup}} \{y_{i1}, ..., y_{in}\},$$

where $\{y_{i1}, ..., y_{in}\}$, with $y_{i1} = y_1, ..., y_{in} = y_n$ *in each free factor* $B_i = B$, *for all* $i \in \mathbb{N}$. *Under rearrangement, let's denote*

$$Y = \{y_j\}_{j \in \mathbb{N}_0}.$$

Then we have $(\mathcal{B}, \nu) \overset{\text{free-iso}}{=} \mathfrak{X}_\varphi$, *and it induces the* C^**-probability space* $(\mathscr{B}, \tau) \overset{\text{free-iso}}{=} \mathscr{X}_\tau$, *generated by the free random variables followed by the semicircular law.*

The aforementioned theorem demonstrates that there are sufficiently many free random variables followed by the semicircular law.

Theorem 25.8 *Let* $x_{j_1}, ..., x_{j_N} \in X$ *be generating semicircular elements of the* C^**-probability space* $\mathfrak{X}_\varphi$, *and let* $\lambda^{k_1}, ..., \lambda^{k_N} \in \lambda$ *be integer shifts generating the shift algebra* Λ, *where either* $j_1, ..., j_N$, *or* $k_1, ..., k_N$ *are not necessarily distinct in* $\mathbb{Z}$, *inducing the generating free random variables* $u_l = u_{k_l, j_l} \in \mathcal{X}$ *of (25.43) in the shift-semicircular* C^**-probability space* $\mathscr{X}_\tau$, *for* $l = 1, ..., N$, *for* $N \in \mathbb{N}$. *If* $w = \prod_{l=1}^{N} x_{j_l}$ *is a free random variable of* $\mathfrak{X}_\varphi$, *and if* $W_{(r_1,...,r_N)} = \prod_{l=1}^{N} u_l^{r_l}$ *is a free random variable of* $\mathscr{X}_\tau$, *then*

$$\tau\left(W_{(r_1,...,r_N)}\right) = \varphi(w), \tag{25.48}$$

for all $(r_1, ..., r_N) \in \{1, *\}^N$. *In the formula (25.48), the joint free moment* $\varphi(w)$ *is characterized by (25.11).*

Proof 25.10 *Let* $w = \prod_{l=1}^{N} x_{j_l} \in \mathfrak{X}_\varphi$, *satisfying*

$$\varphi(w) = \varpi, \text{ in } \mathbb{C},$$

computed by (25.11). If

$$W_{(r_1,...,r_N)} = \prod_{l=1}^{N} u_l^{r_l} \in \mathscr{X}_\tau,$$

for any $(r_1, ..., r_N) \in \{1, *\}^N$, *then*

$$W_{(r_1,...,r_N)} = \left(\prod_{l=1}^{N} \lambda^{i_l}\right) \otimes \left(\prod_{l=1}^{N} x_{j_l}^{r_l}\right) = \left(\prod_{l=1}^{N} \lambda^{i_l}\right) \otimes w,$$

in $\mathscr{X}_{\tau}$*, with*

$$i_l = \begin{cases} k_l & \text{if } r_l = 1 \\ -k_l & \text{if } r_l = *, \end{cases}$$

in $\mathbb{Z}$*, for all* $l = 1, ..., N$*. Then,*

$$\prod_{l=1}^{N} \lambda^{i_l} = \lambda^{k_{(r_1,...,r_N)}} \in \lambda,$$

in Λ*, for some* $k_{(r_1,...,r_N)} \in \mathbb{Z}$*, and hence,*

$$\tau\left(W_{(r_1,...,r_N)}\right) = \varphi\left(\lambda^{k_{(r_1,...,r_N)}}(w)\right) = \varpi = \varphi(w),$$

since $\lambda^{k_{(r_1,...,r_N)}} \in \lambda$ *is a free-isomorphism on* $\mathfrak{X}_{\varphi}$*. So, the formula (25.48) holds.*

The free-distributional data on our shift-semicircular C^*-probability space $\mathscr{X}_{\tau}$ is characterized by (25.49) with the help of (25.11), since all generating elements $\{u_{k,j}\}_{k,j\in\mathbb{Z}}$ are followed by the semicircular law.

Theorem 25.9 *Let* $u_l = u_{k_l,j_l} \in \mathcal{X}$ *be generating free random variables of* $\mathscr{X}_{\tau}$*, for* $l = 1, 2$*. Then* $j_1 \neq j_2$ *in* $\mathbb{Z}$*, if and only if* u_1 *and* u_2 *are free in* $\mathscr{X}_{\tau}$*.*

Proof 25.11 *Suppose* $j_1 \neq j_2$ *in* $\mathbb{Z}$*, and hence, the generating operators* u_1 *and* u_2 *are distinct in* $\mathcal{X} \subset \mathscr{X}_{\tau}$*. For a "mixed" n-tuple* $(l_1, ..., l_n) \in \{1,2\}^n$*, for* $n \in \mathbb{N}_{>1}$*, we have that*

$$k_n^{\tau}\left(u_{l_1}^{r_1}, ..., u_{l_n}^{r_n}\right) = k_n^{\varphi}\left(x_{j_{l_1}}, ..., x_{j_{l_n}}\right),$$

by (25.47), under the Möbius inversion of [22], for all $(r_1, ..., r_n) \in \{1,*\}^n$*. By the freeness of* x_{j_1} *and* x_{j_2} *in* $\mathfrak{X}_{\varphi}$*,*

$$k_n^{\tau}\left(u_{l_1}^{r_1}, ..., u_{l_n}^{r_n}\right) = 0 = k_n^{\varphi}\left(x_{j_1}, ..., x_{j_m}\right),$$

for all mixed n-tuples $(l_1, ..., l_n) \in \{1,2\}^n$*, for all* $n \in \mathbb{N}_{>1}$*, whenever* $j_1 \neq j_2$*. It shows that* u_1 *and* u_2 *are free in* $\mathscr{X}_{\tau}$*.*

Conversely, if $j_1 = j = j_2$ *in* $\mathbb{Z}$*, and hence,* $u_l = u_{k_l,j} = \lambda^{k_l} \otimes x_j \in \mathcal{X}$ *in* $\mathscr{X}_{\tau}$*, then, for all* $(l_1, ..., l_n) \in \{1,2\}^n$*, for* $n \in \mathbb{N}$*,*

$$k_n^{\tau}\left(u_{l_1}^{r_1}, ..., u_{l_n}^{r_n}\right) = k_n^{\varphi}\left(\underbrace{x_j, x_j, ..., x_j}_{n\text{-times}}\right) = \delta_{n,2},$$

by (25.4) and (25.47), for all $(r_1, ..., r_n) \in \{1,*\}^n$*, showing that*

$$k_n^{\tau}\left(u_{l_1}^{r_1}, u_{l_2}^{r_2}\right) = k_n^{\varphi}\left(x_j, x_j\right) = 1 \neq 0.$$

Thus, if $j_1 = j_2$ *in* $\mathbb{Z}$*, then* u_1 *and* u_2 *are not free in* $\mathscr{X}_{\tau}$*.*

By the aforementioned theorem, the following structure theorem of $\mathscr{X}_\tau$ is obtained.

Corollary 25.1 *The shift-semicircular C^*-probability space $\mathscr{X}_\tau$ satisfies*

$$\mathscr{X}_\tau \overset{*\text{-}iso}{=} \underset{j\in\mathbb{Z}}{\star} \left(C^* \left(\mathcal{X}_j\right)\right),$$

where (25.49)

$$\mathcal{X}_j = \left\{u_{k,j} = \lambda^k \otimes x_j \in \mathcal{X} : k \in \mathbb{Z}\right\}, \ \forall j \in \mathbb{Z},$$

where $C^*(Y)$ is the C^*-subalgebras of $\mathscr{X}_\tau$ generated by a subset Y of $\mathscr{X}_\tau$.

Proof 25.12 *The structure theorem (25.49) is obtained by the aforementioned theorem.*

25.5 A GROUP-DYNAMICAL SYSTEM $(\mathbb{Z}, \mathscr{X}_\tau, \alpha)$

Let $\mathscr{X}_\tau = (\mathscr{X}, \tau)$ be our shift-semicircular C^*-probability space, generated by the free random variables,

$$u_{k,j} \overset{\text{denote}}{=} \lambda^k \otimes x_j \in \mathcal{X}, \text{ for all } k, j \in \mathbb{Z},$$

followed by the semicircular law, where the generator set $\mathcal{X}$ is decomposed by

$$\mathcal{X} = \underset{j\in\mathbb{Z}}{\sqcup} \mathcal{X}_j, \text{ with } \mathcal{X}_j = \left\{u_{k,j} : k \in \mathbb{Z}\right\},$$

where the blocks $\{\mathcal{X}_j\}_{j\in\mathbb{Z}}$ are mutually free from each other in $\mathscr{X}_\tau$, satisfying the structure theorem (25.49).

25.5.1 Dynamics on $(\mathbb{Z}, \mathscr{X}_\tau, \alpha)$

In this section, we study a group-dynamical system $(\mathbb{Z}, \mathscr{X}_\tau, \alpha)$ of the infinite cyclic abelian group $\mathbb{Z} = (\mathbb{Z}, +)$ acting on the shift-semicircular C^*-probability space $\mathscr{X}_\tau$ via a natural group-action α. Recall that, by (25.29), the group $\mathbb{Z}$ is isomorphic to the integer-shift group λ.

For any $n \in \mathbb{Z}$, define a "multiplicative" linear transformation,

$$\alpha(n) \in B\left(\mathscr{X}_\tau\right),$$

by a morphism satisfying (25.50)

$$\alpha(n)\left(T \otimes S\right) = T \otimes \lambda^n\left(S\right), \ \forall n \in \mathbb{Z},$$

for all $T \otimes S \in \mathscr{X}_\tau$, with $T \in \Lambda$ and $S \in \mathfrak{X}_\varphi$, where $\lambda^n \in \lambda \subset \Lambda$, and $B(\mathscr{X}_\tau)$ is the operator space of all bounded linear transformations, or Banach-space operators, on $\mathscr{X}_\tau$, by regarding the C^*-probability space $\mathscr{X}_\tau$ as a Banach space with its C^*-norm (e.g., [13]). By the multiplicativity of the Banach-space operators $\{\alpha(n)\}_{n\in\mathbb{Z}} \subset B(\mathscr{X}_\tau)$ of (25.50), they satisfy that

$$\alpha(n)\left(\prod_{l=1}^{N} u_{k_l,j_l}\right) = \prod_{l=1}^{N} \alpha(n)\left(u_{k_l,j_l}\right) = \prod_{l=1}^{N} u_{k_l,j_l+n}, \tag{25.51}$$

since

$$\alpha(n)\left(u_{k,j}\right) = \alpha(n)\left(\lambda^k \otimes x_j\right) = \lambda^k \otimes \lambda^n(x_j) = \lambda^k \otimes x_{j+n} = u_{k,j+n},$$

for all generating operators $u_{k,j}, u_{k_l,j_l} \in \mathcal{X}$ of $\mathscr{X}_\tau$, for $l = 1, \ldots, N$, for all $N \in \mathbb{N}$. By (25.51), for $n \in \mathbb{Z}$, $t \in \mathbb{C}$, and $u_{k,j} \in \mathcal{X} \subset \mathscr{X}_\tau$,

$$\alpha(n)\left((tu_{k,j})^*\right) = \alpha(n)\left(\bar{t}u_{-k,j}\right) = \bar{t}\alpha(n)\left(u_{-k,j}\right) = \bar{t}u_{-k,j+n} = \left(tu_{k,j+n}\right)^* = \left(\alpha(n)\left(tu_{k,j}\right)\right)^*,$$

by (25.51), implying that

$$\alpha(n)\left(W^*\right) = \left(\alpha(n)\left(W\right)\right)^*, \ \forall W \in \mathscr{X}_\tau, \tag{25.52}$$

for all $n \in \mathbb{Z}$.

Notation. In the following text, we denote $\alpha(n)$ of (25.50) simply by α_n, for all $n \in \mathbb{Z}$. □

Lemma 25.1 *The operators $\{\alpha_n\}_{n\in\mathbb{Z}} \subset B(\mathscr{X}_\tau)$ are $*$-isomorphisms on $\mathscr{X}_\tau$, i.e.,*

$$\{\alpha_n\}_{n\in\mathbb{Z}} \subset Aut\left(\mathscr{X}_\tau\right), \tag{25.53}$$

where $Aut(B)$ is the automorphism group of a C^*-algebra B, consisting of all $*$-isomorphisms on B, equipped with the multiplication (or, the composition) of $*$-isomorphisms.

Proof 25.13 *For any $n \in \mathbb{Z}$, the multiplicative Banach-space operator $\alpha_n = \alpha(n)$ satisfies (25.52), and hence, it is a well-defined $*$-homomorphism. It is not hard to check that it is invertible on $\mathscr{X}_\tau$ with its inverse operator $\alpha_{-n} = \alpha(-n)$. Indeed, for any generating operators $u_{k,j} \in \mathcal{X}$ of $\mathscr{X}_\tau$,*

$$\alpha_n\alpha_{-n}\left(u_{k,j}\right) = \alpha_n\left(u_{k,j-n}\right) = u_{k,(j-n)+n} = u_{k,j},$$

and

$$\alpha_{-n}\alpha_n\left(u_{k,j}\right) = \alpha_{-n}\left(u_{k,j+n}\right) = u_{k,(j+n)-n} = u_{k,j},$$

implying that

$$\alpha_n\alpha_{-n}\left(W\right) = W = \alpha_{-n}\alpha_n\left(W\right), \ \forall W \in \mathscr{X}_\tau,$$

by (25.50) and (25.51), equivalently,

$$\alpha_{-n}\alpha_n = I = \alpha_n\alpha_{-n} \Longleftrightarrow \alpha_n^{-1} = \alpha_{-n},$$

on $\mathscr{X}_\tau$. Thus, the $*$-homomorphism α_n is bijective, i.e., it is a $*$-isomorphism on $\mathscr{X}_\tau$. Therefore, the set-inclusion (25.53) holds.

On the subset $\{\alpha_n\}_n$ of $Aut(\mathscr{X}_\tau)$ of (25.53), one has

$$\alpha_{n_1}\alpha_{n_2} = \alpha_{n_1+n_2} \text{ in } Aut\left(\mathscr{X}_\tau\right),$$

because (25.54)

$$\alpha_{n_1}\alpha_{n_2}\left(u_{k,j}\right) = \alpha_{n_1}\left(u_{k,j+n_2}\right) = u_{k,j+(n_1+n_2)} = \alpha_{n_1+n_2}\left(u_{k,j}\right),$$

for all generating operators $u_{k,j} \in \mathcal{X}$ of $\mathscr{X}_\tau$.

Lemma 25.2 *The system $\{\alpha_n\}_{n\in\mathbb{Z}}$ of (25.54) forms a subgroup of $Aut(\mathscr{X}_\tau)$, isomorphic to $\mathbb{Z} = (\mathbb{Z},+)$. i.e.,*

$$\{\alpha_n\}_{n\in\mathbb{Z}} \overset{Group}{=} \mathbb{Z} \text{ in } Aut\left(\mathscr{X}_\tau\right). \tag{25.55}$$

Proof 25.14 *The subgroup-relation (25.55) is shown by (25.54). Indeed, one can define a group-isomorphism,*

$$\Phi : \{\alpha_n\}_{n\in\mathbb{Z}} \to \mathbb{Z},$$

defined by

$$\Phi\left(\alpha_n\right) = n, \ \forall n \in \mathbb{Z}.$$

By (25.53) and (25.55), one can verify that there is a group-action α of $\mathbb{Z}$,

$$\alpha : \mathbb{Z} \to Aut\left(\mathscr{X}_\tau\right),$$

defined by (25.56)

$$\alpha(n) = \alpha_n, \text{ for all } n \in \mathbb{Z},$$

where $\alpha(n) = \alpha_n$ are in the sense of (25.50).

Theorem 25.10 *The mathematical triple,*

$$\Gamma = (\mathbb{Z}, \mathscr{X}_\tau, \alpha), \tag{25.57}$$

is a group(-C^*)-dynamical system of the infinite cyclic abelian group $\mathbb{Z}$ acting on the shift-semicircular C^*-probability space $\mathscr{X}_\tau$ via a group-action α of (25.56).

Proof 25.15 *The morphism α of (25.55) is indeed a well-defined group-action, by (25.53), (25.54), (25.55) and (25.56).*

25.5.2 The Crossed Product $C^* - Algebra\ \mathscr{X}[\Gamma]$ of Γ

Let $\Gamma_G = (G, B, \beta)$ be a group-dynamical system of a group G acting on a C^*-algebra B via a group-action β of G. i.e., the images,

$$\beta\left(G\right) = \{\beta(g) = \beta_g : g \in G\} \overset{\text{Group}}{=} G \subset Aut\left(B\right),$$

form $*$-isomorphisms on B. From this group-dynamical system Γ_G, one can define the corresponding crossed product C^*-algebra,

$$B_G \overset{\text{denote}}{=} B \rtimes_\beta G, \tag{25.58}$$

by the C^*-subalgebra of the tensor product C^*-algebra $B \otimes B(H_G)$, where $H_G = l^2(G)$, satisfying the β-relation:

$$(b_1, g_1)(b_2, g_2) = (b_1 \beta_{g_1}(b_2),\ g_1 g_2),$$

and (25.59)

$$(b, g)^* = (\beta_g(b^*),\ g^{-1}),$$

for all $b_1, b_2, b \in B$ and $g_1, g_2, g \in G$, where g^{-1} is the group-inverse of g in G, regarded as the unitary operator, satisfying $g^* = g^{-1}$.

Let $\Gamma = (\mathbb{Z}, \mathscr{X}_\tau, \alpha)$ be our group-dynamical system (25.57). Then, as in the general case (25.58), we have the crossed product C^*-algebra,

$$\mathscr{X}[\Gamma] = \mathscr{X}_\tau \rtimes_\alpha \mathbb{Z},$$

with its α-relation: (25.60)

$$(W_1, n_1)(W_2, n_2) = (W_1 \alpha_{n_1}(W_2),\ n_1 + n_2),$$

and

$$(W, n)^* = (\alpha_n(W^*), -n),$$

by (25.59), for all $W_1, W_2, W \in \mathscr{X}_\tau$, and $n_1, n_2, n \in \mathbb{Z}$.

On this crossed product C^*-algebra $\mathscr{X}[\Gamma]$ of (25.60), define a linear functional τ_o by a morphism satisfying

$$\tau_o((W, n)) = \tau(\alpha_n(W)), \tag{25.61}$$

for all $(W, n) \in \mathscr{X}[\Gamma]$, with $W \in \mathscr{X}_\tau$ and $n \in \mathbb{Z}$, where α_n is in the sense of (25.50).

Definition 25.8 *The crossed product C^*-algebra $\mathscr{X}[\Gamma]$ of (25.60), generated by the group-dynamical system Γ of (25.57), is called the ($\mathbb{Z}$-)dynamical shift-semicircular (C^*-)algebra. The corresponding C^*-probability space,*

$$\mathscr{X}[\Gamma] \overset{\text{denote}}{=} (\mathscr{X}[\Gamma],\ \tau_o),$$

is called the ($\mathbb{Z}$-)dynamical shift-semicircular (C^*-)probability space, where τ_o is the linear functional (25.61).

We here focus on studying free-distributional data on this dynamical shift-semicircular probability space $\mathscr{X}[\Gamma]$, in terms of the linear functional τ_o of (25.61). They will illustrate how our $\mathbb{Z}$-depending dynamics affect the free probability on $\mathscr{X}_\tau$.

Let $w^m_{k,j:n}$ be a generating operator,

$$w_{k,j:n}^{m} \overset{\text{denote}}{=} \left(u_{k,j}^{m}, n\right) \in \mathscr{X}[\Gamma], \tag{25.62}$$

for all $k, j, n \in \mathbb{Z}$, and $m \in \mathbb{N}$.

If $w_{k_l,j_l:n_l}^{m_l} \in \mathscr{X}[\Gamma]$ are the generating operators (25.62), for $l = 1, ..., N$, for $N \in \mathbb{N}$, then

$$\begin{aligned} w_{k_1,j_1:n_1}^{m_1} w_{k_1,j_1:n_1}^{m_2} &= \left(u_{k_1,j_1}^{n_1}, n_1\right)\left(u_{k_2,j_2}^{m_2}, n_2\right) \\ &= \left(u_{k_1,j_1}^{m_1} \alpha_{n_1}\left(u_{k_2,j_2}^{m_2}\right),\ n_1 + n_2\right) \\ &= \left(u_{k_1,j_1}^{m_1} u_{k_2,j_2+n_1}^{m_2},\ n_1 + n_2\right), \end{aligned} \tag{25.63}$$

and hence, we have

$$\begin{aligned} \prod_{l=1}^{N} w_{k_l,j_l:n_l}^{m_2} &= \left(u_{k_1,j_1}^{m_1} u_{k_2,j_2+n_1}^{m_2} u_{k_3,j_3+n_1+n_2}^{m_3} \dots u_{k_N,j_N+\sum_{l=}^{N-1} n_l}^{m_N},\ \sum_{l=1}^{N} n_l\right) \\ &= \left(\prod_{l=1}^{N} u_{k_l,j_l+\sum_{i=1}^{l-1} n_i}^{m_l},\ \sum_{l=1}^{N} n_l\right), \end{aligned}$$

in $\mathscr{X}[\Gamma]$, with axiomatization: (25.64)

$$\sum_{i=1}^{0} n_i = 0,$$

by the induction on (25.63).

Lemma 25.3 *Let* $w_{k_l,j_l:n_l}^{m_l} = \left(u_{k_l,j_l}^{m_l}, n_l\right) \in \mathscr{X}[\Gamma]$ *be generating operators, for* $l = 1, ..., N$, *for* $N \in \mathbb{N}$. *Then*

$$\prod_{l=1}^{N} w_{k_l,j_l:n_l}^{m_l} = \left(\prod_{l=1}^{N} u_{k_l,j_l+\sum_{i=1}^{l-1} n_i}^{m_l},\ \sum_{l=1}^{N} n_l\right). \tag{25.65}$$

Proof 25.16 *The formula (25.65) is obtained by (25.64).*

Observe that, for $w_{k_l,j_l:n_l} \in \mathscr{X}[\Gamma]$ of (25.62), for $l = 1, ..., N$, for $N \in \mathbb{N}$,

$$\tau_o\left(\prod_{l=1}^{n} w_{k_l,j_l:n_l}^{m_l}\right) = \tau_o\left(\prod_{l=1}^{N} u_{k_l,j_l+\sum_{i=1}^{l-1} n_i}^{m_l},\ \sum_{l=1}^{N} n_l\right)$$

by (25.65)

$$= \tau\left(\alpha_{\sum_{l=1}^{N} n_l}\left(\prod_{l=1}^{N} u_{k_l,j_l+\sum_{i=1}^{l-1} n_i}^{m_l}\right)\right)$$

by (25.61)

$$= \tau\left(\prod_{l=1}^{N} u_{k_l,j_l+\sum_{i=1}^{l-1} n_i+\sum_{l=1}^{N} n_l}^{m_l}\right)$$

by (25.51)

$$=\varphi\left(\prod_{l=1}^{N} x^{m_l}_{j_l+\sum_{i=1}^{l-1} n_i+\sum_{l=1}^{N} n_l}\right), \tag{25.66}$$

since all generating element $u_{k,j} \in \mathcal{X}$ of $\mathscr{X}_\tau$ are followed by the semicircular law. Note that the final quantity in (25.66) is determined by (25.11).

Theorem 25.11 *Let $w^{m_l}_{k_l,j_l:n_l}$ be the generating operators (25.62) of the dynamical shift-semicircular space $\mathscr{X}[\Gamma]$, for $l = 1, ..., N$, for $N \in \mathbb{N}$. Then*

$$\tau_o\left(\prod_{l=1}^{N} w^{m_l}_{k_l,j_l:n_l}\right)=\varphi\left(\prod_{l=1}^{N} x^{m_l}_{j_l+\sum_{i=1}^{l-1} n_i}\right), \tag{25.67}$$

where the quantity (25.67) is characterized by (25.11).

Proof 25.17 *The free-distributional data (25.67) on $\mathscr{X}[\Gamma]$ is obtained by (25.66). Indeed, we have*

$$\tau_o\left(\prod_{l=1}^{N} w^{m_l}_{k_l,j_l:n_l}\right)=\varphi\left(\prod_{l=1}^{N} x^{m_l}_{j_l+\sum_{i=1}^{l-1} n_i+\sum_{l=1}^{N} n_l}\right),$$

by (25.66), and

$$\varphi\left(\prod_{l=1}^{N} x^{m_l}_{j_l+\sum_{i=1}^{l-1} n_i+\sum_{l=1}^{N} n_l}\right)=\varphi\left(\prod_{l=1}^{N} x^{m_l}_{j_l+\sum_{i=1}^{l-1} n_i}\right),$$

by (25.11), because two free random variables,

$$\prod_{l=1}^{N} x^{m_l}_{j_l+\sum_{i=1}^{l-1} n_i+\sum_{l=1}^{N} n_l}, \text{ and } \prod_{l=1}^{N} x^{m_l}_{j_l+\sum_{i=1}^{l-1} n_i},$$

of the C^*-probability space $\mathfrak{X}_\varphi$ have the corresponding noncrossing partitions of (25.10), having the same patterns; since they are induced by the $N\left(\sum_{l=1}^{N} m_l\right)$-tuples of (25.5),

$$(I_1, I_2, ..., I_N), \text{ and } (J_1, J_2, ..., J_N),$$

with

$$I_l=\left(\underbrace{j_l+\sum_{i=1}^{l-1} n_i+\sum_{l=1}^{N} n_l, ..., j_l+\sum_{i=1}^{l-1} n_i+\sum_{l=1}^{N} n_l}_{m_l\text{-times}}\right),$$

and

$$J_l=\left(\underbrace{j_l+\sum_{i=1}^{l-1} n_i, ..., j_l+\sum_{i=1}^{l-1} n_i}_{m_l\text{-times}}\right),$$

for all $l = 1, ..., N$. Since the aforementioned two free random variables induce the same patterned noncrossing partitions of (25.10), they have the same free moments by (25.11). Therefore, the free-distributional data (25.67) holds by (25.11) and (25.66).

25.5.3 Free-Distributional Data on $\mathscr{X}[\Gamma]$

In this section, we consider refined results of the general free-distributional data (25.67) on the dynamical shift-semicircular space,

$$\mathbb{X}[\Gamma] = (\mathscr{X}[\Gamma], \tau_o),$$

where $\mathscr{X}[\Gamma] = \mathscr{X}_\tau \rtimes_\alpha \mathbb{Z}$ of (25.60), and τ_o is the linear functional (25.61). Let

$$w^m_{k,j:n} = \left(u^m_{k,j}, n\right) = \left(\lambda^{km} \otimes x^m_j,\ n\right)$$

be a generating operator (25.62) of $\mathscr{X}[\Gamma]$, for $k, j, n \in \mathbb{Z}$, and $m \in \mathbb{N}$. Then, by the α-relation of (25.60), we have

$$\left(w^m_{k,j:n}\right)^* = \left(u^m_{k,j}, n\right)^* = \left(\alpha_n\left(u^{m*}_{k,j}\right), -n\right) = \left(u^m_{-k,j+n}, -n\right),$$

since

$$u^{m*}_{k,j} = \left(\lambda^{km} \otimes x^m_j\right)^* = \lambda^{-km} \otimes x^m_j = \left(\lambda^{-k} \otimes x_j\right)^m = u^m_{-k,j},$$

in $\mathscr{X}_\tau$, i.e.,

$$\left(w^m_{k,j:n}\right)^* = w^m_{-k,j+n,-n} \text{ in } \mathscr{X}[\Gamma]. \tag{25.68}$$

Also, by (25.65), one has that

$$\left(w^m_{k,j:n}\right)^N = \left(u^m_{k,j} u^m_{k,j+n} ... u^m_{k,j+n(N-1)},\ nN\right),$$

i.e.,

$$\left(w^m_{k,j:n}\right)^N = \left(\prod_{l=1}^{N} u^m_{k,j+(l-1)n},\ nN\right) \text{ in } \mathscr{X}[\Gamma], \tag{25.69}$$

for all $N \in \mathbb{N}$. By (25.68) and (25.69), we have

$$\left(\left(w^m_{k,j:n}\right)^*\right)^N = \left(w^m_{-k,j+n,-n}\right)^N = \left(\prod_{l=1}^{N} u^m_{-k,(j+n)+(l-1)n}, -nN\right),$$

i.e.,

$$\left(\left(w^m_{k,j:n}\right)^*\right)^N = \left(\prod_{l=1}^{N} u^m_{-k,j+ln},\ -nN\right), \tag{25.70}$$

in $\mathscr{X}[\Gamma]$.

Lemma 25.4 *Let $w_{k,j:n}^{m} = \left(u_{k,j}^{m}, n\right) \in \mathscr{X}[\Gamma]$ be a generating operator (25.62), for $k, j, n \in \mathbb{Z}$, and $m \in \mathbb{N}$. Then*

$$\left(w_{k,j:n}^{m}\right)^{*} = w_{-k,j+n:-n}^{m},$$

and

$$\left(w_{k,j:n}^{m}\right)^{N} = \left(\prod_{l=1}^{N} u_{k,j+(l-1)n}^{m},\ nN\right),$$

and (25.71)

$$\left(\left(w_{k,j:n}^{m}\right)^{*}\right)^{N} = \left(\prod_{l=1}^{N} u_{-k,j+ln}^{m},\ -nN\right),$$

for all $N \in \mathbb{N}$.

Proof 25.18 *The operator-equalities of (25.70) are proven by (25.68), (25.69) and (25.70), respectively.*

By (25.71), we obtain the following free-distributional data as special cases of (25.67).

Theorem 25.12 *Let $w_{k,j:n}^{m} = \left(u_{k,j}^{m}, n\right) \in \mathscr{X}[\Gamma]$ be a generating operator (25.62). Then*

$$\tau_o\left(w_{k,j:n}^{m}\right) = \omega_m c_{\frac{m}{2}} = \tau_o\left(\left(w_{k,j:n}^{m}\right)^{*}\right), \tag{25.72}$$

and

$$\tau_o\left(\left(w_{k,j:n}^{m}\right)^{N}\right) = \left(\omega_m c_{\frac{m}{2}}\right)^{N} = \tau_o\left(\left(\left(w_{k,j:n}^{m}\right)^{*}\right)^{N}\right), \tag{25.73}$$

for all $N \in \mathbb{N}$.

Proof 25.19 *By (25.67) and (25.71), one has that*

$$\tau_o\left(w_{k,j:n}^{m}\right) = \tau\left(\alpha_n\left(u_{k,j}^{m}\right)\right) = \tau\left(u_{k,j+n}^{m}\right) = \varphi\left(x_{j+n}^{m}\right) = \omega_m c_{\frac{m}{2}},$$

and (25.74)

$$\tau_o\left(\left(w_{k,j:n}^{m}\right)^{*}\right) = \tau_o\left(w_{-k,j+n:-n}^{m}\right) = \varphi\left(x_{j}^{m}\right) = \omega_m c_{\frac{m}{2}},$$

since $u_{k,j+n}, u_{k,j+n-n} = u_{k,j} \in \mathscr{X}_{\tau}$ are followed by the semicircular law induced by $x_{j+n}, x_j \in \mathfrak{X}_{\varphi}$, respectively. Therefore, the free-distributional data (25.72) is obtained by the formulas of (25.74). Also, we obtain that, for all $N \in \mathbb{N}$,

$$\tau_o\left(\left(w_{k,j:n}^{m}\right)^{N}\right) = \tau_o\left(\left(\prod_{l=1}^{N} u_{k,j+(l-1)n}^{m},\ nN\right)\right)$$

by (25.71)

$$= \tau\left(\alpha_{nN}\left(\prod_{l=1}^{N} u_{k,j+(l-1)n}^{m}\right)\right) = \varphi\left(\prod_{l=1}^{N} x_{j+(l-1)n}^{m}\right),$$

by (25.67); and (25.75)

$$\tau_o\left(\left(\left(w_{k,j:n}^m\right)^*\right)^N\right) = \tau_o\left(\prod_{l=1}^{N} u_{-k,j+ln}^m, \ -nN\right)$$

by (25.71)

$$=\tau\left(\alpha_{-nN}\left(\prod_{l=1}^{N} u_{-k,j+ln}^m\right)\right) = \varphi\left(\prod_{l=1}^{N} x_{j+ln}^m\right).$$

Now, let

$$W_1 = \prod_{l=1}^{N} x_{j+(l-1)n}^m, \text{ and } W_2 = \prod_{l=1}^{N} x_{j+ln}^m,$$

in $\mathfrak{X}_\varphi$. *Then these two free random variables* W_1 *and* W_2 *induce the noncrossing partitions* π_{W_1} *and* π_{W_2}, *whose patterns are same, because the* (mN)*-tuples,*

$$I_{W_1} = (I_1, ..., I_N), \text{ and } I_{W_2} = (J_1, ..., J_N),$$

with (25.76)

$$I_l = \left(\underbrace{j+(l-1)n, ..., j+(l-1)n}_{m\text{-}times}\right),$$

and

$$J_l = \left(\underbrace{j+ln, ..., j+ln}_{m\text{-}times}\right),$$

have the same patterns. It implies that

$$\varphi(W_1) = \varphi(W_2), \text{ in } \mathfrak{X}_\varphi, \tag{25.77}$$

by (25.11). Thus,

$$\tau_o\left(\left(w_{k,j:n}^m\right)^N\right) = \varphi(W_1) = \tau_o\left(\left(\left(w_{k,j:n}^m\right)^*\right)^N\right), \tag{25.78}$$

by (25.77), where $W_1 \in \mathfrak{X}_\varphi$ *is in the sense of (25.76).*

Now, let's concentrate on computing the joint free moment,

$$\varphi(W_1) = \varphi\left(\prod_{l=1}^{N} x_{j+(l-1)n}^m\right),$$

on $\mathfrak{X}_\varphi$, *determined by (2.11). As we have seen in (25.76), the free random variable* $W_1 \in \mathfrak{X}_\varphi$ *induces the* (mN)*-tuple* I_{W_1}, *with*

$$j + (l_1 - 1)n \neq j + (l_2 - 1)n \text{ in } \mathbb{Z},$$

whenever $l_1 \neq l_2$ *in* $\{1, ..., N\}$. *i.e., the sub-sequence* I_{l_1} *of (25.76) are consisting of all identical entries* $j + (l_1 - 1)n$, *which are distinct from the entries* $j + (l_2 - 1)n$ *of other* *sub-sequence I_2, whenever $l_1 \neq l_2$. So, this (mN)-tuple I_{W_1} induces the corresponding noncrossing partition $\pi_{W_1} \in NC([I_{W_1}])$, which is equivalent to a noncrossing partition $\pi \in NC(mN)$,*

$$\pi = \{(1, ..., m), (m+1, ..., 2m), ..., (m(N-1)+1, ..., mN)\}.$$

Thus, by (25.11),

$$\varphi(W_1) = \varphi\left(\prod_{l=1}^{N} x_{j+(l-1)n}^{m}\right) = \prod_{l=1}^{N} \varphi\left(x_{j+(l-1)n}^{m}\right),$$

in $\mathfrak{X}_{\varphi}$. Therefore, by the semicircularity of $\{x_{j+(l-1)n}\}_{l=1}^{N}$,

$$\varphi(W_1) = \left(\omega_m c_{\frac{m}{2}}\right)^N. \tag{25.79}$$

So, the free-distributional data (25.73) are obtained by (25.78) and (25.79).

Suppose now $w_{k,j:n}^{m} \in \mathscr{X}[\Gamma]$ are generating operators of dynamical shift-semicircular space $\mathscr{X}[\Gamma]$, for $k, j, n \in \mathbb{Z}$, and $m \in \mathbb{N}$. Define free random variables $X, Y \in \mathscr{X}[\Gamma]$ by

$$X = \left(w_{k,j:n}^{m}\right)\left(w_{k,j:n}^{m}\right)^* = \left(w_{k,j:n}^{m}\right)\left(w_{-k,j+n:-n}^{m}\right),$$

and (25.80)

$$Y = \left(w_{k,j:n}^{m}\right)^*\left(w_{k,j:n}^{m}\right) = \left(w_{-k,j+n:-n}^{m}\right)\left(w_{k,j:n}^{m}\right),$$

in $\mathscr{X}[\Gamma]$, and hence,

$$X = \left(u_{k,j}^{m}, n\right)\left(u_{-k,j+n}^{m}, -n\right) = \left(u_{k,j}^{m} u_{-k,j+2n}^{m}, 0\right),$$

and (25.81)

$$Y = \left(u_{-k,j+n}^{m}, -n\right)\left(u_{k,j}^{m}, n\right) = \left(u_{-k,j+n}^{m} u_{k,j-n}^{m}, 0\right),$$

in $\mathscr{X}[\Gamma]$, by the α-relation (25.60). Then, one can get that

$$X^N = \left(\prod_{l=1}^{N}\left(u_{k,j+(l-1)\cdot 0}^{m} u_{-k,j+2n+(l-1)\cdot 0}^{m}\right), N \cdot 0\right),$$

and

$$Y^N = \left(\prod_{l=1}^{N}\left(u_{-k,j+n+(l-1)\cdot 0}^{m} u_{k,j+(l-1)\cdot 0}^{m}, N \cdot 0\right)\right),$$

by (25.81), and hence,

$$X^N = \left(\left(u_{k,j}^{m} u_{k,j+2n}^{m}\right)^N, 0\right),$$

and (25.82)

$$Y^N = \left(\left(u_{-k,j+n}^{m} u_{k,j-n}^{m}\right)^N, 0\right),$$

in $\mathscr{X}[\Gamma]$.

Theorem 25.13 *Let $X, Y \in \mathscr{X}[\Gamma]$ be the free random variables (25.80) induced by the generating operator $w_{k,j}^{m} \in \mathscr{X}[\Gamma]$ of (25.80). Then*

$$\tau_o\left(X^N\right) = \left(\omega_{mN} c_{\frac{mN}{2}}\right)\left(\omega_m c_{\frac{m}{2}}\right)^N = \tau_o\left(Y^N\right), \tag{25.83}$$

for all $N \in \mathbb{N}$.

Proof 25.20 *Let $X \in \mathscr{X}[\Gamma]$ be an operator of (25.80). Then, by (25.82),*

$$X^N = \left(\left(u_{k,j}^{m} u_{-k,j+2n}^{m}\right)^N, 0\right) \text{ in } \mathscr{X}[\Gamma],$$

for all $N \in \mathbb{N}$, and hence,

$$\begin{aligned}\tau_o\left(X^N\right) &= \tau_o\left(\left(\left(u_{k,j}^{m} u_{k,j+2n}^{m}\right)^N, 0\right)\right)\\ &= \tau\left(\alpha_0\left(\left(u_{k,j}^{m} u_{k,j+2n}^{m}\right)^N\right)\right) = \tau\left(\left(u_{k,j}^{m} u_{k,j+2n}^{m}\right)^N\right)\\ &= \varphi\left(\left(x_j^m x_{j+2n}^m\right)^N\right) = \varphi\left(x_j^m x_{j+2n}^m x_j^m x_{j+2n}^m \cdots x_j^m x_{j+2m}^m\right),\end{aligned} \tag{25.84}$$

by (25.67), where the last quantity is characterized by (25.11) in $\mathfrak{X}_\varphi$.

From (25.84), take a free random variable $W_X = \left(x_j^m x_{j+2n}^m\right)^N \in \mathfrak{X}_\varphi$, and the $(2mN)$-tuple I_{W_X} of (25.5),

$$I_{W_X} = \left(J_1, J_2, ..., J_{2N-1}, J_{2N}\right),$$

with

$$J_{2l-1} = \left(\underbrace{j, j, j, ..., j}_{m\text{-times}}\right),$$

and (25.85)

$$J_{2l} = \left(\underbrace{j+2n, j+2n, ..., j+2n}_{m\text{-times}}\right),$$

for all $l = 1, ..., N$, and it has its corresponding noncrossing partition $\pi \in NC\left(\left[I_{W_X}\right]\right)$,

$$\pi = \left\{U_1, U_2, ..., U_{m+1}\right\},$$

with its blocks (25.86)

$$U_1 = \bigcup_{i=1}^{N} \left\{J_{2i-1}\right\},$$

and

$$U_s = \left\{J_{2s}\right\}, \text{ for all } s = 2, 3, ..., m+1,$$

by (25.85), where $\{Z\}$ mean the blocks of π induced by $Z \subset [I_{W_X}]$. So,

$$\varphi\left(\left(x_j^m x_{j+2n}^m\right)^N\right) = \prod_{l=1}^{m+1}\left(\omega_{|U_l|} c_{\frac{|U_l|}{2}}\right),$$

i.e.,

$$\varphi\left(\left(x_j^m x_{j+2n}^m\right)^N\right) = \left(\varphi\left(x_j^{mN}\right)\right)\left(\varphi\left(x_{j+2n}^m\right)\right)^N, \tag{25.87}$$

by (25.11), where

$$|U_1| = mN, \text{ and } |U_s| = m,$$

for all $s = 2, 3, ..., m+1$, by (25.86). Therefore, the first equality of (25.83) holds by (25.87).

Similar to (25.87), we have that

$$\tau_o\left(Y^N\right) = \tau_o\left(\left(\left(u_{-k,j-n}^m u_{k,j+n}^m\right)^N, 0\right)\right)$$

by (25.82)

$$= \tau\left(\alpha_0\left(\left(u_{-k,j-n}^m u_{k,j+n}^m\right)^N\right)\right) = \varphi\left(\left(x_{j-n}^m x_{j+m}^m\right)^N\right)$$

by (25.67)

$$= \left(\varphi\left(x_{j-n}^{mN}\right)\right)\left(\varphi\left(x_{j+n}^m\right)\right)^N = \left(\omega_{mN} c_{\frac{mN}{2}}\right)\left(\omega_m c_{\frac{m}{2}}\right)^N, \tag{25.88}$$

as in (25.87), for all $N \in \mathbb{N}$. Therefore, the second equality of (25.83) holds by (25.88).

The aforementioned theorem also provides spacial cases of the general free-distributional data (25.67), induced by a single generating operator $w_{k,j:n}^m$of $\mathscr{X}[\Gamma]$.

25.5.4 Discussion: $\mathbb{Z}$-Dynamics on $\mathscr{X}_\tau$

Let $\mathscr{X}_\tau = (\Lambda \otimes \mathfrak{X}_\varphi, \tau)$ be our shift-semicircular C^*-probability space, generated by the free random variables,

$$\mathcal{X} = \left\{u_{k,j} = \lambda^k \otimes x_j : k, j \in \mathbb{Z}\right\},$$

followed by the semicircular law, where $\lambda^k \in \lambda$ are the integer shifts, generating the integer-shift operator algebra Λ, and $x_j \in X$ are the generating semicircular elements of $\mathfrak{X}_\varphi$.

In Sections 5.1, 5.2, and 5.3, by acting as a discrete group, $\mathbb{Z} = (\mathbb{Z}, +)$, on $\mathscr{X}_\tau$ canonically, we obtain the group-dynamical system,

$$\Gamma = (\mathbb{Z},\ \mathscr{X}_\tau,\ \alpha),$$

of (25.75), and it induces the corresponding crossed-product C^*-probability space,

$$\mathscr{X}[\Gamma] = (\mathscr{X}_\tau \rtimes_\alpha \mathbb{Z},\ \tau_o),$$

of Definition 34. And the free-distributional data (25.76) give the general free-probabilistic information on $\mathscr{X}[\Gamma]$. Some interesting free-distributional data induced by a single generating operator,

$$w_{k,j:n}^{m} = \left(u_{k,j}^{m},\ n\right) \in \mathscr{X}[\Gamma],$$

are considered; see (25.85), (25.86), and (25.87). In particular, one can realize that the dynamics of $\mathbb{Z}$ deforms the free probability on $\mathscr{X}_{\tau}$, and the deformation is characterized by the free probability on $\mathscr{X}[\Gamma]$. For instance, if we have

$$u_{k,j} = \lambda^{k} \otimes x_{j} \in \mathcal{X} \text{ in } \mathscr{X}_{\tau},$$

then

$$\tau\left(u_{k,j}^{2}\right) = \varphi\left(x_{j}^{2}\right) = c_{\frac{2}{1}} = c_{1} = 1,$$

meanwhile, if

$$w_{k,j:n}^{1} = \left(u_{k,j}, n\right) \in \mathscr{X}[\Gamma],$$

for $n \in \mathbb{Z}$, then

$$\tau_{o}\left(\left(w_{k,j:n}^{1}\right)^{2}\right) = \tau_{o}\left(\left(u_{k,j}u_{k,j+n}, 2n\right)\right) = \varphi\left(x_{j}x_{j+n}\right),$$

satisfying (25.89)

$$\tau_{o}\left(\left(w_{k,j:n}^{1}\right)^{2}\right) = \begin{cases} \varphi\left(x_{j}^{2}\right) = 1 & \text{if } n = 0 \\ \varphi\left(x_{j}\right)\varphi\left(x_{j+n}\right) = 0 & \text{otherwise,} \end{cases}$$

by (25.67), for $n \in \mathbb{Z}$. It illustrates that our dynamics of $\mathbb{Z}$ on $\mathscr{X}_{\tau}$ deforms the free probability on $\mathscr{X}_{\tau}$.

Observation. The discrete dynamics of $\mathbb{Z}$ on $\mathscr{X}_{\tau}$ distorts the free probability on $\mathscr{X}_{\tau}$ in general, e.g., see (25.89). However, such deformations of free-distributional data are manageable, or characterizable. e.g., see (25.67), (25.72), (4) and (25.83). □

25.6 MORE ABOUT FREE-DISTRIBUTIONAL DATA ON $\mathscr{X}[\Gamma]$

In this section, we keep considering special cases of the general free-distributional data (25.67) on the dynamical shift-semicircular probability space $\mathscr{X}[\Gamma]$. From the main results of Section 5, we consider certain types of free random variables of $\mathscr{X}_{\tau}$ and those of $\mathscr{X}[\Gamma]$.

25.6.1 Free Random Variables of $\mathscr{X}[\Gamma]$ Followed by the Circular Law

In this section, we consider a certain type of free random variables of $\mathscr{X}_{\tau}$ and those of $\mathscr{X}[\Gamma]$, whose free distributions are followed by the circular law in a certain sense.

25.6.1.1 Free Random Variables Followed by The Circular Law

In an arbitrary topological $*$-probability space (B, ψ), let $y_1, y_2 \in (B, \psi)$ be two, free, semicircular elements. Define a free random variable $y_{1,2}$ of (B, ψ) by

$$y_{1,2} = \frac{1}{\sqrt{2}} (y_1 + iy_2), \text{ with } i = \sqrt{-1} \in \mathbb{C}. \tag{25.90}$$

Definition 25.9 *A free random variable $y_{1,2} \in (B, \psi)$ of (25.90) is called the circular element induced by two, free, semicircular elements y_1 and y_2. The free distribution of $y_{1,2}$ is called the circular law.*

Note that circular elements are not self-adjoint. Indeed, if $y_{1,2} \in (B, \psi)$ is a circular element (25.90), then

$$y_{1,2}^* = \frac{1}{\sqrt{2}} (y_1 - iy_2) \neq y_{1,2},$$

in (B, ψ). Since $y_{1,2}$ is not self-adjoint in B, to study the free distribution of $y_{1,2}$, one needs to compute the joint free moments or the joint free cumulants of $\{y_{1,2}, y_{1,2}^*\}$. It is well known that the circular law of $y_{1,2}$ is characterized by the only non-zero following joint free cumulants of $\{y_{1,2}, y_{1,2}^*\}$.

Proposition 25.3 *(See e.g., [17]) Let $y_{1,2} \in (B, \psi)$ be a circular element (25.90). Then the only non-zero joint free cumulants of $\{y_{1,2}, y_{1,2}^*\}$ are*

$$k_2^{\psi} \left(y_{1,2}, y_{1,2}^*\right) = 1 = k_2^{\psi} \left(y_{1,2}^*, y_{1,2}\right). \tag{25.91}$$

where $k_{\bullet}^{\psi}(.)$ is the free cumulant on B in terms of ψ. □

The proof of (25.91) is obtained by the freeness and the semicircularity (25.3) of y_1 and y_2. By the universality of the semicircular law, the circular law is universal, too, in the sense that: all circular elements are identically free-distributed, and their free distributions are characterized by (25.91).

Corollary 25.2 *If $x_{j_1}, x_{j_2} \in X$ are generating semicircular elements of $\mathfrak{X}_{\varphi}$, where $j_1 \neq j_2$ in $\mathbb{Z}$, then*

$$y = \frac{1}{\sqrt{2}} (x_{j_1} + ix_{j_2}) \in \mathfrak{X}_{\varphi},$$

is circular in $\mathfrak{X}_{\varphi}$, and its free distribution is characterized by the only non-zero joint free cumulants of $\{y, y^*\}$;

$$k_2^{\varphi} (y, y^*) = 1 = k_2^{\varphi} (y^*, y).$$

Proof 25.21 *Since $j_1 \neq j_2$ in $\mathbb{Z}$, the semicircular elements x_{j_1} and x_{j_2} are distinct in the free semicircular family X generating $\mathfrak{X}_{\varphi}$, equivalently, they are free in $\mathfrak{X}_{\varphi}$. So, it is shown by (25.91), by the universality of the circular law.*

Now, let $u_{k_l, j_l} \in \mathcal{X}$ be generating free random variables of the shift-semicircular C^*-probability space $\mathscr{X}_{\tau}$, followed by the semicircular law, for $l = 1, 2$, under an additional condition:

$$j_1 \neq j_2 \text{ in } \mathbb{Z}, \tag{25.92}$$

and $k_1, k_2 \in \mathbb{Z}$ are arbitrary. Note that, by the assumption (25.92), u_{k_1,j_1} and u_{k_2,j_2} are free in $\mathscr{X}_\mathcal{T}$ by (25.49). Define a new free random variable $u \in \mathscr{X}_\mathcal{T}$ by

$$u = \frac{1}{\sqrt{2}} \left(u_{k_1,j_1} + i u_{k_2,j_2} \right) \in \mathscr{X}_\mathcal{T}. \tag{25.93}$$

Then this operator $u \in \mathscr{X}_\mathcal{T}$ of (25.93) is not self-adjoint, since

$$u^* = \frac{1}{\sqrt{2}} \left(u_{-k_1,j_1} - i u_{-k_2,j_2} \right) \in \mathscr{X}_\mathcal{T}.$$

So, the free distribution of u is characterized by the joint free cumulants (or, moments) of $\{u, u^*\}$.

Theorem 25.14 *The free random variable $u \in \mathscr{X}_\mathcal{T}$ of (25.93) is followed by the circular law in the sense that: the only "non-zero" joint free cumulants of $\{u, u^*\}$ are*

$$k_2^\mathcal{T}(u, u^*) = 1 = k_2^\mathcal{T}(u^*, u). \tag{25.94}$$

Proof 25.22 *For convenience, we let $u_{k_l,j_l} \overset{denote}{=} u_l$, for $l = 1, 2$. Then, for any arbitrary $(r_1, ..., r_N) \in \{1, *\}^N$, for $N \in \mathbb{N}$, one has*

$$k_N^\mathcal{T}(u^{r_1}, ..., u^{r_n}) = \left(\frac{1}{\sqrt{2}}\right)^N k_N^\mathcal{T}\left((u_1 + iu_2)^{r_1}, ..., (u_1 + iu_2)^{r_N}\right)$$

since the free cumulant $k_\bullet^\mathcal{T}(...)$ is a bimodule map on $\mathscr{X}_\mathcal{T}$ (e.g., [17])

$$= \left(\frac{1}{\sqrt{2}}\right)^N \left(k_N^\mathcal{T}(u_1^{r_1}, ..., u_1^{r_N}) + k_N^\mathcal{T}(i^{r_1} u_2^{r_1}, ..., i^{r_N} u_2^{r_N})\right)$$

by the freeness of u_1 and u_2 (and hence, the freeness of $C^*(\{u_1, u_1^*\})$ and $C^*(\{u_2, u_2^*\})$, e.g., see [17, 22, 23]), where

$$i^{r_l} = \begin{cases} i & \text{if } r_l = 1 \\ -i & \text{if } r_l = *, \end{cases}$$

for all $l = 1, ..., N$, so

$$= \left(\frac{1}{\sqrt{2}}\right)^N \left(k_N^\varphi(x_{j_1}, ..., x_{j_1}) + k_N^\varphi(i^{r_1} x_{j_2}, ..., i^{r_N} x_{j_2})\right)$$

by (25.49)

$$= \left(\frac{1}{\sqrt{2}}\right)^N k_N^\varphi\left((x_{j_1} + i^{r_1} x_{j_2}), ..., (x_{j_1} + i^{r_N} x_{j_2})\right)$$

since x_{j_1} and x_{j_2} are free in $\mathfrak{X}_\varphi$ under (25.92)

$$= k_N^\varphi(y^{r_1}, ..., y^{r_N}),$$

where (25.95)

$$y = \frac{1}{\sqrt{2}}(x_{j_1} + i x_{j_2}) \in \mathfrak{X}_\varphi$$

is a circular element by the aforementioned corollary.

Therefore, there exist only non-zero joint free cumulants,

$$k_2^{\tau}(u^*, u) = 1 = k_2^{\tau}(u, u^*),$$

by (25.91) and (95). So, the free distribution of the free random variable $u \in \mathscr{X}_{\tau}$ of (25.93) is characterized by the only non-zero free cumulants (25.94).

Now, take the generating operators,

$$w_{k_l, j_l:0} = (u_{k_l, j_l}, 0) \in \mathscr{X}[\Gamma], \text{ for } l = 1, 2,$$

with a condition: (25.96)

$$j_1 \neq j_2 \text{ in } \mathbb{Z}, \text{ and } k_1, k_2 \in \mathbb{Z},$$

in the dynamical shift-semicircular space $\mathscr{X}[\Gamma]$. Since $j_1 \neq j_2$ in $\mathbb{Z}$, the generating operators $u_{k_1,j_1}, u_{k_2,j_2} \in \mathcal{X}$ are free in $\mathscr{X}_{\tau}$ by (25.49).

From the generating operators (25.96), define a new free random variable w of $\mathscr{X}[\Gamma]$ by

$$w = \tfrac{1}{\sqrt{2}}(w_{k_1,j_1:0} + i w_{k_2,j_2:0}) \in \mathscr{X}[\Gamma]. \tag{25.97}$$

Theorem 25.15 *Let $w \in \mathscr{X}[\Gamma]$ be a free random variable (25.97) induced by the generating operators (25.96). Then the free distribution of w is followed by the circular law in the sense that: the only non-zero joint free cumulants of $\{w, w^*\}$ are*

$$k_2^{\tau_o}(w, w^*) = 1 = k_2^{\tau_o}(w^*, w). \tag{25.98}$$

Proof 25.23 *Observe that, for any $(r_1, ..., r_N) \in \{1, *\}^N$, for $N \in \mathbb{N}$, we have that*

$$k_2^{\tau_o}(w^{r_1}, ..., w^{r_N}) = \sum_{\pi \in NC(N)} \left(\prod_{V \in \pi} \tau_o \left(\prod_{l \in V} w^{r_l} \right) \right) \mu(\pi, 1_n)$$

by the Möbius inversion

$$= \sum_{\pi \in NC(N)} \left(\prod_{V \in \pi} \tau \left(\prod_{l \in V} u^{r_l} \right) \right) \mu(\pi, 1_n)$$

by (25.67) (and (25.72), (25.73) and (25.74)), where

$$u = \frac{1}{\sqrt{2}}(u_{k_1,j_1} + i u_{k_2,j_2}) \in \mathscr{X}_{\tau},$$

because

$$\alpha_0^s(w) = w \text{ in } \mathscr{X}[\Gamma], \forall s \in \mathbb{N},$$

and hence, it goes to

$$= \sum_{\pi \in NC(N)} \left(\prod_{V \in \pi} \varphi \left(\prod_{l \in V} y^{r_l} \right) \right) \mu(\pi, 1_n)$$

by (25.94) (or, (25.95)), where

$$y = \frac{1}{\sqrt{2}} \left(x_{j_1} + i x_{j_2} \right) \in \mathfrak{X}_{\varphi},$$

implying that

$$= k_N^{\varphi} \left(y^{r_1}, ..., y^{r_N} \right),$$

i.e., (25.99)

$$\begin{aligned} k_N^{\tau_o} \left(w^{r_1}, ..., w^{r_N} \right) &= k_N^{\tau} \left(u^{r_1}, ..., u^{r_N} \right) \\ &= k_N^{\varphi} \left(y^{r_1}, ..., y^{r_N} \right), \end{aligned}$$

for all $(r_1, ..., r_N) \in \{1, *\}^N$, *for all* $N \in \mathbb{N}$.

Note that the free random variable $y \in \mathfrak{X}_{\varphi}$ *in (25.99) is circular by (25.92), and the operator* $u \in \mathcal{X}_{\tau}$ *in (25.99) is followed by the circular law as in (25.94). Therefore, the free distribution of the operator*

$$w = \frac{1}{\sqrt{2}} \left(w_{k_1, j_1 : 0} + i w_{k_2, j_2 : 0} \right) \in \mathcal{X}[\Gamma]$$

of (25.97) has only non-zero joint free cumulants (25.98) of $\{w, w^*\}$, *whenever* $j_1 \neq j_2$ *in* $\mathbb{Z}$.

The aforementioned theorem demonstrates that there are free random variables of $\mathcal{X}[\Gamma]$ induced by $\{w_{k,j:0}\}_{k,j \in \mathbb{Z}}$, whose free distributions are followed by the circular law in the sense of (25.98). Remark that, as we discussed in Section 5.4 (e.g., see (25.89)), if $n \neq 0$ in $\mathbb{Z}$, then

$$\frac{1}{\sqrt{2}} \left(w_{k_1, j_1 : n} + i w_{k_2, j_2 : n} \right) \in \mathcal{X}[\Gamma],$$

are "not" followed by the circular law, even though $j_1 \neq j_2$ in $\mathbb{Z}$.

25.6.2 Free Random Variables of $\mathcal{X}[\Gamma]$ Followed by Free Poisson Distributions

In this section, we consider a different type of free random variables in the dynamical shift-semicircular space $\mathcal{X}[\Gamma]$, whose free distributions are followed by free Poisson distributions.

Definition 25.10 *Let* x *be a semicircular element of a topological* $*$*-probability space* (B, ψ), *and* $y \in (B, \psi)$, *a self-adjoint free random variable. If* x *and* y *are free in* (B, ψ), *then a free random variable,*

$$w = xyx \in (B, \psi), \tag{25.100}$$

is said to be the free Poisson element generated by x *and* y. *The free distribution of* w *is called "a" free Poisson distribution.*

By (25.100), a free Poisson element $w = xyx$ is self-adjoint in (B, ψ), since

$$w^* = x^* y^* x^* = xyx = w.$$

The free Poisson distribution of w is characterized by the following free-distributional data.

Proposition 25.4 *(e.g., see [17]) If $w = xyx \in (B, \psi)$ is a free Poisson element (25.100), then the corresponding free Poisson distribution is characterized by the free cumulants of w,*

$$k_n^{\psi}\left(\underbrace{w, w, \ldots\ldots, w}_{n\text{-times}}\right) = \psi\left(y^n\right), \forall n \in \mathbb{N}. \tag{25.101}$$

□

By (25.101), one can realize that free Poisson distributions are not universal; they depend on the free distributions of self-adjoint free random variables y.

Corollary 25.3 *If $x_{j_1}, x_{j_2} \in X$ are generating semicircular elements of $\mathfrak{X}_{\varphi}$, with $j_1 \neq j_2$ in $\mathbb{Z}$, then the free distribution of $w = x_{j_1} x_{j_2} x_{j_1} \in \mathfrak{X}_{\varphi}$ is characterized by its free cumulants,*

$$k_N^{\varphi}\left(\underbrace{w, w, \ldots\ldots\ldots, w}_{N\text{-times}}\right) = \omega_N c_{\frac{N}{2}}, \ \forall N \in \mathbb{N}. \tag{25.102}$$

Proof 25.24 *Let $w = x_{j_1} x_{j_2} x_{j_1} \in \mathfrak{X}_{\varphi}$, where $x_{j_1} \neq x_{j_2} \in X$ in $\mathfrak{X}_{\varphi}$. Since x_{j_1} and x_{j_2} are free in $\mathfrak{X}_{\varphi}$, the operator w becomes a free Poisson element of $\mathfrak{X}_{\varphi}$ by (25.100). Thus,*

$$k_N^{\varphi}\left(w, \ldots, w\right) = \varphi\left(x_{j_2}^N\right) = \omega_N c_{\frac{N}{2}},$$

for all $N \in \mathbb{N}$, by (25.101), implying the free-Poisson-distributional data (25.102).

Suppose now that $j_1 \neq j_2$ in $\mathbb{Z}$, and $u_{k_1, j_1}, u_{k_2, j_2} \in \mathcal{X}$ is the generating free random variables of $\mathscr{X}_{\tau}$, followed by the semicircular law. Define a free random variable,

$$u = u_{k_1, j_1} u_{k_2, j_2} u_{k_1, j_1} \text{ in } \mathscr{X}_{\tau}. \tag{25.103}$$

Theorem 25.16 *If $u \in \mathscr{X}_{\tau}$ is a free random variable (25.103), then the free distribution of u is followed by the free Poisson distribution of a free Poisson element $w = x_{j_1} x_{j_2} x_{j_1} \in \mathfrak{X}_{\varphi}$, in the sense that*

$$k_N^{\tau}\left(u^{r_1}, \ldots, u^{r_N}\right) = \omega_N c_{\frac{N}{2}}, \tag{25.104}$$

for all $(r_1, \ldots, r_N) \in \{1, *\}^N$, for all $N \in \mathbb{N}$.

Proof 25.25 *Let $u = u_{k_1, j_1} u_{k_2, j_2} u_{k_1, j_1} \in \mathscr{X}_{\tau}$ be an operator (25.103), where $j_1 \neq j_2$ in $\mathbb{Z}$. Then*

$$k_N^{\tau}(u^{r_1}, ..., u^{r_N}) = \sum_{\pi \in NC(N)} \left(\prod_{V \in \pi} \tau \left(\prod_{l \in V} u^{r_l} \right) \right) \mu(\pi, 1_N)$$

$$= \sum_{\pi \in NC(N)} \left(\prod_{V \in \pi} \varphi \left(\prod_{l \in V} w \right) \right) \mu(\pi, 1_N)$$

since $\{u_{k,j}\}_{k,j \in \mathbb{Z}}$ *are followed by the semicircular law, where*

$$w = x_{j_1} x_{j_2} x_{j_1} \in \mathfrak{X}_{\varphi},$$

and hence,

$$= k_N^{\varphi} \left(\underbrace{w, w, ..., w}_{N\text{-times}} \right),$$

i.e., (25.105)

$$k_N^{\tau}(u^{r_1}, ..., u^{r_N}) = k_N^{\varphi}(w, ..., w),$$

(as in (25.98)), for all $(r_1, ..., r_N) \in \{1, *\}^N$*, for all* $N \in \mathbb{N}$*.*

Note that the free random variable w *in (25.105) is a free Poisson element of* $\mathfrak{X}_{\varphi}$ *satisfying (25.102). Therefore, the free-distributional data (25.104) for* $u \in \mathscr{X}_{\tau}$ *holds by (25.105).*

The aforementioned theorem says that the free random variables followed by the semicircular law induce free random variables whose free distributions are followed by the free Poisson distributions, characterized by (25.102).

Now, let $w_{k_1, j_1:0}, w_{k_1, j_2:0} \in \mathscr{X}[\Gamma]$ be generating operators of the dynamical shift-semicircular space $\mathscr{X}[\Gamma]$, where $j_1 \neq j_2$ in $\mathbb{Z}$. Define a new operator $W \in \mathscr{X}[\Gamma]$ by

$$W = w_{k_1, j_1:0} w_{k_2, j_2:0} w_{k_1, j_1:0} \text{ in } \mathscr{X}[\Gamma]. \tag{25.106}$$

Theorem 25.17 *Let* W *be a free random variable (25.106) of* $\mathscr{X}[\Gamma]$*. Then the free distribution of* W *is followed by the free Poisson distribution characterized by (25.102) in the sense that:*

$$k_N^{\tau_o}(W^{r_1}, ..., W^{r_N}) = \omega_N c_{\frac{N}{2}}, \tag{25.107}$$

for all $(r_1, ..., r_N) \in \{1, *\}^N$*, for all* $N \in \mathbb{N}$*.*

Proof 25.26 *Note first that the free random variable* W *is induced by*

$$w_{k_l, j_l:0} = (u_{k_l, j_l}, 0) \in \mathscr{X}[\Gamma], \text{ for } l = 1, 2,$$

and $\alpha_0^s(u_{k_l, j_l}) = u_{k_l, j_l}$*, for all* $s \in \mathbb{N}$*, for all* $l = 1, 2$*. Thus,*

$$k_N^{\tau_o}(W^{r_1}, ..., W^{r_N}) = k_N^{\tau}(u^{r_1}, ..., u^{r_N}),$$

for all $(r_1, ..., r_N) \in \{1, *\}^N$, *for all* $N \in \mathbb{N}$, *by (25.67), under the Möbius inversion, where*

$$u = u_{k_1,j_1} u_{k_2,j_2} u_{k_1,j_1} \in \mathscr{X}_\tau.$$

So, we have that

$$\begin{aligned} k_N^{\tau_o}\left(W^{r_1}, ..., W^{r_N}\right) &= k_N^{\tau}\left(u^{r_1}, ..., u^{r_N}\right) \\ &= k_N^{\varphi}\left(\underbrace{w, w, .., w}_{N\text{-times}}\right), \end{aligned}$$

by (25.105), implying that

$$k_N^{\tau_o}\left(W^{r_1}, ..., W^{r_N}\right) = \varphi\left(x_{j_2}^N\right) = \omega_N c_{\frac{N}{2}},$$

for all $(r_1, ..., r_N) \in \{1, *\}^N$, *for all* $N \in \mathbb{N}$, *by (25.104).*

The aforementioned theorem shows that there do exist free random variables of $\mathscr{X}[\Gamma]$ followed by the free Poisson distributions characterized by (25.102). However, in general, if $n \neq 0$ in $\mathbb{Z}$, then the free random variables,

$$w_{k_1,j_1:n} w_{k_2,j_2:n} w_{k_1,j_1:n} \in \mathscr{X}[\Gamma],$$

are "not" followed by the free Poisson distributions characterized by (25.102).

BIBLIOGRAPHY

[1] M. Ahsanullah, Some Inferences on Semicircular Distribution, *J. Stat. Theo. Appl.*, 15, no. 3, (2016) 207–213.

[2] H. Bercovici, and D. Voiculescu, Superconvergence to the Central Limit and Failure of the Cramer Theorem for Free Random Variables, *Probab. Theo. Related Fields*, 103, no. 2, (1995) 215–222.

[3] M. Bozejko, W. Ejsmont, and T. Hasebe, Noncommutative Probability of Type *D*, *Internat. J. Math.*, 28, no. 2, (2017) 1750010, 30.

[4] M. Bozheuiko, E. V. Litvinov, and I. V. Rodionova, An Extended Anyon Fock Space and Non-commutative Meixner-Type Orthogonal Polynomials in the Infinite-Dimensional Case, *Uspekhi Math. Nauk.*, 70, no. 5, (2015) 75–120.

[5] I. Cho, Semicircular Families in Free Product Banach $*$-Algebras Induced by p-Adic Number Fields over Primes p, *Compl. Anal. Oper. Theo.*, 11, no. 3, (2017) 507–565.

[6] I. Cho, Semicircular-Like Laws and the Semicircular Law Induced by Orthogonal Projections, *Compl. Anal. Oper. Theo.*, 12, (2018) 1657–1695.

[7] I. Cho, Banach-Space Operators Acting on Semicircular Elements Induced by Orthogonal Projections, *Compl. Anal. Oper. Theo.*, 13, no. 8, (2019) 4065–4115.

[8] I. Cho, Acting Semicircular Elements Induced by Orthogonal Projections on von Neumann Algebras, *Mathematics*, 5, no. 74, (2017) DOI:10.3390/math5040074.

[9] I. Cho, and J. Dong, Catalan Numbers and Free Distributions of Mutually Free Multi Semicircular Elements, (2018) Submitted to *J. Probab. Related Fields.*

[10] I. Cho, and P. E. T. Jorgensen, Semicircular Elements Induced by Projections on Separable Hilbert Spaces, *Monograph Ser., Oper. Theory: Adv. & Appl.*, Published by Birkhauser, Basel, (2019) To Appear.

[11] I. Cho, and P. E. T. Jorgensen, Banach $*$-Algebras Generated by Semicircular Elements Induced by Certain Orthogonal Projections, *Opuscula Math.*, 38, no. 4, (2018) 501–535.

[12] I. Cho, and P. E. T. Jorgensen, Semicircular Elements Induced by p-Adic Number Fields, *Opuscula Math.*, 35, no. 5, (2017) 665–703.

[13] A. Connes, *Noncommutative Geometry*, ISBN: 0-12-185860-X, (1994) Academic Press (San Diego, CA).

[14] P. R. Halmos, Hilbert Space Problem Books, *Grad. Texts in Math.*, 19, ISBN: 978-0387906850, (1982) Published by Springer.

[15] I. Kaygorodov, and I. Shestakov, Free Generic Poisson Fields and Algebras, *Comm. Alg.*, 46, no. 4, DOI:10.1080/00927872.2017.1358269, (2018).

[16] L. Makar-Limanov, and I. Shestakov, Polynomials and Poisson Dependence in Free Poisson Algebras and Free Poisson Fields, *J. Alg.*, 349, no. 1, (2012) 372–379.

[17] A. Nica, and R. Speicher, *Lectures on the Combinatorics of Free Probability* (1st Ed.), London Math. Soc. Lecture Note Ser., 335, ISBN-13:978-0521858526, (2006) Cambridge University Press.

[18] I. Nourdin, G. Peccati, and R. Speicher, Multi-Dimensional Semicircular Limits on the Free Wigner Chaos, *Progr. Probab.*, 67, (2013) 211–221.

[19] V. Pata, The Central Limit Theorem for Free Additive Convolution, *J. Funct. Anal.*, 140, no. 2, (1996) 359–380.

[20] F. Radulescu, Random Matrices, Amalgamated Free Products and Subfactors of the C^*-Algebra of a Free Group of Nonsingular Index, *Invent. Math.*, 115, (1994) 347–389.

[21] F. Radulescu, Free Group Factors and Hecke Operators, notes taken by N. Ozawa, Proceed. 24-th Conference in Oper. Theo., Theta Advanced Series in Math., (2014) Theta Foundation.

[22] R. Speicher, *Combinatorial Theory of the Free Product with Amalgamation and Operator-Valued Free Probability Theory* (1998) Amer. Math. Soc. Mem.

[23] R. Speicher A Conceptual Proof of a Basic Result in the Combinatorial Approach to Freeness, *Infinit. Dimention. Anal. Quant. Prob. & Related Topics*, 3, (2000) 213–222.

[24] R. Speicher, and T. Kemp, Strong Haagerup Inequalities for Free R-Diagonal Elements, *J. Funct. Anal.*, 251, no. 1, (2007) 141–173.

[25] R. Speicher, and U. Haagerup, Brown's Spectrial Distribution Measure for R-Diagonal Elements in Finite Von Neumann Algebras, *J. Funct. Anal.*, 176, no. 2, (2000) 331–367.

[26] V. S. Vladimirov, p-Adic Quantum Mechanics, *Comm. Math. Phy.*, 123, no. 4, (1989) 659–676.

[27] V. S. Vladimirov, I. V. Volovich, and E. I. Zelenov, p-Adic Analysis and Mathematical Physics, *Ser. Soviet & East European Math.*, vol 1, ISBN: 978-981-02-0880-6, (1994) World Scientific.

[28] D. Voiculescu, Aspects of Free Analysis, *Jpn. J. Math.*, 3, no. 2, (2008) 163–183.

[29] D. Voiculescu, Free Probability and the Von Neumann Algebras of Free Groups, *Rep. Math. Phy.*, 55, no. 1, (2005) 127–133.

[30] D. Voiculescu, K. Dykemma, and A. Nica, Free Random Variables, CRM Monograph Series, vol 1., ISBN-13: 978-0821811405, (1992) Published *by Amer. Math. Soc.*

Index